37th Riso International Symposium on Materials Science 2016

IOP Conference Series: Materials Science and Engineering
Volume 139

Riso, Denmark
5 – 8 September 2016

Editors:

B. Madsen
A. Biel
Y. Kusano
H. Lilholt
L.P. Mikkelsen
L. Mishnaevsky Jr.
B.F. Sorensen

ISBN: 978-1-5108-3053-0
ISSN: 1757-8981

Printed from e-media with permission by:

Curran Associates, Inc.
57 Morehouse Lane
Red Hook, NY 12571

Printed by Curran Associates, Inc. (2016)

For permission requests, please contact the Institute of Physics at the address below.

Institute of Physics
Dirac House, Temple Back
Bristol BS1 6BE UK

Phone: 44 1 17 929 7481
Fax: 44 1 17 920 0979

techtracking@iop.org

Additional copies of this publication are available from:

Curran Associates, Inc.
57 Morehouse Lane
Red Hook, NY 12571 USA
Phone: 845-758-0400
Fax: 845-758-2633
Email: curran@proceedings.com
Web: www.proceedings.com

Table of contents

Volume 139

37th Risø International Symposium on Materials Science
5–8 September 2016, Risø, Denmark

Accepted papers received: 8 July 2016
Published online: 2 September 2016

Contributed papers

IOP Conf. Series: Materials Science and Engineering **139** (2016) 011001 doi:10.1088/1757-899X/139/1/011001

05 July 2016

Preface for the 37th Risø Symposium Proceedings, IOP publication

PREFACE

Proceedings of the 37th Risø International Symposium on Materials Science: *Understanding performance of composite materials – mechanisms controlling properties*

B. Madsen, A. Biel, Y. Kusano, H. Lilholt, L.P. Mikkelsen, L. Mishnaevsky Jr., B.F. Sørensen

The theme of the 37th Risø International Symposium is the scientific basis for understanding the performance of composite materials. Specifically, the Symposium focuses on the underlying mechanisms that control the properties of composites, which will lead to better and more reliable model predictions. This will serve to support and further promote the technological development of composite solutions.

The present state of composite development was founded by the first scientific studies on composites in the 1960s, establishing fundamental models for fibre reinforcement in composites. Technological challenges have required continued development of more advanced materials models and experimental techniques to study the underlying mechanisms and phenomena for specific composite properties and performance, both the interactions between fibres and matrix, and the more complex loading modes such as fatigue and fracture behaviour. Further work has addressed the development of new fibre and matrix materials, improved composite manufacturing techniques, and recently, recycling routes for composites.

The present Symposium Proceedings cover fundamental studies on the properties and performance of a range of fibre composite systems, such as polymer matrix composites, biocomposites, nanocomposites, and hybrid composites. Examples of matters addressed are stress-strain behaviour of fibres, cure kinetics and residual stresses in matrices, chemistry and mechanical evaluation of fibre/matrix interfaces, microstructure and micromechanics of composites, damage evolution and ductility in composites, effects of composite manufacturing, and recycling opportunities. The Proceedings contain 10 invited papers, and 40 contributed papers.

The 37th Risø International Symposium is organised by the Section of Composites and Materials Mechanics, Department of Wind Energy, Technical University of Denmark (DTU), at the Risø Campus. We would like to thank all those at DTU Wind Energy who assisted in the preparation of the Symposium. We appreciate additionally the support from the international advisory committee consisting of: E. Greenhalgh, UK; F. Paris, Spain; T. Peijs, UK; K. Pickering, New Zealand; K. Schulte, Germany; R. Talreja, USA; J. Thomason, UK; J. Varna, Sweden. We gratefully acknowledge financial support from the following foundations: Civilingeniør Frederik Leth Christiansens Almennyttige Fond, Knud Højgaards Fond, Otto Mønsteds Fond, Danish Center for Applied Mathematics and Mechanics (DCAMM).

Peer review statement

All papers published in this volume of *IOP Conference Series: Materials Science and Engineering* have been peer reviewed through processes administered by the proceedings Editors. Reviews were conducted by expert referees to the professional and scientific standards expected of a proceedings journal published by IOP Publishing.

IOP Conf. Series: Materials Science and Engineering **139** (2016) 012001 doi:10.1088/1757-899X/139/1/012001

Development of novel composites through fibre and interface/interphase modification

A Bismarck[1,2,*], JJ Blaker[3], DB Anthony[2,4], H Qian[2,4], HA Maples[1], P Robinson[5], MSP Shaffer[4], ES Greenhalgh[5]

[1] Polymer & Composite Engineering (PaCE) Group, Institute of Materials Chemistry and Research, Faculty of Chemistry, University of Vienna, Währinger Str. 42, A-1090 Vienna
[2] Department of Chemical Engineering, Imperial College London, South Kensington Campus, London, SW7 2AZ, UK
[3] University of Manchester, School of Materials, MSS Tower, Manchester, M13 9PL, UK
[4] Department of Chemistry, [5] The Composites Centre, Department of Aeronautics, Imperial College London, South Kensington Campus, London, SW7 2AZ, UK

E-mail: alexander.bismarck@univie.ac.at

Abstract. We show how fibre/matrix interface (or interphase) modification can be used to develop a range of novel carbon fibre reinforced polymer (CFRP) composites that open up new applications far beyond those of standard CFRPs. For example, composites that undergo pseudo-ductile failure have been created through laser treatment of carbon fibres. Composites manufactured with thermo-responsive interphases can undergo significant reductions in stiffness at elevated temperatures. Additionally, structural supercapacitors have been developed through a process that involves encapsulating carbon fibres in carbon aerogel.

1. Introduction

Composite materials have a wide range of applications due to their excellent specific strength and stiffness. These composite properties are determined by the fibre type, the fibre volume fraction and fibre alignment. The role of the matrix is to hold the fibres in place, protect them from damage. The matrix determines the composite transverse properties, in relation to the primary fibre axis. Load is transferred from the matrix to the fibres through the fibre/matrix interface. Tailoring the fibre/matrix interface allows control over the final composite properties but clever design of composite interphases also offers new opportunities to introduce additional functionalities, such as active stiffness control or multifunctionality, into composite materials. Three fibre modification concepts are discussed, which include i) the creation of fibres with modulated properties and shapes as a potential route to short fibre composites which form in-situ upon loading, ii) thermo-responsive interphases that allow for active stiffness control in carbon fibre composites and iii) the storage of electrical energy in structural carbon fibre composites.

2. Interfaces for Ductility: Carbon Fibres with Modulated Shape and Properties

It is possible to create a pseudo-ductile response in composites by allowing discontinuous fibre fragments to move within the matrix. Prepregs with discontinuities formed by cut plies have been studied in detail. Fibre fragments are required to be sufficiently small to suppress unstable propagation of shear from the discontinuities before overall tensile failure. It was shown both theoretically and experimentally for cementitious composites that fibres with slip strain hardening fibre/matrix interfaces can undergo multiple fracture events, spreading fibre/matrix debonding and fibre pull-out across large volumes, allowing for significant deformation before a crack localises and the composite fails [1]. More recently, it was shown that machined steel fibres having conical ends [2] with taper angles of 5° dissipated much more energy than straight fibres when being pulled out from a cured epoxy resin. Considering such fibres embedded in composites it should, therefore, be possible to bridge cracks and plough through the matrix, which should contribute to the energy dissipated during the failure process. There is a growing body of evidence from synthetic composite materials [3, 4, 5, 6], which shows that tapered fibre ends do result in an increased work of pull-out, delocalise inelastic deformation and induce strain-hardening. However, it remains challenging to produce outward tapered features and/or an unduloid fibre shape in continuous carbon fibres, which are the preferred reinforcements in high performance composites. From a manufacturing point it is desired to use continuous fibres to produce high fibre volume faction composites, in which the fibres break *in situ* during loading to form short fibre segments with tapered ends. Such a composite would have a high initial stiffness.

Lasers are being increasingly used to machine high performance unidirectional carbon fibre composites [7, 8] and for cutting composite plies at predefined points to produce highly aligned discontinuous composites mainly aiming to improve manufacturing ductility [9]. Carbon fibres cut via laser ablation have been reported to feature expanded fibre ends near the heat-affected zones of laser-drilled holes [7, 8]. Controlled laser exposure of carbon fibres should present an opportunity to modify the shape of continuous carbon fibres along their lengths.

In order to modify carbon fibres, individual unsized, polyacrylonitrile (PAN) based AS4 fibres (Hexcel Ltd., Cambridge, UK) were fixed on paper frames. These frames allowed for positioning and handling of the fibres for laser treatment and subsequent characterisation. The substrates were positioned underneath the laser beam, such that the laser was run across the fibres perpendicular to the fibre axis. A nanosecond pulsed Nd:YVO_4 laser with a maximum average power of 17.5 W at λ = 1064 nm operating with an average power of 1.1 - 17.5 W, a repetition rate in the range of 50 - 100 kHz and beam scanning speeds in the range of 50 - 200 mm s^{-1} was used to treat the fibres. The laser beam was focused to a spot size of ~ 70 μm in diameter. The laser conditions were tuned and resulted in three different fibre shapes (Figure 1), causing: i) local expansion of the fibre at the irradiation site; ii) expansion and ablation of the fibres, and iii) expansion and cutting of the fibres [10] (Fig. 1). Laser parameters 1.1 W of average output power, frequency 100 kHz, and speed 100 mms^{-1} resulted in approximately 20% of fibres being cut; at 17.5 W, frequency 50 kHz and speed 50 mms^{-1} all fibres were cut. Parameters 1.1 W, frequency 100 kHz and speed 200 mms^{-1} consistently resulted in intact fibres. The laser operated at 1.1 W of average power had an energy fluence of 0.3 J cm^{-2} (energy per pulse 11 μJ). 1D temperature modelling has been conducted for these conditions with an estimated temperature increase in the area irritated by a focused beam in excess of 2900 K. The vaporisation temperature of PAN-based carbon fibres is reported to be 3500 K [11]. The same average power but scanning speeds lower than 100 mms^{-1} resulted in cutting of the fibres.

The laser treated fibres consistently exhibited an affected region along their length. The diameter of laser treated fibres increased from the average diameter of the control fibres 7.0 μm to a maximum of 9.7 μm, corresponding to a diameter increase of 37% (Expanded fibres). The average affected length of these Expanded carbon fibres after a single laser pass was ~140 μm, resulting in a taper angle about 1°. Expanded/Ablated fibres exhibited slightly larger affected lengths with narrower midsections, flanked by expanded regions, resulting in taper angles of 1.6°. Cotton-bud end fibres were completely cut and had average affected lengths of ca. 110 μm and taper angles of 1.8° (Fig. 1). The maximum diameter expansion observed was 53%, after which ablation reduced the expanded fibre diameter, eventually cutting the fibre, resulting in characteristic cotton-bud shaped fibre ends.

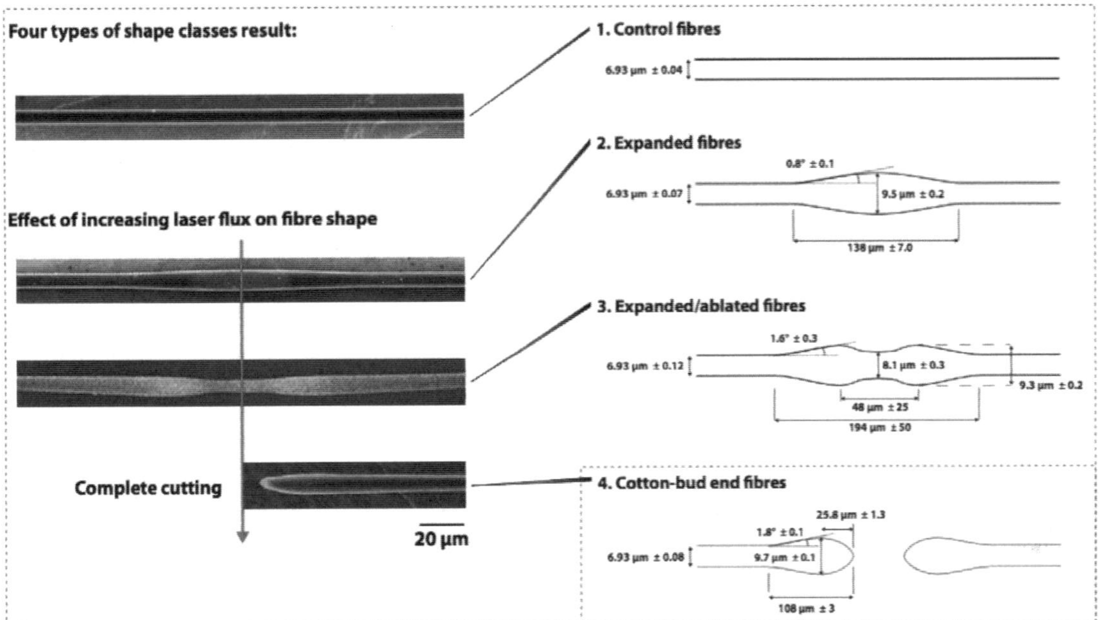

Fig. 1: Characteristic scanning electron micrographs of laser treated of laser treated carbon fibres; depending on the laser parameters, average power, repetition rate and beam scanning speeds, these typical fibre morphologies were observed: Control, Expanded, Expanded/Ablated and Cotton-bud end fibres. With permission of the American Chemical Society (ACS).

The tensile properties of Expanded and Expanded/Ablated carbon fibres were significantly reduced in comparison to control AS4 fibres (Fig. 2). All fibres exhibited a linear elastic behaviour with brittle fracture. As expected the control carbon fibres failed catastrophically, whereas the laser treated fibres broke into two, at their pre-defined weak point (expanded or ablated regions) resulting in outwardly tapered fibre ends (for the expanded carbon fibres). The fibre tensile strength (Fig. 2) was reduced by ~69% for Expanded fibres and 75% for Expanded/Ablated fibres. The tensile modulus of the laser treated fibres (Fig. 2), however, was reduced only by a modest ~17% (averaged over all treated fibres) in comparison to the control fibres. In practice, this reduction in fibre strength is still useful in the right context, for example; fibre manipulation is retained and therefore, it should be possible to use such laser treated fibres to produce unidirectional composites, which should still possess high initial stiffness. A significant reduction in strength is, in fact, a prerequisite to ensure reliable local failure sites to create highly aligned short fibre composites contained cotton bud shaped fibres. Whilst the strength reduction is greater than necessary, in the current guise, a distribution of pre-weakened points can be implemented throughout the composite due to controlled nature of the modification, which can be used to minimise the reduction in overall composite strength.

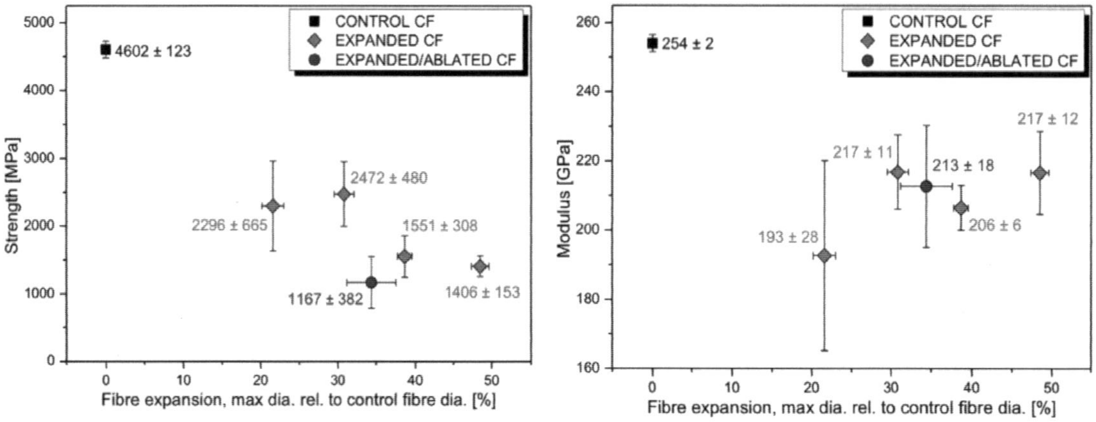

Fig. 2: Tensile strength and modulus of Control and laser treated Expanded and Expanded/Ablated fibres carbon fibres as function of fibre expansion. The fibre tensile strength was normalised to control fibre diameter. With permission of the ACS.

Fibre pull-out tests performed on cotton-bud terminated carbon fibres from a transparent silicone rubber matrix were conducted to visualise the effect of the modified fibre ends on the pull-out behaviour over a long embedded length. Significant ploughing was observed for Cotton-bud end fibres through the matrix in contrast to the control fibres, which were simply pulled from the matrix (Fig. 3). The work of pull-out of the Cotton-bud end fibres increased dramatically. Taper angles of 1-2° are already sufficient to cause cavitation in the matrix along with sliding. The apparent interfacial shear strength determined from the maximum pull-out force normalised by the embedded fibre area increased for cotton-bud fibres by a factor of 5 compared to Control fibres.

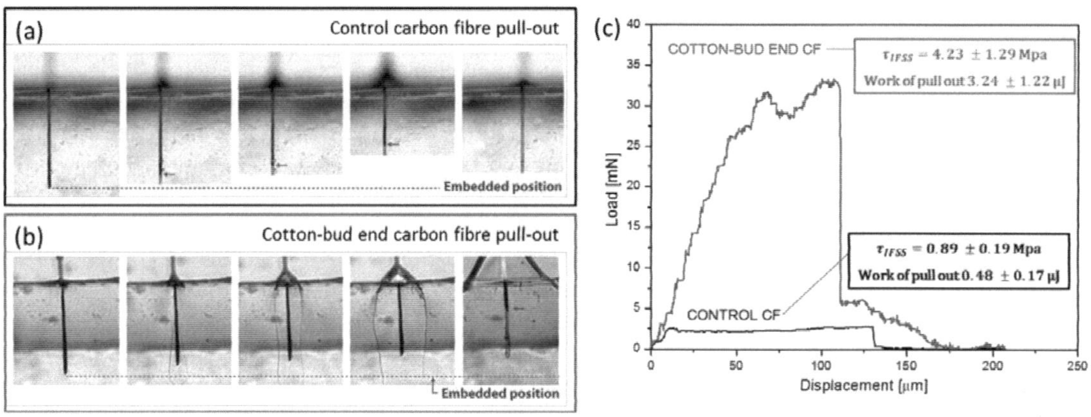

Fig. 3: Optical micrographs taken during the pull-out test of (a) Control (AS4) carbon fibre and a laser treated carbon fibre with a Cotton-bud end carbon fibre. All images have been positioned such that embedded position is maintained (blue dotted line), fibre ends are highlighted with an arrow where appropriate. Characteristic load-displacement curves of the fibre pull-out of the Control and Cotton-bud end carbon fibre with inset proving the average interfacial shear strength (τ_{IFSS}) and work of pull-out. With permission of the ACS.

High power laser treatment on structural carbon fibres was systematically used to create expanded regions at predetermined points; the fibre strength was locally reduced whilst much of their stiffness was preserved. Laser treated fibres consistently fractured in the laser affected regions, producing fibre segments with outwardly tapered ends possessing taper angles of 1 - 2°. The effectiveness of the laser modification method to modify pull-out characteristics was successfully demonstrated using fibres with 'cotton-bud' ends, which resulted in a significant improvement of the work of pull-out and interfacial shear strength. Modified continuous carbon fibres could be used to form aligned high stiffness unidirectional composites with predefined (and distributed) weak regions. Under load fibre fragmentation will occur resulting in the formation of a well-aligned short fibre composite, in which a high modulus close to that of common unidirectional composites is predicted, but with tapered fragment ends resulting in strain hardening composite interfaces.

3. Composite Interfaces/Interphases for Multifunctionality

Multifunctional materials are specifically developed to improve system performance by integrating subsystem components and functions [12]. Multifunctional composites ought to combine high specific mechanical properties with additional functionalities, such as sensing, actuation, repair or energy storage. Composite interfaces/-phases play a crucial role in order to realise such multifunctional composites, which will be shown using the following two examples.

3.1 Responsive Fibre Coatings: A Route to Composites with Stiffness Control

Recently, the development of composites with controllable stiffness has received considerable interest [13, 14, 15]. Such composites are of particular interest for use in morphing and deployable structures as they offer high stiffness with increased compliance when required [16]. A route to allow fibres within a polymer matrix composite to slide is to switch off the fibre/matrix adhesion. The fibre/matrix adhesion is dramatically above the glass transition temperature (T_g) of the composite matrix. In order to allow carbon fibres to slide within a thermosetting high T_g matrix, all fibres could be coated with a thin low T_g thermoplastic coating.

Flexible interlayers between reinforcing fibres and matrix were actively researched over 20 years ago as a means to improve the energy absorption characteristics of composites [17]. Such flexible interlayers promote toughness by interfacial debonding, fibre pull-out and post-debonding friction as well as stress redistribution through localised large scale elastic and plastic deformations [18]. However, the presence of a ductile coating promoting toughness can lead to a reduction in strength and modulus of a composite. Various fibre coating methods, such as electropolymerisation, have been used to deposit polymeric coatings around reinforcing fibres. The carbon fibres within the composite can be used as Joule heaters by passing a current through them. This heats the fibres and also the surrounding thermoplastic coating to temperatures above its T_g until the coating softens, which enables the fibres to slide within the high T_g composite matrix (Fig. 4). Manifestation of this effect has shown to reduce greatly the storage modulus (Fig. 5) and flexural stiffness of the composite (by up to 88 %) [19]. This process is reversible as the stiffness returns to its original value once the coating is allowed to cool below its T_g. The choice of fibre coating depends primarily on the thermal properties of the polymer. It should have a T_g or melting temperature (T_m) below that of the thermosetting or thermoplastic composite matrix but well above the maximum service temperature of the final composite part. Successful demonstration of this concept by electrocoating carbon fibres with polyacrylamide (PAAm) using a continuous electrocoating method is reported [19]. The T_g of PAAm can be tailored by adjusting the moisture content of the coating. PAAm, when partially hydrated (20 % moisture content), has a T_g of approximately 84 °C. However, water soluble fibre coatings like PAAm are not suitable for use in high performance applications and for this reason carbon fibres were coated with non-water soluble polymers, such as polymethylmethacrylate (PMMA) and polystyrene [20]. As only a very thin thermoresponsive coating is needed, the coating can be heated quickly by passing a current through the carbon fibres. This means the stiffness can be controlled on demand. It is important that the matrix remains stiff when the carbon fibre coating has softened. This ensures that the composite can withstand loading at elevated temperatures and prevents fibre misalignment. Below the softening temperature of the fibre coating the epoxy composites have a very similar flexural stiffness compared to those containing unsized carbon fibres, which suggests they are suitable for high performance applications. The major disadvantage of using these composites is that the maximum deflection at elevated temperatures is limited to the maximum strain of the matrix material.

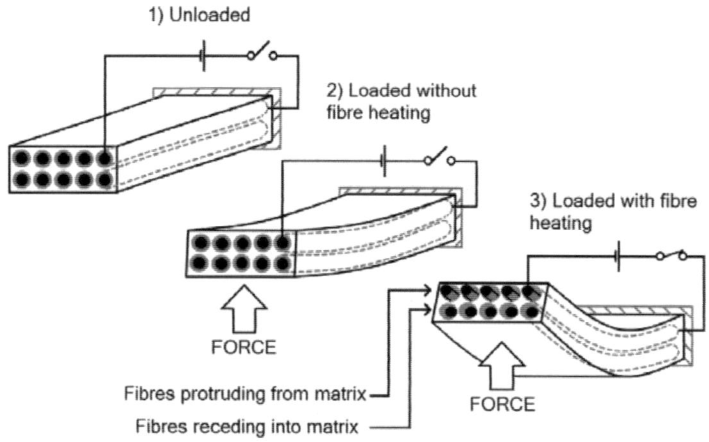

Fig. 4: Concept of the controllable stiffness composite. With permission from the ACS.

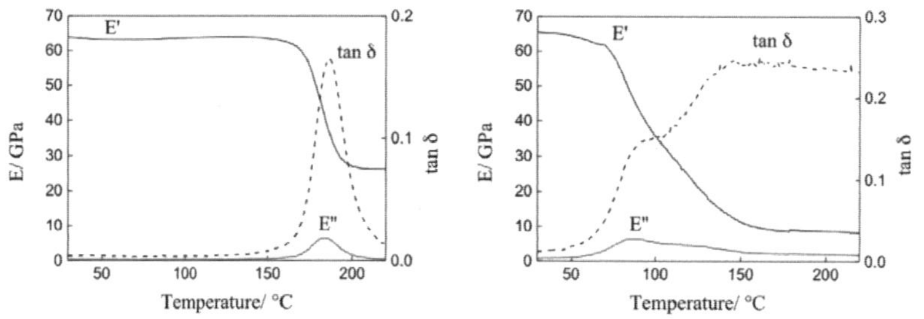

Fig. 5: Storage modulus (E'), loss modulus (E") and damping (tan δ) of an epoxy composite containing unsized (left) and hydrated PAAm coated carbon fibres (right) as a function of temperature showing the stiffness loss when (left) the epoxy matrix passes T_g or the PAAm fibre coating (right) passes through T_g allowing the carbon fibres to slide within the coating.

3.2 High Surface Aero Composite Interfaces for Structural Energy Storage Devices

The combination of structural properties with electrochemical energy storage in a single material, to provide multifunctional structural energy storage devices [21], offer the prospect to decrease the weight and volume of systems, which otherwise do require independent energy storage devices [22], such as batteries or supercapacitors, such electric/hybrid vehicles. The easiest route to a multifunctional structure is to embed energy storage devices into conventional fibre-reinforced composites. Alternatively, truly multifunctional materials could be developed that simultaneously have both structural and electrochemical energy storage capabilities [23, 24]. Show here is the possibility to create multifunctional structural supercapacitor composites. Figure 6 shows a design for a structural supercapacitor consisting of two modified structural carbon fibre fabric electrodes, separated by a structural glass fibre fabric infused with a multifunctional polymeric electrolyte. The critical components for such a device are the electrolyte, which has to provide sufficient mechanical support while being an ionic conductor, and the multifunctional electrode having both a high energy storage capability and good mechanical properties. In order to achieve this goal structural carbon fibres were encapsulated into a carbon aerogel (CAG). Carbon aerogels (CAGs) are widely used for their high surface area and electrical conductivity, particularly in the context of electrochemical electrodes and devices. The most common carbon aerogels are formed by condensation, stabilisation and subsequent carbonisation of resorcinol-formaldehyde (RF) mixtures, to yield a network structure of amorphous carbon.

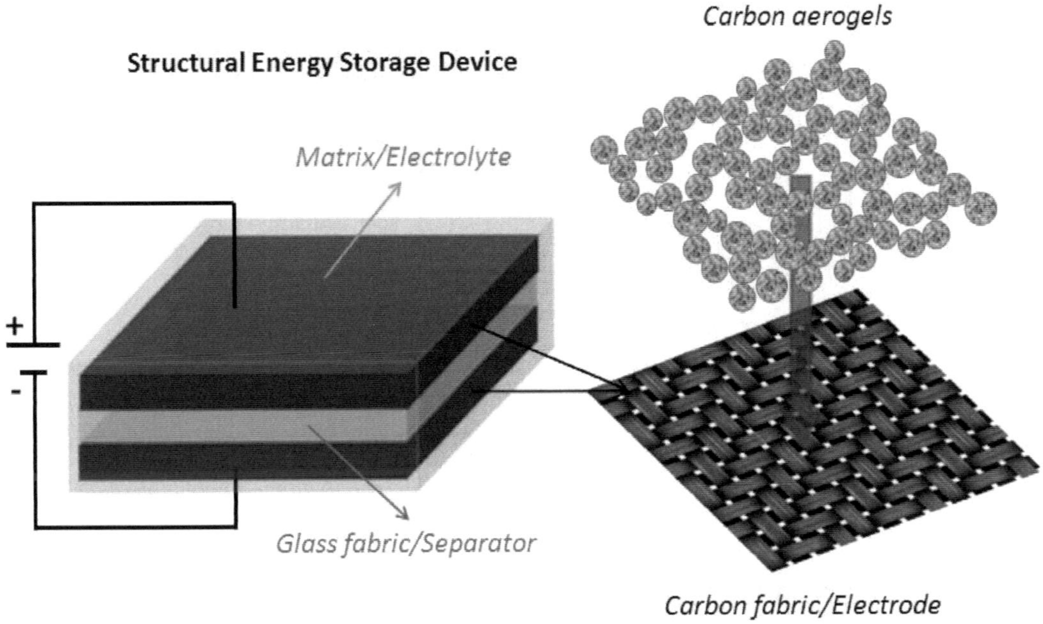

Fig. 6. Schematic of a structural supercapacitor concept based on CAG-modified structural carbon fibre fabrics as electrodes, structural glass fabrics as separator impregnated with a polymer-based electrolyte. With permission of the ACS.

Exploring this idea further using a variety of woven structural carbon fibres, semistructural supercapacitor laminates can be prepared [23]. Scaled production of such laminates is achieved by adapting conventional composite processing techniques. The multifunctional electrode material developed was prepared by resin infusion under flexible tooling (RIFT) of a plain weave carbon fibre fabric (3k, 5-satin-harness weave, 283 gm^{-2}, Cytec Engineered Materials) with resorcinol-formaldehyde (AX2000, INDSPEC Chemical Corporation) diluted with formaldehyde solution and using sodium hydroxide as catalyst. The specimens prepared were both cured at room temperature, 50°C and 90°C with a period of 24 h at each temperature. The dried specimens were then carbonised at 800°C for 30 min in a furnace under flowing nitrogen (Fig. 7). The carbon fibre supported CAG forms a hierarchical reinforcement preform (Fig. 8) consisting of a bicontinuous nanoporous matrix, uniformly surrounding reinforcing carbon fibres. The CAG contributes around 10 wt% to the sample; the surface area of the CAG component alone is typical of such materials at around 700-800 m^2g^{-1}, with pore sizes of a few tens of nanometres. This preform is then infused via a second RIFT cycle with the desired polymeric matrix resin system to form a dense composite. To create a structural supercapacitor the carbon aerogel-carbon fibre (CAG-CF) preform was infused with a non-structural resin, polyethylene glycol diglycidal ether (PEGDGE), which has a low modulus of about 6 MPa but was shown to have reasonable ionic conductivity [23] when combined with a suitable electrolyte. To examine whether a CAG network can be a useful reinforcement, the same types of CAG-CF preforms were infused with a conventional epoxy matrix (Gurit PRIME 20ULV, modulus of about 3 GPa) to generate pure structural systems.

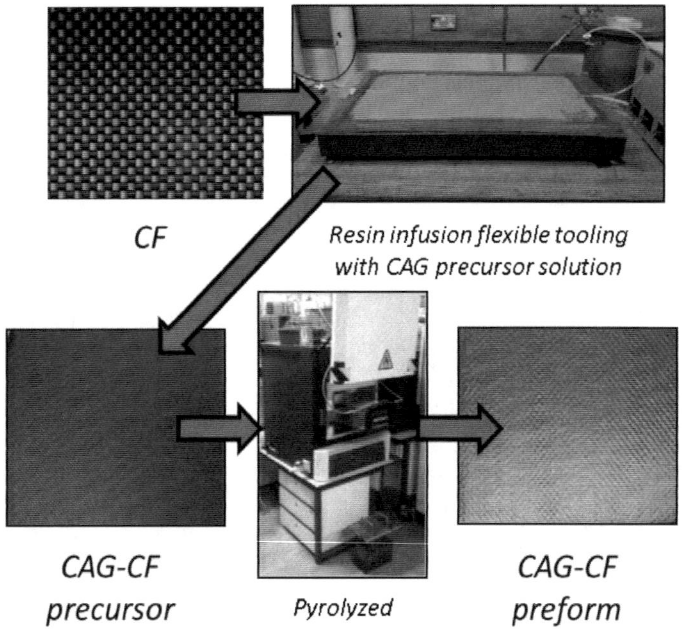

Fig. 7: Schematic flow diagram illustrating the synthesis of carbon aerogel-carbon fibre preforms.

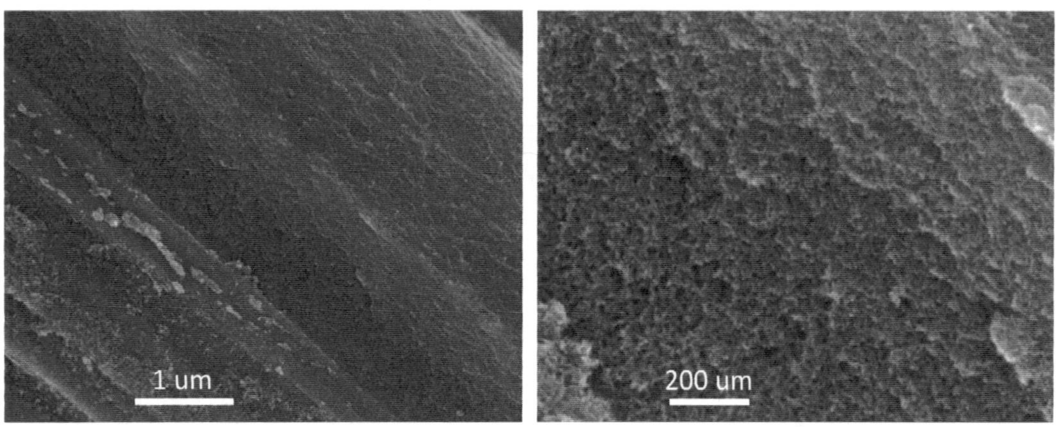

Fig. 8: Scanning electron micrographs of CAG-CF, infused with 40 wt% RF solution, showing carbon fibres embedded in CAG (left) and detail of the CAG nanostructure (right).

The mechanical performance of the hierarchical CF-CAG composites was first examined using ±45° shear tests (ASTM Standard D3518) in order to investigate the response of the CAG reinforced matrix (Table 1). The shear strength and the shear modulus of both multifunctional and structural epoxy matrices increased. Predictably, the properties of the non-structural PEGDGE matrix system are most dramatically improved but surprisingly even the properties of the structural epoxy are apparently improved. To explore whether these improved matrix characteristics translate into better composite properties composites were tested in compression (Table 2). Again, the compression strength and stiffness of the soft CF-CAG PEDGE composite increased significantly. Also, the modulus of CF-CAG composites with the structural epoxy matrix increased but a significant decrease in compression strength was observed in this case. This observed decrease in strength the CF-CAG composites with the structural matrix was due to the reduced adhesion between the CAG modified carbon fibres and the matrix as indicated by interlaminar shear strengths (ILSS), obtained by short beam shear tests (ASTM Standard D2344); the ILSS decreased from 60 MPa to 43 MPa. The heat treatment involved in carbonizing the CAG precursor, can be expected to remove both sizing and surface oxidation from the CFs. The resulting lack of chemical interaction with the subsequently infused resin may be the source of the degraded fibre/matrix interface, and hence mechanical properties.

Table 1: In-plane shear response tensile test of ±45° laminates on monolithic laminates [23] with woven carbon fibre fabrics, ASTM D3518.

Reinforcement	Matrix	Shear Strength (MPa)	Shear Modulus (MPa)	Volume fraction of fibre reinforcement (vol%)
As-received	DGEBA-based epoxy	25.9 ± 2.2	4380 ± 60	45.0
CAG-modified		26.2 ± 0.5	5050 ± 210	40.7
As-received	PEGDGE-based epoxy	5.83 ± 0.14	201 ± 10	47.2
CAG-modified		8.88 ± 0.12	911 ± 60	42.0

Table 2: Longitudinal compression on monolithic laminates in a 0°/90° lay-up with woven carbon fibre fabrics [25], ASTM D3410.

Reinforcement	Matrix	Compressive Strength (MPa)	Compressive Modulus (GPa)	Volume fraction of fibre reinforcement (vol%)
As-received	PEDGE-based epoxy	50.4 ± 3.8	16.6 ± 1.5	57.1
CAG-modified		174.8 ± 9.2 (3.47-fold)	60.5 ± 3.6 (3.65-fold)	55.7
As-received	PRIME-20ULV epoxy	535.7 ± 52.2	65.4 ± 5.1	56.3
CAG-modified		345.2 ± 29.2 (35.6% decrease)	80.6 ± 4.5 (23.2% increase)	56.5

4. Conclusions

We introduced a novel fibre modification concept to produce carbon fibres with tapered shapes in-situ in a composite during loading and two fibre modification methods which highlight the importance of composite interfaces for the development of truly multifunctional composites. We showed that coating carbon fibres with thin thermoplastic polymers and using such fibres as reinforcement for structural thermosetting resins allows for active stiffness control; the thermoplastic coating can be softened when heating it to temperatures above its glass transition temperature, which has to be below that of the composite matrix allowing the fibres to slide through the matrix when the composite is loaded in bending resulting in large deformations. Furthermore, modifying carbon fibres with a carbon aerogel not only allows for the development of functional structural supercapcitors but also improves the mechanical properties of structural polymer electrolytes.

Acknowledgements

This work was partially funded under the UK Engineering and Physical Sciences Research Council (EPSRC) Programme Grant EP/I02946X/1 on High Performance Ductile Composite Technology in collaboration with University of Bristol, and by Defence Science and Technology Laboratory under MAST 2.2.2, and an MBDA EPSRC CASE Award. Supporting data can be requested from Prof Bismarck, access to supporting data will be granted subject to retrospective consent being requested and granted from the original project participants.

References

[1] Z Lin and VC Li, 1997 Crack bridging in fibre reinforced cementitious composites with slip-hardening interfaces. *J. Mech. Phys. Solids* **45** pp. 763-787.
[2] H Humburg, D Zhu, S Bezina and F Barthelat, 2012 Bio-inspired tapered fibres for composites with superior toughness. *Comp. Sci. Tech.* **72** pp. 1012-1019
[3] IJ Beyerlein, YT Zhu and S Mahesh 2001 On the influence of fiber shape in bone-shaped short-fibre composites. *Comp. Sci. Tech.* **61** pp. 1341-1357.
[4] YT Zhu, JA Valdez, N Shi, ML Lovatom, MG Stout, SJ Zhou, DP Butt, WR Blumenthal and TC Lowe 1998 A composite reinforced with bone-shaped short-fibres. *Scripta Materialia* 38 pp 1321-1325.
[5] RC Wetherhold and FK Lee 2001 Shaped ductile fibres to improve the toughness of epoxy-matrix composites. *Comp. Sci. Tech.* **61** pp. 517-530.
[6] RM Bagwell, JF McManaman and RC Wetherhold 2006 Short shaped copper fibres in epoxy matrix: Their role in a multifunctional composite. *Comp. Sci. Tech.* 66 pp. 522-530
[7] CF Cheng, YC Tsui and T.W. Clyne, 1998 Application of a three-dimensional heat flow model to treat laser drilling of carbon fibre composites. *Acta Mater.* **46** pp. 4273-4285.
[8] L Romoli, F Fischer and R. Kling, 2012 A study on UV laser drilling of PEEK reinforced with carbon fibres. *Optics and Lasers in Engineering* **50** pp. 449-457.
[9] RA Ford 2002 Thermo-ductile composites: new materials for 21st century manufacturing - micro-perforation symmetry. *Materials and Design*, **23** pp. 431-439.
[10] JJ Blaker, DA Anthony, G Tang, SR Shamsuddin, G Kalinka, M Wienrich, A Abdolvand, MSP Shaffer and A Bismarck, A. 2016 Property and shape modulation of carbon fibres using lasers. *ACS Appl. Mater. Interfaces* DOI: 10.1021/acsami.6b05228.
[11] CF Cheng, YC Tsui and TW Clyne, 1998 Application of a three-dimensional heat flow model to treat laser drilling of carbon fibre composites. *Acta Mater.* **46** pp. 4273-4285.
[12] L Christodoulou and J Venables, Multifunctional material systems: The first generation. 2003 *J. Min. Metals Mater. Soc.* **55** pp. 39-45.
[13] C Thill, J Etches, I Bond, K Potter and P Weaver 2008 Morphing skins. *The Aeronautics Journal,* **112** pp. 117-39.

IOP Conf. Series: Materials Science and Engineering **139** (2016) 012001 doi:10.1088/1757-899X/139/1/012001

[14] IK Kuder, AF Arrieta, WE Raither and P Ermanni 2013 Variable stiffness materials and structural concepts for morphing applications. *Progress in Aerospace Sciences,* **63** pp. 33-55.

[15] AYN Sofla, SA Meguid, KT Tan and WK Yeo 2010 Shape morphing of aircraft wing: Status and challenges. *Materials and Design,* **31** pp. 1284-92.

[16] W Raither, A Bergamini, F Gandhi and P Ermanni 2012 Adaptive bending-twist coupling in laminated composite plates by controllable shear stress transfer. *Composites Part A-Applied Science and Manufacturing* **43** pp. 1709-16.

[17] JK Kim and YW Mai 1998 Engineered Interfaces in Fiber Reinforced Composites, Amsterdam: Elsevier.

[18] L Mascia, J Dhillon and JF Harper 1993 The influence of fibre/matrix interlayers based on silane functionalised EPDM and UHMWPE on the mechanical properties of glass fibre/epoxy composites *Plast. Rubber Comp. Process. Appl.* 19 pp. 22

[19] C Tridech, HA Maples, P Robinson and A Bismarck 2013 High performance composites with active stiffness control. *Applied Materials and Interfaces* **5** 9111-19.

[20] C Tridech. Smart Fibre Coatings: Stiffness Control in Composite Structures. 2011 Ph.D. Thesis, Imperial College London.

[21] N Shirshova, H Qian, MSP Shaffer, JHG Steinke, ES Greenhalgh, PT Curtis, A Kucernak, A Bismarck 2012 Structural composite supercapacitors. *Composites Part A: Applied Science and Manufacturing* **46** pp. 96-107.

[22] RF Gibson, 2010 A review of recent research on mechanics of multifunctional composite materials and structures. *Compos. Struct.* **92** pp. 2793-2810.

[23] H Qian, A Kucernak, ES Greenhalgh, A Bismarck and MSP Shaffer 2013 Multifunctional structural supercapacitor composites based on carbon aerogel modified high performance carbon fibre fabric. *ACS Appl. Mater. Interfaces* **5** pp. 6113–6122.

[24] N Shirshova, H Qian, M Houlle, JHG Steinke, ARJ Kucernak, QPV Fontana, ES Greenhalgh, A Bismarck and MSP Shaffer 2014 Multifunctional structural energy storage composite supercapacitors. *Faraday Discuss* **172** pp. 81-103.

[25] H Qian, S Nguyen, DB Anthony, A Singh, S Xu, ES Greenhalgh, A Bismarck and MSP Shaffer 2015 Stiff monolithic aerogel matrices for structural fibre composites. *20th International Conference on Composite Materials Proceedings* Copenhagen 5212-4

Causes and remedies for porosity in composite manufacturing

G Fernlund, J Wells, L Fahrang, J Kay and A Poursartip

Department of Materials Engineering, The University of British Columbia, 309-6350 Stores Rd., Vancouver, BC, Canada, V6T 1Z4

E-mail: goran.fernlund@ubc.ca

Abstract. Porosity is a challenge in virtually all composite processes but in particular in low pressure processes such as out of autoclave processing of prepregs, where the maximum pressure is one atmosphere. This paper discusses the physics behind important transport phenomena that control porosity and how we can use our understanding of the underlying science to develop strategies to achieve low porosity for these materials and processes in an industrial setting. A three step approach is outlined that addresses and discusses: gas evacuation of trapped air, volatiles and off-gassing, and resin infiltration of evacuated void space.

1. Introduction

Our inability to foresee the technical challenges associated with producing large integrated composite structures within tight engineering tolerances, such as for the F 35, the Boeing 787 and many other recent development programs, has contributed to schedule delays and cost overruns [1]. One reason why we are poor at product and process scaling in composite manufacturing is that industrial composite manufacturing is largely based on heuristics and not on science. We have developed processes that work based on experience and trial-and-error, but we don't fully understand how and why they work, which makes it challenging to scale them up in size and in production volume.

Most composite parts have multiple engineering requirements they must meet in order to be put into service. These often include a maximum porosity level, geometric dimensions, fibre alignment and other process outcomes. Requirements typically come in the form of engineering specifications that give the allowable range within which a component is deemed satisfactory. When making multiple nominally identical parts, the process outcome will be variable due to variability in incoming materials and in process conditions. Components that are outside engineering tolerances are deemed defective and have to be scrapped, or re-worked so that they meet requirements. Figure 1 illustrates this concept with two processes, process A and process B, where process A has a randomly variable outcome that is largely within engineering tolerances and produce acceptable parts, whereas process B has a variable outcome that does not meet engineering tolerances and produces defective parts.

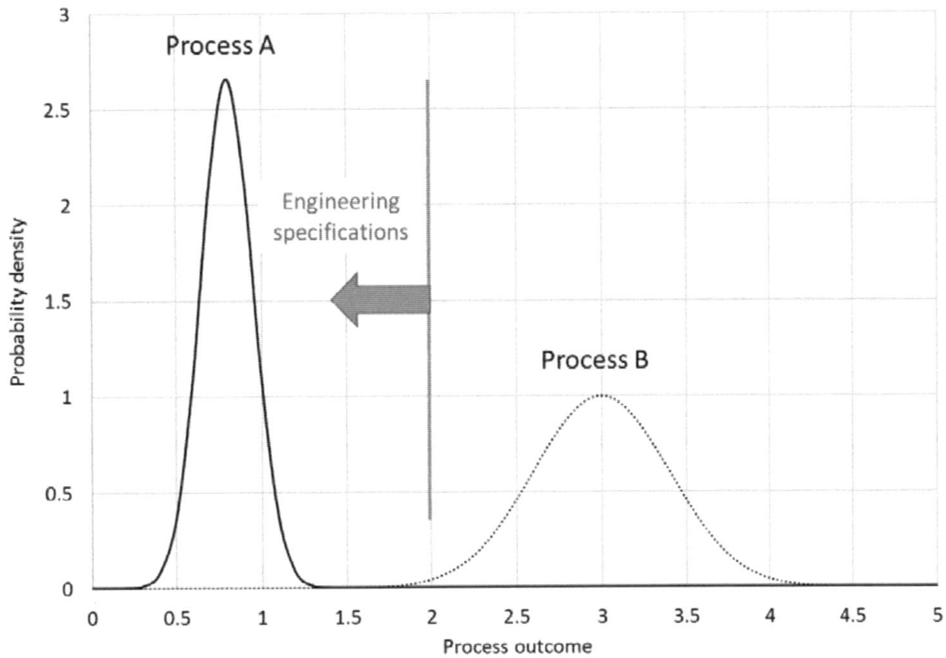

Figure 1. Examples of variable process outcomes (e.g. % porosity) for two processes, A and B. In this case the engineering specifications require a porosity level below 2%.

Although Figure 1 is highly simplified it illustrates two important concepts. To get a process to produce parts that conform to engineering specifications (such as process A), two things needs to be addressed. First, the process needs to be designed or adjusted so that the mean value of the outcome is centered within engineering specifications. Second, the process needs to be controlled so that the variability in the outcome does not take it outside specifications. This means that to produce high quality components that are within tight engineering tolerances we need to understand the physics behind how process outcomes depend on process parameters, and in particular how we can design or adjust our processes so that both the mean outcome and the variability is acceptable.

1.1 Process-induced defects

There are three process outcomes that regularly cause composite fabricators problems and often lead to defective parts: porosity or voids within the composite material, wrinkling or misalignment of fibres in the components, and process induced shape distortions. The outcome of a process can be viewed from a system's perspective, Figure 2, which illustrates a composite factory with equipment, tooling & consumables, a part to be made, and materials & processes [2]. All these objects will together determine the outcome of the process. In addition to ensuring that the manufacturing system is able to make parts within engineering tolerances, i.e. process A in Figure 1, we need to develop and design robust processes that can meet engineering tolerances with ease. Thus, we need to aim for high producibility and design processes that are insensitive to typical variations in process parameters.

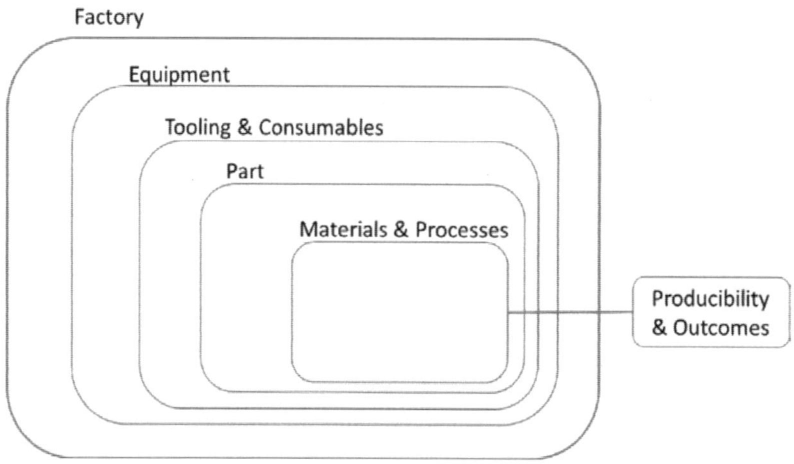

Figure 2. A systems view of process outcomes and producibility. Adapted from [2].

2. Porosity

Porosity is defined as the volume fraction of (small) voids in a material. In a composite, a void is space that is not occupied with resin or fibre. When individual voids get large, they are no longer treated at a continuum level but as discrete objects with a size, shape and specific location within the material. There are many causes for porosity in composite parts and it is very difficult to achieve zero porosity. A good approach to porosity management is to view it as a balance of void sources and void sinks, and strive to design processes where the sinks dominate the sources [3]. Void sources are features or mechanisms that increase porosity such as trapped air, volatiles and off-gassing, tool and bag leaks. Void sinks are features or mechanisms that reduce porosity, such as vacuum evacuation, elevated resin pressure, and bubble mobility. The relative importance of different sources and sinks are dependent on the process and the materials used. The focus of this paper is on out of autoclave prepregs which are challenging to process with low porosity levels due to the low pressures applied.

Out of autoclave prepregs are generally partially impregnated with resin in order to create an interconnected porous network for gas removal [4]. Figure 3 shows photos of a carbon/epoxy MTM 45-1 5HS (five harness satin fabric) OOA prepreg with a resin poor and a resin rich side. MTM45-1 is designed for high performance applications and flexible curing temperatures. It can be cured at temperatures as low as 80°C to allow use of inexpensive tooling with a 180°C post cure heat treatment to maximize mechanical properties [5]. The fibre mat consists of 6K IM7 carbon fibre tows measuring 2 mm in width woven in a 5-harness satin pattern (5 tows per crimp). When layers of partially impregnated prepreg are laid up on top of each other a complex porous gas network that facilitates gas transport is formed. Figure 4 shows a micrograph of an un-debulked and uncured MTM 45-1 5HS laminate that illustrates the nature of the complex porous network with interconnected voids [6].

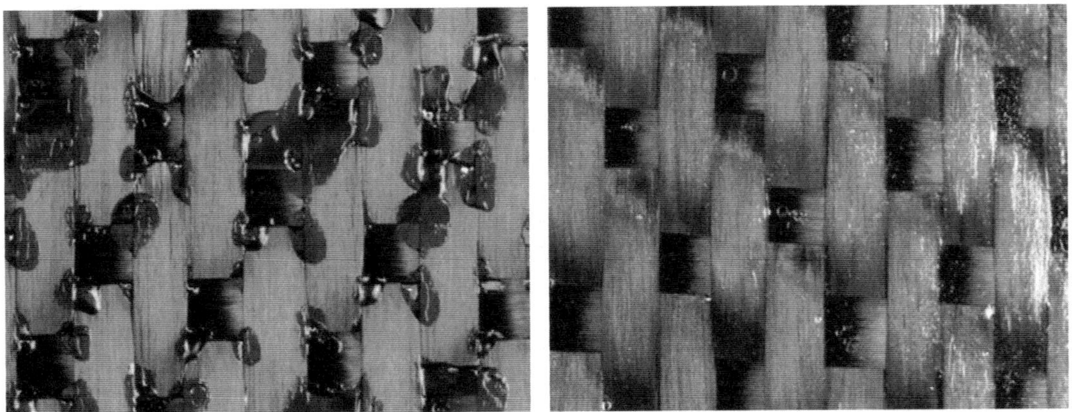

Figure 3. Top view of uncured MTM45-1/CF2426A 5HS prepreg. Resin poor side (left) and resin rich side (right) [6].

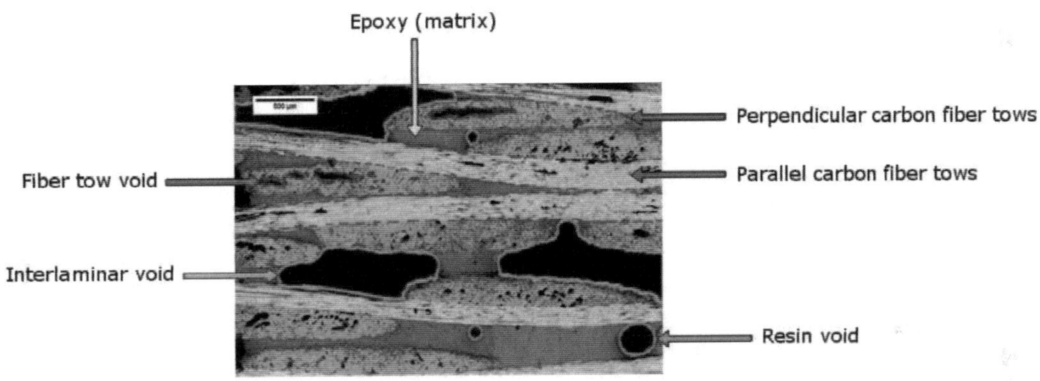

Figure 4. Cross-section of an un-debulked and uncured MTM 45-1 5HS prepreg laminate showing different types of voids [7].

Figure 4 shows three distinct types of voids: voids within partially impregnated fibre tows (red), interlaminar voids between plies (green), and voids fully enclosed by resin (blue). The first two types of voids make up the interconnected porous network and image analysis has shown that the porosity of the inter connected porous network for this prepreg can be as high as 30% before vacuum is applied to debulk the material [7]. When vacuum is applied to the laid up laminate, the laminate compacts but the viscosity of the resin is high enough so that the porous network remains relatively intact during a room temperature debulk and thereby facilitating vacuum evacuation of trapped air. The amount of inter-connected porosity in an uncured laminate of this material that has been under vacuum at room temperature for several hours was measured to be 15-20% [7]. For primary aerospace structures there is often a requirement that the porosity in finished parts is less than 2% and if we start with close to 30% porosity we need to ensure that the available void sinks are powerful enough to overcome this large initial porosity. A simple three step approach to achieve low porosity in OOA prepreg processing is as follows:

1. *Vacuum evacuation of trapped air*. Conduct a sufficiently long room temperature vacuum debulk to remove most of the trapped air (and perhaps some dissolved moisture).

2. *Keep volatiles in solution.* Heat the laminate up to cure it, but keep the resin pressure sufficiently high to keep any dissolved volatiles in solution until the resin gels.
3. *Resin infiltration.* Ensure that there is sufficient time (and resin pressure) for the resin to infiltrate any evacuated voids space between plies and within fibre tows.

The challenge is to turn these three qualitative steps into a robust scalable process recipe that reliably and repeatedly can produce low porosity parts with different sizes and characteristics. The following section will address the physics and timescales associated with advective removal of gas from a porous media by use of a vacuum system, which is the primary porosity sink for trapped air. The presented theory applies to partially impregnated prepregs, such as out of autoclave prepregs, as well as breathers and other porous materials in prepreg processing.

2.1 Vacuum evacuation of trapped air

Consider one-dimensional gas transport through a porous gas permeable medium, Figure 5. The medium can either be a partially impregnated prepreg, a breather, or any other medium that allows for gas transport in composite processing. We assume that there are gas channels, or an inter-connected "vascular network", that allows gas to permeate though the medium and the permeability and volume of the network are constant during the evacuation process.

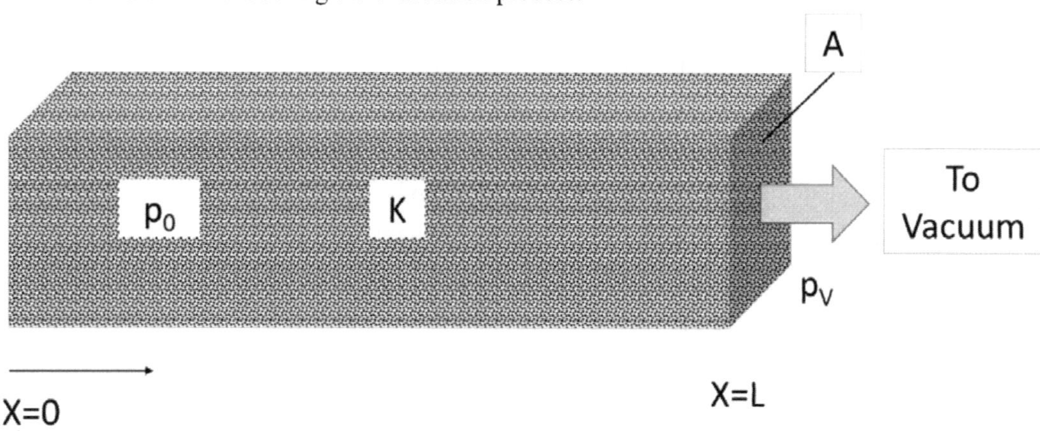

Figure 5. Schematic of the 1D gas transport problem with flow through a porous media.

If we assume that the gas is an ideal gas and that the flow follows Darcy's law, the governing differential equation in terms of gas pressure p can be written

$$\phi \frac{\partial p}{\partial t} - \frac{K}{\mu} \cdot \frac{\partial}{\partial x}\left(p \cdot \frac{\partial p}{\partial x} \right) = 0 \tag{1}$$

where ϕ = porosity, p= pressure [Pa], t = time [s], x = distance [m], K = gas permeability [m^2], μ= dynamic viscosity [Pa·s]. To solve eq. (1) for a specific case we can non-dimensionalize the equation with respect to distance, pressure and time as

$$\frac{\partial p^*}{\partial \tau} - \frac{\partial}{\partial \xi}\left(p^* \cdot \frac{\partial p^*}{\partial \xi}\right) = 0; \qquad \textit{where}$$
$$\xi = \frac{x}{L}; \qquad p^* = \frac{p}{p_0}; \qquad \tau = \frac{Kp_0}{\phi \mu L^2} t \tag{2}$$

ξ is non-dimensional distance, p^* is non-dimensional pressure, and τ is non-dimensional time. The non-dimensional factors in equation (2) give some interesting scaling laws for gas evacuation. For example, the time t it takes to reach a new pressure state is proportional to the viscosity of the gas, μ, the porosity ϕ, the length of the laminate squared, L^2, and is inversely proportional to the initial pressure p_0 and the permeability of the medium K. Thus a laminate with twice the permeability will evacuate twice as fast, and a laminate with twice the distance to the vacuum system will take four times longer to evacuate.

We can now analyze one-dimensional gas evacuation of a laminate. The whole laminate is initially at pressure p_0 (ambient pressure), and vacuum pressure p_V is applied at x=L at time t=0 (Figure 5). The volume of gas in the porous laminate is assumed fixed and the vacuum applied at x=L evacuates the gas from the laminate via Darcy flow. In this case, eq. (2) needs to be solved for the following boundary condition

$$\left.\frac{\partial p^*}{\partial \xi}\right|_{\xi=0} = 0, \quad p^*(1,\tau) = p_V \tag{3}$$

Equation (2) with boundary conditions (3) has no known analytic solution and was solved numerically using the Mathematica software. The numerical solution was used to develop an approximate general expression for the amount of gas removed as a function of time during gas evacuation. By integrating the pressure distribution along the length of the laminate at different times, the mole fraction of gas remaining in the porous media as a function of time was calculated and the following simple relationship for the mole fraction of gas remaining n/n_0 as a function of dimensionless time τ is obtained

$$\frac{n}{n_0} \approx e^{-0.9\tau^{0.6}} \tag{4}$$

Equation (4) is plotted in Figure 6 to illustrate the relationship between mole fraction of gas remaining and dimensionless time τ. The main reason for the strong exponential decay in eq. (4) is that the velocity of the gas is proportional to the pressure gradient in Darcy's law and the more gas that is removed, the smaller that driving force becomes. Hence, it is not practical to strive to remove all the gas from the system but we need to determine what is sufficient to reach a specific porosity target.

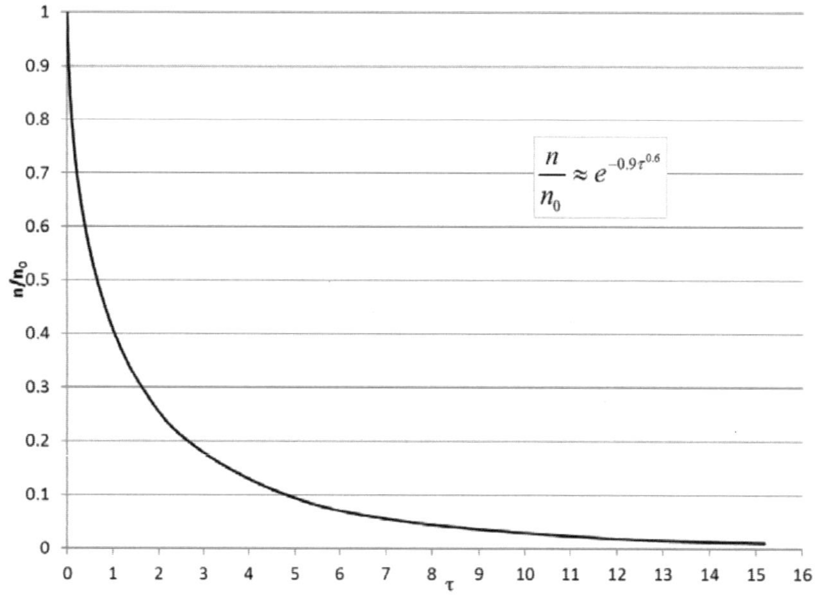

Figure 6. Mole fraction gas remaining in the porous media n/n_0 as a function of dimensionless time τ.

By determining the dimensionless time τ it takes to reach a specific remaining gas fraction n/n_0 from eq. (4), the actual time t can be calculated based on the length L, permeability K, initial pressure p_0, interconnected porosity ϕ and gas viscosity μ of the gas using the relationship between t and τ in eq. (2).

$$t = \frac{\mu}{p_0}\frac{\phi L^2}{K}\left[-\frac{1}{0.9}\ln\left(\frac{n}{n_o}\right)\right]^{\frac{1}{0.6}} \tag{5}$$

To determine how much gas we need to remove during a room temperature debulk of an out of autoclave prepreg to obtain a part with acceptable porosity we can make the following argument. During a room temperature debulk of an out of autoclave prepreg, the porosity level is approximately constant as the resin is designed not to flow at room temperature to keep the breathing network open [4]. Therefore, during debulk, the number of moles of gas in the porous network decreases but the volume is approximately constant. After the room temperature debulk, the prepreg is heated up and the resin flows into the partially evacuated porous network until pressure equilibrium between the resin and any remaining gas is established. Thus if we want a final porosity level ϕ_F and begin with a porosity level ϕ and assume that the resin pressure remains constant during the process, the ideal gas law gives the following condition of how much gas we need to remove

$$\phi\frac{n}{n_o} = \phi_F \tag{6}$$

By combining eqs. (5) and (6) we get the following approximate expression for the minimum required debulk time t^* to reach a specific porosity level ϕ_F for a prepreg with an initial porosity ϕ.

$$t^* = \frac{\mu}{p_0}\frac{\phi}{K}\left[-\frac{1}{0.9}\ln\left(\frac{\phi_F}{\phi}\right)\right]^{\frac{1}{0.6}} L^2 \qquad (7)$$

Gas permeability testing has been performed on the MTM 45-1 5HS prepreg to determine the gas permeability during debulk [7]. The initial porosity during debulk has also been measured [7] and the other properties in eq. (7) are readily known. Table 1 show a summary of the values used to calculate the minimum required debulk time for the MTM 45-1 5HS prepreg.

Table 1. Properties and values used for calculation of the minimum required debulk time t*, eq. (7) for the MTM 45-1 5HS prepreg [7]

Property:	μ (Pa*s)	p_0 (Pa)	K (m^2)	ϕ (–)
Value:	$1.82*10^{-5}$	101,000	$6*10^{-14}$	0.15

Figure 7 shows a plot of the calculated minimum debulk time t^* as a function of evacuation length L. Note that the evacuation length L is typically half the total laminate length as we generally evacuate from all sides and the problem solved is one-sided.

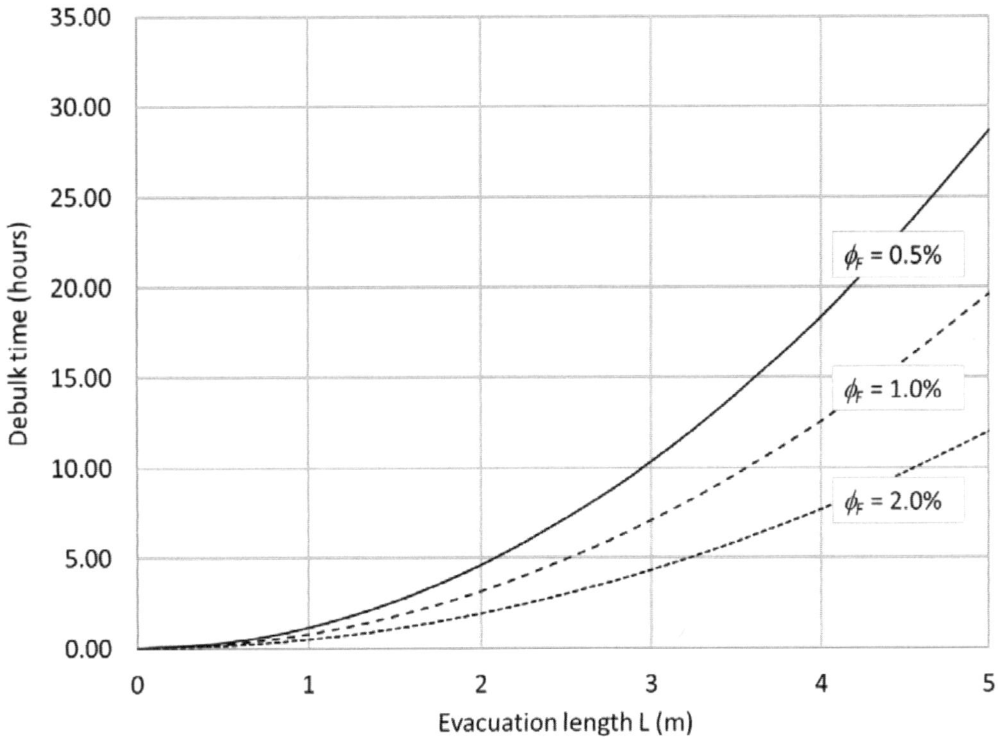

Figure 7. Minimum required room temperature debulk time to achieve different final porosity levels ϕ_F with a perfect vacuum $p_V = 0$.

Figure 7 shows that for long laminates the minimum debulk times get large, which is in accordance with the prepreg manufacturers recommendation of an approximately 20 hour debulk for large laminates. It also shows explicitly how the gas evacuation process scales with size and with other process parameters, which is very important when attempting to scale a process.

It is known empirically that a good (hard) vacuum is required to achieve low porosity and high laminate quality in OOA prepreg processing. The presented analysis clearly shows why. In the analysis above it was assumed that a perfect vacuum $p_V = 0$ was applied at the edge of the laminate during the debulk phase. If the vacuum level is not perfect, $p_V > 0$, we can reformulate eq. (4) as follows

$$\frac{n}{n_0} \approx e^{-0.9\tau^{0.6}} + \frac{p_V}{p_0} \tag{8}$$

If $p_V > 0$ then the mole fraction gas after infinite debulk time would equal that expected from the ideal gas law with a porous network at uniform pressure p_V. Using eq. (8) we can develop an expression for the minimum required debulk time t^* in the case of imperfect vacuum, $p_V > 0$

$$t^* = \frac{\mu}{p_0}\frac{\phi}{K}\left[-\frac{1}{0.9}\ln\left(\frac{\phi_F}{\phi} - \frac{p_V}{p_0}\right)\right]^{\frac{1}{0.6}} L^2 \tag{9}$$

From eq. (9) we see that there is a limit on the maximum absolute vacuum pressure p_V allowed to achieve a fixed final porosity level ϕ_F, even after infinite debulk time

$$p_V < p_0 \frac{\phi_F}{\phi} \tag{10}$$

From the data in Table 1, this gives that $p_V < 0.067$ atm to have a chance to achieve a final porosity of less than 1% for the current prepreg material. With $0 < p_V < 0.067$ atm, 1% final porosity is possible, but the increased absolute vacuum pressure will increase the required debulk time. Eq. (9) is plotted in Figure 8 for a few different vacuum levels p_V to illustrate the effect imperfect vacuum has on the minimum required debulk time t^*.

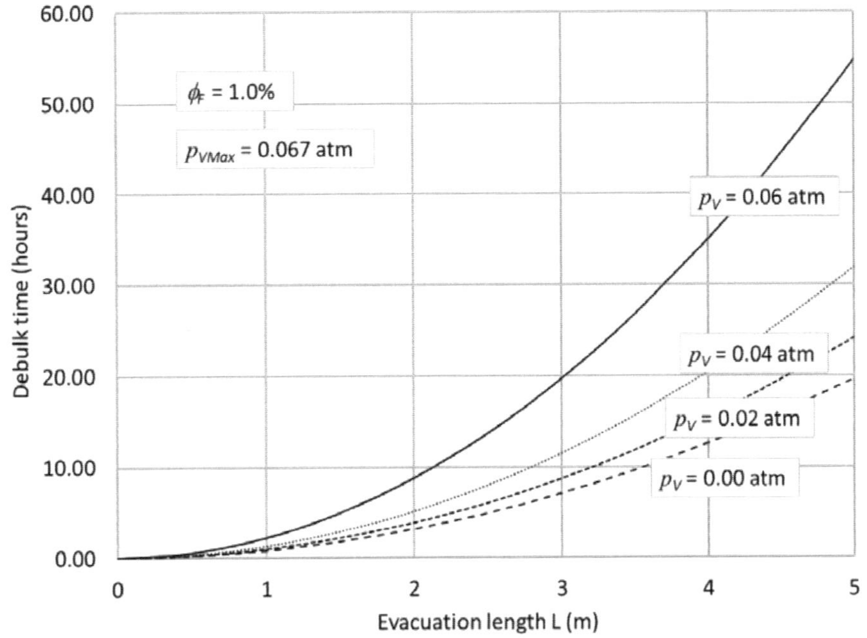

Figure 9. Effect of imperfect vacuum on the minimum required debulk time to achieve 1% porosity for a MTM 45-1 5HS prepreg laminate.

Figure 9 shows how sensitive effective vacuum evacuation is to imperfect vacuum and therefore producibility and the robustness of this process step. The presented analysis gave quantitative guidelines for the first step in the process to achieve low porosity, vacuum evacuation. The second step is to keep any dissolved volatiles in solution while heating the material to cure it as to not generate gases that may cause porosity.

2.2 Keep volatiles in solution

Once sufficient amount of trapped gas has been removed before cure commences we want to avoid that more gas in produced by volatiles and off-gassing before the resin gels and the void morphology is locked it. Up to the point of gelation, when the resin transforms from a liquid to a solid material, voids can be generated, expanded or shrunk because of the fluidity of the resin. After gelation, the void geometry is essentially fixed. Previous generations of prepreg were often solvent impregnated and contained dissolved volatiles in the resin. Modern OOA prepregs are manufactured using either the hot melt or resin filming process, which do not require the use of solvents and therefore the amount of dissolved volatile species in the resin is significantly reduced [8]. However, epoxy resin is hygroscopic and will absorb moisture from the environment and volatiles may be generated during the cure process. Moisture can vapourize from the resin causing existing voids to grow resulting in porosity in the final part. Prepreg exposed to relative humidity (RH) above ~70% RH has been shown to develop porosity in OOA processing but not in autoclave processing [9] due to the higher applied pressure.

Kardos developed a criterion for moisture driven bubble growth which is based on the premise that moisture concentration gradients drive evaporation and bubble growth [10]. Resin that has equilibrated with surrounding moist air has shown to have a concentration of dissolved moisture in the resin C (g/cm^3) that is proportional to the square of the relative humidity RH [11]

$$C = \rho_{\text{resin}} k_1 (RH)^2 = \rho_{\text{resin}} k_1 \left(\frac{p_{H2O}}{p^*_{H2O}}\right)^2 \qquad (11)$$

ρ_{resin} is the resin density (g/cm^3), k_1 is an experimentally determined fitting parameter = 0.00558 (at room temperature), p_{H2O} is the partial pressure of moisture vapour in the surrounding humid air (atm) and p^*_{H2O} is the equilibrium partial pressure of moisture vapour (atm). The relative humidity *RH* is defined as the ratio of the partial pressure of moisture in the surrounding air divided by the equilibrium partial pressure. This relationship was found to agree very well with the measured concentration of moisture in the MTM 45-1 resin used in the current study [6]. At equilibrium, the resin at any resin/air interface will have the same moisture concentration *C* as in the bulk resin. However, if the relative humidity *RH* of the air decreases by increasing the temperature, there will be a higher concentration of moisture in the bulk of the resin than at the resin/air interface which will promote diffusion, void growth and evaporation of moisture at the resin/air interface. This is shown schematically in Figure 10.

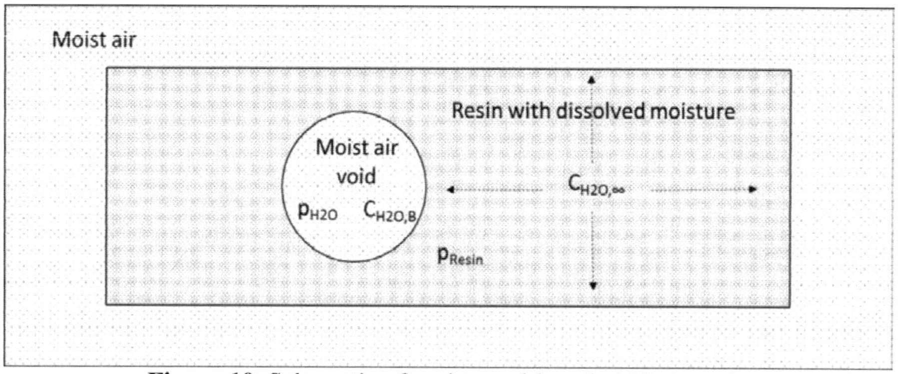

Figure 10. Schematic of moisture driven bubble growth.

Growth of the resin void will occur if bulk moisture diffuses to the resin void interface, requiring $C_{H2O,B} < C_{H2O,\infty}$, which using eq. (11) can be expressed

$$p_{H2O} < p^*_{H2O}(RH_0) \qquad (12)$$

where RH_0 is the relative humidity of the moist air that the resin was equilibrated with. the Ignoring surface tension effects, assuming that $P_{Resin} \approx p_{H2O}$ at equilibrium, and that the equilibrium vapour pressure is described by the Clausius-Clapeyron relation we get the following criterion for moisture driven bubble growth

$$P_{Resin} < P_0 \exp\left(\frac{\Delta H_{Vap}}{RT_0}\right) \exp\left(\frac{-\Delta H_{Vap}}{RT}\right)(RH_0) \qquad (13)$$

ΔH_{Vap} is the enthalpy of vapourization, P_0 and T_0 are the Clausius-Clapeyron reference pressure and temperature respectively. Inserting moisture values for the constants (Table 2) and equating the sides yields the critical resin pressure below which moisture can cause resin voids to grow by diffusion and vapourization [10].

$$P_{R,Critical} = 4962 \exp\left(\frac{-4892}{T}\right)(RH_0) \qquad (14)$$

Table 2. Values used in the Clausius-Clapeyron relation for calculating the equilibrium vapour pressure of water, eq. (13).

Variable	Value
$\Delta H_{Vap}/R$	4892 K
P_0	1 atm
T_0	373 K

In out of autoclave processing, the maximum possible resin pressure is 1 atm but the resin pressure is often slightly lower than that because some of the applied pressure is carried by the fibre bed [12]. If we set $P_{R,Critical}$ to 1 atm in eq. (14) and solve for the temperature for different relative humidity levels RH_0, we can calculate the maximum temperature we can expose the resin to, prior to gelation, before we run into problems with moisture driven bubble growth, see Figure 11.

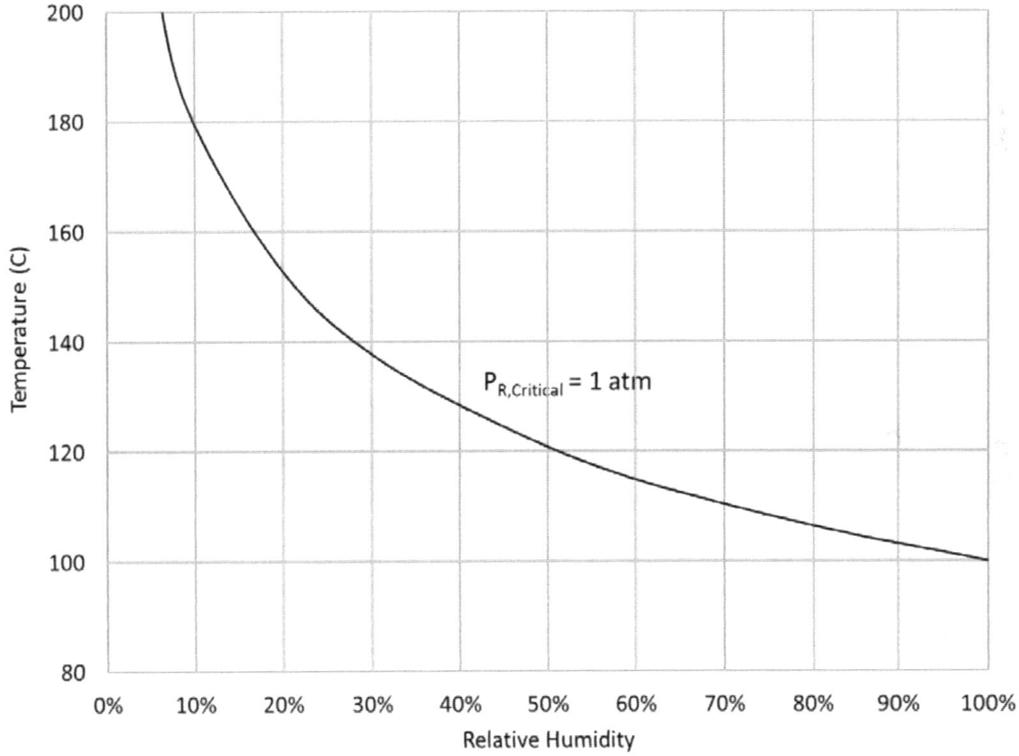

Figure 11. Temperature at which the critical resin pressure for moisture driven bubble growth equals 1 atmosphere for the MTM 45-1 resin system.

Based on Figure 11, we see that for resins equilibrated with air at a RH_0 of less than about 10% we can heat up and gel the material directly to 180°C without causing moisture driven bubble growth. For resins equilibrated at higher relative humidity levels, the figure suggests that we need to put an intermediate hold in the cure cycle, at which the material gels before it is heated up to 180°C for final cure. Using the cure kinetics for the MTM 45-1 resin we can calculate how long these holds have to be [13]. Figures 12-14 present cure cycles for different relative humidity levels that were been designed using the RAVEN composite process simulation software [14] to ensure that moisture driven bubble growth does not occur until the resin gels. Gelation for this resin is assumed to occur at a degree of cure of 50% in the presented analysis.

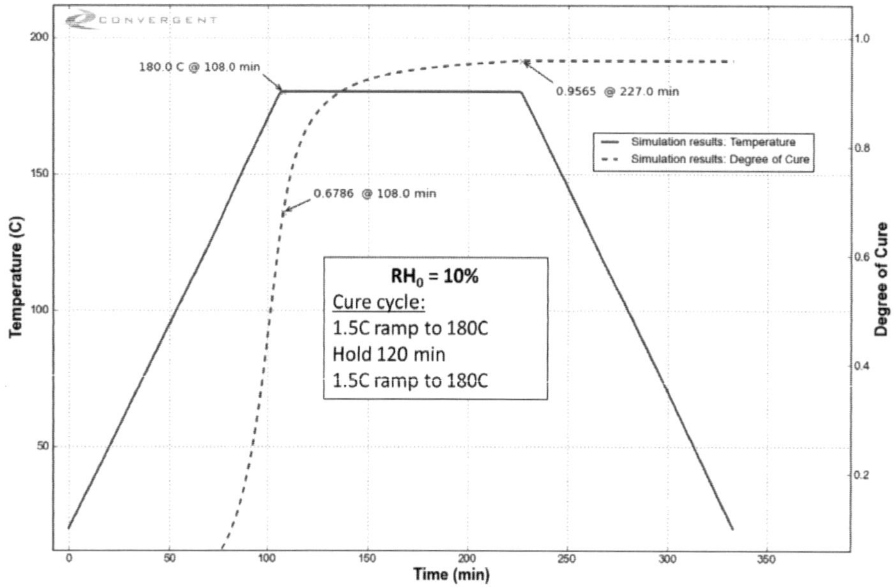

Figure 12. Default cure cycle for RH_0 = 10% or less. The critical temperature from Figure 11 is 180°C and there is no need for an intermediate hold.

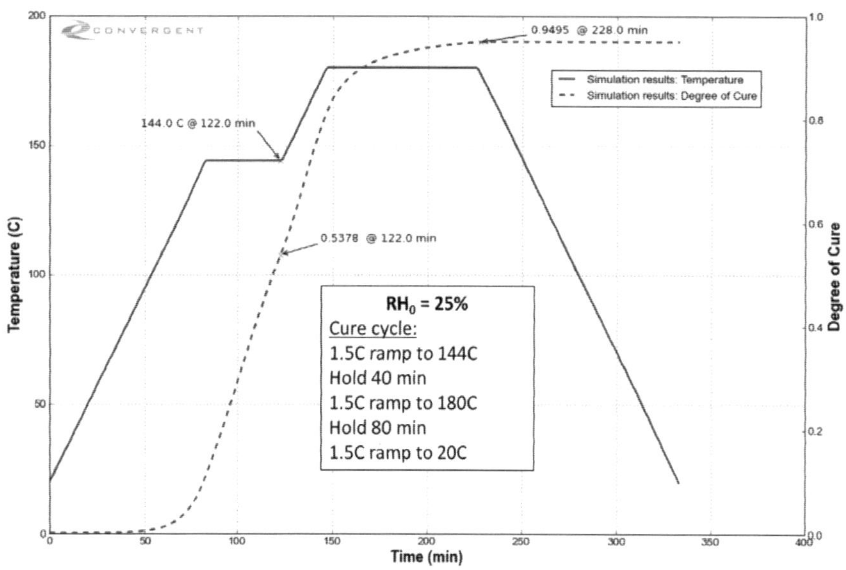

Figure 13. Cure cycle for RH_0 = 25%. The critical temperature from Figure 11 is 144°C and an intermediate hold for 40 mins at 144°C has been added to gel the resin before further heating is performed.

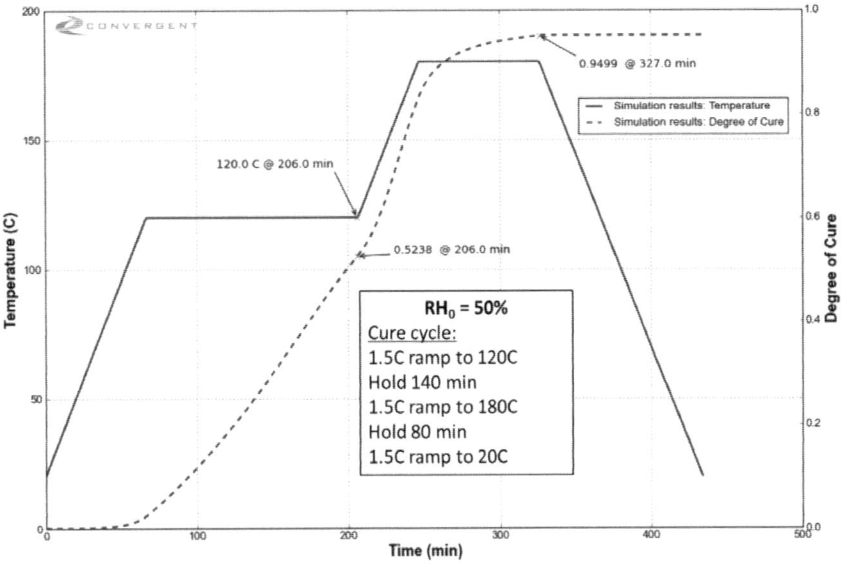

Figure 14. Cure cycle for $RH_0 = 50\%$. The ritical temperature from Figure 11 is 121°C and an intermediate at 120°C hold for 140 mins has been added to gel the resin before further heating is performed.

Figures 12-14 show cure cycles that are designed to supress moisture driven bubble growth for the MTM 45-1 based on the Kardos criterion and the assumption that the resin pressure can be maintained at 1 atmosphere during cure. In all three cases the temperature of the intermediate hold is based on the Kardos criterion, Figure 11. The length of the hold is based on the cure kinetics of the resin [13] so that the resin gels and transforms from a liquid to a solid at the hold temperature. The length of the final hold at 180°C was designed using the RAVEN software to ensure that the resin is fully cured before cool-down. The final degrees of cure for the three cure cycles are 0.96, 0.95 and 0.95, which ensures fully developed physical and mechanical properties of the resin.

2.3 Resin infiltration

Once sufficient amount of trapped gas has been removed according to the principles outlined in section 2.1 and moisture driven bubble growth has been prevented using the methods shown in section 2.2 we need to ensure that the resin pressure, resin viscosity and time are sufficient to fill any evacuated void space within fibre tows or between plies. This is typically not difficult to achieve but approximate analysis for this has been presented elsewhere [15] and will not be discussed in detail here.

3. Discussion

The proposed three step approach to achieve low porosity in OOA prepreg processing: 1. Vacuum evacuation of trapped air; 2. Keep volatiles in solution; 3. Resin infiltration of evacuated void space does not account for all void sources and sinks. It assumes that there are no additional void sources such as bag or tool leaks. It also ignores void sinks such as drying of the resin: diffusion, evaporation and moisture removal through the engineered vacuum channels during room temperature debulk and heat up; and bubble mobility: movement of bubbles through the resin and removal through the vacuum channels [16]. However, provided that there are no bag and tool leaks, the presented approach should be conservative as the ignored voids sinks will only further reduce the porosity.

Individual mechanisms for porosity formation have been addressed in the past but it is important to view this as systems problem with multiple sources and sinks, establish the physics for the individual mechanisms and link them together so that comprehensive porosity predictions on real structures and processes can be performed. By taking a comprehensive science based approach, we can improve the design of our processes and achieve better producibility.

In the introduction, the notion of a variable outcome for industrial processes, such as the level of porosity, was introduced (Fig 1). The science base presented here is deterministic, for example eq. (9) describing vacuum evacuation of trapped air and eq. (14) describing the resin pressure required to keep moisture in solution as a function of temperature and relative humidity. With appropriate material constants, these equations can guide us in determining appropriate nominal process conditions to achieve acceptable mean porosity levels. By measuring or estimating the variability in material constants and process variables we can also get a sense of the expected variability in the process outcome which is essential to achieve low re-work and scrap rates.

Figure 2 presented a system's view of a composite factory consisting of equipment, tooling & consumables, part, materials & processes. We can now relate the developed science base back to these objects so that we better understand how they affect the process outcome. For example, the parameters that govern the minimum required debulk time in eq. (9) can be related to the different factory objects. Equipment: vacuum pressure p_V; Tooling & consumables: presence of vacuum leaks; Part: length L; Materials & Processes: initial porosity ϕ, permeability K, initial pressure p_0, air viscosity μ. By establishing these relationships, we can make better decisions regarding the best corrective actions for a process that produces parts outside engineering specifications (e.g. process B in Fig. 1). For example, we can quantify the return on investment of a high performing vacuum system (lower p_V) on the required minimum debulk time and expected porosity level.

Acknowledgments
We would like to thank the Natural Sciences and Engineering Research Council of Canada (NSERC) for their financial support of this research. We would also like to acknowledge current and past colleagues at The University of British Columbia as well as our industrial collaborators, particularly Michael Thompson at Boeing Research and Technology for sharing his vast knowledge in this area.

References

[1] National Defense Industrial Association 2011 21st Century Manufacturing Modeling & Simulation Research and Investment Needs.
[2] Fabris J, Fernlund G, Poursartip A 2016. Introducing a knowledge framework to integrate composites manufacturing science and industrial practice. In preparation.
[3] Mobuchon C, Keulen C, Hsiao K, Fernlund G, Poursartip A 2014. Development of a structured approach to porosity management in composites manufacturing. *SAMPE Technical Conference*, Seattle, US.
[4] Ridgard C 2009. Out of Autoclave Composite Technology for Aerospace, Defense and Space Structures. *SAMPE Technical Conference*, Baltimore, US.
[5] MTM45-1 Matrix Resin Product Data Sheet, Advanced Composites Group (ACG Inc.), Vol. PDS1104/09.06/4.
[6] Wells J 2016. Behaviour of resin voids in out-of-autoclave prepreg processing. MASc Thesis The University of British Columbia, Canada.
[7] Farhang L 2014. Void Evolution during Processing of Out-of-Autoclave Prepreg Laminates. Ph.D. Thesis The University of British Columbia, Canada.
[8] Campbell FC. 2004. Manufacturing Processes for Advanced Composites. Elsevier, New York.
[9] Grunenfelder LK, Nut SR 2010. Void Formation in Composite Prepregs – Effect of Dissolved Moisture. *Composites Science and Technology*, **70**, 2304-2309.

[10] Kardos JL 2000. Void Growth and Dissolution, in Processing of Composites, Edited by R.S. Dave and A.C. Loos, Hanser Publishers, Munich, 182-206.

[11] Brand RA, Brown GG, McKague EL 1984. Processing Science of Epoxy Resin Composites. Materials Laboratory, Air Force Wright Aeronautical Laboratories.

[12] Roy M, Kay J, Fernlund G, Poursartip A 2015. Porosity in Configured Structures. *SAMPE Technical Conference*, Baltimore, US.

[13] Shahkarami A. VanEee D, Poursartip A 2009. Material Characterization for Processing: ACG MTM45-1 Material Model Development. NCAMP Final Project Wrap-Up.

[14] Convergent Manufacturing Technologies. www.convergent.ca

[15] Centea T, Hubert P 2012. Modelling the effect of material properties and process parameters on tow impregnation in out-of-autoclave prepregs. *Composites Part A* **43** 9 1505–1513

[16] Gangloff Jr, JJ, Hwang WR, Advani SG 2014. Characterization of Bubble Mobility in Channel Flow with Fibrous Porous Media Walls. *International Journal of Multiphase Flow* **60** 76-86.

Moisture induced softening and swelling of natural cellulose fibres in composite applications

E K Gamstedt

Uppsala University, Department of Engineering Sciences,
Box 534, SE-75121 Uppsala, Sweden

E-mail: kristofer.gamstedt@angstrom.uu.se

Abstract. Composites based on natural cellulose fibres are susceptible to moisture. The fibres as well as the composite will inevitably soften and swell as moisture is absorbed. The intention of the present paper is to shed some light on the mechanisms behind softening and swelling. Also references to modelling work are made, to predict the moisture-induced dimensional stability. Characterisation techniques and models of such kind can be useful in choosing suitable fibres for improved moisture resistance, and identifying the main controlling parameters which affect the engineering consequences of moisture absorption. Understanding of the mechanisms and the main contributions to swelling can rationalise materials development. The examples shown in this review attempt to show the benefits by experimental mechanics and modelling in development of moisture resistant cellulose composites.

1. Introduction

1.1. Background

The use of natural cellulose fibres in composite materials has increased substantially during the last few decades. High-end applications with strict demands on mechanical performance typically advocate glass or carbon fibres. For high-volume applications, where low cost is more important than the mechanical performance, cellulose fibres can serve as an alternative to e.g. glass fibres. The main drawback of cellulose fibres is their affinity to moisture [1], which limits the outdoor applications, at least in climates which are not consistently dry [2]. In a short-time perspective, the moisture sensitivity of cellulose is caused by the abundance of accessible hydrogen bonds between hydroxyl end groups [3]. Of course, the moisture sensitivity can be reduced by chemical modification of the cellulose fibres, immobilizing and preventing the hydroxyl groups to adsorb moisture excessively [4, 5]. Any such chemical modification, is however going to lead to higher costs, and reducing the green features of this renewable and biodegradable raw material [6]. Many studies have focused on different strategies to chemically modify cellulose and characterising moisture absorption, swelling or reduction in tensile properties [7-9]. Such empirical studies are of course necessary in the quest to develop cost-effective and environmentally friendly alternatives to mitigate the disadvantages caused by the presence of moisture. In parallel to such empirical studies, it is also of interest to analyse the moisture absorption from a mechanics point of view. Suitable models can then be derived to link the moisture induced-deformation on the fibres to that of the composite [10]. The transfer of the stiffness drop due to moisture absorption from fibre to composite can also be studied [11, 12]. The purpose of the present article is to show how micromechanics can prove itself useful in linking effects of moisture on the fibre level to the composite level. The main engineering properties in this regard is stiffness and dimensional stability. Swelling and elastic properties are amenable to micromechanical modelling, which is typically more complicated for e.g. strength properties, controlled by local inhomegeneities. The advantage of micromechanics and correlations with experiments is that simulations can be made how to constrain swelling and softening phenomena by e.g. changing the microstructure and composition of the composite. Such exercises can provide some advice in how cellulose composite with improved moisture resistance can be made. All models should, if possible, be based on observed physical mechanisms. Although strength is presently not the main focus, figure 1 below shows how experimental observations can be used to identify fibre pull-out and slippage at the fibre-matrix interface as key mechanism in the reduced strength of a wood-fibre composite.

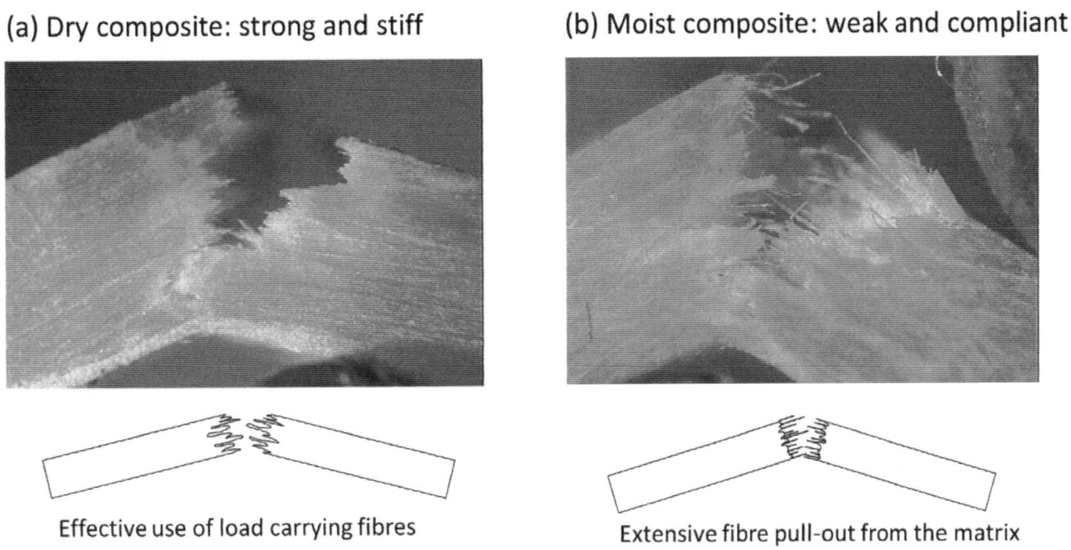

Figure 1. Appearance of fracture in a dry or moist wood-fibre reinforced composite [13]: (a) tortuous crack growth around fibres in the dry composite, and (b) effortless pull-out of fibres from the matrix, indicative of a weakened fibre-matrix interface of moist composite.

1.2. Moisture transport

In regard of effects of moisture on cellulose composites, we will focus only softening and swelling within suitable working climates. For extended time, higher temperatures and consistently high ambient relative humidity, mould formation is a limiting factor for the use of cellulose-based materials. For these limiting cases, the reader is referred to other resources [14-15]. Furthermore, the transient regimes are not considered here. Only steady-state equilibrium conditions are taken into account. For relatively thin composite components and slow changes of the surrounding relatively humidity, the diffusions processes will be fast compared with the cycles of moisture content. In e.g packaging and furniture applications, the component dimensions and time scales are thought to be such that equilibrium conditions can be assumed for most part of the lifetime. For bulkier components or faster changes in surrounding climate, it is worthwhile to investigate the transport phenomena. If the moisture diffusion is slow, it is possible that there is not enough time for the material to absorb a sufficient amount of moisture to result in any noticeable degradation in stiffness or dimensional changes. Assuming Fickean diffusion, it is quite straightforward to characterise the diffusivity of composites with time-lapse weight measurements [16-18]. A particularity of composite diffusion is that diffusion is inherently anisotropic [19, 20], with higher in-plane diffusion rates where the fibres (typically the most hydrophilic phase) are normally oriented [21]. In practice, diffusion is typically characterised only in the out-of-plane direction for relatively thin composites plates with sealed edges to promote essentially one-dimensional moisture transport. However, in composite components in service, only one-dimensional through-thickness diffusion cannot be assumed. It is likely that exposed fibrous edges can accelerate the rate of moisture uptake. In general, an impermeable coating and sealed edges, can slow down the moisture uptake and prevent undesired softening and swelling [22, 23].

1.3. Vapour sorption

Diffusion processes show non-linear behaviour in the sense that the diffusivity depends on the local moisture content [24] and the two-phase transport of free and bound moisture, respectively [25]. Even under equilibrium conditions, non-linear behaviour presents itself in terms of a non-linear relationship between ambient relative humidity and the moisture content of the material. An example for a wood species is shown in figure 2. Dynamic-vapour sorption isotherm curves (equilibrium moisture content vs. surrounding relative humidity) show non-linearity in particular for high levels of relative humidity, close to the fibre saturation point where a noticeable amount of free moisture precipitates in the voids of the material [26]. Another feature in the relative humidity-moisture content relation is that absorption and desorption curves are not identical. There is a limited but undeniable hysteresis during climate cycling due to dissipation processes in moisture transport [27, 28]. Nevertheless, in the present context of moderate climate conditions significantly below the saturation level, the sorption isotherm can be regarded as linear and the difference between sorption and desorption can be neglected in practical micromechanical models.

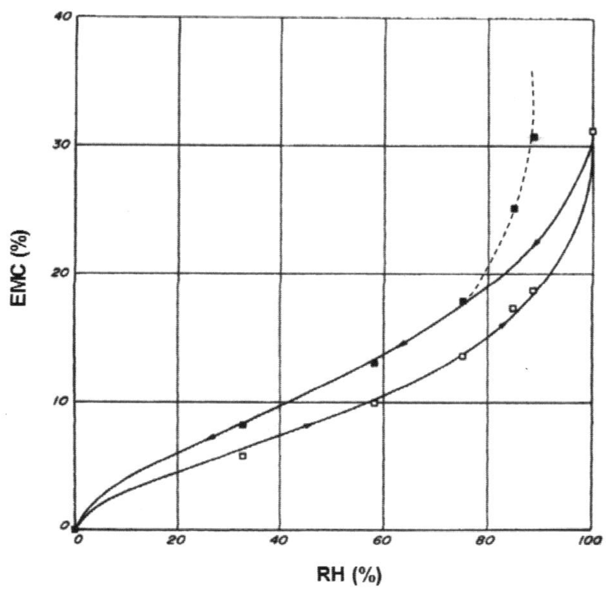

Figure 2. Non-linear sorption isotherm curve of wood measured by dynamical vapour sorption test, with equilibrium moisture content against relative humidity of the ambient atmosphere [29]. The measurements show a saturation point at about 80% relative humidity, and an absorption-desorption hysteresis.

1.4. Temperature effects

The physical climate affecting moisture content also includes the temperature. The temperature is known to markedly affect moisture content even though the relative humidity is kept constant [30]. Most studies on moisture effects of cellulose composites are carried out at room temperature, around 20-25°C, which is acceptable for composites intended for indoor applications. Essentially, there is a lack of experimental data for the temperature dependency of the properties of moist cellulose composites. This is especially the case of high processing temperatures, where extrapolation of a few published measurements may be necessary [31, 32]. In the present work, however, we consider working conditions at constant ambient temperatures, although the hygroelastic models could be generalised to include also effects of the temperature.

1.5. Relative humidity vs. moisture content

Moisture is typically either characterised in terms of relative humidity in surrounding the specimen under investigation or specifically the moisture content (relative moisture mass) in the specimen itself. The two measures are intimately linked, and the relationship is characterised by dynamic vapour sorption (DVS) isotherms. The relative humidity is easily characterised by hygrometers in the ambient air, whereas the moisture content necessitates weighing of sample and comparison with its dry weight. The relative humidity thus has the advantage of straightforward measurement, but provides only and indirect moisture-level characterisation. The moisture content, in contrast, offers a direct intrinsic measure of hygroscopic state of the material, although it is not always practical to quantify experimentally. In the literature, both relative humidity and moisture content are used to describe the state of the material. Since they can be linked by DVS mesurements, micromechanical hygroelastic models based on the two different measures are mutually transferable, which applies also to the methods presented in this paper.

1.6. Natural cellulose fibre composites

Natural cellulose fibres in composites come in many forms. On an industrial scale, the largest class in terms of material is WPC (wood polymer composites). Strictly speaking, the reinforcing phase is constituted by wood particles acting as fillers rather than reinforcing fibres. WPC components are manufactured by extrusion or injection moulding with a polyolefin matrix, such as polypropylene or polyethylene. The limited aspect ratio of the wood particles, i.e. a length to width ratio of about 2-5, is a limiting factor in the mechanical properties of WPC, in particular regarding tensile strength [34]. In fact, for particulate WPC possessing relatively poor fibre-matrix interfaces, the composite strength shows a strong correlation with the interfacial strength, and composite strength can be used to rank the efficiency of fibre surface treatment for improved interfacial stress transfer [35]. For applications with higher demands on the mechanical performance, it is better to use high-aspect ratio reinforcements based on long plant fibres or unbroken wood fibres. When processed in by high-volume techniques, such as injection moulding, these slender fibres are however likely to degrade in dimensions, and eventually reach small aspect ratio like WPC particles with limited reinforcing capabilities [35]. Compression moulding for thermoplastic matrix composites [36] and resin-transfer moulding for thermoset matrix composites [37] are better options for retained fibre slenderness and improved mechanical properties. The fibre-length degradation caused by high shear flow along the screws in injection moulding machine or extruder [38] are then avoided. At the cost of processability, choosing longer fibres usually mean clear improvements in mechanical properties for load-carrying applications, such as stiffness, strength and dimensional stability. For each specific application, a cost-performance assessment must be made in order to select the fibre type and a suitable manufacturing technique. The environmental impact is in the balance pan of cost, which is an important driving force for development of cellulose-fibre composites. In other words, a transition from petroleum-based materials to renewable bio-based materials is afflicted with additional costs, at least during a transition period. From a mechanics point of view, the use of longer fibres from wood (forestry) or plants (agriculture) only has an influence on the geometry of the fibres and their mechanical properties. The wood cells are shorter and have inferior stiffness and strength, due to a lower cellulose content and a higher off-axis orientation of the cellulose microfibils in the cell wall, as compared with plant fibres from annual crops. A comparison of wood and plant fibres for composite applications has been compiled by Madsen and Gamstedt [39].

1.7. Materials modification and design

From a micromechanical perspective, natural cellulose fibre composites are composed of a continuous phase, i.e. a polymer matrix, in which discrete reinforcements are embedded, i.e. the cellulose fibres. Moisture effects of such composites are mainly attributable to the fibre constituent, since the cellulose fibres are usually more hygroscopic than the polymer matrix. In some specific examples, such as cellulose-reinforced starch, the softening and swelling could stem from the very hydrophilic matrix [13]. Composites based on unmodified starch matrix are however not a candidate for load-carrying applications. For cellulose fibres in general, it is safe to assume that the moisture effect is predominantly caused by the moisture uptake by the fibres. This is also the case for most biopolymer matrix materials, such as polylactide, which are still less hydrophilic than the fibres, at least in the transverse direction [10]. Modifications to control or suppress moisture effects are therefore mostly focused on the cellulose fibres. In contrast to man-made synthetic fibres, such as glass and carbon fibres, the natural fibres are not inert and offer many chemical routes to modify their properties. The fibres are also porous on the nanometre scale [40], which facilitate access of low-molecular weight chemicals.

Focussing on the main component, i.e. the natural fibre, it can also be regarded as a composite material on the cell-wall level [41]. Natural cellulose fibres from plants and wood are composed of crystalline cellulose microfibrils embedded in an amorphous biopolymer matrix composed of hemicellulose and lignin. On this ultrastructural level, the crystalline cellulose microfibrils are relatively inert, whereas the hemicellulose-lignin matrix is the moisture-sensitive component. Hemicellulose has more accessible hydrogen bonds and is less fatty than lignin, and is hence the more hydrophilic of these two matrix biopolymers [42]. Pulp fibres can be cooked to remove the hydrophilic biopolymers while retaining most of the crystalline cellulose, leading to higher moisture resistance. This is the case for dissolving pulp consisting almost entirely of cellulose, which can be processed to form highly consolidated stiff and moisture-resistant composites [43]. Chemical cross-linking of remaining hemicellulose improves the moisture resistance by creating a tighter molecular network. Although the hydroxyl groups are active functionalities in such cross-linking, there is such an abundance of accessible hydrogen bonds that far from all groups are used in for cross-links. The hemiscellose still remains hydrophilic, but to a lesser extent than the unmodified material [44].

From a practical engineering top-down approach, one of the first issues to address is which fibre to choose with regard to dimensional stability and softening of the composite. The micromechanics of the cell wall, its ultrastructure and targeted chemical modification schemes are then not of primary concern. Presently, we will only address natural cellulose fibres and composites containing such fibres with regard to their hygroelastic behaviour. Models and experiments on this length scale could potentially be used in materials development in ranking different fibre candidates [12], and provide improvements in experimental design [45].

2. Moisture-induced softening

2.1. Driving forces

So-called green composites, i.e. composite materials at least partially made from renewable resources, contain almost always lignocellulosic biopolymers, typically from annual plants or wood. These have been processed and treated chemically, at least to some extent. Their 'eco-friendliness' may thus vary, although they should be better than synthetic composites from petrochemical resources regarding carbon dioxide emissions from a total life-cycle perspective. The lignocellulosic biopolymers are rich in hydroxyl groups, which form hydrogen bonds as one of the most important secondary bonds along the polymer chains. Apart from the interior of the cellulose crystallites, these hydrogen bonds are relatively accessible to water molecules which can diffuse into the biopolymer. Absorption of water molecules at the hydrogen sites will lead to weakening of the strong hydrogen bonds, and macroscopically a lower stiffness. Decrease of stiffness, typically characterised by a Young's modulus measured in a tensile test, is known as moisture-induced softening, and is generally an undesired property of load carrying cellulose-fibre composites. Since the fibre is moisture sensitive constituent compared with the stress-transferring matrix, the softening of the fibres is of interest, and how the softened fibres contribute to the softened stiffness of the overall composite. In general, if the stiffness of fibres in a moist state of interest is known, then the least softened fibre type can be identified. It is also this fibre type that will lead to the least softened composite. Techniques to measure stiffness of fibres, in try and moist states, are therefore useful. The minuteness and variability of natural cellulose fibres make it cumbersome and time-consuming to characterise individual fibres [46], and many fibres must be tested to gain a statistically representative average.

2.2. *Softening of fibres*

Stiffness testing of fibres under moist conditions is essentially the same as under standard ambient condition. The only difference is that the climate is controlled in the room where the test is carried out, or that the test rig is mounted in a climate chamber with a controlled relative humidity. The fibres should be in equilibrium state, i.e. that all transients of moisture transport have vanished and the fibres are in a steady state with respect to their moisture content. The small dimensions of the natural fibres usually mean that equilibrium state is attained in a short period of time, comparable or even shorter than the preparation and mounting time before testing. However, to be confident that this is the case, it is recommended to weight the fibres, normally a larger fibre mass to reach a practical total weight, and check that the weight remains constant over time in the targeted climate.

Direct testing of single fibres would give the best actual value of the fibre stiffness along the axial direction. In addition to required dexterity, patience and sensitive load cells and deformation measurements, there is also a challenging large variation in stiffness which can be explained natural variability in fibre geometry and the orientation of the microfibrillar angle, i.e. the orientation of the aligned cellulose crystallites in the main cell wall layer with regard to the fibre axis [47]. The properties of the main cell-wall layer, the S2 layer, can be regarded as a material property, at least relatively speaking as compared with the properties of the fibre itself. However, for practical purposes the engineering stiffness on the fibre level is more suitable. In contrast to e.g. glass fibres, natural cellulose fibres are significantly anisotropic, with considerably higher compliance transverse to the axial direction of the composite. Recognising the fibre anisotropy is important in composites, where the fibres are typically multidirectional. In fact, the transverse properties have a higher impact on the stiffness of the in-plane stiffness of a randomly oriented composite, as predicted by the Tsai-Pagano relation [48]. Furthermore, it can be shown that the shear properties of the fibres have considerable influence on the macroscopic in-plane stiffness of a randomly-oriented fibre material [49]. The simplest description of the elastic anisotropy of natural fibres would be transversely isotropic elasticity, quantified by 5 elastic constants, of which 4 are independent. The major and minor Poisson ratios are assumed to have less influence on the overall composite stiffness than the Young's moduli in the axial and transverse direction, and the shear modulus. It is only the axial Young's modulus that is relatively straightforward to characterise experimentally. Estimation of the transverse and shear modulus usually requires an indirect test method since it is not directly feasible to apply pure shear lading or transverse loading to single fibres. In practice, the off-axis elastic parameters, i.e. all but the axial Young's modulus, are estimated by assuming a fixed ration to the axial stiffness [41]. These assumptions should, if possible, be based on physical reasoning, related experimental results or micromechanical modelling. This illustrates that there is a lack of reliable experimental data on the properties of the cellulose fibres.

An alternative method to single fibre tensile test, is to characterise the longitudinal fibre stiffness by compression of fibre mats, and back-calculate the fibre stiffness. The fibre network can be interpreted an in-plane assembly of Bernoulli-Euler beams, which are deformed between contact points of adjacent fibres. The average distance between contact points necessarily decreases during elastic compaction. Toll and Månson [50] have developed a model which predicts the Young's modulus of cylindrical glass fibres from compaction of in-plane randomly oriented fibre mats in the thickness direction. Their model can be modified to describe compaction of wood fibres, provided the contact points are sufficiently far away from one another (low-density fibre mats), and that the variation in second moment of area of are used in the expressions of the bending stiffness [51]. From load-displacement data of compression tests of fibre mats, the volume fraction of fibres V_f in the mat can be determined with respect to the applied pressure P, if the grammage, initial thickness and cell-wall densities are given. For cylindrical fibres, it can be shown that the applied pressure can be expressed as [50]

$$P_c = \frac{512}{5\pi^4} E_{11} f^4 V_f^5 \tag{1}$$

where f is an integration of a weighted autocorrelation function of the orientation distribution. The dimensions of the fibre cross-section play an important role, since the fibre segments are bent between the contact points as the fibre assembly is compressed. Although the variation of cross-sectional geometry is relatively small, it can have a measurable impact on the compaction behaviour [51 52]. From compression test and the expression in equation (1) or statistical counterparts [51], the Young's modulus of the fibre can be determined, and e.g. compared with results from the tensile tests briefly presented above. The stiffness of flax fibres has been estimated both by compaction of fibre mats and tensile testing of fibres [53], with comparable results. Fair agreement was obtained for the first compaction of a pristine fibre mat relatively low density and well-separated fibres.

Testing of moist fibres or fibre mats is more difficult than testing of dry fibres, given the softened state of the fibres. This can lead to difficulties in clamping the fibre ends for single fibre tensile tests. For the compaction test, the softened fibres may induce buckling of the hollow natural cellulose fibres [54]. Despite the difficulties to handle and test soft and moist fibres, the characterisation techniques are the same as for drier fibres.

2.3. Softening of composites

The stiffness of moist composites is measured in the same way as composites at normal ambient laboratory conditions. The stiffness of composites with moist softened composites can be predicted by micromechanical models and laminate analogy, similar to non-moistened composites [41]. Unless the matrix is an extremely hydrophilic biopolymer, the composite softening can be attributed directly to the softening of the cellulose fibres by a micromechanical model for a hypothetical unidirectional lamella [55] and laminate analogy to account for the in-plane fibre orientation distribution [41]. Empirical micromechanical models based on Cox' network model can also be used to relate the microstructure, the fibre stiffness and that of the composite [56]. For the contribution of a unidirectional lamella, the simple rule of mixtures can be used for the longitudinal elastic properties. For transverse properties, the Halpin-Tsai model [57] or models from Hashin [58], can be used to estimate the stiffness parameters. From elastic properties of a unidirectional lamella, a laminate analogy can be used to scale up to the composite plate, as shown schematically in figure 3. Given the fibre orientation distribution for a short fibre composite or mat, the global stiffness matrix [$\overline{Q}$] can be described as

$$\left[\bar{Q}\right]=\frac{1}{\pi}\int_{-\frac{\pi}{2}}^{\frac{\pi}{2}}\left[T(\theta)\right]^{-1}\left[Q\right]\left[T(\theta)\right]p(\theta)\,\mathrm{d}\theta \tag{2}$$

where $T(\theta)$ is the transformation matrix for a rotation angle of θ radians, and $[Q]$ is the local stiffness matrix. It is assumed that $p(\theta)$ is symmetric, which is typically the case for simple geometries like plates used to manufacture test coupons. This means that the material is globally orthotropic, and the stiffness matrix is described by five elastic constants, namely the longitudinal and transverse Young moduli, the major and minor Poisson ratios and the shear modulus. The components in the global stiffness matrix can be determined from standardised macroscopic testing, and the orientation distribution can readily be found by image analysis of scanned sections of the fibre mat [59] or image analysis of X-ray micro computed tomography scans [60]. From the stiffness matrix, the Young's modulus of the composite can be found as a function of the elastic properties of the fibres. Such a model can be used both to investigate the effect of softened fibres on the stiffness of the composite, or reversely to identify the fibre stiffness based on macroscopic measurements of the composite stiffness, provided that the interrelation of the four independent transversely isotropic elastic properties of the fibre has been set [41, 61].

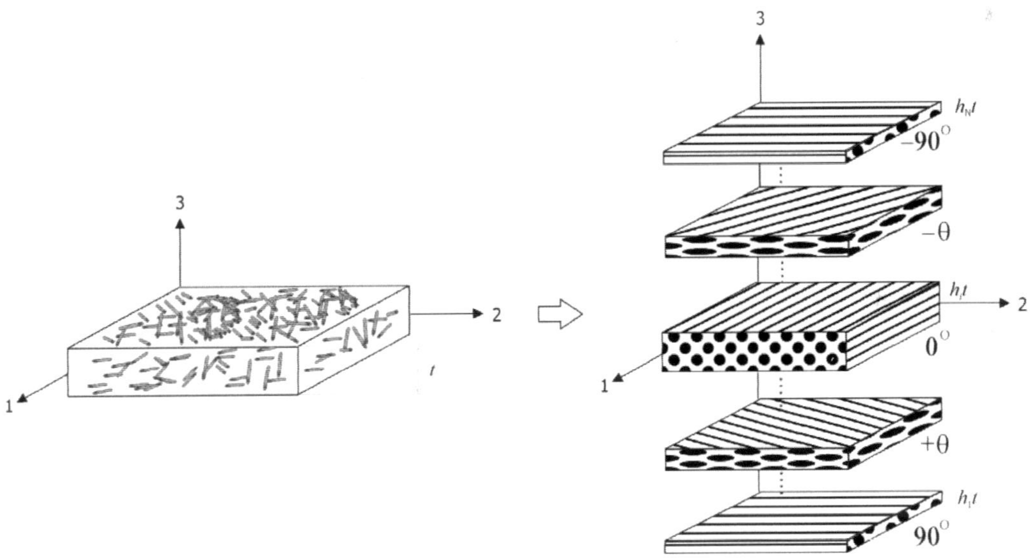

Figure 3. Schematic illustration of a laminate analogy for a short-fibre composite, i.e. with finite length natural fibre reinforcement. The relatively slender fibres have an in-plane random distribution given by the fibre orientation distribution $p(\theta)$, which defines the relative proportion of a small orientation interval in the laminate model.

It should be noted that it is not only the fibres that contribute to the softening of the composite, even if the matrix is relatively hydrophobic. The fibre-matrix interface can be susceptible to moisture uptake. If there the interfacial compatibility is limited, e.g. by a hydrophobic polymer matrix, like most polyolefins thermoplastics, the wetting of the fibres during manufacturing can be incomplete resulting in gaps between the reinforcement and the surrounding matrix. Capillary effects can then promote moisture uptake along the interfacial gaps. The high concentration of water can then promote breakage of hydrogen bonds at the interface, leading to a decrease in stiffness due to reduced fibre-matrix stress transfer [62]. The contribution of a weakened interface to a reduced stiffness is expected to be particularly important if fibre-matrix is susceptible to moisture adsorption. This is expected if the adhesion is poor, meaning capillary channels along the interfaces, and a hydrophilic composite in general, i.e. moisture is readily transported into the composite interfaces.

3. Moisture-induced swelling

3.1. Driving forces

Swelling is here defined as increase in dimensions as moisture is absorbed. The mechanisms are essentially the same as those for softening, i.e. increase of compliance as moisture is absorbed; water molecules diffuse into the material from the surroundings, and break tight hydrogen bonds, thus creating additional space between molecular chains of the lignocellulosic fibres. Therefore, swelling and softening go hand in hand, sometimes also quantitatively [61]. When the hygroexpansion coefficient is reduced by a material modification, the compliance of the material is typically reduced by approximately the same relative amount. Among the constituents of the cell wall of natural fibres, the cellulose crystallites do not swell, since their tight crystalline packing does not allow moisture uptake. Adsorption on the surface crystallites is possible, or in the amorphous or disordered cellulose regions interspacing the crystallites in the cellulose microfibrils. Lignin and in particular hemicellulose, on the other hand, have a large amount of accessible hydroxyl groups and breakable hydrogen bonds [42]. Any attempt to chemically modify the fibres to reduce the dimensional instability should be directed to these biopolymers. An example of cross-linking pulp fibres with butantetracarboxylic acid (BTCA) to increase the resistance to swelling is presented by Almgren et al. [10].

3.2. Swelling of fibres

Moisture-induced swelling is considerably larger in the radial or transverse direction compared with that of the axial direction. This is due to the microfibril direction which is relatively close to the axis of the fibre. Since the swelling is large in the hemicellulose and lignin matrix surrounding the relatively inert cellulose microfibrils, the swelling of the fibres will be predominantly in the transverse direction where the constraint effect of the stiff microfibrils is very limited. The swelling is normally quantified by the hygroexpansion coefficient which can be defined as the change in strain per change in moisture content, or alternatively per change in relative humidity in the surrounding medium at equilibrium conditions. As for fibre stiffness, the best way to obtain a direct measure of the fibre welling is to measure it directly on single fibres. This is however even more cumbersome for fibre swelling than fibre tensile testing, since the fibres generally twist and bend during moisture uptake due to the off-axis orientation of the microfibrils and the non-prismatic and irregular shape of the cross-section along the fibre. Joffre et al. [63] investigated the dimensional changes of a single native spruce fibre subjected to climate changes in relative humidity with X-ray micro computed tomography at high resolution in a synchrotron. The experimental results were compared with finite element simulations accounting for large deformations, with the hygroexpansion coefficients in the longitudinal and transverse directions as the only fitting parameters. The elastic properties needed to be corrected to account for the influence of the changing environment, since moist fibres are more compliant and more prone to deformation when subjected to hygroscopic internal stresses. This approach is schematically outlined in figure 4. The irregular and varying individual shape of natural fibres require that the detailed geometry of specific fibres need to be taken into in such calculations. Only compression softwood fibres [61] or plant fibres with a near prismatic geometry [64] allow analytical modelling. The example shown in figure 4 with a mixed numerical-experimental approach where a finite-element updating scheme was used to simulate the swelling of an earlywood spruce fibre going from the experimental fibre geometry at 47% relative humidity to the predicted geometry of the fibre in the moist state at 80% relative humidity at equilibrium conditions. The hygroexpansion coefficients were identified by comparing the predicted and the experimental three-dimensional fiber geometry in the wet state. The obtained values are 0.17 strain per change in relative humidity transverse to the microfibrils in the cell wall, and 0.014 along the microfibrils. It should be noted that this value is for untreated fibres in the green state. Any further treatment like pulping, hornification from drying (leading to microfibril aggregation) or chemical processing would most likely result in reduced hygroexpansion, since the free swelling is restrained by further cross-linking in the cell wall. The hygroexpansion of the helical cell wall, although at a moderate microfibril angle of 5-10° for wood fibres and even less for most plant fibres, is not the same as that of the fibre. Considering the lumen and the tubular helix structure, the fibre swelling can be calculated from the cell-wall swelling [61].

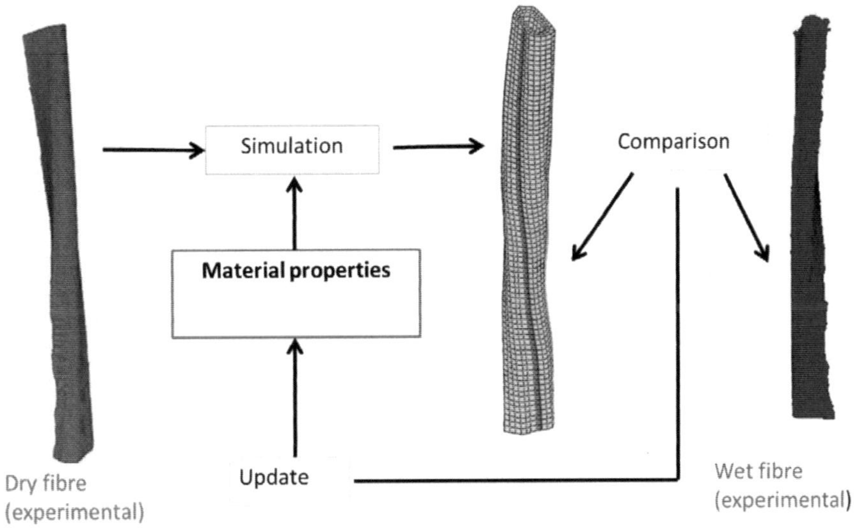

Figure 4. Comparison of 3D geometries of swollen single fibres from experimental measurements (synchrotron X-ray computed tomography images) and from numerical simulations (finite elements). The fitting parameter and the hygroelastic properties of the fibre.

3.3. Swelling of composites

The swelling of the composite is mainly attributed to the swelling of the hydrophilic fibres. Since the fibres in the composite are generally embedded in a relatively hydrophobic matrix, the surrounding matrix should restrain the swelling of the fibres. A micromechanical model has been established to predict the swelling of embedded fibres based on experimentally characterised microstructural parameters and hygroelastic properties of the constituents [60], and the predicted swelling was found to be in concert with direct measurement of various wood-pulp fibre composites by means of three-dimensional X-ray microtomographic images. The internal stresses developed during a change of moisture content can be sufficiently large to induce damage or cracking in the material. This is particularly the case of the local fibre dispersion results in tensile stresses normal to weak fibre-matrix interfaces. Such a situation is however more likely in drying than in wetting.

The model based on classical laminate theory used by Almgren et al. [10] has been used to link the fibre swelling to the out-of-plane swelling of in-plane oriented composite plates. The swelling in the thickness direction is larger than the in-plane swelling of composites (see figure 5), since the fibres tend to orient themselves in the plane of manufactured plates. This seems to be the case for both injection moulded [60] and compression moulded composites [12, 59]. For unsymmetric fibre orientation distributions with regard to the composite mid-plane, not only swelling in the thickness direction, but also a significant plate bending can take place [59], which can also be related to the fibre swelling. In order to avoid such obvious shape deformation, a symmetric lay-up is recommended.

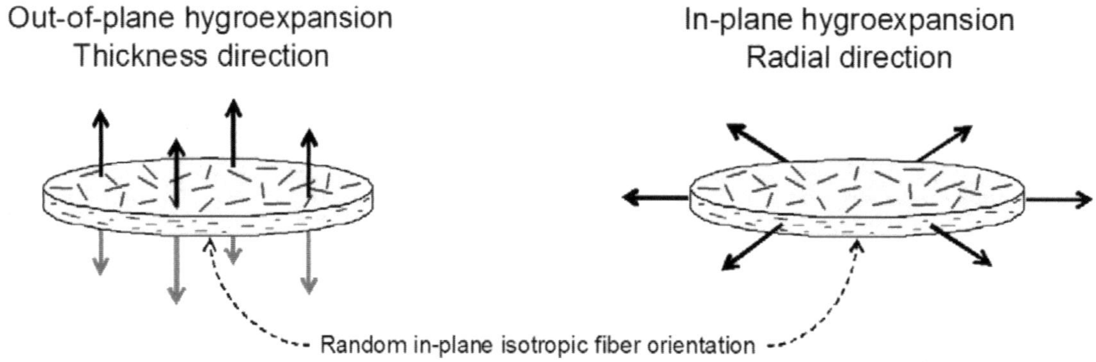

Figure 5. Out-of-plane swelling (left) is typically larger than the in-plane swelling (right) due to the predominantly in-plane fibre orientation.

4. Final remarks

It can be summarised that the inevitable moisture induced softening and swelling of natural cellulose fibre composites is a nuisance in most engineering applications. In parallel to the development of new processing routes and modification schemes to suppress these undesired effects, it is useful to gain better understanding of the underlying mechanisms and develop models that can predict the reduction in stiffness and dimensional instability. The intention of this review paper is to provide some examples of such attempts, using a variety of experimental techniques and numerical and analytical composite mechanics models. The success of a quantitative mechanics-based approach in facilitating and rationalising the materials development process for improved moisture resistance lies in integrating the chemical/processing approach with methods derived from experimental mechanics. Multidisciplinarity and openness to collaborate over scientific and engineering fields are keywords in such a process. The mechanics based methods should the preferably be straightforward to use and pedagogically explained.

Acknowledgments

The author wishes to thank previous collaboration partners in research projects on hygromechanical behaviour of wood-fibre composites, namely Drs. Cristian Neagu, Karin Almgren, Thomas Joffre and Gabriella Josefsson.

References

[1] Saheb DN and Jog JP 1999 *Adv. Polym. Tech.* **18** 351-363
[2] Ellis WD 2000 *Mol. Cryst. Liquid Cryst.* **353** 75-84
[3] Kaplan DL 1998 *Introduction to Biopolymers from Renewable Resources* (Berlin Heidelberg: Springer
[4] Khalil HA, Issam AM, Shakri MA, Suriani R. and Awang AY 2007 *Ind. Crops Prod.* **26** 315-323
[5] John MJ and Anandjiwala RD 2008 *Polym. Compos.* **29** 187-207
[6] Clark JH 2006 *Green Chem.* **8** 17-21
[7] Bisanda ETN and Ansell MP 1991 *Compos. Sci. Technol.* **41** 165-178
[8] Bledzki AK, Reihmane S and Gassan J 1996 J. Appl. Polym. Sci. **59** 1329-1336
[9] Almgren KM, Gamstedt EK, Berthold F and Lindström M 2009 *Polym. Compos.* **30** 1809-1816
[10] Almgren KM, Gamstedt EK. and Varna J. 2010 *Polym. Compos.* **31** 762-771
[11] Salmén L, de Ruvo A and Carlsson L 1984 *A Treatise on the Elastic and Hygroexpansional Properties of Paper by a Composite Laminate Approach* (Polymer Processing and Properties V) ed Nicolais L and Astarita G (New York: Springer) 389-398

[12] Neagu RC, Gamstedt EK and Berthold F 2006 *J. Compos. Mater.* **40** 663-699
[13] Duanmu J, Gamstedt EK and Rosling A 2007 *Compos. Sci. Technol.* **67** 3090-3097
[14] Dawson-Andoh B, Matuana LM and Harrison J 2004 *J. Vinyl Add. Tech.* **10** 179-186
[15] Re GL, Morreale M, Scaffaro R and La Mantia FP 2013 *J. Appl. Poly. Sci.* **129** 3198-3208
[16] Espert A, Vilaplana F and Karlsson S 2004 Compos. A **35** 1267-1276
[17] Pupure L, Doroudgarian N and Joffe R 2014 *Polym. Compos.* **35** 1150-1159
[18] Doroudgarian N, Pupure L and Joffe R 2015 *Polym. Compos.* **36** 1510-1519
[19] Aronhime MT, Neumann S. and Marom G 1987 *J. Mater. Sci.* **22** 2435-2446
[20] Grace LR and Altan MC 2012 Compos. A **43** 1187-1196
[21] Carman ED 2015 *Diffusion Property Prediction Methods for Natural Fibre Kenaf Composite Materials* (Johannesburg: University of the Witwatersrand, dissertation)
[22] Gamstedt EK, Neagu RC and Lindström M 2003 *Proc. 7th Int. Conf. Woodfiber-Plastic Composites* (Madison: Forest Products Society) pp 127-149
[23] Du Y, Zhang J, Wang C. Lacy Jr TE, Xue Y, Toghiani H, Horstemeyer MF, Pittman Jr CU 2010 *Forest Prod. J.* **60** 366-372
[24] Svagan AJ, Hedenqvist MS and Berglund L 2009 *Compos. Sci. Technol.* **69** 500-506.
[25] Engelund ET, Thygesen L, Svensson S and Hill CA 2013 *Wood Sci. Technol.* **47** 141-161
[26] Venkateswaran, A. 1970 *Chem. Rev.* **70** 619-637
[27] Patera A, Derome D, Griffa M and Carmeliet J 2013 J. Struct. Biol. **182** 226-234
[28] Willems W 2014 *Wood Sci. Technol.* **48** 499-518
[29] Djolani B 1970 Notes Rech. Dep. Exploit. Util. Bois Univ. Laval **8**
[30] Skaar C 2012 *Wood-Water Relations* (Berlin: Springer)
[31] Kamke FA and Kutnar A 2010 Wood Fiber Sci. **42** 377–387
[32] Muzamal M, Gamstedt EK and Rasmuson A 2014 *Wood Sci. Technol.* **48** 353-372
[33] Hietala M, Samuelsson E, Niinimäki J and Oksman K 2011 *Compos. A* **42** 2110-2116
[34] Nyström B, Joffe R and Långström R 2007 *J. Reinf. Plast. Compos.* **26** 579-599
[35] Joffre T, Miettinen A, Berthold F and Gamstedt EK 2014 *Compos. Sci. Technol.* **105** 127-133
[36] Lundquist L, Marque B, Hagstrand PO, Leterrier Y and Månson J-A 2003 *Compos. Sci. Technol.* **63** 137-152
[37] Sèbe , Cetin NS, Hill CA and Hughes M 2000 *Appl. Compos. Mater.* **7** 341-349
[38] Franzén B, Klason C, Kubát J and Kitano T 1989 *Compos. 20* 65-76
[39] Madsen B, Gamstedt EK 2013 *Adv. Mater. Sci. Eng.* **2013** 564346
[40] Persson PV, Hafrén J, Fogden A, Daniel G and Iversen T 2004 *Biomacromol.* **5** 1097-1101
[41] Neagu RC and Gamstedt EK 2007 *J. Mater. Sci.* **42** 10254-10274
[42] Cave ID 1978 *Wood Sci. Technol.* **12** 75-86
[43] Nilsson H, Galland S, Larsson PT, Gamstedt EK and Iversen, T. 2012 *Cellulose* **19** 751-760
[44] Zoldners J and Kiseleva T 2013 *Holzforschung* **67** 567-571
[45] Gamstedt EK, Bader TK and de Borst K 2013 *Wood Sci. Technol.* **47** 183-202
[46] Eder M, Arnould O, Dunlop JW, Hornatowska J and Salmén L 2013 *Wood Sci. Technol.* **47** 163-182
[47] Wilhelmsson D, Neagu RC, Bardage S and Gamstedt EK 2006 *Proc. of the 5th Plant Biomech. Conf.* (Stockholm: STFI-Packforsk AB) pp 181-186
[48] Hull D and Clyne TW 1996 *An Introduction to Composite Materials* (Cambridge: Cambridge University Press)
[49] Josefsson G, Chinga-Carrasco G and Gamstedt EK 2015 *RSC Adv.* **5** 58091-58099
[50] Toll S and Månson, JAE 1995 *J Appl Mech.* **62** 223-226
[51] Alkhagen M and Toll S 2009 *J. Appl. Mech.* **76** 041014
[52] Neagu RC, Gamstedt EK and Lindström M 2006 *Wood Fib. Sci.* **36** 95-111
[53] Gamstedt EK, Bommier E and Madsen B 2014 *Meas. Sci. Technol.* **25** 035601
[54] Lundquist L, Willi F, Leterrier Y and Månson JAE 2004 *Polym. Eng. Sci.* **44** 45-55
[55] Marklund E, Varna J, Neagu RC and Gamstedt EK 2008 *J. Compos. Mater.* **42** 2377-2405

[56] Madsen B, Thygesen A and Lilholt H 2009 *Compos. Sci. Technol.* **69** 1057-1069
[57] Halpin JC and Kardos JL 1976 *Polym. Eng. Sci.* **16** 344-352
[58] Hashin Z 1979 *J. Appl. Mech.* **46** 543-550
[59] Neagu RC, Gamstedt EK and Lindström 2005 *Compos. Part A* **36** 772-788
[60] Joffre T, Wernersson EL, Miettinen A, Hendriks CLL and Gamstedt EK 2013 *Compos. Sci. Technol.* **74** 52-59
[61] Joffre T, Neagu RC, Bardage SL and Gamstedt EK 2014 *J. Struct. Biol.* **185** 89-98
[62] Almgren KM and Gamstedt EK 2010 *Compos. Interf.* **17** 845-861
[63] Joffre T, Isaksson P, Dumont PJJ, Rolland du Roscoat S, Sticko S, Orgéas L and Gamstedt EK 2016 *Exp. Mech.* 56 723-733
[64] Aslan M, Chinga-Carrasco G, Sørensen B F and Madsen B 2011 *J. Mater. Sci.* 46 6344-6354

IOP Conf. Series: Materials Science and Engineering **139** (2016) 012004 doi:10.1088/1757-899X/139/1/012004

Improving fiber/matrix interfacial strength through graphene and graphene-oxide nano platelets

A Pegoretti[1], H Mahmood[1], D Pedrazzoli[2] and K Kalaitzidou[3,4]

[1] Department of Industrial Engineering, University of Trento, via Sommarive 9 - I38123, Trento, Italy

[2] Department of Macromolecular Science and Engineering, Case Western Reserve University, 2100 Adelbert Road, Cleveland (OH), 44106-7202, USA

[3] G. W. Woodruff School of Mechanical Engineering—Georgia Institute of Technology, Atlanta, GA 30336, USA

[4] School of Materials Science and Engineering, Georgia Institute of Technology, Atlanta, GA 30322, USA

E-mail: alessandro.pegoretti@unitn.it

Abstract. Fiber/matrix interfacial shear strength (ISS) is a key factor determining the mechanical properties of structural composites. In this manuscript the positive effects of both graphene and graphene oxide in improving the ISS value of glass-fiber reinforced composites are experimentally demostrated. Two strategies will be presented: i) uniform dispersion of the nanofillers in the polymer matrix or ii) selective deposition of the nanofillers at the fiber/matrix interface. Both thermoplastic (polypropylene) and thermosetting (epoxy) matrices are investigated and the effects of nanoparticles on the fiber/matrix interface are determined through micromechanical tests on single-fiber composites. Finally, the beneficial effects of the investigated nanofillers on both mechanical and functional (strain monitoring) properties of multiscale macrocomposites are experimentally proved for the cases of polypropylene-based composites reinforced with short glass fibers and on epoxy-based composites reinforced with continuous unidirectional glass fibers.

1. Introduction

The fiber/matrix interphase is a fundamental issue in composites due to its major role in determining the mechanisms of stress transfer from the matrix to the 'load-bearing' fibrous phase [1]. Several approaches have been proposed in the last 15 years to improve the fiber/matrix adhesion and to design the interphase of polymer composites [2]. The main strategies can be distinguished between (i) interphase tailoring via sizing/coating on fibers, (ii) creation of hierarchical fibers by nanostructures, (iii) fiber surface modifications by polymer deposition and (iv) interphase formation by matrix modifications [3]. In more recent years, a remarkable attention have been devoted to the usage of nanofillers for the interphase engineering of fiber-reinforced polymer composites [4]. In fact, nanofillers may improve the stress transfer ability at the fiber/matrix interface and allow the creation of (multi)functional interphase in polymer composites. Added functionalities provide sensing, self-healing and damping properties [5].

Depending on their morphology, nanofillers can be classified as i) isodimensional (3D, nanoparticles with a characteristic dimension less than 100 nm), ii) bi-dimensional (2D, nanofibres or nanotubes with a diameter less than 100 nm), iii) mono-dimensional (1D, lamellar nanoplatelets with a thickness less than 1 nm). Among 1D lamellar nanoplatelets a remarkable case is represented by graphene [6]. Graphene sheets, one-atom-thick two-dimensional layers of sp^2-bonded carbon, manifest a range of unusual properties. Their thermal conductivity and mechanical stiffness may reach the remarkable in-plane values of graphite, i.e. 3000 $Wm^{-1}K^{-1}$ and 1060 GPa, respectively; their fracture strength should be comparable to that of carbon nanotubes for similar types of defects. Moreover, recent studies have shown that individual graphene sheets have extraordinary electronic transport properties [7].

One possible way to harness these properties for applications would be to incorporate graphene sheets in a polymer matrix. The manufacturing of such composites requires not only that graphene sheets are produced in a sufficient quantity, but that they also are homogeneously distributed in the polymer matrix [7-9]. The current available methods for the mechanical exfoliation of graphite are not suitable for large scale production. On the other hand, chemical oxidation of graphite into graphite oxide may offer an easy path to obtain graphene oxide (GO) in a large quantity that can be eventually reduced (chemically, electrochemically or thermally [10]) into graphene. The bulk production of GO and reduced graphene oxide (rGO) has created opportunities to explore this flat structure of carbon with polymer and nano particles in composites.

The aim of this work is to demonstrate how both graphene and graphene-oxide nanofillers can be used to improve the fiber/matrix adhesion of glass-fiber reinforced composites, as assessed by micromechanical (single-fiber fragmentation and microdebonding) tests. To prove this concept, two cases studies will be presented: i) uniform dispersion of graphene nanoplateletes in the matrix of polypropylene/glass fiber and epoxy/glass fiber composites or ii) selective deposition of graphene oxide and reduced graphene oxide at the fiber/matrix interface in epoxy/glass-fiber composites. Moreover, the positive role of both GO and rGO on both mechanical and functional (electrical conductivity) properties of (real) composites is investigated on polypropylene-based composites reinforced with short glass fibers composites and on epoxy-based composites reinforced with continuous glass fibers.

2. Experimental methods

2.1. Materials

*2.1.1. Polypropylene and epoxy matrices.*The polypropylene (PP) matrix used in this work was an isotactic homo-polypropylene matrix (code PPH-B-10-FB) produced by Polychim Industrie (Loon-Plage, France) and kindly provided by Lati Industria Termoplastici (Varese, Italy). This PP matrix was characterized by a melt flow index value of 6.9 g/10' (190 °C, 2.16 Kg). The mechanical properties of neat PP compression moulded plaques are reported in Table 1. Fusabond_ P M-613-05 maleic anhydride modified polypropylene (PPgMA) was supplied by DuPont™ de Nemours (Geneva, Switzerland).

Two types of bicomponent epoxy resins were used: i) epoxy_1, consisting of an epoxy base EC 252 and a hardener W241 mixed at a weight ratio of 100:40, was provided by Elantas Italia S.r.l.; while ii) epoxy_2, consisting of a mixture of 635 thin epoxy and a 556 slow aminic hardener at a weight ratio of 200:100, was supplied by US Composites. Epoxy_1 resin was cured at room temperature for 3 h followed by 15 h at 60 °C, while epoxy_2 was cured at T = 80 °C for 1 h, followed by post-curing at T = 100 °C for 4 h. The physical properties of both epoxy resins are reported in Table 1.

2.1.2. Glass fibers. Depending on the polymer matrix involved, two types of glass fibers (GF) were used: i) PP-compatible RO99 P319 GF supplied by Saint-Gobain - Vetrotex (average diameter 15.3 ± 1.5 μm), named as GF_1; ii) epoxy-compatible XG 2089 GF supplied by PPG Industries (average diameter 16.0 ± 0.1 μm), named as GF_2. The scale and shape parameters of the Weibull strength distribution of GF are summarized in Table 1. Continuous glass fibers were chopped in short-glass fibers (average length of 6.50 ± 0.44 mm) by using a chopper gun CDA-08 provided by GlasCraft (Graco®, Bury, England).

Table 1. Thermal and mechanical properties of the composite constituents: tensile modulus (E), tensile strength (σ_b), Weibull scale (σ_0) and shape (m) parameters, glass transition temperature (T_g) and melting temperature (T_m).

Material	E (MPa)	σ_b (MPa)	σ_0 (MPa)	m (L_0=5 mm)	T_g (°C)	T_m (°C)
PP	1546 ± 24	37.1 ± 0.11	-	-	9.9	165.1
Epoxy_1	795 ± 28	26.1 ± 1.1	-	-	28.0	-
Epoxy_2	2917 ± 37	59.6 ± 0.7	-	-	64.8	-
GF_1 - (PP-compatible)		-	3206	6.2	-	-
GF_2 - (Epoxy-compatible)			3551	4.4	-	-

2.1.3. Graphene and graphene oxide. Exfoliated graphite nanoplateletes (GNP) type M-5 with an average diameter of 5 μm and thickness in the range of 10–20 nm were supplied by XG Sciences Inc. (East Lansing, USA). Details on the exfoliation process as well as on the morphology of GP can be found elsewhere [11].

Graphene oxide (GO) was synthesized using an approach derived from the Hummer's method [12]. Briefly, 1 g graphite powder was added into 46 ml of H_2SO_4 cooled in an ice bath, followed by the addition of 1 g of $NaNO_3$ and stirred for 15 min. In the next step, 6 g of $KMnO_4$ were slowly added in order to avoid a spontaneous exothermic reaction. The mixture was then stirred for at least 24 h at 35 °C. Finally, excess of distilled water was added to the above mixture while the temperature was kept under 80 °C. In the end, 30% H_2O_2 was added to the mixture to stop the reaction. The resulting suspension was thoroughly washed using HCl solution and distilled water to remove Mn ions and acid respectively. The obtained brown solution was dried in a vacuum oven at 50 °C for at least 36 hours to obtain GO powder.

2.1.4. Electrophorectic deposition of graphene oxide on glass fiber. A stable suspension is the key for uniform deposition of graphene on GF. Initially, graphite oxide powder was added in water at a concentration of 1 mg/ml and the dispersion was subjected to bath-sonication for 1 h. Since GF are non-conductive materials, two copper plates were used as electrodes in the EPD process. The GF (fixed on a window frame) were placed in contact with the anode. Since GO display negative potential due to functionalities attached during the oxidation reaction, during the EPD process GO migrated towards the anode and deposited on the GF. EPD was carried out at various applied voltages up to 10 V/cm with a constant deposition time of 5 min and a gap between the electrodes of 2 cm. A second EPD cycle was performed under the same conditions while reversing the window frame with GF, so that a homogenous deposition could be achieved on the fiber surface. The coated samples were dried in a vacuum oven at 40 °C for 12 h. A schematic of the EPD process used to deposit GO nanosheets on GF_2 is depicted in Figure 1 [13]. To chemically reduce the GO coating into rGO, the dried fibers were exposed to hydrazine hydrate vapors at 100°C for 24 hours, as reported elsewhere [14].

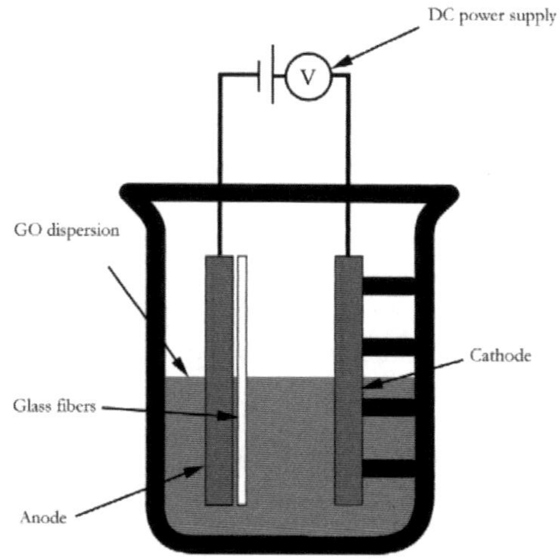

Figure 1. Schematic of the process for the electrophoretic deposition of GO nanosheets on glass fibers.

2.2. Processing techniques

2.2.1. Preparation of polypropylene -based composites. Square sheets (thickness of around 0.7 mm) of neat PP and PP-GNP nanocomposites were prepared by melt mixing in a Thermo Haake internal mixer (temperature = 190 °C, rotor speed = 50 rpm, time = 10 min) followed by compression moulding in a Carver laboratory press (temperature = 190 °C, pressure = 0.76 MPa, time = 10 min). Thin (70–80 μm) matrix films used for the preparation of the microcomposites were obtained by a further hot pressing stage (temperature = 200 °C, pressure = 3.4 MPa, time = 10 min).

Single-fiber microcomposite samples for the single fiber fragmentation tests were prepared by the following procedure [15]. About 10 fibers were aligned between two films of the selected PP matrix, sandwiched between two Mylar sheets (thickness of 0.5 mm) and two aluminum plates. The mould was placed in a vacuum oven at a temperature of 165 °C and at a pressure of about 10 kPa for about 20 min and then it was let to cool in air. The specimens were obtained by cutting strips containing one single fiber longitudinally aligned in the centerline. The dimensions of microcomposites were roughly 5 mm in width and 25 mm in length.

Single-fiber microcomposite samples for the microdebonding test were prepared by depositing a molten PP microdrop onto a single fiber filament supported on a paper tab [16]. In order to avoid the formation of asymmetric droplets with respect to the filament, a PP fiber was tied around the filament prior to heating [17]. PP microdrops were distributed symmetrically around the filament during melting in a hot-stage equipment (Instec HCS302, Boulder, CO, USA) while observing under an optical microscope Leica DMRM (Leica Microsystems Inc., Buffalo Grove, USA). Prior to testing, the microbond samples were examined using the optical microscope in order to determine the fiber diameter (d), embedded fiber length (L), and the maximum droplet diameter (D).

Composites reinforced with short GF_1 (up to 30 wt%) with and without GNP (up to 7 wt%) were made by melt mixing and injection moulding using a vertical, co-rotating, bench-top twin-screw microextruder (DSM Micro 15 cm^3 Compounder) connected to a micro-injection moulding unit (DSM). The compound was mixed for 3 min, at 190 °C and 250 rpm. After the polymer compound had melted and homogenized, short GF_1 strands were directly added to the melt and further mixed for 2 min before injection moulding. The mould temperature was 80 °C and the injection pressure was 800 kPa. ASTM D638 dumbbell specimens were obtained.

2.2.2. Preparation of epoxy-based composites. GNP nanoplatelets were dispersed in isopropanol by sonication using a Misonix S-4000-010 for 1 h (30% amplitude, 8 W power) equipped with a probe of 12.5 mm diameter. Once the isopropanol was filtered away, the GNP powder was mixed with the epoxy_2 at 800 rpm and T = 60 °C for 40 min using a magnetic stirring plate.

The composites were made as follows: first the GNP were dispersed in isopropanol by sonication using the same procedure as mentioned above. GF_1 were then added to the solution and stirring was continued for 20 min followed by addition of the curing agent and subsequent stirring at 800 rpm for 30 min. The mixture was degassed in a vacuum oven and cured with the thermal cycle described in par. 2.1 for epoxy_2.

Microcomposite samples for microdebonding tests were prepared by depositing a microdrop of epoxy_2 (and corresponding nanocomposites) onto a single GF_1 filament supported on a paper tab. Epoxy microdrops were distributed symmetrically around the filament while observing under an optical microscope. The drops were cured with the thermal cycle described in par. 2.1 for epoxy_2. Prior to testing, the samples were examined using an optical microscope (Leica DMRM, Buffalo Grove, IL, USA) in order to determine the fiber diameter (d), embedded fiber length (L), and the maximum droplet diameter (D).

Single GO or rGO coated GF_2 were axially aligned in a silicon mould and epoxy_1 resin was poured into the mould to fabricate single fiber model composites for the single fiber fragmentation tests. The samples were cured according to the thermal cycle described in par. 2.1 for epoxy_1.

Unidirectional composites consisting of rGO coated glass fibers were created by a hand lay-up method. Briefly, strands of rGO coated GF_2 were aligned in a silicon mould in layers and each layer was infused with epoxy_1 resin and cured according to the thermal treatment described in par. 2.1 for epoxy_1. The processed composites manifested an approximate fiber volume fraction of 34% as evaluated on the basis of the measured (by displacement in water) density of the composites. On the cured specimens (100 mm × 13 mm × 2 mm), silver paste was applied on the cross section at a distance of 30 mm and thin aluminum sheet was also covered over the silver paste to avoid ohmic resistance of electrical contacts.

2.3. Testing methods

2.3.1. Single-fiber fragmentation test. Single fiber fragmentation tests (SFFT) tests were performed at room temperature by using a custom-made apparatus represented by a small tensile tester (Minimat, by Polymer Laboratories) placed under a polarized optical stereomicroscope (Wild M3Z by Leica). A schematic of the device is reported in Figure 2.

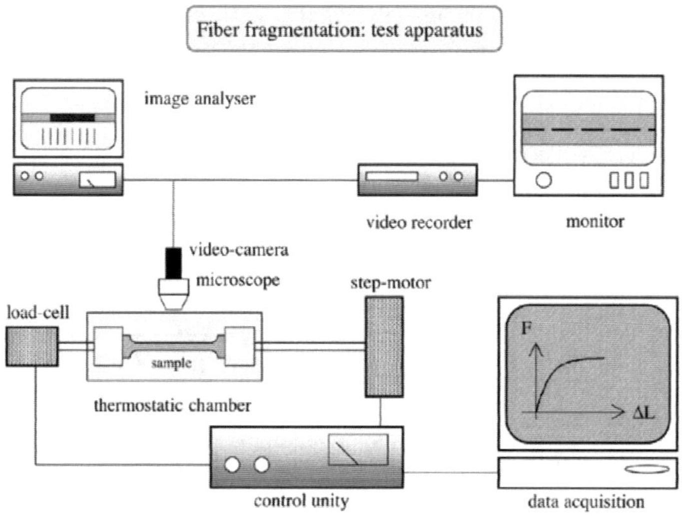

Figure 2. Schematic of the apparatus for the fragmentation test.

Samples of single-fiber microcomposites were 5 mm in width and 25 mm in length. A cross-head speed of 10 mm/min was applied up to a strain of 10%, necessary to assure the saturation of the fragmentation process. The mean fiber length at saturation, LS, was measured by an image analysis software (Image J). The value of critical fiber length, L_C, was considered to be equal to 4/3 of L_S. ISS values were evaluated in accordance to the simplified micromechanical model proposed by Kelly and Tyson. The static equilibrium between the tensile force acting on a fiber and the shear force transferred through the fiber-matrix interface provides an average value of ISS according to the following equation:

$$ISS = \frac{\sigma_{fb}(L_c)\,d}{2\,L_c} \tag{1}$$

where d is the fiber diameter and $\sigma_{fb}(L_c)$ is the fiber tensile strength at the critical length. This latter value can be estimated by assuming a Weibull distribution for the fiber strength, i.e.:

$$\sigma_{fb}(L_c) = \sigma_0 \left(\frac{L_c}{L_0}\right)^{-\frac{1}{m}} \Gamma\left(1+\frac{1}{m}\right) \tag{2}$$

where Γ is the Gamma function, while σ_0 and m are the scale and shape parameters of the Weibull strength distribution (reported in Table 1) at the reference length L_0, respectively.

2.3.2. Microdebonding test. Microdebonding tests were conducted at a crosshead speed of 1 mm/min by an Instron 33R 4466 (Norwood, USA) tensile tester equipped with a 500 N load cell. During testing the paper tab attached to one end of the glass fiber was slowly pulled up, while the droplet was constrained by a shearing plate, which was fixed on a stationary support. A nominal interfacial shear strength (ISS) value was computed by using the following expression:

$$ISS = \frac{F_{max}}{\pi d L} \quad (3)$$

where F_{max} is the maximum applied load recorded during the test, d the fiber diameter, and L the embedded fiber length. When a critical load is reached, the fiber-matrix interface fails and the load dramatically decreases.

2.3.3. Mechanical tests on composites. Tensile tests were performed according to ASTM D638 with an Instron model 33R 4466 tensile tester equipped with a 10 kN load cell at a crosshead speed of 5 mm/min. Each data point is an average of at least five measurements. Axial strain was recorded by using a resistance extensometer Instron_ model 2630-101 with a gauge length of 10 mm. The elastic modulus was measured as a secant value between longitudinal deformations of 0.05% and 0.25%.

2.3.4. Electrical resistivity and strain monitoring measurements. Two different resistivity measurement methods were employed depending on the electrical behaviour of the investigated materials. For specimens having resistivity levels exceeding 10^6 Ωcm, the electrical resistivity was measured using a Keithley 8009 resistivity test chamber coupled with a Keithley 6517A high-resistance meter. In the other case of more conductive samples, a 6-1/2-digit electrometer/high resistance system (Keithley model 6517A) was used and a 2-points electrical measurement was chosen as test configuration.

The possibility to monitor the strain in epoxy_1 composites containing rGO coated fibers by measuring the variation of electrical resistivity was investigated under various loading conditions such as tensile quasi-static (ramp) and dynamic (loading-unloading). Tests were performed by an Instron 5969 testing machine equipped with a 50 kN load cell simultaneously recording the applied strain (by an electrical extensometer Instron model 2620) and the variations of electrical resistivity by a electrometer/high resistance system (Keithley model 6517A) [18, 19].

2.3.5. Electron microscopy. Scanning electron microscopy (SEM) observations of the fracture surfaces of the composites were obtained by a Phenom G2 Pro (Phenom-World BV) microscope, at an acceleration voltage of 5 kV. A thin gold coating was applied onto the surface by plasma sputtering to minimize the charging effects.

A field emission scanning electron microscope (FESEM, Zeiss SUPRA 40) was also utilized to observe the morphology of coatings of GO and rGO nanosheets on glass fibers. Approximately 5 nm thick layer of platinum was deposited on samples prior to FESEM observations.

3. Results and Discussion

3.1. Effects of graphene homogeneously dispersed in the polymer matrix

3.1.1. Polypropylene-based composites. Both single-fiber fragmentation test [15] and microdebond tests [20] clearly evidenced how a homogeneous dispersion of GNP in the PP matrix improve the fiber-matrix adhesion . In fact, as evidenced in Figure 3, the ISS values increases with the GNP content in the polymer matrix. It is worthwhile to note that both micromechanical testing methods provided comparable results and that ISS value is improved by a factor of about 6 when 7 wt% of GNP is dispersed in the polymer matrix.

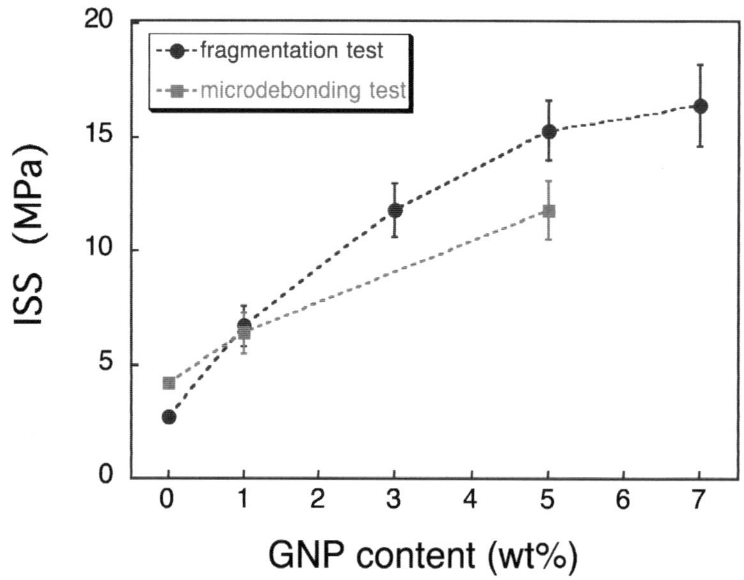

Figure 3. ISS values as determined by both the fragmentation and microdebonding tests on PP-based microcomposites containing GNP homogeneously dispersed in the matrix.

The observed effects of GNP on the interfacial adhesion are comparable to those induced by the addition of more common compatibilizers used to improve the fiber/matrix adhesion in PP/glass fiber composites, such as maleic anhydride modified polypropylene (PPgMA). In fact, as documented in Figure 4, the fiber/matrix adhesion levels reached by dispersing GNP in the PP matrix are quite similar to those observed when PPgMA is dispersed in the same polymer matrix.

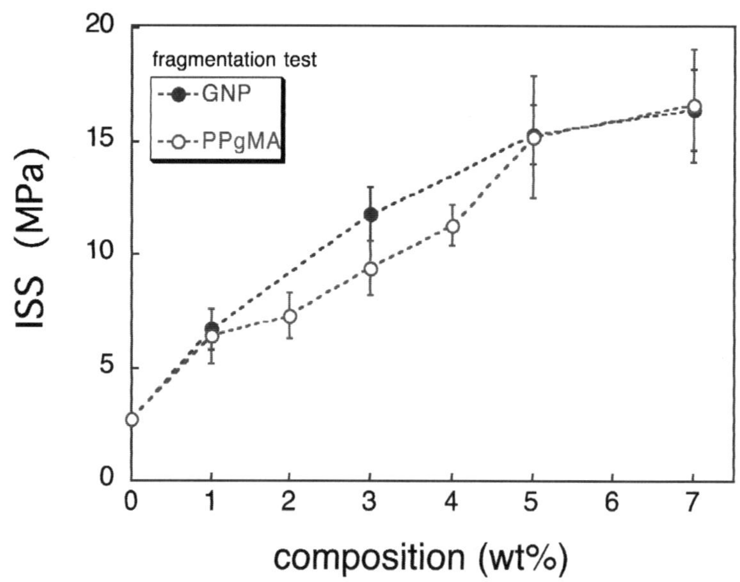

Figure 4. ISS values as determined by the fragmentation test on PP-based microcomposites containing GNP nanoplatelets or PPgMA compatibilizer homogeneously dispersed in the matrix.

A possible reason for the positive role of both GNP and PPgMA on the fiber/matrix interfacial interaction can be explained by analyzing the energetics of the surfaces involved. In fact, as reported by Pedrazzoli and Pegoretti [15], wettability measurements in two different liquids (water and ethylene glycol) evidenced that the thermodynamic work of adhesion of the polymer matrix with respect to glass fiber, was improved by the presence of GNP or PPgMA. In particular, the polar component of the matrix surface tension (γ_p) increased considerably due to the addition of PPgMA, probably due to the presence of hydrophilic maleic anhydride groups with high surface energy. PP/GNP systems manifested a similar increase in γ_p likely because of the possible presence of functional groups at the edges of GNP [21]. The dispersive component (γ_d) is slightly higher for all nanocomposites with respect to unfilled PP and increases proportionally to the content of PPgMA or GNP [15].

It is interesting to investigate how the positive effects of GNP on the fiber/matrix interfacial interactions can reflect on the properties of PP-composites reinforced with large quantities of short glass fibers. For this reason, PP-based injection moulded composites reinforced with various amounts of short GF_1 and GNP were mechanically tested in tension. First of all, as manifested in Figure 5 the appearance of the fracture surfaces of the composites with and without GNP appeared to be markedly different. SEM micrographs of fracture surfaces for PP composites reinforced with 10 wt% GF_1 are shown in Figure 5(a). The GF appeared to be quite homogeneously dispersed in the PP matrix but several fibers were pulled out from the matrix. Interfacial debonding resulted to be the dominant failure mechanism, indicating a rather low adhesion level. This observation is consistent with the outcome of the micromechanical tests. On the other hand, a different situation can be noticed for hybrid PP composites containing 5 wt% GNP and 10 wt% GF whose fracture surface is reported in Figure 5(b). In this case, very few debonded fibers can be observed, thus indicating a significantly better fiber–matrix adhesion. Furthermore, the improved fiber/matrix compatibility is documented by the presence of matrix fragments on the fiber surface after pull-out.

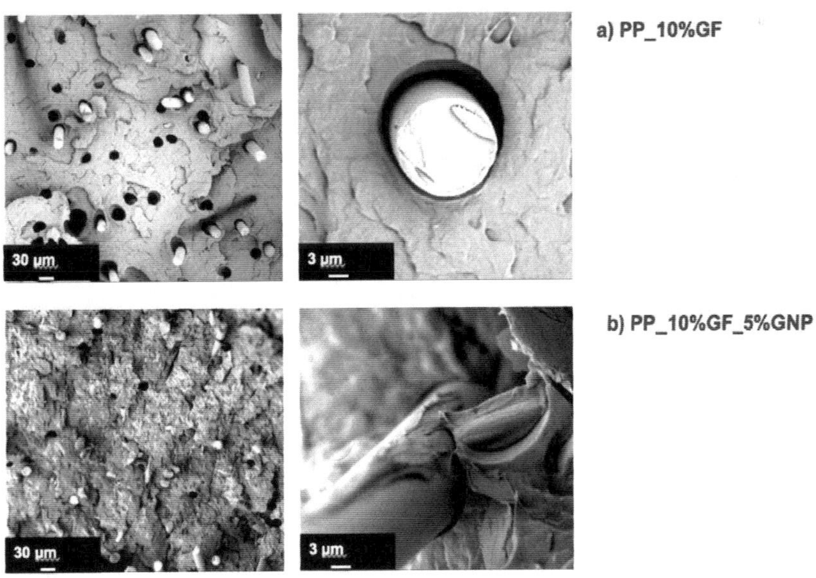

Figure 5. SEM micrographs of fracture surfaces of a) PP composites reinforced with 10 wt% of short glass fibers, b) PP composites reinforced with 10 wt% of short glass fibers and 5 wt% of GNP.

The tensile modulus of composites with 10, 15 and 30 wt% of short GF_1 and various amounts of GNP is plotted in the following Figure 6. The presence of GNP in the polymer matrix plays a beneficial effect of PP-based composites reinforced with 10 wt% of short GF_1, whose elastic modulus continuously increased until an increment of about 38% was reached when 7 wt% of GNP were added. It is also worthwhile to note that the addition of 5 wt% of GNP in a PP-based composite reinforced with 15 wt% of short glass fibers generated a material whose elastic modulus was comparable to that of a composite loaded with 30 wt% of short GF_1. I the inserts of Figure 6 some additional properties measured on the investigated composites are reported, such as the tensile strength σ_b, the Izod impact strength and the density values [20]. The addition of GNP improved the strength values of composites reinforced with 10 and 15 wt% of short GF_1 at levels higher than that of a composite reinforced with 30 wt% of short GF_1 without GNP. Moreover, the Izod impact strength was preserved and the density decreased. In fact the density of the composites with 10 wt% of short GF_1 and 7 wt% of GNP was about 12% lower than that of the composites reinforced with 30 wt% of short GF_1 without GNP. Therefore, the presence of GNP lead to improved multiscale composites both in term of mechanical properties and lightness.

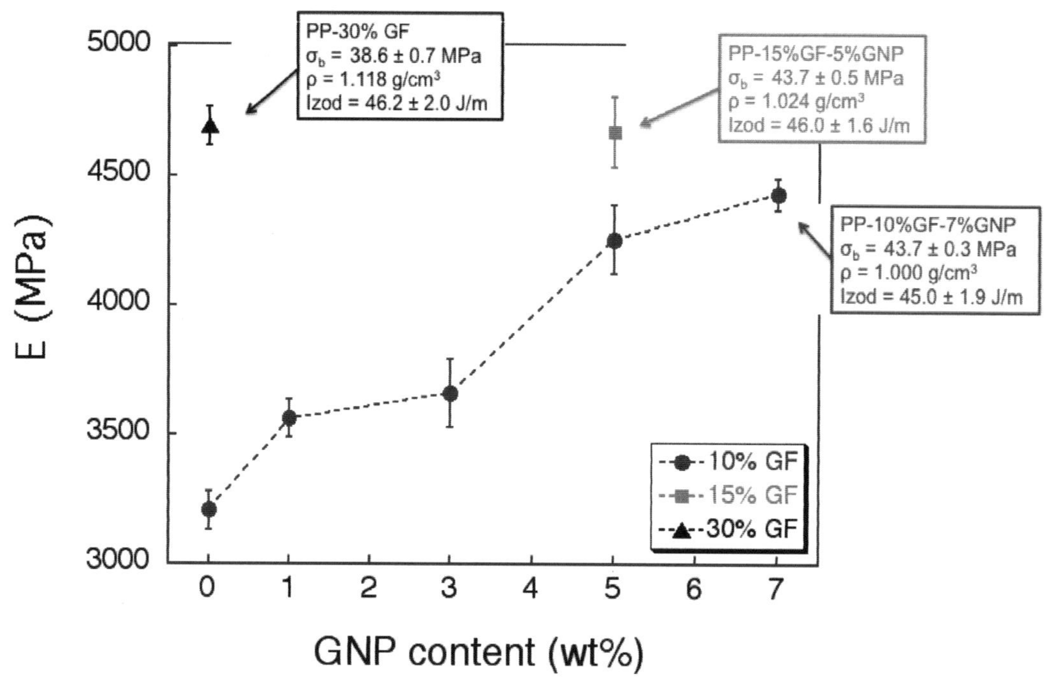

Figure 6. Tensile modulus of PP-based composites with various amounts of short glass fibers (i.e. ● 10 wt%, ■ 20 wt% and ▲ 30 wt%) and GNP.

3.1.2. Epoxy-based composites. The beneficial effect of the homogeneous dispersion of 5 wt% of GNP in epoxy_2 was tested by microdebonding test on microcomposites in which small epoxy droplets are deposited on GF_1 fibers (see Figure 7a), and the obtained results are reported in Figure 7b.

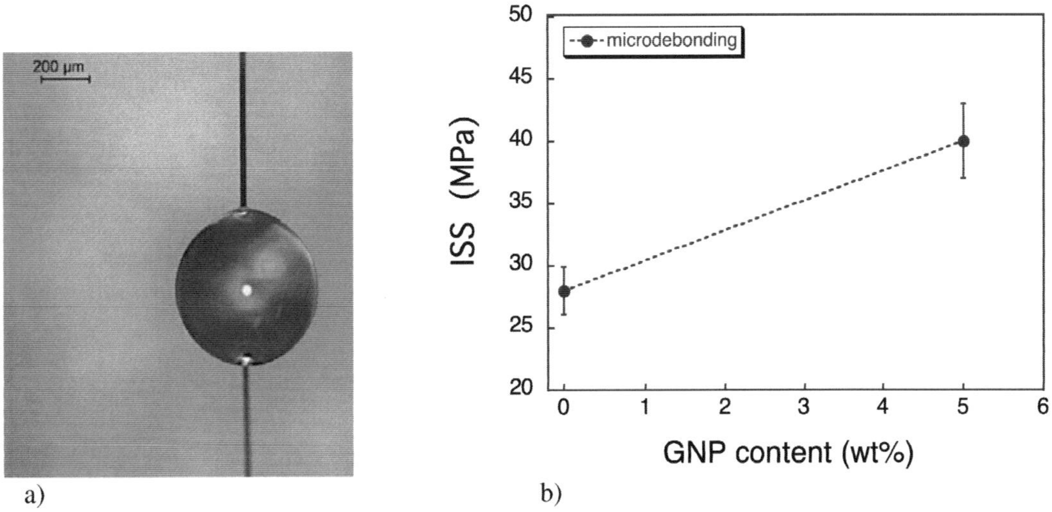

Figure 7. a) Epoxy-glass single-fiber specimen for microdebonding; b) effect of GNP on the interfacial shear strength in epoxy-glass microcomposites.

Similarly to what was observed for the PP-based composites, even for epoxy/glass composites a positive effect of GNP on the fiber/matrix interfacial shear strength was clearly detected. In fact, the dispersion of 5 wt% of GNP in the polymer matrix induced an increase of ISS values by about 40%.

Concurrently, the mechanical properties of the neat epoxy_1 resin resulted to be improved by the presence of GNP, as documented by the data reported in Figure 8. In fact, for a GF content of zero (corresponding to the neat epoxy), an improvement of tensile modulus (E, in Figure 8a), tensile strength (σ_b, in Figure 8b) was observed, while the Izod impact strength (in Figure 8c) was practically not affected. From the data reported in Figure 8, it can be observed that improvement of E, σ_b and also of the impact strength are caused by the presence of GNP in epoxy_1 composites with various amounts of short glass fibers. In particular, it can be observed that a composite filled with 15 wt% of short GF_1 (PP-15%GF-5%GNP) exhibited tensile modulus and impact strength values similar to the corresponding values of a epoxy_1 composite containing 30 wt% of short GF_1 (PP-30%GF). At the same time, for the hybrid composite PP-15%GF-5%GNP an higher tensile strength value (70.4 MPa) with respect to the sample PP-30%GF (67.8 MPa) was measured. In the plots of Figure 8 the density of each composite tested is also reported. It is worthwhile to note that the hybrid composite PP-15%GF-5%GNP had a density lower by ~ 7% compared to the density of 30 wt% of short GF_1/epoxy.

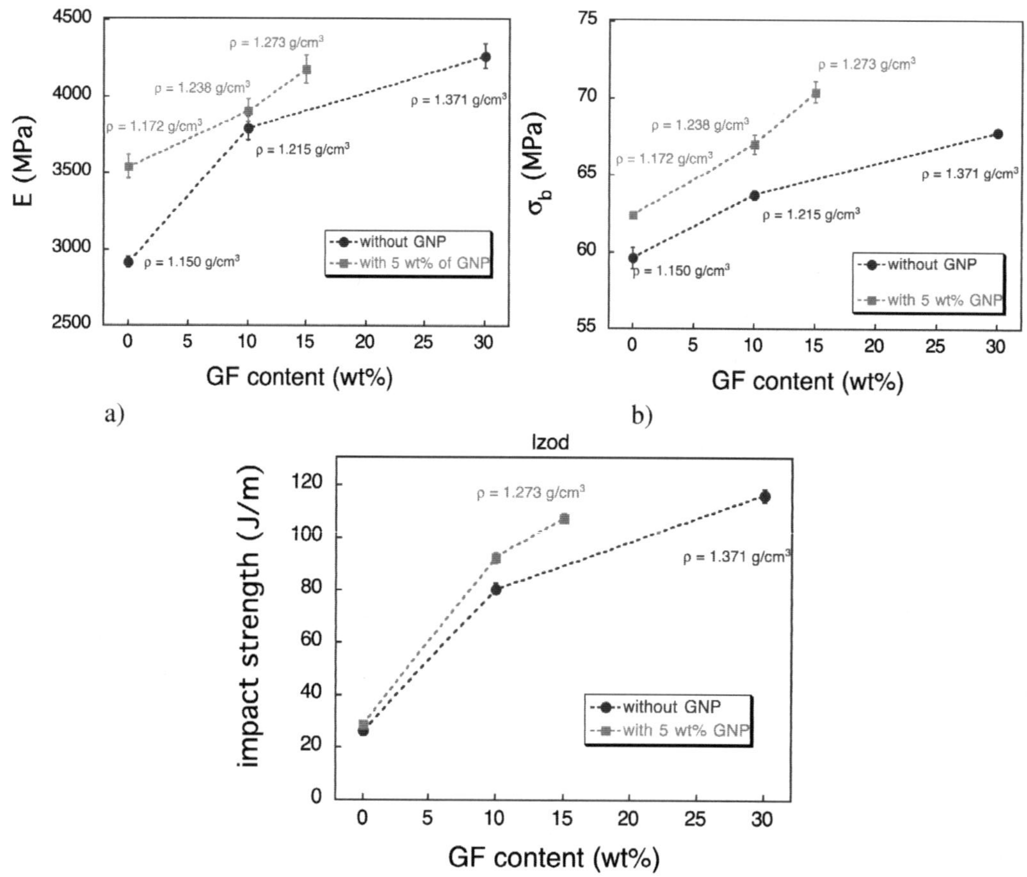

Figure 8. Effect of 5wt% of GNP on a) the tensile modulus, b) the tensile strength and c) the Izod impact strength of epoxy_1 composites reinforced with various amounts of short GF_1.

3.2. Effects of graphene-oxide and reduced graphene oxide deposited on the glass fibers

The electrophorectic deposition (EPD) process of graphene oxide on GF_2 fibers lead to a compact and quite homogeneous coating. The appearance of GF_2 fibers coated with both GO at two different deposition voltages can be observed in the SEM pictures reported in Figure 9. Figure 9c the appearance of fiber coated with reduced graphene oxide can also be observed.

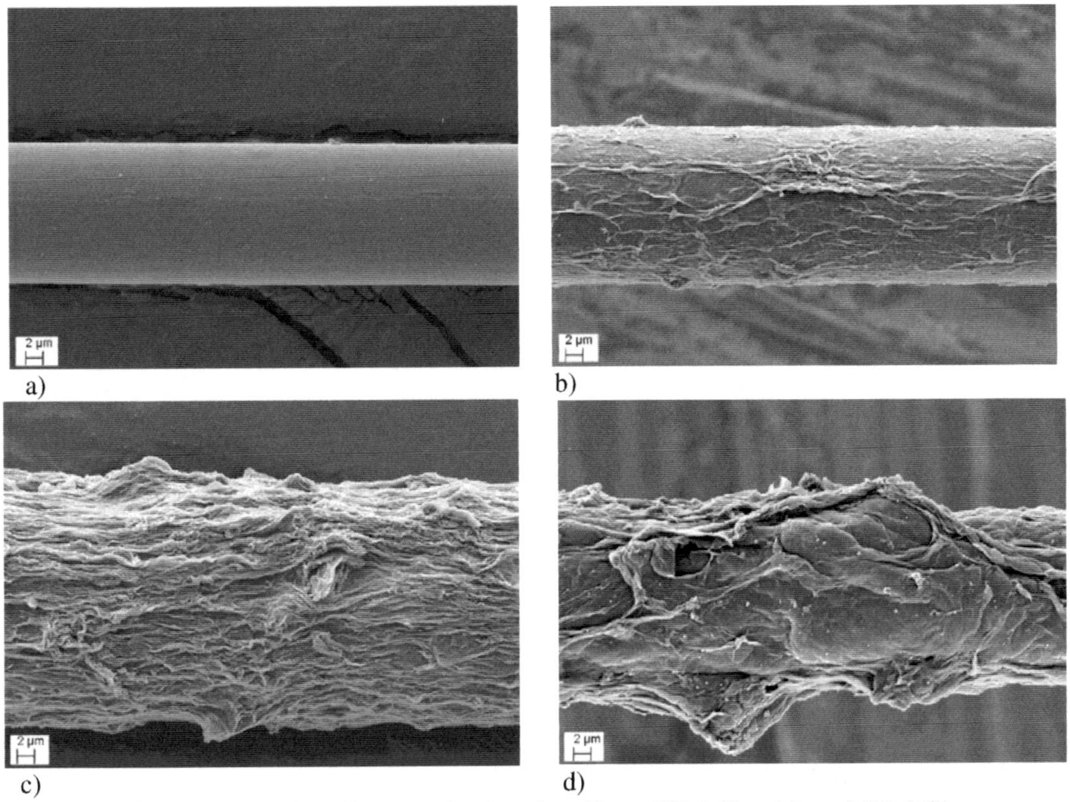

Figure 9. Scanning electron micrographs of bare GF_2 fiber (a), and GF_2 fibers electrophoretically coated with GO deposited at 7.5 V/cm (b) or 10 V/cm (c) and subsequently reduced to rGO (d).

The thickness (weight) of the deposited coatings resulted to linearly increase with the intensity of the applied voltage adopted during the electrophorectic deposition process [13].

The adherence of the GO coating on the glass fiber surface was investigated by tribological tests performed at a nanoscale level by atomic force microscopy (AFM) [13]. Calibrated tip was slid from bare GF to GO covered region at a fixed normal force (FN). This method was capable of measuring substrate-coating adhesion energy by debonding the deposited GO. The analysis and modeling of the obtained force data allowed us to assign a value of about 130 MPa to the shear stress required for the debonding of GO coating from the glass fiber. A reported in Figure 9c, even after the chemical reduction process the rGO coating appears to be tightly bounded on the glass fiber surface.

The single-fiber fragmentation test indicated a positive effect of the GO coating on the stress transfer ability at the fiber/matrix interface. In fact, as reported in Figure 10, the ISS values improved as the intensity of the voltage applied for the GO deposition increased. It is noteworthy to observe that for GO-coated fiber at the maximum EPD voltage of 10 V/cm an increment of the ISS value of about 220% was obtained with respect to the uncoated fibers. The effect of a chemical reduction of GO coating into rGO on the ISS values was also investigated by single-fiber fragmentation tests whose results are reported in Figure 10. It is interesting to observe that the rGO coating electrophoretically deposited on the glass fibers played a positive effect on the fiber/matrix ISS values which displayed an improvement of about 70% at the maximum EPD voltage of 10 V/cm. At the same time, it is clear that the beneficial effect of rGO coating was lower than that observed for GO coating. This difference can be attributed to the much lower amount of oxygenated functional groups in rGO with respect to GO, as experimentally proven by Fourier transformation infrared spectroscopy (FTIR) and X-ray photoelectron spectroscopy (XPS) [22].

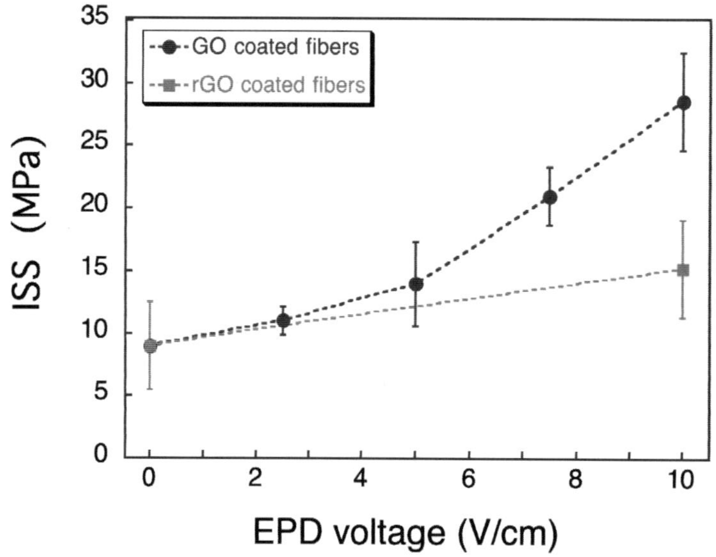

Figure 10. ISS values as determined by the fragmentation test performed on microcomposites containing GF_2 fiber coated with GO or rGO applied by electrophoretic deposition (EPD) under various applied voltages.

Electrical resistivity measurements on unidirectional composites containing GO or rGO coated glass fibers are reported in Figure 11 and compared with the case of uncoated glass fibers. It clearly emerges that the GO coating did not significantly modify the electrical resistivity of epoxy/glass composites. On the other hand, the electrical conductivity value of composites containing rGO coated glass fibers markedly decreased (by 12 orders of magnitude) with respect to the resistivity value of composites with uncoated fiber.

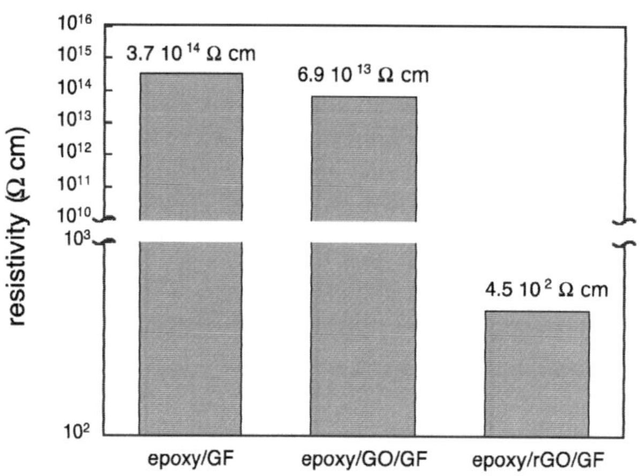

Figure 11. Electrical resistivity of composites reinforced with continuous uncoated glass fibers (epoxy/GF), continuous glass fibers coated with GO (epoxy/GO/GF) and continuous glass fibers coated with rGO (epoxy/rGO/GF).

Therefore, composites containing rGO coated fibers were subjected to various (both static and dynamic) mechanical loading conditions, and the strain was monitored (by an axial extensometer) simultaneously to electrical resistance variations along axial direction. For shake of brevity, only two cases are reported on Figure 12, regarding the obtained results under quasi static and the dynamic loading conditions. It is interesting to note that the variations in the electrical resistivity of the investigated composites quite carefully reflected the applied strain histories under both loading conditions. This experimental result clearly indicates that the rGO coating can extend to epoxy/glass composites the possibility of a strain monitoring based on the control of the electrical resistance variations.

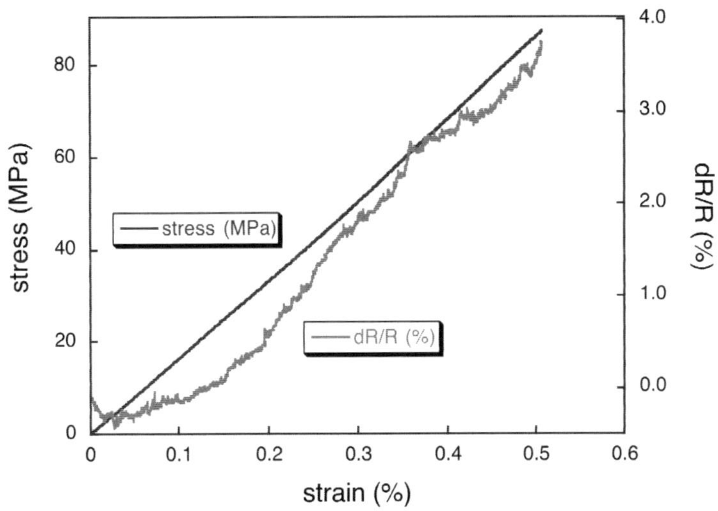

a)

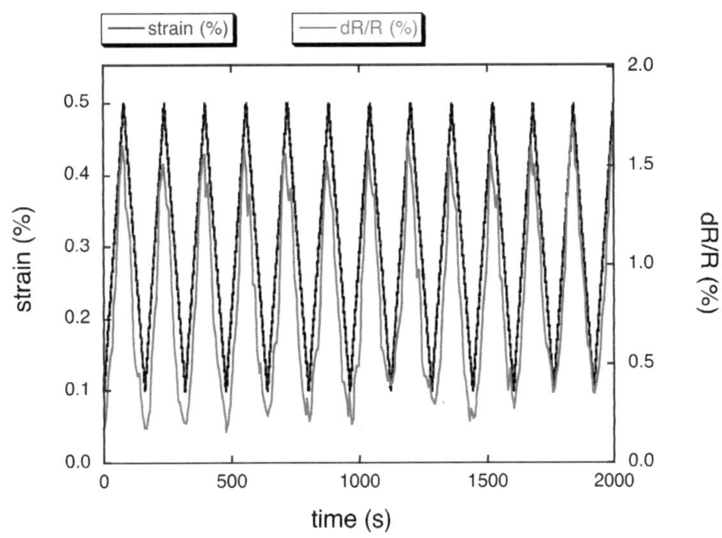

b)

Figure 12. Electrical resistivity variations of composites reinforced with continuous rGO coated glass fibers under a) quasi static and b) dynamic loading conditions.

4. Conclusions

In this manuscript an overview is presented on some possible strategies for improving the mechanical and functional properties of glass fiber reinforced composites through graphene and graphene-oxide nano platelets.

A homogenous dispersion of graphene nanoplatelets in a thermoplastic polymer such as polypropylene induced a remarkable improvement in the fiber/matrix adhesion with interfacial shear strength values improved up to a factor of about 6 when 7 wt% of GNP is dispersed in the polymer matrix. As a consequence, the presence of GNP played also a role in determining the mechanical properties of short-glass fiber reinforced PP. For instance, in PP-based composites reinforced with 10 wt% of short glass fibers an elastic modulus increment of about 38% was observed when 7 wt% of GNP were added. Moreover, the addition of 5 wt% of GNP in a PP-based composite reinforced with 15 wt% of short glass fibers generated a material whose elastic modulus was comparable to that of a composite loaded with 30 wt% of short glass fibers, with a significative reduction of the material specific weight.

The beneficial effects on the fiber/matrix adhesion of a the dispersion of GNP in the matrix resin had been also experimentally proved by microdebonding test on epoxy-based microcomposites. In fact, the dispersion of 5 wt% of GNP in the matrix resin induced an increase of ISS values by about 40%. Concurrently, improvements of tensile modulus, tensile strength and Izod impact strength were promoted by the presence of GNP in epoxy composites reinforced with various amounts of short glass fibers.

Another investigated strategy for the fiber/matrix interphase engineering was the electrophorectic deposition of graphene oxide on the glass fiber surface and the chemical reduction of the obtained coating into rGO. It is noteworthy to observe that ISS value for GO-coated fiber at the maximum EPD voltage of 10 V/cm an increment of the ISS value of about 220% was obtained with respect to the uncoated fibers. Noteworthy, the chemical reduction of the GO coating in rGO allowed to increase the electrical conductivity of unidirectional epoxy/glass composites by 12 orders of magnitude with respect to the typical values of composites with uncoated glass fibers. The piezoresistive response of composites reinforced with rGO coated glass fibers was monitored by measuring the electrical resistance change on specimens subjected to mechanical loading under quasi static and dynamic loading conditions. In this way, the possibility of strain monitoring due to the presence of an engineered fiber/matrix interphase was experimentally proved.

References

[1] Kim J-K and Mai Y-M 1998 *Engineered interfaces in fiber reinforced composites* (Amsterdam, The Netherlands: Elsevier)

[2] Pegoretti A and Karger-Kocsis J 2015 Interphase engineering in polymer composites: Challenging the devil… *Express Polymer Letters* **9** 838

[3] Karger-Kocsis J, Mahmood H and Pegoretti A 2015 Recent advances in fiber/matrix interphase engineering for polymer composites *Progress in Materials Science* **73** 1-43

[4] Karger-Kocsis J, Kékic S, Mahmood H and Pegoretti A in press *Interfacial and Adhesion Aspects in Polymer Nanocomposites,* ed K Mittal and A Netravali: Scrivener Publishing)

[5] Pegoretti A 2015 Editorial corner – a personal view. Where micro- and nano-worlds meet: multiscale polymer composites *Express Polymer Letters* **9** 329

[6] Geim A K 2009 Graphene: status and prospects *Science* **324** 1530-4

[7] Stankovich S, Dikin D A, Dommett G H, Kohlhaas K M, Zimney E J, Stach E A, Piner R D, Nguyen S T and Ruoff R S 2006 Graphene-based composite materials *Nature* **442** 282-6

[8] Bianco A, Cheng H-M, Enoki T, Gogotsi Y, Hurt R H, Koratkar N, Kyotani T, Monthioux M, Park C R, Tascon J M D and Zhang J 2013 All in the graphene family – A recommended nomenclature for two-dimensional carbon materials *Carbon* **65** 1-6

[9] Singh V, Joung D, Zhai L, Das S, Khondaker S I and Seal S 2011 Graphene based materials: Past, present and future *Progress in Materials Science* **56** 1178-271
[10] Traina M and Pegoretti A 2012 In situ reduction of graphene oxide dispersed in a polymer matrix *Journal of Nanoparticle Research* **14** 801_1-_6
[11] Fukushima H 2003 Graphite nanoreinforcements in polymer nanocomposites, PhD thesis. Department of Chemical Engineering and Materials Science, Michigan State University.
[12] Hummers W S and Offeman R E 1958 Preparation of Graphitic Oxide *Journal of the American Chemical Society* **80** 1339-
[13] Mahmood H, Tripathi M, Pugno N and Pegoretti A 2016 Enhancement of interfacial adhesion in glass fiber/epoxy composites by electrophoretic deposition of graphene oxide on glass fibers *Composites Science and Technology* **126** 149-57
[14] Watcharotone S, Dikin D A, Stankovich S, Piner R, Jung I, Dommett G H B, Evmenenko G, Wu S-E, Chen S-F, Liu C-P, Nguyen S T and Ruoff R S 2007 Graphene–Silica Composite Thin Films as Transparent Conductors *Nano Letters* **7** 1888-92
[15] Pedrazzoli D and Pegoretti A 2014 Expanded graphite nanoplatelets as coupling agents in glass fiber reinforced polypropylene composites *Composites Part A: Applied Science and Manufacturing* **66** 25-34
[16] Pedrazzoli D 2014 Understanding the effect of nanofillers on the properties of polypropylene and glass fiber/polypropylene multiscale composites Department of Industrial Engineering, University of Trento.
[17] Herrerafranco P J and Drzal L T 1992 Comparison of methods for the measurement of fiber matrix adhesion in composites *Composites* **23** 2-27
[18] Pedrazzoli D, Dorigato A and Pegoretti A 2012 Monitoring the mechanical behaviour of electrically conductive polymer nanocomposites under ramp and creep conditions *Journal of Nanoscience and Nanotechnology* **12** 4093-102
[19] Pedrazzoli D, Dorigato A and Pegoretti A 2012 Monitoring the mechanical behaviour under ramp and creep conditions of electrically conductive polymer composites *Composites Part A - Applied Science and Manufacturing* **43** 1285-92
[20] Pedrazzoli D, Pegoretti A and Kalaitzidou K 2015 Synergistic effect of graphite nanoplatelets and glass fibers in polypropylene composites *Journal of Applied Polymer Science* **132**
[21] Laskowska A, Marzec A, Zaborski M and Boiteux G 2014 Reinforcement of carboxylated acrylonitrile-butadiene rubber (XNBR) with graphene nanoplatelets with varying surface area *Journal of Polymer Engineering* **34**
[22] Mahmood H, Vanzetti L, Bersani M and Pegoretti A in preparation Strain monitoring in glass-epoxy composites with graphene-coated fibers

Applications for carbon fibre recovered from composites

SJ Pickering, Z Liu, TA Turner and KH Wong

Composites Research Group, Faculty of Engineering, The University of Nottingham, University Park, Nottingham, NG7 2RD, United Kingdom

E-mail: stephen.pickering@nottingham.ac.uk

Abstract. Commercial operations to recover carbon fibre from waste composites are now developing and as more recovered fibre becomes available new applications for recovered fibre are required. Opportunities to use recovered carbon fibre as a structural reinforcement are considered involving the use of wet lay processes to produce nonwoven mats. Mats with random in-plane fibre orientation can readily be produced using existing commercial processes. However, the fibre volume fraction, and hence the mechanical properties that can be achieved, result in composites with limited mechanical properties. Fibre volume fractions of 40% can be achieved with high moulding pressures of over 100 bar, however, moulding at these pressures results in substantial fibre breakage which reduces the mean fibre length and the properties of the composite manufactured. Nonwoven mats made from aligned, short carbon fibres can achieve higher fibre volume fractions with lower fibre breakage even at high moulding pressure. A process for aligning short fibres is described and a composite of over 60% fibre volume fraction has been manufactured at a pressures up to 100 bar with low fibre breakage. Further developments of the alignment process have been undertaken and a composite of 46% fibre volume fraction has been produced moulded at a pressure of 7 bar in an autoclave, exhibiting good mechanical properties that compete with higher grade materials. This demonstrates the potential for high value applications for recovered carbon fibre by fibre alignment.

1. Introduction

Carbon fibre is finding increasing numbers of applications in lightweight structures due to its high specific mechanical properties. High growth rates are already being observed in several major civil aerospace programmes, such as Boeing 787 Dreamliner and Airbus A350. Optimistic levels of growth are also predicted for the automotive, industrial and power generation sectors. Current levels of carbon fibre usage are in excess of 100,000 tonnes per annum with growth forecast to be between 10% and 20% per annum. Automotive remains an interesting potential future user of considerable amounts of fibre with global production now in excess of 90,000,000 vehicles per annum. Even a small amount of fibre in a small fraction of total production would make a significant impact on demand.

Several processes are now operating commercially for the recovery of carbon fibres from end of life components, such as aircraft parts and tooling, as well as in-process scrap such as out of shelf-life prepreg and ply cuter offcuts. The high energy use associated with the production of virgin carbon fibre leads to a high market value and gives the potential for the production of recovered fibre at a significantly lower cost than that associated with virgin fibre. Process scale is of course a dominant factor with virgin fibre production lines showing significantly increased efficiency through improved process monitoring and hardware, whereas recovered fibre production is typically smaller scale thus struggling to make use of the efficiencies of larger scale operation.

Currently it is difficult to understand the impact of legislation on the recovered carbon fibre market; increasing pressure to manufacture fuel efficient vehicles drives car manufacturers towards lightweighting methodologies but often these are somewhat at odds with consumer desires which ultimately drive production. The End of Life Vehicle Directive serves to put pressure on vehicle makers but often has the effect of prohibiting the use of materials which will save energy over the use phase of the product lifecycle. No corresponding regulation exists in aerospace where the use phase dominates and lightweighting through the use of carbon fibre is overwhelmingly compelling. Landfill tax remains as a motivating factor for fibre recovery with increasing difficultly experienced with the disposal of both uncured and cured waste and a growing impetus for cost saving at a business level.

Although fibres may be recovered at a cost which is lower than the corresponding grade of virgin fibre, the form of the recovered fibres is inevitably different from virgin and this factor has to-date limited the penetration of recovered fibres into virgin fibre markets. The challenge therefore is to change the material into a form which competes with virgin materials (whether on the basis of cost or performance) without incurring so much cost that the price is not competitive with virgin fibres.

1.1. The nature of recycled carbon fibre

In pyrolysis or thermal fluid recycling processes, the physical form of the fibres recovered is basically that of the material fed in with the polymer removed. If scrap prepreg materials are processed it may be possible to recover the carbon fibre in the form of a textile, in which the fibres are continuous, ready for re-impregnation. However, in most cases the scrap being processed would be in a variety of physical forms, as shown in figure 1, and so a short fibre product could be produced by either chopping the waste before processing or by chopping the carbon fibre recyclate after recycling. The fibre products are in a fluffy form as there is no longer any size holding the fibres bundles together. The fluidized bed process also produces a recyclate with a fluffy form, as shown in figure 2, in which the carbon fibre is in the form of individual filaments with a distribution of fibre length. All the processes produce a clean fibre recyclate with no substantial residue on the surface, as shown in figure 3.

Figure 1. End of life laminate scrap after primary size reduction (left) and secondary size reduction (right).

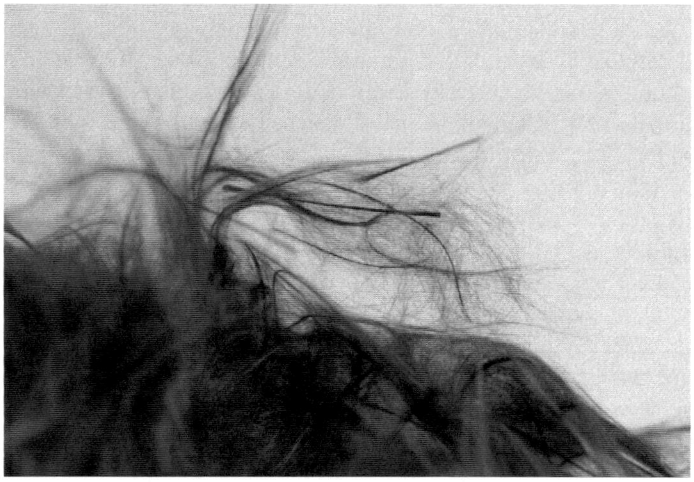

Figure 2. Recycled fibre in raw form showing fluffy, discontinuous, 3D random and highly entangled structure with areas of higher fibre packing due to retained bundle integrity.

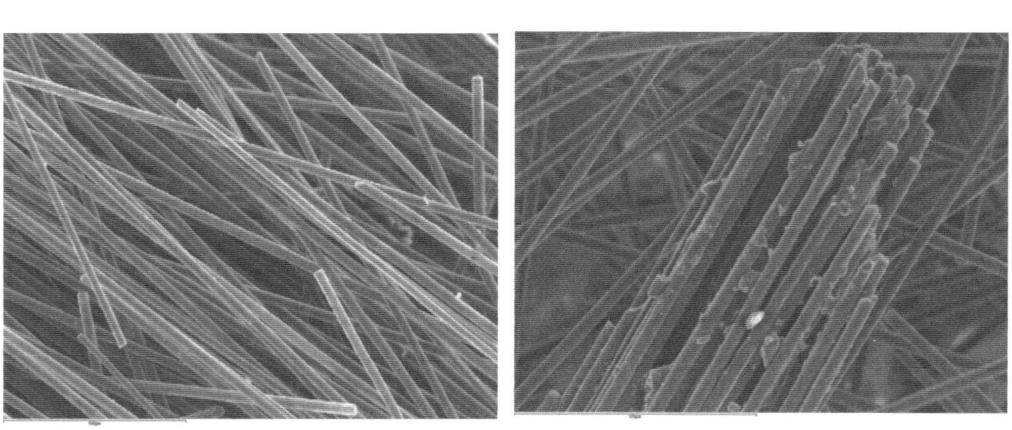

Figure 3. Recycled fibres, showing a clean surface free from polymer residue (left) and with some char remaining (right) causing bundle integrity to be maintained.

The mechanical properties of recycled carbon fibre are generally good and compare well with virgin fibre. Measurements of stiffness show that the recycling processes generally gives recycled fibres with a stiffness similar to that of virgin fibre. However, there is some degradation in tensile strength. The thermal fluid processes only show a reduction of a few percent. The pyrolysis processes are reported to show slightly more strength reduction with up to 10% loss in strength. However, the fluidized bed process shows a loss of strength of up to 50%. This may be due to the increased mechanical agitation in the process and also the effect of the oxidizing atmosphere. Though process is particularly suitable for contaminated end-of-life waste which may not be appropriate for other recycling processes.

The electrical conductivity of the recycled fibre has been found to be similar to that of virgin fibre and analysis of the surface chemistry shows that after recycling there are still active oxygenated species on the surface and the recycled fibres have been found to bond well to epoxy resin.

1.2. Applications for recycled carbon fibre

There are a many potential applications for recycled carbon fibre and it can be used as a structural reinforcement in composites with a wide range of mechanical properties and hence value. Broadly there are several approaches for reuse:

1. Direct competition with virgin materials (composite or metallic)
   - Reduced mechanical performance may be offset by reduced cost
   - At what scale should the recycled materials be compared with other materials: tow, mat, or compound?
2. Development of or penetration into new markets
   - New material forms may be appropriate for use of recycled fibre (e.g. commingled with other fibres)
   - Replacement for glass as well as virgin carbon fibre.
3. Increased functionality
   - Discontinuous fibres may be used in applications to give enhanced drape properties for easier manual or automated forming
   - Exploiting the conductivity of the recycled fibre in applications where reduced mechanical strength is of secondary importance.
   - Manufacturing low fibre areal mass materials, in which discontinuous recycled fibre may be preferred.

Currently the outlook is unclear and many routes for market development are under consideration. Milled fibre is an obvious end use application but it has low mechanical properties as a reinforcement and the market is not significant when compared to the volumes of fibre that may become available. Pellets for injection moulding processes are one potential avenue and highly aligned mat is another area showing promise.

1.3. The case for high value applications for recycled carbon fibre

The key driving force for future recycled carbon fibre-based materials will be a business case whereby the fibre recovery provider can sell the fibres to an end user or intermediate product manufacturer. A high value end use application enables a more viable supply chain for recovered fibres rather than a low value, high volume re-use route. Penetration into the automotive market is seen as a significant enabler for carbon fibre recycling in general and the costs proposed by automotive vehicle manufacturers provide a good target for recovered carbon fibre-based materials. The challenge is to manufacture materials from the recovered fibre without significantly increasing the cost. Aluminium is a good mechanical property target as the automotive industry already performs high levels of research into weight saving through the use of this material. These two cost and performance requirements therefore define the area of applicability for recovered fibre products.

2. Achieving high fibre volume fraction with fluffy discontinuous fibre

Composite structural performance depends largely on the mechanical properties and volume content of the reinforcement and how these structural properties can be attained will define the market that the composite will be targeted toward. At an initial level, the mechanical property of virgin and recycled carbon fibres don't appear very different: both of them have insignificant difference in tensile stiffness and interfacial shear strength; in the case of tensile strength, it depends on the chosen recycling process and the commercially available pyrolysis processes can offer fibre with high degree of strength retention. However, a marked difference is in their physical form: virgin fibre is continuous and can readily be weaved into the trademark woven pattern that marks it out as belonging to high grade applications, for instance found in superlight bicycles, Formula One and aerospace structure; whereas recycled fibre is short and discontinuous, without sizing and in a fluffy form, preventing it from being a direct substitute for virgin fibre and requiring further conversion processes to improve handling and processability. What they have in common is that both are valuable materials that can be processed specifically for specific applications. For recycled fibre, having economically viable conversion processes is pivotal for them to be established away from markets associated with a perception of lower quality, such as being used as a filler in injection moulding, and to reach out to wider markets offering higher grade applications.

Using milled recycled fibre in injection moulding is the earliest conversion process to enable the recycled fibre to be used in high volume basis. This application will have a long term presence but wider routes are required as the market will become saturated. Attention is focusing on seeking out established technology in which recycled fibre can be used directly, and the papermaking industry may be able to offer not only a process for direct use of recycled fibre but also an economically viable process for turning the fluffy fibre into a nonwoven form. Work is still ongoing to optimise the process and to establish markets for the nonwovens. Meanwhile, the recycling industry is continuing to develop new processes for higher value usage and fibre alignment is considered one an important conversion options as it not only offers higher fibre volume fractions to be achieved with lower moulding pressure, but it will also reduce high pressure fibre-to-fibre contact points and preserve fibre length for much needed structural application development. The following two sub-sections report the use of nonwoven mats in reinforcing thermoset composites, and consider the pros and cons as well identifying the need to lead to the development of an aligned fibre structure.

2.1. Use of random wet lay non-wovens for composites manufacture

In a paper presented at the SAMPE conference in 2009 [1], the feasibility of using recycled fibre in a nonwoven form for a flame-retardant epoxy composite was reported. The fibre was of grade T300 from Toray and was recycled via a pyrolysis process by ELG Carbon Fibre Ltd, Coseley, UK (formerly Recycled Carbon Fibre Ltd) . The average input fibre length before moulding was 12 mm and it was converted into a 100 g/m^2 nonwoven mat 520 mm wide using a papermaking process at Technical Fibre Products Ltd, Kendal, UK. The nonwoven mats were collated with 200 g/m^2 resin film, MTM56-2FRB, supplied by Cytec (formerly Advanced Composites Group), Heanor, UK and compression moulded at 120 °C under a pressure varying from 2 MPa to 14 MPa to make a composite of fibre volume content ranging from 20% to 40%. The moulding pressure required to achieve a required fibre volume fraction was determined by performing a compression test on a stack of dry non-woven mats between two platens attached to a universal testing machine with distance between the platen and load were being constantly recorded. To evaluate the mechanical performance of the composite, tensile and flexural tests were performed on at least 5 specimens for each fibre loading. The change in fibre length due to the moulding pressure was measured by matrix burn-off, fibre separation, digital image capture via a microscope and finally length measurement via the Image Pro Plus image analysis software, provided by MediaCybernetics, Cambridge, UK.

Figure 4 and figure 5 show that the nonwoven mat has been successful in enhancing the tensile and flexural moduli of the composite and a linear correlation between the moduli and fibre content can be observed. However, for the case of composite strength, such a linear correlation only applies for fibre content up to 30% as beyond which the composite performance starts to drop. Several factors can affect the strength of a fibre reinforced composite, including the inherent fibre strength, shape, orientation and interface with the matrix. All these factors are considered common between all samples tested and the drop in strength is attributed to the degradation in fibre length distribution and increase in composite void content and these are shown in figure 6and figure 7 respectively. For full utilisation of the fibre reinforcement potential, the fibre must be longer than L_c, the critical fibre length, which is given by $L_c = r_f\sigma_f/\tau_c$, where r_f, σ_f and τ_c are the fibre radius (3.6 μm), fibre ultimate tensile strength (4.16 GPa) and composite shear strength respectively. Shear strength varies with the chosen type of carbon fibre and for a typical epoxy composite reinforced with recycled carbon fibre, the value varies between 60.0 MPa and 80.0 MPa [2]. Thus, the estimated critical length is between 0.19 mm and 0.25 mm. It can be seen in figure 6 that at 20% fibre content, about 70% of the fibre population is longer than 0.25 mm. However, at 40% fibre content, there is only about 10% and thus the fibre reinforcing potential is greatly reduced. As mentioned earlier, another factor that could contribute to the decrease in strength is the presence of high a void content, as depicted in figure 7.

Nonwoven mat has a high degree of loft as it comprises filamentised fibres with random planar orientation and a wide range of length distribution. This suggests a very irregular flow channels between the fibres and the permeability very much depends on the availability of interstitial free space. It is hypothesized that under high moulding pressure, the creation of short fibres promotes nesting and reduces the interstitial free space, disrupting or locally blocking up the flow channels, resulting in higher void in the resultant composite.

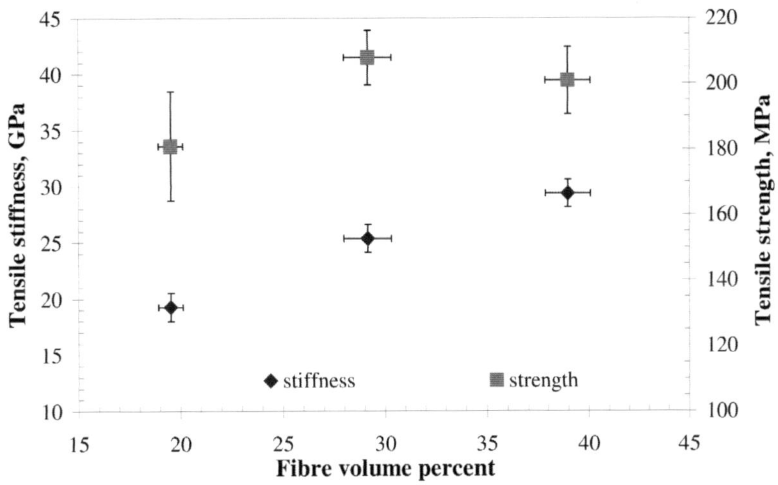

Figure 4. Tensile properties of composite reinforced with recycled carbon fibre nonwoven mats at various fibre volume content [1].

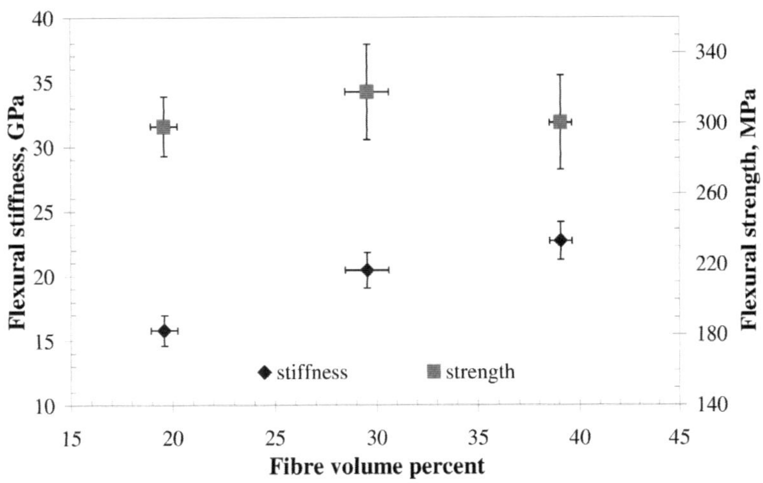

Figure 5. Flexural properties of composite reinforced with recycled carbon fibre nonwoven mats at various fibre volume content [1].

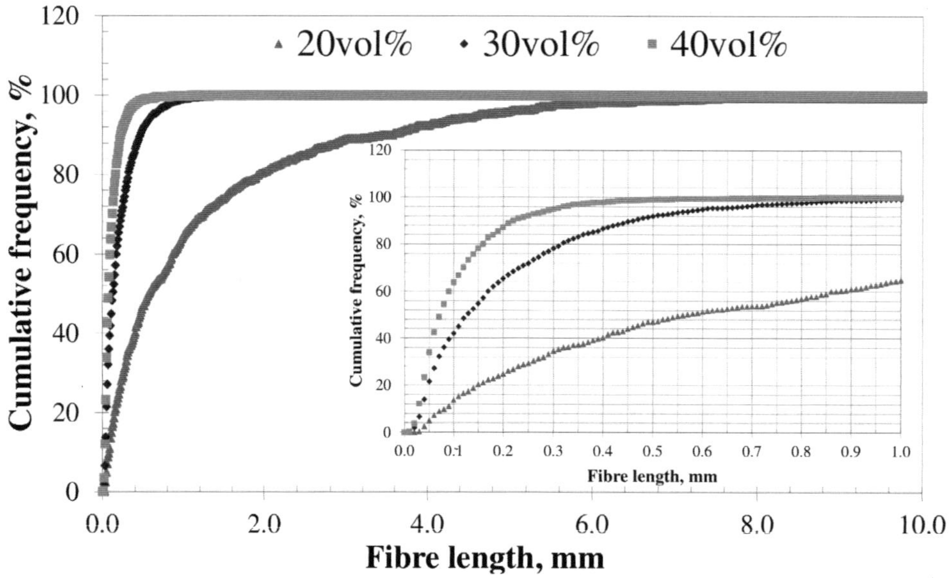

Figure 6. Fibre length distribution from composites at various fibre volume contents [1].

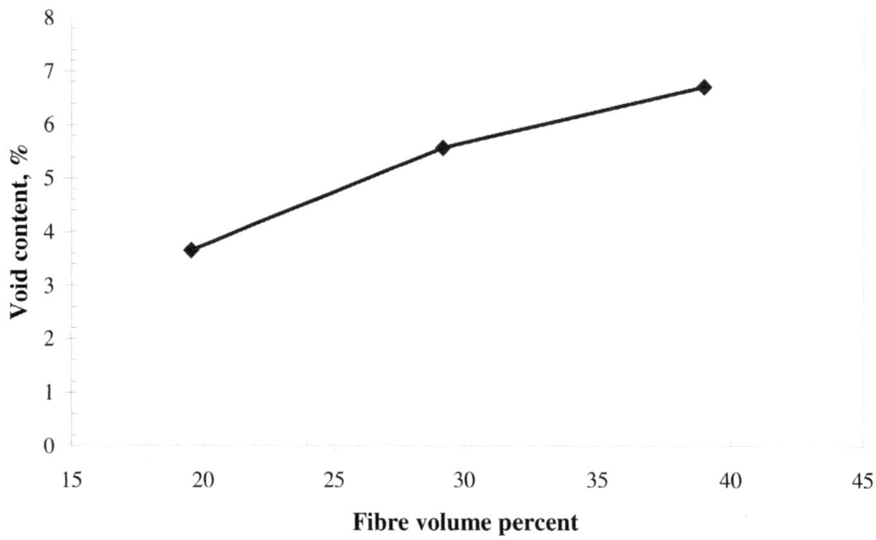

Figure 7. Composite void content [1].

2.2. The need for fibre alignment

Research and development in nonwoven mat manufacture continues to be of importance to the composite industry as it is built on an established know-how and is capable of lending itself to existing moulding technology, such as resin film infusion and compression moulding for semi-structural applications. Thermoplastic material, whether in the form of powder, filament or film, can potentially be included in the nonwoven mat, producing a commingled structure ready for thermoforming. However, with a random, filamentised fibre architecture, the maximum practical volume content is about 30% in order to minimize excessive length degradation. Higher volume fraction can only be realised through fibre alignment, a conversion process to arrange individual filament to a preferential direction to form a close-packed structure.

3. Technology for fibre alignment

Much work has been done in the past to control orientation of discontinuous fibre. Conventional textile processes such as carding, combing and gilling have been successful in improving orientation for cotton yarn, which is generally long, tough and flexible. Unfortunately, for the recycled carbon fibre, which is unsized, entangled and brittle, these intense mechanical actions are likely to cause severe damage to the fibre itself. Injection moulding is a process capable of introducing anisotropy to composite mechanical property by affecting the fibre orientation. A 3-layer structure is generally produced across the thickness of moulded sample with two skin layers having fibre preferentially aligned to the flow direction, but fibre at the core section has major orientation distribution normal to the flow. Besides the high shear stress inside the melting and dispersion zones of the moulding machine, the interaction between fibres and machine parts have a significant effect on the resultant fibre length of the moulded part. Practically, 35 vol% is the maximum fibre content this process can offer without causing too much damage to the fibre length. Ideally, interaction between fibres should be kept to minimum, or in other words, fibres are allowed to rotate and translate freely toward the preferential direction and this implies that alignment is best taken place with low fibre volume content and then the aligned fibres are densified to obtain a close-packed structure. These requirements can be met by wet hydrodynamic processes, which subject a dilute or semi-dilute fibre suspension toward a convergent device for alignment and after which the liquid medium is removed to obtain a layer of well aligned fibres. Further details of the hydrodynamic processes and their process parameters that govern the alignment efficiency are reported in the subsequent section.

3.1. Techniques for alignment of discontinuous fibre

Several approaches have been developed using different means to align discontinuous fibre. The diamagnetic anisotropic property of carbon fibre has been exploited under magnetic field for alignment development but this is limited to fibre length less than 500 micron [3, 4]. Electric fields have also been used for orientation control but generally for carbon nanotube [5] or polymer fibres [6]. Another approach that has drawn attention is via hydrodynamic force created by fluid flow. Of these approaches, the hydrodynamic process is considered to the most practical as the working principal is based on well-established and mature hydrodynamic technology and it also offers versatility for pilot scale research and can readily be scaled up for commercial application. Between the late sixties and the seventies, the Explosives Research and Development Establishment (ERDE), UK devised and developed three patented hydrodynamic alignment techniques based on extrusion [7], filtration [8] and centrifugal [9] process.

The extrusion process was designed for aligning asbestos and silicon carbide whiskers in an ammonium alginate solution [10]. The fibre suspension was extruded through an orifice of 2mm diameter into an acid precipitating bath. The extrudate gelled immediately in the bath and hence the alignment was maintained. After reeling up on a winding drum, it was then washed and dried. Then the residual alginic acid was burned off. In the pilot plant, the nozzle traversed across a winding drum to build up several complete layers of extrudate sheets. The whole sheet was then cut across to give a large flat sheet. The level of alignment was not presented but high volume packing up to 50% was reported. However, this process was deemed to be impractical and uneconomical as the ammonium alginate itself was not recyclable and it required a lengthy combustion stage for complete removal of the alginic acid [11].

The filtration approach is the most widely published ERDE fibre alignment process. It started from a batch scale, which was capable of making aligned fibre sheets up to 40 cm x 30 cm. It was later developed into a continuous process for a 152 mm wide sheet roll [11, 12]. Discontinuous fibre was dispersed in glycerine to form a suspension. A good dispersion quality was obtained if there were no undispersed bundles in the bulk suspension and this could be achieved with the selection of short fibre length, low fibre loading and the presence of a high shear rate within the viscous medium. Air entrapment should be kept to minimum as fibres deviating around air bubbles would affect the alignment quality. The suspension was then pumped gently toward a convergent device with a narrow slit exit. Fibres were aligned inside the convergent device through a velocity gradient across the suspension streamlines. The device was located above a fine conveying mesh and travelled repeatedly depositing suspension over a region on the mesh. A vacuum was applied under the mesh to promote draining. The damp mat was then washed with a gentle water spray and binder applied. The degree of alignment attained from this process was good and more than 90% of the fibre population was within ±15° of the desired direction [13]. However, when the mat thickness increases, it's through-thickness permeability would decrease, resulting in a longer dwell period to allow completion of glycerine drainage before the next deposition was applied and leading to an economically unacceptable production rate [14]. In addition, due to the reciprocating motion of the convergent slit, a lower degree of alignment was seen at the end regions of the reciprocating path.

The centrifugal process was built on the strength of the filtration approach but equipped with two significant modifications i.e. the narrow slit exit was replaced with a series of convergent nozzle as the latter offered an all-round converging contour along the nozzle axis and the conveying mesh was replaced with a rotating drum lined with a fine mesh internally as the drainage could be accelerated with the aid of centrifugal force. It had been claimed that a better alignment was achieved with higher throughput rate [15].

3.2. Initial development of high fibre volume fraction composites from recycled carbon fibre

A bench-top centrifugal rig was built according to the principal outlined by the ERDE process [9]. The capability of the rig was investigated via a $2^{(6-1)}$ fractional factorial experiment design to provide a better insight into the effect of operating conditions on alignment quality. The six factors under consideration were fibre length, fibre loading in the slurry, glycerine viscosity, nozzle diameter ratio, aligned mat areal density and relative speed between the rotating drum and slurry deposition rate. Orientation of the fibre within the aligned mat was measured through image analysis of the fibre elliptical cross-sectional geometry and this orientation data was used as an output response to assess the effect of changes from the six input factors. Virgin Toho Tenax HTA 5131 carbon fibre with its sizing removed by heating at 550 °C for 10 minutes inside a furnace was used. Once the best operating condition was identified from the test, a 3mm long recycled Toray T800 carbon fibre, supplied by ELG Carbon Fibre Ltd, was then used for making the mats, which was later compression moulded with epoxy resin films into a composite for mechanical testing. A summary of a more detailed analysis [16] of the test result is presented here. A schematic representation of the alignment rig is shown in figure 8. After the deposition process, the rotational speed of the perforated drum was set to 1500 rpm and this speed was maintained for 1 minute in order to further remove glycerine from within the mat. The mat was then placed on a vacuum tray for removal of the residual glycerine through water spraying and later was followed with application of a general grade PVA binder, sourced from Hobbycraft, Dorset, UK. The mat was then dried at room temperature overnight.

From the results of the test, it was found that there were four main effects: alignment quality improved with a higher relative speed ratio and nozzle diameter ratio but there was a negative correlation with mat areal density and fibre length. The only interaction effect was between the fibre length and veil area density. Figure 9 shows the fibre orientation distribution of the best and the worst aligned mats from the tests. The former had 94% of the fibres laid within ±10° from the preferred orientation, whereas the latter case only 78%.

The results of the test allowed the formulation of the best operating conditions for aligning recycled carbon fibre. It is found that a fibre slurry with 0.1 vol% of recycled fibre was required in order to prevent entanglement of the fibre filaments. The glycerine was diluted with 5 wt% of tap water to lower its viscosity in order to accelerate filtration rate during the fibre deposition stage. A higher relative speed was selected for getting a better alignment by further stretching the extruded fibre slurry before it reached the rotating mesh. The designated areal density for the recycled fibre mats was 35g/m^2. The mats produced were collated with resin film and pressed to 100 bar 3 times to increase the packing by a nesting effect. Then a moulding pressure of 80 bar was applied at 120 °C to fabricate a composite of thickness of around 2 mm and with a targeted fibre volume content of 60%. The mechanical properties of the composite were evaluated by performing a 3-point bending test along and normal to the alignment direction and the result is shown in table 1. It can be seen that the flexural strength measured along the alignment direction is a magnitude higher than the value perpendicular to the alignment and similarly for the flexural stiffness. This clearly indicates that a good fibre alignment has been achieved and this is further supported by referring to the micrograph shown in figure 10, which presents a close-packed fibre architecture. Image analysis of these micrographs revealed that the local fibre volume content was between 60% and 63%, which was in agreement with the targeted value of 60%. The void content of the composite was found to be around 1%. The fibre length distribution of the composite was determined by matrix burn off to allow individual fibre lengths to be measured digitally. The length distribution based on number fraction is shown in figure 11and a peak can be seen near the 3.1 mm region. Weight average and number average of the fibre lengths are also included in the figure 11 and when compared to the corresponding input fibre length, no significant degradation in fibre length is observed, despite the mats had been subjected to 100bar pressure in the early stage of the moulding process. The outcome of this test clearly highlighted the advantages of having a well aligned fibre structure compared to the nonwoven mat reported early.

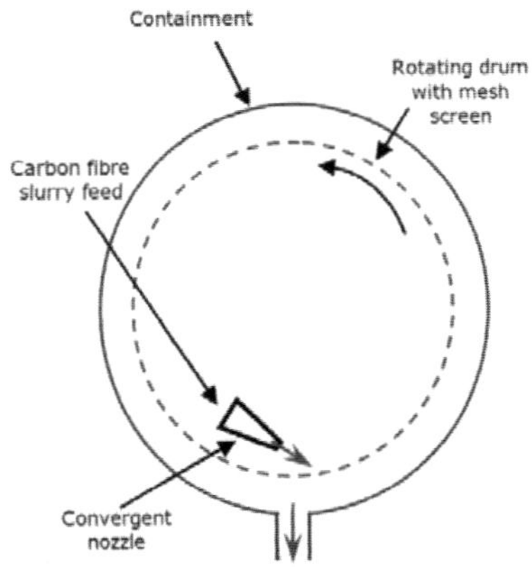

Figure 8. A schematic representation of the fibre alignment rig [16].

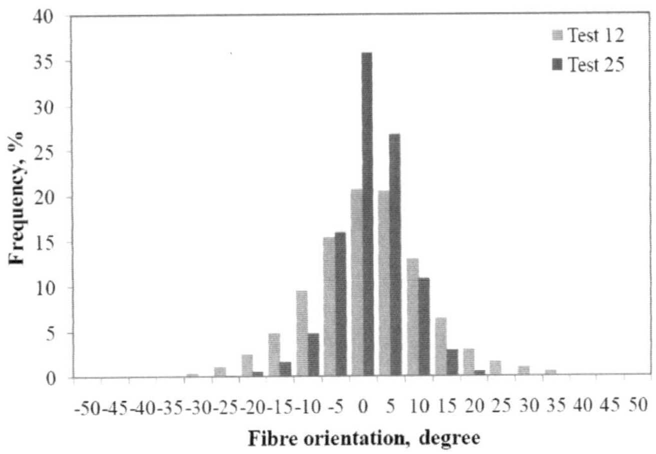

Figure 9. Fibre orientation distribution obtained from the best and worst samples from the factorial test [16].

Table 1. Average flexural properties of composite containing recycled and aligned carbon fibre [17].

Specimen	Flexural strength, MPa	Flexural stiffness, GPa
Along the alignment direction	1240 ± 69	81.8 ± 5.0
Normal to the alignment direction	107 ± 2	7.2 ± 0.2

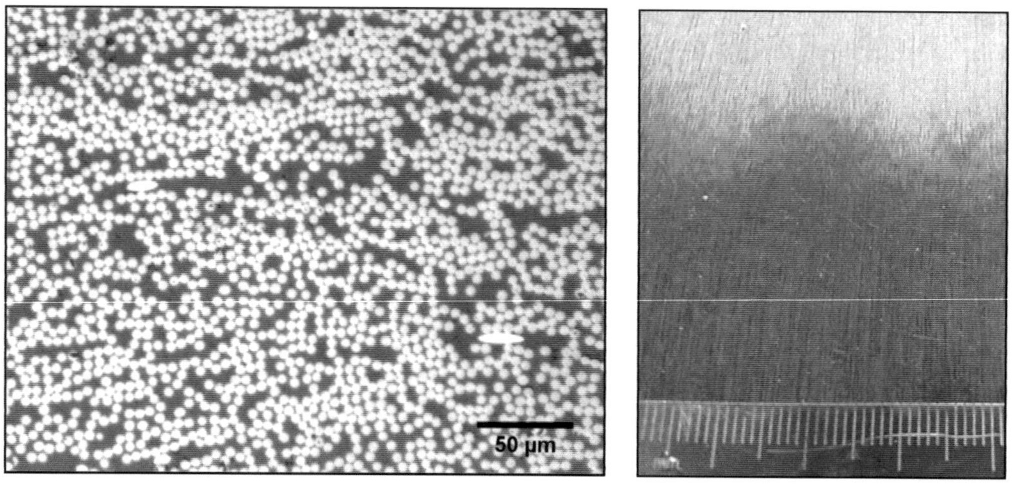

Figure 10. Micrograph of a composite containing 60% (vol) recycled, aligned fibre mat (left) and a top view of the mat before moulding (right) [16].

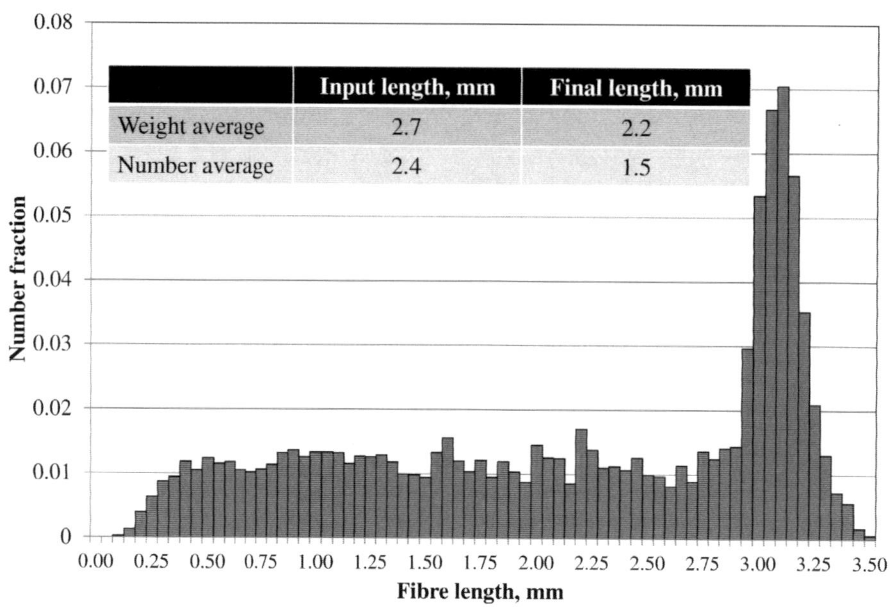

Figure 11. Length distribution of fibre obtained from composite reinforced with recycled, aligned carbon fibre with 60% volume fraction [16].

4. High fibre volume fraction composites moulded at lower pressures

Within the composites industry, the best mechanical properties are achieved in composites manufactured from unidirectional continuous fibre where a fibre volume fraction of about 60% can be achieved by moulding a pressures of less than 10 bar in an autoclave. Whilst a composite fibre volume fraction of 60% (vol) from aligned recycled carbon fibre has been achieved, processing pressures of up to 100 bar were required. Moulding at high pressure is inevitably more expensive and further work is currently being undertaken to gain a better understanding of the process of fibre alignment with the aim of being able to achieve better alignment at lower moulding pressures. In this section, work on liquid jet geometry and its dependency on fluid properties and suspended fibre is reported. Modifications were then made to the test rig and this resulted in the production of an aligned mat able to give a higher fibre content at a lower moulding pressure. Full size composite coupons were then fabricated and the tensile properties are presented.

4.1. Improving fibre alignment – understanding convergent jet behaviour

This section introduces the use of ink jet technology to provide an insight into the understanding of factors that governing the geometry of jet emerging from a given nozzle.

4.1.1. Jet geometry influencing alignment quality. The fundamental principle of the hydrodynamic alignment process is that short fibres are aligned by the velocity gradient along the flow direction in a convergent nozzle. Then, the jet of fibre suspension is deposited on a nylon mesh and the liquid drained to form an aligned fibre mat. The alignment process has been further developed through experience gained from the paper manufacturing industry. It was found through careful observation of the fibre suspension jet that an unstable jet (figure 12a) which splashes during deposition process will reduce fibre alignment. A varicose jet and air bubbles (figure 12– b, c) can misalign fibres.

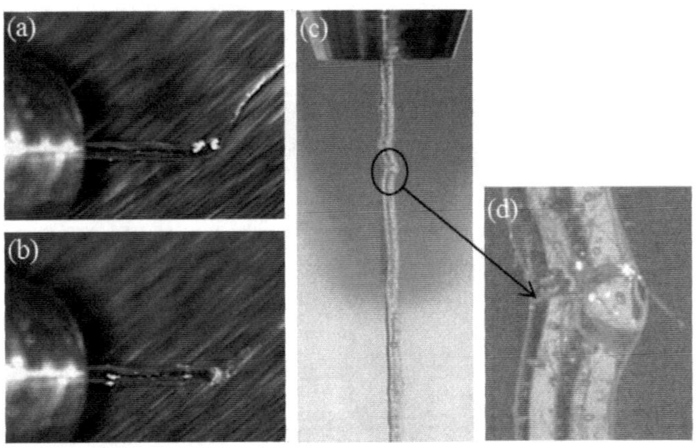

Figure 12. Photos of unsatisfactory fibre suspension jets – (a) unstable jet; (b) varicose jet; (c) jet containing air bubbles.

4.1.2. Investigating influencing factors on fibre suspension jet stability. Liquid jets are known to break up because of the interfacial forces between the jet and the surrounding air. When a liquid jet is discharged into air, disturbances on the jet surface will be augmented by the aerodynamic interactions between the jet and the surroundings. The most commonly accepted jet disintegration classification in fluid mechanics literature was developed by Ohnesorge (1936) who combined the Reynolds and Weber numbers and proposed a dimensionless Ohnesorge number Z (or Oh) [19].

$$Z = \frac{\mu_L}{\sqrt{\rho_L \sigma d_j}} = We_L^{0.5} Re_L^{-1} \qquad (1)$$

where We_L is liquid's Weber number, Re_L is liquid's Reynolds number and d_j is the liquid jet diameter at nozzle exit. Based on Ohnesorge's work, Reitz suggested that there are four liquid jet breakup regimes which are presented in figure 13and table 2 [19] and Derby stated that the criterion for a drop to possess sufficient kinetic energy to be ejected from the nozzle is $We_{crit} \geq 4$ or $Re \leq 2/Z$ [20]. Therefore, a stable jet can be developed when $4 \leq We_L \leq 8$.

Table 2. Expressions for breakup regimes [19].

Regime	Range
Rayleigh (varicose)	$We_L \geq 8$ and $We_g < 0.4$
First wind-induced (sinuous)	$1.2+3.41Z^{0.9} < We_g < 13$
Second wind-induced	$13 < We_g < 40.3$
Atomization	$We_g > 40.3$

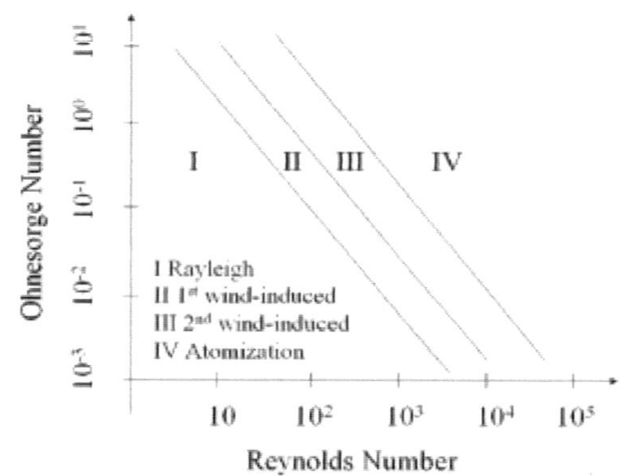

Figure 13. Classification of the modes of liquid jet disintegration [19].

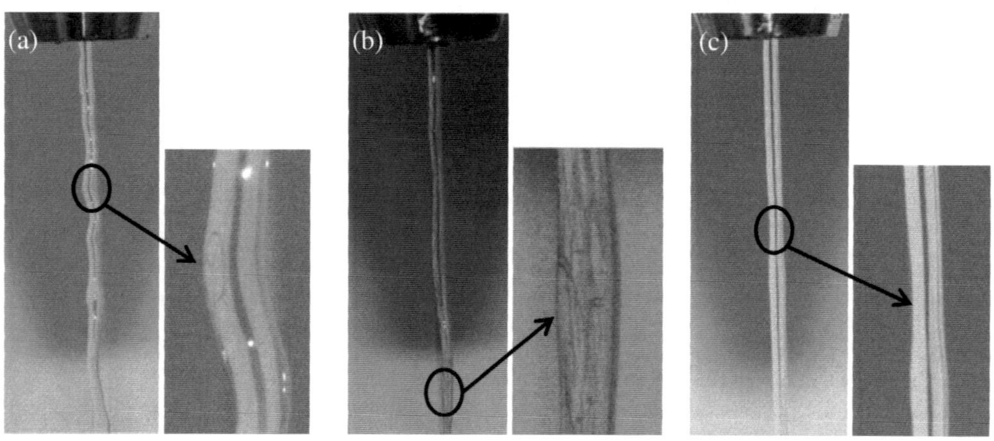

Figure 14. Images of fibre suspension jets and glycerine/water jets all at the same Weber number: (a) 12mm length fibre_0.05% volume fibre concentration jet – the fibre becomes disturbed in the wavy jet; (b) 3mm length_0.15% volume fibre concentration - fibre suspension jet and fibres are aligned with flow direction; (c) glycerine/water solution jet behaves varicose slightly.

In a two phase flow jet (liquid/fibre), the jet breakup regimes are likely to be different to the liquid jets described in figure 13 and table 2. Figure 14 shows that the fibre–liquid interaction and fibre–fibre interaction give jets different unstable effects although the Weber numbers are the same. Future work will focus on exploring the effects of different fibre lengths and fibre volume concentrations on suspension jet stability.

4.2. Exploring hydrodynamic alignment process - experimental investigations
This section reports the mechanical properties of composite, which contained aligned mats that were produced frm the modified alignment process. A Design of Experiment (DoE) approach was implemented to investigate correlation between factors that affected the alignment quality.

4.2.1. Materials. Tenax®-A HT C124 carbon fibre staples coated with a water-soluble sizing, was supplied by Toho Tenax Europe GmbH. The fibre lengths were 3 mm and 12 mm and the individual fibre nominal diameter is 7 µm. Fibre sizing was removed by placing the staples inside an ashing furnace at 550 °C for 15 mins. Oleon Glycerine 4810 with glycerol content ≥99.5% was supplied by Univar UK. The glycerine was diluted with tap water until a viscosity of around 400 mPas was measured at room temperature (15 °C) using a Brookfield LVDVII viscometer. The diluted glycerine was used as a dispersion medium.

4.2.2. Manufacture of aligned mat and Design of Experiment (DoE). The hydrodynamic alignment process, as depicted in figure 8, was used for this study. During the alignment process, the fibre dispersion is pumped into a pressure pot and is pressurized to create a steady flow to a concentric conical nozzle that is located above a rotating drum. A nylon mesh positioned inside the drum is employed as a filter to separate fibre from dispersion medium. A vacuum suction is applied under the mesh to increase the liquid draining rate. The nozzle is fixed to a linear actuator and continuously moves forward and backward to create a certain width fibre mat. The filtered glycerine is separated from air by a cyclone and is pumped back to a storage tank for reuse. Once the target veil areal density has been met, the slurry deposition process will be halted. The mat is then washed with warm water to remove the remaining dispersion medium and an epoxy based binder is applied to maintain the aligned fibre orientation.

A 2-level full factorial experimental work plan was designed to explore the effects of processing factors on alignment quality of discontinuous carbon fibre in hydrodynamic alignment process. Details of the factors and levels are listed in table 3. The dimensions of the aligned mats made were 135mm in width by 900mm in length.

Table 3. Details of full factorial DoE.

Fibre length, mm	Fibre volume concentration in suspension, %	Wire / Jet transvers velocity ratio	Dewatering vacuum level	Number of replicates
3	0.05	1.5	Low	2
12	0.15	5	High	2

4.3. Dry fibre mats compaction results
A compaction test rig for measuring carbon fibre mat thickness (used to determine the average fibre volume fraction) under different compression pressure is shown in figure 15. The rig essentially contains two flat platens; the bottom platen is fastened to the base of a 5969 universal testing machine from Instron; the upper platen is connected to a 50 kN load cell, which is mounted to the machine crosshead. System compliance of the entire rig was first determined by closing the two platens together at a crosshead speed of 1mm/min until a compression pressure of 10 bar was registered by the

load cell. The crosshead location was automatically logged by the machine's control software during the test. The data was used to measure the distance between the two platens. Two linear variable differential transducers (LVDTs) were also mounted on the top platen to provide further readings of the distance between the platens. To measure the thickness and volume fraction of the mats, test specimens 35mm length square were obtained by stamping carbon fibre mat using a cutting template. Each mat was weighed before the test. Then 4 samples were stacked together and carefully transferred to the bottom platen and were compressed at 1 mm/min until 10 bar pressure was attained. The thickness and fibre volume fraction of the mats were determined by taking average of the readings from the two LVDT units.

Figure 15. Dry fibre mat compaction testing apparatus.

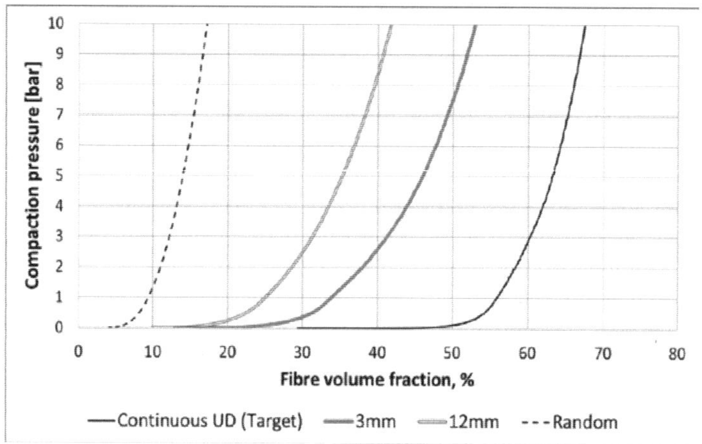

Figure 16. Dry fibre mat compaction test results.

The compaction result is presented in figure 16, includes two benchmarks: a 2D planar random mat and a unidirectional continuous fibre (UD) mat as well as two aligned mats of different fibre lengths obtained from the full factorial test. Under 10bar compaction pressure, the fibre volume fraction of UD mat is 68% though the random mat can only give 18% fibre volume content at the same pressure. This clearly indicates that an aligned fibre orientation improves fibre volume fraction. For discontinued short fibre (especially the fluffy recycled fibre), to achieve good mechanical properties, high fibre volume fraction is required. Thus, alignment is a necessity. It can be seen that with 0.1% of fibre concentration in the suspension, the 3 mm fibre aligned mat can achieve a volume content of 47.7% at 10 bar, which is 32.5% higher than the 12 mm fibre mats and this is attributed to having a better fibre suspension quality, which is free of fibre bundles for the shorter fibre. Further improvement was made to the rig and process operating parameters and this allows better alignment performance to be attained from a suspension with higher fibre concentration. At 10 bar pressure, the achievable fibre volume content for the 3 mm and 12 mm mats are 52.8% and 41.6% respectively.

4.4. Manufacture of aligned fibre composite in an autoclave and tensile testing

Laminates were made from discontinuous aligned fibre mats (details given in table 4) and MTM57 resin films supplied by Cytec Industrial Materials, UK. The materials were vacuum bagged to remove trapped air. The compacted materials were then moulded at 120 °C, 7 bar pressure, for 1 hour. Finally, it was cooled down to around 60 °C under pressure before demoulding.

Table 4. Specifications of aligned preforms and composite samples.

Fibre length	Fibre mat areal density	Fibre volume fraction	No. of layers fibre mat	MTM57 resin film areal density	No. of layers resin film	Moulded laminate dimensions
mm	g/m^2/layer	%	/	g/m^2/layer	/	$l \times w \times t$(mm^3)
3	72	46	24	83	16	280×135×1.87
12	71	43	20	83	19	280×135×1.93

Tensile tests were performed with a 250 kN load cell which was attached to the crosshead (figure 15), and a test speed of 1 mm/min was applied for all specimens, i.e. 4 specimens from each laminate. Extensometer was attached on each specimen to record strain. Micrographs of the cross-sectional view of the laminates are shown in figure 17 and a close-packed fibre structure is observed. Image analysis software was used on figure 17 and showed that the fibre volume content of 3 mm fibre composite is 46% and the fibre volume content of 12 mm fibre composite is 43%.

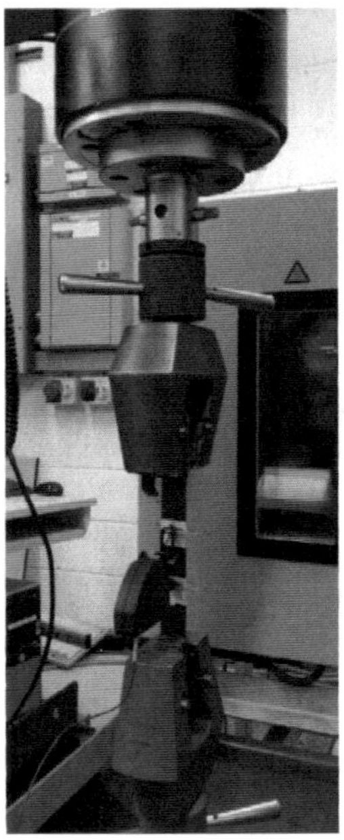

Figure 15. Tensile testing setup.

Figure 16. Fracture cross section of 3mm fibre laminate composite.

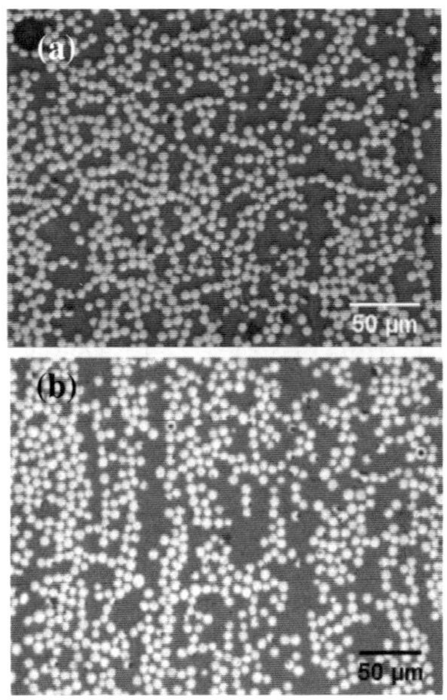

Figure 17. Micrograph of aligned composite: (a) 3mm_46% fibre volume fraction; (b) 12mm_43% fibre volume fraction.

4.5. Results

The results obtained from the tensile test are presented in figure 18. It can be seen that with 46% fibre volume fraction, 3 mm fibre laminate composite gave a 15.15 GPa higher tensile modulus than the laminate made with 12 mm fibre although the fibre length is much smaller. This indicates that good fibre alignment has been achieved and the aligned fibre structure can make a significant improvement to the mechanical properties. The void fraction in the composites was estimated, by image analysis, to be less than 2% for the composite made with 3mm fibre and less than 0.5% for the composite made with 12 mm fibre. The properties achieved for the 46% volume fraction composite give a specific tensile modulus of 0.057 GPa/kg/m^3 and a specific tensile strength of 0.42 MPa/kg/m^3, given a composite density of 1.5 kg/m^3. These specific properties are greater than alternative competing structural materials with the exception of carbon fibre composites and show the potential of fibre alignment for the manufacture of high value composites from recovered carbon fibre.

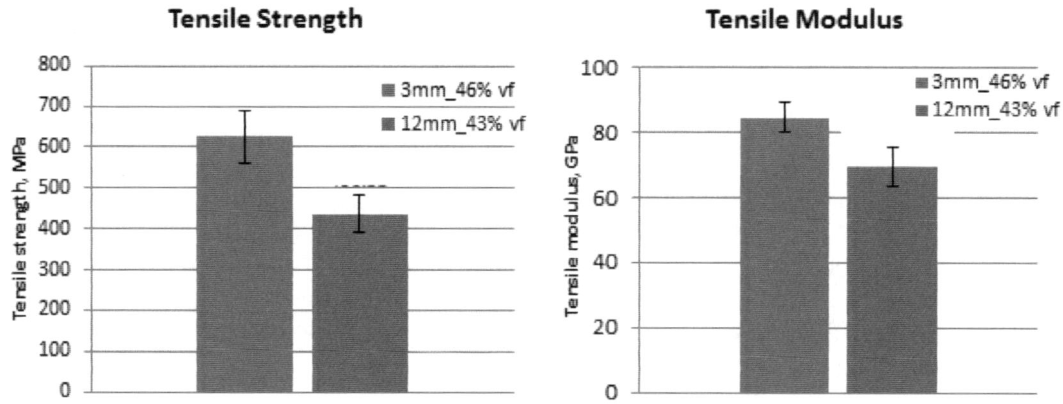

Figure 18. Tensile properties of composite reinforced with 3mm and 12mm aligned fibre mats.

5. Conclusions and future directions

New applications for increasing quantities of recovered carbon fibre produced by emerging commercial recycling activities are sought and use as structural reinforcement utilises the excellent mechanical properties of this material. Wet lay processing to produce nonwoven mats lend themselves to discontinuous recovered carbon fibre that typically has a fluffy form. However, random nonwoven mat structures cannot be processed into composites with mechanical properties better than readily available low cost metallic and composite materials. Nonwoven mats with aligned discontinuous fibres allow significantly better mechanical properties to be achieved, through higher fibre volume fractions in the composite and reduced fibre breakage during moulding processes. A process for achieving high levels of fibre alignment has been developed, based on existing technology, and this demonstrates the potential of fibre alignment for the production of high value composites that give specific mechanical properties better than other widely used low costs metallic and composite materials and approaching those of virgin carbon fibre composites.

Future work must now focus on developing alignment processes further, considering in particular process capital and operating costs, so that commercially viable processes are available to support future expansion of the emerging carbon fibre recycling industry.

Acknowledgment

The support of the Boeing Company in undertaking the research presented in Section 4 of this paper is gratefully acknowledged.

References

[1] Wong K, Pickering S, Turner T and Warrior N 2009 SAMPE '09 Spring Symposium Conference Proceedings, Baltimore, MD, 2009.
[2] Jiang G, Pickering S, Lester E, Turner T, Wong K and Warrior N 2009 *Comps Sci Tech* **69** 192-8
[3] Timbrell V 1972 *J App Phys* **43** 4839-40
[4] Matthews M, Dresselhaus M, Dresselhaus G, Endo M, Nishimura Y, Hiraoka T and Tamaki N 1996 *Appl Phys Let* **69** 430-2
[5] Park C, Wilkinson J, Banda S, Ounaies Z, Wise K, Sauti G, Lillehei P and Harrison J 2006 *J Pol Sci Part B: Pol Phys* **44** 1751-62
[6] Seyam A, Cai Y, Wendisch B and Kim Y 2008 *J Text Inst* **99** 375-83
[7] Parratt N 1968 GB 1128321

[8] Bagg G, Dingle L, Jones R and Pryde A 1971 US 3617437
[9] Bagg G, Cook J, Dingle L, Edwards H and Ziebland H 1977 US4016031
[10] Parratt N 1969 , *Comps* **1** 25-7
[11] Bagg G, Evans M and Pryde A 1969 *Comps* **1** 97-100
[12] Bagg G, Dingle L, Edwards H, Evans M and Ziebland H 1970 British Plastics Federation Seventh International Reinforced Plastics Conference, Brighton.
[13] Kacir L, Narkis N and Ishai O 1975 *Polym Eng Sci* **15** 525-31
[14] Edwards H and Evans N 1980 *Econ Comput Econ Cyb* **2** 1620-35
[15] Salariya A and Pittman J 1980 *Polym Eng Sci* **20** 787-97
[16] Wong K, Pickering S, Turner T and Warrior N 2011 SAMPE Spring Technical Conference and Exhibition - State of the Industry: Advanced Materials, Applications, and Processing Technology, Long Beach, CA
[17] Liu Z, Wong K, Thimsuvan T, Turner T and Pickering S 2015 20th International Conference on Composite Materials, Copenhagen Denmark
[18] Ohnesorge W 1936 ZAMM – *J Appl Math Mech* **16** 355-8
[19] Reitz R and Bracco F 1986 Encyclopedia of Fluid Mechanics: Gas Fluid Flows 233-49
[20] Derby B 2010 *Annu Rev Mater Res* **40** 395-414

IOP Conf. Series: Materials Science and Engineering **139** (2016) 012006 doi:10.1088/1757-899X/139/1/012006

A damage-based approach for the fatigue design of composite structures

Marino Quaresimin

Department of Management and Engineering (DTG), University of Padova, Stradella San Nicola 3, Vicenza, Italy

marino.quaresimin@unipd.it

Abstract. The paper illustrates a fatigue design strategy, based on the physics of the damage evolution, under development by the Composite Group at DTG-University of Padova. After a brief introduction, where the motivations of the work are analysed, examples of damage mechanisms at the microscopic scale are discussed. Then the procedures for the quantitative description of these mechanisms are illustrated.

1. Introduction

Composite materials are excellent candidates for the development of reliable lightweight components and structures complying with the needs, for instance, of decreasing fuel consumptions in the transportations field or increasing the specific energy production in the wind energy field.

Most of structural components manufactured with composite materials are subjected during their in-service life to cyclic loadings, which might lead to a progressive damage, a consequent loss of stiffness and residual strength and, eventually, to the final failure. The "design against fatigue" is therefore fundamental to improve the reliability of composite structural parts.

To meet the demand of tools for designing against fatigue of composites in many industrial fields, the author's group is working since several years to the development of a design framework suitable to predict the initiation of damage, its evolution and the final failure of composite laminates under cyclic loadings.

As a starting point of the discussion, it is important to clarify the concept of "design against fatigue", which can be meant in different ways depending on the requirements of the part under design:

i) design against crack initiation (no damage);
ii) design against stiffness degradation (damage tolerant design);
iii) design against final failure.

In some applications (like for instance fuel and pressure vessels or fuel rails) avoiding the onset of damage is a safety requirement. In this case the capability to predict the life spent for the initiation of the first crack is essential. In other cases like automotive composite frames, turbine blades, bicycle cranks or composite rims, the global stiffness can be the design driver. Due to the several fatigue damage mechanisms, structural composite components can lose up to 30-40% of their initial stiffness (depending on lay-up and load conditions) much before the final failure. Therefore, for a more reliable and cost-effective design, it is important in these cases to estimate the stiffness degradation under the specific loading conditions.

When only the load bearing capability of the part is of interest, the final failure (separation into two or more pieces) is the design target. However, the prediction of the fatigue response of a composite part is not an easy task, particularly when variable amplitude cyclic loadings applied in multiple directions are considered

As discussed later on in more details, the development of damage in composite laminates is characterised by multiscale and hierarchical processes, from the damage initiation at the microscopic scale to the onset and propagation of macro-cracks and delaminations leading to the final failure. These mechanisms are also dependent on the materials adopted, the lay-up and the loading type and multiaxial conditions [1].

The only way to deal with such a complicated problem and provide reliable design methods is to develop models based on the damage mechanisms actually occurring at the different length scales [1] and link them in a multiscale and multi-mechanism design procedure.

The next sections illustrate, with representative examples, the main mechanisms and damage modes leading a laminate to its degradation and failure under cyclic loadings. Afterwards the modelling strategy under development is presented.

2. Observations of damage mechanisms under fatigue loadings

In this section the damage mechanisms characterising the fatigue life of generic composite laminates are discussed, taking advantage of experimental results and observations collected within recent research programs by the Group and by other researchers in the literature. As already mentioned, the damage evolution in composite materials is a multiscale and hierarchical process, involving several length scales. Therefore the damage mechanisms will be presented at two different scales, i.e. macro- and micro-scales. In this case, the macro-scale is that related to the ply and laminate level, whereas the micro-scale is relevant to the fibres and the inter-fibre spacing.

2.1 Macro-scale mechanisms

If a unidirectional (UD) laminate is tested under a uniaxial off-axis fatigue load, it will fail suddenly, without any indication of damage development, as the number of cycles for the initiation of the first off-axis crack (macro-crack) is reached [2-4]. Here, with the term off-axis crack (or macro-crack) we mean a crack involving the entire thickness of a ply and propagating in a plane containing the fibres, along the fibres direction. Therefore, no off-axis cracks are detectable before the final failure and no stiffness degradation is measured. This does not implies necessarily the absence of damage occurring and evolving before the final failure. In fact, this occurs at a lower scale and does not influence significantly the global stiffness. Conversely, in multidirectional laminates the 90° UD plies are constrained between tougher 0° layers, which prevent the laminate separation when the first off-axis crack initiates in the weakest position. This leads to the initiation of further cracks and their stable propagation in the fibres direction.

Figure 1. Example of fatigue damage evolution and failure in a $[0/60_2/0/-60_2]_s$ laminate [5].

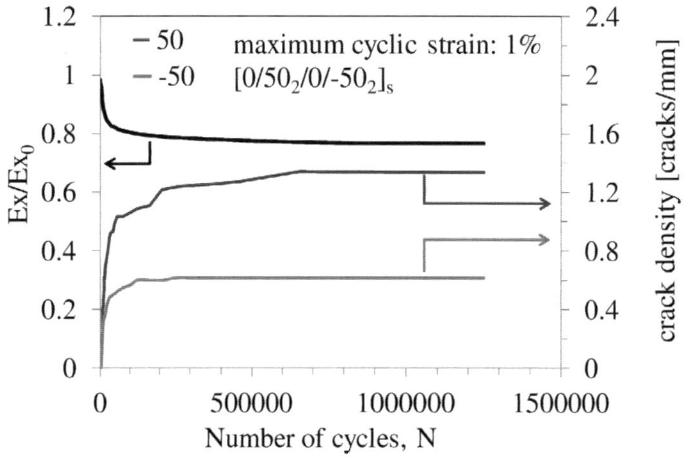

Figure 2. Crack density evolution and stiffness degradation in a $[0/50_2/0/-50_2]_s$ laminate [5].

This scenario is documented, as an example, by Figure 1 for a $[0/60_2/0/-60_2]_s$ glass/epoxy laminate [5]. In this case, several off-axis cracks had already initiated after 5000 cycles in both the 60° and -60° layers. Then, at 10000 cycles, the existing cracks had propagated and new ones initiated. The crack density increases until a saturation condition is reached. This is usually followed, or preceded, by the initiation of delaminations, which then propagate for the remaining part of the fatigue life. Even if the initiation of multiple off-axis cracks and delaminations is not critical for the final failure, these mechanisms lead to the degradation of the global elastic properties and can trigger the fibre failure.
Figure 2 shows the trend of the elastic modulus of a $[0/50_2/0/-50_2]_s$ laminate and the crack density evolution in the off-axis plies. Most of stiffness loss occurs in the earlier stages of fatigue life, where the crack density increases steeply. Then when the crack density reaches a saturated value, the rate of stiffness degradation becomes very low. This is a proof that off-axis cracks are the main responsible for the loss of stiffness, the initiation and propagation of delaminations providing, typically, a limited contribution in this sense.

A schematic of the typical stiffness trend for multidirectional laminates is shown in Figure 3.

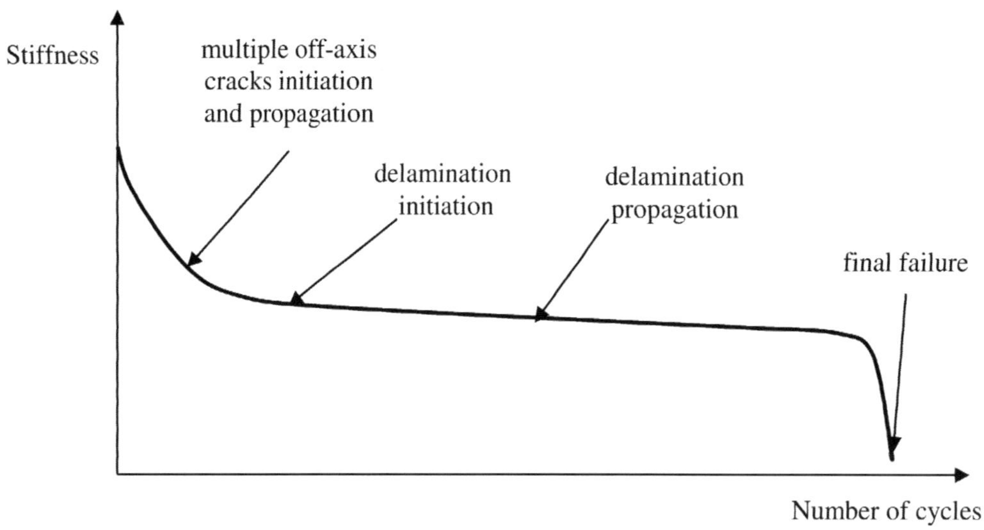

Figure 3. Schematic of the typical stiffness trend in multidirectional laminates under a fatigue load.

2.2 Micro-scale mechanisms

As discussed in the previous paragraph, the first damage event visible at the macro-scale in composite laminates is the initiation of off-axis cracks. In the case of off-axis UD laminates this event leads to the unstable propagation and sudden failure, whereas in multidirectional laminates the damage evolution process is more complicated, involving several damage mechanisms.

In both cases the initiation of an off-axis crack is preceded by the initiation and evolution of the damage at a lower scale, i.e. the micro-scale. In fact, recognising the presence of irreversible processes at the micro-scale is necessary for justifying the sensitivity of composite materials to a fatigue [6].

With the aim of identifying the microscopic damage mechanisms leading to the initiation of a macro-crack, glass/epoxy $[45/-45/0]_s$ laminates were tested under a uniaxial cyclic load [7]. The top surface of the 45° layer was accurately polished in order to make the fibres and the matrix clearly visible. At regular intervals the specimens were removed from the machine and their top surface observed under a microscope. It was observed that the first event of damage at the micro-scale, before the initiation of a visible off-axis crack, was the initiation of micro-cracks in the matrix, inclined of a certain angle with respect to the fibres direction (Figure 4a). The accumulation and coalescence of these micro-cracks led then to the initiation of an off-axis crack propagating in the fibres direction. It was shown that also the propagation phenomenon occurred with a similar mechanism, i.e. by the initiation and coalescence of micro-cracks in a region in front of the macro-crack tip [7].

This scenario gives reason of the fracture surfaces observed on glass/epoxy tubes tested under combined tension-torsion loadings, leading to the presence of the transverse (σ_2) and in-plane shear (σ_6) stresses [8]. In fact, it was shown that, if the shear stress was high enough with respect to the transverse stress, the fracture surfaces were characterised by a wide presence of shear cusps (figure 4b), reasonably due to the initiation of inclined micro-cracks as those observed on the 45° ply of the $[45/-45/0]_s$ laminates.

This scenario was observed only if the σ_6/σ_2 ratio (λ_{12} from now on) is higher than a certain threshold value, which, for the material tested, was about 0.5.

In fact, the tubes tested with $\lambda_{12}=0$ (pure tension) and 0.5 did not show the presence of shear cusps and the fracture surfaces were quite regular and smooth, as those characterising a mode I driven failure [8].

Figure 4. Damage mechanisms at the microscopic scale: a) micrograph of a 45° ply in a $[45/-45/0]_s$ laminate under tension-tension loadings (frontal view) [7] and b) SEM image of a fracture surface of a tube under combined tension/torsion [8].

After identifying the micro-scale damage modes leading to the initiation and propagation of an off-axis crack, the delamination and fibre failure phenomena were analysed from the microscopic point of view. Fatigue tests were carried out on glass/epoxy cross-ply laminates. Again, multiple transverse cracks were seen to initiate and propagate. The crack density saturation was slightly preceded by the initiation of delaminations. Micrographs were taken on the specimens edges, revealing as a typical scenario that shown in Figure 5. As expected, the delaminations were triggered by the presence of the transverse cracks and then propagated between the resin layer at the interface and the 90° fibres.

The role that transverse cracks and delaminations play on the failure of the fibres is clear from Figure 5. In fact, 0° fibres breaks are concentrated in the vicinity of the transverse crack tip and above the delamination. It is therefore reasonable to assume that the fibre failure phenomenon occurs according to the scenario sketched in Figure 6 for a cross-ply laminate. After the first load application, some randomly distributed fibre breaks occur in the weakest positions. As a transverse crack initiates, it causes a strong stress concentration in the 0° ply, promoting the failure of the fibres in the crack tip neighbourhood. Then, as a delamination initiates and propagates, a progressively larger region of the 0° ply is subjected to a higher value of the longitudinal stress, thus leading to further fibre breaks in that region, until a critical condition is reached, which leads to the laminate separation.

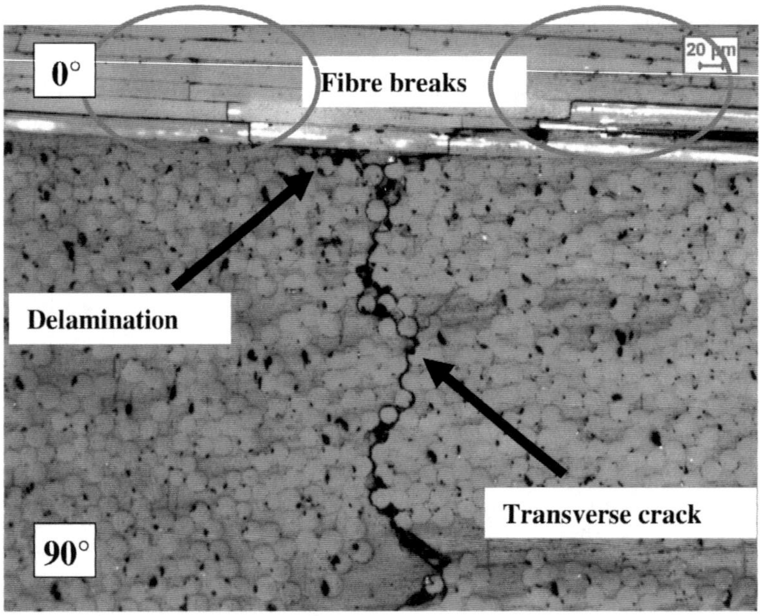

Figure 5. Example of delamination from a transverse crack tip with broken fibres.

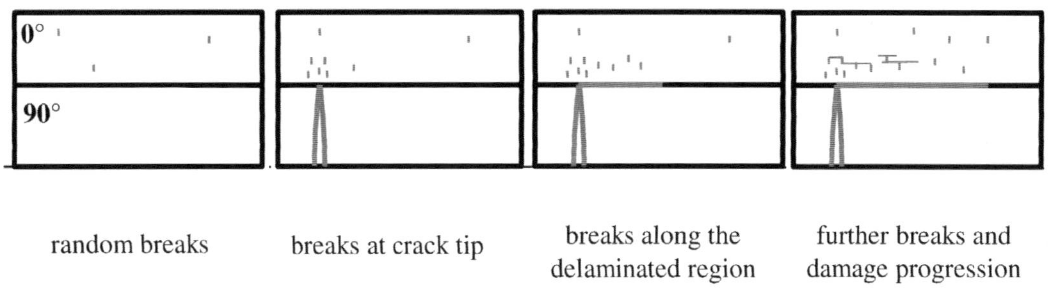

Figure 6. Schematic of micro-scale delamination- and fibre-related damage.

3. A damage based modelling strategy

According to the experimental evidences deduced from the damage observation at the macro- and micro-scales, the fatigue life of a composite laminate is divided into different phases, each characterized by a dominating damage mode (Figure 7):

- In the first stages, the damage initiates and accumulates at the microscopic scale in the form of micro-cracks, until an off-axis crack forms in the weakest position ($N_{i,c}$);
- Then multiple cracks initiate and propagate causing a steep stiffness reduction, until, thanks to the stress re-distribution between cracks, a saturation condition is reached ($N_{p,c}$);
- Typically, slightly before or after the crack density saturation state, delaminations initiate at the tips of the off-axis cracks ($N_{i,d}$);
- The remaining part of the fatigue life is spent for the propagation of the delaminations ($N_{p,d}$);
- During the entire life, fibre breaks occur, first randomly distributed in the weakest positions, then driven by the critical stress state due to the presence of off-axis cracks and delaminations, until a critical condition occurs and leads to the final failure (N_f).

According to the design requirement, for a safe and efficient design against fatigue it is necessary to predict:

i) The fatigue life spent for crack initiation (no damage);
ii) The fatigue life required to produce a certain crack density/delamination length associated to a specific stiffness reduction (damage tolerant approach);
iii) The fatigue life to the laminate failure.

As the final failure is driven by the presence of cracks and delaminations, the last point includes the prediction of the initiation and propagations of off-axis cracks and delaminations and their effect on the development of the fibre breaks.

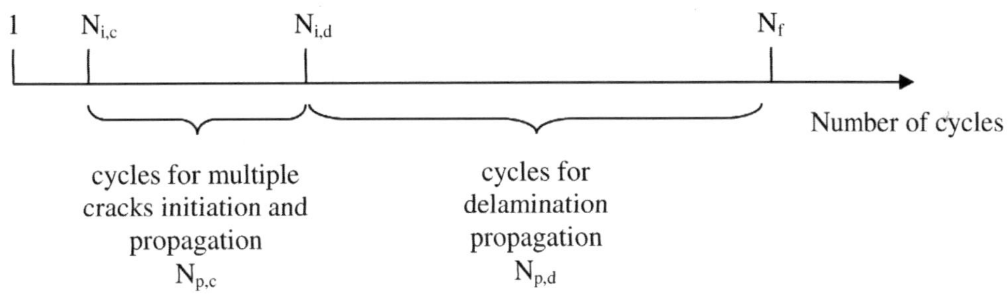

Figure 7. Typical sequence of damage mechanisms in multidirectional laminates under fatigue.

In the following sections the models developed by the authors for the prediction of crack initiation, crack density evolution, stiffness degradation and delamination propagation are briefly introduced.

4. A damage based criterion for crack initiation

To predict the initiation of the first crack, as well as the crack density evolution, it is necessary to define a crack initiation criterion, suitable to deal with multiaxial stress states. Such a tool was developed by Carraro and Quaresimin [9] and is briefly recalled in the following. As already mentioned, at a microscopic level, a progressive and irreversible damage process takes place from the very early stages of the fatigue life. Thus, the number of cycles spent for the initiation of an off-axis crack is controlled by the damage evolution occurring at the microscopic level. As a consequence, the definition of a crack initiation criterion consists in the identification of the driving force for the damage evolution at the microscopic scale. To this aim, the concept of the *local nucleation plane* is introduced in the following. This is defined as the plane in which the micro-cracks initiate in the matrix, as shown in Figure 4a. The authors have proved that the orientation of this plane is normal to the direction of the maximum principal stress in the matrix, between the fibres [7].

Accordingly, the stress component to be considered as the driving force for this kind of damage is the Local Maximum Principal Stress (LMPS) in the matrix.

When a ply is subjected to a pure transverse stress, the *local nucleation plane* is normal to the transverse direction. In fact, in this case no shear cusps can be observed in the matrix between the fibres [8]. In addition, Asp et al. [10] proved that in the case of pure transverse tension the local stress state at or close to the fibre-matrix interface is nearly hydrostatic. Therefore, the damage initiation occurs in the form of cavitation-induced matrix cracking and debonding. Asp et al. showed that a suitable criterion for predicting the static failure in the case of pure transverse stress is based on reaching a critical value for the dilatational energy density expressed in Eq. (1):

$$U_v = \frac{1-2\nu}{6E} I_1^2 \tag{1}$$

where I_1 is the first invariant of the local stress tensor.

If we assume to extend this finding also to the case of cyclic loadings, a change has to be expected in the leading damage mode, moving from a loading condition near the pure transverse stress to another one, characterized by the presence of a high enough shear stress component. In the former case, the micro-scale damage evolution is assumed to be driven by the Local Hydrostatic Stress in the matrix (LHS = $I_1/3$), in the latter by the LMPS.

On these basis, the estimation of the life to crack initiation under multiaxial fatigue can be made by using the S-N curves expressed in terms of two different, alternative parameters:

- the peak of the Local Hydrostatic Stress (LHS = $I_1/3$), for nearly pure transverse tension case;
- the peak of Local Maximum Principal Stress (LMPS) for high enough shear stress component.

As the LHS and LMPS are local parameters, representative of the driving forces for the microscopic damage evolution, they have to be expressed in terms of local stresses (or micro-stresses) acting in the matrix and at the fibre-matrix interface. A multiscale approach, to link the applied stresses σ_1, σ_2 and σ_6 to the local stress fields, is therefore adopted (Figure 8).

If a regular fibres array of the square type is assumed for a UD ply, a unit cell as that shown in Figure 8 is representative of the homogenised lamina in terms of average mechanical behaviour. According to the micromechanics theory, the global stresses σ_1, σ_2 and σ_6 are the average stresses acting on the faces of the unit cell.

Regular fibres distributions are indeed not representative of the real microstructure in composite plies. In spite of this, it was proved that the unit cell provides reliable results for the application of the proposed initiation criterion [7, 9] even in the case of real random fibre distributions.

The micro-stresses can be calculated by means of FE analyses of a fibre-matrix unit cell subjected to the average (or macroscopic) stresses σ_1, σ_2 and σ_6. In addition, residual stresses due to the different thermal expansion of the two phases during the cooling process must be accounted for. In the FE code thermal loads have been applied as a uniform temperature gap $\Delta T = T_c - T_r$, where T_c is the curing temperature and T_r is the room temperature. Since tensile loading conditions are analysed here and the epoxy behaviour is usually linear elastic in that range, linear elastic FE analyses have been carried out with the software ANSYS 11® using 20 node solid elements.

At each point P of the unit cell the mechanical micro-stresses in polar coordinates (r, φ, z) can be defined in terms of stress concentration factors $k_{i,jl}$ relating the macro-stress σ_i to the local stress σ_{jl}, as in equation (2). Finally the thermal stresses are related to the temperature gap ΔT by means of thermal concentration factors h_{jl}, as in Eq. (2).

$$\begin{Bmatrix} \sigma_{rr} \\ \sigma_{\varphi\varphi} \\ \sigma_{zz} \\ \sigma_{r\varphi} \\ \sigma_{\varphi z} \\ \sigma_{rz} \end{Bmatrix}^P = \begin{bmatrix} k_{1,rr} & k_{2,rr} & 0 \\ k_{1,\varphi\varphi} & k_{2,\varphi\varphi} & 0 \\ k_{1,zz} & k_{2,zz} & 0 \\ k_{1,r\varphi} & k_{2,r\varphi} & 0 \\ 0 & 0 & k_{6,\varphi z} \\ 0 & 0 & k_{6,rz} \end{bmatrix}^P \begin{Bmatrix} \sigma_1 \\ \sigma_2 \\ \sigma_6 \end{Bmatrix} + \Delta T \begin{Bmatrix} h_{rr} \\ h_{\varphi} \\ h_{zz} \\ 0 \\ 0 \\ 0 \end{Bmatrix}^P \tag{2}$$

The peak values of the LMPS and LHS have been found to be always at the points A or B (mostly at point A) of Figure 8 where, due to the symmetry, some stress components vanish, allowing one to adopt the following simple expressions for the LMPS and LHS (valid in the segment AB):

$$LMPS = \frac{1}{2}\left[\sigma_{rr} + \sigma_{zz} + \sqrt{\sigma_{rr}^2 + 4\sigma_{rz}^2 - 2\sigma_{rr}\sigma_{zz} + \sigma_{zz}^2}\right] \tag{3}$$

$$LHS = \frac{\sigma_{rr} + \sigma_{\varphi\varphi} + \sigma_{zz}}{3} \tag{4}$$

In Eqs. (3) and (4) the local stresses obtained from Eq. (2) must be substituted.

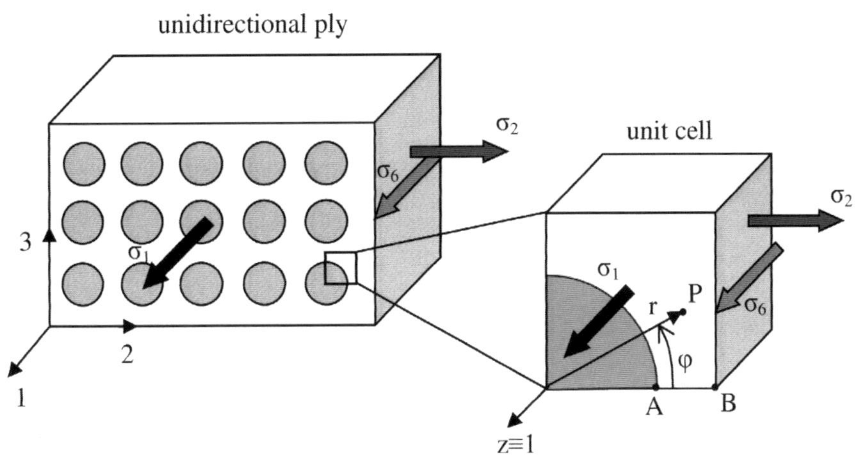

Figure 8. Definition of the fibre-matrix unit cell for stress analysis art the microscopic scale.

As a validation of the proposed criterion, the S-N curves for the first crack initiation in glass/epoxy tubes subjected to tension/torsion loadings are considered [8]. These are shown in Figure 9a in terms of the maximum cyclic value of the transverse stress, for four values of the biaxiality ratio λ_{12}. Four distinct curves are obtained in terms of σ_2, but they are collected in only two scatter bands when plotted in terms of the LHS and LMPS parameters. Therefore, when designing against crack initiation with this material, under a general multiaxial stress state, one of these two curves can be adopted, depending on the fact that the ratio λ_{12} is higher or lower than the threshold value 0.5. The criterion validation through a bulk of experimental results from the literature can be found in Ref. [9].

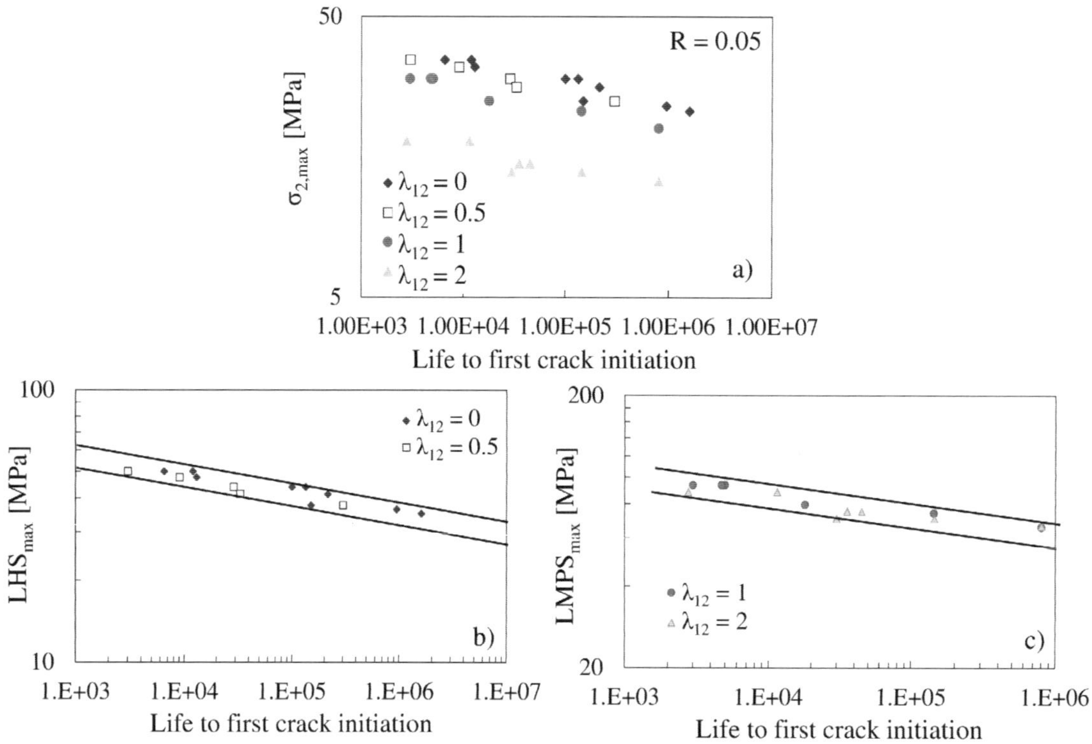

Figure 9. First crack initiation results for glass/epoxy tubes [8]: maximum a) cyclic transverse stress, b) LMPS and c) LHS against the number of cycles for first crack initiation.

5. Crack density prediction model

The crack initiation criterion described in the previous section, combined with the classical laminate theory (CLT), is enough for predicting the first crack initiation within a ply in a laminate, according to a given probability of survival. The prediction of the crack density evolution is instead a much more complicated task, which was treated according to the following basic ideas [11-13]:

- The initiation of multiple cracks can be predicted, again, using S-N curves for crack initiation in terms of the local parameters LHS and LMPS. The statistical distribution of the fatigue strength has now to be introduced, as it is the only explanation for having crack initiations at different number of cycles;
- The CLT is not suitable for calculating the stress distributions in the plies of a laminate in the presence of a high number of cracks. To overcome this limitation, Carraro and Quaresimin developed a suitable stress re-distribution model [14];
- The propagation of off-axis cracks along the fibres direction can be treated with a Paris-like curve, linking the crack growth rate to the energy release rate.

A schematic of the multiscale procedure for the prediction of the crack density evolution is shown in Figure 10, and it consists in the simulation of the fatigue life by progressively increasing the number of cycles by steps ΔN.

For a given step of fatigue cycles ΔN, and for every off-axis layer, the analysis is carried out as follow:

1) Calculation of ply stresses in each layer, with the possible presence of off-axis cracks, and therefore considering the stress re-distribution.
2) Calculation of micro-stresses (or local stresses) by means of a fibre matrix unit cell subjected to periodic boundary conditions (see the stress concentration factors defined previously).
3) Calculation of the local parameters LHS and LMPS.

4) According to the multiaxial condition, calculation of the total density of nucleated cracks, by using the LHS of LMPS master curves for the material considered. The statistical distribution of fatigue strength and the sequence effect, due to the stress re-distribution, must be accounted for.
5) Analysis of the crack propagation phase to compute the length of all the nucleated cracks.
6) Increasing the number of cycles of a quantity ΔN and repeat steps 1)-6) until the required number of cycles or crack density saturation is reached.

It is worth mentioning that, for the calculation of the local stresses at the step 2) of the procedure, one does not need to carry out a FE analysis of the unit cell every time. In fact the micro-stresses can be simply calculated with Eq.(2), once the stress and thermal concentration factors are calculated at the beginning of the analysis.

Once the crack density is predicted in all the plies of a laminate, it can be used as an input for the stiffness degradation model developed by Carraro and Quaresimin [14]. This model is capable of dealing with generic laminates with cracks in any number of plies, also accounting for the interaction between cracks in different layers. This is important in view of avoiding underestimations of the stiffness loss due to the presence of off-axis cracks in multiple layers [14].

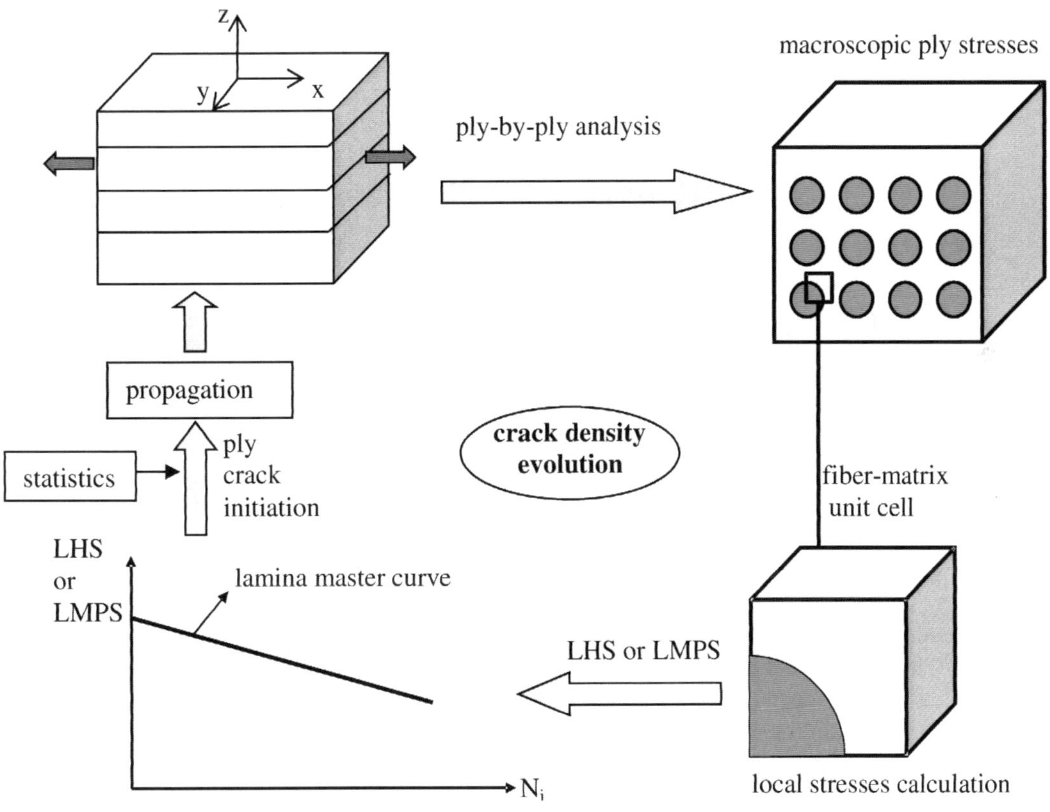

Figure 10. Procedure for the prediction of the crack density evolution.

6. Prediction of the delamination propagation

It has been observed that the propagation of delaminations plays an important role in the fibres failure process. Therefore, treating this phenomenon is of great importance for the prediction of the final failure. With the aim of characterising the delamination initiation and propagation, the authors have recently conducted a dedicated experimental investigation. Fatigue tests were conducted on glass/epoxy cross-ply laminates, focusing the attention on the initiation of multiple transverse cracks and also on the initiation and propagation of delaminations from the nucleated transverse cracks. It was observed that the crack density saturation and the initiation of delaminations occurred at a very small percentage of the total fatigue life (1-2%), so that the cycles spent for these phenomena can be even neglected if the final failure is the target of the analysis.

The delamination length was measured during the fatigue tests (figure 11a), the relevant values of the mode I and II Energy Release Rate (ERR) were computed with Finite Element analyses and the VCCT technique, and eventually related to the delamination growth rate. It has been confirmed that the propagation of the delaminations can be treated by means of Paris-like curves relating the growth rate to the mode II energy release rate (the mode I contribution is very limited or absent, at least under tensile loadings) (figure 11b).

This tool allows the delamination length to be predicted as a function of the number of cycles. In addition it is possible to estimate the stiffness loss due to the delaminations, by means of easy analytical tools as for instance a shear lag model.

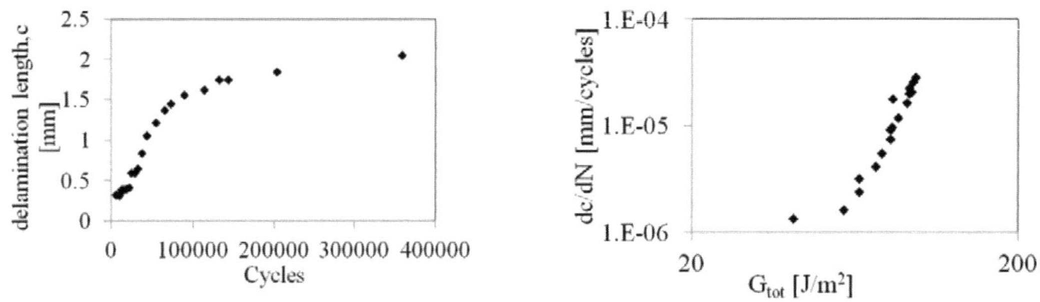

Figure 11. a) Example of delamination length evolution and b) relevant Paris-like curve.

7. Conclusions

The fatigue behavior of composite laminates under fatigue loading is characterized by a progressive and complex damage evolution which involves different length scales. It was shown that, in the presence of a high enough shear stress, the first event of damage is the initiation of micro-cracks in the matrix between the fibres. The accumulation and coalescence of these cracks lead to the initiation of an off-axis crack, involving the entire ply thickness and propagati G_{II} ng the fibres direction. It was proved that this kind of micro-scale damage evolution is driver. .ie Local Maximum Principal Stress in the matrix. Conversely, when the shear stress is very low, the microscopic damage evolution is driven by the Local Hydrostatic Stress (LHS). According to these findings, a fatigue crack initiation criterion was proposed by the authors, suitable to deal with multiaxial stress states, based on the use of these two local parameters. This criterion can be used to predict the initiation of the first cracks in laminates, but also the entire evolution of the crack density, through a model which accounts also for the statistical distribution of the fatigue strength and the stress re-distribution in the presence of cracks. The predicted crack density can be used as in input for a stiffness degradation model, to estimate the laminate stiffness loss as a function of the number of cycles. The developed tool is fundamental in a damage tolerant design against fatigue.

To predict the cycles spent for the final separation of a laminate, the subsequent mechanisms of the initiation ad propagation of delaminations must be addressed. An experimental campaign on cross-ply laminates was carried out for this purpose, showing that the cycles spent for initiating the delaminations and saturating the crack density is about 1-2% of the total fatigue life. It was proved that a Paris-like curve is a suitable tool for predicting the propagation of delamination, which is another fundamental step for the estimation of the total life. To reach this final goal, the influence of the presence of off-axis cracks and delamination on the development of fibre-related damage must be modelled. The authors are currently working in this direction, to eventually provide a damage based approach of industrial applicability for the prediction of damage initiation, evolution and the final failure of composite laminates.

References

[1] Quaresimin M, Susmel L, Talreja R, 2010, Fatigue behaviour and life assessment of composite laminates under multiaxial loadings, *Int J Fatigue* **32** 2-16

[2] Quaresimin M, Carraro PA, 2013, On the investigation of the biaxial fatigue behaviour of unidirectional composites, *Compos Part B-Eng* **54** 200-208

[3] El-Kadi H, Ellyin F, 1994, Effect of stress ratio on the fatigue of unidirectional fibre glass-epoxy composite laminae, *Composites* **25** 917-924

[4] Awerbuch J, Hahn HT, 1981, Off-axis fatigue of graphite/epoxy composite. Fatigue of fibrous composite materials, ASTM STP 723, *American Society for Testing and Materials* 1981: 243-273

[5] Quaresimin M, Carraro PA, Mikkelsen LP, Lucato N, Vivian L, Brøndsted P, Sørensen BF, Varna J, Talreja R, 2014, Damage evolution under internal and external multiaxial cyclic stress state: a comparative analysis, *Compos Part B-Eng* **61** 282–290

[6] Talreja R, 1981, Fatigue of composite materials: Damage mechanisms and fatigue-life diagrams, *Proceedings of the Royal Society of London. Series A, Mathematical and Physical Sciences* **378** 461-475

[7] Quaresimin M, Carraro PA, Maragoni L, 2016, Early stage damage in off-axis plies under fatigue loading, *Compos Sci Technol* **128** 147-154

[8] Quaresimin M, Carraro PA, 2014, Damage initiation and evolution in glass/epoxy tubes subjected to combined tension-torsion fatigue loading, *Int J Fatigue* **63** 25–35

[9] Carraro PA, Quaresimin M, 2014, A damage based model for crack initiation in unidirectional composites under multiaxial cyclic loading, *Compos Sci Technol* **99** 154-163

[10] Asp LE, Berglund LA, Talreja R, 1996, Prediction of matrix initiated transverse failure in polymer composites, *Compos Sci Technol* **56** 1089-1097

[11] Carraro PA, Quaresimin M, 2014, Prediction of crack density evolution for multidirectional laminates under fatigue loading, In: *Proceedings of 16th European Conference on Composite Materials*, Sevilla, Spain, 22-26 June 2014

[12] Carraro PA, Maragoni L, Quaresimin M, 2014, A tool for the simulation of fatigue damage evolution in multidirectional laminates, In: *Proceedings of 16th European Conference on Composite Materials*, Sevilla, Spain, 22-26 June 2014

[13] Carraro PA, Quaresimin M, Maragoni L, Prediction of crack density evolution in multidirectional laminates under fatigue loading, to appear

[14] Carraro PA, Quaresimin M, 2015, A stiffness degradation model for cracked multidirectional laminates with cracks in multiple layers, *Int J Solids Struct* **58** 34–51

IOP Conf. Series: Materials Science and Engineering **139** (2016) 012007 doi:10.1088/1757-899X/139/1/012007

The micro-mechanics of strength, durability and damage tolerance in composites: new insights from high resolution computed tomography

S Mark Spearing and Ian Sinclair

Faculty of Engineering and the Environment, University of Southampton, SO17 1BJ, UK

Abstract. Recent work, led by the authors, on impact damage resistance, particle toughening and tensile fibre failure is reviewed in order to illustrate the use of high-resolution X-ray tomography to observe and quantify damage mechanisms in carbon fibre composite laminates. Using synchrotron and micro-focus X-ray sources resolutions of less than 1 μm have been routinely achieved. This enables individual broken fibres and the micromechanisms of particle toughening to be observed and quantified. The data for fibre failure, cluster formation and overall tensile strength are compared with model predictions. This allows strategies for future model development to be identified. The overall implications for using such high-resolution 3-D measurements to inform a "data-rich mechanics" approach to materials evaluation and modeling is discussed.

1. Introduction

It has been more than fifty years since the development of carbon fibres and the introduction of high performance composite materials [1]. Material properties have developed remarkably over that period, with a more than doubling in each of the useful tensile strength, strain to failure and functional impact damage tolerance since the first generation of carbon fibre composite materials. Perhaps surprisingly this has been achieved without a strong physically-based understanding of the underlying mechanisms and also an ability to predict the resultant structural response, particularly with regard to failure, damage resistance, damage tolerance and durability. This is exemplified by the work done over the past twenty years in a series of "World wide failure exercises" [2] which have been conducted in order to assess the predictive capability for the failure of composite materials. The results have indicated that the overall modeling capability continues to relatively poor, at least compared to that for the metallic materials which composites are often used to replace. The difficulty in predicting failure is that composites typically fail by multiple interacting damage mechanisms, which occur in three dimensions. Furthermore it is usually difficult to infer the underlying damage mechanisms solely by inspection of the surface of the structure or test coupon. The response of the modeling and design communities has been to rely on extensive testing for any safety critical parts in order to supplement, or in some case, completely replace any reliance on modeling. These issues are further compounded when issues of durability are to be addressed, and where comprehensive test programmes are particularly time consuming and expensive.

Significant work has been conducted over the past fifty years to understand the impact damage tolerance [e.g 3,4], role of particle toughening e.g [5,6] and tensile fibre failure e.g [7,8], however there has been a lack of detailed experimental data to allow direct validation of the assumptions and predictions of these models. In order to address this deficiency we have been conducting research over the past ten years into the use of high-resolution computed tomography (HRCT) to elucidate damage mechanisms in composite materials [9-20]. This is motivated by the strong belief that in order to make real progress in developing superior materials and to be able to predict the material and structural response it is essential to understand the underlying mechanisms and to incorporate these into predictive models. In particular we have found the ability to identify features in composites under load and to measure their relative displacements in order to infer strains or to calculate crack opening and shear displacements is a very powerful means by which to interrogate and validate mechanics models for composite failure processes. The use of HRCT, in combination with other techniques allows a few careful experiments to yield large, very valuable data sets which can be used to inform, validate and challenge modeling approaches. This approach has the potential to change radically the approach to composite modeling and mechanical testing. This paper provides an overview of our current and recent work in this area and an assessment of the potential for HRCT to contribute further to the understanding and modeling of composite failure.

2. Testing methods

The details of the testing and imaging procedures are described more fully elsewhere [9, 10, 17, 19]. High resolution tomographic imaging was conducted on beam line ID19 at the European Synchrotron Radiation Facility (ESRF) in Grenoble and the TOMCAT beamline at the Swiss Light Source in Villigen, Switzerland. In these facilities resolutions down to 0.3 μm can be achieved. In all cases a compromise must be considered between the resolution and field of view. The synchrotrons produce a highly coherent, monochromatic X-ray source, which allows for both absorbtion and phase contrast imaging, which is particularly important for allowing crack surfaces and damage feature to be identified. Lower resolution imaging was conducted using the microfocus CT beamlines available at the μVis facility at Southampton.

For *in situ* tensile loading one of two fixtures was used, and these are described elsewhere [9, 19]. These have load capacities of up to 5kN and are designed to allow 360° X-ray access with isotropic absorbtion. The fixtures are particularly suited to use in synchrotrons, with their high resolution capabilities, but can be used in conjunction with micro-focus sources. The load frames allow damage development to be tracked as a function of load, as well as allowing damage to be imaged in the loaded and unloaded states, which allows for the separation of the effect of residual stresses and applied loads on crack opening and shear displacements. *In situ* loading has been performed to generate tensile damage in double edge notched specimens with a total width of 4 mm, with two semi-circular notches of nominally 1mm radius/depth. This leaves a net section between the notches of 2 mm. In other work, wedge-opened double cantilever beam specimens have been used to study mode 1 fracture and end notched flexure tests loaded in three point bending to study mode 2 fracture [17].

3. Results

3.1. Damage tolerance

Quasi-isotropic laminates, approximately 4 mm thick, cut into panels 150 mm x 100 mm were tested under low velocity impact conditions. Compression after impact tests were subsequently conducted. The experimental details are described elsewhere [13-15]. Five material systems were investigated, identified as UT and T1 to T4, with UT referring to an untoughened epoxy matrix and T1 to T4 representing toughened systems with increasing levels of damage resistance. Figure 1 shows the damage area as a function of the incident impact energy. The untoughened material exhibits about three times the damage area as material T4 for a given impact energy. Figure 2 shows two rendered computed tomography images of the damage in the untoughened material and one of the toughened systems. It is clear that the principal effect of toughening is to reduce the extent of delamination under the impact site. There is no noticeable effect on the extent of intralaminar damage.

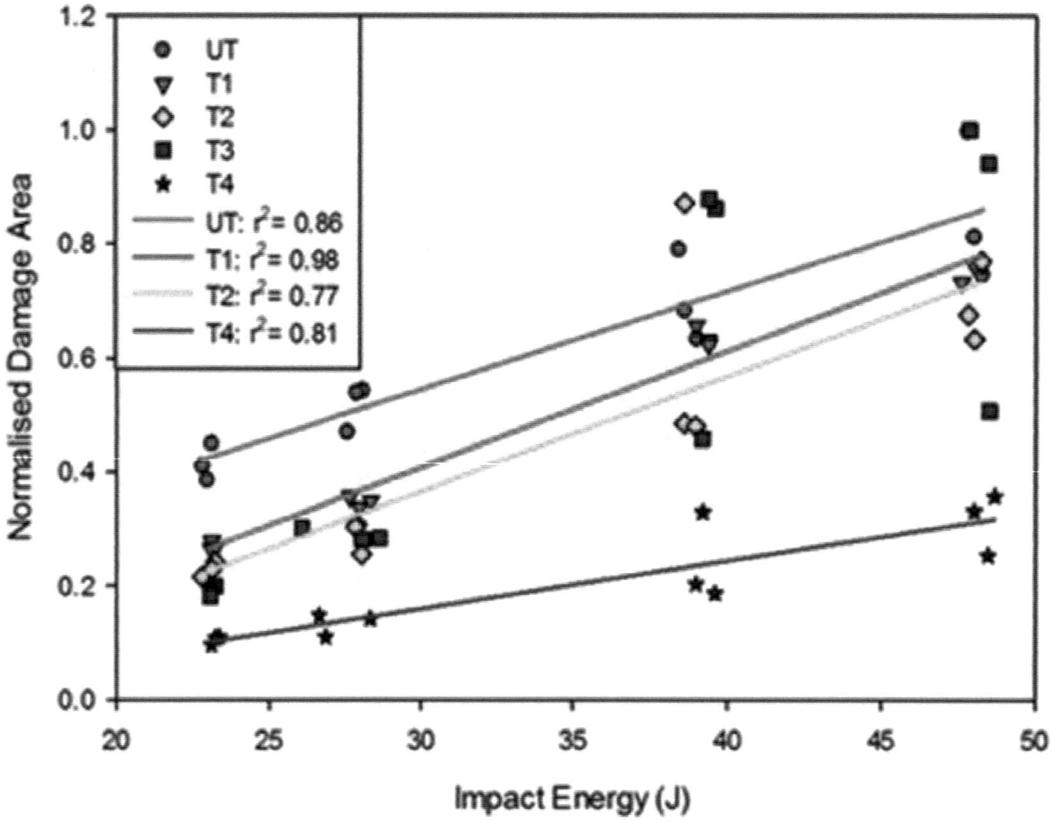

Figure 1. Impact damage resistance as a function of incident impact energy after [14].

Figure 3 shows higher resolution CT-generated cross sections of interlaminar cracks in each of the materials studied. Analysis of the full three dimensional image file allows the performance of the material to be correlated with aspects of the crack microstructure, such as crack path tortuosity, crack surface roughness, area fraction of bridging ligaments and bridging ligament size. Figure 4 shows the data for bridging ligament interconnectivity (area fraction) and bridging ligament size for the five materials as a function of position, measured from the crack tip. It is clear that the order of toughness of the materials correlates strongly with the bridging ligament interconnectivity close to the crack tip. The evidence from this study indicates that toughening strategies, which promote near-tip ligament formation are likely to have a greater effect on improving impact damage resistance.

3.2. Micro-scale toughening and durability

The observation that bridging ligament formation is important for determining the functional toughness and damage resistance of toughened composite systems motivates a closer look at the micro-mechanisms of toughening. This requires higher resolution imaging, and benefits greatly from in situ loading. SRCT was used to obtain the images shown in Figure 5, which show the behaviour observed in a relatively tough material system. Toughening is achieved by the incorporation of an interply region containing polymer particles. These particles partially debond ahead of the crack tip. This debonding process has the effect of maintaining the crack in the interply region and of promoting the creation of bridging ligaments in the crack wake. Figure 6 shows a delamination propagating at the ply interface, following individual fibre-matrix interfaces. In this case the particles provide no direct contribution to the toughness of the material, and there is greatly reduced opportunity for the formation of bridging ligaments.

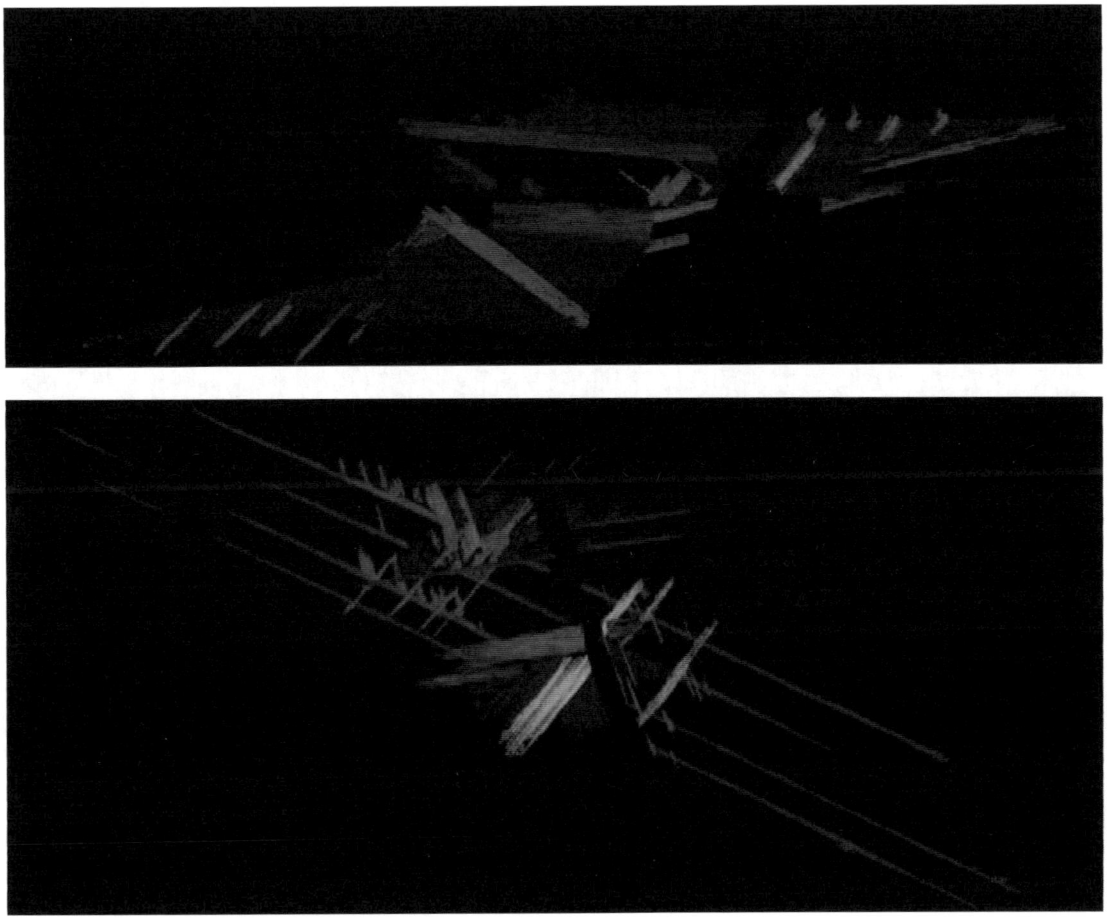

Figure 2. A three-dimensional rendering of impact damage in a) an untoughened material and b) a toughened material. The impact energy in b) is twice that in a). Note the reduced extent of delamination (blue) in the toughened material. After [13].

The dependence on bridging ligaments for toughening and damage resistance raises questions as to the effect of fatigue loading on such toughened systems. Work conducted by Garcea et al [18, 19] has shown that the mechanisms of matrix crack advance are microscopically similar in fatigue as for quasi-static loading. Figure 7 shows two SRCT cross-section of the same region of a split emanating from a notch under macroscopic tensile loading, with an increment of 100 cycles between the two images. Overall crack advance occurs by the linking together of discrete debond events ahead of the continuous crack tip, as for the quasi-static case. In addition, ligaments in the rear of the crack tip fail, removing this load transfer mechanism between the crack faces and facilitating further crack advance. In this particular case, fatigue loading, even at quite high loads, is not detrimental to the overall specimen residual strength and the resulting S-N curves are quite shallow or even flat.

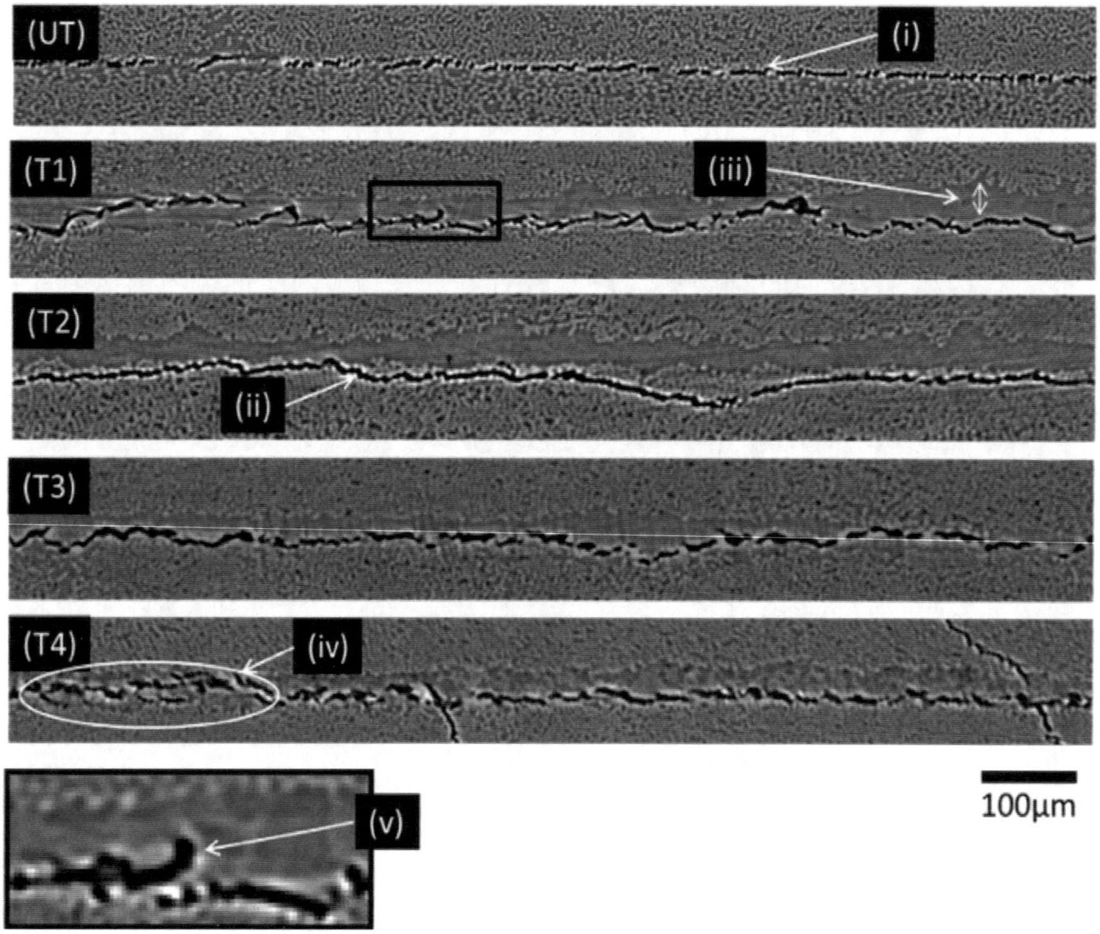

Figure 3. CT cross sections of delaminations in the five material systems. Bridging ligaments are visible to varying extents in all five materials (e.g region "v") after [13].

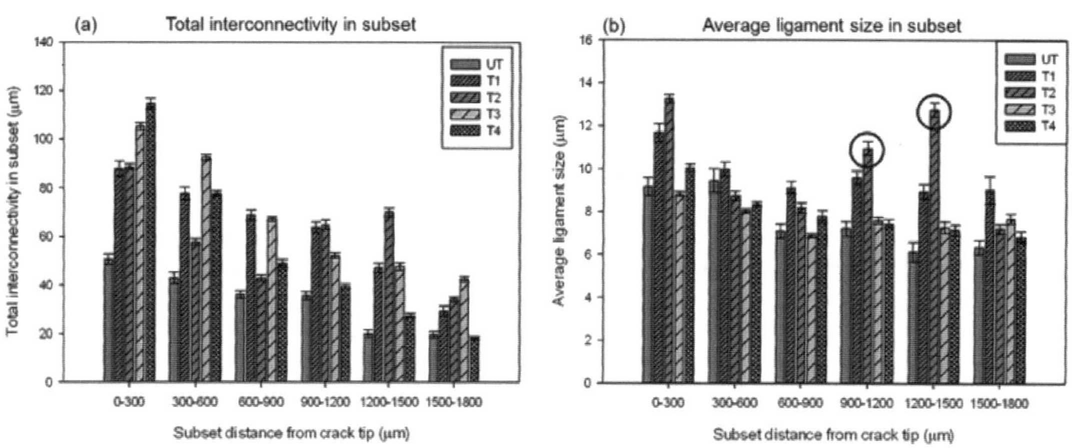

Figure 4. Data obtained by analysis of CT image volumes for a) bridging ligament interconnectivity and b) bridging ligament size as a function of position behind the crack tip. After [15].

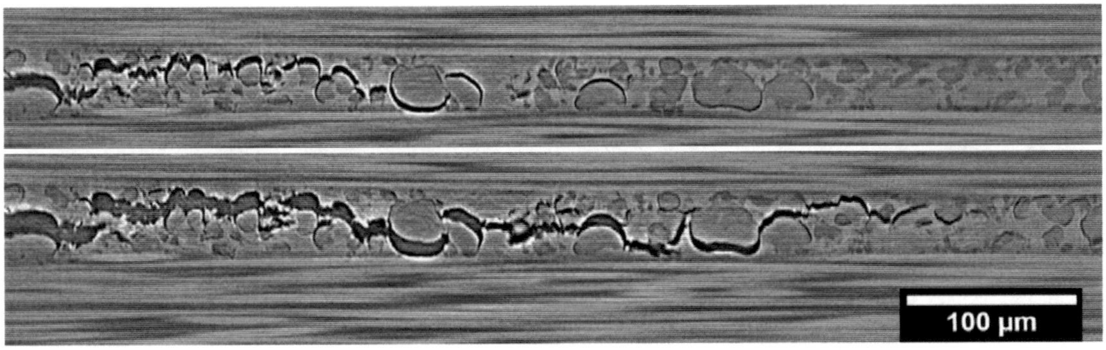

Figure 5. SRCT cross-sections of a delamination in a particle toughened interlayer. The same region is viewed at two different load levels, the crack has grown between the load steps. After [17].

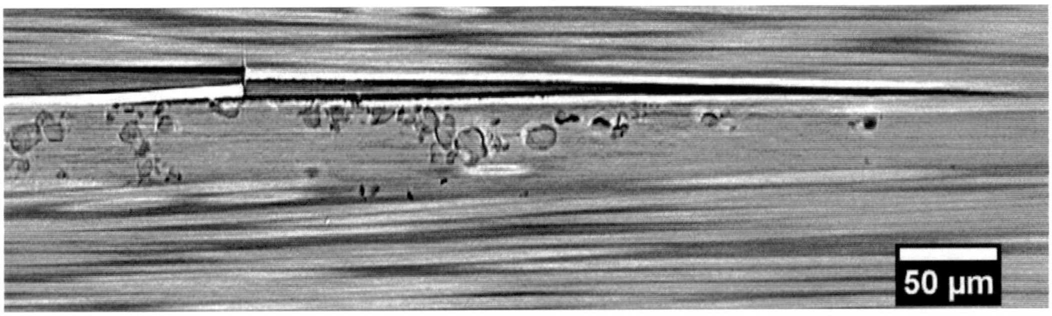

Figure 6. Mode 1 fracture at the interface between the ply and the toughened region after [17].

3.3. Tensile strength and fibre fracture

Fibre fracture is the key failure mode in tensile loading of 0° dominated laminates. Perhaps surprisingly, more than 50 years after the creation of the first continuous carbon fibre composites [1], it is still unclear as to how the tensile strength of individual fibres, in combination with the matrix and fibre-matrix interface, affects the strength of a composite. Figure 8 shows data for composites made with several similar intermediate modulus carbon fibres, combined with epoxy matrices to form composites. Despite very little variation in fibre strength, very significant variation in composite tensile strength results. High-resolution CT imaging combined with in situ loading provides a unique means to observe the development of broken fibres as a function of load. In particular the technique offers the opportunity to correlate broken fibres with microstructural features as well as the tendency for broken fibres to develop as clusters. Figure 9 shows images of clusters of broken fibres in specimens loaded in tension [20]. Using such images, combined with in situ loading capabilities, it is possible to obtain data such as that shown in Figure 10, for the evolution of broken fibres with applied tensile load, for the same set of materials examined in Figure 8. Very wide variation in broken fibre accumulation is apparent between ostensibly quite similar materials systems. In order to understand these variations, with a view to providing guidance for improving material strength, existing models for tensile strength have been applied [21, 22]. In both modeling strategies applied to date, comparisons similar to that shown in Figure 11 are obtained. The prediction of fibre break density and of isolated broken fibres is reasonably accurately modeled. However, the formation of clusters consisting of multiple broken fibres (as shown in figure 8) is not well modeled, as presented in figure 12. It is expected that cluster formation plays a critical role in determining the point of failure, so this suggests that further work is required to develop a capable modeling framework.

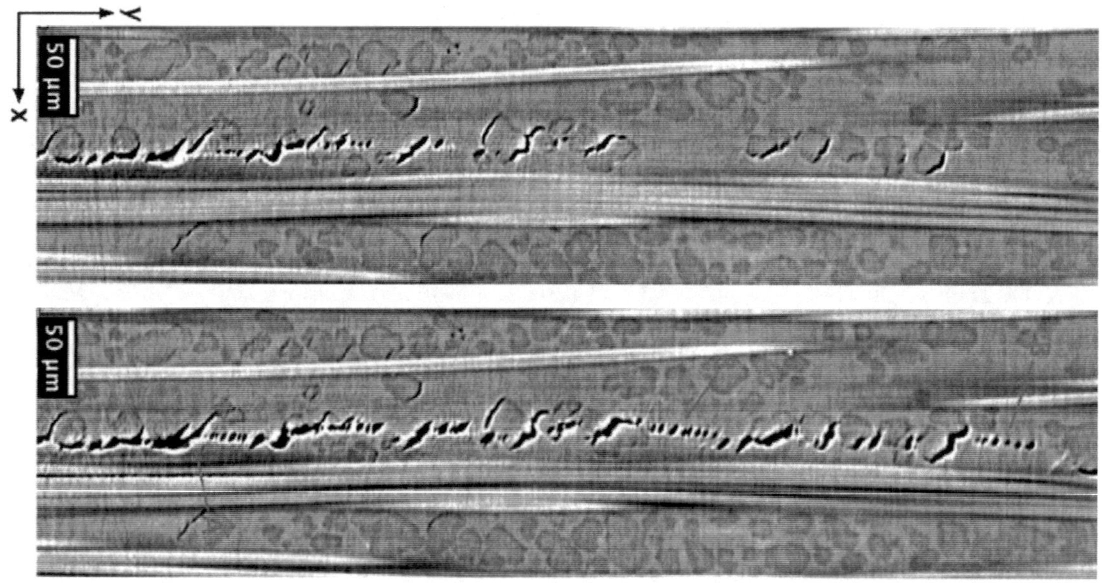

Figure 7. Fatigue damage propagation with 100 cycles between image a) and image b), after [19]

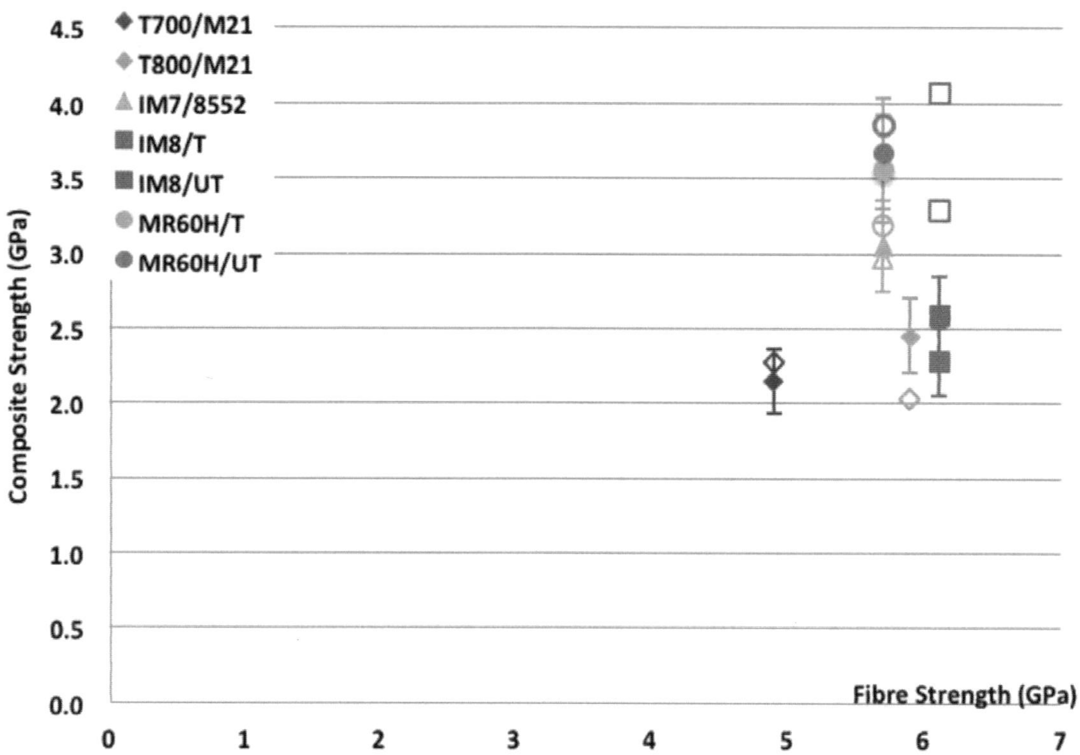

Figure 8. Plot of composite tensile strength vs. mean fibre strength for seven material systems all with intermediate modulus fibres but different matrices and fibre/matrix interfaces. After [22].

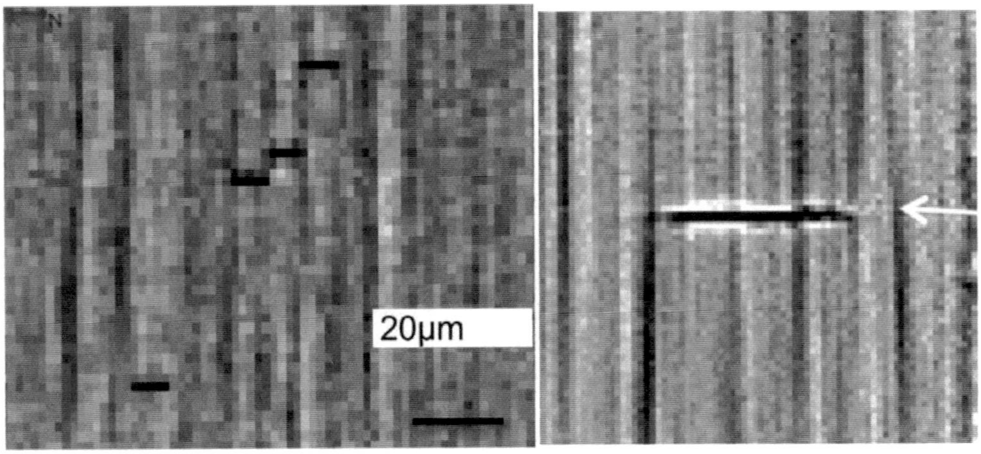

Figure 9. SRCT images of a) a diffuse cluster of broken fibres and b) a coplanar cluster of broken fibres after [20].

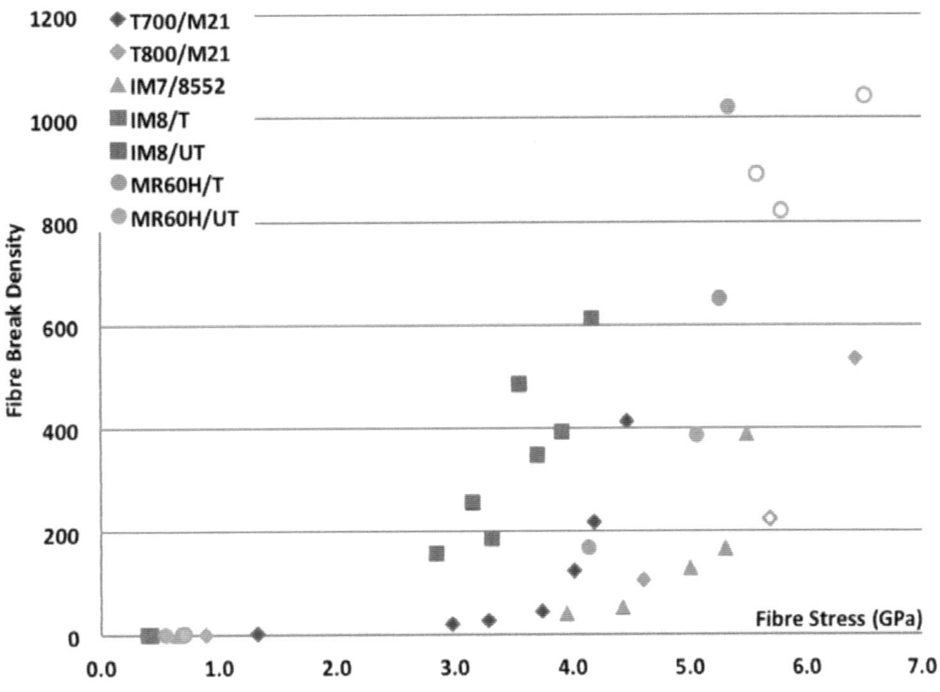

Figure 10. Data obtained via SRCT observations of fibre break density as a function of the nominal fibre stress for the same set of materials examined in Figure 8. After [22].

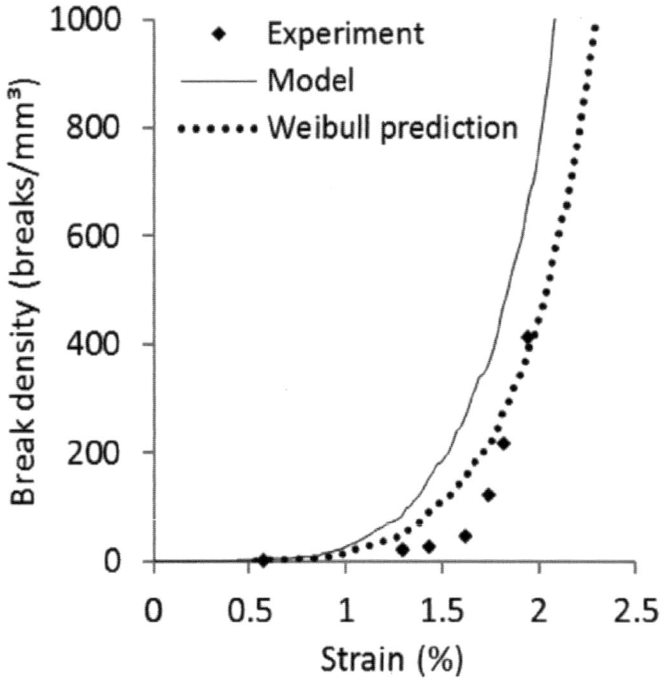

Figure 11. Comparison of experimental data for fibre break density with applied strain, and a model [22]. A comparison with the Weibull prediction based on individual fibre tests is provided.

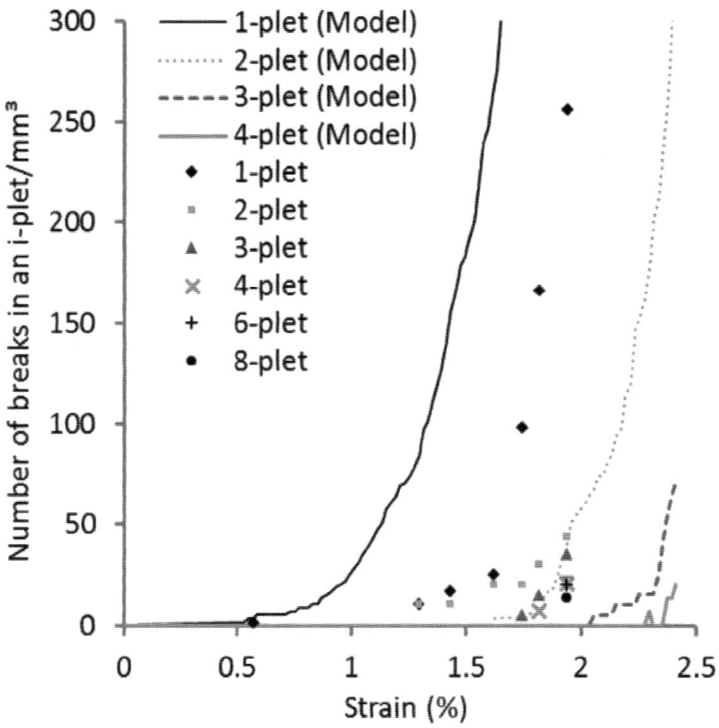

Figure 12. Fibre fracture evolution with applied strain compared to model predictions. After [22]

4. Discussion

In the three areas of investigation described in this paper, a number of common themes can be identified. It is clear that the very high resolution imaging capability afforded by computed tomography, particularly when using synchrotron generated radiation, allows for insights to be obtained that were not possible more than a decade ago. As the equipment, the image processing and data handling becomes more routine, the accessibility of these techniques will continue to grow. When applied to composite materials these techniques allow the interaction of the relevant damage mechanisms to be understood in 3-D, which is often of critical importance. There is particular value when looking at the mechanisms affecting toughening, fatigue and damage tolerance, and fibre fracture and tensile strength. These areas have been particularly amenable to investigation due to the match between the relevant lengthscales of the mechanisms and the resolution and field of view available by high resolution computed tomography. The extension via in situ loading to examine the micro-mechanisms of damage formation, particularly in fatigue offers a high degree of novelty and great potential to advance our understanding in the field. Even a 100 cycles of loading at an intermediate load level results in a few 10's of μm of crack advance, which would be difficult to detect using conventional techniques such as radiography, ultrasound or visual inspection. However, at SRCT resolution it allows the individual micro-mechanisms to be identified very clearly. This capability also offers the potential to provide a high quality benchmark for structural health monitoring and non-destructive evaluation techniques. Computed tomography is unlikely to be useful for anything but the highest criticality and highest value inspection of manufactured and in service parts, but as a means of calibrating and verifying lower cost techniques to be used in such situations, it has considerable potential.

The most important element of the work described in this paper, and often conducted with collaborators over the past decade is the use of the extensive data files obtained from high resolution computed tomography and other techniques to inform, calibrate and challenge models. Elsewhere this has been termed "Data Rich Mechanics", the particular example of fibre failure and tensile strength has been highlighted in this work, but other examples have been explored for damage propagation from notches [12, 23]. The ability for high resolution CT, often in combination with in situ loading rigs, to yield large, comprehensive data sets, of damage, displacements and strains is unparalleled. When combined with models there is a real opportunity to develop fundamental, mechanism-based understanding and predictive capabilities. Such models have an important role to play in guiding material development and material and structural design. There is a need and an opportunity to continue to improve the predictive capability for composite materials and to move away from the empirical approaches to material development and structural design, as well as to achieving better material utilization and increased economic and environmental sustainability as a result.

In order to achieve this ambition of a data rich mechanics framework to enable an improved model-based predictive capability for composites a number of challenges remain to be overcome. The highest resolution imaging is necessarily restricted to limited fields of view. This results in a challenge of ensuring that observations made at small scales are representative and relevant to those in larger components and structures. Multiscale imaging, using micro-focus sources as well as synchrotron sources has been important in this regard, but there is still considerable progress required. The extraction of damage from data sets is still a largely manual/visual process defined by human operators. There is considerable scope for improved automated feature recognition software to allow automated data reduction and comparison with models. For damage such as fibre failure and impact damage formation, the processes are inherently dynamic. Imaging hardware, data acquisition and image reconstruction software has developed significantly over the past decade to allow for relatively fast scanning, with useful CT images being obtained within durations of 1-10s. This is still many times longer than the timescale for some of the critical damage events, so either further development is needed, or more likely, augmentation with other experimental and modeling techniques. The data sets provided by CT imaging are relatively large; a typical image file is about 4 Gb. This creates significant challenges for long term data storage, manipulation and provision of access to other users for comparison with models. While computer speeds, data storage capacity and transmission bandwidth are all improving, it is not clear whether current approaches will prove to be sufficient. Finally, while good progress has been made in using CT data sets to inform model development, hitherto the full 3-D potential of the data sets available has not been exploited. This will require increased sophistication in the comparison of models with experimental data in 3-D, including developing statistical techniques to evaluate the quality of agreement between experiments and models and visual techniques for representing the results.

5. Conclusions

In situ and ex situ observations have been made using high-resolution computed tomography of damage mechanisms in carbon fibre composites. Work has been conducted on impact damage resistance and damage tolerance, interlaminar fracture toughness and fibre fracture and tensile strength. These experimental studies have allowed key damage modes to be identified and quantified, which has allowed novel insights to be achieved regarding the relative importance of the individual damage modes and their relationship to microstructural features. Such data sets are particularly powerful when used to inform, calibrate and challenge models for composite performance. It is critical that there is an open and honest dialogue between those conducting experiments, those developing models and the user community.

Acknowledgements

The authors wish to thank many collaborators, postdoctoral researchers and doctoral students who have contributed to the work described herein. Funding from multiple sources, including the UK's Engineering and Physical Sciences Research Council is gratefully acknowledged.

References

[1] Philips L N 1987 Carbon fibre comes of age *Endeavour*, **11** 127-132
[2] Hinton M J, Kaddour A S Soden P D 2013 The background to the third world-wide failure exercise", *J.Comp. Mater.* **47** 2417-2426
[3] Cartie D D R Irving P. E. 2002 Effect of resin and fibre properties on impact and compression after impact performance of CFRP *Composites Part A* **33** 483-493
[4] Cantwell W J Morton J 1989 Comparison of the low and high velocity impact response of CFRP *Composites* **20** 545-551
[5] Evans A G Ahmad Z B Gilbert D G Beaumont P W R 1986 Mechanisms of toughening in rubber toughened polymers *Acta Metallurgica* **34** 79-87
[6] Sela N and Ishai O 1989 Interlaminar Fracture-Toughness and Toughening of Laminated Composite Materials – A review *Composites* **20** 423-435
[7] Rosen B W 1964 Tensile failure of fibrous composites, *AIAA Journal* **2** 1985-1991
[8] Harlow D G Phoenix S L 1981 Probability distributions for the strength of composite materials 2. A convergent sequence of tight bounds *International Journal of Fracture* **17** 601-630
[9] Wright P Fu X Sinclair I Spearing S M 2008 Ultra high resolution computed tomography of damage in notched carbon fiber-epoxy composites *J. Comp. Mater.* **42** 1993-2002
[10] Moffat A J Wright P Buffiere J Y Sinclair I Spearing S M 2008 Micromechanisms of damage in 0 degrees splits in a [90/0]s composite material using synchrotron radiation computed tomography *Scripta Mater* **59** 1043-1046
[11] Moffat A J Wright P Helfen L Baumbach T Johnson G Spearing S M Sinclair I 2010 *In situ* Synchrotron Computed Laminography of damage in Carbon Fibre Epoxy (90/0)s Laminates *Scripta Materialia*, **62** 97-100
[12] Wright P Moffat A Sinclair I Spearing S M 2010 High-resolution tomographic imaging and modeling of notch tip damage in a laminated composite *Comp. Sci. Tech.* **70** 1444-52

[13] Bull D J Spearing S M Sinclair I Helfen L 2013 Three-dimensional assessment of low velocity impact damage in particle toughened composite laminates using micro-focus X-ray computed tomography and synchrotron radiation laminography *Composites Part A* **52**, 62-69
[14] Bull D J Scott A E Spearing S M Sinclair I 2014The influence of toughening-particles in CFRPs on low velocity impact damage resistance performance *Composites Part A* **58** 47-55
[15] Bull D J Spearing S M Sinclair I 2014 Observations of damage development from compression-after-impact experiments using ex situ micro-focus computed tomography *Comp. Sci. Tech.* **97**, 106-114
[16] Bull D J Spearing S M Sinclair I 2015 Investigation of the response to low velocity impact and quasi-static indentation loading of particle-toughened carbon-fibre composite materials, *Composites Pt. A*, **74**, 38-46
[17] Borstnar G Mavrogordato M N Helfen L Sinclair I Spearing S M Interlaminar fracture micro-mechanisms in toughened carbon fibre reinforced plastics investigated via synchrotron radiation computed tomography and laminography", Composite Pt. A, 71, 176-183, 2015
[18] Garcea S C Mavrogordato M N Scott A E Sinclair I Spearing S M 2014 Fatigue micromechanism characterisation in carbon fibre reinforced polymers using synchrotron radiation computed tomography *Comp. Sci. Tech.* **99** 23-30
[19] Garcea S C Sinclair I Spearing S M 2015 *In situ* synchrotron tomographic evaluation of the effect of toughening strategies on fatigue micromechanisms in carbon fibre reinforced polymers *Comp. Sci. Tech.* **109** 32-39
[20] Scott A E Mavrogordato M Wright P Sinclair I Spearing S M 2011 *In situ* fibre fracture measurement in carbon fibre-epoxy laminates using high resolution computed tomography *Comp. Sci. Tech.* **71** 1471-77
[21] Blassiau S Thionnet A and Bunsell A R 2009 Three-dimensional analysis of load transfer micro-mechanisms in fibre/matrix composites *Comp. Sci. Tech.* **69** 33-37
[22] Swolfs Y Morton H Scott A E Gorbatikh L Reed P A S Sinclair I Spearing S M Verpoest I 2015 Synchrotron radiation computed tomography for experimental validation of a tensile strength model for unidirectional fibre-reinforced composites *Composites Pt. A,* **77** 106-113
[23] Yang Q D Schesser D Niess M Wright P Mavrogordato M N Sinclair I Spearing S M Cox B N 2015 On crack initiation in notched, cross-plied polymer matrix composites *J.Mech. Phys. Solids* **78** 314-32

Tensile failure of hybrid composites: measuring, predicting and understanding

Yentl Swolfs[1,2], Ignaas Verpoest[1] and Larissa Gorbatikh[1]

[1]Department of Materials Engineering – KU Leuven, Kasteelpark Arenberg 44 bus 2450, 3001 Leuven, Belgium
[2]Department of Aeronautics – Imperial College London, South Kensington Campus, SW7 2AZ, London, United Kingdom

Yentl.Swolfs@mtm.kuleuven.be

Abstract. Fibre-hybrid composites are attracting an ever-increasing interest from academia and industry. It is therefore vital to develop a solid understanding of their basic mechanical properties. Measuring and predicting the tensile failure of hybrid composites however remains a challenging task. This paper describes how failure develops in unidirectional (UD) hybrid composites, and how this can be predicted using fibre break models. It also provides recommendations for experimental measurements of the hybrid effect, which is a synergetic increase of the failure strain of low elongation fibres when hybridised with higher elongation fibres. Finally, limitations of our understanding of the tensile failure of hybrid composites are discussed and recommendations for future research are proposed.

1. Introduction

Fibre-reinforced polymer composites are composed of fibres in a polymeric matrix. The fibres impart stiffness and strength, whereas the matrix provides structural integrity. The low density of the matrix and fibres combined with excellent mechanical properties makes composites a preferred choice in many lightweight applications. The two most common reinforcement fibres are carbon and glass. Carbon fibres are typically used in applications where the highest performance is needed, such as sports and aerospace industries. Glass fibres, which are 5-10 times cheaper than carbon fibres, are more common in applications where cost is the main driver. Cost is however not straightforward to define for a subcomponent, and should also be evaluated on the global level. The increased use of carbon fibre composites in airplanes is a good illustration of this point. In recent airplanes like the Boeing 787 Dreamliner and Airbus A350, carbon fibre composites have replaced a large part of the advanced metal alloys, even though such alloys are cheaper. Nevertheless, carbon fibre composites lead to lighter parts, which reduce fuel consumption and eventually lead to cost savings over the lifetime of the airplane.

One of the greatest benefits of fibre-reinforced composites over other material families is the potential they offer for design optimisation. Being able to choose the fibre type, matrix type, the layup and the preform gives designers a myriad of options. Not being restricted to a single fibre type expands this design envelope even further. The development of fibre-hybrid composites, like carbon/glass hybrids, is therefore a logical evolution towards even more design freedom and hence more possibility for optimisation and cost reduction. Recently, researchers are also starting to explore other types of fibres, such as carbon/polypropylene [1], carbon/polyamide [2] or glass/natural fibres [3]. This further expands the potential design freedom that can be achieved through fibre-hybridisation.

Since carbon and glass fibres are the dominant reinforcement fibres, carbon/glass hybrids have also been the dominant hybrid combination. Carbon/glass hybrids were initially developed to reduce the cost of carbon fibre components in the sixties [4]. This triggered a large body of research in the seventies and the eighties [5-12] to explore their mechanical properties. The typical stress-strain diagram of a hybrid composite (see Figure 1) shows two distinct peaks, which are associated with failure of the low (carbon) and high elongation (glass) fibres respectively. The transition between these two peaks can be sharp as in Figure 1, or can consist of a plateau [13].

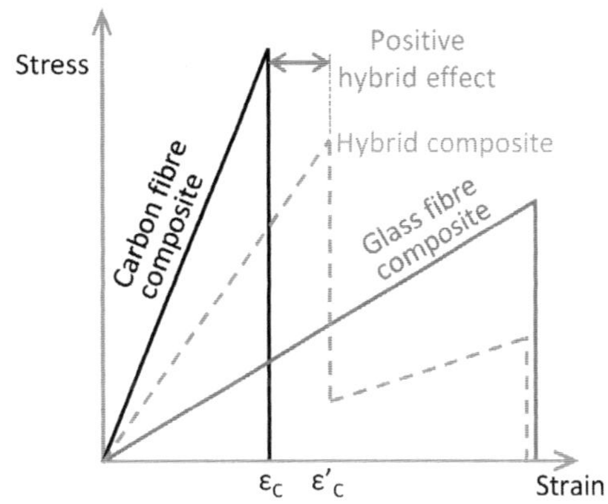

Figure 1: Schematic stress-strain diagram of a hybrid composite and its two reference composites. The hybrid shows two peaks, which are linked to failure of the carbon and glass fibre composite respectively.

In 1972, Hayashi [14] reported a remarkable effect, which later turned out to be the subject of intense scientific discussions [5]. By sandwiching a carbon fibre layer in between glass fibre layers, Hayashi improved the tensile failure strain of the carbon fibre layer by 40%. In terms of Figure 1, this means that the first peak shifted from ε_C to ε'_C, which is an increase of 40%. The term "hybrid effect" was coined to describe this synergetic effect. It is defined as the relative increase of the failure strain of the carbon fibre composite/layer in a hybrid composite relative to the failure strain in an all-carbon fibre composite. With reference to Figure 1, the hybrid effect can be calculated as:

$$Hybrid\ effect = \frac{\varepsilon'_C - \varepsilon_C}{\varepsilon_C} \quad (1)$$

The origin of the hybrid effect was initially unclear, which greatly contributed to the initial controversy. Later, three hypotheses were coined: (1) thermal residual stresses [8,9], (2) changes in the damage development leading to final failure [8,15], and (3) dynamic stress concentrations [16]. Many authors have reported on the hybrid effect, and a collection of the experimental data is presented in Figure 2.

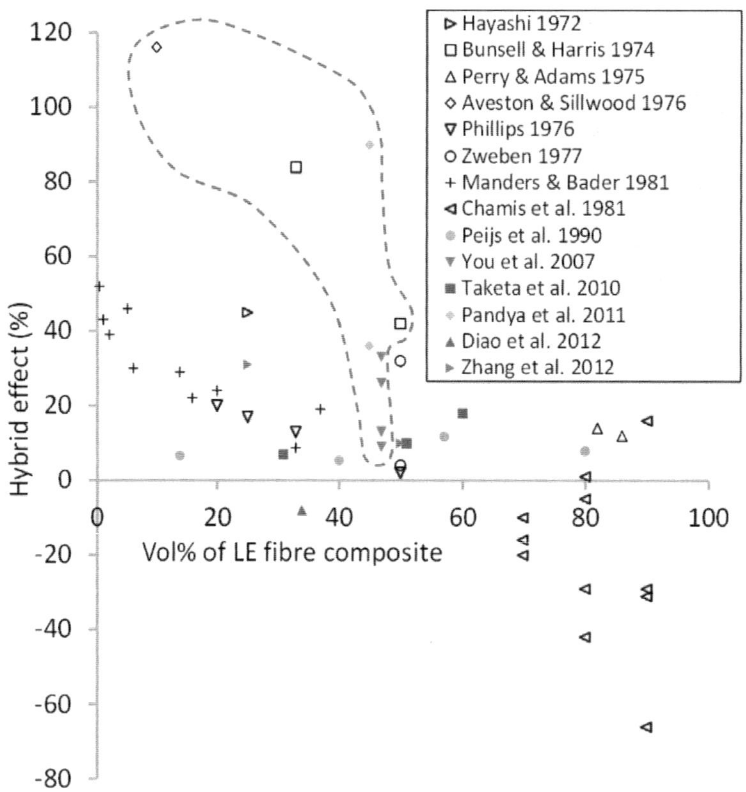

Figure 2: Overview of experimentally measured hybrid effects in carbon fibre-reinforced hybrids, with LE indicating the low elongation fibre, i.e. carbon fibre. Data from before 1987 are in black, whereas more recent data are in colour. The red dashed line encircles data points that are doubtful. (reprinted from Swolfs et al. [17] with permission from Elsevier).

Until a few years ago, there was no model available to quantitatively compare the experimental data against. These models were all based on 1D packings [8,18,19], which implies a single row fibres. The failure development in such simplified packings is significantly different from that in 2D packings. 1D packings also fail to properly capture fibre dispersion, which is well recognised to have a drastic influence on the hybrid effect [17]. Based on such models, it was therefore difficult to judge how realistic the data in Figure 2 is. Recent modelling evidence however, revealed that the hybrid effect at its best is up to 20% [20,21]. Such high values could be achieved for very thin carbon fibre layers or for very well-dispersed hybrids only. The data in Figure 2 seem to reveal an upwards trend towards 40% with decreasing carbon fibre content. This is significantly higher than the highest modelling predictions, and seems unrealistically high in light of the rather low degrees of dispersion that were achieved.

A part of the discrepancies in Figure 2 can be explained by the inherent difficulties in measuring the hybrid effect. The hybrid effect is normally measured by comparing the failure strain of a carbon fibre composite to that of a hybrid composite, both of which contain only 0° fibres. Tensile tests on fully unidirectional (UD) carbon fibre composites are however challenging to perform for three reasons. Firstly, the grips induce stress concentrations [22,23], which increases the probability of failure at the grips. Secondly, the grips limit the Poisson contraction within the grips, while such contraction is unconstrained outside of the grips. This creates a tendency for premature onset of splitting just outside of the grips [24]. Finally, the high stiffness and strength of UD carbon fibre composites lead to a very high energy release when they fail. This creates additional damage and failure, thereby disguising the location of the initial failure [24]. It is therefore difficult to establish whether failure started within the gauge length or near the grips. These three problems can lead to underestimations of the reference carbon fibre failure strain. Additionally, hybridisation with glass fibre will tend to reduce the severity of the first and third problem. All of this combined will tend to lead to overestimations of the hybrid effect.

This paper will investigate the tensile failure of hybrid composites, and attempt to discern best practices for measuring and predicting it. This will be necessary to advance our understanding of the tensile failure of hybrid composites.

2. Predicting the hybrid effect for initial failure strain

2.1. Failure development

The failure of unidirectional fibre-reinforced composites is controlled by fibre breaks. Since fibre strength follows a Weibull distribution, some fibres will have a relatively low strength. These fibres fail first, and locally lose their load transfer capability. The length over which they lose this capability, is called the ineffective length. Over this length, the broken fibres will shed their load to the nearby fibres, which causes stress concentrations [25]. The stress concentrations will increase the failure probability of the nearby fibres, thereby causing a tendency to create clusters of fibre breaks. These clusters cause even larger stress concentrations [26]. Eventually, a critical cluster will develop that will propagate rapidly and cause the entire composite to fail.

The failure development in hybrid composites essentially follows the same steps, but with a few additional complications. At a given strain, the high elongation fibres are less likely to fail than the low elongation fibres. This has two consequences. Firstly, this reduced failure probability of the high elongation fibres hinders the development of clusters of fibre breaks, as the high elongation fibres limit the number of pathways for the cluster to grow into. Secondly, in a given volume of composite material, there are now less low elongation fibres. This reduces the probability of finding a collection of relatively weak fibres that will cause the formation of a cluster, and eventually the critical cluster. These aspects will contribute to increasing the strain at which the low elongation fibres or plies fail.

A key implication of this failure process is that correctly capturing the microstructure is essential to accurate predictions of the failure development [18]. Swolfs et al. [21] have recently revealed how important this could be (see Figure 3). Alternating hybrid carbon/glass microstructures with different layer thicknesses were modelled, and the hybrid effect was found to increase significantly with decreasing layer thickness. Having layers of a single fibre thick led to a hybrid effect of 16%, which was larger than that of a randomly dispersed configuration. For layers that were 8 fibres thick, the hybrid effect reduced to about 2%. It should be noted though that most commercial plies are much thicker.

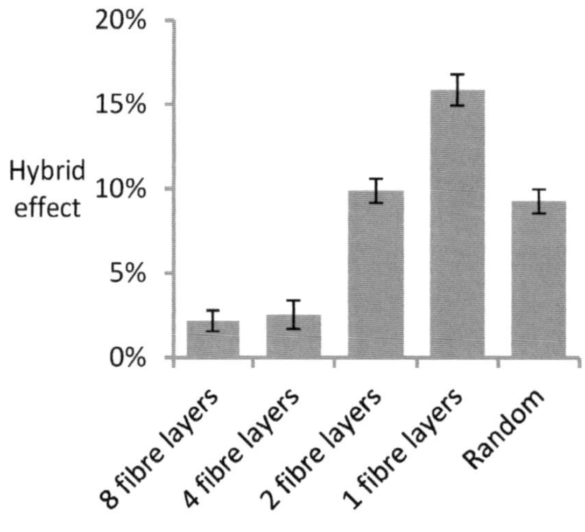

Figure 3: The hybrid effect as a function of the microstructure (reprinted from Swolfs et al. [21] with permission from Elsevier).

2.2. State-of-the-art models

The development of strength models for UD hybrid composites started with Zweben's model in 1977 [8]. Zweben's model was based on a 1D packing, which is basically a single row of parallel fibres. Zweben's hybrid packing was a single row of alternating carbon and glass fibres. Such a model hence does not capture the microstructure of a real hybrid composite. Nevertheless, this model was a breakthrough, as it was the first one proving the hypothesis that adding high elongation fibres delays the development of fibre breaks. Given the initial confusion and discussions on the hybrid effect in the early seventies [5], this model was a landmark for future model developments.

The reader is referred to the original work [8] for more details, but several key conclusions can be drawn from his work:

- Small Weibull moduli, which correspond to a large strength scatter, should lead to large hybrid effects.
- The ineffective length and stress concentrations have a minor influence on the hybrid effect.
- The largest hybrid effects are achieved if the two fibre types have a large difference in their average failure strains.

Conclusion #1 confirms that the hybrid effect is essentially a result of the scatter in the fibre strength of the low elongation fibre. This is well captured in the statement of Manders [27]: "*the hybrid effect arises from a failure to realise the full potential strength of the fibres in all-carbon fibre composites, rather than from an enhancement of their strength in the hybrids*". Conclusions #2 and #3 require further validation with more advanced models, as the assumptions and simplified geometry of Zweben's model may have tainted these conclusions.

The next step in the model development for hybrid composites was achieved by Fukuda [28] in 1984. He addressed some of the shortcomings of Zweben's model, but still used the same basic geometry. Fukuda's model revealed that the failure strain of the high elongation fibre does not influence the hybrid effect. This invalidates conclusion #3 from Zweben's model. While Zweben's model perhaps exaggerated the influence of the high elongation failure strain, Fukuda's model underestimated it. More refined models with realistic microstructures are needed to assess the importance of the high elongation failure strain.

After the models of Zweben and Fukuda however, the progress in this area was relatively slow. There was some progress in later years [18,29], but these were still limited for relatively simple microstructures or packings. Models for non-hybrid composites were developing at a much faster rate. Only in recent years, the models for hybrid composites are catching up. Swolfs et al. [30] developed a model based on global load sharing, which was further developed by Rajan and Curtin [31]. While global load sharing theory inherently ignores the local microstructure, this approach does capture several of the basic phenomena:

- The hybrid effect increases when more high elongation fibres are added, which confirms the overall trend in Figure 2.
- A small Weibull modulus for the low elongation fibre leads to a larger hybrid effect, which confirmed conclusion #1 from Zweben's model.
- A small Weibull modulus for the high elongation fibre leads to a smaller hybrid effect.
- Larger differences between the failure strains of both fibre types increase the hybrid effect, which confirms Zweben's conclusion #3. This effect is however relatively small, and when a certain ratio of both failure strains has been exceeded, no further benefit is obtained.

More advanced models were developed simultaneously by Swolfs et al. [20,21]. The first model was based on very local load sharing in a hexagonal packing [21]. This approach simplified several aspects, such as:

- Assuming both fibre types have the same radius.
- Assuming only the six nearest neighbours carry the stress concentrations.
- Using hexagonal packings instead of the more realistic random packings.
- Using a linear stress recovery in the broken fibre instead of the actual stress recovery profile.

A second model improved all these aspects by using a random fibre packing, which also allowed to include fibres with different radiuses [20].

Figure 3 revealed the key conclusion from the first model: the microstructure is essential in accurately predicting the hybrid effect. Having alternating layers of a single fibre thick led to a hybrid effect of 16%, which was larger than the 10% achieved for a random dispersion. While this is a reasonably large hybrid effect, it is still significantly lower than the 40% found by Hayashi. Compared to the other literature data in Figure 2, this may seem realistic at first. However, most of this data was obtained for much poorer dispersion. In some cases, the layer thicknesses were several 100 μm, in which case, the model predictions would predict a hybrid effect close to 0%. There is hence a significant discrepancy between modelling predictions and the experimental measurements. The question therefore is: is this discrepancy caused by inaccurate models or by inaccurate measurements? Let us first turn to potential issues with the modelling predictions.

As with any model, models for hybrid composites require several simplifications. Additionally, a model is only as reliable as its input data. There are several issues with the input data that have been identified. The most prominent issue is the reliability of the Weibull distribution for fibre strength [32]. Accurately determining this distribution is challenging from an experimental point of view, as it requires hundreds of accurate tests and a time-consuming preparation. As explained before, the Weibull distribution for the high elongation fibre is not crucial for the hybrid effect predictions. The Weibull distribution for the low elongation fibre however has an important effect on the predictions. While the literature on this specific topic is not exhaustive, there are strong indications that a small Weibull modulus for the low elongation fibre will increase the hybrid effect (see Figure 4). This also makes sense in light of Manders' quote: a small Weibull modulus implies that the all-carbon fibre composites is far away from reaching the potential strength of the fibres. Adding a high elongation fibre allows the hybrid composite to get closer to this potential strength.

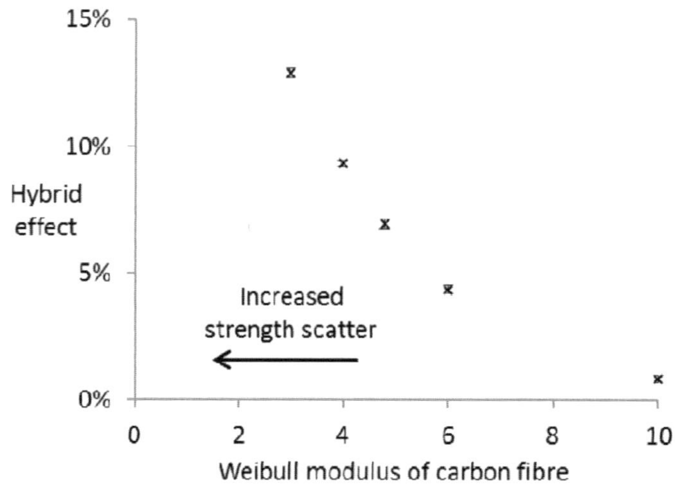

Figure 4: The hybrid effect decreases strongly with increasing Weibull modulus of the low elongation fibre (reprinted from Swolfs et al. [20] with permission from Elsevier).

The next question is how the results in Figure 4 could help to explain the large discrepancy between modelling predictions and experimental results. Close inspection of Figure 2 reveals that the majority of the data points with a large hybrid effect were obtained in the seventies and the eighties. Carbon fibre production in those days was not as well developed as it is at the moment. It is not unreasonable to assume that the lower quality of the carbon fibres also resulted in a large fibre strength scatter and hence a lower Weibull modulus. While this could bring the modelling predictions closer to the experimental values in Figure 2, the influence seems too small to explain the complete discrepancy.

2.3. Experimental validation

As any strength model for UD composites has inherent assumptions, the discrepancy could also be explained by the fact that models for UD hybrid composites are not capturing all relevant mechanisms. Experimental validation is therefore a crucial step to confirm the validity of strength models. KU Leuven and the University of Bristol have recently teamed up for such an experimental validation. Thin carbon fibre plies of 29 μm were sandwiched in between thicker glass plies. Every carbon fibre plies contained an average of just three fibres through the thickness. By changing the number of carbon fibre plies in the middle, different values for the hybrid effect could be obtained, as predicted by Figure 3. If sufficient carbon fibre plies are used, the hybrid effect should revert to zero, which means that this failure strain can be used as the reference failure strain for calculating the hybrid effect. Changing the number of carbon fibre plies from 1 to 4 was enough for achieving this objective.

Figure 5 presents the decrease in the hybrid effect with increasing number of carbon fibre plies. The hybrid effect decreases to 0% when 3-4 plies have been used, which means that the hybrid effect is absent for layer thicknesses above 100 μm. For layer thicknesses of 25-30 μm however, hybrid effects of up to 12% (for the model) and 18% (for the experiments) could be achieved. While the model parameters were fitted to achieve the correct reference failure strain (see dashed line in Figure 5), the predictions for the hybrid effect capture the correct trend. This is an encouraging result, as it demonstrates that the model captures the most important micromechanical phenomena.

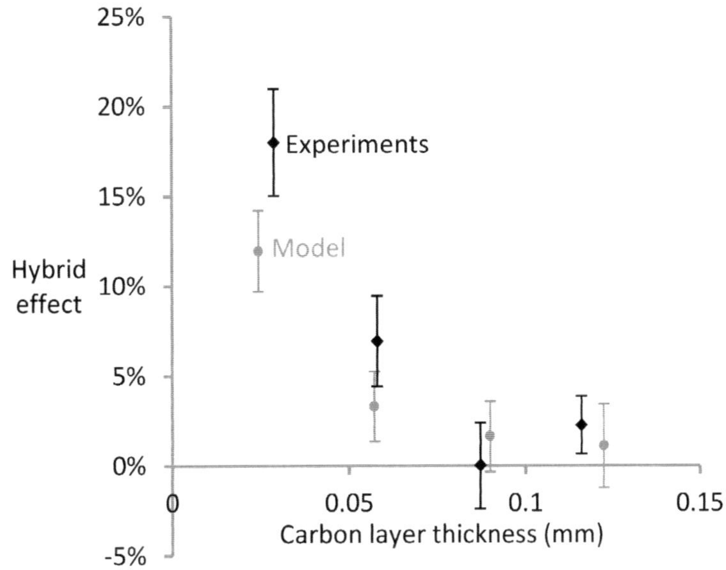

Figure 5: The model predicted the experimentally measured failure strain of hybrid composites well. The presence of thermal residual stresses were taken into account. (data based on Wisnom et al. [33])

What was so unique about this set of experiments that it matched well with the modelling predictions? The main improvement lies in the way that the reference failure strain was determined. This value is typically obtained by measuring the failure strain of the carbon fibre plies in an all-carbon fibre composite. That approach however is flawed, as testing non-hybrid UD composites nearly always leads to underestimations of the strength and failure strain. The introduction revealed three main issues in tensile testing of UD composites: (1) stress concentrations at the grips, (2) Poisson contractions near the grips, and (3) difficulties in establishing the origin of the failure. Stress concentrations at the grips have been widely discussed in literature [22,23]. They can be somewhat reduced by the use of suitable end tabs, but completely avoiding them seems to be impossible. For UD composites, an additional problem arises due to the Poisson contractions near the grips. Within the grips, the Poisson contractions are prevented, while outside the grips the sample can freely contract. This creates a tendency to initiate splitting near the end tabs. A final problem is that UD composites release a large amount of strain energy when they fracture. This rapidly creates secondary damage, thereby obscuring the initial failure location. Without a clear determination of the initiation location, it is impossible to determine whether failure started sufficiently far from the grips to qualify as a good test.

All three issues mentioned above can be resolved by placing a relatively thick layer of carbon fibre plies in between glass fibre plies:

- The glass fibre plies shield the carbon fibre plies from the stress concentrations at the grips.
- The glass fibre plies have a lower tendency for splitting, and they will also help to prevent splitting of the carbon fibre plies if their volume fraction is sufficiently high.
- By limiting the amount of carbon fibre in the hybrid composite, the amount of energy released upon failure is significantly reduced and the glass fibre continues to carry load. This allows an easy determination of the initiation location.

One additional issue that can arise is the occurrence of thermal stresses. As glass fibres have a larger coefficient of thermal expansion than carbon fibres, the proposed hybrid layup results in compressive stresses on the carbon fibre layer(s). These stresses can however be predicted and accounted for in the calculations, as was also done by Wisnom et al. [33].

Having all of this information, the literature can be re-examined. Examining the references cited in Figure 2 in more detail, it becomes clear that they did not describe how their samples failed or how failure at the grips was prevented or monitored. This makes it rather likely that the large hybrid effects they achieved, were caused by an unrealistic low value for the failure strain of their low elongation fibre composite. The proposed approach would hence be to sandwich a sufficiently thick layer of the reference material in between two layers with a higher failure strain.

3. Limitations in our understanding

Predictive models have greatly helped to advance our understanding of the tensile failure of hybrid composites. Being able to predict the hybrid effect for failure initiation was a large step forward. It is however not the end point, as there are still several open questions and areas for further improvement.

3.1. Improvements for initial failure

One of the open questions for modelling the initial failure is the importance of dynamic effects. When a fibre fails, it releases strain energy and springs back, thereby causing dynamic stress concentrations on the nearby fibres. The existence of dynamic stress concentrations is undisputed in the literature, but its importance remains unclear. While several authors have indicated that these dynamic effects can be up to 50% larger than the static ones [34-36], it remains unclear how large the influence on the failure would be. No one has included dynamic stress concentrations in a strength model yet, and the understanding for hybrid composites is even more limited. Only one study has looked at dynamic stress concentrations for hybrid composites, and this study seemed to indicate that it could contribute to the hybrid effect [16]. This particular study was however based on a simple geometry consisting of a row a carbon fibres on top of a row of glass fibres. It also limited itself to stress concentrations, and did not include them in a strength model. It hence remains unknown how large the influence of dynamics on the hybrid effect is.

There are also a number of open questions with respect to modelling the strength of unidirectional composites in general. The main issue is linked to measuring the Weibull distribution for fibre strength [32]. These problems are however not linked to hybrid composites specifically, and are outside of the scope here.

3.2. Improvements for final failure

The models that can accurately capture the initial failure do not yet capture the final failure. Some models do capture the entire stress-strain diagram, but they do this by ignoring individual fibre breaks [37,38] and/or the local microstructure [30,31]. The accuracy in predicting how the final failure strain shifts by hybridisation is therefore limited. In general, it would be expected the final failure strain is slightly reduced by the presence of the low elongation fibres (see Figure 1). The fibre breaks in the low elongation fibres trigger more fibre breaks in the high elongation fibres than in the reference composite with only high elongation fibres. This should in principle lead to earlier failure of the high elongation fibres. Furthermore, the thermal residual stress that contribute to delaying the initial failure should lead to an earlier onset of final failure. In a carbon/glass hybrid, the thermal residual stresses put the carbon fibres in compression, but they put the glass fibres in tension. These tensile stresses in the glass fibres will be added on top the externally applied stresses, which will lead to an earlier onset of final failure.

3.3. Improvements for the transition between initial and final failure

In recent years, there has been significant progress in tuning the transition the between initial and final failure. This is mainly driven by the interest in achieving pseudo-ductility. Jalalvand et al. [37,38] for example has developed modelling approaches to predict how different layups lead to different transitions between the initial and final failure. A plateau region can be achieved if a suitable layer thickness and overall volume fraction of the low elongation fibre is chosen. If done correctly, the low elongation fibre layer will fragment, instead of causing an unstable delamination or immediate fracture of the high elongation fibre layers. A crucial parameter in achieving this, is the mode II interlaminar fracture toughness. This parameter determines the amount of energy required to propagate a delamination, and should be higher than the strain energy released.

The plateau region is in principle not completely horizontal, but should increase slightly. The low elongation fibre layer does not have a constant strength, but a strength varying over the length of the layer. The first fracture hence occurs in the region where the layer has the lowest strength. When this happens, the stresses and strains redistribute themselves. To cause a second failure elsewhere in the layer and start fragmentation, the stress needs to sufficiently high to cause the second weakest location to fail. This implies that the stress needs to be higher than the stress at which the first location failed. There are two inherent issues associated with this:

1. The slope of the plateau region depends on the scatter in the strength of the low elongation ply. Measuring this strength distribution is however rather difficult, as it is often affected by surface damage and other experimental variations.
2. The length of the delamination around the failure locations will determine how the stresses and strains redistribute themselves. This length is however created dynamically when the low elongation ply fails, making it challenging to predict.

Estimates for the Weibull strength distribution of plies do exist, but such estimates inevitably depend on the exact carbon fibre type and ply thickness. One powerful approach would be to use visual observation of the fragmentation in combination with models to fit the strength distribution of the plies. A similar approach could be used for the length of the delamination immediately after a fracture. This length could in principle be measured experimentally and then used in the model [39], but that would limit the usefulness of the model for blind predictions. Progress is clearly needed on both fronts to advance our capacity to predict the full tensile behaviour of hybrid composites.

Acknowledgments
YS wishes to acknowledge the European Commission for his Marie Skłodowska-Curie Individual European Fellowship at Imperial College London and FWO Flanders for his postdoctoral mandate at KU Leuven.

References

[1] Swolfs Y, Shi J, Meerten Y, Hine P, Ward I, Verpoest I and Gorbatikh L 2015 The importance of bonding in intralayer carbon fibre/self-reinforced polypropylene hybrid composites *Compos Part A: Appl Sci Manuf* **76** 299-308
[2] Hine PJ, Bonner M, Ward IM, Swolfs Y, Verpoest I and Mierzwa A 2014 Hybrid carbon fibre/nylon 12 single polymer composites *Compos Part A: Appl Sci Manuf* **65** 19-26
[3] Jawaid M and Khalil H 2011 Cellulosic/synthetic fibre reinforced polymer hybrid composites: A review *Carbohydr Polym* **86** 1-18
[4] Phillips LN 1969 Improving racing-car bodies *Composites* **1** 50-1
[5] Phillips LN 1976 The hybrid effect - does it exist? *Composites* **7** 7-8
[6] Phillips LN 1976 On the usefulness of glass-fibre-carbon hybrids *British Plastics Federation Congress* Brighton United Kingdom
[7] Phillips MG 1981 Composition parameters for hybrid composite materials *Composites* **12** 113-6
[8] Zweben C 1977 Tensile strength of hybrid composites *J Mater Sci* **12** 1325-37
[9] Manders PW and Bader MG 1981 The strength of hybrid glass/carbon fibre Composites Part 1 Failure strain enhancement and failure mode *J Mater Sci* **16** 2233-45
[10] Short D and Summerscales J 1980 Hybrids - a review: Part 2. Physical properites *Composites* **11** 33-8
[11] Short D and Summerscales J 1979 Hybrids - a review: Part 1. Techniques, design and construction *Composites* **10** 215-21
[12] Bunsell AR and Harris B 1974 Hybrid carbon and glass fibre composites *Composites* **5** 157-64
[13] Czél G and Wisnom MR 2013 Demonstration of pseudo-ductility in high performance glass/epoxy composites by hybridisation with thin-ply carbon prepreg *Compos Part A: Appl*

Sci Manuf **52** 23-30

[14] Hayashi T 1972 On the improvement of mechanical properties of composites by hybrid composition *Proc. 8th Intl. Reinforced Plastics Conference* 149-52

[15] Manders PW and Bader MG 1981 The strength of hybrid glass/carbon fibre composites Part 2 A statistical model *J Mater Sci* **16** 2246-56

[16] Xing J, Hsiao GC and Chou TW 1981 A dynamic explanation of the hybrid effect *J Comp Mater* **15** 443-61

[17] Swolfs Y, Gorbatikh L and Verpoest I 2014 Fibre hybridisation in polymer composites: a review *Compos Part A: Appl Sci Manuf* **67** 181-200

[18] Fukunaga H, Chou TW and Fukuda H 1989 Probabilistic strength analyses of interlaminated hybrid composites *Compos Sci Technol* **35** 331-45

[19] Fukunaga H, Chou TW and Fukuda, H 1984 Strength of intermingled hybrid composites *J Reinf Plast Comp* **3** 145-60

[20] Swolfs Y, Verpoest I and Gorbatikh L 2016 Maximising the hybrid effect in unidirectional hybrid composites *Mater Design* **93** 39-45

[21] Swolfs Y, McMeeking RM, Gorbatikh L and Verpoest I 2015 The effect of fibre dispersion on initial failure strain and cluster development in unidirectional carbon/glass hybrid composites *Compos Part A: Appl Sci Manuf* **69** 279-87

[22] De Baere I, Van Paepegem W and Degrieck J 2009 On the design of end tabs for quasi-static and fatigue testing of fibre-reinforced composites *Polym Composite* **30** 381-90

[23] Portnov GG, Kulakov VL and Arnautov AK, 2007 A refined stress-strain analysis in the load transfer zone of flat specimens of high-strength unidirectional composites in uniaxial tension - 2. Finite-element parametric analysis *Mech Compos Mater* **43** 29-40

[24] Fuwa M, Bunsell AR and Harris B 1975 Tensile failure mechanisms in carbon fibre reinforced plastics *J Mater Sci* **10** 2062-70

[25] Swolfs Y, Gorbatikh L, Romanov V, Orlova S, Lomov SV and Verpoest I 2013 Stress concentrations in an impregnated fibre bundle with random fibre packing *Compos Sci Technol* **74** 113-20

[26] Swolfs Y, Morton H, Scott AE, Gorbatikh L, Reed PAS, Sinclair I, Spearing SM, Verpoest I 2015 Synchrotron radiation computed tomography for experimental validation of a tensile strength model for unidirectional fibre-reinforced composites *Compos Part A: Appl Sci Manuf* **77** 106-13

[27] Manders PW 1979 The strength of mixed fibre composites PhD thesis University of Surrey

[28] Fukuda H 1984 An advanced theory of the strength of hybrid composites *J Mater Sci* **19** 974-82

[29] Zeng QD 1994 A statistical analysis of the tensile failure and hybrid effect of an intraply hybrid composite *Int J Fract* **68** 351-62

[30] Swolfs Y, McMeeking RM, Rajan VP, Zok FW, Verpoest I and Gorbatikh L 2015 Global load sharing model for unidirectional hybrid fibre-reinforced composites *J Mech Phys Solids* **84** 380-94

[31] Rajan VP and Curtin WA 2015 Rational design of fiber-reinforced hybrid composites: A global load sharing analysis *Compos Sci Technol* **117** 199-207

[32] Swolfs Y, Verpoest I and Gorbatikh L 2015 Issues in strength models for unidirectional fibre-reinforced composites related to Weibull distributions, fibre packings and boundary effects *Compos Sci Technol* **114** 42-9

[33] Wisnom MR, Czel G, Swolfs Y, Jalalvand M, Gorbatikh L and Verpoest I 2016 Hybrid effects in thin ply carbon/glass unidirectional laminates: accurate experimental determination and prediction *Compos Part A: Appl Sci Manuf* **88** 131-9.

[34] Hedgepeth JM 1961 Stress concentrations in filamentary structures *NASA TN* **D-882** 1-36

[35] Sakharova EN and Ovchinskii AS 1984 Influence of dynamic effects accompanying rupture of fibers and separation of fibers from the matrix on interaction between failure micromechanisms of composite materials *Mech Compos Mater* **20** 323-7

IOP Conf. Series: Materials Science and Engineering **139** (2016) 012008 doi:10.1088/1757-899X/139/1/012008

[36] Ji X, Liu XR and Chou TW 1985 Dynamic stress concentration factors in unidirectional composites *J Comp Mater* **19** 269-75
[37] Jalalvand M, Czél G and Wisnom MR 2014 Numerical modelling of the damage modes in UD thin carbon/glass hybrid laminates *Compos Sci Technol* **94** 39-47
[38] Jalalvand M, Czél G and Wisnom MR 2015 Damage analysis of pseudo-ductile thin-ply UD hybrid composites - a new analytical method *Compos Part A: Appl Sci Manuf* **69** 83-93
[39] Swolfs Y, Crauwels L, Van Breda E, Gorbatikh L, Hine P, Ward I and Verpoest I 2014 Tensile behaviour of intralayer hybrid composites of carbon fibre and self-reinforced polypropylene *Compos Part A: Appl Sci Manuf* **59** 78-84

IOP Conf. Series: Materials Science and Engineering **139** (2016) 012009 doi:10.1088/1757-899X/139/1/012009

Fiber pull-out test and single fiber fragmentation test - analysis and modelling

B F Sørensen[1] **and H Lilholt**

Section of Composites and Materials Mechanics, Department of Wind Energy, The Technical University of Denmark, Risø Campus, Frederiksborgvej 399, DK-4000 Roskilde, Denmark

[1]e-mail: bsqr@dtu.dk

Abstract. A mathematical model is developed for the analysis of the fiber debonding phase of a pull-out experiment where the matrix is supported at the same end as the fiber is loaded in tension. The mechanical properties of the fiber/matrix are described in terms of two parameters, a fracture energy for fiber/matrix debonding and a frictional sliding shear stress. Results for the debond length and fiber debond displacement are compared with results from similar models for single fiber pull-out experiments where the specimen is gripped at the end opposite to the end where the fiber is pulling-out and with results for a single fiber fragmentation test.

1. Introduction

The fiber/matrix interface plays an important role in controlling the macroscopic mechanical properties of fiber composites [1]. A number of tests involving specimens with a single fiber have been developed, such as single fiber pull-out tests, single fiber fragmentation tests and fiber push-out tests [2-4]. Yet it still remains a challenge to characterize the mechanical properties of the fiber/matrix interface for several reasons. First, the practical side. The manufacture of specimens with a single fiber involves handling of individual thin fibers which can be difficult and the testing usually requires special testing devices and measurement equipment. Second, a part of the challenge is from the theoretical side. There is a lack of agreement in which parameters to use for the characterization of the mechanical behavior of the fiber/matrix interface and there is a need for suitable theoretical models and approaches for the extraction of interface parameters from experimental data.

Historically, the mechanical properties of fiber/matrix interfaces in composites were first described in terms of a maximum interfacial shear stress [5], representing yielding of a ductile interface or interface strength in case of a brittle interface. This idea led to the development of the single-fiber fragmentation test (SFFT) in which the saturated distribution of spacings between positions of multiple fiber breaks is used to for the calculation of an interfacial shear strength [5]. More recently, it has been proposed to characterize the fiber/matrix interface in terms of a debond energy and a frictional shear stress or Coulomb friction [6-9]. Such models are motivated by observations made during SFFT testing, indicating that fiber/matrix debonding takes place progressively during monotonic loading [2]. Another test type for fiber/matrix interface characterization is fiber pull-out tests. There are (at least) two ways of performing single fiber pull-out tests. One is to clamp the specimen end opposite to the one where the fiber is loaded in tension - we will denoted this test for fiber pull-out Type 1 (PO1). Another configuration is where the specimen is supported in the matrix material in the same end as the fiber is loaded - pull-out Type 2 (PO2). In both cases, we denote the (un-deformed) length of the free fiber by L_f and the embedded length by L.

A number of micromechanical models exist for the analysis of pull-out and SFFT tests. Several of these models are quite advanced and incorporate e.g. Poisson's effects of fiber and matrix as well as residual stresses [8-12]. Unfortunately, due to mathematical complexity many of these models require a numerical implementation to be used. It is not always easy to see how model parameters are to be extracted from experiments.

In this paper, we develop a relative simple analytical shear-lag model for the analysis of fiber pull-out Type 2 (PO2). The complete pull-out experiment, from the start to the final separation of fiber and matrix, consists of three stages: the initial *debonding and sliding* phase, the *load drop* at maximum fiber stress and the *sliding and pull-out* phase to complete pulling-out of the fiber. The model describes the fiber/matrix interface in terms of two parameters, a fiber/matrix debond energy, $\mathcal{G}_c^i$, and a constant interfacial frictional a sliding shear stress, τ_s. The model incorporates residual stresses, but neglects Poisson´s effects. This allows us to make a relative simple 1-dimensional model. We describe an approach on how the interfacial parameters are determined from the debonding/sliding phase of PO2 pull-out experiments. Unlike most previous approaches, we utilize the fact that polymer matrix composites are optical transparent so that the debond length can be measured experimentally e.g. using optical miscopy during the experiment and used as an input parameter to extract model parameters from the model.

We also compare the mathematical structure of the model with other similar models for pull-out experiments where the other end is held fixed (clamped) and with the SFFT. These test methods are closely related, see Figure 1 for a comparison.

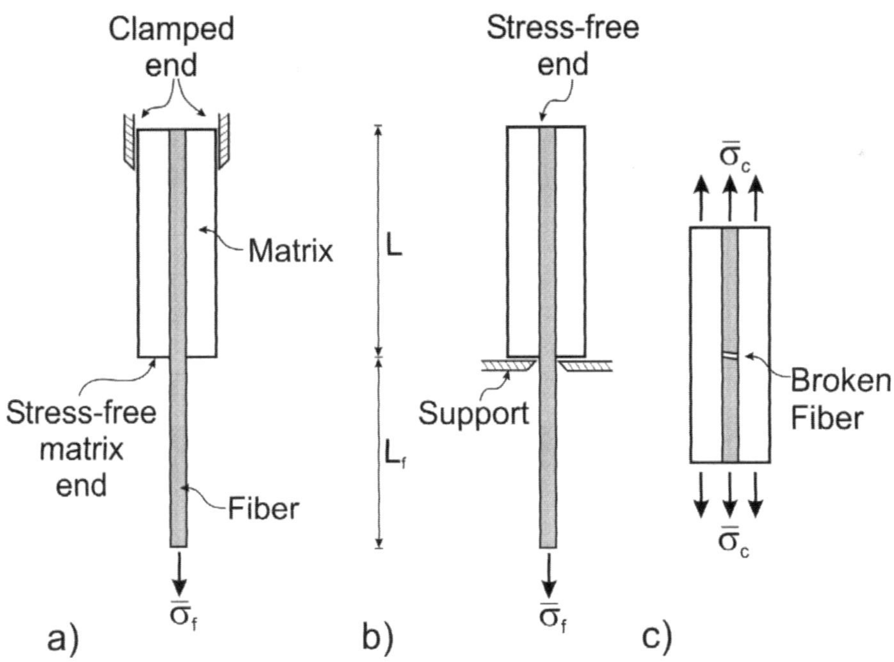

Figure 1. Comparison of testing methods (a) Pull-out Type 1 (PO1), (b) pull-out Type 2 (PO2), and (c) the single fiber fragmentation tests (SFFT).

The paper is organized as follows. First, Section 2 describes the assumption and development of the mathematical model. Next, in Section 3, we discuss the key results and explain how the interfacial parameters can be determined from experiments. Section 4 discusses the basic assumptions and limitations of the model, compares the mathematical structure of the model results with other, similar models for pull-out experiments PO1 and models for the SFFT, which are closely related, see Figure 1 for a comparison. Section 5 summarizes the major conclusions of this work.

2. Theory - model development

The section describes the basic assumption of the model, defines the geometry and boundary conditions of the model, and then the major steps in the model derivation are described.

2.1. Basic assumptions for debonding and sliding during pull-out Type 2 (PO2)

The problem we analyze is a single circular fiber embedded in a matrix cylinder. The fiber radius is r and the outer radius of the matrix cylinder is R and the fiber volume fraction is denoted V_f. The fiber has debonded a length, ℓ_d. A z-coordinate system with origin at the debonded end of the specimen is used, see Figure 2. The interfacial fracture energy is denoted $\mathcal{G}_c^i$ and the frictional sliding shear stress is τ_s. The fiber is loaded by a tensile stress, $\overline{\sigma}_f$, applied at the debonded fiber end ($z = 0$). The specimen is supported at the matrix at $z = 0$, inducing a traction $\overline{\sigma}_m$ that is in force balance with $\overline{\sigma}_f$ (note the we have defined the traction $\overline{\sigma}_m$ positive in the direction of the z-axis, opposite to the positive direction of the applied fiber stress $\overline{\sigma}_f$). As a result of the interfacial debonding and sliding, the fiber and matrix have a relative displacement δ at $z = 0$; δ is defined positive when the fiber slips out. The geometry and boundary conditions are shown in Figure 2. The fiber and matrix are taken to possess linear-elastic behavior described in terms of a Young's modulus of the fiber and matrix, denoted E_f and E_m respectively.

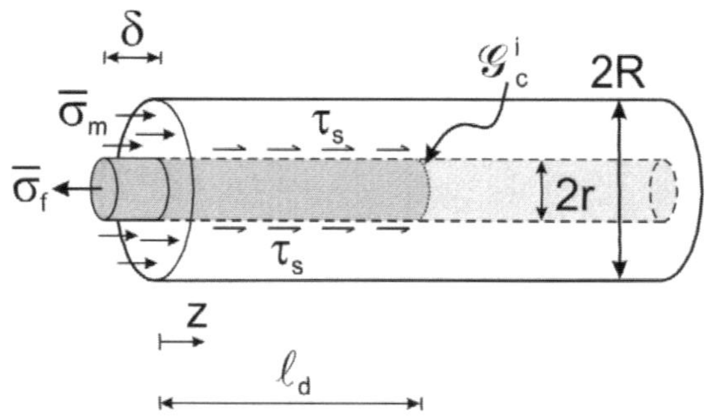

Figure 2. Geometry and loading of the model for pull-out Type 2 (PO2).

2.2. Residual stresses

The composite is taken to be stress-free at the processing temperature, T_0. However, residual stresses can be present in the unloaded composite when cooled down to the temperature, T, where the pull-out test is performed. In the following, we will express the residual stresses in terms of a mismatch strain, $\Delta\varepsilon^T$. The mismatch strain is the strain difference that would exist between fiber and matrix at the testing temperature if the fiber and matrix were allowed to expand and contract freely following the processing history. Thus, $\Delta\varepsilon^T$ can be a result of shrinkage due to cross linking of polymers chains during curing, contraction due to thermal expansion mismatch during cool-down, etc. In the following, we will calculate the mismatch strain in case it originates solely from a mismatch in thermal expansion coefficients. With α_f and α_m being the thermal expansion coefficients of fiber and matrix, the residual stresses in the fiber and matrix induced by thermal expansion mismatch are found as (see Appendix A)

$$\frac{\sigma_f^{res}}{E_f} = -\Delta\varepsilon^T (1-V_f)\frac{E_m}{E_c} \qquad \frac{\sigma_m^{res}}{E_m} = \Delta\varepsilon^T V_f \frac{E_f}{E_c}, \tag{1}$$

where the mismatch strain is defined as

$$\Delta\varepsilon^T = (\alpha_f - \alpha_m)(T - T_0), \tag{2}$$

and E_c is the Young's modulus of the composite, given by

$$E_c = (1-V_f)E_m + V_f E_f . \tag{3}$$

In principle, residual stresses can originate from other phenomena than a mismatch in the thermal expansion coefficients. However, irrespective of their cause, residual stresses can be represented in terms of a mismatch strain, $\Delta\varepsilon^T$. Therefore the model is formulated such that $\Delta\varepsilon^T$ is considered being an unknown quantity that must be determined experimentally. Appendix A describes how to account for a pre-stress applied to the fiber during specimen manufacturing. In the present paper, we will treat three microscale parameters as unknowns that must be determined from experiments: $\mathcal{G}_c^i$, τ_s and $\Delta\varepsilon^T$.

2.3. Stresses and strains in the debonded zone

With the applied stresses at $z = 0$ being $\bar{\sigma}_f$, force balance of the specimen gives

$$\bar{\sigma}_m = \bar{\sigma}_f \frac{V_f}{(1-V_f)}. \tag{4}$$

The stresses in the debonded zone are found by force balance, accounting for the stress transfer from the fiber to the matrix via the interfacial shear stress, τ_s, which acts in the direction opposite to the slip direction at the fiber/matrix interface. The stress in the fiber at position z is:

$$\sigma_f^-(z) = \bar{\sigma}_f - 2\tau_s \frac{z}{r} \qquad for\ 0 \le z \le \ell_d, \tag{5}$$

(superscript "-" indicates the stress in the wake of the crack tip) and the stress in the matrix cylinder at position z is

$$\sigma_m^-(z) = \frac{V_f}{(1-V_f)}\left(2\tau_s \frac{z}{r} - \bar{\sigma}_f\right) \quad for\ 0 \le z \le \ell_d. \tag{6}$$

The strain in the fiber at position z within the debonded zone is obtained from Hooke's law:

$$\varepsilon_f^-(z) = \frac{\bar{\sigma}_f}{E_f} - 2\frac{\tau_s}{E_f}\frac{z}{r} + \alpha_f \Delta T \quad for\ 0 \le z \le \ell_d. \tag{7}$$

Likewise, we find the strain in the matrix as:

$$\varepsilon_m^-(z) = \frac{V_f}{(1-V_f)}\left\{2\frac{\tau_s}{E_m}\frac{z}{r} - \frac{\bar{\sigma}_f}{E_m}\right\} + \alpha_m \Delta T \quad for\ 0 \le z \le \ell_d. \tag{8}$$

2.4. *Stresses far ahead of the crack tip*

Far ahead of the debond crack tip (superscript "+" indicates position ahead of the debond crack tip), the stresses are independent of z -position and equal to the residual stresses:

$$\sigma_f^+(z) = \sigma_f^{res} = -\Delta\varepsilon^T(1-V_f)\frac{E_m E_f}{E_c} \qquad \text{and} \qquad \sigma_m^+(z) = \sigma_m^{res} = \Delta\varepsilon^T V_f \frac{E_m E_f}{E_c}. \tag{9}$$

2.5. *Fiber debond displacement*

The fiber debond displacement at position $z = 0$, δ, is obtained by integration of the strain difference along the sliding part of the interface, assuming that there is no displacement difference (slip) between the fiber and matrix at the crack tip ($z = \ell_d$):

$$\delta = \int_0^{\ell_d} \left[\varepsilon_f^-(z) - \varepsilon_m^-(z)\right] dz. \tag{10}$$

The debond displacement in (10) is defined to be positive when the fiber slips out of the matrix. Inserting $\varepsilon_f^-(z)$ and $\varepsilon_m^-(z)$ from (7) and (8) into (10) and performing the integration leads to:

$$\frac{\delta}{r} = \left(\frac{E_c}{(1-V_f)E_m}\frac{\bar{\sigma}_f}{E_f} + \Delta\varepsilon^T\right)\left(\frac{\ell_d}{r}\right) - \frac{E_c}{(1-V_f)E_m}\frac{\tau_s}{E_f}\left(\frac{\ell_d}{r}\right)^2. \tag{11}$$

Eq. (11) is the first major result of our analysis. Eq. (11) can be re-written as follows:

$$\Delta\varepsilon^T = \left(\frac{\delta}{\ell_d}\right) + \frac{E_c}{(1-V_f)E_m}\left[\frac{\tau_s}{E_f}\left(\frac{\ell_d}{r}\right) - \frac{\bar{\sigma}_f}{E_f}\right]. \tag{12}$$

Eq. (12) enables the determination of $\Delta\varepsilon^T$ from experimental measurements of ℓ_d and δ as a function of the applied fiber stress, $\bar{\sigma}_f$, if the interfacial shear stress τ_s is determined independently. Note that Eq. (12) is independent of $\mathcal{G}_c^i$.

2.6. *Determination of debond length*

In order to establish equations for the determination of $\mathcal{G}_c^i$ and τ_s, we proceed to determine the debond length ℓ_d as a function of the applied fiber stress, $\bar{\sigma}_f$. To do so, we utilize the potential energy approach of Budiansky, Hutchinson and Evans [13]. This approach is developed to calculate the potential energy loss accounting for large-scale frictional sliding. More precisely, the approach calculates the potential energy differences of two states of a cracked body with fixed surface tractions as it undergoes further cracking and frictional sliding. In State I, the body is subjected to surface tractions and cracks are present and frictional slippage has occurred along internal interfaces in the body. The body now undergoes further debonding and frictional slipping to State II. During this transition, frictional shear stresses perform further work and thus dissipates energy. Utilizing the principle of virtual work, Budiansky *et al.* [13] were able to eliminate the work of the applied tractions and express the potential energy difference between the two states as follows:

$$\Pi_I - \Pi_{II} = \frac{1}{2}\int_V (\sigma_I - \sigma_{II}) : M(\sigma_I - \sigma_{II}) dV + \xi_F, \tag{13}$$

where Π_I and Π_{II} are the potential energy of State I and State II. In (13), V is the volume of the body, σ_I and σ_{II} are the stresses associated with State I and State II, M indicates an elastic operator (Hooke's law) and ξ_F is the frictional energy dissipation, defined as

$$\xi_F = \tau_s \int_{S_F} |\Delta u| \, dS . \tag{14}$$

In (14), $|\Delta u|$ denotes the relative frictional slip between the two states (assumed to occur monotonically during the transition from State I to State II) and S_F is the surface area at which frictional sliding occurs.

The potential energy loss will be available for the energy absorption and dissipation during the transition from State I to State II. Therefore, a criterion for debond crack growth can be stated as follows: The debond crack propagation will occur only when the potential energy loss, eq. (13), is equal to the energy absorption by the debonding of the fiber/matrix crack tip and the frictional energy dissipation. This can be written as:

$$\Pi_I - \Pi_{II} = \mathcal{G}_c^i \, 2\pi r \, \Delta\ell_d + \xi_F . \tag{15}$$

The right hand side of Eq. (15) is the energy absorption by the debonding of the fiber/matrix crack tip and the frictional energy dissipation. Inserting (13) into the left hand side of (15), we note that the frictional energy dissipation term, ξ_F, will appear on both sides in (15) and thus will cancel out. This is a major advantage of the approach since it simplifies the calculations; the frictional energy dissipation and (it turns out) the potential of the applied tractions do not need to be calculated.

We first identify State I as the problem of a debond length ℓ_d and State II as the situation where the debond crack length has increased by a small distance, $\Delta\ell_d$. We note that with the tractions at the external boundaries held fixed, the State II stresses, $\sigma_f^-(z)$ and $\sigma_m^-(z)$, in the region $0 < z < \ell_d$, are exactly the same as for State I. In State II, the length of the slip zone has increased by $\Delta\ell_d$, so the equations (4) and (5) for $\sigma_f^-(z)$ and $\sigma_m^-(z)$ are now valid also for the new (State II) slip zone, $0 \le z \le \ell_d + \Delta\ell_d$. The upstream stresses, σ_f^+ and σ_m^+, are unchanged for $z > \ell_d + \Delta\ell_d$. Then, for the transition from State I to State II, the stress state changes only for $\ell_d < z < \ell_d + \Delta\ell_d$. Therefore, (13) becomes

$$\Pi_I - \Pi_{II} = \frac{1}{2} \int_{\ell_d}^{\ell_d + \Delta\ell_d} \left\{ \pi r^2 \frac{\left(\sigma_f^+ - \sigma_f^-(z)\right)^2}{E_f} + \pi\left(R^2 - r^2\right) \frac{\left(\sigma_m^+ - \sigma_m^-(z)\right)^2}{E_m} \right\} dz + \xi_F . \tag{16}$$

Omitting details (the analyses presented here follows that of Sørensen [14] who analyzed fiber failure and progressive debonding of a SFFT using a similar approach), we insert σ_f^+ and σ_m^+ from (9) and $\sigma_f^-(z)$ and $\sigma_m^-(z)$ from (5) and (6) into (16). After having performed the integration, we neglect higher order terms of $\Delta\ell_d$ (since $\Delta\ell_d \ll \ell_d$), and insert the result for the potential energy loss into the left hand side of (15). Then, $\Delta\ell_d$ appears on all terms on both sides and cancels out. We end up with an equation for the determination of the debond length, ℓ_d. For $\tau_s \neq 0$ we get

$$\frac{\ell_d}{r} = \frac{E_f}{2\tau_s} \left(\frac{\bar{\sigma}_f}{E_f} + \frac{(1-V_f)E_m}{E_c} \Delta\varepsilon^T \right) - \frac{E_f}{\tau_s} \sqrt{\frac{(1-V_f)E_m}{E_c} \left(\frac{\mathcal{G}_c^i}{E_f r} \right)} . \tag{17}$$

Eq. (17) is our second major result. From (17) we note that the debond length depends on all three unknown parameters, $\mathcal{G}_c^i$, τ_s and $\Delta\varepsilon^T$.

3. Further analysis

In this section we re-write and combine equations (11) and (17) to forms that are particularly suited for use in connection with experimental determination of the unknown parameters, $\mathcal{G}_c^i$, τ_s and $\Delta\varepsilon^T$.

3.1. Re-arranging equations

We can rewrite (17) to give $\bar{\sigma}_f$ as a function of ℓ_d:

$$\frac{\bar{\sigma}_f}{E_f} = 2\sqrt{\frac{(1-V_f)E_m}{E_c}\left(\frac{\mathcal{G}_c^i}{E_f r}\right)} - \frac{(1-V_f)E_m}{E_c}\Delta\varepsilon^T + 2\frac{\tau_s}{E_f}\left(\frac{\ell_d}{r}\right). \tag{18}$$

We define the fiber stress corresponding to debond initiation as the value of $\bar{\sigma}_f$ for $\ell_d = 0$:

$$\frac{\bar{\sigma}_f^i}{E_f} = 2\sqrt{\frac{(1-V_f)E_m}{E_c}\left(\frac{\mathcal{G}_c^i}{E_f r}\right)} - \frac{(1-V_f)E_m}{E_c}\Delta\varepsilon^T, \tag{19}$$

so that (18) can be written as

$$\frac{\bar{\sigma}_f}{E_f} = \frac{\bar{\sigma}_f^i}{E_f} + 2\frac{\tau_s}{E_f}\left(\frac{\ell_d}{r}\right). \tag{20}$$

Eq. (20) shows that $\bar{\sigma}_f$ is related linearly to ℓ_d. This suggests an experimental way to determine the interfacial frictional shear stress from the slope of the $\bar{\sigma}_f$ - ℓ_d relationship. Differentiation of (20) leads to:

$$\tau_s = \frac{r}{2}\left(\frac{d\bar{\sigma}_f}{d\ell_d}\right). \tag{21}$$

Thus, by recording the relationship between $\bar{\sigma}_f$ and ℓ_d we can determine τ_s from (21). Eq. (21) is our third major result.

We now insert ℓ_d from (17) into (11). The result is

$$\frac{\delta}{r} = \frac{1}{4}\frac{(1-V_f)E_m}{E_c}\frac{E_f}{\tau_s}\left(\frac{E_c}{(1-V_f)E_m}\frac{\bar{\sigma}_f}{E_f} + \Delta\varepsilon^T\right)^2 - \frac{\mathcal{G}_c^i}{\tau_s r}. \tag{22}$$

Rewriting (19), we obtain

$$\frac{\mathcal{G}_c^i}{E_f r} = \frac{1}{4}\frac{(1-V_f)E_m}{E_c}\left(\frac{\bar{\sigma}_f^i}{E_f} + \frac{(1-V_f)E_m}{E_c}\Delta\varepsilon^T\right)^2. \tag{23}$$

This result, which is our fourth important result, is independent of τ_s but requires the knowledge of $\Delta\varepsilon^T$. Eq. (23) thus provides one equation to determine $\mathcal{G}_c^i$ from measured values of $\bar{\sigma}_f^i$ and $\Delta\varepsilon^T$, i.e. one result per test specimen.

Inserting $\Delta\varepsilon^T$ from (12) into (23) gives

$$\frac{\mathcal{G}_c^i}{E_f r} = \frac{1}{4}\frac{E_c}{(1-V_f)E_m}\left(\frac{\bar{\sigma}_f - \bar{\sigma}_f^i}{E_f} + \frac{(1-V_f)E_m}{E_c}\left(\frac{\delta}{\ell_d}\right) + \frac{\tau_s}{E_f}\left(\frac{\ell_d}{r}\right)\right)^2. \tag{24}$$

This, our fifth main result, is independent of $\Delta\varepsilon^T$. Unlike the result (23), Eq. (24) enables the determination of multiple values of $\mathcal{G}_c^i$ from multiple sets of data ($\bar{\sigma}_f$, ℓ_d and δ) from one test specimen. This approach is thus likely to provide a more accurate estimate of $\mathcal{G}_c^i$ that the result based on a single data set, Eq. (23).

3.2. *Approximations for $V_f \to 0$*

A further simplification of results can be obtained by considering $V_f \to 0$, which is practically the case for specimens consisting of a single fiber cast into a matrix specimen. First we note that $E_c \to E_m$ for $V_f \to 0$, so that $(1-V_f)E_m/E_c \to 1$ for $V_f \to 0$. Then (12) reduces to

$$\Delta\varepsilon^T = \left(\frac{\delta}{\ell_d}\right) + \left[\frac{\tau_s}{E_f}\left(\frac{\ell_d}{r}\right) - \frac{\bar{\sigma}_f}{E_f}\right], \tag{25}$$

while (19) simplifies to

$$\frac{\bar{\sigma}_f^i}{E_f} = 2\sqrt{\frac{\mathcal{G}_c^i}{E_f r}} - \Delta\varepsilon^T, \tag{26}$$

while eq. (22) becomes

$$\frac{\delta}{r} = \frac{1}{4}\frac{E_f}{\tau_s}\left(\frac{\bar{\sigma}_f}{E_f} + \Delta\varepsilon^T\right)^2 - \frac{\mathcal{G}_c^i}{\tau_s r}, \tag{27}$$

and (24) reduces to

$$\frac{\mathcal{G}_c^i}{E_f r} = \frac{1}{4}\left(\frac{\bar{\sigma}_f - \bar{\sigma}_f^i}{E_f} + \left(\frac{\delta}{\ell_d}\right) + \frac{\tau_s}{E_f}\left(\frac{\ell_d}{r}\right)\right)^2. \tag{28}$$

Finally, eq. (23) can be written as

$$\frac{\mathcal{G}_c^i}{E_f r} = \frac{1}{4}\left(\frac{\bar{\sigma}_f^i}{E_f} + \Delta\varepsilon^T\right)^2. \tag{29}$$

Eq. (29) is simply another form of (26).

3.3. *Approach for parameter determination*

We will now describe an approach for the determination of the interface parameters $\mathcal{G}_c^i$ and τ_s as well as the strain mismatch, $\Delta\varepsilon^T$, from measurement from a continuous pull-out Type 2 (PO2) test. The procedure follows a similar procedure proposed earlier for the pull-out Type 1 (PO1) [15]. We assume that during the single fiber pull-out experiments the parameter $\bar{\sigma}_f$, δ and ℓ_d are recorded simultaneously. We propose the following approach to determine the three parameters in three steps:

<u>Step 1</u>: $\bar{\sigma}_f$ is plotted as a function of ℓ_d. Since, according to (20), $\bar{\sigma}_f$ should depend linearly on ℓ_d, a straight line can be fitted to the data. This allows the determination of τ_s using (21). A sketch of such a plot is shown in Figure 3a.

<u>Step 2</u>: Having determined τ_s, another plot is made in accordance with (12) or (25) using the experimental values of $\bar{\sigma}_f$, δ and ℓ_d. The result should be a constant value, $\Delta\varepsilon^T$. An average value of $\Delta\varepsilon^T$ can be determined from such data. A sketch of such a plot is shown in Figure 3b.

Step 3: Knowing τ_s, the value of $\mathcal{G}_c^i$ can now be determined by the use of (24) or (29), from a plot showing $\mathcal{G}_c^i$ as a function of $\bar{\sigma}_f$ or ℓ_d using the associated values of δ. A sketch of such a plot is shown in Figure 3c.

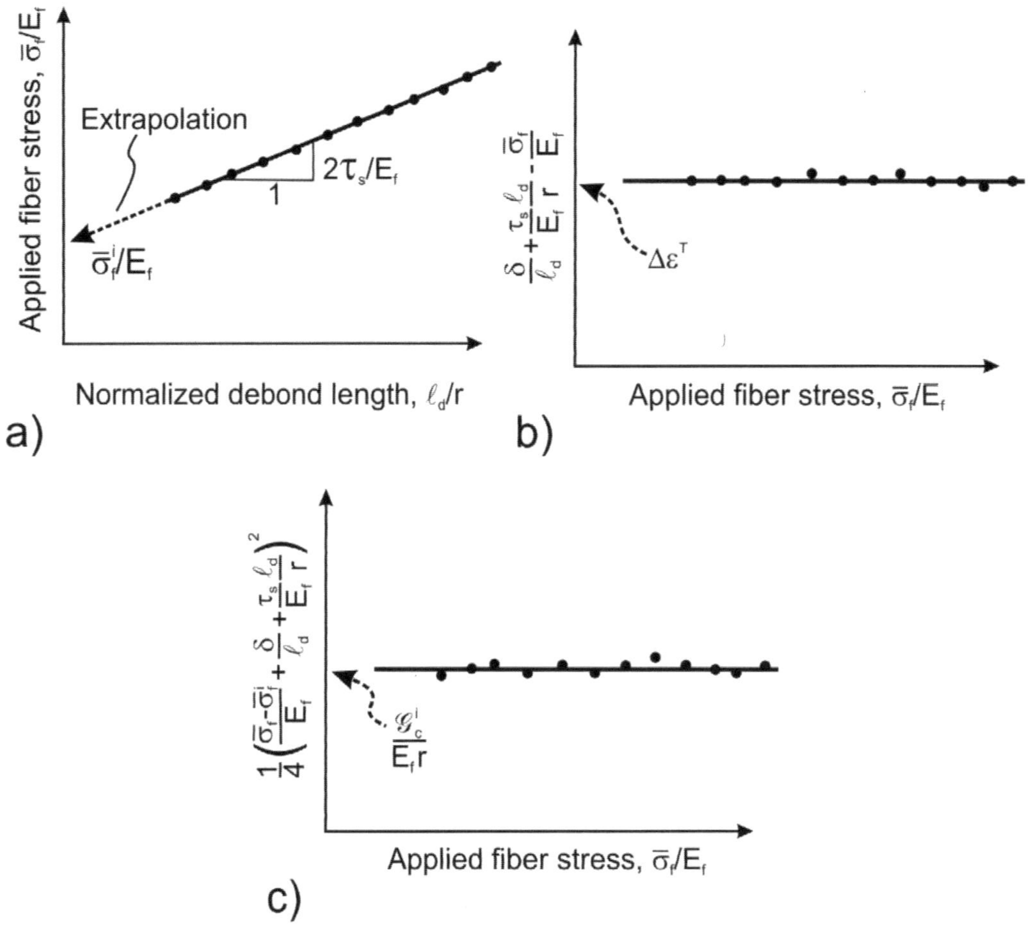

Figure 3. Sketches of plots for the determination of parameters τ_s, $\Delta\varepsilon^T$ and $\mathcal{G}_c^i$ from data from pull-out experiments. (a) A plot of $\bar{\sigma}_f$ as a function of ℓ_d for the determination of τ_s according to (21). (b) Plot of experimental values of $\bar{\sigma}_f$, δ and ℓ_d combined according to Eq. (25) to give $\Delta\varepsilon^T$. (c) Values of $\bar{\sigma}_f$, δ and ℓ_d combined according to (28) to give $\mathcal{G}_c^i$.

An independent check of $\mathcal{G}_c^i$ can be made by (23) or (28) where first $\bar{\sigma}_f^i$ is determined from a plot showing $\bar{\sigma}_f$ as a function of ℓ_d (as Figure 3a) by extrapolation to $\ell_d = 0$, and then using $\bar{\sigma}_f^i$ and $\Delta\varepsilon^T$ in (23). It should be noted that $\bar{\sigma}_f^i$ cannot be obtained directly as the experimental stress value where debonding is found to initiate in the experiments, because the model is not really valid for initiation. Having determined all the parameters, $\mathcal{G}_c^i$, τ_s and $\Delta\varepsilon^T$, we can make a another check of the accuracy of the parameters by plotting $\bar{\sigma}_f$ as a function of δ, using eq. (22), and compare the predicted graph with the original experimental data for $\bar{\sigma}_f$ as a function of δ.

4. Discussion

In the following, we will compare the equations of the present analysis for PO2 with similar equations for PO1 and SFFT. We also discuss the pull-out part of the experiment and the major assumptions of the present model.

4.1. Comparison with results from other pull-out and SFFT problems

The three basic test problems, PO1, PO2 and SFFT (Figure 2) can be considered as special cases of a more general problem, where there are three applied stresses, $\bar{\sigma}_f$ and $\bar{\sigma}_m$ applied to the fiber and matrix at $z = 0$ and a stress $\bar{\sigma}_c$ applied to the uncracked end of the composite specimen at $z = L$, see Figure 4. Care needs to be taken on the direction of sliding and pull-out and the sign (direction) of the frictional shear stress - note that the sliding direction of the SFFT is opposite to the pull-out tests. Force equilibrium gives

$$\bar{\sigma}_c - V_f\bar{\sigma}_f + (1-V_f)\bar{\sigma}_m = 0. \tag{30}$$

The three cases PO1, PO2 and SFFT correspond to the following conditions for the applied stresses:

$\bar{\sigma}_m = 0$ corresponds to PO1

$\bar{\sigma}_c = 0$ corresponds to PO2

$\bar{\sigma}_f = 0$ corresponds to SFFT

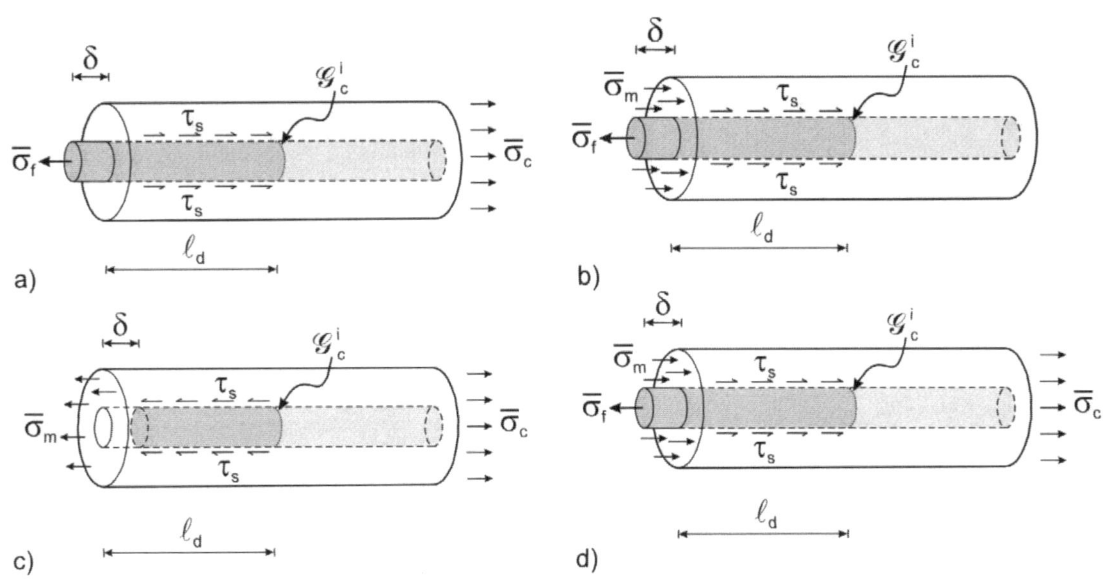

Figure 4. Models for (a) pull-out Type 1 (PO1), (b) pull-out Type 2 (PO2), (c) SFFT and (d) general model.

The conditions enter the equations for the stresses $\sigma_f^-(z)$, $\sigma_m^-(z)$, σ_f^+, σ_m^+, corresponding to (5), (6) and (9) for case PO2, and thus affects the results for the debond length ℓ_d and the relative displacement, δ. Let us compare the equation for $\bar{\sigma}_f$ versus ℓ_d for the two pull-out cases.

For PO1, the equation for $\bar{\sigma}_f$ as a function of ℓ_d is given by Hutchinson and Jensen [8]:

$$\frac{\bar{\sigma}_f^{PO1}}{E_f}=2\frac{E_c}{(1-V_f)E_m}\sqrt{\frac{(1-V_f)E_m}{E_c}\left(\frac{\mathcal{G}_c^i}{E_f r}\right)}-\Delta\varepsilon^T+2\frac{\tau_s}{E_f}\left(\frac{E_c}{(1-V_f)E_m}\right)\left(\frac{\ell_d}{r}\right). \tag{31}$$

Superscripts indicate test method. For PO2, the result is given by Eq. (18), reproduced below:

$$\frac{\bar{\sigma}_f^{PO2}}{E_f}=2\sqrt{\frac{(1-V_f)E_m}{E_c}\left(\frac{\mathcal{G}_c^i}{E_f r}\right)}-\frac{(1-V_f)E_m}{E_c}\Delta\varepsilon^T+2\frac{\tau_s}{E_f}\left(\frac{\ell_d}{r}\right). \tag{32}$$

It is seen that the result for PO2 is simply the result of PO1 multiplied by $(1-V_f)E_m/E_c$. This implies that for the same ℓ_d the stresses scales according to:

$$\sigma_f^{PO2}=\frac{(1-V_f)E_m}{E_c}\bar{\sigma}_f^{PO1}. \tag{33}$$

Next we compare the equations for δ as a function of $\bar{\sigma}_f$ for the two pull-out cases. For PO1 we have [8]:

$$\frac{\delta^{PO1}}{r}=\frac{1}{4}\frac{(1-V_f)E_m}{E_c}\frac{E_f}{\tau_s}\left(\frac{\bar{\sigma}_f^{PO1}}{E_f}+\Delta\varepsilon^T\right)^2-\frac{\mathcal{G}_c^i}{\tau_s r}, \tag{34}$$

while for PO2 we have eq. (22):

$$\frac{\delta^{PO2}}{r}=\frac{1}{4}\frac{(1-V_f)E_m}{E_c}\frac{E_f}{\tau_s}\left(\frac{E_c}{(1-V_f)E_m}\frac{\bar{\sigma}_f^{PO2}}{E_f}+\Delta\varepsilon^T\right)^2-\frac{\mathcal{G}_c^i}{\tau_s r}. \tag{35}$$

We see that these equations are identical if the applied stress terms in the brackets are equal, i.e. under the condition

$$\bar{\sigma}_f^{PO1}=\frac{E_c}{(1-V_f)E_m}\bar{\sigma}_f^{PO2}. \tag{36}$$

This is the same relationship as above, eq. (33). Thus, for the same ℓ_d and same δ the stresses scale by the factor $(1-V_f)E_m/E_c$. This factor is very close to, but slightly lower than unity for a single fibre composite specimen because from (3), $E_c\rightarrow E_m$ for $V_f\rightarrow 0$. This shows that there is very little difference in terms of stresses and pull-out displacement for the two pull-out cases, PO1 and PO2.

A similar comparison of equations for ℓ_d and δ for the SFFT case [14] gives the following scaling factors, noting that during a SFFT experiment it is the composite stress $\bar{\sigma}_c$ that is recorded, since in the SFFT the fiber is broken and thus stress free where debonding occurs.

$$\bar{\sigma}_c^{SFFT}=-\frac{(1-V_f)E_m}{E_f}\bar{\sigma}_f^{PO1}. \tag{37}$$

Again, if we consider $V_f\rightarrow 0$ we obtain

$$\bar{\sigma}_c^{SFFT} = -\frac{E_m}{E_f}\bar{\sigma}_f^{PO1}, \tag{38}$$

so that, with all parameters (including ℓ_d and δ) held fixed, the stresses are scaled approximately as the ratio of the stiffnesses of the matrix and fiber. This can be a significant difference for a polymer matrix composite where the Young's modulus of the matrix is usually significantly lower than that of the fiber. The comparison reveals that the applied composite stress of the SFFT - for the same ℓ_d - is much lower than the fiber stress of the PO1 experiment. It might be of interest to design experiments to test this relationship.

4.2. Comments regarding the full pull-out experiment

PO2 pull-out experiments are usually conducted in "displacement control", i.e. the position of the free end of the fiber is displaced in the negative z-direction at a constant rate. Such experiments can be considered to take place in two phases. In the first debonding and sliding phase, the applied load to the fiber increases and the debond crack tip propagates along the fiber towards the un-cracked end of the specimen ($z = L$) and the applied fiber stress increased. In the sliding/pull-out phase, the fiber is fully debonded and slides along the entire fiber/matrix interface, and the stress carried by the fiber is given by the frictional shear stress and the embedded fiber length, ℓ_s. The following considerations are valid for experiments being conducted in "displacement control", i.e. when the position of the free end of the fiber is displaced in the negative z-direction at a constant rate.

For the first debonding/sliding phase, our equations for the stress-displacement (22) is expected to be valid until the crack tip stress field approaches the end of the specimen, i.e. until $\ell_d \approx L$. The maximum applied fiber stress $\bar{\sigma}_f^c$ can thus be estimated from by setting $\ell_d = L$ in (20). The result is

$$\frac{\bar{\sigma}_f^c}{E_f} = \frac{\bar{\sigma}_f^i}{E_f} + 2\frac{\tau_s}{E_f}\left(\frac{L}{r}\right), \tag{39}$$

and the corresponding debond displacement is found by inserting $\bar{\sigma}_f^c$ from (39) into (11), giving

$$\frac{\delta^c}{r} = \frac{E_c}{(1-V_f)E_m}\left[\frac{\bar{\sigma}_f^i}{E_f}\left(\frac{L}{r}\right) + \frac{\tau_s}{E_f}\left(\frac{L}{r}\right)^2\right] + \Delta\varepsilon^T\left(\frac{L}{r}\right). \tag{40}$$

Just as the debond crack tip reaches the end of the specimen, the crack tip stress field vanishes and during the remainder of the experiment, the stress field is independent of $\mathcal{G}_c^i$ and $\Delta\varepsilon^T$. The stress field is now controlled by the interfacial friction and the experiments starts its sliding/pull-out phase. As this happens, the applied fiber stress drops to a value denoted $\bar{\sigma}_f^d$. The value of $\bar{\sigma}_f^d$ depends on τ_s and the initial free fiber length, L_f (see definition of L_f in Figure 1a).

In the sliding phase, the stress carried by the fiber is controlled by the frictional shear stress and the embedded fiber length, ℓ_s, along which the interfacial sliding shear stress τ_s acts. A simple force balance gives:

$$\bar{\sigma}_f = 2\tau_s\frac{\ell_s}{r}. \tag{41}$$

Clearly, the embedded fiber length ℓ_s decreases with increasing pull-out. However, the exact value of ℓ_s depends on the initial length of the free part of the fiber, i.e. the initial distance from the gripping point of the fiber to the start of the specimen, L_f. This is because the applied load level $\bar{\sigma}_f$

decreases during the sliding phase of the pull-out experiment as the embedded sliding length ℓ_s decreases (see (41). Then, the matrix-free part of the fiber unloads too and thus contracts elastically. Being griped at the loaded fiber end ($z = -L_f$), the fiber pulls out of the matrix and thus decreases the embedded length, ℓ_s. The longer the initial free length, the larger will the elastic contraction and thus the larger shortening of ℓ_s and thus the lower will $\bar{\sigma}_f^d$ be. The smallest possible load drop will be to consider $L_f = 0$, corresponding to that the fiber is gripped at $z = 0$. Then, the sliding length just after complete debonding $\ell_s^d \approx L$. An upper bound for $\bar{\sigma}_f^d$ is thus

$$\bar{\sigma}_f^d = 2\tau_s \frac{L}{r}. \tag{42}$$

From (39) and (42) it follows that this stress drop is $\bar{\sigma}_f^i$. The fiber stress $\bar{\sigma}_f$ will decrease linearly with ℓ_s in accordance with (41) and $\bar{\sigma}_f$ will vanish for $\delta = L$ corresponding to $\ell_s = 0$. Figure 5 shows a sketch of the entire $\bar{\sigma}_f - \delta$ relationship.

Relation (41) has been widely used to characterize the interfacial shear stress of pull-out experiments. It is clear from our model perspective that the use of relations like (41) is only relevant for characterization of the frictional sliding resistance of the fiber/matrix interface. It will not be useful to characterize the chemical bonding between fiber and matrix, because the bond rupture occurs in the first phase (debonding/sliding) of the experiment where the applied fiber stress is rising.

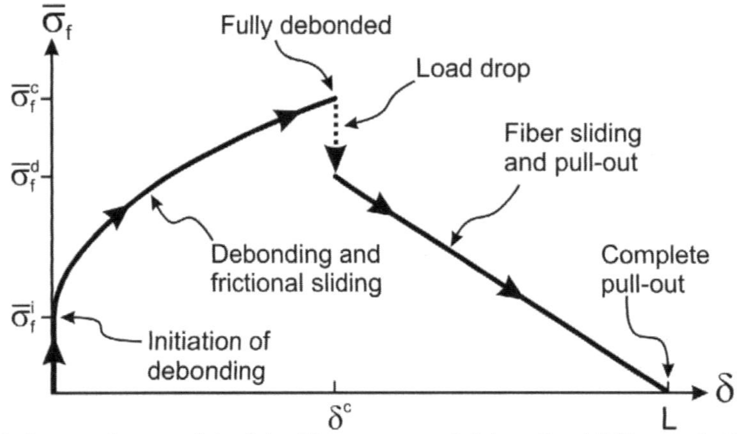

Figure 5. Sketch (not made to scale) of the fiber stress and debonding/sliding and sliding/pull-out displacement behavior of a pull-out type 2 (PO2) experiment.

4.3. Traditional analysis of pull-out experiments

An "apparent" average shear stress, calculated from the maximum applied force, $P_{\max}$, has sometimes been used to characterize the fiber/matrix interface. In connection with our analysis we obtain

$$\tilde{\tau}_s = \frac{P_{\max}}{2\pi r L} = \frac{\bar{\sigma}_f^c r}{2L} = \frac{\bar{\sigma}_f^i}{2}\frac{r}{L} + \tau_s. \tag{43}$$

It is clear that such an approach does not enable a clear separation of the interface parameters, $\mathcal{G}_c^i$ and τ_s, since $\bar{\sigma}_f^c$ depends on both parameters ($\mathcal{G}_c^i$ through $\bar{\sigma}_f^i$), see (39). Moreover, in case of non-zero $\bar{\sigma}_f^i$, the use of (43) will result in different values for $\tilde{\tau}_s$ for different values of L. This is because $\bar{\sigma}_f^c$ is linear but not directly proportional to L, see (39); the term $\bar{\sigma}_f^i$ depends on both $\mathcal{G}_c^i$ and $\Delta\varepsilon^T$, see (19). This can be seen by considering Figure 6, which shows the theoretical relationship between $\bar{\sigma}_f$ and ℓ_d, eq (20). Consider three specimens having different initial fiber embedded length, L_1, L_2 and L_3. The associated maximum fiber stresses of the three specimens, $\bar{\sigma}_f^{c,1}$,$\bar{\sigma}_f^{c,2}$ and $\bar{\sigma}_f^{c,3}$ are indicated. The different slopes of the dashed lines indicate that the apparent shear stress $\tilde{\tau}_s$ calculated using (43) would come out differently for different values of L.

The constant term in (39), $\bar{\sigma}_f^i$, can be positive or negative depending on the values of $\mathcal{G}_c^i$ and $\Delta\varepsilon^T$. If $\bar{\sigma}_f^i$ is positive (as shown in Figure 6), the apparent shear stress $\tilde{\tau}_s$ will decrease with increasing embedded fiber length and, conversely, if $\bar{\sigma}_f^i$ is negative (in case the fiber is pre-debonded), $\tilde{\tau}_s$ will increase with increasing embedded length.

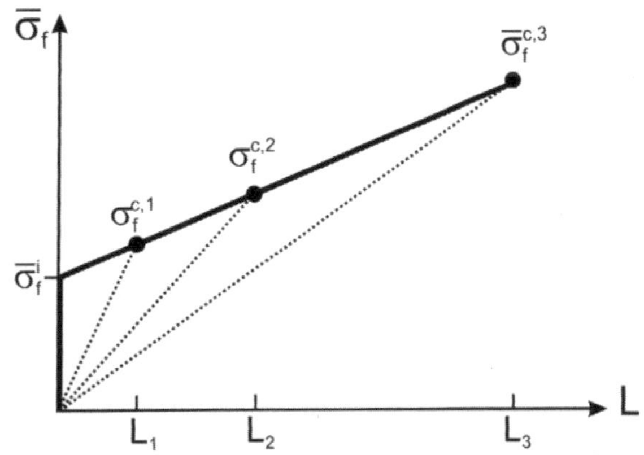

Figure 6. Sketch of the stress-debond length relationship, eq. (39), of a pull-out Type 2 (PO2) experiments with specimens of different lengths. Slopes of dashed lines indicate the apparent shear stress $\tilde{\tau}_s$.

Eq. (43) suggest a consistent way to analyze old data from traditional pull-out experiments and obtain τ_s from measurements of $\tilde{\tau}_s$ as a function of L. We note from (43) that $\tilde{\tau}_s \to \tau_s$ for $L \to \infty$ or $r/L \to 0$. Thus, if $\tilde{\tau}_s$ is plotted as a function of r/L, we can read off the true frictional shear stress τ_s at $r/L = 0$. The slope of the plot will be $\sigma_f^i/2$. According to (19), the parameter $\bar{\sigma}_f^i$ depends on both $\mathcal{G}_c^i$ and $\Delta\varepsilon^T$. Having determined $\bar{\sigma}_f^i$ from the slope of the plot, we cannot determine $\mathcal{G}_c^i$ and $\Delta\varepsilon^T$ independently but only in the combination given by (19).

4.4. Model approximations

The present model builds on a number of assumptions, such as taking fiber and matrix as linearly-elastic materials and neglecting Poisson's ratio's effects, effectively creating a one-dimensional model. Most fibers used in engineering composites, such as glass fibers and carbon fibers possess rather small non-linearity in their stress-strain behavior. The neglecting of Poisson's effects requires a few comments. First, for some fibers such as carbon fibers the Poisson's ratio is not readily known. Secondly, the Poisson's effects would induce a lateral contraction in fibers in pull-out experiments, possibly changing (reducing) the interfacial normal stress acting normal to the fiber/matrix interface. Considering Coulomb friction, a decrease in the normal stress across the interface could change (decrease) the frictional shear stress. However, there are also experiments that indicate that roughness asperities of the fiber surface affects the frictional sliding behavior of fibers in a composite material [3, 16]. With these complexities in mind, a friction law based on a constant frictional shear stress can be seen as a crude approximation. We think, however, that the most important point is that we characterize the fiber/matrix interface in terms of two parameters that represent two distinctive physical phenomena, the debond process, in which breakage of chemical bonds are the main feature, and frictional sliding resistance. Each of the two phenomena can be modelled in various ways, but how this should be done should be based on careful physical observations.

The assumption that the mechanical behavior of the fiber/matrix interface can be adequately be described in terms of $\mathcal{G}_c^i$ and τ_s can to some extend be checked with the proposed approach outlined in Section 3.2 in which parameters are determined from combinations of measured experimental data ($\bar{\sigma}_f$ as a function of ℓ_d) which according to (20) should result in linear relationships, as outlined in Figure 3a. In case the plots show that the experimental data come out with non-linear relationships, we must conclude that the models do not represent the experiments well. Then, more advanced interface model concepts should be explored.

Finally a remark about the use of shear-lag model for the present problem. This is expected to be a fairly good approximation when the debond crack length is long. Hutchinson and Jensen [8] showed by comparing results from a shear-lag model with results from a more accurate numerical model that shear-lag models can be good approximations once the debond length is longer than about 2 -3 times r. This implies that the model is best suited for composites with relatively low values for the interface shear stress and debond energy.

5. Concluding remarks

An analytical model was developed for debonding and sliding of a single fiber specimen supported at the matrix at the end where the fiber is loaded in tension. The model takes the fracture energy for fiber/matrix debonding, the interfacial sliding fraction and a strain mismatch as unknowns that must be determined from experimental measurements. An approach is presented for sequential identification of these parameters. Equations for debond length and fiber debond displacement were found to have similar form as equations derived for single fiber pull-out testing where the specimen is gripped at the opposite end and for single fiber fragmentation tests.

Acknowledgements

This study was funded by the "Danish Centre for Composite Structure and Materials for Wind Turbines (DCCSM)", fund no. 09-067212 from the Danish Council for Strategic Research.

Nomenclature

ℓ_d	debond length
ℓ_s	embedded (sliding) length during pull-out phase
r	fibre radius
z	coordinate axis
E_c	Young's modulus of the composite
E_f	Young's modulus of the fibre
E_m	Young's modulus of the matrix material

$\mathcal{G}_c^i$	interfacial fracture energy
L	length of embedded part of the fibre (debond and sliding phase)
L_f	length of free fibre
R	radius of matrix cylinder
S_F	surface area where frictional slip occurs
T	temperature at which pull-out experiment is conducted
T_0	temperature at specimen is free of residual stresses
V	volume of test specimen
V_f	fibre volume fraction
α_f	thermal expansion coefficient of fibre
α_m	thermal expansion coefficient of matrix
δ	relative displacement between fibre and matrix at $z = 0$
δ^c	relative displacement between fibre and matrix at end of debonding phase ($\ell_d = L$)
$\Delta\varepsilon^T$	mismatch strain
Δu	relative slip between fibre and matrix along the fibre matrix interface
ε_f^-	strain in the fibre in the debonded/slipping zone ($z < \ell_d$)
ε_m^-	strain in the matrix in the debonded/slipping zone ($z < \ell_d$)
σ_I	stresses associated with State I
σ_{II}	stresses associated with State II
$\bar{\sigma}_c$	applied composite stress at the un-cracked end ($z = L$)
$\bar{\sigma}_f$	applied stresses at the fibre end ($z = 0$)
$\bar{\sigma}_f^c$	applied fibre stress at end of debonding phase ($\ell_d = L$)
$\bar{\sigma}_f^d$	applied fibre stress at end of beginning of the pull-out phase
$\bar{\sigma}_f^i$	applied fibre stress at initiation of fibre/matrix debonding and sliding
σ_f^-	stress in the fibre in the debonded/slipping zone ($z < \ell_d$)
σ_f^+	stress in the fibre ahead of the debonded/slipping zone ($z > \ell_d$)
σ_f^{res}	residual stress in the fibre in the bonded zone ($z > \ell_d$)
$\bar{\sigma}_m$	traction applied to the matrix at $z = 0$
σ_m^-	stress in the matrix in the debonded/slipping zone ($z < \ell_d$)
σ_m^+	stress in the matrix ahead of the debonded/slipping zone ($z > \ell_d$)
σ_m^{res}	residual stress in the matrix in the bonded zone ($z > \ell_d$)
τ_s	interfacial frictional sliding shear stress
$\tilde{\tau}_s$	apparent interfacial shear stress
ξ_F	frictional energy dissipation

Appendix A

In the following we wish to calculate the residual stresses at temperature T accounting for a pre-stress, σ_f^0 , applied to the fiber during the processing at temperature T_0. The pre-stress is applied to the fiber only, see Figure A-1. The matrix remains stress-free at temperature T_0. Thus, the resulting strain in the fiber and matrix after the application of the pre-stress to the fiber are thus given by Hooke's law

$$\varepsilon_f^0 = \frac{\sigma_f^0}{E_f} \quad \text{and} \quad \varepsilon_m^0 = 0. \tag{A-1}$$

Following the application of the pre-stress the matrix is cured and the single fiber composite is cooled down to temperature T. The fiber is taken to possess a constant thermal expansion coefficient α_f. We assume that the matrix solidifies and obtains its full stiffness, given in terms of its Young's modulus E_m, during the curing. The matrix may undergo shrinkage before consolidation, but this will not be considered in the following, since residual stresses are not expected to build-up as long as the matrix is in a viscous state. The strain that the stress-free matrix material would undergo after solidification (usually contraction) is described in terms of a constant thermal expansion coefficient α_m.

As a part of the consolidation, the matrix is assumed to form a chemical bond to the fiber. Thus, the matrix cannot contract freely but must undergo the same strain change as the fiber. This constrain induces a resulting stress in the fiber, denoted σ_f^1 and a residual stress in the matrix, σ_m^1. Following consolidation and cool down to temperature T, the strain in the fiber ε_f^1 and the matrix ε_m^1 are given by

$$\varepsilon_f^1 = \alpha_f(T - T_0) + \frac{\sigma_f^1}{E_f} \quad \text{and} \quad \varepsilon_m^1 = \alpha_m(T - T_0) + \frac{\sigma_m^1}{E_m}. \tag{A-2}$$

The strain changes are obtained from (A-1) and (A-2) as

$$\Delta\varepsilon_f^1 = \varepsilon_f^1 - \varepsilon_f^0 = \alpha_f(T - T_0) + \frac{\sigma_f^1}{E_f} - \frac{\sigma_f^0}{E_f} \quad \text{and} \quad \Delta\varepsilon_m^1 = \varepsilon_m^1 - \varepsilon_m^0 = \alpha_m(T - T_0) + \frac{\sigma_m^1}{E_m}. \tag{A-3}$$

Taking the thermal expansion coefficients as know material properties we can determine σ_f^1 and σ_m^1 by the two conditions: The strain changes after matrix consolidation have to be identical, $\Delta\varepsilon_f^1 = \Delta\varepsilon_m^1$, and the stresses σ_f^1 and σ_m^1 must be in force balance with the applied stress σ_f^0. This leads to

$$\sigma_f^1 = \sigma_f^0 - \frac{(1 - V_f)E_m}{E_c} E_f \Delta\varepsilon^T \quad \text{and} \quad \sigma_m^1 = \frac{V_f E_m}{E_c} E_f \Delta\varepsilon^T, \tag{A-4}$$

where

$$\Delta\varepsilon^T = (\alpha_f - \alpha_m)(T - T_0). \tag{A-5}$$

is the mismatch strain.

Finally, the pre-stress is released. This is equivalent to adding a pressure (a negative stress) to the free fiber end. This creates a stress change to both the fiber and the matrix, denoted $\Delta\sigma_f^2$ and $\Delta\sigma_m^2$, respectively. Again the two conditions must be fulfilled: force balance and the fiber and matrix undergo the same strain change: $\Delta\varepsilon_f^2 = \Delta\varepsilon_m^2$, where the strain changes are given by Hooke's law

$$\Delta\varepsilon_f^2 = \frac{\Delta\sigma_f^2}{E_f} \quad \text{and} \quad \Delta\varepsilon_m^2 = \frac{\Delta\sigma_m^2}{E_m}. \tag{A-6}$$

Solving for $\Delta\sigma_f^2$ and $\Delta\sigma_m^2$ we find:

$$\Delta\sigma_f^2 = -\frac{V_f E_f}{E_c}\sigma_f^0, \quad \text{and} \quad \Delta\sigma_m^2 = -\frac{V_f E_m}{E_c}\sigma_f^0. \tag{A-7}$$

The final stress state, the residual stresses in the fiber and matrix, σ_f^{res} and σ_m^{res}, are obtained by adding the two stress fields (assuming linearity), Eq. (A-4) and (A-7):

$$\sigma_f^{res} = \sigma_f^1 + \Delta\sigma_f^2 \quad \text{and} \quad \sigma_m^{res} = \sigma_m^1 + \Delta\sigma_m^2. \tag{A-8}$$

The result is:

$$\sigma_f^{res} = \frac{(1-V_f)E_m}{E_c}\left\{\sigma_f^0 - E_f\Delta\varepsilon^T\right\} \quad \text{and} \quad \sigma_m^{res} = -\frac{V_f E_m}{E_c}\left\{\sigma_f^0 - E_f\Delta\varepsilon^T\right\}. \tag{A-9}$$

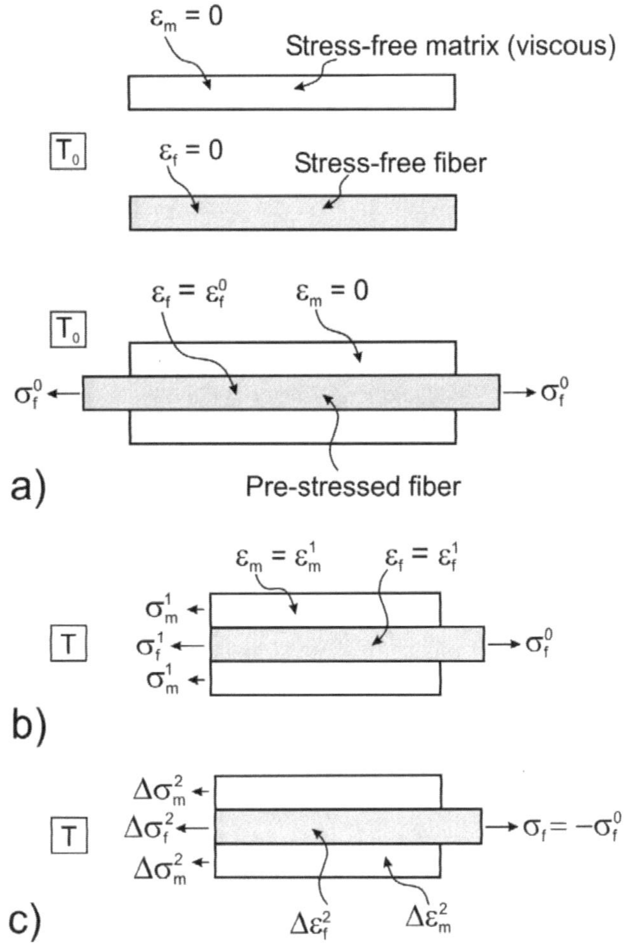

Figure A-1. Sketch of stress and strain during the processing of single fiber composite specimen with a pre-stressed fiber. (a) Pre-stress is applied at temperature T_0, (b) the stresses in balance at temperature T, (c) additional stresses induced as negative pre-stress is applied.

In the absence of pre-tension, (A-9) reduces to

$$\sigma_f^{res} = -\frac{(1-V_f)E_m}{E_c} E_f \Delta\varepsilon^T \qquad \text{and} \qquad \sigma_m^{res} = \frac{V_f E_m}{E_c} E_f \Delta\varepsilon^T . \tag{A-10}$$

Obviously, (A-9) and (A-10) have very similar forms. It follows that experiments conducted with pre-tension ($\sigma_f^0 \neq 0$) can be analyzed as a problem without pre-tension by substituting

$$\Delta\tilde{\varepsilon}^T = -\frac{\sigma_f^0}{E_f} + \Delta\varepsilon^T , \tag{A-11}$$

i.e., treating $\Delta\tilde{\varepsilon}^T$ as an "apparent" mismatch strain in the data analysis. When parameters e.g. ($\mathcal{G}_c^i$, τ_s and $\Delta\tilde{\varepsilon}^T$) are identified from experimental data, the true mismatch strain can be obtained from

$$\Delta\varepsilon^T = \Delta\tilde{\varepsilon}^T + \frac{\sigma_f^0}{E_f} . \tag{A-12}$$

Finally, it follows from (A-9) that the residual stresses σ_f^{res} and σ_m^{res} vanish when the pre-stress are chosen as:

$$\sigma_f^0 = E_f \Delta\varepsilon^T . \tag{A-13}$$

References

[1] Feih S, Wei J, Kingshott P and Sørensen BF 2005 *Composites part A* **36** 245
[2] Kim BW, Nairn JA 2002 *Journal of Composite Materials* **36** 1825
[3] Parthasarathy TA, Jero PD, and Kerans RD 1991 *Scripta Metall. Mater* **25** 2457
[4] Tripathi D, Jones F R 1998 *J. Mater. Sci.* **33** 1
[5] Kelly A and Tyson WR 1965 *Journal of the Mechanics and Physics of Solids* **13** 329
[6] Gao YC, Mai YW and Cotterell B 1988 *J. Appl. Mathematics and Physics (ZAMP)* **39** 550
[7] Hsueh CH 1990 *Materials Science and Engineering* **A123** 1
[8] Hutchinson JW and Jensen HM 1990 *Mechanics of Materials* **9** 139
[9] Graciani E, Mantic V, Paris F and Varna J 2009. *Comp. Sci. Techn.* **69** 2514
[10] Kessler H, Schüller T, Beckert W and Lauke B 1999 *Comp. Sci. Techn.* **59** 2231
[11] Nairn JA 2000 *Advanced Composite Letters* **9** 373
[12] Wu W, Verpoest I and Varna J 2000 *Comp. Sci. Techn.* **60** 351
[13] Budiansky B, Hutchinson JW and Evans AG 1986 *J. Mech. Phys. Solids* **34** 167
[14] Sørensen BF 2016, *Mechanics of Materials*, submitted for publication
[15] Prabhakaran DRT, Gupta M Lilholt H Sørensen BF and Mahajan P 2013 Experiments and analyses for determining fibre/matrix interface parameters in composites *34th Risø International Symposium on Materials Science, "Processing of fibre composites – challenges for maximum materials performance*. eds. B Madsen, H Lilholt, Y Kusano, S Fæster and B Ralph, Department of Wind Energy, Technical University of Denmark, Roskilde, Denmark, pp. 341-8
[16] Marshall D B, Shaw MC and Morris WL 1992 *Acta Metall. Mater.* **40** 443

Mechanisms to create high performance pseudo-ductile composites

M R Wisnom

Advanced Composites Centre for Innovation and Science, University of Bristol, UK

E-mail: M.Wisnom@bristol.ac.uk

Abstract. Current composites normally fail suddenly and catastrophically, which is an undesirable characteristic for many applications. This paper describes work as part of the High Performance Ductile Composite Technology programme (HiPerDuCT) on mechanisms to overcome this key limitation and introduce pseudo-ductility into the failure process.

1. Introduction

Composite materials have excellent specific strength and stiffness, but a major drawback is their lack of ductility. The HiPerDuCT programme is a collaboration between the University of Bristol and Imperial College to address this challenge by developing new materials and architectures that give a more gradual failure.

Fibre reinforced composite materials normally exhibit sudden brittle failure, with linear elastic response to failure. Although some analyses have suggested the possibility of gradual degradation as a result of successive ply failures, such behavior is rarely observed experimentally, and failure is normally catastrophic, with little warning.

Ideally, ductile failure is desirable, with no loss of modulus on reloading. A more achievable target using currently available materials is pseudo-ductility, where non-linearity or "pseudo-yielding" is achieved via damage. Although this results in a loss of modulus on reloading, it still allows load redistribution around stress concentrations, potentially making the material less notch sensitive and more damage tolerant. A long, flat plateau is desirable, preferably with a positive "work hardening" slope to avoid localization of failure.

Three fundamental mechanisms are considered in this paper which enable pseudo-ductility under tensile loading: fibre reorientation, hybridisation, and interfacial slip in discontinuous fibre composites. Pseudo-ductile strain has been proposed as a key measure to assess progress, defined as the difference between the final failure strain, and the elastic strain at the same stress based on the initial modulus E_0, figure 1. In some cases a sharp "yield" point is observed, but in others there is a gradual non-linearity. To account for this, the stress σ_y at an offset "plastic strain" ε_p is used, analogous to the proof stress in metals, with corresponding pseudo-yield strain ε_y.

Design strains would be expected to be within the initial linear region to avoid issues with fatigue, with the pseudo-yielding catering for overloads and localized damage and stress concentrations. Although the pseudo-yield strains are lower than the ultimate failure strains of the baseline materials, they are still potentially higher than current design strains which are very low due to concerns about brittle failure, and issues such as notched strength and impact.

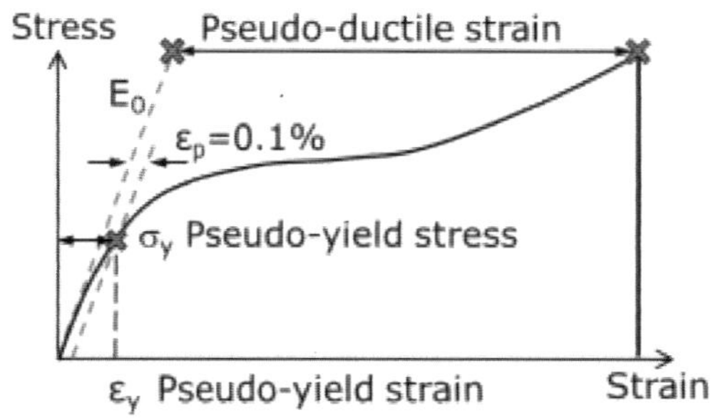

Figure 1. Definition of pseudo-ductile strain

2. Ductility via fibre re-orientation

It has been successfully demonstrated that using Skyflex carbon/epoxy plies of only 0.03 mm thickness, matrix cracking and delamination can be completely suppressed in angle-ply laminates, allowing the fibres to rotate under tensile loading, creating additional strain and pseudo-ductility [1]. Angle plies of (±45) layup can produce strains of over 20% and necking behaviour despite the brittle nature of the matrix (see figure 2). There is a trade-off between the stresses and strains that can be achieved depending on the angle, and this has been investigated in modelling studies [2]. A good balance of properties has been achieved for example with thin ply (±25) carbon/epoxy laminates that gave a pseudo-ductile strain of 1.23% and a maximum stress of 927 MPa, figure 3 [1].

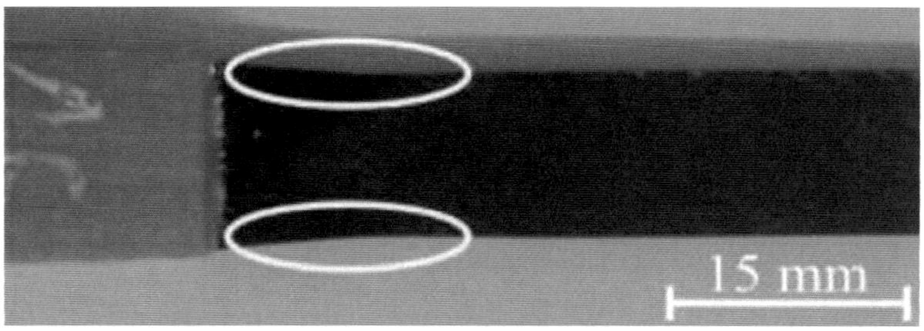

Figure 2. Necking of (±45) thin ply carbon/epoxy angle ply specimen

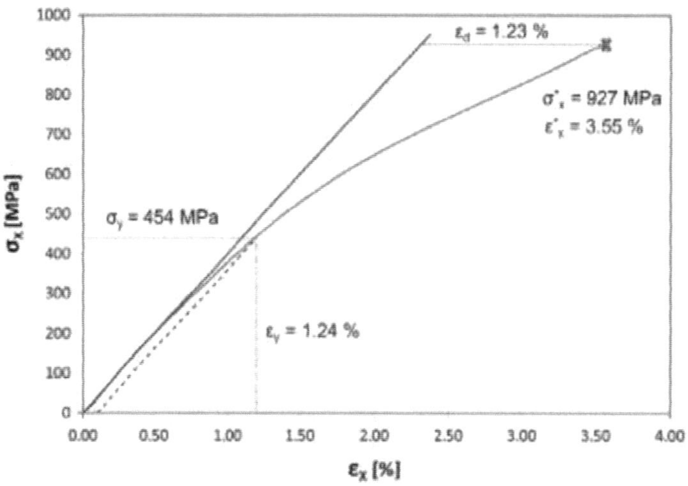

Figure 3. Pseudo-ductile response of thin ply (±25) carbon/epoxy laminate

Tests involving loading, unloading and then reloading have shown that the initial modulus is fully recovered, and so these laminates may be considered as ductile rather than pseudo-ductile, figure 4 [3].

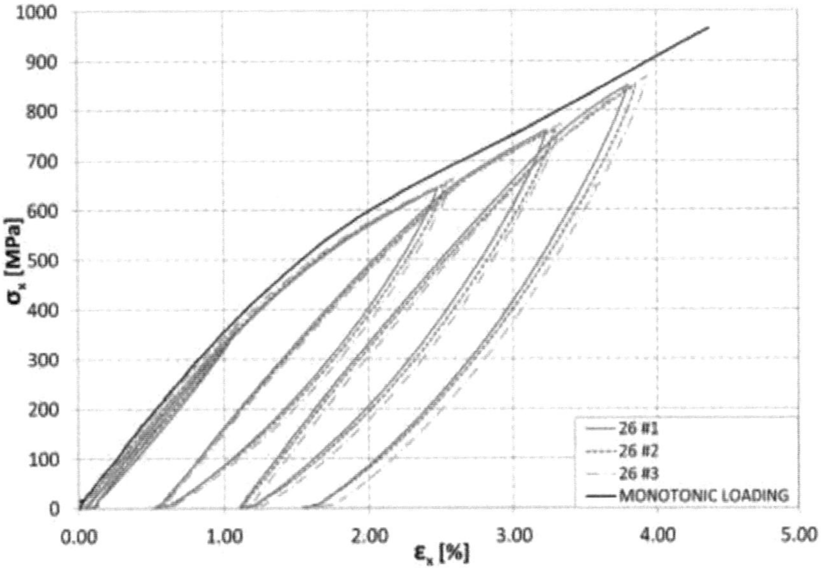

Figure 4. Cyclic loading of $(\pm26_5)_s$ carbon/epoxy laminate

3. Ductility via fragmentation in thin-ply hybrids

Suppressing delamination in hybrid laminates can also be achieved by using thin plies [4]. Modelling has shown that both the relative thickness (i.e. proportion of carbon) and absolute thickness of the carbon plies are important [5-7]. Damage mode maps can be produced such as Figure 5, with the different failure mode regions in this case indicated approximately based on FE analysis [5]. However analytical models allow the boundaries to be calculated explicitly [6]. With the correct thicknesses, premature brittle failure of the whole hybrid plate and catastrophic delamination can be avoided. This leads to the new failure mechanisms of fragmentation of the stiffer ply, and stable pull-out of the fragments,

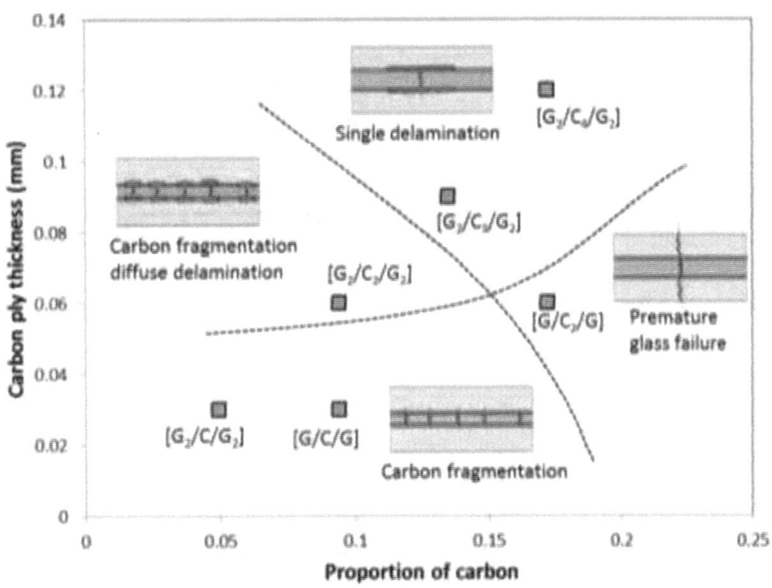

Figure 5. Damage mode map for E glass/thin carbon hybrid composite [5]

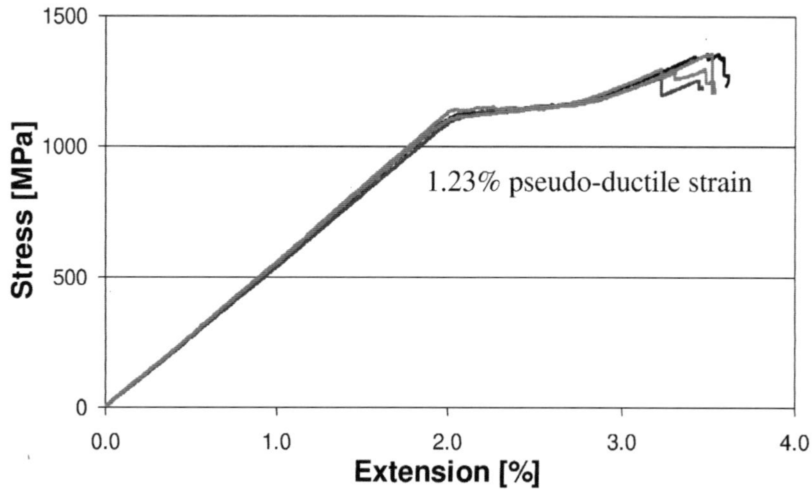

Figure 6. Pseudo-ductile response of S-glass/high strength carbon hybrid [8]

producing a pseudo-ductile stress-strain response. There is a trade-off between pseudo-ductility and yield stress. Figure 6 shows a typical response for two 21 g/m^2 plies of high strength TR30 carbon sandwiched between single 190 g/m^2 S-glass epoxy plies on either side. A range of different glass-carbon hybrid configurations has been evaluated, and pseudo-ductile strains of up to 2.66% have been obtained with a plateau stress of 520 MPa, or 0.86% pseudo-ductile strain with a plateau stress of over 1300 MPa [8].

Similarly pseudo-ductile response has been demonstrated with hybrids with different grades of carbon fibres. Figure 7 shows the response of hybrid specimens made from thin plies of Granoc XN80 ultra-high modulus carbon between layers of T1000 intermediate modulus carbon fibres, both prepregs supplied by North Thin Ply Technology [9]. It has also been shown that pseudo-ductile behaviour is obtained with quasi-isotropic laminates made from similar unidirectional carbon fibre hybrid sub-laminates [10].

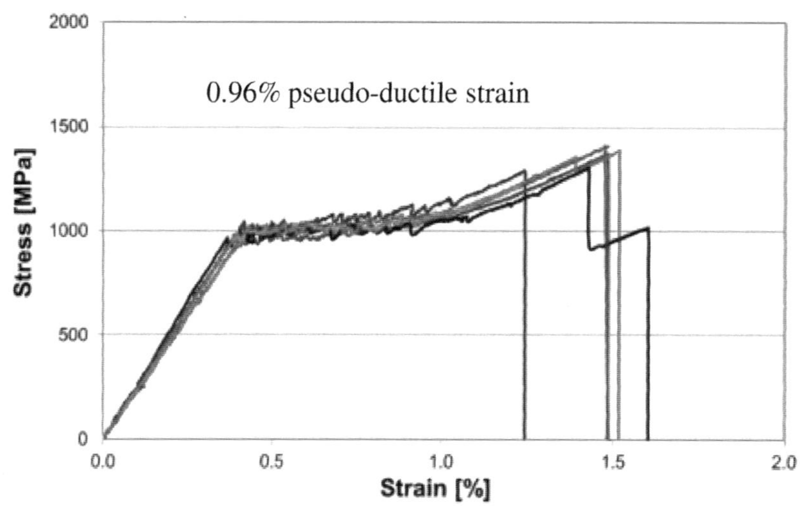

Figure 7. Stress-strain response for [$T1000_2/XN80_2/T1000_2$] carbon hybrid

Damage mechanisms have been investigated and it has been shown that acoustic emission can be used to detect fragmentation, with a direct correlation between acoustic emission and fragmentation events, allowing the technique to be used to detect fragmentation in opaque carbon/epoxy laminates [11].

Loading-unloading-reloading tests show a reduction in initial modulus due to the damage, and so these laminates do not show true ductility [12].

When the low strain plies in the hybrid laminates are very thin, there is an enhancement in the strain to failure due to the hybrid effect [13]. This only occurs for plies less than 0.1 mm thick, and can be as high as 20% for a single ply 0.03 mm thick. This effect can be modelled, and has been shown to be due to the constraint on forming critical clusters of fibre breaks [13]. This means that as well as producing pseudo-ductile response, these hybrid laminates are able to take greater advantage of the intrinsic properties of the carbon fibres.

Initially it was thought that this mechanism would only be operative in tension, but it has been found that fragmentation also occurs in thin ply M55 carbon/ S glass laminates in compression on the surface of specimens loaded in bending [14]. The carbon fibres fracture, and then some relative movement between the fracture surfaces allows further loading and fracture elsewhere, leading to a similar fragmentation behaviour and associated change in stiffness to that observed in tension.

4. Ductility via fragmentation in thin angle plies with 0° plies

The two previous mechanisms can be combined by replacing the lower modulus glass plies in a glass/carbon hybrid with carbon fibre angle plies [15]. This allows the fragmentation mechanism exhibited by the unidirectional hybrid composites to occur in the 0° plies together with the fibre rotation of the angle plies. For example $[\pm 26_5/0]_S$ laminates of Skyflex thin carbon/epoxy gave the response shown in figure 8, with a pseudo-ductile strain of 2.2%. On reloading, these laminates do show some loss of initial modulus due to the fragmentation of the 0° plies, and so are pseudo-ductile [3].

Different materials can be used for the 0° and angle plies. For example a (25/-25/0/-25/25) layup using thin ultra-high modulus carbon for the 0° ply but standard modulus thin carbon for the angle plies allows a pseudo-ductile sub-laminate to be produced which is only 0.15 mm thick and has a modulus of 135 GPa [16].

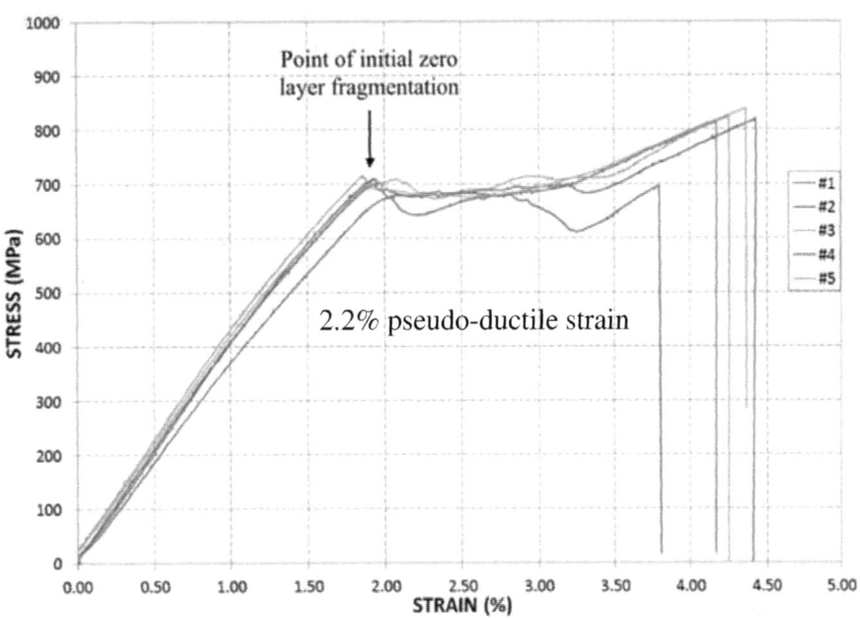

Figure 8. Pseudo-ductile response of $[\pm 26_5/0]_S$ laminate

5. Ductility in discontinuous fibre composites

Another mechanism for pseudo-ductility is slip at the interfaces between discontinuous fibres or plies. This has been demonstrated in model systems of discontinuous carbon/epoxy prepreg where the plies have been cut through the thickness prior to layup. The effect of ply thickness, cut spacing and alignment on the response have been investigated both numerically and experimentally [17]. For example specimens of IM7/8552 carbon/epoxy with 0.25 mm thick discontinuous ply blocks (0.125 mm for surface plies) and overlap length of about 8 mm were tested, (see figure 9). Significant non-linearity was obtained, providing a clear indication of damage, with a modest pseudo-ductile strain of 0.25%.

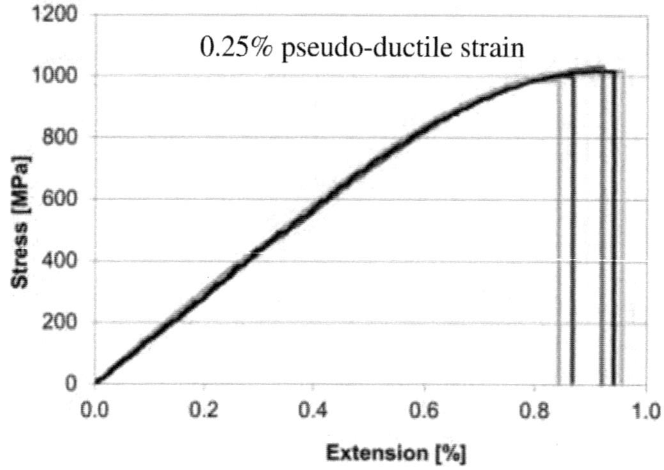

Figure 9. Non-linear response of discontinuous carbon/epoxy laminate

Non-linear tensile behaviour can also be produced in short fibre composites. The HiPerDiF method allows manufacture of high volume fraction well aligned unidirectional short fibre composites [18]. When high-modulus carbon and glass fibres are mixed, non-linear response and pseudo-ductility can be obtained similar to that achieved with thin plies. For example different relative volume fractions of Granoc XN90 ultra-high modulus carbon and E-glass fibres produce the stress-strain responses shown in figure 10 [19], with the relative carbon ratio indicating the volume of carbon compared with the total fibre volume.

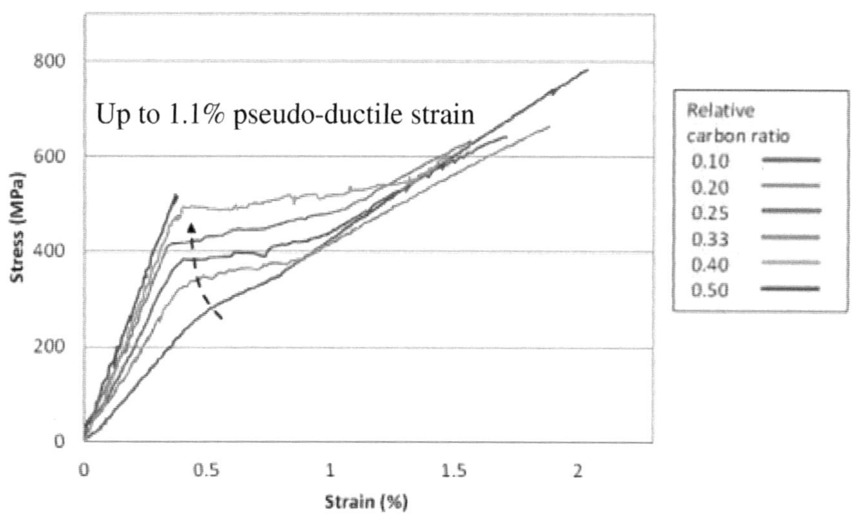

Figure 10. Non-linear response of hybrid XN90/E-glass short fibre composites

In this case the non-linearity is due to fragmentation, and the mechanism is similar to that obtained with hybrid laminates, but at the fibre rather than ply scale. However if the fibres are shorter than the critical length, a different mechanism occurs, with slip between fibres, similar to the behaviour demonstrated with the discontinuous prepreg. For example well aligned composites with 3 mm long standard modulus carbon fibres (C124, Toho Tenax) have been produced with a polypropylene matrix. These materials give the response shown in figure 11 [20]. Whilst there is some non-linearity, the amount of pseudo-ductility is modest, and the composite tends to fail at a relatively low strain due to localisation.

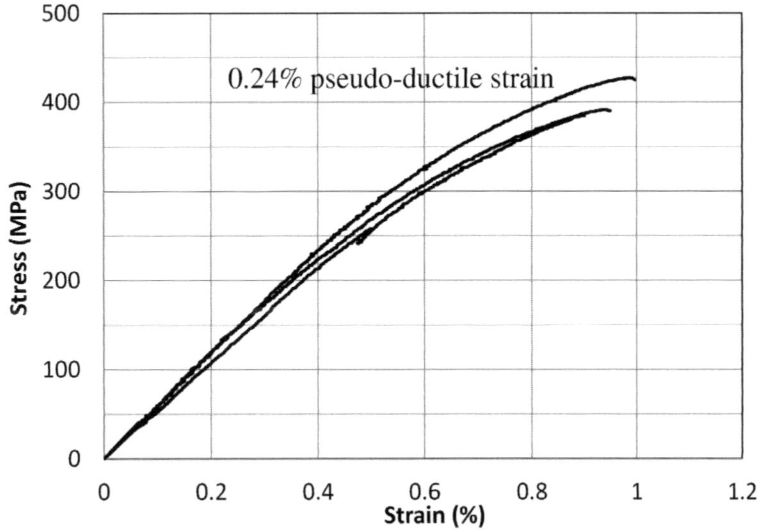

Figure 11. Non-linear response of short carbon fibre/polypropylene composite

6. Notched response of pseudo-ductile composites

The pseudo-ductility produced due to fragmentation and the associated non-linear response should allow load redistribution to occur at stress concentrations, resulting in notch-insensitive response, in a similar way to stress redistribution due to plasticity in ductile metals. To investigate this, notched quasi-isotropic pseudo-ductile laminates made from T1000/XN80 hybrid carbon sub-laminates similar to the unnotched unidirectional specimens presented in section 3 were tested [10]. Specimens 16 mm wide with 3.2 mm holes drilled were tested in open hole tension. Figure 12 shows the response, indicating that the

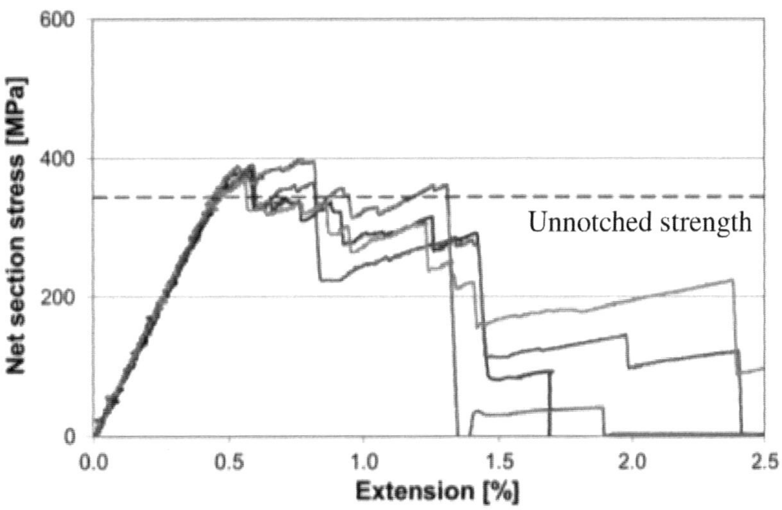

Figure 12. Open-hole tension of QI hybrid carbon specimens

specimens have actually exceeded the net section unnotched strength of the laminate, showing complete notch insensitivity. In addition, after the net section strength had been reached, they did not immediately fail catastrophically, but still showed some residual load carrying capacity, gradually reducing with further increasing strain.

Similarly modelling studies have shown that angle ply laminates with fragmenting 0° plies can also show notch insensitivity [21] and this has subsequently been demonstrated experimentally.

7. Conclusions

A number of different mechanisms have been demonstrated by which the sudden catastrophic failure of conventional composites in tension can be overcome, and a pseudo-ductile, gradual failure achieved. Using very thin plies can suppress matrix cracking and delamination, allowing fibre rotation and large strains to be obtained in angle-ply laminates. Thin ply hybrids can fail gradually by in-situ ply fragmentation. These mechanisms can be combined to produce laminates that demonstrate significant pseudo-ductile strain by both fragmentation and rotation. More limited pseudo-ductility has also been demonstrated with discontinuous carbon/epoxy plies and short fibres. Pseudo-ductile hybrid laminates have been shown to be notch insensitive. These results open up new possibilities for creating high performance composites that fail in a more gradual manner.

Acknowledgements

This work was funded under the UK Engineering and Physical Sciences Research Council (EPSRC) Programme Grant EP/I02946X/1 on High Performance Ductile Composite Technology in collaboration with Imperial College, London.

References

[1] Fuller J D and Wisnom M R 2015 Pseudo-ductility and damage suppression in thin ply CFRP angle-ply laminates *Composites Part A* **69** pp 64–71 (doi:10.1016/j.compositesa.2014.11.004)

[2] Fuller J D and Wisnom M R 2015 Exploration of the potential for pseudo-ductility in thin ply CFRP angle-ply laminates via an analytical method *Composites Science and Technology* **112** pp 8–15 (doi:10.1016/j.compscitech.2015.02.019)

[3] Fuller J D and Wisnom M R 2016 Ductility and pseudo-ductility of thin ply angle-ply CFRP laminates under quasi-static cyclic loading *Composites Part A,* submitted

[4] Czél G and Wisnom M R 2013 Demonstration of pseudo-ductility in high performance glass-epoxy composites by hybridisation with thin-ply carbon prepreg *Composites Part A: Applied Science and Manufacturing* **52** pp 23–30 (doi:10.1016/j.compositesa.2013.04.006)

[5] Jalalvand M, Czél G and Wisnom M R 2014 Numerical modelling of the damage modes in UD thin carbon/glass hybrid laminates *Composites Science and Technology* **94** pp 39–47 (doi:10.1016/j.compscitech.2014.01.013)

[6] Jalalvand M, Czél G and Wisnom M R 2015 Damage analysis of pseudo-ductile thin-ply UD hybrid composites - A new analytical method *Composites Part A: Applied Science and Manufacturing* **69** pp 83–93 (doi:10.1016/j.compositesa.2014.11.006)

[7] Jalalvand M, Czél G and Wisnom M R 2015 Parametric study of failure mechanisms and optimal configurations of pseudo-ductile thin-ply UD hybrid composites *Composites Part A: Applied Science and Manufacturing* **74** pp 123–131 (doi:10.1016/j.compositesa.2015.04.001)

[8] Czel G, Jalalvand M and Wisnom M R 2016 Design and characterization of advanced pseudo-ductile unidirectional thin-ply carbon/epoxy-glass/epoxy hybrid composites *Composite Structures* **143** pp 362–370 (doi:10.1016/j.compstruct.2016.02.010)

[9] Czel G, Jalalvand M and Wisnom M R 2015 Pseudo-ductile carbon/epoxy hybrid composites *ICCM20 20th International Conference on Composite Materials* (Copenhagen 19-24 July 2015)

[10] Czél G, Rév T, Jalalvand M, Fotouhi M and Wisnom M R 2016 Demonstration of pseudo-ductility in quasi-isotropic laminates comprising thin-ply UD carbon/epoxy hybrid sub-laminates *ECCM17 17th European Conference on Composite Materials* (Munich 26-30 June 2016)

[11] Fotouhi M, Suwarta P, Jalalvand M, Czel G and Wisnom M R 2016 Detection of fibre fracture and ply fragmentation in thin-ply UD carbon/glass hybrid laminates using acoustic emission *Composites Part A: Applied Science and Manufacturing* **86** pp 66–76 (doi:10.1016/j.compositesa.2016.04.003)

[12] Wisnom M R, Fuller J D, Suwarta P and Czél G 2015 Repeated tensile loading of thin-ply pseudo-ductile laminates *American Society for Composites 30th Technical Conference on Composite Materials* (Michigan 28-30 Sept 2015)

[13] Wisnom M R, Czél G, Swolfs Y, Jalalvand M, Gorbatikh L and Verpoest I 2016 Hybrid effects in thin ply carbon/glass unidirectional laminates: accurate experimental determination and prediction *Composites Part A,* in press

[14] Czél G, Jalalvand M and Wisnom M R 2015 Hybrid specimens eliminating stress concentrations in tensile and compressive testing of unidirectional composites *Composites Part A*, submitted

[15] Fuller J D, Jalalvand M and Wisnom M R 2016 Combining fibre rotation and fragmentation to achieve pseudo-ductile CFRP laminates *Composite Structures* **142** pp 155–166 (doi:10.1016/j.compstruct.2016.01.073)

[16] Fuller J D, Jalalvand M and Wisnom M R 2016 A pseudo ductile angle-ply sub-laminate approach for multidirectional thin ply cfrp laminates *ECCM17 17th European Conference on Composite Materials* (Munich 26-30 June 2016)

[17] Czél G, Pimenta S, Wisnom M R and Robinson P 2015 Demonstration of pseudo-ductility in unidirectional discontinuous carbon fibre/epoxy prepreg composites *Composites Science and Technology* **106** pp 110–119 (doi:10.1016/j.compscitech.2014.10.022)

[18] Yu H, Potter K D and Wisnom M R 2014 A novel manufacturing method for aligned discontinuous fibre composites (High Performance-Discontinuous Fibre method) *Composites Part A* **65** pp 175–185 (doi:10.1016/j.compositesa.2014.06.005)

[19] Yu H, Longana M L, Jalalvand M, Wisnom M R and Potter K D 2015 Pseudo-ductility in intermingled carbon/glass hybrid composites with highly aligned discontinuous fibres *Composites Part A: Applied Science and Manufacturing* **73** pp 35–44 (doi:10.1016/j.compositesa.2015.02.014)

[20] Yu H, Longana M L, Grail G, Pimenta S, Robinson P, Wisnom M R, Potter K D 2015 Aligned short fibre composites with nonlinear behaviour *ICCM20 20th International Conference on Composite Materials* (Copenhagen, 19-24 July 2015)

[21] Wu X, Fuller J D and M R Wisnom 2016 Open-hole response of pseudo-ductile thin-ply angle-ply laminate *ECCM17 17th European Conference on Composite Materials* (Munich 26-30 June 2016)

IOP Conf. Series: Materials Science and Engineering**139** (2016) 012011 doi:10.1088/1757-899X/139/1/012011

Nanofibre distribution in composites manufactured with epoxy reinforced with nanofibrillated cellulose: model prediction and verification

Yvonne Aitomäki[1], Mikael Westin[1], Jani Korpimäki[2] and Kristiina Oksman[1]

[1]Luleå University of Technology, 971 87 Luleå, Sweden,
[2]CSI Composites, 35700 Vilppula, Finland

Email: yvonne.aitomaki@ltu.se

Abstract. In this study a model based on simple scattering is developed and used to predict the distribution of nanofibrillated cellulose in composites manufactured by resin transfer moulding (RTM) where the resin contains nanofibres. The model is a Monte Carlo based simulation where nanofibres are randomly chosen from probability density functions for length, diameter and orientation. Their movements are then tracked as they advance through a random arrangement of fibres in defined fibre bundles. The results of the model show that the fabric filters the nanofibres within the first 20 μm unless clear inter-bundle channels are available. The volume fraction of the fabric fibres, flow velocity and size of nanofibre influence this to some extent. To verify the model, an epoxy with 0.5 wt.% Kraft Birch nanofibres was made through a solvent exchange route and stained with a colouring agent. This was infused into a glass fibre fabric using an RTM process. The experimental results confirmed the filtering of the nanofibres by the fibre bundles and their penetration in the fabric via the inter-bundle channels. Hence, the model is a useful tool for visualising the distribution of the nanofibres in composites in this manufacturing process.

1. Introduction

Nanofibres from different cellulose sources have been shown to reinforce polymers, even at low volume fractions [1], [2]. One possible way of exploiting this is reinforcing the matrix of a polymer composite with the aim of increasing the properties of the matrix and thus the properties of the entire composite. For such matrices to be used effectively, it is important to understand their behaviour when used in typical composite manufacturing methods so that appropriate processes can be found and modified if necessary. Resin transfer moulding (RTM) is a typical composite manufacturing method where a dry preform is held in a matched, closed mould and the resin is forced into the mould by pressurising the resin at the inlet. The advantages of this process are that it is a closed process, so chemicals are not released into the working environment, the pressure applied can be high to allow rapid impregnation, the volume fraction can be controlled by compressing the fibre mat into the mould and both surfaces of the part have a good finish. However, if a matrix reinforced with nanofibres is used in this process, the nanofibres are likely to be filtered by the fabric. This filtering will depend on a number of parameters such as the volume fraction and the size of the nanofibres. Numerical models that capture the complex particulate flow in RTM processes, and the result this has on the distribution of particles have been developed but requires complex modelling of the flow both in and between the bundles [3], [4]. Evaluated here is whether a simple scattering model, that is easily modified, can be used as a visual aid to give a good initial understanding of the distribution of nanofibres when these types of particulate matrices are used in RTM. The proposed model is based on a Monte-Carlo model of photon transport in a scattering media [5] but modified substantially to reflect the movement of nanoparticles and their interaction with different surfaces.

The results of the model are a distribution of the nanofibres in the composite. Thus to verify these results a means of measuring this distribution is required. The proposed approach is to infuse a preform of glass fabric of known geometry and permeability with a resin containing stained nanofibres using RTM. The composite is then cut and the cross-section examined for the presence of nanofibres.

2. Method

2.1. Theory

The Monte-Carlo model has the following key elements:

- The geometry of the fabric to be impregnated can be modified to represent the fibre arrangement of the materials of interest.
- The nanofibres are randomly distributed across the resin inlet and the inlet can be adjusted to fill the whole surface of the composite.
- The length and width of the nanofibres are randomly selected from a probability density function (PDF) and these PDFs are the input parameters to the model.
- The nanofibres are randomly oriented.
- The nanofibres progress in the direction of the flow unless they encounter an obstacle. The nanofibres reflect off the surface of the obstacle at an angle based on the gradient of the surface at that point. The obstacles are the fibres in the fabric that is being impregnated.
- The gradient of the surface of the fabric fibres can be set to represent the hydrodynamic "surface". Hence at slow speeds, the fabric fibres are set to a circular aspect whereas at higher flow velocities, the fibre is distorted to have an elliptical shape with the major axis aligned in the direction of the flow.

2.2. Implementation

The model has been implemented in Matlab (R2014b) and the fibres are represented as ellipses, randomly arranged in bundles (see Figure 1), hence the image represents a cross-section of the fibre mat. The markers show the position of different points on the nanofibres but are not drawn to scale so that their position can be seen even on the larger scale images. The size of the fibres, the major and minor axis of the ellipse, the number of fibres and the number of bundles are set as input parameters.

(a) (b) (c)

Figure 1. Nanofibres of width 70 nm and length of 3000 nm in a fabric where the single fibre diameter was 7 μm and the bundle size was 50μm. (a) Shows the arrangement of fibres (black ellipses) within bundles (b) nanofibres hitting the surface of a single fabric fibre (c) a close up of a nanofibre showing it is made up in this case of 7 pixels representing 70 nm.

The fibres in the fabric are created as circles on an image using the equation of an ellipse such that

$$\frac{x^2}{a^2}+\frac{z^2}{b^2}=r \tag{1}$$

where r is the radius, a and b are the radius fraction on the x and z axis respectively and x and z are the position on the image where the fibre is located. The centre for the fibre is randomly positioned on the image under the condition that the ellipse created does not overlap with another. The fibre bundles are created by simply using ellipses of larger size, superimposed over the image such that areas outside of the large ellipses are cleared of fibres. The radius is defined and the position of these bundles calculated based on this diameter and the gap between the fibre bundles. The size of the fibre bundles and the size of the fibres can be measured from the fabric or obtained from datasheets.

The positions of the nanofibres are randomly distributed across the image and represent the random distribution of the nanofibres across the upper surface of the composite. The current size of the image is set to 100 μm. The fibre arrangement is produced at this scale then is enlarged by a scale currently set to 4 to allow the gradient at the point of contact of the nanofibres to be calculated with reasonable accuracy. This also allows a computationally efficient reproduction of a part of the fibre fabric. As stated earlier, the length and width of the nanofibres are randomly selected from a PDF. The currently implemented PDF is simply a normal distribution. The nanofibres are plotted on the image of the fabrics and a resolution is used such that each pixel represents 10 nm. A close up of a nanofibre on the surface of a fabric fibre is shown in Figure 1.

As well as being randomly positioned, the nanofibres are randomly oriented and this is implemented by calculating the shape of a nanofibre using vectors from a single point on its surface. The direction of the vector defines the orientation of the nanofibre.

The nanofibres move stepwise in the z-axis and their movement is defined by a directional vector associated with each nanofibre. The velocity of the flow is calculated based on a Darcian flow such that

$$v_i=-\frac{K_{ij}}{\mu}\frac{\partial p}{\partial x_j} \tag{2}$$

where v_i is the superficial velocity vector, K_{ij} the permeability tensor μ, the dynamic viscosity, p the pressure and x_i are the coordinates. The subscripts i and j denote the free and summation indices, respectively. The superficial velocity is the local flow rate per unit area including both the solid and pore space and is thus related to the averaged interstitial velocity according to $v=u_{\text{int}}\epsilon$ where ϵ is the porosity. Hence for the one-dimensional case and assuming a constant pressure gradient and with L as the length of the flow path, the velocity is

$$u_{int}=-\frac{K}{\epsilon\mu}\frac{\Delta p}{L}. \tag{3}$$

It is assumed that the nanoparticles move with this velocity in the z-direction. The repercussions of this assumption are that the particles undergo an increase in speed once a collision has occurred which is not possible. Issues surrounding this will be dealt with in later versions of the model, if deemed an important source of error. The velocity is not reflected directly in the current simulations although a timestamp for each point is calculated, thus it can be used to calculate position with respect to fill-time. The velocity is used indirectly to estimate the aspect ratio and thus the gradient of the surface from which the nanoparticles will be reflected.

As discussed earlier, the model was initially based on photon scattering, thus only representing the interaction with surfaces rather than of position throughout the fabric. The advantage with this is that it

IOP Conf. Series: Materials Science and Engineering **139** (2016) 012011 doi:10.1088/1757-899X/139/1/012011

rapidly gives the resulting distribution of nanofibres with little memory usage. However, to aid in the visualisation an additional function was added so that the progress of the nanofibres could be seen.

2.3. Experimental setup
The nanofibres used were fibrillated from Kraft Birch pulp, SCA, Munksund using an ultrafine grinder (Masuko Sangyo Co., Ltd., Japan). The dimensions of the nanofibres, as well the larger residual fibrous particles in the suspension were measured from dried samples using AFM (Veeco Multimode Scanning Probe, USA) and optical microscopy. The epoxy was Prime 20ULV (Gurit, UK) that has an initial viscosity of 139 mPas at 25°C. The procedure for the preparation of the stained nanofibre/epoxy is as follows: the colouring agent, Carmine, was added to the aqueous nanofibre suspension mixed and then a solvent exchange to acetone was done. In the process unbound carmine was removed. The mixture was then diluted to 1.5 wt.% nanofibres.

Epoxy/nanofibre resin was prepared by adding the now stained nanofibres to the epoxy such that final resin would contain 0.5 wt.% nanofibres. The nanofibre, epoxy and acetone mixture was thoroughly mixed and degassed after which the mixture was stirred and kept at a temperature of 60°C for 48 hours to remove the majority of the acetone. The residual acetone was estimated to be 5.1 wt.%.

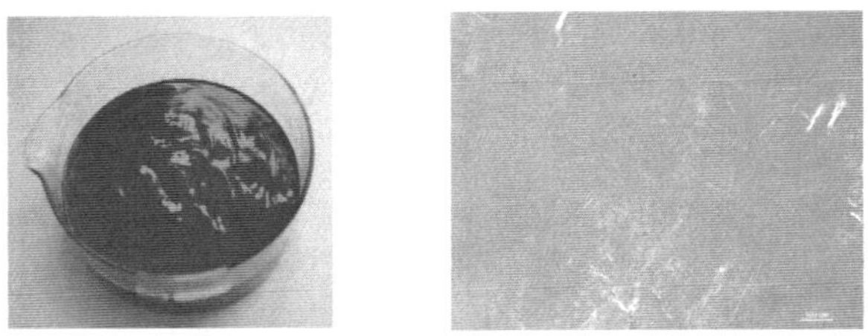
(a) (b)

Figure 2. (a) Photograph of the stained nanofibres (b) microscope image of stained nanofibres.

Figure 2 shows the stained nanofibres in epoxy. With the hardener, the epoxy system has a viscosity of 393 mPas at 25°C, making it suitable for processing via RTM. This resin was used to infuse a dry glass fibre preform of $[0/90]_4$ made up of two types of biaxial fabric, a core of weight 600 gm^{-2} and outer layers of 450 gm^{-2}. The stack was placed in the mould and a release film and flow mesh was put on the top before the mould was closed. The nanofibre/epoxy resin was degassed for 10 min before injected into a heated mould (35°C). A pressure of 0.2 MPa was used at the inlet and a vacuum pressure was applied to the outlet. After 15 min, the outlet was clamped and the inlet pressure increased to 0.5 MPa for 90 min. The part was cured for 10 hours at 35°C then post cured at 70°C for a further 10 hours. The composite was then removed from the mould, trimmed and samples were cut and polished for evaluation under a zoom stereomicroscope (Nikon, Japan). The glass fibres and bundles sizes were measured from the polished surface of the composite.

3. Results
Table 1 shows the distribution of the size of particles in the nanofibrillated suspension. Although a large portion of the fibres lie in the 500nm class, this is thought to be due to the drying process in the AFM sample preparation and hence the aqueous suspension is more likely to be made up of nanofibre bundles in the class 18 nm. Reassigning 50% of group in the 500 nm to the 18 nm class, gives an average nanofibre size of 172 nm. This does not include the larger fibre sizes (25μm and 2μm) as their behaviour would be quite different. It is noted that a better approach would be to use a PDF for a binomial distribution of the fibres and nanofibre, instead of the normal distribution for the nanofibre sizes that is currently implemented.

Table 1. Distribution of size of particles in the dried samples of the nanofibrillated suspension.

Diameter size class (μm)	Area of coverage (%)
25	6
2	16
0.5	34
0.018	27

The average glass fibres were 10±1 μm, the average fibre bundle was 1307± 286 μm (n=5) in width and 212±13 nm (n=10) in height and the inter fibre bundle spacing was 210±66 μm (n=18). The volume fraction of the composite, V_f, was 40%.

Frames taken at different points in the video of a simulation are shown in Figure 3 when the option to trace the nanofibres is on. As can be seen, use is made of Matlab's scaling feature that means the red circles indicate the position of the nanofibres and not their actual size. The random arrangement of the nanofibres on the top surface can be seen and their position at different time intervals. It can also be seen that these simulations do not capture a continuous flow of nanofibres but rather shows the final positions of all nanofibres injected into the area represented by the image.

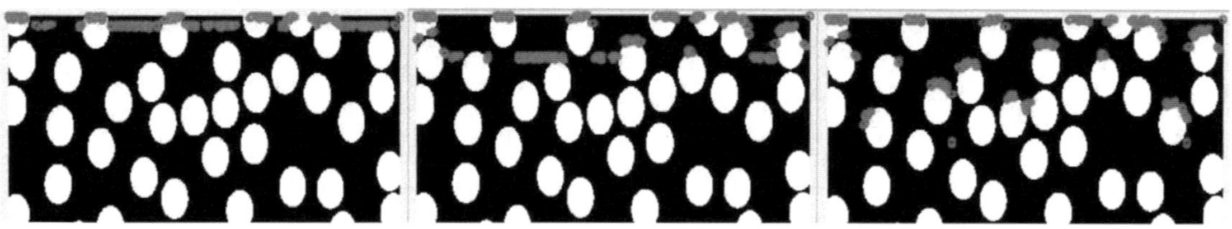

Figure 3. Time sequenced frames from the video of the simulated flow of nanofibres through a fabric.

3.1. Simulation

The following selected scenarios were run in order to illustrate effects of changing different parameters and with this guide processing of nanofibre filled epoxy by RTM. In different scenarios, each time a simulation is run the position of the fibres in the fabric changes as well as the starting point of the nanofibres, their length, width and orientation. Running the simulations several times allows for a greater reliance on the results. Where stated, n represents the number of simulations run.

3.1.1. Scenario 1: Effect of increase flow velocity. When the flow velocity increases the nanofibres are more likely to be carried around the fibre by the flow. This behaviour is mimicked by decreasing the size of the fibres and increasing the gradient of the fibre surface that the nanofibres are reflecting off. Figure 4 show the resulting distributions of nanofibres in the glass fibre fabric. In Figure 4(a) an aspect ratio replicating the true geometry of the fabric fibres is used and in Figure 4(b) the aspect ratio is adjusted to a = 0.7 (Equation 1) in order to emulate the hydrodynamic shape of the fabric fibres

Comparing Figure 4(a) to Figure 4(b) the predominating effect on the distribution is the differences in the number of fibres in any one area i.e. the fibre volume fraction. This was confirmed by running the simulation again whilst maintaining the same volume fraction, which showed there is little effect of the surface gradient on the nanofibre distribution.

Figure 4. Effect of velocity represented by change in aspect ratio of the glass fibres on the nanofibre distribution.

3.1.2. Scenario 2: Effect of using shorter and thinner nanofibres. Nanofibres coming from different sources and isolated with different processing treatments have different geometries. Taking an extreme example, nanocrystals can be extracted from cellulose and are shorter and thinner e.g. 200nm in length and widths of 5-10 nm [6] . To see the effect of changing geometry, the simulation were run with nanofibres with a width of 10 nm and length of 200nm. We also test the effect of having more homogeneous distribution of nanofibres with widths of 70nm x 3000nm. The distribution of nanofibres in the glass fibre fabric resulting from the simulation of this scenario is shown in Figure 5(a) which can be compared to Figure 5(b) where the nanofibres are 3000 by 75 nm and to Figure 4(a) where the nanofibres were 3000 by 175 nm. The results show that the crystal sized particles can penetrate the bundles, even cross the entire simulated region (marked red in Figure 6(a)), however, a large proportion are still filtered out. Increasing the size to that of homogeneous nanofibre, reduced the depth of penetration to approximately < 40 μm in the bundles, which is similar to that achieved by the larger nanofibres modeled in Figure 4(a).

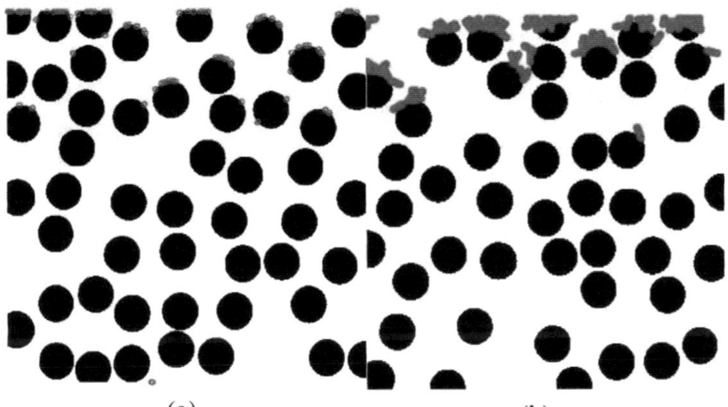

Figure 5. Nanofibre distribution with (a) nanofibres of 200nm by 10nm (b) nanofibres of 3000nm by 75nm.

3.1.3. Scenario 3:Effect of inter-bundle channels. The effect of inter-bundle channels where simulated by using an inter-bundle channel width of 100μm rather than the actual 200 μm channels that are in the glass fibre mat. Since this channel width is much greater than the length of the nanofibres, this will not introduce a difference of behaviour and allows a more rapid simulation of the behaviour. Also tested is the effect of having smaller fibre bundles (400 μm) and channels (60 μm) but maintaining the same fibre to bundle ratio as in the glass fabric. The results for this scenario are shown in Figure 6(a) and it can be seen that nanofibres reach the end of the image, which represent a flow path of 100 μm.

The nanofibres are still however trapped in the upper part of the fabric when the flow path is through the fibre bundle. The results of having inter-bundle channels, suggest that this would allow good penetration of the nanofibres into the fabric though possibly results in a lack of homogeneity of the nanofibre distribution. Placement of several layers may however disrupt these channels.

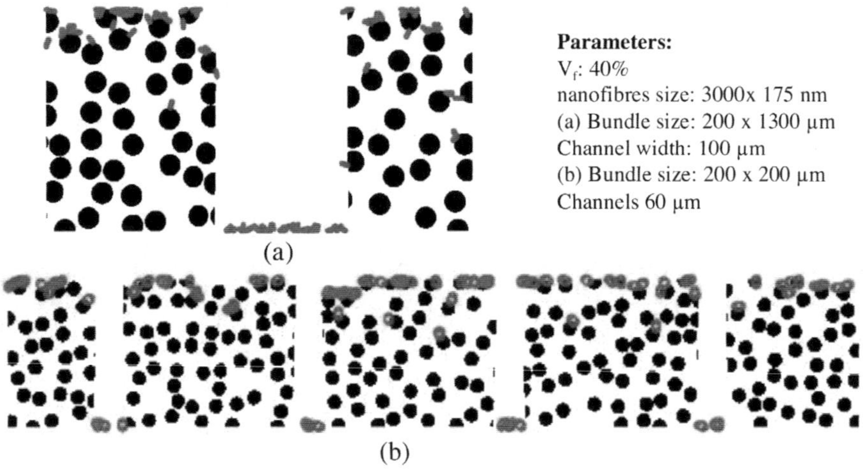

Figure 6. Effect of inter-bundle channels on the nanofibre distribution (a) with large of 100 μm and (b) with narrow channels and fibre bundles.

The nanofibres form long tangled networks on the fabric fibres at the surface (inlet) and that this high concentration of nanofibre in the matrix will have a gradual transition to matrix without nanofibre in the main part of the composite plate. It is possible that this integrated layer would then provide good properties in terms of surface hardness and add rigidity to the composite. It is also clear that the impregnation should be made in the thickness direction to have any effect from the reinforcement in the resin on the composite since otherwise the nanofibres will be filtered out in the trim, before the composite part as been reached.

3.2. Experimental results
Microscope images of the polished cross-section of the composite are show in Figure 7(a) and (b) alongside a schematic of the glass fibre layup in Figure 7(c).

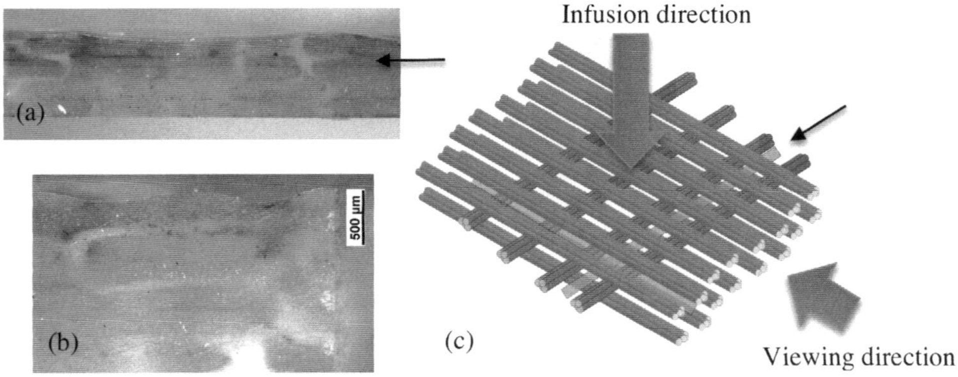

Figure 7. (a)Cross-section of composite showing nanofibres in the channels (b) higher magnification showing little impregnation the fibre bundles by the stained nanofibres and (c) schematic of the glass fibre layup.

The microscope images show that the nanofibres are concentrated in the channels and this is represented by the pink colour in the schematic diagram. Despite staining the nanofibres, due to their size, it is not possible to see individual or even low amounts of nanofibres. It is clear then that only low amounts nanofibre, if any, have penetrated the fibre bundles.

It is difficult to verify the model with precision but the overall result of the filtering effect of the nanofibres by the glass fibre is evident. Higher amounts of nanofibres can be seen in the sides of the bundles, which are not seen in the model. The most likely explanation for this is that the fibre volume fraction at these bundle edges is less than elsewhere as the fibre here can move into the inter-bundle space when compacted. The model shows that volume fraction has a large influence on penetration of nanofibres, and hence would support this explanation. The layer of fibres running perpendicular to the view direction (shown by the arrow in Figure 7(a)) appears to be pink. The likely cause of this is that the nanofibre/epoxy has filled the channel behind the transparent glass fibres (marked by the black arrow in Figure 7(c)). The model does not currently capture that behaviour as it does take into account transverse flow in these channels.

4. Conclusion

A simple scattering model has been developed and used to predict the location of nanofibres in a composite manufactured by RTM using a matrix reinforced with nanofibres. The results show clearly the effect of filtering by the fabric fibres and these are backed by the experimental results. Inter-bundle channels allow the nanofibres to penetrate the composite but are stopped by fibre bundles. The model shows that fibre volume concentration plays a major role in the penetration depth, and this is supported by penetration into the sides of the fibre bundles where the fibre volume fraction is lower. In conclusion, the model is a useful tool for showing the distribution of the nanofibres in composites though more work is needed to confirm the rest of the simulation results.

Acknowledgments

This work is funded by the European Commission under INCOM EC FP7, Grant Agreement No: 608746. The authors would also like to thank Dr. Maxime Noël for his work on the fibre size distribution.

References

[1] Hassan M L, Fadel S M, El-Wakil N and Oksman K 2012 *J Appl. Polym. Sci.* **125**, 216-222
[2] Bondeson D, Syre P and Niska K O 2007 *J. Biobased Materials and Bioenergy* **1**(3) 367-371
[3] Frishfelds V and Lundström T 2011 *Plast. Rubber Compos.* **40** 65-69
[4] Lundström T S and Frishfelds V 2013 *J.Compos. Mater.* **47** 1907-1915
[5] Wang L, Jacques S L and Zheng L 1995 *Comput. Methods Programs Biomed.* **47** 131-146
[6] Siqueira G, Bras J and Dufresne A 2010 *Polymers* **2** 728-765

IOP Conf. Series: Materials Science and Engineering **139** (2016) 012012 doi:10.1088/1757-899X/139/1/012012

Dissipating mechanical mechanisms and their link to morphology in polyamide toughened carbon-epoxy UD laminates

Emmanuel Baranger

LMT Cachan, ENS Cachan, CNRS, Université Paris-Saclay, 61 avenue du Président Wilson, 94235 Cachan cedex, France

E-mail: baranger@lmt.ens-cachan.fr

Abstract. Toughened carbon/epoxy laminates include thermoplastic particles. Depending of the curing cycle, the morphology of the thermoplastic phases can differ. This strongly impacts the degradation scenario of the composite. In the present paper, such reinforced laminates are evaluated. Different morphologies are studied. Classical experiments are used to evaluate the impact of the morphology on the damage and plasticity evolution laws at the scale of the ply. Some elements to model the toughening mechanism at the inter-ply scale are given and limitations are discussed.

1. Introduction

Carbon/epoxy UD laminates have relatively limited performances regarding damage tolerance as they are prone to delamination. The toughness of epoxy matrices being relatively low regarding thermoplastics, some authors proposed to improve their performances by adding thermoplastic phases such as PES [1] or PA [2]. These two examples modify the composite morphology at different scales, PES is used at the submicron scale while PA is used at inter-plies and therefore for scale-lengths about 10 microns. This kind of intrinsic toughening strategy [3, 4] does not necessary lead to improved macroscopic properties as shown in [2, 5]. For example, different scenarios of degradations are observed during mode I or mode II macroscopic delamination tests. This extrinsic part of the toughening strategy may play a major role and be modified by the processing conditions of the material. In this work, we focus on the curing cycle of prepregs. This study has been conducted during the French ANR project: PROMORPH. The objective of this paper is to give some insight of the toughening scenarios and their relation to the morphology of toughened carbon/epoxy UD laminates. For that, an extensive literature survey has been conducted associated to micro-mechanical models. Based on that understanding, the modelling of the toughening mechanisms by PA at interfaces is then studied at the inter-ply scale. The local constitutive relations and their identification as well as their use in a finite element framework describing a simplified geometry are detailed. In particular, limitations of the different parts of the model are tackled such as localization. This modelling study is accompanied with an experimental campaign at the mesoscale conducted on different materials. In fact, modifying curing conditions can lead to different phase morphologies at the scale of the thermoplastic reinforcement. Five morphologies have been produced, the associated mechanical properties

are compared using the mesomodel developed at LMT-Cachan [6, 7]. An understanding of the underlying mechanisms is proposed to explain the differences observed.
The material manufactured in this paper is relatively close to the one used by [2] i.e. a carbon/epoxy laminate toughened by poly-amide (PA) particles of about $20\mu m$. Depending on the impregnation and curing cycle parameters, several morphologies have been obtained:

1 - un-toughened material;
2 - intra-ply particle dispersion, melted particles;
3 - intra-ply particle dispersion, phase separated particles;
4 - inter-ply particle dispersion, melted particles;
5 - inter-ply particle dispersion, phase separated particles.

The first three configurations have been mainly evaluated for the moment.

2. Literature survey

A wide literature exists on the toughenability of polymers [Argon and Cohen, 2003] using different methods. For exemple, [1] shows the influence of PES modification on epoxy toughening by phase separation, modifying the crack path. This phase separation introduces inclusions of characteristic size of 1 micron. It also modifies the viscosity of the matrix and therefor the processability of the laminate. These studies on pure polymers are not always relevant. [8] shows that, the addition of fibers to the polymers may modify the polymer blend morphology and thus lead to an unexpectedly poor toughening. Thus, the composite should be studied. This as been done extensively for non toughened composite in [9] but also by Kinloch [10]. In [9], many composites were characterized whatever the matrix, the fibers, the sizing, or the process. The objective of this study is to show the correlation of the raw matrix properties on the composite ones regarding delamination. Two regimes are demonstrated driven by the size of the plastic zone at the crack tip versus the thickness of the inter-ply region. Note also that adjacent plies modify the local stress field leading to a shielding effect (the stress wavelength in front of the crack tip is longer and smoother). [11] studied the formation of hackles in an interphase. The conditions for a crack in an interphase to remain in this inter-ply zone has been studied by[12] for elastic isotropic adhesive between two substrates. In mode I (macroscopic), the pre-stress plays a major role and destabilize the crack that will deviate from the interfacial zone (even in presence of toughening particles [13]). Crack fiber bridging (for interfaces between O^o plies) may be involved as mentioned by most of the authors and visible on increasing R-curves. Crack Fiber bridging is related to the local intraply delamination. [14] observed this phenomenon. The plastic zone is assumed to extend in the ply region (very locally because plies are stiff). This plastic zone being at constant stress level (perfectly plastic behavior), the weakest link triggers the delamination which may occur in the ply. Thus crack fiber bridging is related to the matrix toughness via plasticity as shown in the second regime in [9].

[5, 13] shows an increase in KIc and KIIc for PA particles additions at composite inter-ply zones. These particles work as crack bridging when the crack remains in the inter-ply zone, as noted in the second paper. [13] also discussed the differences observed between static and fatigue loadings. Concerning Mode I, [5] proposes a model leading to a good correlation with experimental results in the case of a crack remaining at the interface. In [2], PA particles are much larger and the authors play on many different parameters to study the influence on mode II toughness. A process zone modification is demonstrated. [15, 16] also shows this action using high resolution X ray computed tomography. [5] has developed a model for mode I based on particles pinning the crack. The particle partial debonding and drawing can also be described using a simple model of a spherical particle in a cracked domain as in [17, 18]. This model doesn't describe the evolution of the debonding of the particle. Contrary to mode I, the toughening

mechanism for mode II is deployed ahead of the crack tip. First, plasticity develops in the inter-ply zone more or less depending on the epoxy blend properties enforcing the PA deformation. For a sufficient loading, the formation of a hackle network can be observed. This network as been modeled by [11] using a pure mode I crack propagation criterion or by [19] for the global GIIc estimation. These hackles do not extend on the entire width of the specimen but on dimensions related to the size of the fibers or the thickness of the interface as shown by [2]. Some of these cracks especially develops close to the PA particles because of stress concentration. The crack reaching the particles are deviated as in [20, 21] or [22] for thick interphases. A particle/matrix debonding then occurs which allows the particle to develop plasticity by drawing if the debonding is not total. Finally, the debonding or the failure of the particles leads to the macroscopic crack front increase.
Note that more recent references exist on the characterization of this kind of materials as some are already used in the aeronautical industry. However, they do not focus on the understanding of the role of the microstructure.

3. Experimental results

To evaluate the impact of the morphology on the mechanical behaviur of the laminate, the meso-model developed by [6] has been chosen.

3.1. Ply characterization

Classical tests associated to the characterization of the ply have been conducted. Regarding tension tests on $[0_8]$ coupons, the force/strain curves are the same but not the stress/strain curves. The addition of particles modifies the thickness of the laminate leading to different definitions of stresses for the same volume of fiber. Regarding tension tests on $[\pm 45]$ laminates to characterize the shear behavior, the damage evolution law as well as the plasticity evolution law are not changed as shown on Figure 1 and 2.

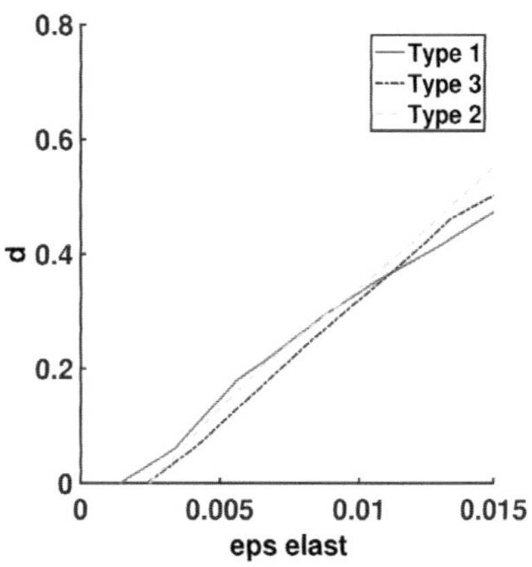

Figure 1. Shear damage evolution vs. shear elastic strain.

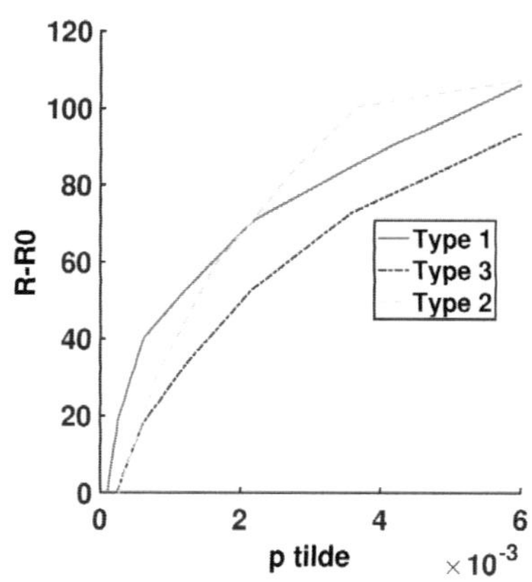

Figure 2. Shear hardening evolution vs. cumulated effective plasticity strain.

For $[0/90_2]_s$ coupons, the evolution of the crack density versus the applied strain is plotted on Figure 3. The crack density is defined as the ratio of the mean crack spacing and the 90 plies thickness. The addition of particules leads to an early transverse crack creation. Their evolution speed remain comparable while the saturation seems to be delayed by the presence of particles. Considering that delamination at the transverse crack tips is delayed by particles, the transverse crack density can be higher.

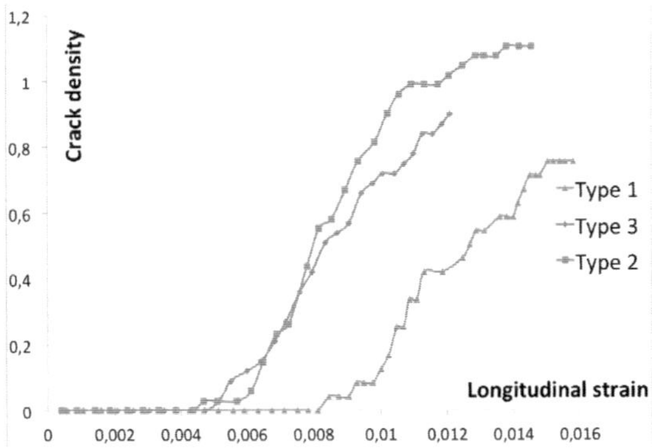

Figure 3. Transverse crack density vs. applied strain.

3.2. Inter-ply characterization
DCB tests have been performed on purely unidirectional coupons. The measured energy release rates are greater of about 50% for the toughened composites. Note that bridging is observed for the toughened coupons, this may explain part of the meso-scopic improved properties. Regarding mode II delamination, 3ENF tests have been performed. The results are presented in Table 1. Two configurations are treated related to crack propagation from an existing crack (and the associated process zone) or not. For the same kind of materials, the results are significantly different, showing the impact of the curing cycle.

Table 1. Critical energy release rates for 3ENF tests (J/m^2).

Type	Mean, non pre-cracked	Mean, pre-cracked
1	1303	728
2	1535	999
3	1061	879
4	533	459

4. Numerical model

The experimental results shown before are not intuitive. The use of analytical or numerical model is therefor of great interest. At the scale of the fiber, [23] uses the elasto-plastic model from [24] to describe the mechanical behavior of the matrix. This model, implemented in Abaqus, as been preferred to the model from [25] in a first time. The model is calibrated using data from the literature. The behavior is supposed to be elastic brittle in tension. In compression, the behavior is identified regarding shear. A compact tension test is performed to identify the post-peak behavior of the material to obtain the good macroscopic energy release rate. For the sake of simplicity, first simulations have been conducted with pure matrix inter-plies. An inter-ply zone is considered (thickness about 10 microns), it is composed of pure epoxy matrix between two stiff plies. The fiber/matrix debonding is modeled using cohesive elements. The applied boundary conditions on the upper and lower plies are extracted from a maroscopic beam analysis to get a relevant representation of the local loading. First a DCB test is performed as shown on Figure 4. The initial crack deviates from the inter-ply region to rich the fiber/matrix region.

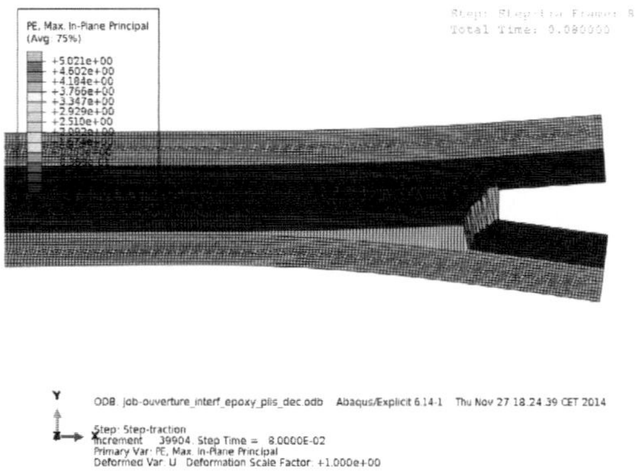

Figure 4. Crack propagation in inter-ply region for a DCB test.

In mode II, a CLS test has been simulated. On Figure 5, hackles can be observed. Their inclination is about 60° while the maximum traction direction is at 45°. This leads to a large crack spacing. The associate process zone is not compatible with fracture mechanics. In fact, it is impossible, from the local kinematics to define an equivalent crack front. The localization orientation is not well predicted, it is directly linked to the model and not tunable. Another model such as [25] has to be used. Also note that this last model should be able to describe the plasticity development at the crack tip that occurs at this scale. One pure matrix coupons, the plastic zone ahead of the crack tip is about 10 to 20 microns. Inside the inter-ply region, such a zone can't develop totally, the model must take this into account.

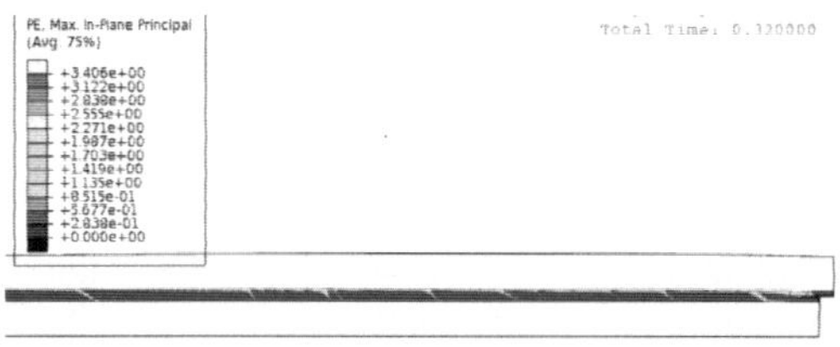

Figure 5. Crack propagation in inter-ply region for a CLS test.

The next step is to be able to introduce particles in the inter-ply region. For that, a 3D model is necessary. The PA particle has been modeled using [24] and an interface between the particle and the epoxy matrix described by a cohesive zone model. An explicit algorithm is used to overcome convergence difficulties while the kinetic energy is controlled to remain negligible. Note that the application of relevant boundary conditions is not straight-forward as the model is of very limited size.

5. Conclusion

This paper presents several ingredients towards the understanding of the link between toughened composite morphologies and mesoscopic mechanical properties. Different morphologies have been tested (mainly 3). The in-plane shear properties remain unchanged while the transverse cracking evolutions are impacted by the toughening. Transverse cracks appear earlier for toughened composite and their saturation is higher. Regarding delamination, critical energy release rates are generally improved due to fiber bridging for the tested configurations between 0° plies. Regarding simulations, a first model from the literature has been used to describe the mechanical behavior of the matrix. The localization properties of this model leads to slightly

Figure 6. Particle partial debonding simulation for a DCB test.

mis-oriented crack networks and to the difficulty to identify mesoscopic associated quantities. A first 3D model of an inter-ply reinforced by a particle has been developed. The application of relevant boundary conditions remaines an issue.

Acknowledgement
This work has been found by the ANR PROMORPH project which is greatly acknowledged.

References

[1] Kinloch A, Yuen M and Jenkins S 1994 *Journal of materials science* **29** 3781–3790
[2] Groleau M, Shi Y, Yee A, Bertram J, Sue H and Yang P 1996 *Composites science and technology* **56** 1223–1240
[3] Argon and Cohen 2003 *Polymer* **44** 6013–6032
[4] Ritchie R O 2011 *Nature materials* **10** 817–822
[5] Cardwell B and Yee A 1998 *Journal of materials science* **33** 5473–5484
[6] Ladeveze P and LeDantec E 1992 *Composites Science and Technology* **43** 257–267
[7] Allix O, Leveque D and Perret L 1998 *Composites Science and Technology* **58** 671–678
[8] Varley and Hodgkin 1996 *polymer papers* **38** 1005–1009
[9] Bradley W L 1991 *Key Engineering Materials* **37** 161–198
[10] Kinloch A 1985 *Epoxy Resins and Composites I* (Springer) pp 45–67
[11] Fleck N 1991 *Proceedings of the Royal Society of London. Series A: Mathematical and Physical Sciences* **432** 55–76
[12] Fleck N, Hutchinson J and Zhigang S 1991 *International Journal of Solids and Structures* **27** 1683–1703
[13] Hojo M, Matsuda S, Tanaka M, Ochiai S and Murakami A 2006 *Composites Science and Technology* **66** 665–675
[14] Johnson W S and Mangalgiri P 1986 Investigation of fiber bridging in double cantilever beam specimens. Tech. rep. DTIC Document
[15] Wright P, Fu X, Sinclair I and Spearing S 2008 *Journal of Composite Materials* **42** 1993–2002
[16] Moffat A, Wright P, Helfen L, Baumbach T, Johnson G, Spearing S and Sinclair I 2010 *Scripta Materialia* **62** 97–100
[17] Gang B and Chung-Yuen H 1990 *International journal of solids and structures* **26** 631–642
[18] Sigl L S, Mataga P, Dalgleish B, McMeeking R and Evans A 1988 *Acta Metallurgica* **36** 945–953
[19] Lee S M 1997 *Journal of materials science* **32** 1287–1295
[20] Ming-Yuan H and Hutchinson J W 1989 *International Journal of Solids and Structures* **25** 1053–1067
[21] Leguillon D, Lacroix C and Martin E 2000 *Journal of the Mechanics and Physics of Solids* **48** 2137–2161
[22] Lacroix C, Leguillon D and Martin E 2002 *Composites science and technology* **62** 519–523

[23] Mora D F, Gonzalez C, Lopes C, Naya F and Llorca J 2014 *ECCM 16 -16th European Conference on Composite Materials* (Séville, Spain)
[24] Lee J and Fenves G L 1998 *Journal of engineering mechanics* **124** 892–900
[25] Boyce M C, Parks D M and Argon A S 1988 *Mechanics of Materials* **7** 15–33

IOP Conf. Series: Materials Science and Engineering **139** (2016) 012013 doi:10.1088/1757-899X/139/1/012013

Investigations on the sintering response of steel-ceramic composites

C Baumgart[1], C Weigelt[2], L Krüger[1] and C G Aneziris[2]

[1] Institute of Materials Engineering, Technische Universität Bergakademie Freiberg, Germany

[2] Institute of Ceramics, Glass and Construction Materials, Technische Universität Bergakademie Freiberg, Germany

E-mail: christine.baumgart@iwt.tu-freiberg.de

Abstract: Purpose of this article is the evaluation of the influence of sintering parameters on the microstructure evolution and mechanical properties of pressureless sintered metal matrix composites consisting of metastable 16Cr7Mn7Ni-steel with 0 or 5 vol.% magnesia partially stabilized zirconia (Mg-PSZ) particles. The materials were prepared from powder raw materials via extrusion at ambient temperature. Three different temperatures between 1280 °C and 1380 °C and two varying dwell times of 40 min and 120 min at maximum temperature were applied. Both, tensile and compression tests are conducted at quasi-static strain rates for comparison of strength level, deformability and energy absorption capability. The results are discussed with regard to the porosity of the specimens, the interface between steel and ceramic, the TRansformation Induced Plasticity (TRIP)-effect occurrence and the failure behavior.

1. Introduction

It is well known that the reinforcement of a ductile and tough metal by stiffer and stronger phases such as ceramics results in a new kind of material with advanced mechanical properties. Considerable are the adjustable high stiffness and specific strength [1, 2] as well as the wear resistance [3, 4] which are not achievable with monolithic alloys. In the past a variety of manufacturing processes have been developed to obtain metal matrix composites (MMC) gaining significant importance for high performance applications in fields of aerospace and transportation industry [5, 6]. A criterion for classification provides the condition of aggregation so that a distinction in liquid, solid and solid-liquid state manufacturing is feasible [7]. The powder metallurgical (PM) processing route belongs to the solid or solid-liquid techniques which depends on the raw materials and the applied temperature regime. Thus, PM possesses numerous advantages compared to casting technologies. A fine and homogenous microstructure is achievable due to the insertion of fine-grained metal and ceramic powders and the lower processing temperatures. Besides, a near-net-shape fabrication becomes possible increasing the material utilization and decreasing the tooling costs which are in case of MMCs and their abrasive reinforcements typically quite high. For exploiting the unique material profile, a fundamental understanding of the influencing process parameters on the microstructure and the mechanical properties is essential for every material system [8, 9].

In the present article the sintering response of a MMC consisting of Cr16Mn7Ni7-steel with 0 or 5 vol.% magnesia partially stabilized zirconia will be analyzed in dependence of the sintering temperature and dwell time. Additional to the potential strengthening-effect generated by the Mg-PSZ particles, a transformation of the metastable austenitic steel into ferromagnetic α'-martensite or ε-martensite known as the TRansformation Induced Plasticity (TRIP)-effect can contribute to an enhancement of strength during plastic deformation without a loss of ductility [10]. So far, it is not reported in literature which effect both sintering temperature and time have on the occurrence of the TRIP-effect in pressureless sintered MMC-compacts deformed under tensile and compressive load.

2. Materials and methods

The manufacturing process of the investigated TRIP-steel (0Z) and TRIP-matrix-composite (5Z) samples bases on a ceramics-derived cold extrusion process. Therefore, a gas-atomized spherically-shaped steel powder (d_{90} < 40 μm) was mixed with 3.3 wt.% magnesia partially stabilized zirconia powder (d_{90} value of 30 μm). The chemical composition of the initial powders is revealed in table 1. In a second step, deionized water, plasticizers and binders were added to the steel/ceramic blend. A further mixing process led to a homogenous plastic paste which was subsequently pressed using a de-airing single-screw extruder to obtain strands with a circular cross section of 13.5 mm. All extruded test samples were stepwise dried and cut into 120 mm long rods before debindering at 450 °C for 30 min in air. Afterwards pressureless sintering in a 99,999 % argon-atmosphere at varying temperatures (1280 °C, 1350 °C, 1380 °C) and dwell times (40 min, 120 min) took place. For every condition three strands were produced in order to machine at least two samples for quasi-static compression (cylindrical: 6x6) and tensile (B6x30 DIN 50125) tests as well as microstructure investigations. Compression behavior was investigated according to DIN 50106 using a servohydraulic universal testing machine whereas tensile properties were determined referring to DIN EN ISO 6892-1 using an electronic screw driven testing machine. The microstructure characterization of initial state and deformed samples was realized by light optical microscopy and scanning electron microscopy (SEM) combined with energy and wavelength dispersive X-ray spectroscopy (EDX/WDX). Furthermore, ferromagnetic phase content was determined by the use of a Fluxmeter 480 (Lake Shore) connected with a MSAT-system (Metis Instruments and Equipment NV). Measurements according to Archimedes (DIN EN 993-1) served to determine porosity and the density of the samples.

Table 1. Chemical composition of steel and ceramic powder at initial state in wt.%.

steel	Cr	Mn	Ni	C	Si	S	N	Mo	Nb
initial state	16.30	7.20	6.60	0.03	1.00	0.007	0.088	0.007	0.001
ceramic	ZrO_2	HfO_2	MgO	SiO_2	Al_2O_3	CaO	TiO_2	Fe_2O_3	Y_2O_3
initial state	92.88	1.85	3.25	0.10	1.58	0.060	0.130	0.019	0.130

3. Results and discussion

In order to assess the influence of the varying process parameters on the microstructure evolution, characteristic light optical micrographs of the unreinforced samples are depicted in figure 1. Principally, the increase of dwell time from 40 min to 120 min at a constant temperature level causes a decrease of closed porosity, as exemplary shown for 1280 °C in figure 1 a)/b). The Archimedes-measurements for evaluation of total porosity confirm these observations. The results reveal an average difference of 3.5 % for short and long time sintering at 1280 °C. Comparable results are recordable at 1350 °C where the porosity decreases from 14.4 ± 0.7 % to 10.4 ± 0.7 %. Nevertheless, the reduction from 11.1 ± 0.4 % to 10.3 ± 0.6 % at 1380 °C is less pronounced. Under consideration of pore quantity evolution for a constant dwell time, an increase of temperature effects a decrease of porosity. The described tendencies are also valid for the composite samples. However, the absolute porosity values range for the low and medium temperature levels above those of the corresponding 0 vol.% Mg-PSZ samples. Only after sintering at 1380 °C the composite porosity amounts 10.6 ± 0.7 % (40 min)/9 ± 0.3 % (120 min) and is therefore comparable with the unreinforced compacts.

Essentially, sintering is a diffusion controlled mechanism where the reduction of free surface operates as driving force. Material transfer processes from the particles center to the points of contact introduce the initial stage of sintering denoted as neck formation. In the following intermediate stage an interconnection of the interstices takes place so that a great amount of densification can be achieved. The curve progress of relative density as a function of sintering follows a logarithmic shape. For this reason, relative long sintering times in comparison to the other stages become necessary to remove the isolated pores in the third sintering interval [11, 12]. Through the microstructure images of 1280 °C samples (figure 1 a)/b) it is apparent that an insufficient diffusion for both 40 min and 120 min sintering occurred due to the low process temperature. The shape of the former steel particle boundaries are still apparent and narrow sinter necks have formed indicating sintering stage one. At 1350 °C and 1380 °C the selected time of 120 min (figure 1 c)/d)) is sufficient to reach mechanism stage three. For obtaining a dense microstructure without any porosity sintering times much longer than 120 min have to be chosen. However, grain growth would start being adverse for the strength properties. Besides, further temperature increase will lead to the melting of the steel particles which is not desirable in terms of the compacts form stability. The presence of Mg-PSZ seems to constrain the densification process for temperatures of 1280 °C and 1350 °C. Rahimian et al. [13] explain this phenomenon with the reduction of compressibility at initial state as well as the inhibiting effect of the reinforcement in the rearrangement of the particles during sintering.

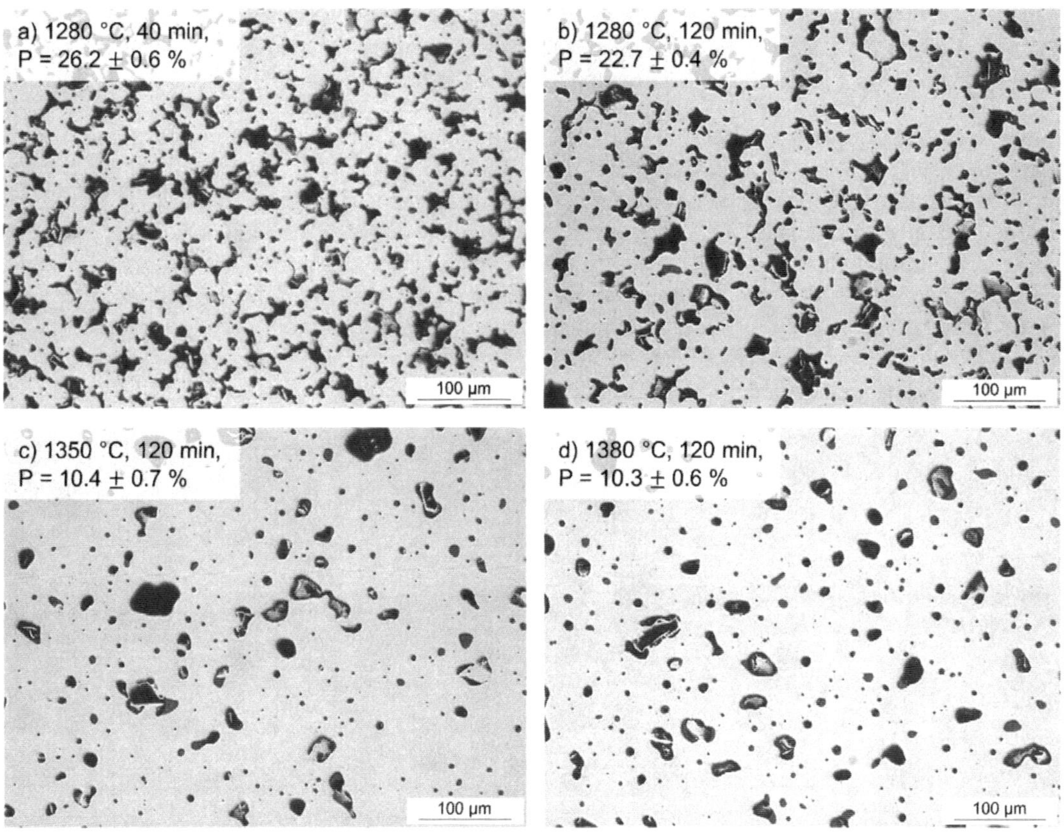

Figure 1. Microstructure evolution as a function of sintering temperature and dwell time.

In figure 2 the true stress-strain-curves of the composites recorded under tensile load are displayed. It becomes obvious that the increase of process temperature and time results in an increase of yield strength as well as fracture strain and thus leads to a higher energy absorption capability. Besides, a change of curve progress appears from an almost linear to a sigmoidal shape at 1380 °C/120 min indicating the formation of deformation induced martensite (TRIP-effect) [14].

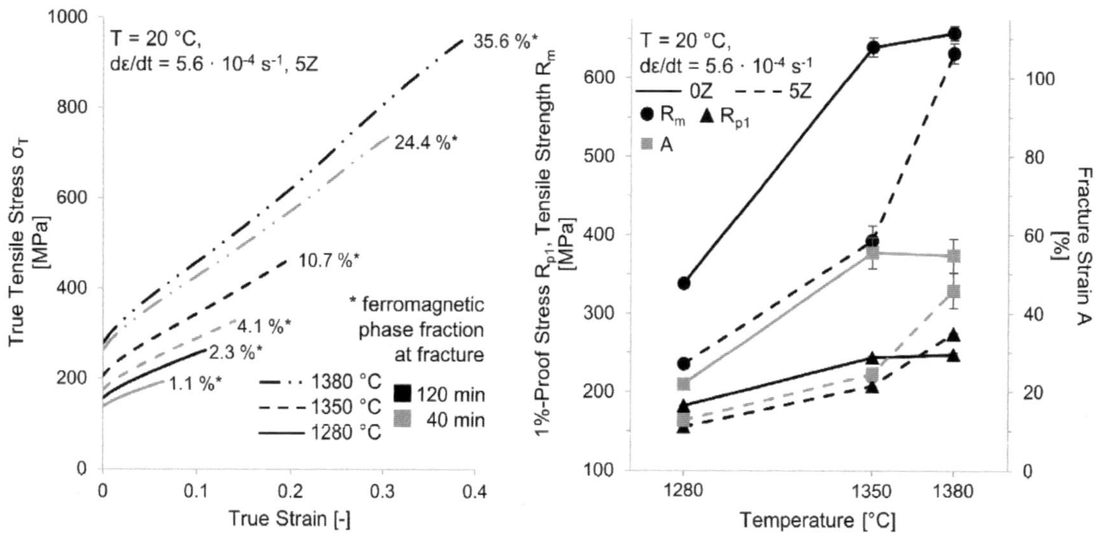

Figure 2. Influence of sintering temperature and time on tensile properties of 16Cr7Mn7Ni-steel reinforced with 5 vol.% Mg-PSZ (5Z) and comparison with unreinforced state (0Z).

Important for the uniform distribution of stress and thus isotropic mechanical properties is the homogeneous dispersion of the reinforcements. This requirement was confirmed by optical microscopy for the investigated compacts. Reason for the low load bearing capacity and deformability of the samples manufactured at low and medium temperatures is rather the already mentioned porosity. The load transfer from one particle to the other one is limited due to the weak sinter necks promoting an easy pore coalescence and crack initiation along the previous interfaces of steel/steel or steel/ceramic particles. Besides, a high number of pores exhibit an edged shape leading to stress concentration fields. Under consideration of the fracture surfaces depicted in figure 03 which are also representative for the 40 min sintered samples, an intergranular failure mechanism can be confirmed for 1280 °C and 1350 °C.

Figure 3. Fracture surfaces of 5 vol.% reinforced 16Cr7Mn7Ni steel samples sintered at varying temperatures for 120 min.

In contrast to this, the SEM-image of the 1380 °C sample displays a dimpled surface indicating a ductile failure mechanism. Through the centered position of the Mg-PSZ within the dimples, it can be concluded that during deformation the failure starts with the detachment of the steel matrix from the Mg-PSZ followed by the consolidation of cavities. Thereby, a better load transfer is possible and more energy is needed to initiate the described processes leading to an increase of energy absorption capability.

However, noticeable is the distinct higher strength and ductility level of the 1380 °C/120 min composites despite that after 40 min and 120 min sintering a similar porosity is meassureable. Although the applied sintering process belongs to the solid-state processes where often the absence of an extensive reaction is mentioned as an advantage, an interfacial layer between steel and Mg-PSZ is visible for the samples sintered at 1350 °C and 1380 °C (see figure 4). The WDX-profiles show that the formed phases which are located at the interfaces steel/ceramic and ceramic/ceramic mainly consist of oxygen, manganese, magnesium and silicon. It is comprehensible that the formation of this layer is also diffusion controlled so that a longer sintering time at a constant temperature leads to an enhanced formation. Obviously, the greater layer thickness implies a better connection to the steel matrix and thus contributes advantageously to a higher tensile strength and deformability.

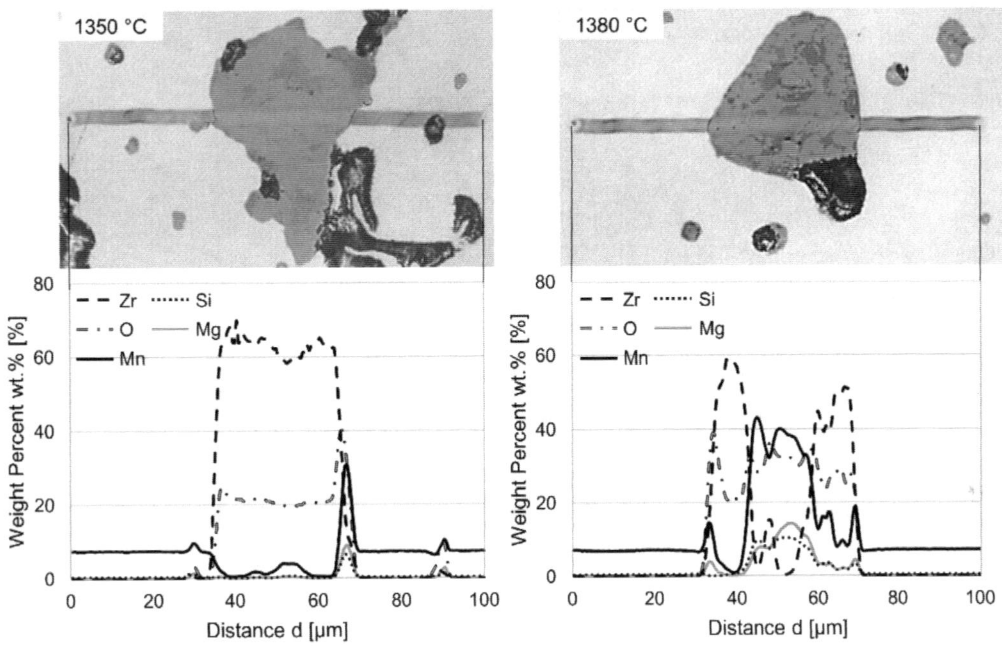

Figure 4. WDX-profiles of Mg-PSZ particles surrounded by 16Cr7Mn7Ni steel matrix sintered for 120 min at varying temperatures.

The results of MSAT-measurements included in figure 2 clarify that in every composite sample a deformation induced transformation from paramagnetic austenite into ferromagnetic α'-martensite occurred indicating the TRIP-effect. Furthermore, the amount of ferromagnetic phase fraction increases with increasing sintering temperature and time. In summary it is concluded, that the high strength and ductility at 1380 °C arise from a lower porosity as well as a superior connection to the steel matrix (interfacial layer) inducing a better load transfer and thus providing an greater amount of mechanical work for the phase transformation of the TRIP-effect.

Under consideration of tensile properties (figure 2), the proof stress of the composite sintered at 1380 °C/120 min exceeds the value of the unreinforced state whereas the tensile strength and the fracture strain are comparable. For lower process temperatures the performance of the steel samples is distinct superior. By comparing the present results with the tensile properties of particulate reinforced samples reported in literature, it becomes obvious that the achieved strength and failure strains of the investigated TRIP-matrix-composites are unobtainable with these ferritic steel and aluminum MMC systems [8, 15].

Figure 5 reveals the stress-strain behavior of the 5 vol.% Mg-PSZ samples under compressive load. Consistent with the tensile flow curves, the yield strength increases with increasing sintering temperature and time. However, appropriate to the strength differential (S-D) effect reported in literature for metallic materials [16] a higher stress level for all investigated samples is visible under compression. Furthermore, the shape of the curves changes from linear at 1280 °C to a parabolic progress at 1380 °C. As under tensile load, the 1280 °C samples provide a low resistance against plastic deformation due to the high porosity values. The mechanical densification of the vacancies contribute a high amount of deformation to the overall compression strain so that minor stresses are generated in the steel particles and a lower strain hardening takes place. Therefore, the TRIP-effect is less pronounced as indicated by lower ferromagnetic phase fractions after deformation. The almost similar ferromagnetic contents of the samples sintered at 1350 °C and 1380 °C can be explained by exceeding the saturation value which is dependent on the chemical composition, test temperature and strain rate. Nevertheless, through the strong connection of the steel matrix and the Mg-PSZ generated by the interfacial layer (figure 4) a better load transfer becomes possible so that also under compressive load the mechanical properties of the 1380 °C composite samples are preferable to the 0 vol.% Mg-PSZ samples (figure 5).

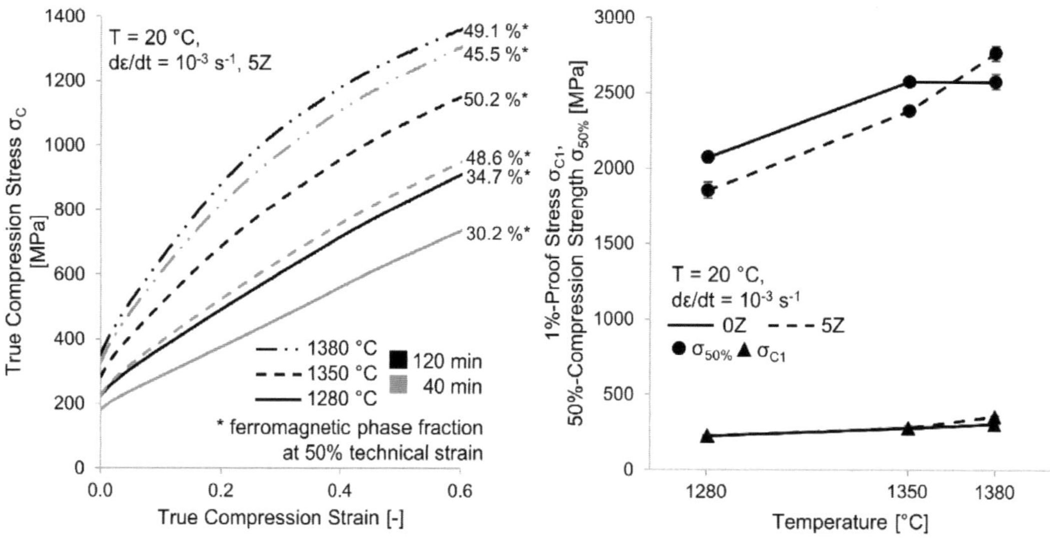

Figure 5. Influence of sintering temperature and dwell time on compression properties of 16Cr7Mn7Ni-steel reinforced with 5 vol.% Mg-PSZ (5Z) and comparison with unreinforced state (0Z).

4. Conclusion

The microstructural investigations as well as mechanical testing on TRIP-steel and TRIP-matrix-composites manufactured by pressureless sintering with varying process parameters led to the following findings:

- An increase of sintering temperature and time results in an enhancement of strength under tensile and compressive load.
- In all samples the deformation induced transition from paramagnetic austenite into ferromagnetic martensite is recordable indicating the occurrence of the TRIP-effect.
- Under tensile deformation the amount of ferromagnetic phase increased with increasing processing temperature and time whereas under compressive deformation a comparable amount (saturation value) for 1350 °C and 1380 °C samples could be measured.
- WDX-investigations reveal the formation of an interfacial layer during sintering which contributes to a better connection between steel and Mg-PSZ.

- By comparing un- and reinforced state, the composite samples manufactured at 1380 °C/120 min exceed the mechanical properties of the 0 vol.% Mg-PSZ samples and thus possess a greater energy absorption capability.

Acknowledgment

The authors thank the German Research Foundation for the financial support of the Collaborative Research Center "TRIP-Matrix-Composite" (CRC 799) in which the presented results were obtained as well as Dr. Dietrich Heger (Institute of Materials Science, Technische Universität Freiberg) for WDX-measurements.

References

[1] Kok M 2005 *J. Mater. Process. Technol.* **161** 381
[2] Chawla N and Shen Y 2001 *Adv. Eng. Mater.* **3** 357
[3] Tjong S and Lau K 1999 *Mater. Lett.* **41** 153
[4] Ruiz-Navas EA, Gordo E and Garcia R 2005 *Mater. Res.* **8** 1
[5] Rawal SP 2001 *JOM* **53** 14
[6] Chawla N and Chawla KK 2006 *JOM* **58** 67
[7] Ibrahim IA, Mohamed FA and Lavernia EJ 1991 *J. Mater. Sci.* **26** 1137
[8] Panda SS, Upadhyaya A and Agrawal D 2007 *J. Mater. Sci.* **42** 966
[9] Umasankar V, Anthony Xavior M and Karthikeyan S 2014 *J. Alloys Compd.* **582** 380
[10] Oettel H and Martin U 2006 *Int. J. Mater. Res.* **97** 1642
[11] Cahn RW and Haasen P 1996 *North-Holland*
[12] Kang SL 2005 *Elsevier Butterworth-Heinemann*
[13] Rahimian M, Ehsani N, Parvin N and Baharvandi HR 2009 Mater. Des. **30** 3333
[14] Jahn A, Kovalev A, Weiβ A, et al. 2009 *ESOMAT*
[15] Kamat SV, Hirth JP and Mehrabian R 1989 *Acta Metall.* **37** 2395
[16] Drucker DC 1973 *Metall. Trans.* **4** 667

IOP Conf. Series: Materials Science and Engineering **139** (2016) 012014 doi:10.1088/1757-899X/139/1/012014

On the nature of interface of carbon nanotube coated carbon fibers with different polymers

Harpreet Singh Bedi, Srikant S Padhee and Prabhat K Agnihotri

Mechanics of Advanced Materials Laboratory, School of Mechanical, Materials and Energy Engineering, Indian Institute of Technology Ropar, Rupnagar, Punjab, India-140001.

E-mail: harpreet.bedi@iitrpr.ac.in, prabhat@iitrpr.ac.in

Abstract. Experimental investigations are carried out to analyse the wetting behaviour of carbon nanotube (CNT) coated carbon fiber to determine their suitability to process carbon nanotube coated carbon fiber/polymer multiscale composites for structural applications. To overcome the problem of agglomeration, CNTs are grown directly on the surface of carbon fibers as well as fabric using thermal chemical vapour deposition (CVD) technique. The term multiscale is used because different reinforcement mechanisms operate at the scale of long fibers and CNTs which are of few micrometers in length. The load carrying capacity of these multiscale composites critically depends on the efficiency and extent of load transfer from low strength matrix to high strength fiber which in turn depends on the interfacial strength between CNT coated carbon fiber and polymer matrix. A systematic analysis of wetting behaviour of CNT coated carbon fiber with epoxy and polyester matrix is carried out in this study. It is shown that CNT coated carbon fibers as well as fabric show better wettability with epoxy matrix as compared to polyester matrix. This results in stronger interface of CNT coated carbon fiber with epoxy as compared to polyester in multiscale composite system. A similar observation is made in nanoindentation testing of single fiber multiscale composites processed with epoxy and polyester matrix. In addition, it is observed that wettability, interfacial strength and average properties of CNT coated carbon fiber/polymer composites are a function of CNT density on the surface of carbon fibers.

1. Introduction

Advances in aviation and automobile industry, military and sports equipment demands better carbon fiber reinforced polymer (CFRP) composites with improved strength to weight ratio. To serve this purpose, nano-materials such as carbon nanotubes (CNTs) have proven to be an unmatched candidate [1]. Tensile modulus of the order of 1 TPa, tensile strength of 100-600 GPa and low density of 1.3 g/cm^3 along with high thermal and electrical conductivity [2] are some of the excellent properties of carbon nanotubes (CNTs) that may be harnessed into CFRP composites to improve their average properties. However, agglomeration of CNTs in polymer matrix [3] poses a difficult challenge for their use in polymer composites. To address this issue, CNTs are grown directly on the surface of carbon fibers [4] rather than being mixed with polymer resin. CNT coated carbon fiber/polymer multiscale composites show improvement in interfacial shear strength [5], enhanced fracture toughness [6] and tensile strength [7]. However, the observed improvement in overall properties of the composite is not as much as expected. One of the possible reasons could be the lower load/stress transfer due to weaker adhesion at CNT coated carbon fiber-polymer interface. Since, the quality of fiber-polymer interface and the extent of adhesion critically depends on the wetting of fiber surface by polymer resin [8]; an attempt is made in this work to characterize the effect of CNTs on the wetting behaviour of CNT coated carbon fiber with different polymer resins. Interfacial strength and average properties are studied by nanoindentation and uniaxial tensile testing of single carbon fiber/polymer multiscale composites processed with epoxy and polyester matrix. It is shown that epoxy matrix is better suited as compared to polyester matrix to process CNT coated carbon fiber/polymer multiscale composites.

2. Experimental

2.1. Raw materials and catalyst coating

Hindoostan Technical Fabrics, India, supplied the unidirectional carbon fiber (CF) fabric that is preheated at about 550°C for 30 minutes to remove any polymer sizing applied to it. These sizing free (will be referred as unsized from here onwards) CFs are then coated with nickel catalyst using dip coating technique [4] (refer table 1). The catalyst solution is prepared by dissolving nickel sulphate, sodium hypophosphite, sodium succinate and succinic acid in distilled water. Variation in number of nucleation sites on the surface of carbon fiber is achieved by dipping CFs in the catalyst solution for different time varying from 3 to 10 minutes. Catalyst coated carbon fibers are then heated in an oven at 120°C for 20 minutes to remove any moisture content. While, 1wt% of accelerator and hardener is used to cure the polyester resin, the cross linking in epoxy resin is facilitated by mixing 60wt% of hardener in uncured epoxy.

Table 1. Composition of the acidic dip coating solution [4].

Nickel sulphate [$NiSO_4.6H_2O$] (g/l)	Sodiumhypophosphite [$NaH_2PO_2.H_2O$] (g/l)	Sodium succinate [$Na_2C_4H_4O_4.6H_2O$] (g/l)	Succinic acid [$C_4H_6O_4$] (g/l)	Coating temperature (°C)
30	15	15	0.36	75

2.2. CNT synthesis on catalyst coated carbon fibers

Growth of CNTs on the surface of catalyst coated CFs is facilitated inside a CVD reactor at an initial vacuum of the order of 10^{-2} Torr using CNT growth protocol shown in figure 1. Argon (Ar) helps in achieving an inert atmosphere and acts as a carrier of hydrogen (H_2) and acetylene (C_2H_2) gases. H_2 is used for dissociating the NiP film deposited on CF surface into Ni or NiP particles which act as CNT growth sites. C_2H_2 dissociates at 750°C providing carbon for the formation of CNTs at nucleation sites.

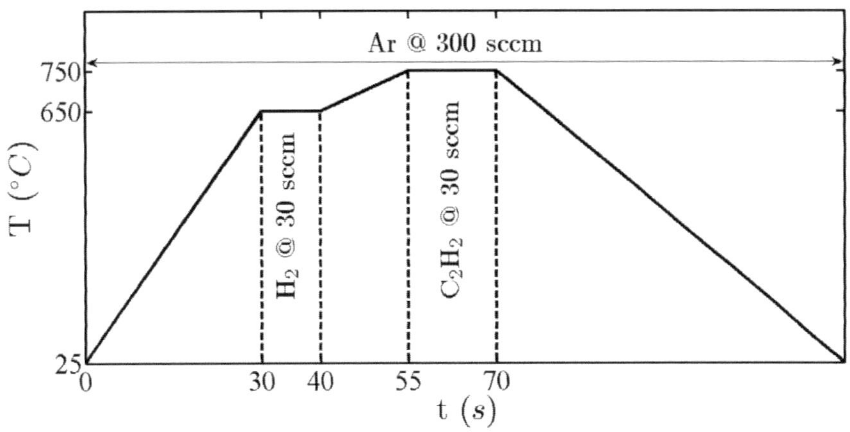

Figure 1. CNT growth protocol used in chemical vapour deposition (CVD) reactor.

3. Results and discussion

3.1. Effect of catalyst coating time on the growth of CNTs

Scanning electron microscopy (SEM, JEOL 6610 LV, Japan) of CF with and without sizing layer (figure 2) reveals their average size of 7-10 µm. Figure 3(a-c) shows the SEM micrographs of catalyst coated CFs with different coating times displaying an increase in thickness of catalyst film on fiber surface with increasing coating time. This leads to a higher number of nucleation sites for CNTs on reducing the catalytic layer during CNT growth process. Energy dispersive spectroscopy (EDS) of catalyst coated carbon fibers predicts a rise in the amount of Ni from 0.3wt % to 1.25wt% as catalyst coating time is increased from 3 mins to 10 mins. Increase in the number of growth sites leads to denser CNTs on the surface of catalyst coated carbon fiber as observed in SEM micrographs shown in figure 3(d-f).

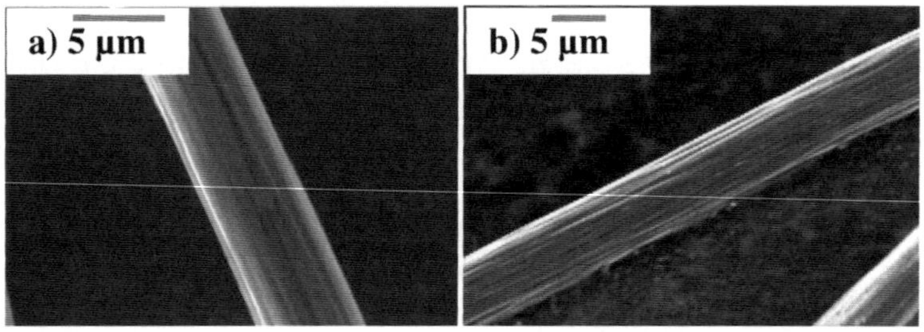

Figure 2. SEM micrograph of (a) as received and (b) unsized (sizing free) carbon fibers.

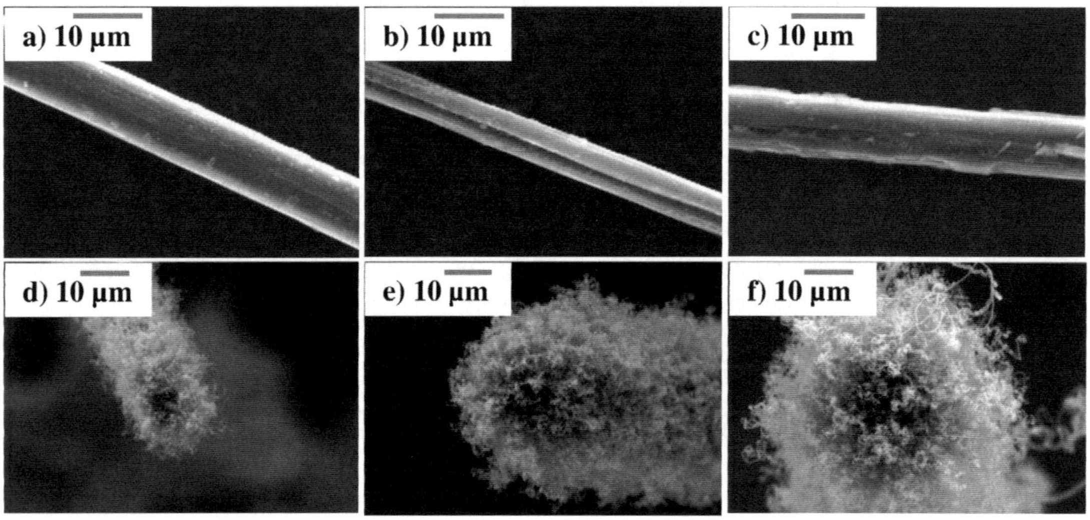

Figure 3. SEM micrograph of catalyst coated carbon fibers with dip coating time of (a) 3 mins (b) 5 mins and (c) 10 mins. Corresponding CNT coated carbon fibers are shown in micrographs 3(d-f).

3.2. Effect of CNT density on the wettability of CNT coated carbon fiber

Effect of CNT growth on the wetting behaviour of CNT coated carbon fiber fabric as well as single carbon fiber with deionized water, polyester and epoxy is investigated by performing contact angle measurements using First Ten Angstroms (FTA) contact angle setup, Virginia, USA.

First, interfacial tension (IFT) of deionized water is measured to check the accuracy of FTA system and is found to be in good agreement with the value reported in literature (see figure 4a). Next, IFT of epoxy and polyester is also determined which will later be used in the calculation of work of adhesion (*Wa*).

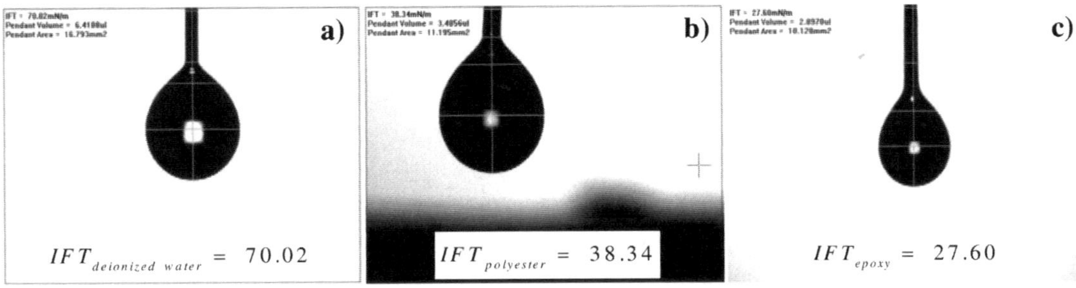

Figure 4. Interfacial tension (IFT) of various liquids in mN/m.

3.2.1 Wettability analysis of CNT coated carbon fiber fabric with deionized water. Liquid droplets are placed on CF-fabric and average contact angle (θ) is measured using freely available imageJ software (http://imagej.nih.gov/ij/download.html). θ for deionized water droplets on various CNT coated CF-fabric is found to be $\geq 123.78^{\circ}$ (see figure 5), showing hydrophobic nature of deionized water with CNT coated CF-fabric. Moreover, hydrophobicity strengthens with increase in CNT density on the fabric surface (see figure 5(a to c)). This behavior is not surprising as the increase in catalyst coating time increases the CNT density on the fiber surface which in turn increases the surface roughness factor r of carbon fibers. According to Wenzel's equation [9], increase in r magnifies the inherent wetting property (hydrophobicity in this case) of a rough surface.

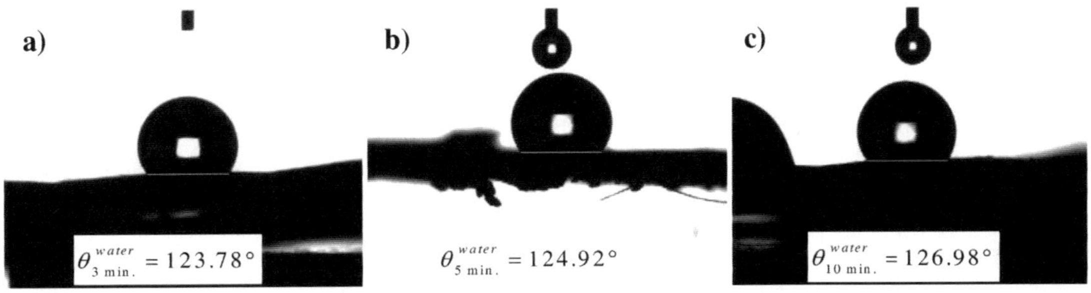

Figure 5. Average contact angle *(θ)* of deionized water droplets on CNT coated carbon fiber-fabric dip coated in Ni catalyst solution for different time durations.

3.2.2. Wettability analysis of CNT coated carbon fiber fabric with polyester. Average contact angle θ for polyester droplet on various CNT coated CF-fabric is found to be $\geq 132.76^{\circ}$ (see figure 6), showing poor wetting of CNT coated CF-fabric with polyester resin. Moreover, θ increases with increasing CNT density on the fiber surface (see figure 6(a to c)) further deteriorating the capacity of polyester to wet the fabric, in accordance with Wenzel's equation.

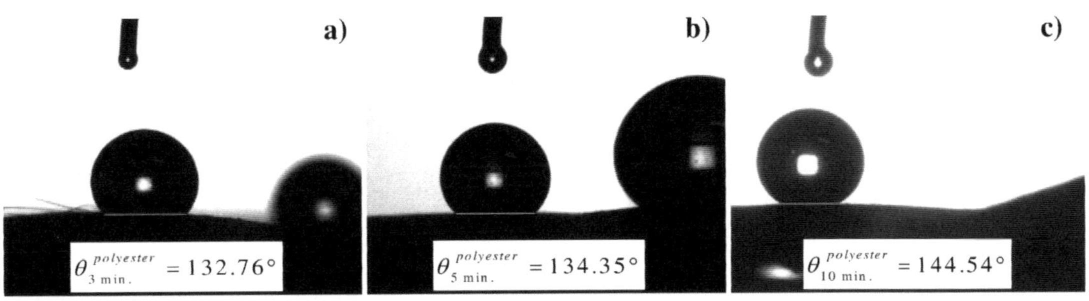

Figure 6. Average contact angle (θ) of polyester droplets on CNT coated carbon fiber-fabric dip coated in catalyst solution for different time durations.

For epoxy droplets θ cannot be measured effectively due to their instant spreading on all types of CNT coated carbon fiber fabric samples. This shows good wetting of CNT coated carbon fibers with epoxy matrix. To quantify the wetting behavior of epoxy resin, shapes of polymer droplets on single carbon fiber filaments are studied in the next section.

3.2.3. Wettability analysis of CNT coated single carbon fibers. To get further insight, the wetting behavior of carbon fibers is characterized by examining the shape of polymer droplets on CNT coated single carbon fiber (SCF) filaments, as shown in figure 7. While epoxy forms axially symmetric barrel type droplets (see figure 7a) indicating good wettability, clam type non-axisymmetric droplets of polyester suggest poor wetting [9] of CNT coated SCF (see figure 7b). This further indicates that probability of having stronger interface is higher with epoxy resin as compared to polyester resin in CNT coated carbon fiber/polymer multiscale composites.

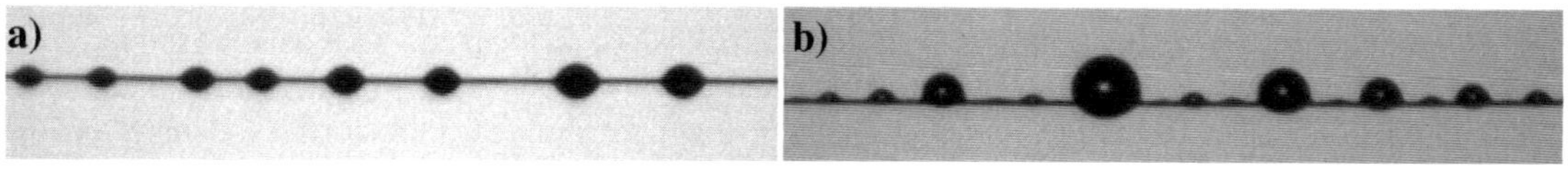

Figure 7. Optical micrographs of (a) barrel type epoxy droplets and (b) clam type polyester droplets on CNT coated single carbon fibers.

3.2.4. Work of adhesion (Wa). Effect of CNT density and hence surface roughness of carbon fiber on wettability is quantified by calculating *Wa* of the polymer using Young-Dupre equation (1) [10].

$$W_a = \gamma_{LA}(1 + \cos\theta) \tag{1}$$

Where, γ_{LA} is the surface energy of the liquid-air interface as determined from IFT value. On various CNT coated CF-fabrics, *Wa* of deionized water and polyester is found to be decreasing with increasing θ (figure 8). This implies that less amount of work is required to separate the liquid-solid phases from one another. Hence, the contact strength between the two phases gets diminished with the increase in density of CNTs on fiber surface which supports our earlier findings, that polyester has poor capacity to wet carbon fibers and wettability degrades with increasing CNT density.

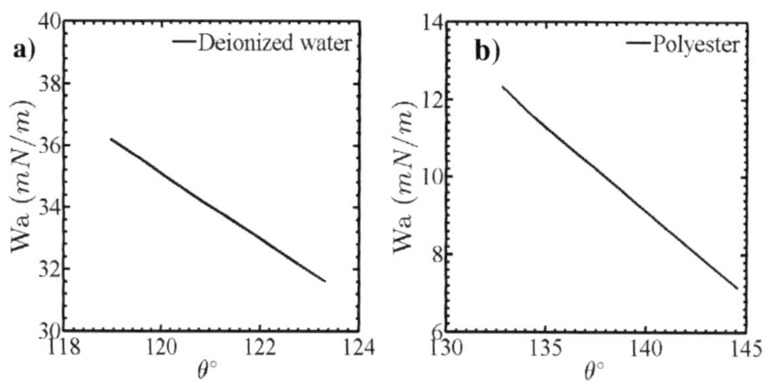

Figure 8. Work of adhesion (*Wa*) of (a) deionized water and (b) polyester with static contact angle θ.

3.3. Effect of CNT density on interfacial stiffness.

Nanoindentation testing on various samples is performed to qualitatively investigate the effect of CNT coating on interfacial stiffness in single carbon fiber composites processed with epoxy and polyester matrix using TI 950 TriboIndenter, Hysitron. To this end, single carbon fibers (SCF) are embedded vertically into epoxy and polyester resins mixed with corresponding hardeners and allowed to cure at room temperature. Circular discs of epoxy and polyester SCF composites, with an approximate diameter of

10 mm and 0.5 mm thickness are then post-cured at temperatures of 55°C and 90°C for 4 hours and 2 hours, respectively. These single fiber composites are then indented with Berkovich indenter with a half angle of 65.35° (see figure 9).

Assuming the fibers to be rigid in comparison to matrix and interface/interphase, the displacement of indenter can be thought of primarily due to the displacement of interace/interphase region surrounding the rigid fiber.

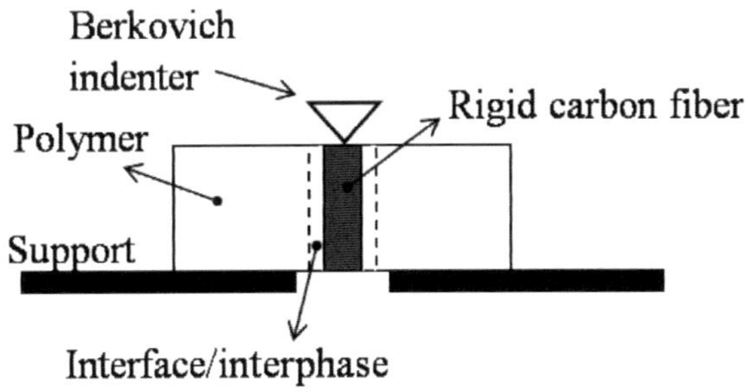

Figure 9. Schematic showing nanoindetation testing of single carbon fiber polymer composite.

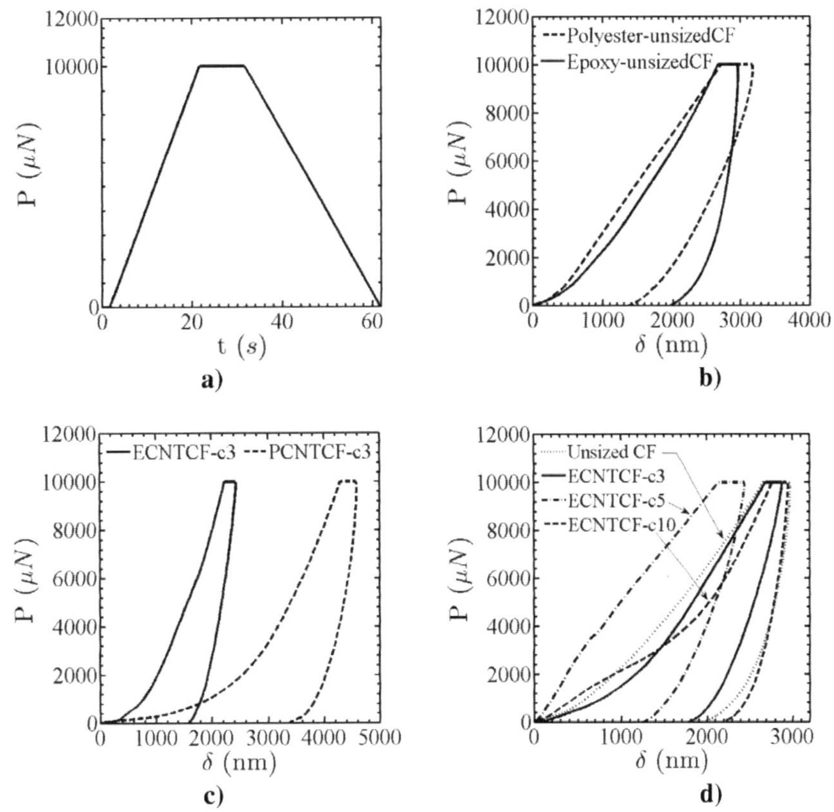

Figure 10. (a) Rate of loading employed in nanoindentation of single carbon fibers (SCF). (b, c) Contrast in load P-displacement δ curve between unsized (b) and CNT coated (c) SCF composites made with epoxy (-) and polyester (--) resins. (d) Comparison of P-δ curves of multiscale SCF epoxy composites with different CNT density on fiber surface.

Thus, the load-displacement (P-δ) curves recorded by varying the load on indenter (see figure 10a) will provide a qualitative measure of interfacial stiffness in various SCF composite samples. At first, P-δ curve of unsized carbon fibers embedded in polyester and epoxy matrix is compared in figure 10b. It is to be noted here that multiple samples are tested in each case and a representative curve is shown here for comparison purpose. It has been found (see figure 10b) that displacement at peak load in unsized carbon fiber/epoxy composite is slightly lower as compared to unsized carbon fiber/polyester composite indicating slightly better interface in unsized carbon fiber/epoxy composite.

Similar analysis is carried out for CNT coated carbon fibers corresponding to 3 minute dip coating time and the comparison of P-δ curves has been shown in figure 10c. Here again, the displacement at peak load in CNT coated carbon fibers is significantly higher in polyester matrix as compared to epoxy matrix (see figure 10c). These observations suggest higher interfacial stiffness and hence stronger interphase in composites processed with epoxy matrix than with polyester matrix.

In the next step, the effect of CNT density on the interfacial stiffness is studied by performing indentation experiments on composites processed with CNT coated single carbon fibers having different catalyst coating time in epoxy matrix. Figure 10d compares the P-δ response for this case. It is observed that the load carrying capacity of CNT coated single fiber composites is improved as the dip coating time is increased from 3 to 5 mins but decreases for 10 mins of catalyst coating time (see figure 10d). This suggests that 5 min. coating time provides optimum density of CNTs on fiber surface to improve the interfacial stiffness of CNT coated carbon fiber/epoxy composites.

3.4. Tensile testing of single carbon fiber epoxy composites.

To characterize the effect of CNT density on the average response of single carbon fiber (SCF)/epoxy composites, various samples are tested under uni-axial tensile loading using Tinius Olsen, UTM. Dogbone shaped samples of 1.5 mm thickness are prepared (see figure 11a) by placing SCF at half the mold depth. The fiber is fixed at the ends using an adhesive; epoxy resin/hardener mixture is then poured in the mold and allowed to cure at room temperature. The cured dogbone samples are further post cured at elevated temperature as described in previous section and are tensile tested at a crosshead speed of 0.5 mm/minute. Figure 11b compares the evolution of load P as a function of crosshead displacement δ for different composite samples.

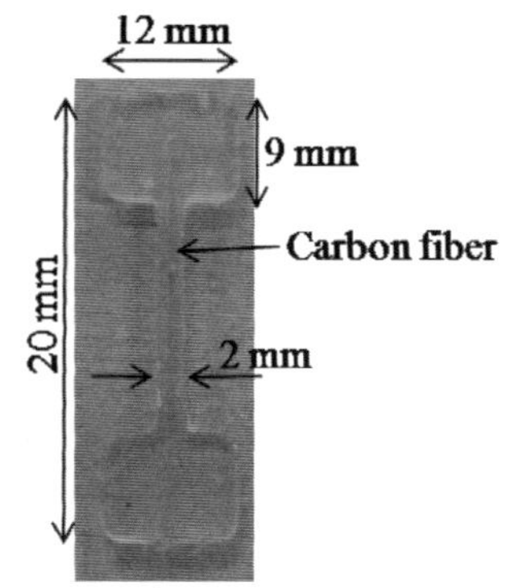

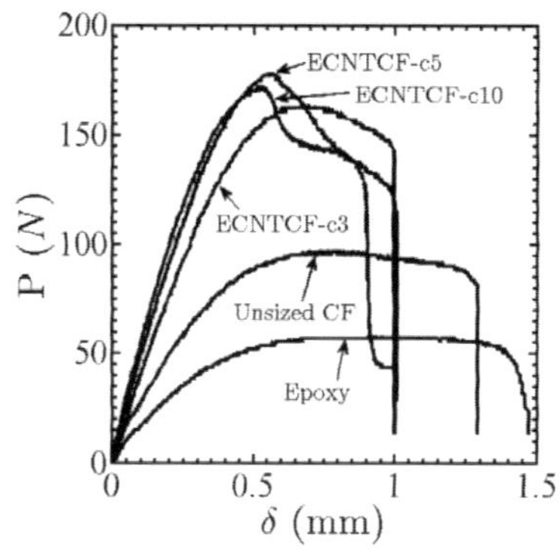

Figure 11. Sample dogbone shaped tensile test specimen (a) and tensile strength of various single carbon fiber dogbone specimens (b).

It is found that ultimate tensile strength of all CNT coated SCF composites is higher than both unsized SCF composites and pure epoxy matrix composite. One of the possible reason for this improvement in tensile strength could be the nucleation effect of CNTs on polymer crystallization [11] leading to microstructural development of polymer nearby CNTs. This results in much better properties of polymer in the interphase around carbon fibers than in the bulk. Moreover, the average response of

CNT coated SCF/epoxy composites show very weak dependence on CNT density on fiber surface and hence catalyst coating time.

4. Conclusions

In this work, a methodology is presented to predict the suitability of a polymer matrix to process CNT coated carbon fiber/polymer multiscale composites. In particular, the effect of CNT density on the wettability, interfacial stiffness and average properties of CNT coated carbon fiber/polymer multiscale composite is investigated. The salient results of the present investigation can be summarized as:

- Clam type droplets, decreasing work of adhesion *Wa* with increasing CNT density (or roughness factor *r*) and hydrophobic nature of polyester shows its poor wetting behavior with CNT coated carbon fibers.
- Axi-symmetric barrel type epoxy droplets on single carbon fibers, instant spreading of epoxy on CNT coated carbon fiber fabric indicates good wettability of CNT coated carbon fiber fabric with epoxy matrix.
- Nanoindentation testing reveals that interfacial stiffness of CNT coated single carbon fiber/epoxy composites is higher as compared to their polyester counterparts. Moreover, the interfacial stiffness is found to be a function of CNT density on the surface of carbon fiber.
- Tensile testing of single carbon fiber epoxy composites indicates an improvement in tensile strength with the use of CNT coated carbon fibers. However, the average tensile response of these composites depends very weakly on CNT density on the fiber surface.

As per the Wenzel's relation [9], increase in roughness of any surface augments its inherent wetting behaviour with a specific liquid. It is observed through SEM analysis (see figure 3) that the growth of CNTs increases the surface roughness of carbon fibers. Hence, consistent with Wenzel's relation, the wettability of carbon fiber fabric as well as single fiber improved with epoxy (see figure 7a) and deteriorated with polyester (see figure 6 and 7b) after growth of CNTs on their surface. The ability or inability of a particular polymer to wet the fiber fabric further gets magnified on increasing the CNT density on the surface of fabric (see figure 6). Moreover, it has been observed that there is an optimum CNT density on the fiber surface to maximize the increase in interfacial stiffness, as observed in [4] as well. The optimum catalyst coating time to achieve this is found to be 5 minutes in the case of CNT coated/epoxy composites studied here. In addition, the average properties as well as interfacial stiffness depends on the type of polymer used to process the multiscale composites (see figure 10 and 11). Hence, the catalyst coating time and the type of polymer matrix needs to be judiciously chosen in order to harness the excellent properties of CNTs in multiscale polymer matrix composites.

Acknowledgments

One of the Author (PKA) would like to acknowledge the financial support provided by Indian Institute of Technology Ropar under Institute Seed Grant to carry out this work.

References

[1] Breuer O, Sundararaj U, 2004, *Polymer composites,* **25(6)**, 630.
[2] Soutis C, 2005, *Progress in Aerospace Sciences*, **41(2)**, 143.
[3] Qian H, Greenhalgh ES, Shaffer MS, Bismarck A, 2010, *J. of Materials Chemistry,* **20(23)**, 4751.
[4] Agnihotri P, Basu S, Kar KK, 2011, *Carbon,* **49(9)**, 3098.
[5] Sager RJ, Klein PJ, Lagoudas DC, Zhang Q, Liu J, Dai, Baur, 2009, *Comp. Sci. Tech.,* **69(7)**, 898.
[6] Kepple KL, Sanborn GP, Lacasse PA, Gruenberg KM, Ready WJ, 2008, *Carbon,* **46(15)**, 2026.
[7] Sharma SP, Lakkad SC, 2011, *Composites Part A: Applied Science and Manufacturing,* **42(1)**, 8.
[8] An F, Lu C, Guo J, He S, Lu H, Yang Y, 2011, *Applied Surface Science,* **258(3)**, 1069.
[9] Song B, Bismarck A, Tahhan R, Springer J, 1998, *J. colloid. and interface science,* **197(1)**, 68.
[10] Schrader ME, 1995, *Langmuir,* **11(9)**, 3585.
[11] Liu Y, Kumar S, 2014, *ACS applied materials & interfaces,* **6(9)**, 6069.

IOP Conf. Series: Materials Science and Engineering **139** (2016) 012015 doi:10.1088/1757-899X/139/1/012015

Delamination initiated by a defect

A Biel and H Toftegaard

Department of Wind Energy, Section of Composites and Materials Mechanics, Technical University of Denmark, P.O. Box 49, Frederiksborgvej 399, DK-4000 Roskilde, Denmark

abie@dtu.dk

Abstract. Composite materials in wind turbines are mainly joined with adhesives. Adhesive joining is preferable since it distributes the stresses over a larger area. This study shows how a defect can influence the fracture behaviour of adhesively joined composite. Repeated experiments are performed using *double cantilever beam* specimens loaded with bending moments. The specimens consist of two 8 mm thick GFRP-laminates which are joined by a 3 mm thick epoxy adhesive. A thin foil close to one of the laminates is used to start the crack. For some of the specimens a defect is created by an initial load-unload operation. During this operation, a clamp is used in order to prevent crack propagation in the main direction. For the specimens without defect, the crack propagates in the middle of the adhesive layer. For the specimens with defect, the crack directly deviates into the laminate. After about 25 mm propagation in the laminate, the crack returns to the adhesive. Compared to the adhesive the fracture energy for the laminate is significantly higher.

1. Introduction

Manufacturing of large structures, e.g. blades for wind turbines, always include defects. Reduction of defects usually increase the manufacturing costs i.e. it is beneficial to some extent to allow for manufacturing defects. Blades for modern wind turbines are made of adhesively joined composites. A recent study has shown that the adhesive joint at the trailing edge is critical [1]. Accordingly knowledge of how cracks initiate and develops from defects, is essential when designing new structures. Material behavior can be derived from specimens with prefabricated defects.

In the current study we use the *double cantilever beam* specimens (DCB) to determine influence from defects. The specimens have, prior to the main experiment, been exposed to a load-unload operation. The operation crates a visible defect, which influences the result. For comparison, experiments without defects are also performed. A similar experimental procedure has previously been used to determine inelastic properties of an adhesive layers between steel sheets [2,3].

2. Method

The DCB-specimen can be loaded by forces or by moments, see figure 1a. The loads gives a non-uniform stress distribution, see figure 1b. If certain conditions are fulfilled the experiments can be evaluated using the path independent J-integral approach [4]. The method has been used in several studies, cf. e.g. [2,5-11]. In order to simplify the experimental setup and the evaluation it is common to load the specimen with either pure forces or pure moments. A short summary of the method is given in the sequel.

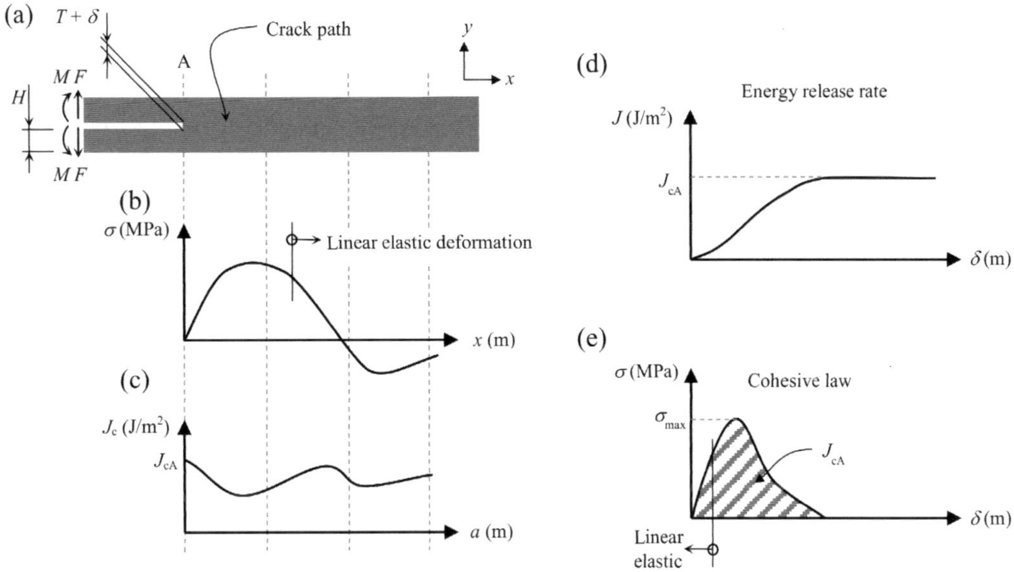

Figure 1. (a) DCB-specimen with used notation, (b) stress distribution along the specimen at beginning of crack propagation, (c) fracture resistance during crack propagation, (d) curve for energy release rate and (e) cohesive laws.

From an external path including the forces and the moments, the *energy release rate* is given by,

$$J = \frac{2F\theta}{B} + \frac{12M^2}{EB^2H^3} \tag{1}$$

where, F is the applied force, θ is the rotation at the loading point, M is the applied bending moment, B is the width of the beam, H is the height of the beam and E is Young's modulus. At crack propagation, the energy release rate equals the fracture energy, e.g. $J = J_c$. The fracture resistance may vary, during the crack propagation, see figure 1c.

By differentiating the energy release rate with respect to a displacement, δ, it is possible to derive the stresses in the beginning of the fracture region, see curves in figure 1d and figure 1e. The stress is given by,

$$\sigma = \frac{\partial J}{\partial \delta} \tag{2}$$

The relation $\sigma(\delta)$ is usually named the *cohesive law*. The cohesive law represents the stress deformation relation of the material. The shape of the cohesive law is dependent on the measurement length T, see figure 1a. Before a macroscopic crack is present, it is common to observe a material softening. For composites, softening can include fibre bridging.

It should be stressed that the method assumes a uniform cohesive law in the loaded region ahead of the crack tip. If the shape of the cohesive law changes along the x-axis, the result using the J-integral approach may be affected [2]. The size of the loaded region is e.g. dependent on the bending stiffness of the beams [8].

3. Specimens and experiments

Twenty-one experiments are performed using three types of specimens. To improve the readability of the paper we introduce three groups, G1, G2 and G3 corresponding to the three types of specimens. The beams of the specimens are made of GFRP-laminates. An insert with a thickness of 12.5 μm is used to create a start for the crack. In order to simplify the manufacturing the insert is attached to one of the beams. Tabs are attached in order to introduce the loads, see figure 2. For two of the groups G1 and G2, the beams are joined with a 3 mm thick epoxy adhesive. Nominally G1 and G2 have the same geometry but for G2 a defect is created by an initial a load-unload operation. For group G3 the adhesive is excluded. The nominal geometry of the specimens are presented in table 1.

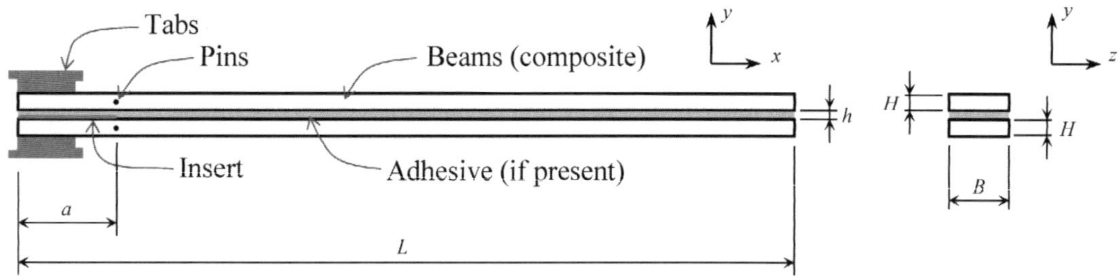

Figure 2. DCB-specimen with used notation

Table 1. Nominal geometry of the used specimens.

Group	Specimens	Defect	H (mm)	h (mm)	B (mm)	a (mm)	L (mm)
G1	8	NO	8	3	30	70	500
G2	7	YES	8	3	30	70	500
G3	6	NO	8	0	30	70	500

The seven specimens in group G2 are exposed to a load-unload operation in order to achieve a defect. During this operation, a clamp with two M8-screws is used in order to prevent main crack propagation, see figure 3a. The load to introduce the defect is applied manually as a moment, see figure 3b. The created defect is visible as small crack across the adhesive, starting from the insert. The direction of the crack is almost perpendicular to the insert. Besides the manually applied load and the loads from the clamping, the direction of the crack may be influenced by residual stresses.

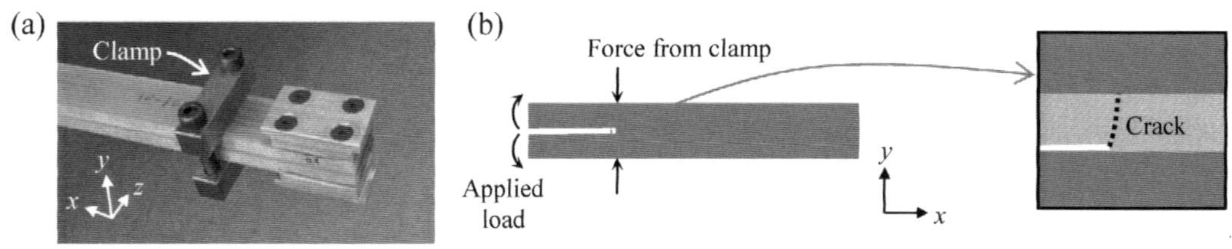

Figure 3. (a) Specimen with applied clamp and (b) forces and crack achieved during the manual load.

By use of a special test frame the DCB-specimens are loaded with pure moments, see figure 4. The setup has been used in several previous studies, cf. e.g. [7,10]. The experiments are loaded quasi statically by constant displacement of the loading point. Each experiment takes about two minutes. The specimen is clamped at the bottom and the moments are applied by use of a wire and two lever arms. The lengths of the lever arms are x_L and x_R. Two load cells are used to measure the force P in the wire and accordingly the applied moment is, $M \approx P(x_L + x_V)/2$. The separation of the beams δ_{ext} is measured with an extensometer which is applied to the specimens by two pins, see figure 2. Measurement for acoustic emission (AE) and shear deformation (LVDT) is also recorded during the experiments. These data are not used in the present paper.

The main loading is mode I however a slightly mixed mode loading is applied to the crack. As mentioned previously the insert is not placed in the centre of the specimen and there is also a small difference in the length of the lever arms, x_L = 0.0835 m and x_R = 0.0825 m. All specimens are performed using the same loading conditions. It is presumed that the influence due to mixed mode is minor. A more detailed analyse of mixed mode loading is e.g. given in [12].

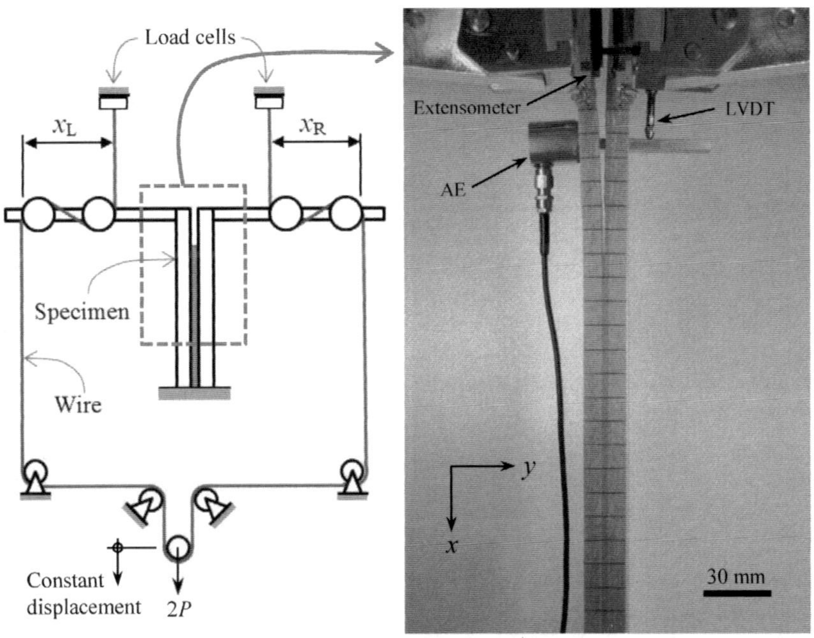

Figure 4. Experimental setup used to load DCB-specimens with pure moments.

4. Results

Significant differences are observed between the three groups. Figure 5a shows the energy release vs. time for all the experiments. The energy release rate is normalized to the average of fracture energy in group G1, J_{CG1}. In the beginning, specimens from group G1 and group G2 follow the same curve. For the specimens in group G3 the energy release rate increases faster compared to G1 and G2. The specimens in G3 are stiffer since the adhesive layer is excluded. At, $t \approx 60$ s a deviation is visible between G1 and G2. For the specimens in G1 the energy release rate suddenly drops due to instable crack propagation in the adhesive. Further loading of the specimens in G1 result in a stick-slip crack propagation in the adhesive. For all specimens in G2 except two the energy release rate increase until, $J/J_{CG1} \approx 3.2$. A sudden drop is observed at $t \approx 90$ s. The energy release rates measured after the drop is similar to the ones measured for the stick-slip behaviour in G1. For the specimens in G3 no main drop is observed. The energy release rate increase until, $J/J_{CG1} \approx 2.3$. Figure 5b shows the energy release vs. separation of the beams. For the experiments in G2 and in G3 some deformation is observed before the main crack propagation. For the experiments in G1, no deformation is observed.

Figure 6 shows photos of representative fracture surfaces and sketches of the crack paths for the different groups. For the specimens in G1 a smooth surface is observed. The crack always propagates in the adhesive close to one of the beams. No significant marks from the stick-slip behaviour is visible.

For the specimens in G2 (with defect) the crack starts in the laminate. After a distance about 25 mm the crack enters the adhesive layer. The sudden drop for the specimens in G2 is a result of the crack entering the adhesive layer. No different crack surfaces are observed for the two specimens in G2 which have lower values of the energy release rate. For the specimens in G3 the fracture surfaces are relatively smooth. Some regions with bridging are observed. Compared to the delamination in G2 the fracture surfaces in G3 are smother.

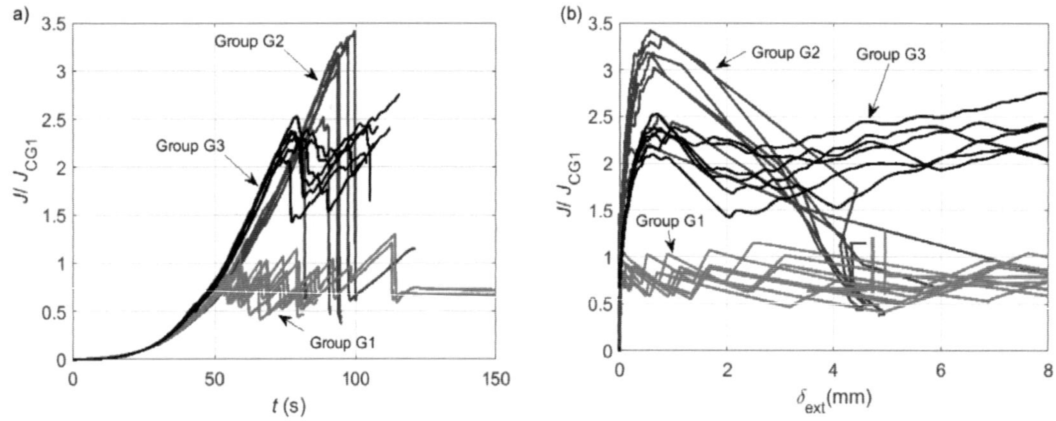

Figure 5. Energy release rate for all experiments (G1, G2, G3) normalized to the average fracture energy for experiments in group G1 vs. (a) time and (b) separation of the beams.

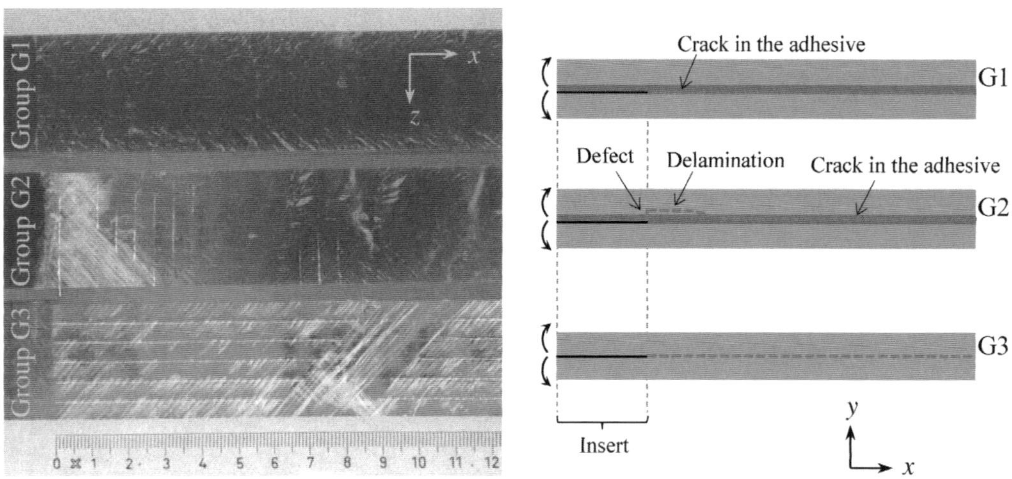

Figure 6. Crack propagation in the specimens; left: Fracture surfaces and right: sketches of the crack paths.

5. Conclusion and discussion

In present paper, it is shown that a defect can change the fracture behavior of a joint. The defect temporary forces the crack to deviate into the laminate. The fracture resistance of the laminate is significantly higher making it harder to propagate the crack. The defect also increases the instability of the joint. During the crack propagation in the laminate, more elastic energy is stored in the specimen. The energy is suddenly released when the crack enters the adhesive layer.

The result from this study can be used to design new wind turbines. However, some care may be advised. In this paper only one type of defect is analysed. The result will not be valid for all kinds of defects. The used method (*J*-integral approach) assumes a uniform cohesive law along the specimen. The instable fracture propagation (stick-slip) is an obvious indicator of variations in the cohesive law. To what extent this influences the result is not evaluated.

References

[1] Eder M and Bitsche R 2015 *Wind Energy* **18** 1007
[2] Biel A and Stigh U 2010 *Int. J. Fract.* **165** 93
[3] Biel A and Stigh U 2011 *Procedia Eng.* **10** 2280
[4] Rice J R 1968 *J. Appl. Mech.* **35** 379
[5] Paris A J and Paris P C 1988 *Int. J. Fract.* **38** R19
[6] Suo Z, Bao G and Fan B 1992 *J. Mech. Phys. Solids* **40** 1
[7] Sørensen B F 2002 *Acta Mat.* **50** 1053
[8] Andersson T and Biel A 2006 *Int. J. Fract.* **141** 227
[9] Stigh U, Alfredsson K S, Andersson T, Biel A, Carlberger T and Salomonsson K 2010 *Int. J. Fract.* **165** 149
[10] Toftegaard H, Rask M, Rasmussen S and Sørensen B F 2013 *6th International Conference on Composites Testing and Model Identification, Department of Mechanical and Manufacturing Engineering, Aalborg University, Denmark* 75
[11] Marzi S, Rauh A, and Hinterhölzl R M 2014 *Comp Struct* **111** 324
[12] Sørensen B F, Jørgensen K, Jacobsen T K and Østergaard R C *Int. J. Fract.* **141** 163

IOP Conf. Series: Materials Science and Engineering **139** (2016) 012016 doi:10.1088/1757-899X/139/1/012016

Advanced fibre reinforced thermoplastic composites with reduced processing times by use of nanoscale fillers

Georg Brockerhoff, Robert Brüll, Klaus Vonberg, Gunnar Seide, Thomas Gries

Institut für Textiltechnik (ITA) der RWTH Aachen University, Otto-Blumenthal-Str. 1 in 52074 Aachen, Germany

Robert.Bruell@ita.rwth-aachen.de

Abstract. The industrial standard for the manufacturing of fibre reinforced thermoplastic composites (FRTCs) is the film stacking method. An alternative to this is commingling thermoplastic fibres with reinforcing fibres into hybrid rovings. These rovings are woven into weaves and consolidated through compression moulding. This paper evaluates the effects of 5 weight percent (wt.-%) titanium dioxide (TiO_2) in commingled polyamide 6 (PA6) on the cycle time during the consolidation process and the mechanical properties. A product representing the industrial standard is used as reference. In order to achieve a good comparability with this product, the film stacking process is also reproduced. Finally, the three plate types are compared regarding their consolidation, tensile and flexural strength. The results show that the hybrid roving FRTC is more consolidated, has better mechanical properties and enables shorter cycle times when compared to the film stacking process.

1. Introduction

The importance of light weight construction in the mobility sector has increased throughout the recent years. Both the automotive industry and the aerospace industry aim to achieve the lowest possible weight, whilst maximizing mechanical properties [1]. A reduced weight leads to higher energy efficiency, thus decreasing operating costs. In order to achieve this goal, fibre reinforced thermoplastic composites (FRTCs) are being developed [2]. FRTCs are made out of two or more components. The reinforcing fibres define the mechanical properties of the composite. Commonly used fibres are carbon, glass, aramid or basalt fibres. The matrix component defines the thermal properties, as well as enables a force distribution between the reinforcing fibres. Typical thermoplastic matrix components are polyamide 6 (PA6), polypropylene (PP) and polyether ether ketone (PEEK). FRTCs can be produced using the film stacking method. This involves stacking the two components into a sandwich structure and consolidating them through compression moulding. Alternatively, FRTCs can be produced by commingling thermoplastic fibres with reinforcement fibres into hybrid yarns or rovings and weaving these into fabrics. The textile structures are subsequently heated and consolidated through compression moulding. For an optimal consolidation, the temperature distribution needs to be homogenous throughout the composite, as excessive temperatures can lead to degradation of the polymer and low temperatures lead to consolidation constraints. The heating and cooling times are an essential component to this process, but are also the cycle time determining factors to the production chain. Shorter cycle times lead to higher outputs, thus reducing the costs per part. Thus the goal is to find the shortest cycle time, without compromising the consolidation quality. During the course of the "NanoOrgano" project for the Federal Ministry of Education and Research (BMBF) of the German government, nanoscale titanium dioxide (TiO_2) fillers were added to the thermoplastic component and a shortening effect on the heating and cooling times of the hybrid yarn based thermoplastic composites was observed [3]. The production chain suggested by this study is shown in Figure 1. The goal of the follow-up project "VIP Organo" is the validation of the innovation potential of this effect. As a first step in the project, studies looking into Polyamide 6 (PA6) compounds containing different concentrations of TiO_2 are being conducted.

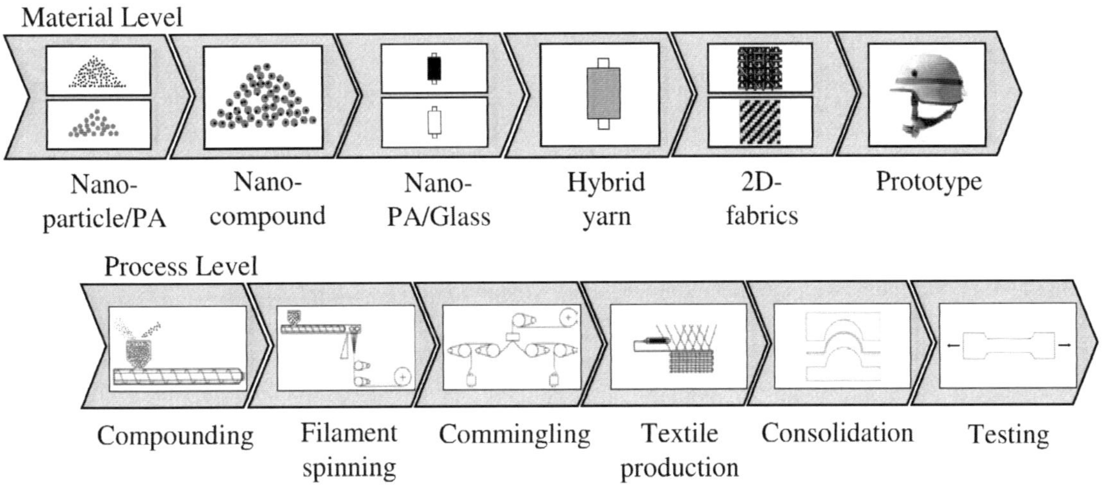

Figure 1. Production chain of the BMBF project NanoOrgano [3]

The modification of PA6 with nanoparticles of TiO_2 has several effects. A small increase in the concentration of titanium dioxide (TiO_2) leads to an increase in the thermal conductivity and a decrease in the heat capacity [4]. This leads to shorter heating and cooling times, which in turn have a significant effect on the cycle time during the consolidation process. In addition, the particle orientation is decreased, which reduces the characteristic thermal shrinking of the thermoplastic compound, and thus enables an easier thermal processing of the material. However, the concentration cannot be increased indefinitely as the TiO_2 particles also lower the strength of the compound [3]. Therefore, a compromise needs to be made. The results of Brüll et al. [4] show that the most promising concentration of TiO_2 is 5 wt.-%. This study therefore focused on the influence of the 5 wt.-% of TiO_2 in commingled PA6 on the mechanical properties, the consolidation and the cycle time during the production of FRTCs.

2. Methods

In order to assess the effect on the cycle time and compare it with the industrial standard this study compares three different FRTCs. The first one is an industrially produced FRTC. This composite is used as a reference and thereby defines the benchmark regarding the material properties of the other two. The second is produced using the commingling method. This composite has the same material properties as the first. However, the PA6 used has been modified with 5 wt.-% of TiO_2. Since the production parameters of the industrially produced FRTC are unknown, a third composite needs to be manufactured. This is done by replicating the production process (film stacking) of the industrially produced FRTC while using the production parameters of the hybrid roving FRTC discovered during pre-tests. This enables a direct comparison between the second (film stacking) and third (hybrid roving) composites and thus makes deductions in regards to the first (industrially produced) composite possible.

2.1. Industrially produced FRTC
The Tepex® Dynalite 102-RG600(10)/47% black Typ B (Tepex) of the Bond-Laminates GmbH, Brilon, Germany is produced using the film stacking method and is chosen to represent the industrial standard. This composite, due to its good mechanical properties, and its production method is widely used by the automotive industry. The material properties are as follows: the glass roving weave is a 2/2 twill weave with a glass roving fineness of 1200 tex. The PA 6 matrix material is added as films. The resulting FRTC has a fibre weight percentage of 63 % and a plate thickness of 5 mm [5][6].

2.2. Hybrid roving FRTC
For the production of the hybrid roving FRTC, a glass roving with a fineness of 1200 tex is used. The PA6 utilised is the nano-modified PA6 with 5 wt.-% TiO_2, which is spun into a roving with a fineness of 800 tex. These components are mixed into a hybrid roving with a fineness of 2000 tex using the commingling method. This method involves feeding the separate rovings into an air jet. In order to process these rough materials, with a combined fineness of 2000 tex, a customized air jet is used. During the commingling process the pressurised air and the resulting turbulences open the two different rovings to their filaments and then entangle them. Ideally, the manufactured hybrid roving has an even distribution of the thermoplastic filaments between the glass filaments. This reduces the distance the PA6 needs to flow during the consolidation process. In order to achieve good properties, pre-tests are conducted. The best parameters are then chosen for the production of the hybrid roving which is afterwards woven into a 2/2 twill weave and consolidated using a heating press.

2.3. Film Stacking FRTC
This composite is produced using the production method (film stacking) of the Tepex plate, while using the process parameters of the hybrid roving process. Therefore, a glass roving 2/2 twill weave and a PA6 films are used. The components are stacked in a sandwich structure in accordance with the reference materials specifications. This enables the production of plates equal in thickness and fibre mass ratio to the Tepex plate. The resulting stacks are consolidated in a heating press. This enables a direct comparison between the two production types and allows inferences to be made regarding the industrial process.

2.4. Consolidation & Analysis
During the consolidation process the material stacks are placed into the heating press at the consolidation temperature (T_c) of 250 °C. This is done in order to eliminate the effect of the heating up stage of the press tools on the consolidation process. At this temperature the pressure in the cavity (p) is maintained at 18 bar for different holding times (t_h) (0 s, 30 s, 60 s). Afterwards the tools are cooled to the extraction temperature (T_e) of 180 °C and the consolidated plate is removed. In the second part of the experiment the effects of cavity pressure are investigated. In order to do this, the holding time (t_h) is kept at 30 s, while the pressure (p) in the cavity is changed (10 bar, 14 bar, 18 bar). These parameter combinations are used in the production of both the hybrid roving FRTC and the film stacking FRTC. The resulting plates for each parameter combination, as well as the Tepex plate, are then compared regarding their consolidation and mechanical properties. In order to evaluate the consolidation of the three different plates, two different analysis methods are used. Firstly, an examination under a light microscope is undertaken. During this process the probes are placed in epoxy resin. The epoxy resin fixates the different components during probe preparation (polishing) and enables equal probe quality in a test series. Afterwards the probes are analysed under the light microscope and the visible consolidation is evaluated. The second method is the micro computed tomography (micro-CT) scan. This method involves scanning a whole FRTC plate with x-rays. This method enables an inspection of the exterior surface, as well as the interior of the complete FRTC.

The mechanical properties are investigated using tensile and flexural tests. This is done in order to determine the parameter combination that enables a short cycle time, while maintaining high mechanical properties of the composite.

3. Results

A suitable commingling process is determined by pre-tests. The resulting hybrid rovings are analysed regarding the maximum force (F_{max}) that can be applied in accordance with the DIN EN ISO 2062 norm. The distribution is evaluated using light microscopy. It is observed that the maximum force which can be applied to the hybrid roving is overall lower than that of the pure glass roving. At the same time a negative correlation between the distribution of the filaments and the maximum force can be detected. The better the distribution becomes, the lower the maximum force appears to be. This enables a choice of commingling parameters for the production of the hybrid roving for the main experiments. For the further tests an average hybrid roving, with medium tensile strength and distribution, is chosen. The consolidated hybrid rovings, together with the film stacking and Tepex plates, are separately investigated in regards to their consolidation, tensile and flexural strength.

3.1. Light microscopic analysis

For each of the parameter combinations, a probe is prepared and viewed under the light microscope. The results show that during the consolidation process of the film stacking FRTC, the PA6 does not flow in-between all the glass filaments. Figure 2 shows the empty gaps as being filled with epoxy resin. Furthermore, distinct layers remain visible. This effect is observed throughout all the tested parameter combinations. However, an increase in the holding time (t_h), as well as an increase in pressure (p), is shown to improve the distribution.

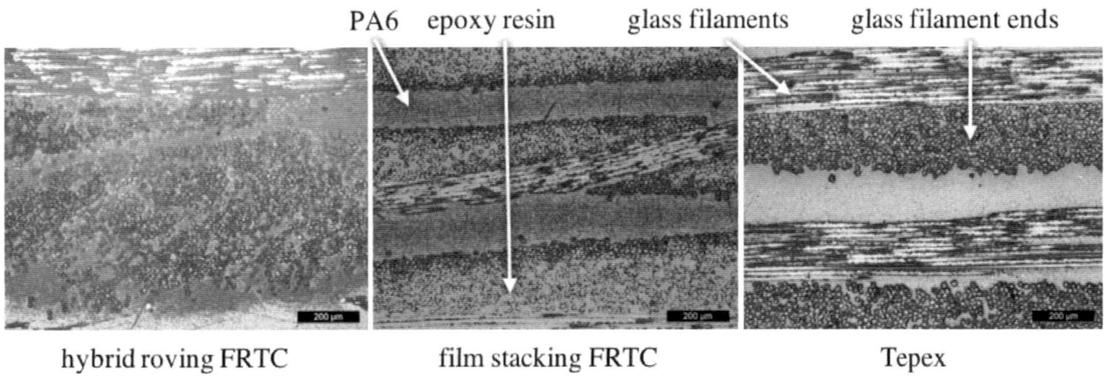

Figure 2. Light microscopic pictures of the parameter combination (30 s, 10 bar)

The Tepex probe is shown to contain the same distinct layers as the film stacking FRTC probes (see Figure 2). However, an inspection of the area around the separate glass filaments shows that the PA6 flows in-between the filaments, while still not achieving an even distribution. The probes of the hybrid roving FRTCs, on the other hand, display a good distribution of the PA6 (see Figure 2). The filaments are well impregnated and there is no distinct layer of PA6 visible between the glass roving weaves. An increase in holding time (t_h) and a decrease in pressure (p) improve the distribution.

3.2. Micro-CT analysis

A micro-CT scan enables a clearer evaluation of the consolidation throughout the composite. However, due to similar radiation absorption of the PA6 and glass rovings, a clear view of the consolidation and impregnation of the filaments is not possible. Figure 3 shows the resulting pictures of the different plate types.

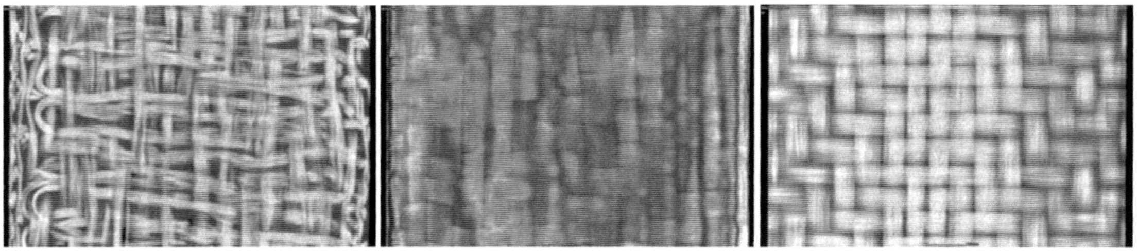

Figure 3. CT-pictures of the three plate types

Although these pictures do not enable an evaluation of the consolidation of the filaments, they do show the different weave structures in the FRTCs. The weaves of the film stacking and the Tepex probe are still clear and well structured. However, the image of the hybrid roving FRTC shows larger gaps and a less clear structure, indicating that the glass fibres are not perfectly aligned.

3.3. Tensile test

The tensile strength (σ_M) is examined in accordance with the DIN EN ISO 527-4 norm. The results can be seen in Figure 4.

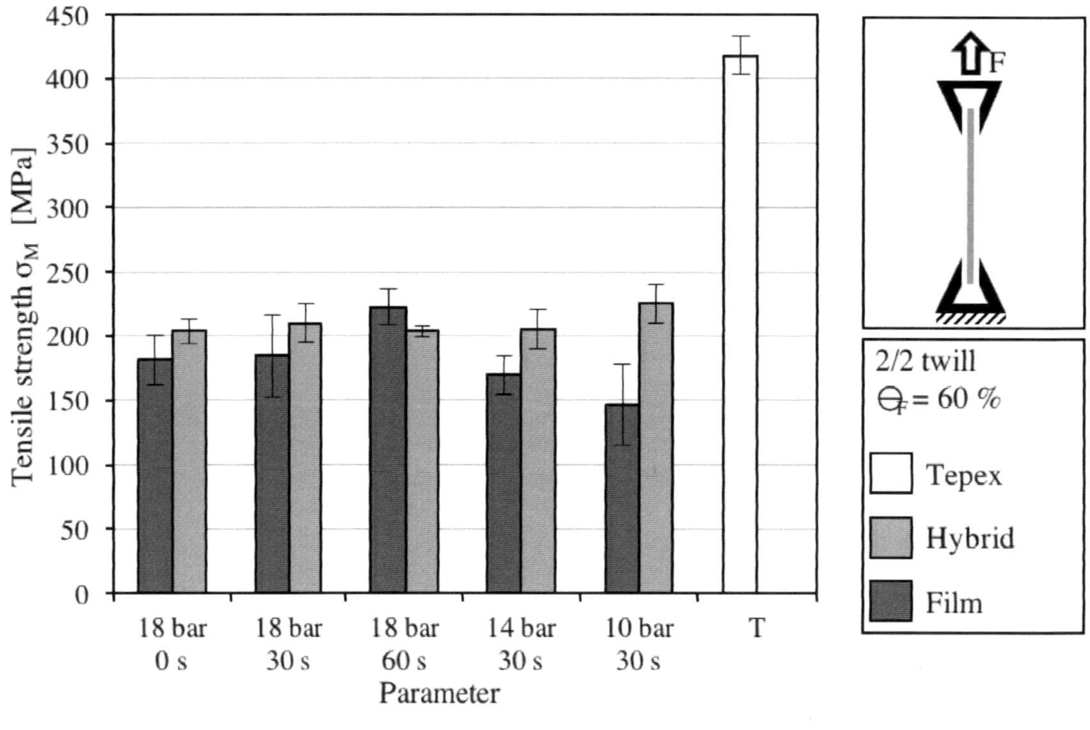

Figure 4. Results of the tensile tests

It is shown that the tensile strength (σ_M) of the film stacking FRTC improves with an increasing holding time (th) and increasing cavity pressure (p). However, the hybrid roving FRTC displays similar strengths even with an increased holding time (th). Interestingly, the strength increases with decreasing cavity pressures (p). Overall the strength is higher than that of the film stacking FRTC with an exception at the sample produced with 18 bar pressure and 60s holding time. The industrially produced Tepex outperforms both plate types for all the parameter combinations.

3.4. Flexural test

The flexural strength (σ_{fM}) is examined in accordance with the DIN EN ISO 14125 norm. The results are shown in Figure 5.

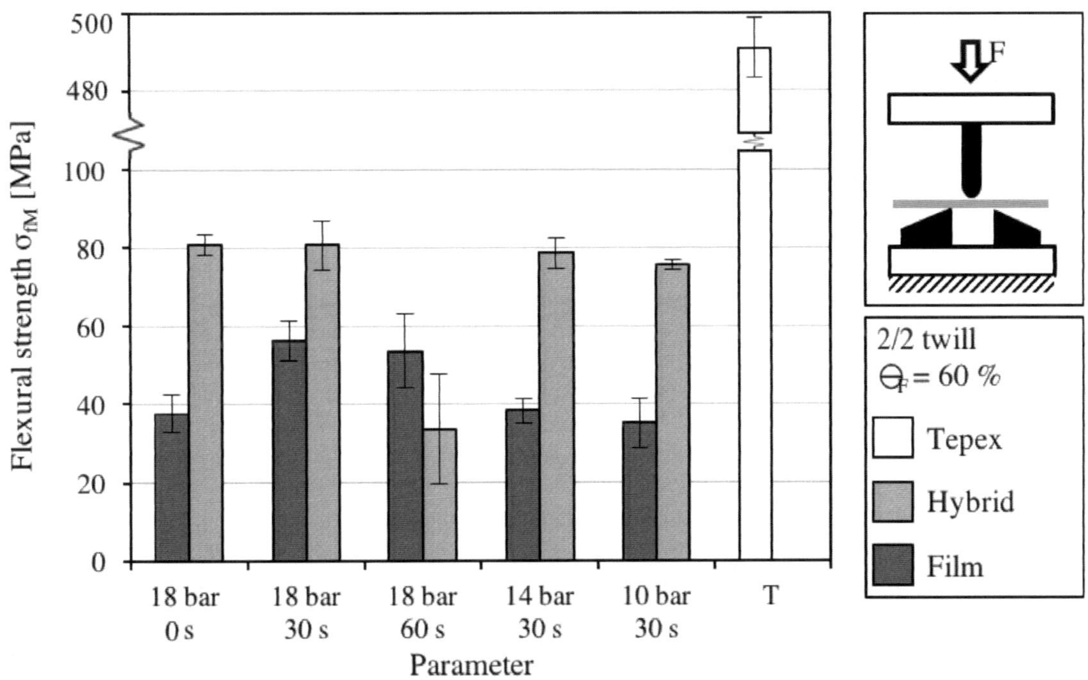

Figure 5. Results of the flexural tests

The flexural strength behaves in a similar way to the tensile strength. It increases for the film stacking FRTC when the holding time and the pressure increase. For the hybrid roving FRTC, the holding time and pressure result in similar flexural strengths. The Tepex plate clearly achieves significantly higher strengths.

4. Discussion

The commingling pre-test illustrate a weaker maximum force can be applied to the hybrid roving in comparison to the pure glass roving. This may be explained with the structure of the hybrid roving. The glass filaments are no longer perfectly straight, but wound and entangled. This leads to an uneven force transmission between the separate filaments. Due to the entanglement with the PA6 filaments, it is to be expected that this effect can be reduced after consolidation. Nevertheless, a trend can be seen. The better the distribution of the different filaments is, the lower the maximum force appears to be. This means a compromise between strength and distribution needs to be made.

A comparison of the consolidation of the different plate types, for all the parameter combinations, shows a clear trend. Even though the consolidation of the film stacking FRTC increases with an increased holding time and pressure, it cannot compete with the consolidation of the hybrid roving FRTC. Not even the industrially produced Tepex plate achieves a comparable distribution of the PA6 between the glass filaments. This may be due to a combination of the nano-modification and the hybrid roving structure produced by commingling the two components. The hybrid roving FRTC is shown to be well consolidated for all the parameter combinations, which suggests that they can be produced with a lower cavity pressure, a lower holding time and thus reduce the cycle time and production costs. This may be due to the nano-modification, which increases the thermal conductivity and decreases the heat capacity. However, in order to clearly distinguish between the impact of the nano-modification and the commingling method, further experiments need to be conducted. It may be helpful to have a study looking into a nano-modified film stacking FRTC, a hybrid roving FRTC without a nano-modification, as well as an investigation into the minimum cavity pressure needed for consolidation.

When evaluating the results of the tensile test it needs to be taken into account that the probe length set by the norm cannot be met. The length is 200 mm instead of 250 mm. The clamping surface size is maintained, while the measuring length is adjusted. However, as the probe dimensions of all the different probe types are the same, a comparison can be made anyway. The results of the tensile tests show that the hybrid roving FRTCs are stronger than the film stacking FRTCs produced with the same parameter combination. This may be due to the better distribution of the PA6 between the glass filaments, which enables a better force transmission to the reinforcing glass roving filaments. This theory is supported by the CT-pictures. The scans show that the glass roving weaves of the Tepex and film stacking plates maintain their tight weave structure, while those of the hybrid roving plates do not. This is likely due to the commingling process during which the components are mixed. The resulting hybrid roving weave is made out of a roving containing both components. In the film stacking process the glass roving weave is added separately and therefore maintains its structure after consolidation. The only exception to these findings can be observed at the samples produced with 18 bar pressure and 60 s of holding time, where the film stacking FRTC show a higher tensile strength then the hybrid roving FRTC. The tensile strength of the film stacked FRTC rises due to a better impregnation of the fibres and surpasses the hybrid roving FRTC due to less fibre damage during production. However, the direct comparison of the two processes shows that the better consolidation of the hybrid roving FRTC outweighs the benefits of a tighter and more aligned structure of the reinforcing glass roving weave achievable using the film stacking method. Interestingly, the strength of the hybrid roving FRTC increases with decreasing cavity pressures. A possible explanation is that higher pressures result in a reduced porosity of the fabric, leading to an uneven distribution of PA6 and contributing to the loss in strength. Therefore, the effects of the cavity pressure on the strength should be further investigated. As the costs of production plants rise significantly with their pressure capacity, using lower pressures also has the potential of significantly lowering investment costs.

The results of the flexural tests show a similar trend to the tensile tests. The film stacking plates are more resilient with increasing pressure and holding time. The hybrid roving plates show a similar strength for different holding times and cavity pressures. However, the results of the flexural tests at 18 bar and 60 s, (see Figure 5) seem to show a decrease in strength. This might be due to increasing cavity pressure resulting in damage to the glass fibres or a long holding time causing the PA6 to decompose. However, as these results are not observed for the film stacking process or in the tensile tests, this is not consistent. The result for this parameter combination may be an anomaly and further studies should elucidate this and investigate possible causes.

In general, when evaluating the flexural test results, it has to be taken into account that the probes do not fracture according to the norm. All the probes, apart from the Tepex one, develope interlaminar fractures. This is reflected in the low flexural strength values displayed by the two self-made probe types. Due to this, a comparison between the two remains possible. This shows that the hybrid roving is stronger for the tested parameter combinations. The interlaminar fractures may result from an inadequate sizing on the glass fibres, as it probably differs between the tested glass fibres and the glass fibres used for the Tepex plate.

The Tepex plate proves to be the strongest in the tensile and flexural tests. As the production parameters are unknown, it is difficult to establish the source of these mechanical properties. However, it suggests that there is scope to adjust parameters further than this study investigates, to achieve a better result. The comparison between the two self-made plate types remains an important finding, as it shows that under the same conditions, nano-particles, combined with commingling, are superior to the traditional film stacking method.

5. Conclusion

From this study it can be surmised that the nano-modified hybrid roving FRTC enables shorter cycle times, while achieving good mechanical properties. The consolidation and impregnation of the glass filaments are clearly better, even in direct comparison with the industrial standard. The tensile and flexural tests highlight the possibility to achieve shorter cycle times, while maintaining high levels of strength. These properties allow a balance between keeping production costs down and maintaining the integral properties. However, further investigation needs to be conducted. In particular, the differentiation between the two overlapping effects of the nano-modification and the hybrid roving need to be examined. Nevertheless, both the process and the material demonstrate promising potential for the future of FRTCs.

Acknowledgments

Special thanks go to the Federal Ministry of Education and Research (BMBF) which funded the research project "NanoOrgano" and the current project "VIP Organo".

References

[1] Witten E, Kraus T and Kühnel M 2014 *Marktentwicklung, Trends, Ausblicke und Herausforderungen* Composite-Marktbericht 2014, AVK, CCeV
[2] Renault T 2014 Latest Developments in Thermoplastic Composites for Automotive Applications *Proc. 2nd Int. Conf. & Exhibition on Thermoplastic Composites* (Bremen, Germany) ed. H Borgmann
[3] Kravaev P, Seide G and Gries T 2012 Drapierfähige Halbzeuge aus nanomodifizierten Hybridgarnen für die Herstellung von faserverstärkten thermoplastischen Bauteilen *Abschlussbericht* (Aachen and Hannover, Germany)
[4] Brüll R, Hütten A, Seide G and Gries T 2015 Using nanoscale fillers to improve the thermal properties of fibre reinforced thermoplastic composites regarding processing times *CAMX Conference Proceedings* (Dallas, TX, USA) p. 1609-1623.
[5] Bondlaminates Tepex® Dynalite. *www.Bondlaminates.com.* Accessed : 13.03.2016
[6] Bondlaminates Automotive. *www.Bondlaminates.com.* Accessed : 13.03.2016

Investigation of out of plane compressive strength of 3D printed sandwich composites

V Dikshit[1,*], Y L Yap[1], G D Goh[1], H Yang[2], J C Lim[2], X Qi[2], W Y Yeong[1] and J Wei[2]

[1] Singapore Centre for 3D Printing, School of Mechanical and Aerospace Engineering, Nanyang Technological University, 50 Nanyang Avenue, Singapore 639798
[2] Singapore Institute of Manufacturing Technology, 71 Nanyang Drive, Singapore 638075

E-mail: vishdixit@ntu.edu.sg

Abstract. In this study, the 3D printing technique was utilized to manufacture the sandwich composites. Composite filament fabrication based 3D printer was used to print the face-sheet, and inkjet 3D printer was used to print the sandwich core structure. This work aims to study the compressive failure of the sandwich structure manufactured by using these two manufacturing techniques. Two different types of core structures were investigated with the same type of face-sheet configuration. The core structures were printed using photopolymer, while the face-sheet was made using nylon/glass. The out-of-plane compressive strength of the 3D printed sandwich composite structure has been examined in accordance with ASTM standards C365/C365-M and presented in this paper.

1. Introduction

Composites sandwich structures are usually comprised of two thin stiff facesheets and a lightweight core structure [1]. Extrusion, expansion and corrugation process are main conventional techniques of manufacturing honeycomb core [2]. However, few new core manufacturing techniques are also available but all are limited to specific simple core design, so complex core manufacturing is not possible by these conventional techniques [3-5].

3D printing is gaining attention of researcher to explore new sandwich structure owing to its design flexibility [6-9]. Composite printing is the emerging technology to revolutionize the composite manufacturing and researchers have become interested to analyse the material behavior of such composite [9]. Due to the high flexural stiffness with overall low core density of composite sandwich structures, it is used by many industries like fashion, automotive, aerospace, marine and etc [6-8].

Many researchers have tested different types of new core design to improve the performance of the core structure, like kagome, kirigami, corrugated or spider web [10-15]. The two new core structures, vertical pillared sine wave corrugated (VPSC) structures and vertical pillared trapezoidal corrugated (VPTC) structures are designed. These types of corrugated structures are known to have high compressive strength due to its specific corrugated design [16, 17]. The 3D inkjet printing technique was selected to fabricate these corrugated core structures because of the high capability of inkjet printing to fabricate accurate features [18]. In the present study the research is focused on compressive behavior of 3D composite sandwich structure having 1.5 mm Kevlar-nylon facesheet and ABS-rubber core.

IOP Conf. Series: Materials Science and Engineering **139** (2016) 012017 doi:10.1088/1757-899X/139/1/012017

2. 3D printed sandwich composites and out of the plane compressive test

In this study, the focus area is to use two different 3D printing technique to print composite sandwich structure. The corrugated structures were produced by the polymer inkjet printing and the composite face sheet was produced by the composite filament fabrication technique.

2.1. Sandwich design

Before the fabrication of the sandwich, core and the facesheets, the design were modelled using computer-aided (CAD) software, SOLIDWORKS 2015, after this the models were exported into STL files to be 3D printed. figure 1a shows the CAD model of the VPSC structure, whereas figure 1b shows the VPTC structure. Table 1 shows the overall dimensions of the 3D printed corrugated sandwich structures.

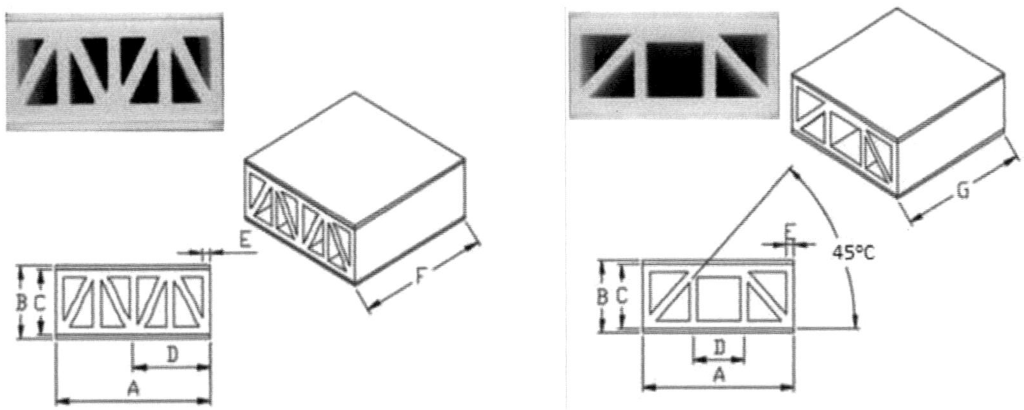

(a) VPSC structures (b) VPTC structures

Figure 1. Design of VPSC and VPTC structures

Table 1. Dimensions of cellular core segments

Designs	Dimensions (mm)						
	A	B	C	D	E	F	G
Vertical pillar corrugated sine wave (VPSC) structure	40	18	16	20	1.45	40	-
Vertical pillar corrugated trapezoidal (VPTC) structure	40	18	16	13	1.5	-	40

2.2. Inkjet core printing

Corrugated structures were fabricated using 3D SystemsProJet 5500X 3D printer, an inkjet 3D printer which is capable of printing multiple materials simulateneously. The UV curable rubbery and rigid polymers could be selectively deposited on the built tray based on the material selected during the printing process setup. RWT-FBK 100 (80% rigid-ABS, 20% flexible-Rubber) semi-rigid material was selected for core material printing. Figure 2a displays the ProJet 5500X 3D printer, whereas figure 2b shows the 3D printing of corrugated structures by ProJet 5500X 3D printer.

(a) Projet 5500X 3D printer (b) Printing of corrugated structures

Figure 2. 3D inkjet printing

2.3. Facesheet printing by composite filament fabrication

The facesheet was fabricated using Mark Forged Mark One 3D printer. The Mark One is a fused deposition modeling technique based 3D printer which has dual nozzles to extrude both the fiber composite filament and pure Nylon filaments to 3D print fibre reinforced composite structures. Kevlar-nylon filament were selected for facesheet fabrication. The printed facesheet thickness is 1.5 mm with 0/90 fiber orientation. Figure 3a shows the Mark One 3D printer, whereas figure 3b shows the 3D printed facesheet.

(a) Mark One 3D printer (b) Composite facesheet

Figure 3. 3D composite printing

Upon completion of both types of printing, the parts were taken out of the building chamber for post-processing. Post-processing of ProJet printed parts involved melting of wax support material that surrounds the final part. On the other hand, there is no separate support material for Mark One 3D printer. A thin layer of nylon was printed as support structure, and it can be peeled off manually from the printed part. After that, the 3D printed core and the facesheet wereassembled to form sandwich beams using M-bond 200 adhesive (AVPG brand).

2.4. Out of plane compressive test

The out of plane compression test was performed by using SHIMADZU universal testing machine with a 10 kN load cell and loading rate is selected 0.50 mm/min, in accordance with ASTM standards C365/C365M. This test was used to find out the ultimate compressive strength, compressive stress at 2 % deflection and compressive chord modulus of the 3D printed composite sandwich structures (refer

Equation 1 to 3). Figure 4a shows the test setup of out of plane compression test of corrugated structures. Figures 4b and 4c show the failed specimen of VPTC structure and VPSC structure, respectively.

$$\sigma^{u} = \frac{P_m}{A} \tag{1}$$

$$\sigma^{fc}_{.02} = \frac{P_{.02}}{A} \tag{2}$$

$$E^{fc} = \frac{(P_{.003} - P_{.001})}{(\delta_{.003} - \delta_{.001})} \frac{t}{A} \tag{3}$$

σ^{u} is the ultimate compressive strength (MPa),

$\sigma^{fc}_{.02}$ is the compressive stress at 2 % deflection (MPa),

E^{fc} is compressive chord modulus (MPa),

P_m is the maximum load before fracture (N),

A is the cross section area of the sandwich (mm^2),

t is the thickness of the specimen,

$\delta_{.02}$ deflection value at $\delta/t\sim$ 0.02,

$\delta_{.001}$ deflection value at $\delta/t\sim$ 0.001,

$\delta_{.003}$ deflection value at $\delta/t\sim$ 0.003,

$P_{.02}$ is applied force corresponding $\delta_{.02}$,

$P_{.001}$ is applied force corresponding $\delta_{.001}$, and

$P_{.003}$ is applied force corresponding $\delta_{.003}$.

(a) Specimen subjected to compression testing

(c) Failed VPTC structure

(c) Failed VPSC structure

Figure 4. Compression test of corrugated structures

3. Result and Discussion

The relative density of bothe type of corrugated structures was retained constant. Five specimens were tested for each type of corrugated structures. Figure 5a shows the average load vs. compressive displacement curve for both types of corrugated structures. Ultimate compressive strength, compressive stress at 2 % deflection and compressive modulus were also calculated from the Equation 1, 2 and 3, respectively. Figure 5b gives a summary of the calculated average compressive modulus, and average ultimate compressive strength for both types of corrugated structures. Table 2 gives a summary of the calculated average ultimate compressive strength, average compressive modulus and average compressive stress for both types of corrugated structures. The standard deviation of test results is also shown in figure 5 and table 2. The test results show that the VPSC structures can withstand more load than the VPTC structures. The ultimate strength of the VPSC structures was found to be 4.9 MPa which is 16.6% higher than the ultimate strength of VPTC structures.

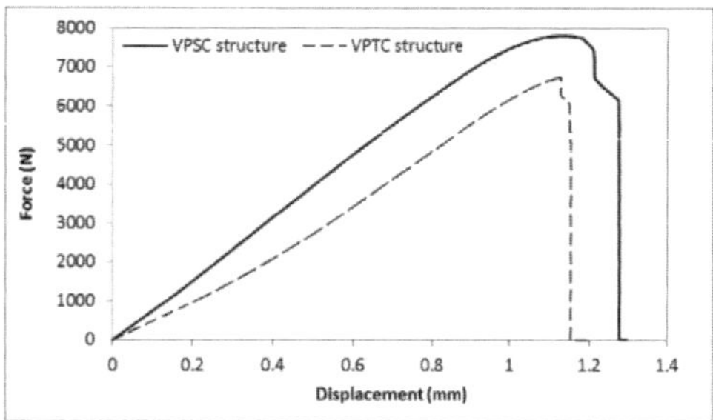

(a) Compressive load vs. displacement curve

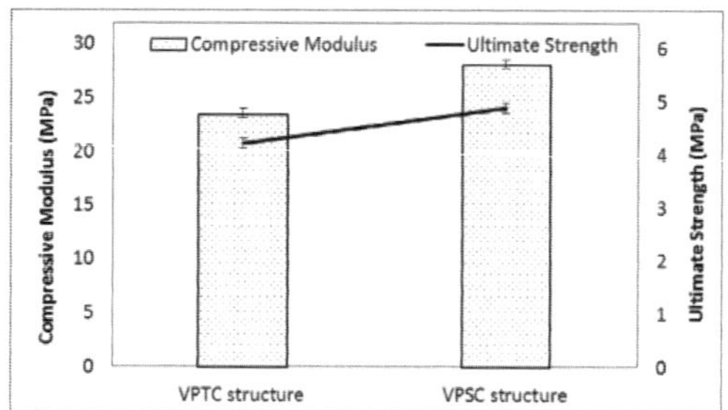

(b) Comparison of compressive modulus and compressive strength

Figure 5. Out-of-plane compression test results

Table 2. Values of ultimate strength, compressive modulus and compressive stress

Samples	Ultimate Strength σ^u (MPa)	Compressive Modulus E^{fc} (MPa)	Compressive stress $\sigma^{fc}_{.02}$ (MPa)
Vertical pillar corrugated sine wave (VPSC) structure	4.9±0.3	28.0±2.2	1.3±0.2
Vertical pillar corrugated trapezoidal (VPTC) structure	4.2±0.2	23.4±1.7	0.8±0.1

The failure behavior of the VPSC structures is slightly different from the VPTC structures during loading as well as in breaking. Both types of specimen failed due to buckling of the sine-wave/trapezoidal walls. In both types of specimens, the initial cracks were developed in the mid cross-section area in the thickness direction, however VPTC structures also showed failure cracks at the top and bottom facesheet. During testing, the vertical walls of the VPSC structures protruded outwards and the sine-wave curve undergoes the inward bending. However, in the trapezoidal structure, both the vertical walls as well as the inclined trapezoidal cell experienced outward protrusion that eventually led to lower load bearing capacity.

4. Summary

The out-of-plane compressive behaviour was studied and the ultimate strength, compressive modulus and compressive stress were also observed. The following conclusions for this research work are summarized:

- The observed results of the out-of-plane compressive tests of VPSC structures have revealed higher compressive strength and modulus as compared to VPTC structures.
- Buckling of the walls is the main failure mode.

Acknowledgement
This work was supported under the A*STAR TSRP – Industrial Additive Manufacturing Programme by the A*STAR Science & Engineering Research Council (SERC) [grant number 1325504105].

References

[1] Gibson L J and Ashby M F 1997 *Cellular solids : structure and properties* (New York : Cambridge University Press, 1997. 2nd ed.: Cambridge)
[2] Bitzer T 1997 *Honeycomb Techonology: Materials, Design, Manufacturing, Applications and Testing* (London: Chapman & Hall: Springer)
[3] Bonnard J C 2004 Thermoplastic honeycomb core: An innovative continuous process *JEC Compos. Mag.* 73-5
[4] Zuhri M Y M, Guan Z W and Cantwell W J 2014 The mechanical properties of natural fibre based honeycomb core materials *Compos Part B: Eng* **58** 1-9
[5] Rashed M G, Ashraf M, Mines R A W and Hazell P J 2016 Metallic microlattice materials: A current state of the art on manufacturing, mechanical properties and applications *Mater. Des.* **95** 518-33
[6] Yap Y L and Yeong W Y 2014 Additive manufacture of fashion and jewellery products: a mini review *Virtual Phys. Prototyping* **9** 195-201
[7] Lee J M, Zhang M and Yeong W Y 2016 Characterization and evaluation of 3D printed microfluidic chip for cell processing *Microfluidics and Nanofluidics* **20** 1-15
[8] Yeong W-Y, Chua C-K, Leong K-F, Chandrasekaran M and Lee M-W 2005 Development of scaffolds for tissue engineering using a 3D inkjet model maker *Virtual modelling and rapid manufacturing-advanced research in virtual and rapid prototyping* 115-8
[9] Wohlers T 2010 Additive manufacturing a new frontier for composites *Compos. Technol.* **16**
[10] Khoo Z X, Teoh J E M, Liu Y, Kai Chua C, Yang S, An J, Leong K F and Yeong W Y 2015 3D printing of smart materials: A review on recent progresses in 4D printing *Virtual Phys. Prototyping*
[11] Yap Y L and Yeong W Y 2015 Shape recovery effect of 3D printed polymeric honeycomb *Virtual Phys. Prototyping* **10** 91-9
[12] Lee B-K and Kang K-J 2009 Compressive strength of tube-woven Kagome truss cores *Scripta Materialia* **60** 391-4
[13] Zhang G, Ma L, Wang B and Wu L 2012 Mechanical behaviour of CFRP sandwich structures with tetrahedral lattice truss cores *Compos Part B: Eng* **43** 471-6
[14] Foo C C, Chai G B and Seah L K 2007 Mechanical properties of Nomex material and Nomex honeycomb structure *Compos. Struct.* **80** 588-94
[15] Zhang G, Wang B, Ma L, Xiong J and Wu L 2013 Response of sandwich structures with pyramidal truss cores under the compression and impact loading *Compos. Struct.* **100** 451-63
[16] Fleck N A and Deshpande V S 2004 The Resistance of Clamped Sandwich Beams to Shock Loading *Journal of Applied Mechanics* **71** 386-401
[17] Rejab M R M and Cantwell W J 2013 The mechanical behaviour of corrugated-core sandwich panels *Compos Part B: Eng* **47** 267-77

[18] Singh R 2011 Process capability study of polyjet printing for plastic components *Journal of mechanical science and technology* **25** 1011-5

An exponential chemorheological model for viscosity dependence on degree-of-cure of a polyfurfuryl alcohol resin during the post-gel curing stage

JC Domínguez[1], M Oliet[1], M V Alonso[1], F Rodriguez[1] and B Madsen[2]

[1]Department of Chemical Engineering, Complutense University of Madrid, Avda. Complutense s/n, 28040 Madrid, Spain.
[2]Department of Wind Energy, Section of Composites and Materials Mechanics, Technical University of Denmark, Risø Campus, Frederiksborgvej 399, DK-4000 Roskilde, Denmark.

E-mail: jucdomin@ucm.es

Abstract. In the present study, the chemorheological behavior of a bio-based polyfurfuryl alcohol (PFA) resin has been determined by rheological isothermal tests at different curing temperatures for the post-gel curing stage of the resin, using three different amounts of catalyst (2, 4 and 6 wt %). Instead of modeling the evolution of the complex viscosity using a widely used chemorheological model such as the Arrhenius model for each tested temperature, the change of the complex viscosity as a function of the degree-of-cure was predicted using a new exponential type model. In this model, the logarithm of the normalized degree-of-cure is used to predict the behavior of the logarithm of the normalized complex viscosity. The model shows good quality of fitting with the experimental data for 4 and 6 wt % amounts of catalyst. For the 2 wt % amount of catalyst, scattered data leads to a slightly lower quality of fitting. Altogether, it is demonstrated that the new exponential model is a good alternative to conventional chemorheological models due to its simplicity and suitability.

1. Introduction

For composite materials, to achieve both optimization of the manufacturing process and tailor-making of the final composite properties, fundamental knowledge of the properties of the constituent materials phases is required. Thus, the study of polymers and fibers employed as matrix and reinforcements of composites is of great importance. In the case of a thermosetting polymer resin, the analysis of its kinetics of curing, and its evolution of properties during crosslinking, including the gelling and vitrification transitions that may occur, are crucial.

The gel point of a thermosetting polymer resin is defined as the point at which the behavior of the resin is more similar to a solid than to a liquid; the point at which the insoluble, infinite-molar-mass network is formed (note that the matrix may still contain unreacted soluble intermediates) [1]. After this point, in the post-gel stage, the kinetics of the curing reactions are commonly changed compared with the pre-gel stage, e.g. the curing rate is typically reduced. The chemorheology of the curing process of a bio-based polyfurfuryl alcohol (PFA) resin during its pre- and post-gel stages has been investigated in previous studies [2,3] employing conventional and widely used chemorheological models, such as the Arrhenius model [1]. The basis of the present study is to use a new exponential model to predict the chemorheological behavior of a PFA resin during the post-gel stage. The degree-of-cure evolution is estimated from rheological variables, such as the complex modulus, which have been determined in a previous study [2]. The new exponential model is shown to be a simple and reliable model.

2. Theory

2.1. Chemorheology

The term used to define the viscoelastic behavior of chemically reacting systems is *chemorheology* [4]. A chemorheological study of a thermosetting polymer resin is typically addressing variations in viscosity due to chemical reactions and process conditions, together with characterization of the molecular network growth and transitions such as gelation and vitrification [5]. Monitoring the variation of rheological properties of a resin during curing allows modelling of the process. The models commonly employed in the literature use viscosity as the selected rheological property, either the steady viscosity (η), or the complex viscosity (η^*), where the latter one is used in the case of oscillatory tests. The steady viscosity is only useful for obtaining chemorheological parameters during the pre-gel stage, since when the resin approaches its gel point, the steady viscosity will be approaching infinity. On the other hand, the complex viscosity is a more suitable parameter to study the curing process of a resin since it can be used for both the pre-gel and the post-gel stages.

2.2. Chemorheological models

The complex viscosity of a thermosetting polymer resin is affected by a number of variables. It can be expressed as a function of temperature (T), pressure (P), shear rate ($\dot{\gamma}$), time (t) and filler properties (F) [5]:

$$\eta^* = f\left(T, P, \dot{\gamma}, t, F\right) \tag{1}$$

The relevant effects on the chemorheological behavior can be separated as: cure effects ($\eta_c^* = f(T,t)$), shear rate effects ($\eta_s^* = f(T, \dot{\gamma})$), and filler effects ($\eta_f^* = f(F,T,t)$). The effect of pressure is commonly very low, and it is typically not taken into account unless for certain manufacturing processes such as high pressure injection molding and resin transfer molding [5]. When studying curing processes of resins, the shear rate is usually fixed to simplify the analysis. Thus, in cases where fillers are not present in the resin, it can be considered that only cure effects are relevant, i.e. $\eta^* = \eta_c^*(T,t)$.

Among a number of models in the literature, the Castro-Macosko model is a frequently employed model to describe the variation of the steady dynamic viscosity of a polymer resin during the pre-gel stage as a function of the degree-of-cure [6]. The Castro-Macosko model is shown in Eq. (2):

$$\frac{\eta}{\eta_0} = \left(\frac{\alpha_{gel}}{\alpha_{gel} - \alpha}\right)^{A+B\cdot\alpha} \tag{2}$$

where η is the steady dynamic viscosity of the resin at a given degree-of-cure (α), η_0 is the initial viscosity (viscosity of the resin before curing), α_{gel} is the degree-of-cure at the gel point, and A and B are empirical model constants.

A simple and reliable model that could predict the change of viscosity during the post-gel stage – in the same way that the Castro-Macosko model does for the pre-gel stage – would be of great interest. The widely used Arrhenius model turns into a complicated model when viscosity is expressed as a function of the degree-of-cure, instead of as a function of time. Taking the Castro-Macosko model as inspiration, the model for the post-gel stage should propose a function that can predict the ratio between the complex viscosity at a certain time after gelation, and the complex viscosity at the gel point, i.e. η^*/η^*_{gel}. Due to the high viscosity values of the resin in the post-gel stage, the natural logarithm of the ratio ($ln(\eta^*/\eta^*_{gel})$ is more appropriate to use. In analogy to the Castro-Macosko model, the proposed relationship should relate to the maximum degree-of-cure ($\alpha = 1$) since the model should predict the change of the complex viscosity for the range $\alpha_{gel} \leq \alpha < 1$. Altogether, the proposed new functional relationship for the post-gel stage is described by Eq. (3):

$$\ln\left(\frac{\eta^*}{\eta^*_{gel}}\right) = f\left(\ln\left(\frac{1-\alpha_{gel}}{1-\alpha}\right)\right) \tag{3}$$

In the literature, a phenomenological sigmoidal model has been successfully applied to predict the chemorheological behavior of resins, such as polyurethane resins [7,8]. However, to the best of the author's knowledge, an exponential empirical model that can forecast the behavior for the post-gel stage has not been used. In the present study, based on the experimental behavior observed for a PFA resin at different curing temperatures, a Box-Lucas like exponential equation will be used for the functional relationship proposed in Eq. (3). Moreover, to properly model the post-gel curing stage of the PFA resin, the parameters of the model will be proposed in a way so that they have a physical meaning.

3. Experimental procedure

3.1. Materials

A bio-based polyfurfuryl alcohol (PFA) resin under development, Furolite 050915A, and an "S type + D" acid catalyst based on p-toluenesulfonic acid dissolved in water (45 % aqueous solution), were supplied by Transfurans Chemicals BVBA (Belgium). According to the supplier's datasheet, the specifications of the PFA resin are: water content 5.40 % and monofurfuryl alcohol content below 1 %.

3.2. Instrumentation

Rheological runs were performed using an AR Rheometer 2000 (TA Instruments©) with a 25 mm diameter upper plate and a Peltier lower plate. Normal force was zeroed and fixed during the tests to prevent contact loss between the sample and plates. Five isothermal curing runs of the PFA resin in the temperature ranges 65 – 85 °C, 55 – 75 °C and 50 – 70 °C (using a 5 °C increment) for 2, 4 and 6 wt % amounts of catalyst, respectively, were carried out for 60 min. Six frequencies equally separated on a logarithm scale were employed in the range from 1 to 10 Hz, and the applied strain was fixed at 0.1 % for all the amounts of catalyst tested. A 20 °C min^{-1} heating ramp was programmed in order to reach the various curing temperatures from a starting temperature of 20 °C. All samples were first stabilized at the starting temperature for 2 min before the heating ramp was applied. All rheological measurements were performed within the linear viscosity region of the PFA resin (strain sweep tests were performed to ensure the applied strain was within the linear viscosity region). Three replicates were used for each experimental setting.

4. Results and discussion

By plotting $ln(\eta^*/\eta^*_{gel})$ vs. $ln((1-\alpha_{gel})/(1-\alpha))$ in Figure 1, it is clear that the change of the logarithm of the normalized complex viscosity can be described by an exponential function. Moreover, by plotting the results in this way, no effect of the changed applied isothermal curing temperatures can be seen. Some scattering is found between the applied curing temperatures, especially for the 2 wt % amount of catalyst, however, the scattering is decreased when the amount of catalyst is increased. This can be explained, as was also done previously for the Arrhenius model [2], by the existence of small gradients of catalyst in the resin (heterogeneity), and the larger impact of these gradients on the behavior of the complex viscosity during the post-gel stage.

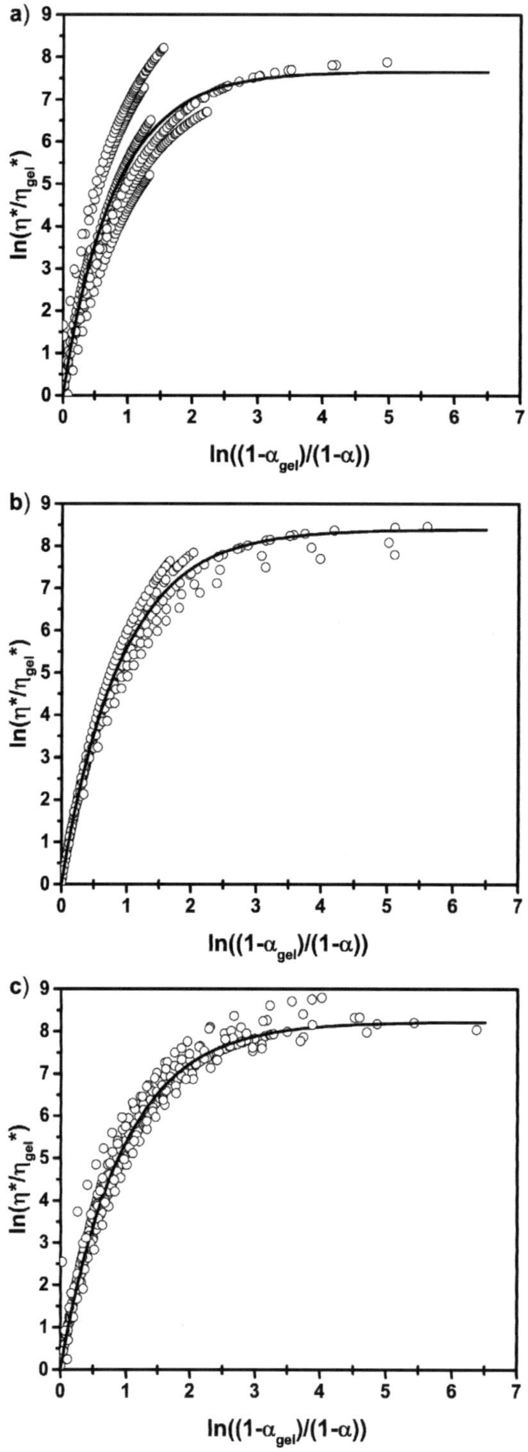

Figure 1. Natural logarithm of the experimental normalized complex viscosity (symbols) and the predicted normalized complex viscosity (lines) vs. natural logarithm of the normalized degree-of cure

for a PFA resin with a) 2 wt % of catalyst; b) 4 wt % of catalyst; c) 6 wt % of catalyst. For each amount of catalyst, results are shown for a range of applied isothermal curing temperatures.

The Box-Lucas type equation ($y = b\cdot[1-exp(-a\cdot x)]$) was chosen to describe the exponential behavior of the PFA resin. Based on this equation, and by giving physical meaningful values to its constants (a and b), an exponential empirical model is proposed to predict the chemorheological behavior of the PFA resin in the post-gel stage. The model is described by Eq. (4):

$$\ln\left(\frac{\eta^*}{\eta^*_{gel}}\right) = \ln\left(\frac{\eta^*_{\alpha=1}}{\eta^*_{gel}}\right)\cdot\left[1-\exp\left(-a\cdot\ln\left(\frac{1-\alpha_{gel}}{1-\alpha}\right)\right)\right] \tag{4}$$

where $\eta^*_{\alpha=1}$ is the complex viscosity of the fully cured resin, and a is a model parameter that describes the complex viscosity rate with respect to the degree-of-cure. The model parameter $ln(\eta^*_{\alpha=1}/\eta^*_{gel})$ represents the logarithm of the maximum reachable ratio between the complex viscosity at any time during the post-gel stage, and the complex viscosity at the gel point. This parameter can either be measured directly, or it can be determined by fitting the model to experimental data. The equation shown in Eq. (4) can be simplified to Eq. (5):

$$\ln\left(\frac{\eta^*}{\eta^*_{gel}}\right) = \ln\left(\frac{\eta^*_{\alpha=1}}{\eta^*_{gel}}\right)\cdot\left[1-\left(\frac{\alpha-1}{\alpha_{gel}-1}\right)^{-a}\right] \tag{5}$$

The proposed new model is a two-parameter model. The applicability of the model is valid for the range of degree-of-cure from the gel point to the fully cured resin.

In Figure 1, the resulting model predictions for the post-gel curing stage of the PFA resin are shown. The established model parameters are exhibited in Table 1. As can be seen from the mean square error values, the quality of the fittings is good for the 4 and 6 wt % amounts of catalyst. For the 2 wt % amount of catalyst, the fitting is however less good due to the more scattered data, as already indicated above, and explained by the gradients of catalyst in the resin. This problem was not observed for the larger amounts of catalyst since the gradients are getting smaller as the amount of catalyst is increased.

Table 1. Established model parameters of the exponential chemorheological model for the post-gel curing stage of a PFA resin.

Amount of catalyst (wt %)	$\ln\left(\frac{\eta^*_{\alpha=1}}{\eta^*_{gel}}\right)$	***a***	**Mean square error**
2	7.7	1.2	0.6
4	8.4	1.1	0.1
6	8.2	1.1	0.1

Next, the suitability of the model is evaluated:

- The model consists of a simple equation that relates the complex viscosity (a ratio) with the degree-of-cure for the post-gel curing stage; therefore, simplicity is a clear benefit of the model. For Arrhenius models, a relatively complicated equation is required when the model is used to predict the change of η^* with respect to the degree-of-cure; in some cases, a six-parameter Arrhenius model is required to obtain reliable predictions.

- For a known complex viscosity, the model can be used to determine the degree-of-cure of a resin in its post-gel stage; in this way, rheology is a fast and easy alternative to other well-known possibilities such as DSC, TMA, and FTIR.
- The results show similar ratios for the parameter $ln(\eta^*_{\alpha=1}/\eta^*_{gel})$ independently of the curing temperature under isothermal tests – notice the ratio is between both complex viscosities at the same temperature –. The overall values of the ratio in Table 1 are similar to the ones that can be determined for each of the applied different temperatures, e.g. for the 6 wt % amount of catalyst, the overall value is 8.2, and it is 8.0 at 65 °C, and 8.3 at 70 °C. Therefore, only a few isothermal tests could be used to determine the parameters of the model – in principle, a single temperature could be enough; however, more than one temperature is recommended –.
- The estimated $ln(\eta^*_{\alpha=1}/\eta^*_{gel})$, obtained by fitting the model to the experimental data, could also be a good indicator of the effect of the amount of catalyst on the cross-linking density of the fully cured resin Therefore, this parameter could be employed to compare the importance of the amount of catalyst added to the system beside the curing rate. In this work, similar values were found for this parameter, 8.40 and 8.23 for 4 and 6 wt % amounts of catalyst, respectively, and this seems to indicate that the same cross-linking density was reached by both amounts of catalyst – 2 wt % amount of catalyst was not considered due to the high scattering found, as mentioned before –; however, this result should be checked in further works. Moreover, this parameter can be easily measured at different curing temperatures and compared with the obtained by the model either to check the validity of the model or to avoid a compensation effect between both parameters of the model.
- In principle, the a parameter of the model can be used as an indicator of how the curing rate is changed by the amount of catalyst. A higher curing rate is expected when the catalyst amount is increased, and the same behavior is expected for the a parameter, i.e. the higher amount of catalyst, the higher value of a. In the present study, the a parameter is determined in the range from 1.1 to 1.2, considering only 4 and 6 wt % amounts of catalyst; however, with no clear effect of the amount of catalyst. Further studies are needed to reveal the real utility of this parameter in the model, which probably will depend on the resin system under study.
- The complex viscosity of the fully cured resin can be predicted at any temperature if the complex viscosity is known at the gel point. Likewise, the complex viscosity of the resin at its gel point can be predicted at any temperature if the complex viscosity is known for the fully cured resin. Further improvements of the model should be focused at finding a simple equation that can predict the change of the complex viscosity at the gel point as a function of temperature, and deploy this equation into the model.

5. Conclusions

A new empirical chemorheological model was proposed based on the experimental behavior found for a bio-based polyfurfuryl alcohol resin cured using three amounts of catalyst under isothermal conditions. Then new chemorheological model was an exponential type equation. It is indicated that the model employed is a suitable chemorheological model for the post-gel curing stage due to the good agreement found with the experimental results when a 4 or 6 wt % amount of catalyst was employed. However, the modeling of the curing process when a 2 wt % catalyst was used could not be conducted satisfactorily; the main reason for the unsuccessful application of the model and the scattering found for the data could be the heterogeneity of the samples due to a insufficient amount of catalyst was employed to keep a constant concentration of it in the whole sample. Further work on the validation of the model under different conditions and for other resins and the development of the model to be employed under non-isothermal conditions is still required.

Acknowledgments

The authors are grateful for the support from "Ministerio de Educación" (Spanish Government) for financial support (Programa Nacional de Movilidad de Recursos Humanos del Plan Nacional de I-D+i 2008-2011). The research has been funded by the European Community's Seventh Framework Programme (FP7/2007-2013) under grant agreement nº 210037 (WOODY).

References

[1] Halley P J and George G A 2009 *Chemorheology of Polymers;* Cambridge University Press
[2] Domínguez J C and Madsen B 2013 *34th Risoe Symposium on Materials Science*, Roskilde, Denmark
[3] Domínguez J C and Madsen B 2014 *Ind. Crop. Prod.* **52** 321-328
[4] Tajima Y A and Crozier D 1983 *Polym. Eng. Sci.* **23** 186-190
[5] Halley P J and Mackay M E 1996 *Polym. Eng. Sci.* **36** 593-609
[6] Castro J M and Macosko C W 1980 *Soc. Plast. Eng.* **26** 434-438
[7] Teyssandier F and Love B J 2010 *Polym. Eng. Sci.* **50** 499-503
[8] Teyssandier F and Love B J 2011 *J. Appl. Polym. Sci.* **120** 1367-1371

Surface, interphase and tensile properties of unsized, sized and heat treated basalt fibres

T Förster, G S Sommer, E Mäder and C Scheffler

Leibniz-Institut für Polymerforschung Dresden e.V. (IPF), Hohe Straße 6, 01069 Dresden, Germany

E-Mail: foerster-theresa@ipfdd.de

Abstract. Recycling of fibre reinforced polymers is in the focus of several investigations. Chemical and thermal treatments of composites are the common ways to separate the reinforcing fibres from the polymer matrices. However, most sizings on glass and basalt fibre are not designed to resist high temperatures. Hence, a heat treatment might also lead to a sizing removal, a decrease of mechanical performance and deterioration in fibre-matrix adhesion. Different basalt fibres were investigated using surface analysis methods as well as single fibre tensile tests and single fibre pull-out tests in order to reveal the possible causes of these issues. Heat treatment in air reduced the fibre tensile strength in the same level like heat treatment in nitrogen atmosphere, but it influenced the wetting capability. Re-sizing by a coupling agent slightly increased the adhesion strength and reflected a decreased post-debonding friction.

1. Introduction

Recent investigations on the strength loss of heat treated basalt fibres were reported by [1] and [2]. Jenkins et al. compared thermal induced strength loss of sized E-glass and basalt fibres. Heat treatment beyond 300 °C led to significant drop of fibre strengths for both fibre types.

After treatment at 600 °C (25 min) the strength loss was about 80% for the basalt fibre sample. Jenkins et al. [1] pointed out that no correlation with the sizing degradation was found and assumed that the progressive oxidation of the fibre could affect the strength level of heat treated basalt fibres. Sabet et al. [2] observed the influence of the treatment temperature and duration on the tensile strength of basalt bundles. The authors determined a strength loss of about 20% after 5 to 15 min exposure at 300 °C. However, a 20 min exposure at 400 °C or 500 °C dropped the strength level drastically. The determined strength loss was about 65% and 89%, respectively. Sabet et al. [2] concluded that the strength loss was attributed to an amorphous-to-crystalline transformation, which is maximized at 450 °C. However, they pointed out that additional experimental studies are mandatory for the study of early crystallisation phenomena. In addition, based on their literature work, they refer to explanations for the basalt fibre degradation, e.g. the distortion of amorphous and local crystalline areas, the sizing burn-off or the corrosion/oxidation of iron.

Our previous studies indicate that the high temperature performance of basalt fibres preferably depends on their chemical composition. We investigated the single fibre tensile strength of unsized basalt fibres and could show that one of the basalt fibre compositions, in particular designated BAS11, remained the strength level whereas the other investigated composition, designated BAS12, showed a strength loss of more than 70% after 5 h heat treatment at a temperature of 600 °C. XPS results indicated that both fibre surfaces were slightly contaminated by carbon. However, considering

Thomason's rule of thumb for evaluating sizing coverage [3], the C/Si ratio of both fibre samples indicated unsized fibres. Therefore, a potential sizing burn-off is negligible. Interestingly, the XPS results reveal also that heat treated basalt fibres show a reduced presence of silica, whereas the calcium content was increased on the fibre surface. The iron content seemed to be not affected [4].

A formation of nano-layers caused by the heat treatment of basaltic glasses was already reported by Smedskjaer et al. [5]. The authors pointed out that the induced diffusion of divalent cations can influence properties like temperature stability as well as chemical durability. However, the impact on fibre-matrix adhesion due to thermal induced nano-layers has been less investigated.

In this study, first results of de-sizing and re-sizing experiments on basalt fibres were reported. Sized and unsized basalt fibres were heat treated in air and in nitrogen atmosphere. The initial tensile strength and the tensile strength after 0.5 h exposure at temperatures in the range of 250 to 600 °C were investigated by single fibre tensile tests. In addition, X-ray photoelectron spectroscopy (XPS), Atomic Force Microscopes (AFM) investigations, contact angle measurements as well as single fibre pull-out tests were carried out. Both initial sized and unsized fibres were investigated comparatively in order to evaluate the impact of the sizing-burn off on the mechanical performance and the surface properties, respectively

2. Experimental

Basalt fibre samples BAS11U and BAS11S were manufactured using an industrial scale pilot plant. The nominal diameter of the basalt fibres is 15 μm and the nominal yarn count is 100 tex. The unsized state is achieved by sizing only with water during the drawing process. BAS11U indicates the unsized sample, whereas the sized specimen is designated BAS11S. The sizing is compatible to EP resin and the amount is < 0.5%.

Unsized as well as sized fibres were heat treated. The heat treatment was managed by fibre storage in an oven (L5/11 Nabatherm) for 0.5 h or 1.0 h at temperatures up to 600 °C in air as well as nitrogen atmosphere. In addition, initially sized fibres were re-sized after the 0.5 h heat treatment at a temperature of 600 °C. Re-sized means the single fibres were dipped in an aqueous silane-surfactant solution. 3- glycidyloxypropyl-trimethoxysilane (GLYMO supplied by Evonik) was used. In addition, a nonionic surfactant was added in order to improve the wetting behaviour. Thus, surface tension of 27 mN/m was determined by Wilhelmy plate technique.

Single fibre tensile tests were conducted under air-conditioning (temperature 23°C, rel. humidity 50%) by using a Favigraph semi-automatic testing device (Textechno, Mönchengladbach, Germany) equipped with a 1 N load cell. The cross head velocity was 25 mm/min and the gauge length was 50 mm. The fineness of each selected fibre was determined by using the vibroscope method in accordance with ASTM D 1577. 50 single fibres were tested to calculate the mean values and standard deviations. A value of 2.8 g/cm^3 is used as density for the evaluation of the tensile tests.

XPS was performed on Axis Ultra photoelectron spectrometer (Kratos Analytical, Manchester, UK). XPS spectra were acquired with a pass energy of 160 eV using a Mono-Al KαX-ray source (hv = 1486.6 eV).

Advancing contact angles were measured with a Krüss K14 tensiometer (Krüss GmbH, Germany) using the Wilhelmy technique. Different liquids (water, ethylene glycol, hexadecane) were used in order to determine the surface energy (γ) according Owens-Wendt-Rabel-Kaelble (OWRK) approach.

Quasi-static single fibre pull-out (SFPO) tests were carried out. Therefore, single fibre model composites made of epoxy resin RIMR135/RIMH137 (Momentive) and treated basalt fibres were prepared by self-made sample preparation equipment and 15-20 force-displacement curves for each sample obtained. A characteristic force-displacement curve is shown in figure 1.

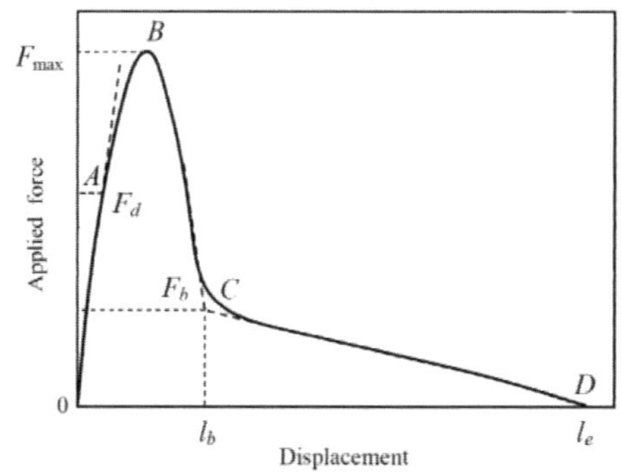

Figure 1. Typical force-displacement curve [6].

In order to get an interfacial shear stress value, the simple approach was applied using the maximum pull-out force, $F_{max.}$, for the calculation of the apparent interfacial shear strength (apparent IFSS, τ_{app}): $\tau_{app} = F_{max} / (\pi\, d_f\, l_e)$, where d_f is the fibre diameter and l_e is the embedded fibre length [7].

3. Results

3.1. Strength-strain behaviour of BAS11 fibre after 5 h heat treatment at 500 °C in air

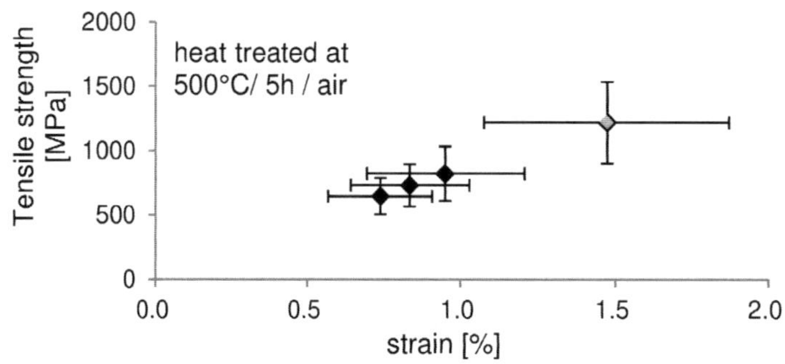

Figure 2. Tensile strength-strain behaviour of heat treated basalt fibre samples; (gray diamond indicates unsized BAS11 sample).

Figure 2 shows the strength-strain behaviour of different basalt fibre samples after a heat treatment at 500 °C for 5 h. The strength level of the unsized BAS11 fibre is advantageous compared to other investigated basalt fibre samples. As shown in [4] a moderate strength level remains after an enhanced heat treatment at a temperature of 500 °C or 600 °C. Thus, we continued experiments with this fibre type varying treatment time, temperatures and atmosphere. Figure 3 and 4 display the mechanical performance of sized BAS11 fibres. However, different samples were used in order to review the reproducibility.

3.2. Sizing burn-off

Figure 3a shows the single fibre tensile strength of sized fibre, BAS11S, after heat treatment at temperatures between 250 and 550 °C. A continuous strength loss with increasing temperatures is noticeable, especially if the fibres are treated at temperatures higher than 300 °C.

TGA/DTG-curve reveals mass loss at temperatures above 200 °C, preferable in the range of 200 to 370 °C in air as well as in nitrogen atmosphere. Thus, the sizing mostly seems to be removed up to a temperature of 450 °C. Considering the strength levels in Figure 3a, the strongest decrease in strength is obtained between 350 and 450 °C heat treatment.

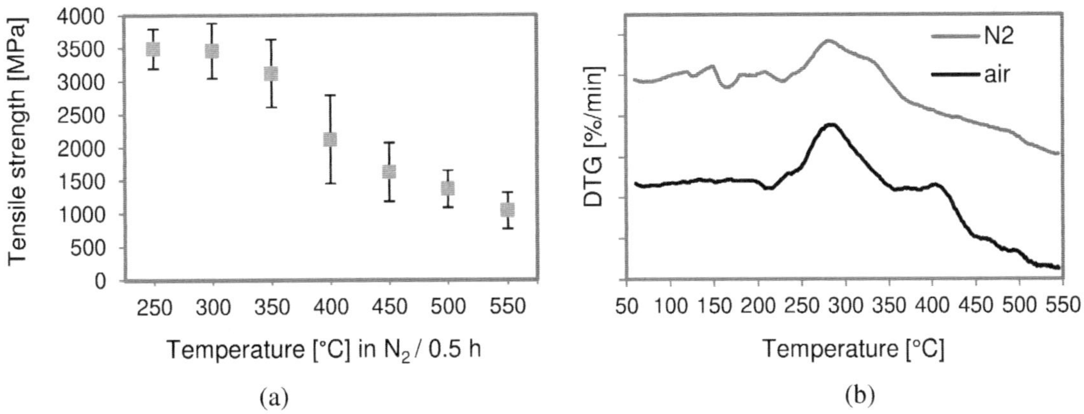

Figure 3. (a) Single fibre tensile strength of sized basalt fibres after 0.5 heat treatment in nitrogen atmosphere (b) TGA/DTG curves of sized basalt fibres.

3.3. Impact of temperature and atmosphere on the single fibre tensile strength

In spite of the advantageous chemical composition, the strength of BAS11S is decreased more than 55 % after treatment at elevated temperatures. Figure 4 compares the mechanical performance of BAS11U and BAS11S.

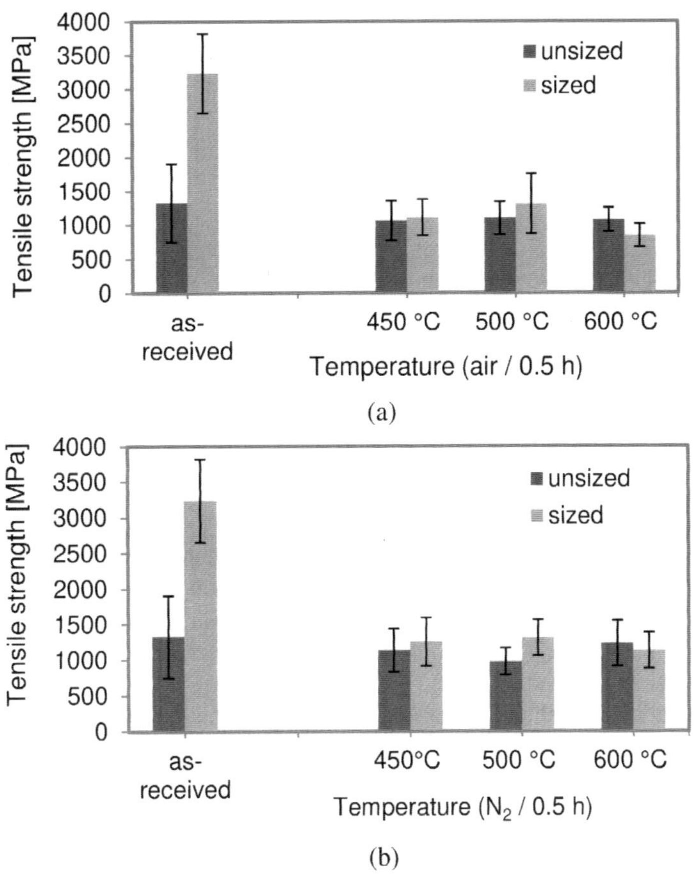

Figure 4. Single fibre tensile strength of unsized and sized basalt fibres after heat treatment in air (a) and in nitrogen atmosphere (b).

The unsized basalt fibre, BAS11U, shows a significantly reduced strength level compared to the sized as-received basalt fibre BAS11S. However, the strength of BAS11U remains independent of the treatment temperature. Heat treatments above 450 °C reduce the strength of BAS11S to the strength level of BAS11U. A significant impact by atmosphere was neither noticed for unsized nor for sized fibres. The results in figure 3a and 4b indicate that both samples taken from BAS11S show comparable strength values after 500-600 °C heat treatments. The remaining tensile strength is about 1000 -1400 MPa.

AFM topography images of the unsized basalt fibre BAS11U indicate a nanostructured surface (Figure 5d). The arithmetic surface roughness (R_a) is 0.37 nm. The surface of sized fibres is less rough due to the nanostructure is covered homogeneously by the sizing (Figure 5a). After 600 °C treatment the nanostructure is visible indicating a significant sizing removal.

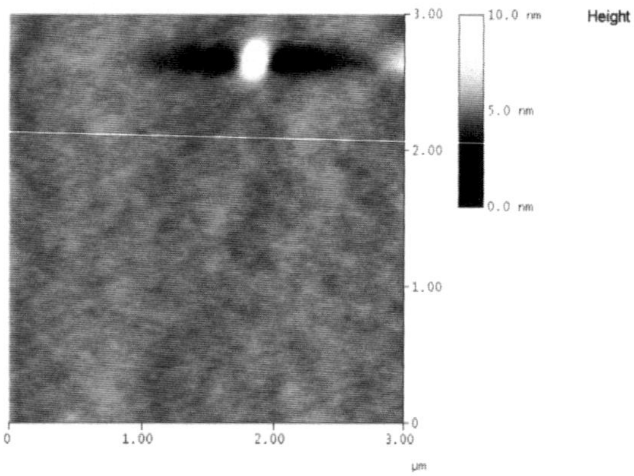

(a) BAS11S – as- received
R_a= 0.28nm

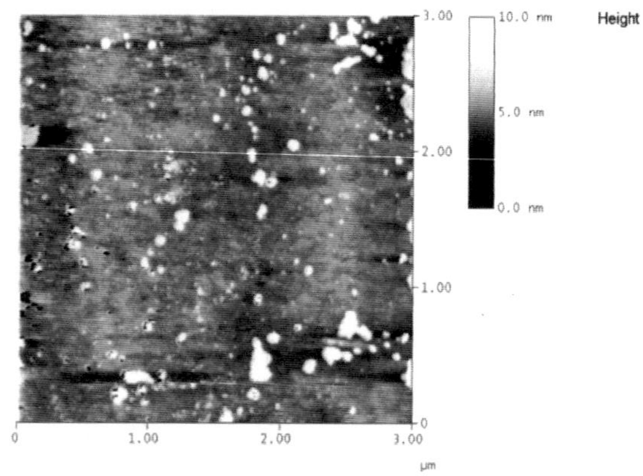

(b) BAS11S – heat treated 450 °C/0.5 h / ai
R_a = 0.56nm

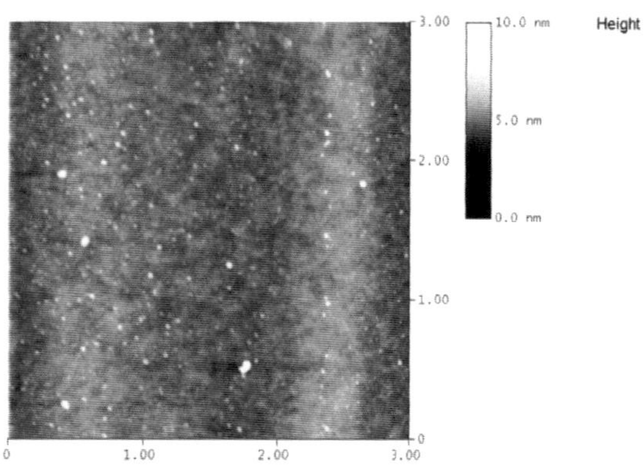

(c) BAS11S – heat treated 600 °C/ 0.5 h/air
R_a=0.31nm

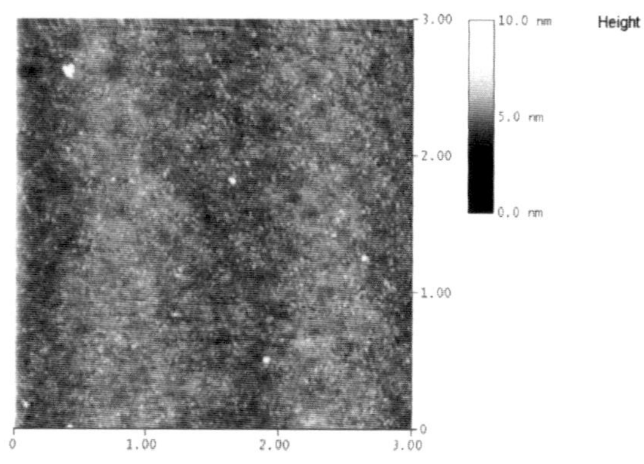

(d) unsized basalt fibre
R_a =0.37nm

Figure 5. Fibre surface and arithmetic surface roughness R_a of as-received (a), heat treated (b), (c) and unsized basalt fibres (d).

The chemical composition of the fibre surfaces was investigated by XPS. The amount of characteristic glass elements increased with the sizing removal. The surface composition of the unsized basalt fibre is shown as reference in table 1. In our previous work, we could show that the chemical composition of basalt fibres is changed after a heat treatment of 600 °C for 5 h [4]. Thus, we also analyzed the surface compositions of unsized basalt fibres after several heat treatments. The ratio of [C] / [Si] indicates high contaminations on the heat treated surfaces. However, an increase of Ca within the surface layer is already noticeable after 0.5 h heat treatment at a temperature of 600 °C (table 1).

Table 2 shows the results of dynamic contact angle measurements and calculated surface energies of as-received and heat treated basalt fibres. Heat treatment in air changes the wettability. The measurement in water leads to a reduced contact angle. Furthermore, the polar-component of surface energy increased slightly.

Interfacial strength parameters of the BAS11S – as-received fibre and re-sized basalt fibres embedded in epoxy resin were determined by quasi-static single fibre pull-out tests. Re-sized fibres show a slightly increased adhesion strength compared to the sized as-received sample. The apparent IFSS were calculated for sized and re-sized fibres for sized fibres as τ_{app} = 35±4 MPa and for re-sized fibres τ_{app} (re-sized) = 45±8 MPa. Furthermore, we observed a decreased post-debonding friction in the case of the re-sized fibres.

Table 1. Surface composition in at %.

	Al2p	**Mg2s**	**Si2p**	**C1s**	**K2p**	**Ca2p**	**O1s**	**Fe2p**	**Na1s**	**Others**	**[C]/[Si]**
BAS11S			0.5	78.9			20.6				156
BAS11S 600°C/0.5h air	3.5	1.0	11.7	40.1		4.0	37.8	0.8	0.4		3.4
BAS11U	4.5	1.3	13.6	29.6	0.4	2.1	46.3	0.8	1.3	0.2	2.2
BAS11U 600°C/0.5h air	2.2	0.8	7.6	42.8		6.8	38.6	0.3	0.6	0.5	5.7
BAS11U 600°C/1.0h air	2.8	0.6	10.2	40.6		6.0	37.2	0.3	0.7	1.6	4.0
BAS11U 600°C/0.5h N2	3.0	0.8	9.0	37.1		7.9	40.3	0.8	0.4	0.6	4.1

Table 2. Contact angles and surface energies of basalt fibres.

	Contact angle [°] Advancing			**Surface energy [mN/m]**		
	Water	**Ethylene glycol**	**Hexadecane**	**total** γ	**polar** γ^p	**disperse** γ^d
BAS11U	65±6					
BAS11S as- received	71±5	31±3	31±4	**38**	13	25
BAS11S 450 °C/0.5/air	75±4					
BAS11S 600 °C/0.5/air	55±7	52±5	27±3	**42**	22	20
BAS11S - 600 °C/0.5/N2	73±4					

4. Discussion and Conclusions

Considering sustainable applications of basalt fibres, there is a lack of knowledge about the performance of heat treated basalt fibres. Thus, we continued heat treatment experiments and carried out first re-sizing trials.

The results substantiate that the strength level of heat treated basalt fibres is dropped significantly. However, the chemical composition as well as the sizing is able to influence the performance. We could show that the sizing burn-off led to a significant strength loss, but the maximum decrease of tensile strength took place at temperatures above 350 °C. However, TGA results indicated that until a temperature of 350 °C a high amount of sizing was already removed. The strength level of the as-received sized basalt fibre, BAS11S, remained almost the same after 0.5 h heat treatments at moderate temperatures. This confirms findings by Jenkins et al. [1] about a 300 °C threshold. Due to the high standard deviation of single fibre tensile tests we could not clearly verify an influence of the atmosphere on the fibre strengths. Here, further experiments and data treatment is necessary.

The AFM investigations of heat treated BAS11S indicated that the nanostructure of basalt fibres is still not unveiled after the heat treatment at 450 °C. Sizing remnants lead to an increase of surface roughness. The increase of surface roughness and the partial exposure of surface flaws accompanied by the possibility of superelevation of flaw depths both are discussable items for the drastic decrease in fibre strength after heat treatments above 300 °C. Interestingly, the strength loss after heat treatments above 450 °C is less significant neither for unsized nor sized basalt fibres.

The contact angle measurements indicate that the wetting properties are affected by the heat treatments. However, here also further investigations are necessary. Especially, XPS should be used to separate the influences of changed surface roughness and changed surface composition due to thermal induced cation diffusion.

The re-sizing experiments seem to be encouraging, but need to be extended and combined with tensile tests. The impact of the surfactant on the observed decrease in post-debonding friction cannot be excluded.

References

[1] Jenkins, P , Riopedre Mendez, S, Saez Rodriguez, E, Yang, L and Thomason, J 2015 Investigation of the strength of thermally conditioned basalt and E-glass fibres *ICCM 20-20th Int. Conf. on Composite Materials (*Copenhagen, Denmark)

[2] Sabet, S M M, Akhlaghi, F and Eslami-Farsani, R 2015 The effect of thermal treatment on tensile properties of basalt fibers *J Ceram Sci Tech* **6** **(3)** 245

[3] Thomason, J and Dwight, D 1999 The use of XPS for characterization of glass fibre coatings. *Compos Part* A *–Appl S* **30** 1401

[4] Förster, T, Mäder, E, Jesson, D A and Watts, J F 2013 Surface analyses of basalt fibres: Tailoring the interphase of "green" fibre reinforced composites *ICCM19-19th Int. Conf. on Composite Materials (*Montreal, Canada)

[5] Smedskjaer, M M Yue, Y, Deubener, J , Gunnlaugsson, H P and Mørup, S 2010 Modifying glass surfaces via internal diffusion *J Non-Cryst Solids* **356** **(6)** 290

[6] Zhandarov, S and Mäder, E 2015 Effect of meniscus on the calculated interfacial interaction parameters in the pull-out and microbond test. *POLYCOMTRIB-2015* (Gomel, Belarus)

[7] Zhandarov, S and Mäder, E 2005 Characterization of fiber/matrix interface strength: applicability of different tests, approaches and parameters *Compos Sci Technol* **65** 149

IOP Conf. Series: Materials Science and Engineering **139** (2016) 012020 doi:10.1088/1757-899X/139/1/012020

Thermal recycling and re-manufacturing of glass fibre thermosetting composites

A Fraisse, J Beauson, P Brøndsted and B Madsen

Section for Composites and Material Mechanics, Department of Wind Energy, Technical University of Denmark, (Risø campus) Roskilde, Denmark

E-mail: antf@dtu.dk

Abstract. The impact of using thermally recycled glass fibre in re-manufactured composites was investigated. A unidirectional glass fibre thermosetting composite laminate was manufactured. The matrix in one part of the laminate was burnt off to recover the glass fibres. These recycled glass fibres were used to manufacture a new composite laminate with the same fibre architecture as the pristine one. The fibres, the matrix and the composite laminates were thoroughly characterised and analysed. The results show that good materials quality was obtained for both laminates. A difference in fibre packing behaviour was observed in the composites with the pristine and the recycled fibres, which lead to a lower fibre volume fraction in the latter one. The Young's modulus of the composites was not changed by the recycling process, if the lower fibre volume fraction is taken into account. However, a marked drop in the maximum stress of the composites was reported, which was found to be related to the loss in maximum stress of the fibres.

1. Introduction

Glass fibre composites are massively used nowadays for their high strength, lightweight and low cost. In 2014, the volume of produced composites was reported to represent 1.04 million tons in Europe [1]. Composites are used in different industry sectors such as transport, electronics, constructions, sports, as well as in the energy sector for wind turbines.

Recycling of wind turbines has become a topic of interest as the first wind farms are currently reaching end of life [2]. The tower, the gearbox, the hub are made of steel, the nacelle is made of aluminium, and these materials have already established recycling routes [3]. However, the blades made of glass fibre reinforced thermosetting polymer composites represent a large volume of the wind turbine, and they are challenging to recycle.

One of the simplest techniques to recycle glass fibre thermosetting composites is to burn off the matrix material and reuse the fibres. The main challenge in this solution is that the properties of burnt glass fibres are dropping with the applied temperature and duration of the recycling process [4,5]. The mechanism for this loss of fibre strength has been intensively studied, and it has been assumed to be due to different aspects, such as surface structural relaxation [6], and water absorption into the glass molecular network leading to formation of defects at the fibre surface [7].

Many studies have characterized the impact of thermal treatments on the properties of glass fibres, however only few have used recycled long continuous glass fibres to manufacture new polymer composites. Manufacturing of composites with recycled glass fibres is challenging, because the fibres are sensitive to mechanical handling.

The goal of this study is to investigate the effect of using recycled glass fibres on the properties of thermosetting composites. For that, a reference composite made with pristine glass fibres and a composite made with recycled glass fibres are manufactured. In order to ease the evaluation of the results, a simple fibre architecture consisting of unidirectional continuous glass fibres is selected. In addition, the selected manufacturing procedure involves minimal fibre handling in order to minimize damage of the fibres, and to preserve the fibre architecture. In order to understand the impact of the recycled glass fibres on the re-manufactured composite, the properties of the fibre and matrix constituents and the composites are thoroughly characterized and analysed.

2. Materials & Methods

2.1. Materials

Figure 1 presents the experimental procedure applied in this study. First, pristine glass fibres are used to manufacture a composite laminate, which will be named pristine laminate. One part of this laminate is burnt and the glass fibres are recovered. These fibres are named burnt fibres and are used to manufacture a new composite laminate, named re-manufactured laminate. In addition, a sample of pristine glass fibres is burnt at 565°C for testing of fibre properties.

The fibres used are Hybon 2026 PPG glass fibres with a silane sizing type which is dedicated to be used in polyester, vinyl ester and epoxy resin systems. The resin used is a Huntsman 1568 epoxy resin system with Huntsman hardener 3489.

The pristine laminate is a unidirectional fibre composite, which was manufactured using filament winding followed by vacuum assisted resin transfer molding. The laminate was cured in two stages in order to minimize the residual stresses in the matrix. The first stage was performed at 40°C for 19 hours, followed by a second stage at 75°C for 5 hours. The laminate dimensions were 300 x 205 x 2 mm^3.

A piece of the pristine laminate with dimensions 150 x 205 mm^2 was placed on a metal plate in a furnace at 565°C. After 10 hours, the fibres were cooled down to room temperature in the furnace. In order to keep them aligned and avoid direct handling; a second metal plate covered with a release foil was placed on the top of the fibres. The two plates were hold together and turned around to have the fibres lying on the infusion plate with the release foil. The fibres were infused and cured with the same conditions as the pristine laminate.

In addition, a pure resin plate was manufactured using the same curing procedure.

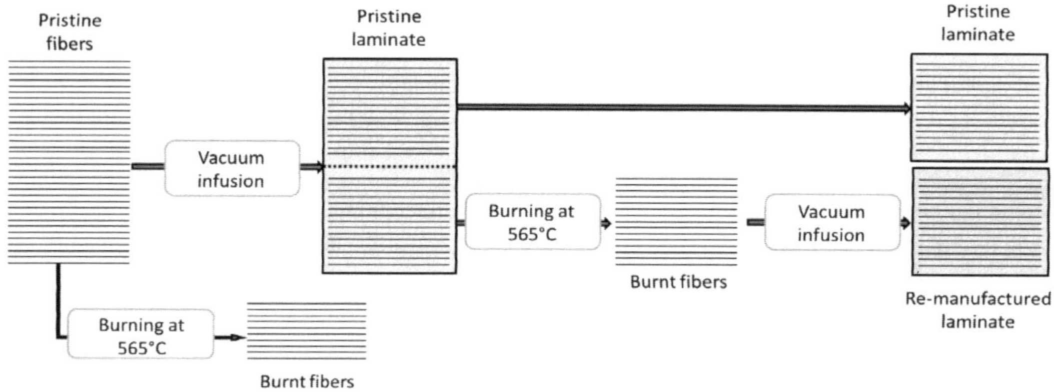

Figure 1. Experimental procedure.

2.2. *Fibre characterization*

2.2.1 Fibre density. To measure the density of the glass fibres, a gas pycnometer Quantachrome Ultrapyc 1200e was used, with nitrogen as the displacement medium. The fibres were rolled around a spool and placed in the pycnometer.
Before the measurements, the fibre samples (12.0 g of pristine fibres and 9.5 g of burnt fibres) were dried for 24 hours to remove moisture which would influence the measurements in the pycnometer. For both types of fibres, 8 repeated density measurements were performed.
2.2.2 Fibre mechanical properties. Fibres were mechanically tested by an automated single fibre tensile machine (Textechno Favimat+ Robot2, Germany). The test procedure consists firstly in a linear density test, in order to determine the fibre cross-sectional area, and secondly in a tensile test. A 50 mm gauge length was used in order to get convergence during the linear density test. Tensile testing speed was 1 mm/min. For both types of fibres, 50 single fibres were prepared to be tested. The fibre Young's modulus is calculated according to the composite materials standards, between 0.05% and 0.25% strain.

2.3. *Matrix characterization*
The mechanical properties of the manufactured pure epoxy plate were measured according to the ISO 527-2:1993(E) standard. Six specimens were tested.

2.4. *Composite characterization*

2.4.1 Volumetric composition. To determine the fibre, matrix and porosity content of the composites, five rectangular samples of dimensions 15 x 15 mm^2 were cut from each composite laminate. The samples were roughly polished, dried overnight, and weighed. The samples were then sealed with paraffin wax to be able to include the correct porosity volume in the determinations. The sealed samples were weighed first in air, and then in water in order to determine the composite density. Then, the epoxy matrix in the samples was burnt off overnight at 565°C. The weight of the remaining glass fibres was recorded in order to determine the fibre weight content. Using standard equations for composite materials [8], and based on the established values of fibre weight content, composite density, fibre density, and matrix density (set equal to 1.15 g/cm^3 [9]), the volume content of fibres, matrix and porosity in the composites was calculated.

2.4.2 Microstructure. To characterize the 2D microstructure of the composites, samples with dimensions 22 x 2 mm^2 were cut from each laminate. The samples were casted in epoxy, and the sample cross-section in the plane normal to the fibre direction was grinded and polished. Observations were made using an optical microscope.

A Xradia 520 from Zeiss was used to investigate the 3D microstructure of the composites by X-ray tomography. One sample with dimensions 15 x 2 x 2 mm^3 was cut from each laminate. To ensure a good scanning quality, the sample width was ensured to be equal to the sample thickness. Since the objective was to describe the composite microstructure, the scans had to cover as large volume as possible, and therefore, scans were stitched together by an image analysis algorithm. The X-ray source was placed at -10 mm and the detector at 25 mm from the sample in order to get a good transmission factor. The accelerating voltage was set to 40 KeV with an exposure time of 5 s using a LE2 filter and 5201 projections to ensure the best quality possible. All these parameters have been chosen to get a voxel size of 1.9 µm^3.

2.4.3 Mechanical properties. The mechanical properties of the composites were measured according to the ISO 527-5:2009(E) standard. The length of the tensile specimens was however shortened to 200 mm (instead of 250 mm) as this was the maximum length to fit in the furnace. The tab material used was 2 mm thick cross-ply glass/epoxy plates produced by Elektro-Isola (Denmark). Seven tensile specimens were tested for each laminate.

2.5. Modelling of composite properties
The composite laminate mechanical properties were analyzed by using the rule of mixtures:

$$E_L = E_f V_f + E_m V_m \quad (1)$$

$$\sigma_L = \sigma_f V_f + \sigma_m^* V_m \quad (2)$$

where E_L, E_f and E_m are the composite laminate, fibre and matrix Young's moduli, respectively, V_f and V_m are the fibre and matrix volume contents, respectively, σ_L and σ_f are the composite laminate and fibre maximum stresses, respectively, and σ_m^* is the stress in the matrix at the composite laminate failure strain.

3. Results and discussion

3.1. Fibres
The density of both pristine and burnt fibres has been measured according to the presented procedure. As shown in Table 1, the pristine fibres have a density of 2.61 g/cm^3, and the burnt fibres have a density of 2.64 g/cm^3. This increase in fibre density is believed to be due to glass densification during the thermal treatment [10].

Regarding the single fibre testing, 50 fibres of each type were prepared; 42 pristine fibres were successfully tested, while only 17 burnt fibres were successfully tested. This difference is due to that the burnt fibres are brittle, and during clamping of the fibres, a high number of them fail. The measured Young's modulus and maximum stress of the fibres are presented in Table 1. The pristine fibres have a lower Young's modulus than the burnt fibres, with values of 82.2 and 87.5 GPa, respectively. However, the pristine fibres have a much higher maximum stress, 2050 MPa, compared to the maximum stress of the burnt fibres, 400 MPa. Thus, the thermal treatment has impacted the glass fibre maximum stress by a reduction of 80%.

Table 1. Density and mechanical properties of pristine and burnt fibres.

Fibre type	Density (g/cm^3)	Young's modulus (GPa)	Maximum stress (MPa)
Pristine	2.61 ± 0.02	82.2 ± 0.7	2050 ± 520
Burnt	2.64 ± 0.01	87.5 ± 0.8	400 ± 180

Increased stiffness and decreased strength of glass fibres after thermal treatment has also been reported elsewhere in the literature. These phenomena are believed to be due to a densification of the glass molecular network [11] and a removal of sizing. During heating glass fibres in normal atmosphere, the water molecules present in the atmosphere are attacking the fibre surface and replacing atoms. These changes are building stresses at the fibre surface, and this is leading to a decrease of the fibre maximum stress [6].

3.2. Matrix
Based on the measured stress-strain curves of the neat epoxy resin plate, the Young's modulus of the matrix was measured to be equal to 3.1 ± 0.02 GPa, and with a maximum stress of 65.1 ± 0.3 MPa.

3.3. Composites

The two composite laminates were manufactured under the same processing conditions. During manufacturing of the laminate with the recycled fibres, a distinct noise could be heard when the vacuum pressure was applied to the bag with the fibres. This noise is believed to be associated with cracking of the glass fibres, due to their brittleness. A similar noise could also be heard when the resin was infused into the bag with the recycled fibres.

3.3.1 Volumetric composition. Table 2 shows the volumetric composition of the pristine and the re-manufactured laminates. The porosity content was measured to be around 2% for both laminates, which corresponds to a good composite quality. After burning off the matrix in the composite laminates, a small amount of ashes from the combustion was observed on the surface of the fibres. This did not seem to influence the quality of the composites. Regarding the fibre volume content, it was measured to be 59.2% for the pristine laminate and 46.8% for the re-manufactured laminate. Since the same amount of fibres was used for the manufacturing of the two laminates, the difference in fibre volume contents must be due to a difference in the composite volume, which can be assessed by the thickness of the laminates. The average thickness of the pristine laminate was 1.89 mm, while the average thickness of the re-manufactured laminate was 2.21 mm. This indicates that under the same processing conditions, the recycled fibres pack less efficiently than the pristine fibres. To control the laminate thickness and thereby to obtain the same fibre volume content, a frame could be placed around the fibres and a higher pressure could be applied during infusion of the resin.

Table 2. Volumetric composition of the pristine and re-manufactured composite laminates.

Composite laminate type	Fiber volume content (%)	Matrix volume content (%)	Porosity content (%)
Pristine	59.2 ± 0.8	39.1 ± 0.8	1.7 ± 0.1
Re-manufactured	46.8 ± 1.1	51.0 ± 1.1	2.1 ± 0.3

3.3.2 Microstructure. Figure 2 shows optical microscopy images of the microstructure of the pristine and re-manufactured composite laminates. In the pristine laminate, the fibres are well distributed across the laminate thickness, with a few porosities. The microstructure is different in the re-manufactured laminate. The fibres are non-uniformly distributed with large matrix-rich regions. This observation supports the hypothesis that recycled fibres pack less efficiently than pristine fibres. This different fibre packing configuration could be due to the removal of sizing from the fibres during the thermal treatment. One important role of sizing is to avoid electrostatic repulsion between fibres and therefore having a more efficient packing.

Figure 3 shows X-ray tomography images of the microstructure of the re-manufactured composite laminate. The fibres can be seen to be well-aligned, and matrix-rich regions can also be seen, similar to the ones seen by optical microscopy. Careful inspections of the scanned 3D volume revealed a number of locations where the fibres are broken, see 2D projections in Figure 4. These observations support the hypothesis of brittle fibres breaking during manufacturing of the re-manufactured laminate.

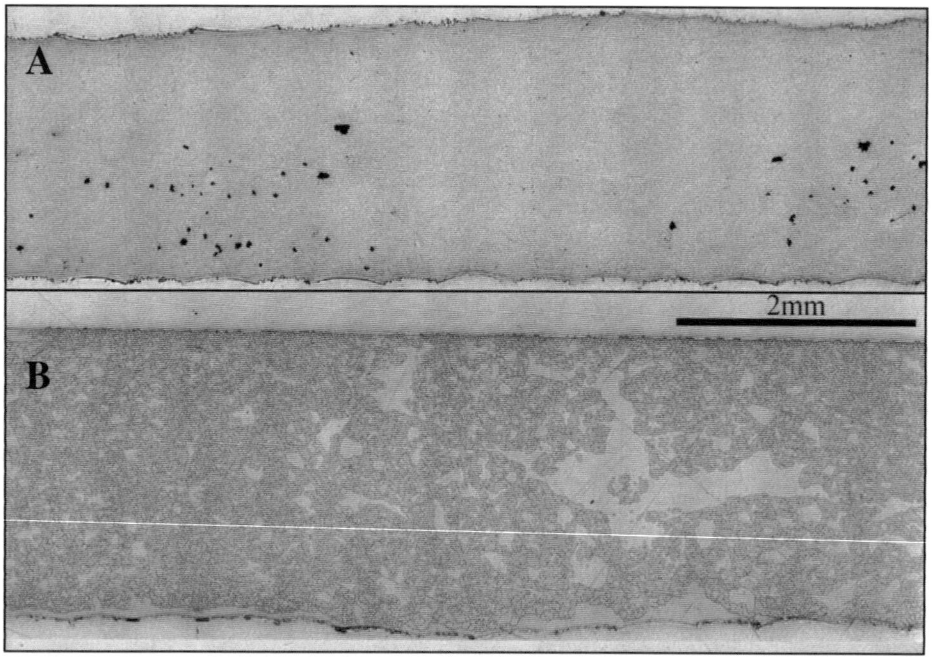

Figure 2. Optical microscopy image of microstructure in (A) pristine composite laminate, and (B) re-manufactured composite laminate.

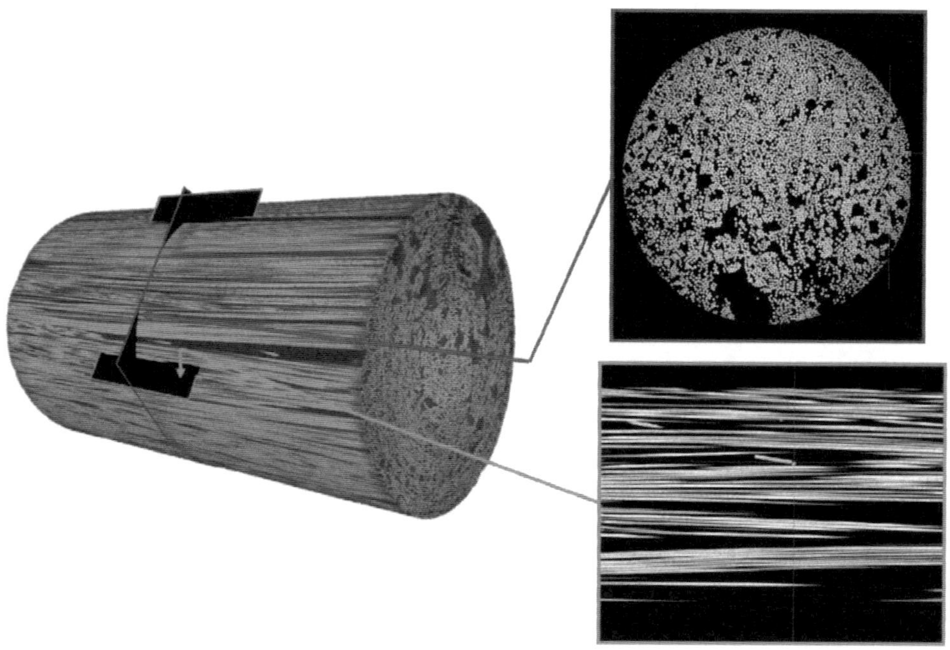

Figure 3. Reconstructed 3D X-ray tomography volume of re-manufactured composite laminate (left), with two examples of 2D projections (right).

IOP Conf. Series: Materials Science and Engineering **139** (2016) 012020 doi:10.1088/1757-899X/139/1/012020

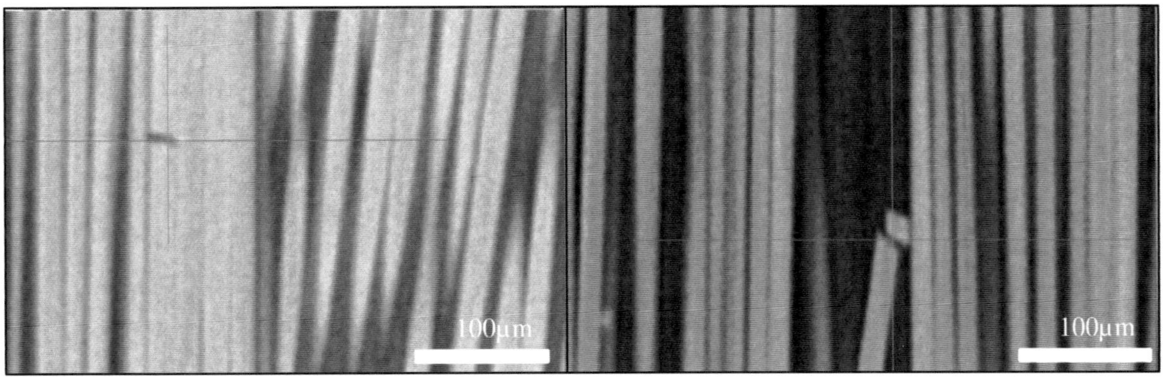

Figure 4. 2D projections of X-ray tomography volume showing broken fibres (marked by crossed lines) in the re-manufactured composite laminate.

3.3.3 Mechanical properties. Results of the tensile tests for the pristine and re-manufactured composite laminate are gathered in Table 3. The pristine laminate has a Young's modulus of 52.8 GPa, whereas the re-manufactured laminate has one of 45.1 GPa. As will be demonstrated later, the main reason for the lower Young's modulus of the re-manufactured laminate is its lower fibre volume fraction (see Table 2). The difference in Young's modulus between the two laminates can also be observed from the stress-strain curves presented in Figure 5, where the curves of the re-manufactured laminate lie slightly under the stress-strain curves of the pristine laminate.

Regarding the maximum stress of the composites, the pristine laminate has a maximum stress of 1210 MPa, whereas the re-manufactured laminate has one of only 120 MPa. This corresponds to a 90% reduction in maximum stress. Static tensile properties of unidirectional composites reflect mostly the fibres properties, and therefore, the low maximum stress of the re-manufactured laminate can be directly related to the low maximum stress of the recycled fibres (see Table 1).

Table 3. Experimental and theoretical mechanical properties of the pristine and re-manufactured composite laminates.

Composite laminate type	Experimental			Theoretical	
	Young's modulus (GPa)	Maximum stress (MPa)	Strain at maximum stress (%)	Young's modulus (GPa)	Maximum stress (MPa)
Pristine	52.8 ± 2.8	1210 ± 70	2.9 ± 0.2	49.9	1240
Re-manufactured	45.1 ± 4.7	120 ± 20	0.3 ± 0.1	42.5	220

The failure behaviour of the tensile specimens can be seen in Figure 6. The specimens from the pristine laminate are showing a chaotic fracture mode with extensive fibre splitting, which is typical for unidirectional glass fibre composites. In contrast, the failure mode of specimens from the re-manufactured laminate is characterised by a single crack across the specimens, and located close to one of the tabs. This difference in failure mode is explained by the difference in maximum stress of the fibres. The pristine fibres are able to carry large stresses before they fail, which then results in a large energy release, leading to a chaotic fracture mode where fibre bundles are split apart in the whole length of the specimens. The recycled fibres are only able to carry small stresses, which does not results in a large energy release at failure, and therefore, the failure mode in these specimens is characterised by a single crack running transverse to the fibres.

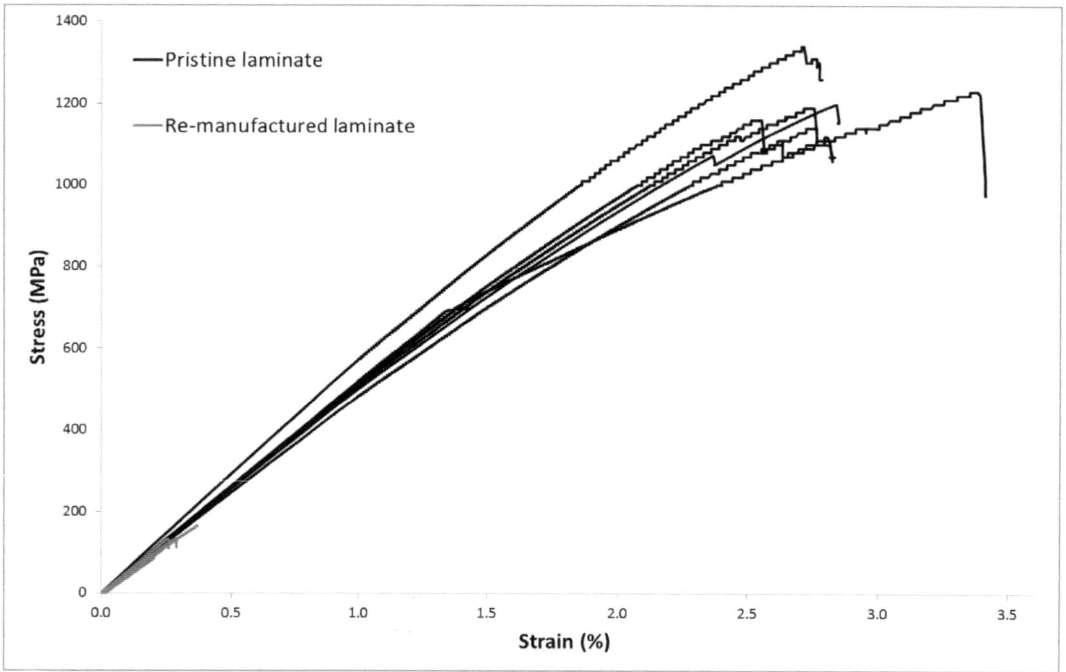

Figure 5. Stress-strain curves of pristine and re-manufactured composite laminates.

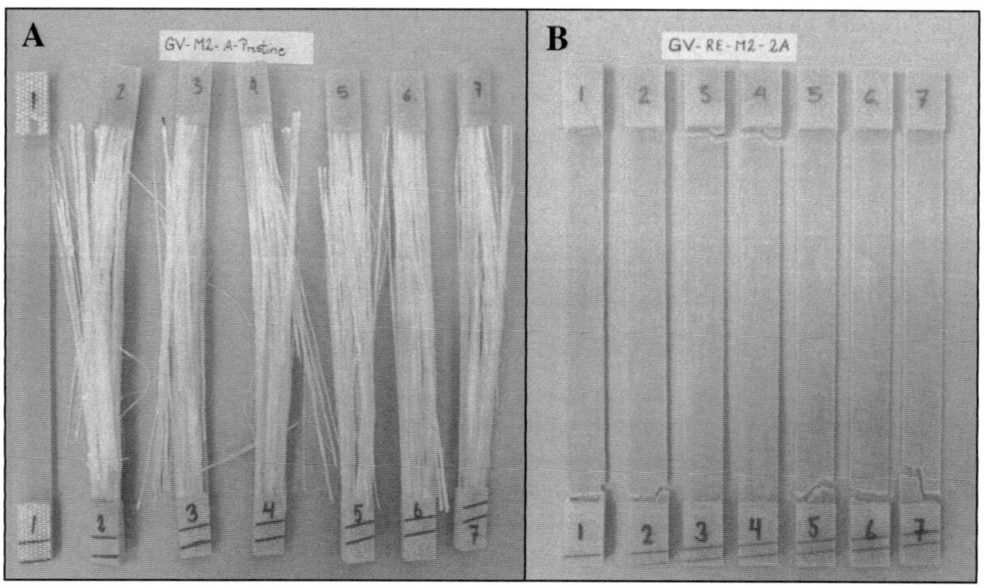

Figure 6. Tensile specimens of (A) pristine and (B) re-manufactured composite laminates after testing.

3.4. Modelling of composite properties

Using the measured volumetric composition of the composites, together with the measured fibre and matrix properties, the theoretical Young's modulus and maximum stress of the pristine and re-manufactured laminates were predicted using the rule of mixtures models, Equations (1) and (2). The results are presented in Table 3.

The theoretical Young's modulus for the pristine laminate is equal to 49.9 GPa, and it is equal to 42.5 GPa for the re-manufactured laminate. As can be observed in Table 3, these results are essentially similar to the measured Young's moduli, taking the scatter (± stdv.) of the measurements into account.

The Young's modulus of the re-manufactured laminate can also be estimated by considering it to have the same fibre volume fraction as found in the pristine laminate, i.e. 59.2 %. Then, the theoretical Young's modulus for the re-manufactured laminate is equal to 53.0 GPa. This is similar to the measured Young's modulus of the pristine laminate on 52.8 GPa. It is however larger than the theoretical Young's modulus of the pristine laminate on 49.9 GPa due to the larger Young's modulus of the recycled fibres.

To predict the maximum stress of the composite laminates, the stress in the matrix at composite failure strain was obtained from the measured stress-strain curve of the matrix. This value is equal to 62.8 MPa for the pristine laminate and 8.4 MPa for the re-manufactured laminate. Then, the maximum stress of the pristine and re-manufactured laminates can be predicted to be 1240 and 190 MPa, respectively. Both predictions are in acceptable agreement with the measured values on 1210 and 120 MPa, respectively. The slightly lower measured maximum stress of the re-manufactured laminate can be explained by the failure of these specimens close to the tabs due to stress concentrations. Thus, if more optimal specimen geometry was used (e.g. dog-bone shaped), the measured maximum stress would be expected to approach 190 MPa.

4. Conclusions

Considering optimal manufacturing conditions and fibre alignment, the present study investigates the effect of using recycled glass fibres on the properties of composites. A unidirectional glass fibre thermosetting composite laminate was manufactured, and part of it was burnt to recycle the fibres. The recycled fibres were then used to re-manufacture a new composite laminate using the exact same processing conditions, and involving minimal handling of the fibres in order to avoid damaging them.

The two types of manufactured composite laminates were shown to be of good quality, with low porosity content. The recycled glass fibres could not be packed as efficiently as the pristine fibres, which lead to a lower fibre volume fraction in the re-manufactured laminate. This was assumed to be due to electrostatic attraction between the recycled fibres where the sizing is removed.

Regarding the performance of the composites, the Young's modulus was not significantly impacted by the recycling process. The recycled glass fibres were found to have a larger Young's modulus than the pristine glass fibres, which theoretically also would lead to a larger Young's modulus of the re-manufactured laminate, in case of similar fibre volume fractions. The maximum stress of the re-manufactured laminate showed a 90% reduction compared to the pristine laminate. This can be attributed to the lower maximum stress of the recycled glass fibres.

Regarding the impact of the manufacturing process on the re-manufactured laminate, it was not possible to avoid it completely. It was shown that the brittle recycled fibres were breaking during the vacuum infusion process.

The work in the present study does not involve measurements of the possible changed adhesion of the recycled glass fibres to the thermosetting matrix. Further studies should be focused on characterizing the fibre/matrix adhesion in re-manufactured glass fibre composites.

Acknowledgements

The authors would like to thank the Danish innovation consortium GenVind ("Nationalt initiativ omkring genanvendelse af plastkompositter", Innovation Fund Denmark) for financial support. Jonas Kreutzfeldt Heininge, Christen Malte Markussen and Tom Løgstrup Andersen are much appreciated for technical support.

References

[1] Witten E, Kraus T and Kühnel M 2014 Composites Market Report 2014, Market developments, trends, challenges and opportunities, pp 44
[2] Larsen K 2009 Recycling of wind turbine blades *Renew. Energy Focus* **9** 70
[3] Papadakis N, Ramirez C and Reynolds N 2010 Designing composite wind turbine blades for disposal, recycling or reuse (Management, recycling and reuse of waste composites) ed V Goodship (Cambridge: Woodhead Publishing Limited) chapter 16 pp 443-457
[4] Feih S, Mouritz A P and Case S W 2015 Determining the mechanism controlling glass fibre strength loss during thermal recycling of waste composites *Compos. Part A* **76** 255
[5] Thomason J L, Yang L and Meier R 2014 The properties of glass fibers after conditioning at composite recycling temperatures *Compos. Part A* **61** 201
[6] Tomozawa M and Hepburn R W 2004 Surface structural relaxation of silica glass: a possible mechanism of mechanical fatigue *J. Non-Cryst. Solids* **345-346** 449
[7] Davis K M and Tomozawa M 1995 Water diffusion into silica glass: structural changes in silica glass and their effect on water solubility and diffusivity *J. Non-Cryst. Solids* **185** 203
[8] Beauson J, Madsen B, Toncelli C, Ilsted Bech J and Brøndsted P 2016 Recycling of shredded composites from wind turbine blades in new thermoset polymer composites [submitted for publication]
[9] Hüther J 2015 The effects of resin curing kinetics and shrinkage on the mechanical properties of glass fibre reinforced plastics used for wind turbine blades MSc. Thesis Karlsruhe Institute of Technology Germany pp 127
[10] Beauson J, Fraisse A, Nørgaard Petersen H and Madsen B 2016 Study of glass compaction after heat treatment and the impact on the glass fibre mechanical properties [in preparation]
[11] Otto W H 1961 Compaction Effects in Glass Fibers *J. Am. Ceram. Soc.* **44** 68

Use of micro-tomography for validation of method to identify interfacial shear strength from tensile tests of short regenerated cellulose fibre composites

A Hajlane[1], A Miettinen[2], B Madsen[3], J Beauson[3] and R Joffe[1]

[1]Luleå University of Technology, Department of Engineering Sciences and Mathematics, SE-971 87 Luleå, Sweden
[2]University of Jyvaskyla, Department of Physics, P.O. Box 35 (YFL) FI-40014 Jyvaskyla, Finland
[3]Technical University of Denmark, Department of Wind Energy, Risø Campus, Frederiksborgvej 399, 4000 Roskilde, Denmark

e-mail: Roberts.Joffe@ltu.se

Abstract. The interfacial shear strength of short regenerated cellulose fibre/polylactide composites was characterized by means of an industry-friendly adhesion test method. The interfacial shear strength was back-calculated from the experimental tensile stress-strain curves of composites by using a micro-mechanical model. The parameters characterizing the micro-structure of the composites, e.g. fibre length and orientation distributions, used as input in the model were obtained by micro-tomography. The investigation was carried out on composites with untreated and surface treated fibres with various fibre weight contents (5wt%, 10wt%, and 15wt% for untreated fibres, and 15wt% for treated fibres). The properties of fibres were measured by an automated single fibre tensile test method. Based on these results, the efficiency of the fibre treatment to improve fibre/matrix adhesion is evaluated, and the applicability of the method to measure the interfacial shear strength is discussed. The results are compared with data from previous work, and with other results from the literature.

1. Introduction

During recent years, cellulosic fibres (natural and manmade) have been considered as feasible replacement of synthetic reinforcement (e.g. glass fibres) for polymer composites designed for long term use in structures. The driving force for such development is the concern regarding impact of production of composites on environment, and the growing demand for sustainable and recyclable materials. This trend is also reflected in the growing number of publications on the subject as shown in literature reviews [1-2] on development of bio-based engineering materials.

One of the weaknesses of cellulosic fibres is their typical poor compatibility with polymers, and the resulting weak fibre/matrix adhesion which strongly affects the mechanical properties of composite materials. This problem can be addressed by modifying the fibre surface to enhance the chemical bonding between composite constituents [3]. The problem can also be addressed by creating hierarchical reinforcement (combining nano- and micro- scales) with larger surface area to promote both, chemical and mechanical, fibre/matrix interaction. Hierarchical reinforcement can be created by grafting of nano-size particles on micro-size fibres, and this is believed to enhance the fibre/matrix adhesion. For example, it has been shown that the grafting of carbon nano-tubes on conventional fibres increases the through thickness performance of composites due to strengthening of fibre/matrix adhesion by improving the fibre surface area, and by creating mechanical interlocking [4]. In another example, composites based on hierarchical cellulosic fibres were successfully achieved by depositing bacterial cellulose on natural fibres [5]. The tensile and flexural properties of the composites of hierarchical sisal fibres embedded in acrylated epoxidised soybean oil resin were significantly improved over neat fibre reinforced composites. In the current study, cellulose nano-whiskers extracted from date palm trees are deposited onto regenerated cellulose fibres.

One of the important parts of the optimization process of the fibre/matrix interface is its characterization to evaluate the efficiency of various fibre treatment methods. The usual way to characterize the fibre/matrix adhesion is by determining the interfacial shear strength (IFSS). There is number of direct and indirect methods to evaluate IFSS, ranging from tests on single fibres (fibre pull-out, micro-droplet test, single fibre fragmentation) to test on full scale composites (e.g. short beam shear). However, these methods, especially those on micro-scale, are time consuming and not easy to perform, and moreover the results may significantly vary between different methods. Therefore, there is a need for a simple, fast, and yet reliable technique to measure IFSS in composites. The method used in the present study has been considered to be the most industry-friendly one [6], and is a modification of a method originally presented in [7-8]. The model scheme is based on analysis of the tensile stress-strain curves of a composite system. Two different strain values are used to obtain values for IFSS and the fibre orientation factor.

The current study follows the method presented in [6-8]. All experimental parameters related to the composite morphology and fibre properties are obtained directly from experiments. This allow more reliable predictions by the micro-mechanical models since input parameters are accurately measured rather than estimated based on materials datasheets and assumptions. The output of the approach proposed in [6-8] is compared with the results obtained based on X-ray tomography data, and possible factors responsible for discrepancies between methods are discussed.

2. Theoretical background

Based on the Kelly-Tyson model for the strength of aligned short fibre composites [9], and by dividing the total fibre volume fraction (V_f) into sub-fractions of fibres with given lengths (L_f), Bowyer and Bader [7] developed a model to predict stress (σ_c) as function of strain (ε_c) in composites with non-aligned fibres:

$$\sigma_c = \eta_o \left(\sum_i \left[\frac{\tau_i \, L_{f(i)} \, V_{f(i)}}{d_f} \right] + \sum_j \left[E_f \, \varepsilon_c \, V_{f(j)} \left(1 - \frac{E_f \, \varepsilon_c \, d_f}{4 \, \tau_i \, L_{f(j)}} \right) \right] \right) + (1 - V_f) \, E_m \, \varepsilon_c \qquad (1)$$

where η_o is the fibre orientation factor, d_f is the fibre diameter, E_f is the fibre stiffness, and τ_i is the interfacial shear strength.

It is assumed that a critical fibre length ($L_{f\,crit\,(\varepsilon)}$) is defined for any given strain level:

$$L_{f\,crit\,(\varepsilon)} = \frac{E_f \, \varepsilon_c \, d_f}{2 \, \tau_i} \qquad (2)$$

Fibres shorter than $L_{f\,crit\,(\varepsilon)}$ contribute with the index i in equation (1), and fibres longer than $L_{f\,crit\,(\varepsilon)}$ contribute with the index j in equation (1).

By selecting two strain levels from a measured stress-strain curve of a given composite system, and by following the described fitting approach in Bowyer and Bader [7], a single value for τ_i (IFSS) and a single value for η_o can be determined.

It should be noted that the term $E_f \cdot \varepsilon_c$ in equations (1) and (2) corresponds to the fibre stress (σ_f). Likewise, the term $E_m \cdot \varepsilon_c$ in equation (1) corresponds to the matrix stress (σ_m). This is assuming linear stress-strain behaviour of fibres and matrix. If stress-strain curves are measured for fibres and matrix, the accurate stress levels (σ_f and σ_m), at a given composite strain level, can be used in the equations.

3. Experimental

This section very briefly presents experimental details about the surface treatment of fibres, manufacturing of composites as well as characterization of fibre and composite properties, including detailed characteristics of the internal structure of the composite materials. More complete information concerning the applied procedures can be found in [10-13].

3.1. Materials

The regenerated cellulose fibre (RCF) Cordenka 700 Super 3, 2440 dtex, Z100 (Z100 refers to twist, 100 rev/meter) was used as received. Fibres were supplied in the form of bundles, each bundle contains 1350 fibres and average diameter of single fibre is 12.5 microns (assuming circular cross-section). Density of fibres is 1.50 g/cm^3.

For the modification of the fibre surface, a number of different chemicals were used. The initiator cerium ammonium nitrate (CAN), nitric acid, NaOH, MPS (organo-functional trialkoxy silane γ-methacryloxypropyl trimethoxysilane) and pure ethanol (99%) were used as received without purification. Distilled water was employed as co-solvent with ethanol for the grafting reaction. Cellulose nano-whiskers (CW) were extracted from date palm tree according to the procedure reported in [11].

Polylactic acid (PLA) biopolymer was used as matrix in composites. The polymer specification is Ingeo biopolymer 2003D from NatureWorks LLC. Density of PLA is 1.24 g/cm^3.

3.2. Fibre surface treatment

The fibre treatment was carried out in two steps. First a silane treatment of fibres was performed and then CW were grafted onto the RCF surface (see [12] for details).

The silane treatment was performed as follows: a 1600 ml mixture of ethanol and water (50/50 v%) was poured into a reactor and heated up to 65°C; after stabilizing the temperature, NaOH and nitric acid solutions were utilized to adjust pH to 7; once pH was stabilized (after approximately 2 hours), 16 grams of fibres, wound on a holder and fixed on a mechanical stirrer, were introduced into the solution (the ratio of fibres and solvent was adjusted to be 1% w/v); the polymerisation (or co-polymerisation) of MPS was carried out by adding first CAN 10^{-3} mol/L in the reactor and then stirring for 30 minutes. Meanwhile, the mixture was purged with nitrogen gas for 15 minutes to remove any dissolved oxygen gas. At this stage, free radicals are supposed to be created at the surface of the cellulose backbone, and are ready to react with vinyl monomers (MPS). To initiate the graft co-polymerisation, 10^{-3} ml/L of MPS was added to the reactor, and the nitrogen gas flow was maintained until the end of the reaction (5 hours) while stirring at medium constant rate. Fibres were then washed two times in ethanol and once in water to remove unreacted products and other impurities trapped in the fibres.

The grafting process of the nano-whiskers was performed by using CW in a 0.2 wt% concentration. Distilled water (1600 ml) was poured in the reactor and pH was adjusted to be between 4 and 5. When pH was stabilized, the MPS-modified fibres were re-placed in the reactor. Prior to sonication, a solution of NaOH (0.1 M) was added to the cellulose nano-whiskers to reduce the amount of sulfonate groups and release the hydroxyl (or alcoholate) groups of CW. The CW suspension was then sonicated for 5 minutes and added into the reactor.

The reaction was maintained at medium stirring rate at room temperature for 5 hours. Afterwards, the fibres were washed three times in distilled water and dried at 110°C for 4 hours under vacuum.

Finally, the treated fibres were stored in a desiccator to protect the fibres from dust.

3.3. Manufacturing of short fibre composites

In order to manufacture short fibre composites with randomly oriented fibres, the continuous RCF were cut to produce 2 mm long bundles. The cutting was done by using an array of 60 razor blades (see [10] for details).

The manufacturing of composite specimens consisted of two stages: 1) homogenization of RCF fibres and PLA granules to produce composite pellets; 2) injection moulding of test specimens. The homogenization of fibres and PLA was carried out in a twin-screw Brabender PLE 650 mixer. The mixing was done for 5 minutes at 215°C with screw rotation of 70 rpm. Directly after mixing, the material was pressed into flat 2 mm thick plates in a hydraulic manually operated press preheated to 80°C. The plates were cut into small pieces to produce composite pellets with a square shape (approximately 2-3 mm in size). Composite pellets were fed into a HAAKE™ MiniJet laboratory injection moulding machine (Thermo Scientific™, Karlsruhe, Germany) to produce tensile specimens according to ASTM D638 (type 5 dog-bone specimens). Detailed information about the manufacturing procedure and conditions can be found in [10].

Four different composite compositions were produced (six specimens for each): three composites of PLA with untreated fibres (5, 10 and 15 wt%), and one composite of PLA with treated fibres (15 wt%).

3.4. Characterization of fibres

In order to evaluate if the fibres are degraded during the composite manufacturing process, fibre bundles were thermally conditioned to simulate the applied composite processing conditions (exposure to 220°C for 15 min). Accordingly, in total, four types of fibres were characterized: 1) untreated and unconditioned (U-UC); 2) untreated and conditioned (U-C); 3) treated and unconditioned (T-UC); 4) treated and conditioned (T-C).

Tensile tests of single fibres were carried out using a Textechno machine (Favimat and Airobot2). The linear density of fibres was obtained by a vibroscopy test. The cross-sectional area of fibres was calculated from its linear density, and the density of cellulose (1.50 g/cm^3). In total 50 fibres of each type were tested with gauge length of 50 mm. Tests were carried out in displacement controlled mode with a speed of 5 mm/min.

3.5. Tensile tests of composites

Tensile tests of composites were carried out on an Instron 3366 electro-mechanical testing machine equipped with a 10 kN load cell and pneumatic grips. Displacement controlled mode was used with a strain rate set to approximately 1 %/min. An Instron 2620-601 extensometer with a gauge length of 12.5 mm was used to register strain. All experiments were carried out at ambient laboratory conditions (T ~ 24°C, RH ~ 42%) without prior conditioning of specimens. At least four samples for each type of composite were tested. A more detailed description of the procedure is presented in [10].

3.6. X-ray micro-tomography

X-ray micro-tomography was used to determine the fibre length and orientation distribution in the composites while the fibre orientation factor is calculated according to Krenchel's model [14]. First, cylindrical samples of approximately 2 mm in diameter were cut from the central part of tensile test specimens of each of the four composite types. The samples were imaged using Xradia MicroCT-400 tomograph with 1.2 μm pixel size, 30 kV acceleration voltage and 3 W X-ray tube power. After reconstruction, the images were filtered with Gaussian blur (standard deviation = 1 pixel) and thresholded, yielding a binary image of fibres. The threshold value was chosen such that the average fibre diameter was 12.5 μm. Regions whose volume was less than 500 pixels were removed from analysis as they were primarily caused by imaging noise. Fibre diameter, fibre length, fibre orientation distribution as well as fibre orientation factor were then determined as described in [13].

4. Results and discussion

This section consists of two parts; first experimental data are reported and discussed, and then the outcome of processing of results is presented with assessment of applicability of models as well as comparison with other available data.

4.1. Experimental results

The typical stress-strain curves for untreated and treated conditioned fibres are presented in figure 1 (left). It is evident that the performance of the fibres is drastically affected by the treatment and conditioning. This is also shown by the analysis of fibre strength distribution as shown in figure 2 (left) by plotting probability of failure P_f as a function of stress at failure σ. The strength distribution was analysed by fitting a Weibull two-parameter statistical distribution to the experimental data (the same procedure as described in [15] is used). The parameters of the distribution (scale σ_0 and shape m) are shown in figure 2 (right). The summary of average fibre properties (diameter, stiffness, strength) along with Weibull parameters are presented in table 1. The results for unconditioned untreated fibres are in line with previously reported results for similar RCF [16-17]. The results in table 1 show that treatment reduces fibre stiffness by 5% and strength by approximately 25%. The thermal conditioning does not affect stiffness of untreated fibres and reduces it for treated fibres by about 7%. However, strength is much more affected by thermal conditioning, it is reduced by 16% for untreated fibres and by 63% for treated fibres. It also should be noted that scatter of strength has significantly increased for treated conditioned fibres (the shape parameter is much lower than for all other fibres). Since during the manufacturing of composites, fibres are subjected to the same temperatures as during the thermal conditioning, it is assumed that it has the same effect on fibre performance in composites. Therefore, in all further calculations, properties of conditioned fibres were used.

The typical stress-strain curves for unreinforced PLA and the four different composites are shown in figure 1 (right). The average mechanical properties of composites are summarized in table 2. The results show that stiffness and maximum stress increases with increase of fibre content in the composites. The performance of composites with 15wt% of treated RCF is lower than that for untreated RCF composites, which is expected since fibre performance was degraded by treatment and thermal conditioning (see table 1). It should be noted that for all of the composites with untreated fibres and for unreinforced PLA, the strain at maximum stress is almost the same (between 2.6% and 2.8%). The results on mechanical performance presented here compare very well to the data for RCF/PLA composites reported in [18] (although fibre length and orientation in that study may differ).

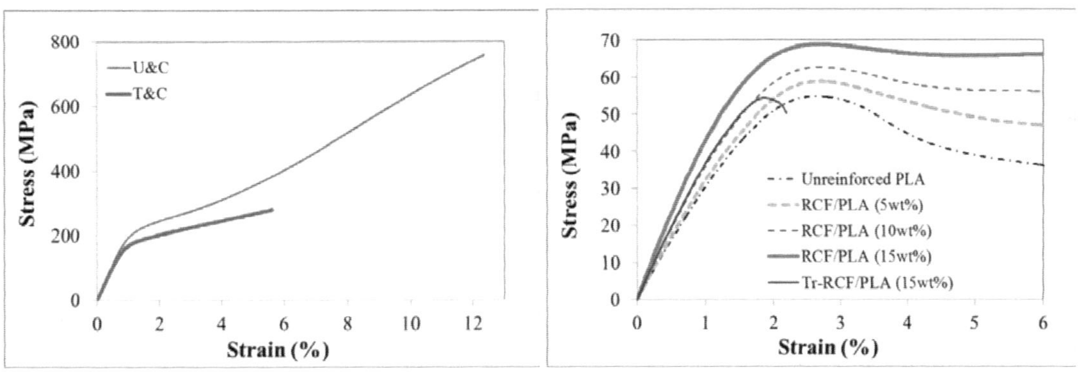

Figure 1. Typical stress-strain curves from single fibre tensile tests of untreated and treated conditioned fibres (left). Typical tensile stress-strain curves for unreinforced polymer and short fibre composites with untreated and treated fibres (right).

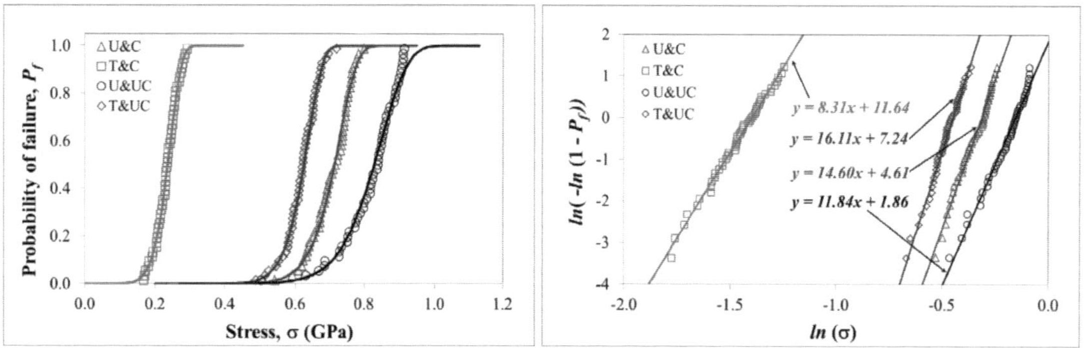

Figure 2. Fibre strength distribution obtained from single fibre tensile tests (symbols in the left plot) and determination of parameters of Weibull distribution (plot on the right). The theoretical Weibull strength distributions are shown as solid lines in the plot on the left.

Table 1. Single fibre tensile test results and parameters of Weibull distributions.

Batch	Fibres	d_f [μm]	Strength [GPa]	Stiffness [GPa]	Shape parameter m	Scale parameter σ_0 [GPa]	$<\sigma>$[a] [GPa]
U-UC	Untreated Unconditioned	12.4 ± 0.2	0.82 ± 0.08	24.04 ± 0.39	11.84	0.86	0.82
U-C	Untreated Conditioned	12.4 ± 0.2	0.69 ± 0.08	23.87 ± 0.45	14.60	0.73	0.70
T-UC	Treated Unconditioned	12.4 ± 0.3	0.62 ± 0.05	22.89 ± 0.45	16.11	0.64	0.62
T-C	Treated Conditioned	12.5 ± 0.2	0.23 ± 0.03	21.50 ± 0.52	8.31	0.25	0.23

[a]Average fibre strength calculated by using parameters of Weibull distribution.

Table 2. Average mechanical properties of neat PLA and RCF reinforced composites.

Batch	Stiffness[a] [GPa]	Max stress [MPa]	Strain at max stress [%]
PLA	3.30 ± 0.13	54.2 ± 0.8	2.64 ± 0.12
RCF/PLA 5wt%	3.52 ± 0.20	57.7 ± 1.4	2.68 ± 0.15
RCF/PLA 10wt%	3.91 ± 0.19	63.9 ± 0.9	2.78 ± 0.15
RCF/PLA 15wt%	4.78 ± 0.13	70.2 ± 1.8	2.66 ± 0.06
Tr-RCF/PLA 15wt%	4.07 ± 0.19	52.5 ± 2.4	1.76 ± 0.10

[a]Measured within strain interval of 0.05%-0.25%.

The determined fibre length and orientation distributions are presented in figure 3 and figure 4, respectively. The results in figure 3 show that fibres were rather significantly reduced in length during the injection moulding process. Initially the fibre bundles were manually cut to 2 mm long pieces, and the final fibre length in the composites is within interval of 0 mm – 1 mm with majority of fibres being around 0.1 mm – 0.3 mm long.

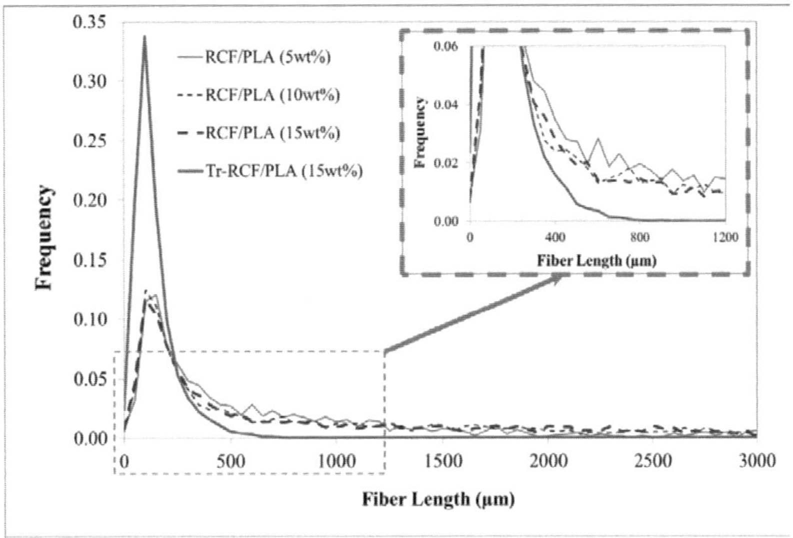

Figure 3. Fibre length distribution obtained from X-ray tomography of composites.

Table 3 summarizes results of the X-ray tomography. It is clear from those data that all untreated fibres have very similar average fibre length (in range 0.266 mm – 0.280 mm) independent of fibre content in the composites, while treated fibres are more than two times shorter. This can be explained by much lower mechanical properties of the treated fibres due to the severe thermal degradation which makes it much easier to break up fibres during composite manufacturing.

The fibre orientation distribution in figure 4 show that fibres are not completely randomly oriented and there is some fibre alignment towards the loading direction (majority of fibres are oriented within 0°-45° with respect to the loading direction). However, the alignment with loading direction is not dominant. It is also worth noticing that treated fibres are less oriented than untreated fibres, which is likely related to their length being shorter.

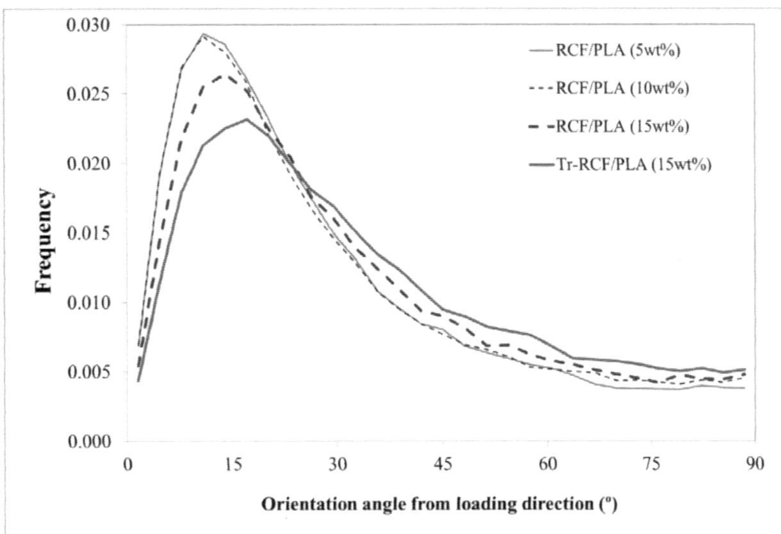

Figure 4. Distribution of angles between fibre direction and loading direction, obtained from X-ray tomography of composites.

The volume fraction of fibres in the composites measured from X-ray tomography agree very well with the values obtained by converting fibre weight fractions into volume fraction by using the density of constituents. Only for composites with 5wt% untreated fibres and 15wt% untreated fibres there is some discrepancy which can be explained by the small volume of composite that is scanned during the X-ray tomography (the fibre distribution may slightly vary along the specimen). Such a good correspondence between measured and theoretically calculated fibre volume fractions indicates that the composites are almost void free. This can be also seen from the images of composite cross-sections obtained from X-ray tomography (see figure 5). In order to avoid error due to the localized measurements, the theoretical fibre volume fractions are used in the further calculations. Inspection of the images in figure 5 also suggests that the fibres are evenly dispersed in the composites.

The fibre orientation factor is also very similar for all composites, being slightly lower for the composite with treated RCF. 3D X-ray tomography images showing fibres in composites are presented in figure 6. Altogether, the data in table 3 and in figures 4 and 6, show that the orientation of fibres is not completely random (orientation factor is higher than that for randomly oriented fibres ~0.4) but it is not clearly unidirectional either (the orientation factor is much smaller than 1).

Table 3. Fibre geometry and content (average values) obtained from X-ray tomography of composites.

Composite	Fibre length L [μm]	Aspect ratio L/d_f	Orientation factor η_o	Fibre volume fraction V_f [%]	Fibre volume fraction, theory[a] V_f^{th} [%]
RCF/PLA 5 wt%	266	21.3	0.62	5.2	4.2
RCF/PLA 10 wt%	281	22.5	0.61	8.6	8.4
RCF/PLA 15 wt%	277	22.1	0.57	12.8	12.7
Tr-RCF/PLA 15 wt%	123	9.9	0.53	11.5	12.7

[a]Calculated from fibre weight fraction and densities of constituents.

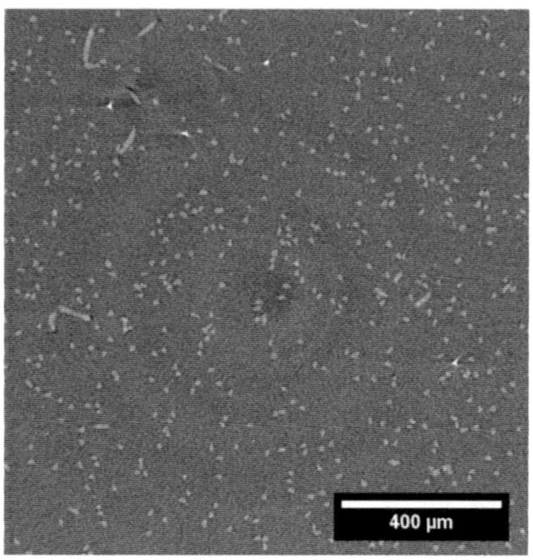

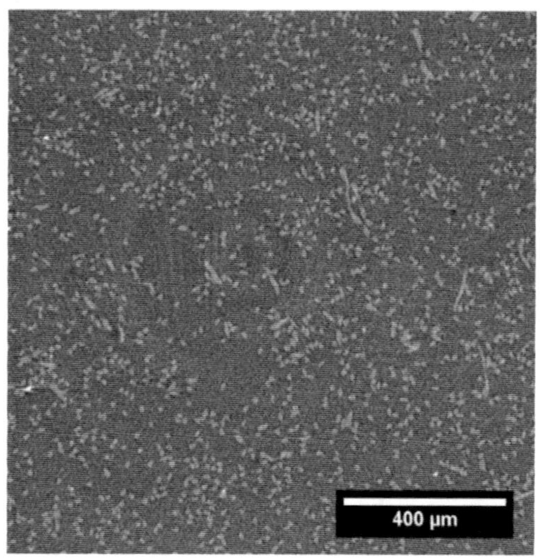

Figure 5. Cross-sections of 5%wt (left) and 15%wt (right) composites obtained from X-ray tomography. The discrete bright regions are cross-sections of individual fibres and the continuous grey region is matrix.

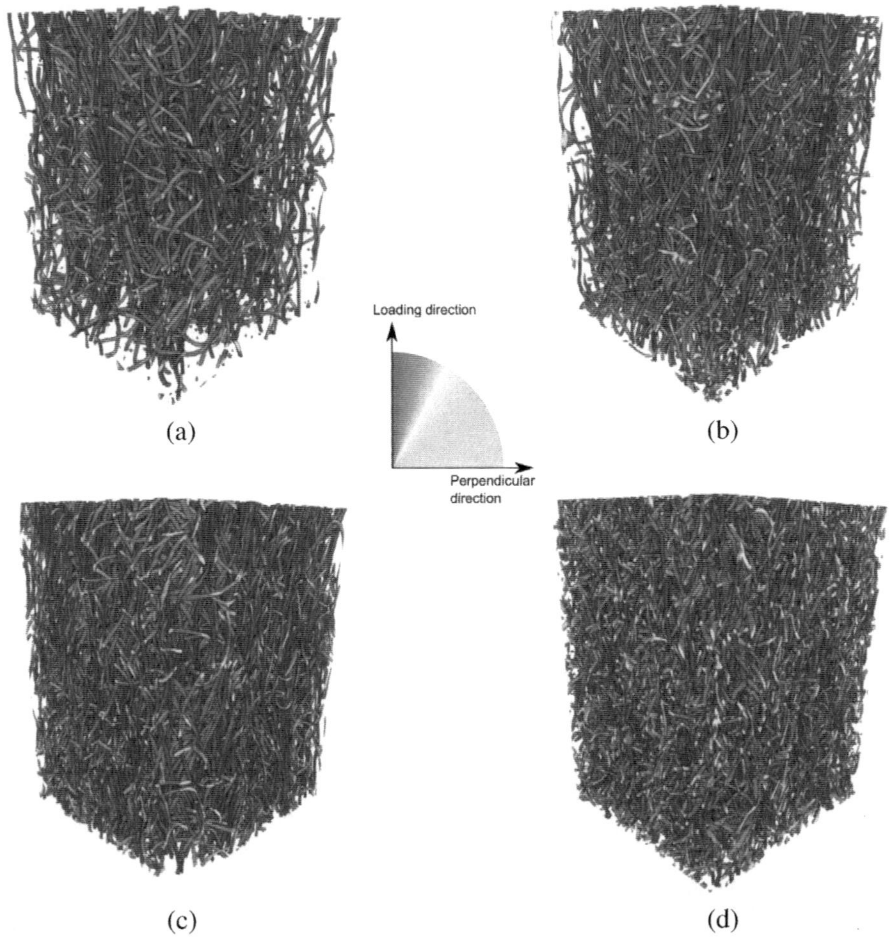

Figure 6. 3D X-ray tomography images of fibres in composites visualizing fibre orientation with colour coding: RCF/PLA (5wt% (a), 10wt% (b), 15wt% (c)) and Tr-RCF/PLA 15wt% (d).

4.2. Calculation of IFSS

The calculation of the IFSS was carried out in a couple different ways by the following procedures:

a) According to method proposed in [6], using two data points from stress-strain curves of composites (figure 1, right) and fitting IFSS and fibre orientation factor according to equations (1) - (2). A linear-elastic behaviour of reinforcement is assumed (using fibre stiffness from table 1 for conditioned fibres). For PLA matrix, the experimental stress-strain curve is used.

b) According to method proposed in [6], but only IFSS is fitted. The fibre orientation factor obtained from X-ray tomography is used (see table 3).

c) The whole stress-strain curve of composites up to the maximum stress value is fitted by the least squares method to obtain IFSS. This approach has been successfully applied to cellulosic fibre composites in [19-20]. The fibre orientation factor from X-ray tomography is used also in this case.

The results of the calculations of IFSS are summarized in table 4, the notation for methods are used according to the list presented above. The data obtained by fitting both IFSS and the fibre orientation factor (method a), show that IFSS is decreasing as the fibre volume fraction is increasing, while the orientation factor is increasing (only untreated fibres are compared). While values for 5wt% and 10wt% are comparable (approximately 15% difference), the biggest change is observed for the composite with 15wt% of fibres. A detailed discussion on such phenomena can be found in [19-20], where it is suggested that friction induced by residual thermal stresses may affect the measured IFSS, and since residual stress is affected by fibre content in the composite so is the measured IFSS. However, even though changes of IFSS (increase or decrease) with fibre content have been observed in short fibre composites, the changes for 15wt% are too dramatic (approximately 3 times compared to 5wt% and 10wt% composites) to be attributed to the frictional forces. Moreover, the fitted value of the fibre orientation factor for the 15wt% composite is very close to 1, which corresponds to perfect fibre alignment, and this is obviously not the case as shown by X-ray tomography (see table 3, and figures 4 and 6). The fibre orientation factors for 5wt% and 10wt% composites are lower than the values obtained from X-ray tomography (by about 20% – 40%), nevertheless these values seem to be more reasonable than the one obtained from fitting for the 15wt% composite. It should be noted that the orientation factor for the 15wt% composite with treated fibres is very close (but lower) to the value from X-ray tomography (only 8% difference).

The results obtained from methods (b) and (c) show a different trend with respect to IFSS than method (a), IFSS is increasing with fibre content (composites with untreated fibres are compared). However, also in this case, values of IFSS obtained from 5wt% and 10wt% composites are fairly comparable (25% - 33% difference), while the value calculated for the 15wt% composite is much higher (2-3 times). Although the IFSS values obtained in this study by the different methods vary rather significantly, it should be noted that the results obtained for the PLA/Cordenka composite system are similar to the IFSS reported in the literature for other cellulosic fibre composite systems. For example, IFSS obtained for Lyocell/PLA from fibre pull-out test [21] is 10.3-12.0 MPa (depending on fibre treatment), for flax/PLAA obtained from microbond, it is 9.9-22.2 MPa [22] (depending on thermal treatment of fibres), and for flax/PLGA, hemp/PLGA, and sisal/PLGA, it is 9.0 MPa, 11.3 MPa, and 14.3 MPa, respectively [23].

Table 4. Summary of the calculations of IFSS and the fibre orientation factor by different methods.

Material	RCF/PLA 5wt%		RCF/PLA 10wt%		RCF/PLA 15wt%		Tr-RCF/PLA 15wt%	
Method	IFSS [MPa]	η_o	IFSS [MPa]	η_o	IFSS [MPa]	η_o	IFSS [MPa]	η_o
a	14.1	0.39	12.2	0.50	4.5	0.93	25.7	0.49
b	4.7	0.62[a]	7.0	0.61[a]	17.7	0.57[a]	19.7	0.53[a]
c	5.0	0.62[a]	6.7	0.61[a]	12.9	0.57[a]	17.6	0.53[a]

[a]From X-ray tomography.

Even though reliability of the obtained IFSS values may be questioned, it is worth noticing that independently on the method used to calculate IFSS, the values obtained for treated fibres are consistently higher than those for untreated fibres. This may indicate that the chemical treatment of the fibres does improve adhesion between RFC and PLA.

As it was shown above (see figure 1), the stress-strain curves of the Cordenka RCF are highly non-linear, and the composites based on these fibres also exhibit non-linear stress-strain behaviour. Thus, it would be logical to use actual stress-strain curve for fibres rather than the assumed linear-elastic behaviour in the calculation procedure of IFSS. This approach was tried out in this study but the values of IFSS obtained from these try-outs were unreasonably high, or it was not possible at all to fit stress-strain curves for composites by the least square method. Therefore, these results are not presented and discussed in this paper, since no reasonable conclusive explanation of this problem has been found yet. Only speculations regarding possible reasons for discrepancy between modelling and experimental results may be offered at this point. It may be possible that the proposed model in [6-8] is not applicable for composite materials which exhibit highly non-linear behaviour (with viscoelastic and viscoplastic phenomena). It is also possible that the use of only axial fibre stiffness in the model is not sufficient, since RFC is expected to show anisotropic properties. Further investigation of abovementioned factors should be carried out in order to develop more appropriate and stable procedure for the characterization of composites containing highly non-linear reinforcement.

5. Conclusions

A comprehensive characterization of regenerated cellulose fibres and their composites based on a PLA matrix was carried out in order to evaluate interfacial shear strength in the composites. The internal structure of composite was characterized by X-ray tomography, and input parameters (e.g. fibre length and orientation distributions) for micro-mechanical models were obtained. The stress-strain curves of short fibre composites with different fibre volume fractions were used to calculate the interfacial shear strength. The obtained results are comparable to the data reported in the literature acquired from other direct test methods (e.g. fibre pull-out or microbond techniques).

The results obtained in this study, however, are somewhat unreliable and depend very much on the particular fitting process of the micro-mechanical model that is employed. Different trends of changes of interfacial shear strength with respect to the fibre volume fraction were obtained depending on the fitting method used. This may be due to the complex stress-strain behaviour of the regenerated cellulose fibres which exhibit highly non-linear behaviour (with viscoelastic/viscoplastic phenomena). Moreover, the regenerated cellulose fibres are expected to show anisotropic properties. Therefore, more advanced model with improved data analysis is required. This however will make the investigated characterization method much less industrially friendly (the initial objective of this work was to validate this approach).

Nevertheless, the determined interfacial shear strength for chemically treated regenerated cellulose fibres was consistently higher than for untreated fibres. This indicates that the proposed chemical treatment of cellulosic fibres is effective, and it does improve adhesion of these fibres to the PLA matrix.

Acknowledgements

Authors affiliated with Luleå University of Technology are thankful to Hassan II Academy of Science and Technology, the Moroccan National Centre for Scientific and Technical Research, the Swedish Research Council (ref. nr. 2009-6433), and EXCEL project (funded by local government Norrbotten, SWEDEN) for their financial supports.

Emma Lindgren, MSc student at Luleå University of Technology (2015), is acknowledged for help with preparation and characterization of composites.

References

[1] Faruk O, Bledzki AK, Fink H-P and Sain M 2012 Biocomposites reinforced with natural fibers: 2000–2010 *Prog Polym Sci* **37** 1552

[2] Shah DU 2013 Developing plant fibre composites for structural applications by optimising composite parameters: a critical review *J Mater Sci* **48** 6083

[3] Belgacem MN and Gandini A 2005 The surface modification of cellulose fibres for use as reinforcing elements in composite materials *Composite Interfaces* **12** 41

[4] Qian H, Greenhalgh ES, Shaffer MS and Bismarck A 2010 Carbon nanotube-based hierarchical composites: a review *J Mater Chem* **20** 4751

[5] Lee K-Y, Ho KKC, Schlufter K and Bismarck A 2012 Hierarchical composites reinforced with robust short sisal fibre preforms utilising bacterial cellulose as binder, *Composites Sci Technol* **72** 1479

[6] Thomason JL 2002 Interfacial strength in thermoplastic composites - at last an industry friendly measurement method? Composites: Part A 33 pp 1283-1288

[7] Bowyer WH and BaderMG 1972 On the reinforcement of thermoplastics by imper-fectly aligned discontinuous fibres *J Mater Sci* **7** 1315

[8] Bader MG and Bowyer WH 1973 An improved method of production for high strength fibre-reinforced thermoplastics *Composites* **4** 150

[9] Kelly A, Tyson WR, 1965 Tensile properties of fibre-reinforced metals *J Mech Phys Solids* **13** 329

[10] Lindgren E 2015 Development and Verification of Indirect Test Method for Characterisation of

Fibre/Matrix Adhesion in Regenerated Cellulose Fibre Based Composite (Luleå: Luleå University of Technology, Thesis Master of Science in Materials Engineering), p 56

[11] Hajlane A, Kaddami H, Joffe R and Wallström L 2013 Design and characterization of cellulose fibers with hierarchical structure for polymer reinforcement *Cellulose* **20** 2765

[12] Hajlane A, Kaddami H, Joffe R 2016 A green route for modification of regenerated cellulose fibres by cellulose nano-crystals *Cellulose* (submitted, currently under review)

[13] Miettinen A, Ojala A, Wikström L, Joffe R, Madsen B, Nättinen K and Kataja M 2015 Non-destructive automatic determination of aspect ratio and cross-sectional properties of fibres *Composites Part A: Applied Science and Manufacturing* **77** 188

[14] Krenchel H 1964 Fibre reinforcement, *dr.techn.thesis* (Copenhagen)

[15] Nordström Y, Joffe R and Sjöholm E 2013 Mechanical characterization and application of Weibull statistics to the strength of softwood lignin-based carbon fibers *Journal of Applied Polymer Science* **130** 3689

[16] Joffe R, Andersons J and Sparniņš E 2009 Applicability of Weibull strength distribution for cellulose fibers with highly non-linear behaviour *Proc. 17th Int. Conf. on Composite Materials (Edinburgh, UK)* p 10

[17] Pupure L, Doroudgarian N and Joffe R 2014 Moisture uptake and resulting mechanical response of biobased composites. I. constituents", *Polymer Composites* **36** 1150

[18] Ganster J and Fink H-P 2011 Man-Made Cellulose Short Fiber Reinforced Oil and Bio-Based Thermoplastics *Cellulose Fibers: Bio- and Nano-Polymer Composites*, ed S Kalia, B S Kaith and I Kaur (Berlin Heidelberg: Springer Verlag) chapter 18 pp 479-506

[19] Modniks J, E. Poriķe, Andersons J and Joffe R 2012 Evaluation of the apparent interfacial shear strength in short-flax-fiber/PP composites *Mechanics of Composite Materials* **48** 571

[20] Andersons J, Modniks J, Joffe R, Madsen B and Nättinen K 2016 Apparent interfacial shear strength of short-flax-fiber/starch acetate composites *International Journal of Adhesion & Adhesives* **64** 78

[21] Graupner N, Fischer H, Gerhard Z and Müssig J, 2014 Improvement and Analysis of Fibre/Matrix Adhesion of Regenerated Cellulose Fibre Reinforced PP-, MAPP- and PLA-composites by the Use of Eucalyptus Globulus Lignin *Composites: Part B* **66** 117

[22] Le Duigou A, Davies P and Baley C 2010 Interfacial Bonding of Flax Fibre/Poly(L-lactide) Bio-composites *Composites Science and Technology* **70** 231

[23] Czigány T, Morlin B and Mezey Z 2007 Interfacial Adhesion in Fully and Partially Biodegradable Polymer Composites Examined with Microdroplet Test and Acoustic Emission *Composite Interfaces* **14** 869

IOP Conf. Series: Materials Science and Engineering **139** (2016) 012022 doi:10.1088/1757-899X/139/1/012022

Local fatigue behavior in tapered areas of large offshore wind turbine blades

Seyed Aydin Raeis Hosseiny and Johnny Jakobsen

Department of Mechanical and Manufacturing Engineering, Aalborg University, Aalborg, Denmark

E-mail: aydin@m-tech.aau.dk

Abstract. Thickness transitions in load carrying elements lead to improved geometries and efficient material utilization. However, these transitions may introduce localized areas with high stress concentrations and may act as crack initiators that could potentially cause delamination and further catastrophic failure of an entire blade structure. The local strength degradation under an ultimate static loading, subsequent to several years of fatigue, is predicted for an offshore wind turbine blade. Fatigue failure indexes of different damage modes are calculated using a sub-modeling approach. Multi axial stresses are accounted for using a developed failure criterion with residual strengths instead of the virgin strengths. Damage initiation is predicted by including available Wöhler curve data of E-Glass fabrics and epoxy matrix into multi-axial fatigue failure criteria. As a result of this study, proper knock-down factors for ply-drop effects in wind turbine blades under multi-axial static and fatigue loadings can be obtained.

1. Introduction

Damage tolerant design of composites attracts attention of industry, owing to significantly reducing the maintenance costs in remote structures. Offshore wind turbine blades are designed to fulfil high endurance limits with minimum repair requirements+. Improving the effective life-time of the blades requires special considerations in areas, where continuous degradation of material characteristics due to high loading cycle is inevitable. Thickness transitions (ply-drops) are critical failure areas, because of local stress concentrations. Failure at ply-drops is largely dominated by interlaminar and transverse shear effects in the matrix [1, 2]. The process of failure can lead to ply delamination at relatively low applied strains under fatigue loading [3].

Among the current fatigue models in effect, fatigue life model and residual strength or stiffness models are formulated in macro-scale, while progressive damage model is defined in micro-scale [4, 5]. Progressive damage model is a computationally expensive method with inherent complexities. Additionally, obtaining the experimental data for generating sufficiently accurate Wöhler (S-N) curves and residual strength models is considerably time consuming [5]. A method of reducing computational and experimental cost in fatigue life prediction of multi-directional laminates through conducting experiments of different stress levels and geometries has been established [6]. By constructing a relationship between fatigue life and stress level for each stage, multi-axial damage index of a static strength failure criterion can predict the fatigue life of a general stacking sequence and ply orientation [5]. A relationship between damage evolution and remaining life of the composites, outlines the process of damage mechanism that would take place in the course of ultimate fracture [7]. Despite the promising applications of these methods, they are mainly experimentally validated against constant amplitude tension fatigue type problems.

A vast amount of damage mechanisms in composite materials, force designers to consider the degraded stiffness and residual strength of blades, subjected to life-time fatigue in their calculations. This is typically done through design codes, where suitable knock-down factors are recommended. A new blade design has to fulfill its reliability under an extreme of a 20 years fatigue, followed by static loading to its ultimate limits. This particular loading scenario has been considered as an example for the present study. Local strength degradation and onset of damage in two selected ply-drop regions of the DTU 10 MW reference wind turbine [8] has been investigated. The blade geometry is provided in the form of an ABAQUS [9] finite element model with conventional layered shell elements (figure 1). Homogeneous anisotropic approximation of material properties are assigned to multidirectional laminates in the model, for simplicity [10, 11]. Composite layup is defined by partitioning the blade into 11 regions circumferentially (figure 2) and 100 regions radially [8]. The reference model considers the blade body as a prismatic beam passing through defined sections and does not consider the effect of tapering throughout the blade [8].

Figure 1. DTU 10 MW wind turbine blade reference model layup partitioning.

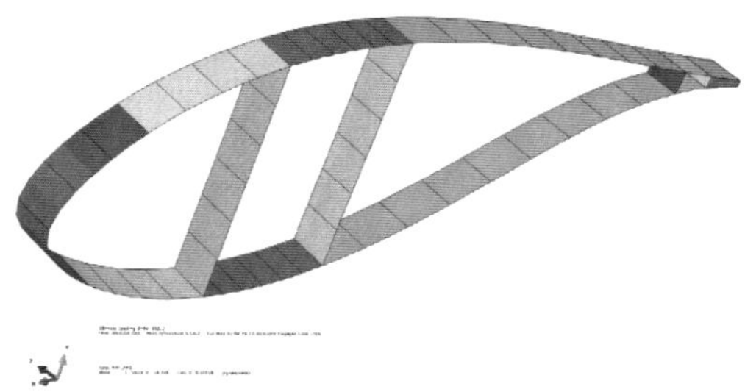

Figure 2. DTU 10 MW wind turbine blade reference model composite layup definition along circumferential partitions.

This study aims to determine the loss of strength in ply-drop areas in a wind turbine blade structure, where material transition discontinuities, abrupt change in bending stiffness, and singular inter-laminar stresses turn them into critical failure prone regions.

2. Methodology

Load combinations outcome have been provided in 27 cross sections along the reference blade, in accordance with IEC 61400-1 Ed.3 [8]. The ultimate load case(1.3)[1] for maximum moment resultant is the prominent design load case. Corresponding forces and moments are distributed over nodal points in cross sections, through defining coupled constraints with respect to nodal distance from the elastic center (figure 3). The resulted displacement field is then applied to a sub-model to investigate the local stress concentrations and damage onset in selected thickness tapered regions as sub-models in ABAQUS (figure 4). Sub-modeling provides finer mesh resolution and use of different set and geometry of elements, while all essential and natural boundary conditions from the main model are inherited over the desired area [13]. Conventional layered shell elements are replaced with solid elements to model the geometrical variations and obtain a local strain distribution in ply-drop areas, especially near the boundaries of the resin

[1] Extreme turbulence model wind conditions with safety factor of 1.35, according to IEC 61400-1 Ed.3.

Figure 3. DTU 10 MW offshore wind turbine blade reference model under ultimate design load case(1.3), according to the Guideline approach [12] with an inherent partial safety factor of 1.35 for flap-wise blade bending moment.

rich pocket area. Biaxial and triaxial laminae have been considered as an assembly of individual uni-directional laminae with their corresponding directions, bonded together in interfaces. Large displacements due to blade movements, have been carefully decoupled from internal displacement field. Linear elastic material properties for the constituent materials are mentioned in tables 1-4. Tensile, compressive, and shear strengths of the material that have not been reported in the reference blade document [8] have been derived from the literature by matching the materials elastic moduli [14] (tables 1 and 2).

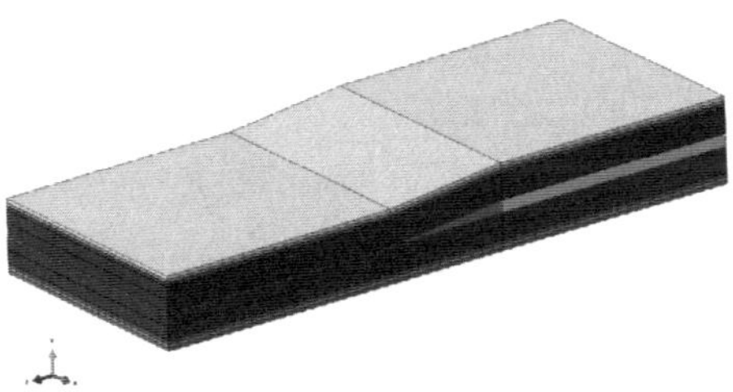

Figure 4. Finite element sub-model of ply-drop effect, to study localized stresses, leading to failure initiation.

Ply stacking of the sub-model which is a 1:10 tapered section of $[[-45/0/+45]/0_5/[+45/0/-45]]$ to $[[-45/0/+45]/0_2]_s$ (figure 4), has to be considered under fatigue loading. Tension and compression fatigue test results from equivalent sample of QQ1[2][3] of the same material with very close mechanical properties is adopted in calculation of residual strength by means of tensile and compressive stress to fatigue life curves in both fiber and transverse directions (figures 5 and 6).

[2] Test specimen series of epoxy and glass fiber with 53% fiber volume with a similar layup to the numerical model of the current study [3].

Table 1. Mechanical properties of the epoxy matrix [8, 14].

Property	Value	Unit
Young's modulus E_m	4.0	GPa
Poisson's ratio ν_m	0.35	...
Shear modulus G_m	1.48	GPa
Mass density ρ_m	1140	kg/m^3
Tensile strength Y_{mt}	0.075	GPa
Compressive strength Y_{mc}	0.150	GPa
Shear strength S_m	0.070	GPa

Table 2. Mechanical properties of E-glass fibers [8, 14].

Property	Value	Unit
Longitudinal Young's modulus E_{f1}	75.0	GPa
Transverse Young's modulus E_{f2}	75.0	GPa
Major Poisson's ratio ν_{12}	0.2	...
In-plane shear modulus G_{f12}	31.25	GPa
Transverse shear modulus G_{f23}	31.25	GPa
Mass density ρ_f	2550	kg/m^3
Longitudinal tensile strength X_{ft}	2.150	GPa
Longitudinal Compressive strength X_{fc}	1.450	GPa

Table 3. Apparent mechanical properties of two uni-directional laminae [8].

Property	Lamina 1	Lamina 2	Unit
Fiber volume fraction V_f	0.50	0.55	...
Young's modulus in fiber direction E_1	39.50	43.05	GPa
Young's modulus in dir. transverse to dir.1 E_2	12.10	13.43	GPa
In-planar shear modulus G_{12}	4.54	5.05	GPa
Out-of-plane shear modulus G_{23}	4.54	5.05	GPa
In-plane Poisson's ratio ν_{12}	0.275	0.268	...
Out-of-plane Poisson's ratio ν_{23}	0.333	0.330	...
Mass density ρ	1845	1915	kg/m^3

A set of modified Hashin-type damage criterion [15], capable of predicting different modes of fatigue failure is employed to model the progressive fatigue damage in the sub model. In order to excess the necessity of performing intensive test numbers, the material properties are considered as function of number of cycles, stress states and stress ratios [6]. A residual strength approach, which considers the material stiffness and strength as a function of time, predicts the remaining fatigue life of a unidirectional ply under a multiaxial state of fatigue stress [6]. Different modes

Table 4. Fiber orientation and apparent mechanical properties of the multi-directional plies [8].

Multi-directional ply	Uniax	Biax	Triax	Unit
Fiber volume fraction V_f	0.55	0.5	0.5	...
Unidirectional lamina	Lamina2	Lamina1	Lamina1	...
0°fibers	95	0	30	%
90°fibers	5	0	0	%
+45°fibers	0	50	35	%
−45°fibers	0	50	35	%
Young's modulus E_1	41.63	13.92	21.79	GPa
Young's modulus E_2	14.93	13.92	14.67	GPa
Shear modulus G_{12}	5.047	11.50	9.413	GPa
Poisson's ratio ν_{12}	0.241	0.533	0.478	...
In-plane shear modulus G_{f12}	31.25	GPa		
Transverse shear modulus $G_{13} = G_{23}$	5.05	4.54	4.54	GPa
Mass density ρ	1915	1845	1845	kg/m^3
Longitudinal tensile failure strain ϵ_1^T	2.10	1.60	2.20	%
Longitudinal Compressive failure strain ϵ_1^C	1.50	1.50	1.80	%

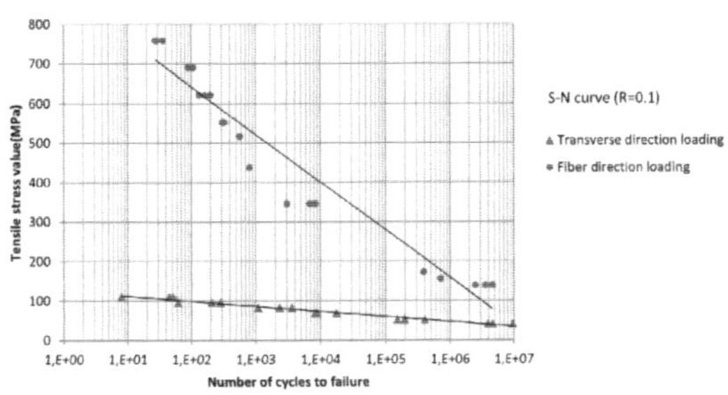

Figure 5. S-N curve for the laminate under different tensile fatigue states(R=0.1) for fiber and transverse directions [3].

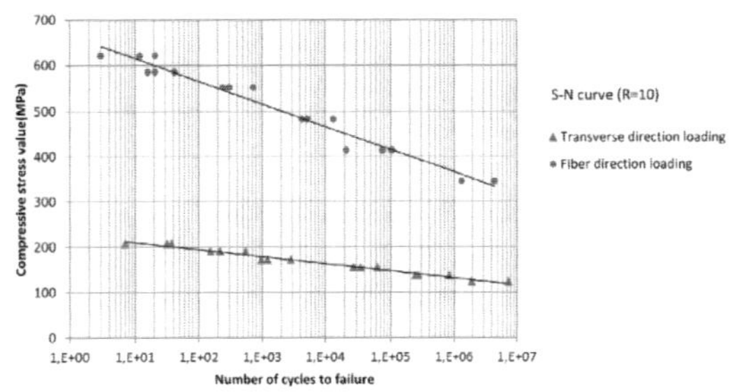

Figure 6. S-N curve for the laminate under different compressive fatigue states(R=10) for fiber and transverse directions [3].

of fatigue failure initiation, propagation, and ultimate damage of a unidirectional ply under uniaxial fatigue loading are calculated by the following criteria [6].

For fiber tension fatigue failure mode ($\sigma_{xx} > 0$), the criterion is:

$$[\frac{\sigma_{xx}}{X_t(n,\sigma,\kappa)}]^2 + [\frac{\frac{\sigma_{xy}^2}{2E_{xy}(n,\sigma,\kappa)} + \frac{3}{4}\delta\sigma_{xy}^4}{\frac{S_{xy}^2(n,\sigma,\kappa)}{2E_{xy}(n,\sigma,\kappa)} + \frac{3}{4}\delta S_{xy}^4(n,\sigma,\kappa)}] + [\frac{\frac{\sigma_{xz}^2}{2E_{xz}(n,\sigma,\kappa)} + \frac{3}{4}\delta\sigma_{xz}^4}{\frac{S_{xz}^2(n,\sigma,\kappa)}{2E_{xz}(n,\sigma,\kappa)} + \frac{3}{4}\delta S_{xz}^4(n,\sigma,\kappa)}] = g_{F+}^2 \quad (1)$$

where $X_t(n,\sigma,\kappa)$, $S_{xy}(n,\sigma,\kappa)$, and $S_{xz}(n,\sigma,\kappa)$ are the fiber residual tensile, in-plane shear, and out of plane shear fatigue strengths, $E_{xy}(n,\sigma,\kappa)$ and $E_{xz}(n,\sigma,\kappa)$ are in-plane and out of plane shear residual fatigue stiffness of a unidirectional ply under uniaxial fatigue loading, and n,σ, κ, and δ account for number of cycles, stress state, stress ratio, and parameter of material nonlinearity, respectively.

For fiber compression fatigue failure mode ($\sigma_{xx} < 0$), the criterion is:

$$[\frac{\sigma_{xx}}{X_c(n,\sigma,\kappa)}] = g_{F-}^2 \tag{2}$$

where $X_c(n,\sigma,\kappa)$ is the fiber compressive residual fatigue strength of a unidirectional ply under uniaxial fatigue loading.

For fiber-matrix shearing fatigue failure mode ($\sigma_{xx} < 0$), the criterion is:

$$[\frac{\sigma_{xx}}{X_c(n,\sigma,\kappa)}]^2 + [\frac{\frac{\sigma_{xy}^2}{2E_{xy}(n,\sigma,\kappa)} + \frac{3}{4}\delta\sigma_{xy}^4}{\frac{S_{xy}^2(n,\sigma,\kappa)}{2E_{xy}(n,\sigma,\kappa)} + \frac{3}{4}\delta S_{xy}^4(n,\sigma,\kappa)}] + [\frac{\frac{\sigma_{xz}^2}{2E_{xz}(n,\sigma,\kappa)} + \frac{3}{4}\delta\sigma_{xz}^4}{\frac{S_{xz}^2(n,\sigma,\kappa)}{2E_{xz}(n,\sigma,\kappa)} + \frac{3}{4}\delta S_{xz}^4(n,\sigma,\kappa)}] = g_{FM}^2 \tag{3}$$

For matrix tension fatigue failure mode ($\sigma_{yy} > 0$), the criterion is:

$$[\frac{\sigma_{yy}}{Y_t(n,\sigma,\kappa)}]^2 + [\frac{\frac{\sigma_{xy}^2}{2E_{xy}(n,\sigma,\kappa)} + \frac{3}{4}\delta\sigma_{xy}^4}{\frac{S_{xy}^2(n,\sigma,\kappa)}{2E_{xy}(n,\sigma,\kappa)} + \frac{3}{4}\delta S_{xy}^4(n,\sigma,\kappa)}] + [\frac{\sigma_{yz}}{S_{yz}(n,\sigma,\kappa)}]^2 = g_{M+}^2 \tag{4}$$

where $Y_t(n,\sigma,\kappa)$ and $S_{yz}(n,\sigma,\kappa)$ are residual transverse tensile and out of plane shear fatigue strengths of a unidirectional ply under uniaxial fatigue loading, respectively.

and finally for matrix compression fatigue mode ($\sigma_{yy} < 0$), we can write:

$$[\frac{\sigma_{yy}}{Y_c(n,\sigma,\kappa)}]^2 + [\frac{\frac{\sigma_{xy}^2}{2E_{xy}(n,\sigma,\kappa)} + \frac{3}{4}\delta\sigma_{xy}^4}{\frac{S_{xy}^2(n,\sigma,\kappa)}{2E_{xy}(n,\sigma,\kappa)} + \frac{3}{4}\delta S_{xy}^4(n,\sigma,\kappa)}] + [\frac{\sigma_{yz}}{S_{yz}(n,\sigma,\kappa)}]^2 = g_{M-}^2 \tag{5}$$

where $Y_c(n,\sigma,\kappa)$ is the transverse compressive residual fatigue strength of a unidirectional ply under uniaxial fatigue loading. In each of the equations 1 to 5, failure occurs when g_* value on the right hand side of the criterion exceeds 1. This means that the material properties under the increasing number of cycles decrease to a level that fail to withstand the applied stresses.

The residual strength in the material by the number of fatigue loadings cycles with random stress ratio is formulated as [6]:

$$R(n,\sigma,\kappa) = [1 - (\frac{\log(n) - \log(0.25)}{\log(N_f - \log(0.25)})^{\beta}]^{\frac{1}{\alpha}}(R_s - \sigma) + \sigma \tag{6}$$

where R_s is static strength, N_f is the fatigue life at σ, and α and β are experimental curve fitting parameters.

A similar equation for the stiffness degradation by number of cycles is rearranged as [16]:

$$E(n,\sigma,\kappa) = [1 - (\frac{\log(n) - \log(0.25)}{\log(N_f - \log(0.25)})^{\lambda}]^{\frac{1}{\gamma}}(E_s - \sigma/\varepsilon_f) + \sigma/\varepsilon_f \tag{7}$$

where E_s is static stiffness, ε_f is average strain to failure, and λ and γ are experimental curve fitting parameters.

3. Results and Discussion

Ply-drop configurations have been modelled on two extreme stress concentration positions throughout the cap sections on pressure and suction sides of the reference wind turbine blade. The sub-model with continuum solid elements of quadratic order inherits the displacement field from the global model and shows the local distribution of different stress states around the dropped thickness area (figures 7 and 8). Average nodal stress components of elements with high stress concentrations in the vicinity of resin pocket (table 5) are input in equations 1-5 to calculate the failure index for each fatigue failure mode. Final failure mechanism and remaining life is estimated by replacing the virgin static strengths in the criteria by residual static strength of the material [17].

Table 5. Average stress values at points of stress for calculating the fiber/matrix fatigue failure index.

Stress component	suction side	pressure side	Unit
σ_{xx}	-83	224	MPa
σ_{xy}	-27	7	MPa
σ_{xz}	-12.6	-24.4	MPa
σ_{yy}	-31.3	-45	MPa
σ_{yz}	-2	-45	MPa
σ_{zz}	29.8	41.06	MPa

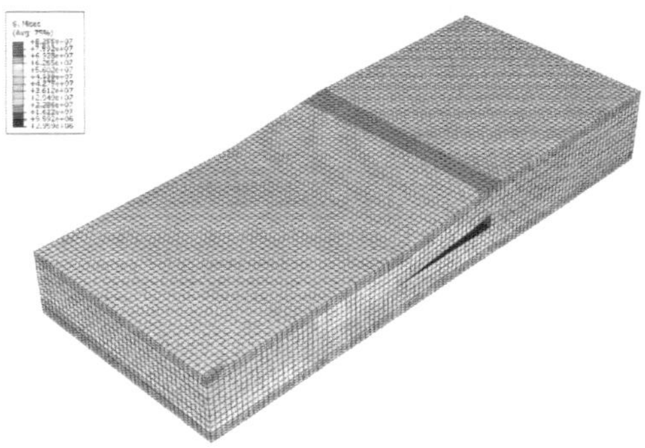

Figure 7. Axial stress distribution in ply-drop sub-model under ultimate tensile loading within pressure side of the cap section.

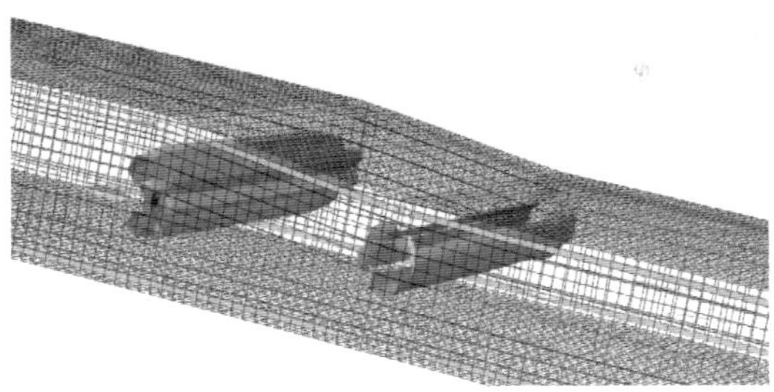

Figure 8. Interlaminar stress concentrations about resin pocket edges in ply-drop area within suction side of the cap section.

Residual strength and stiffness throughout the life-time of blade have been approximated by curve-fitting over existing Wöhler curve data and equations 6-7. Residual strength has been calculated in 20%, 60%, and 80% of the normalized fatigue life and introduced into the fatigue failure criterions (figure 9). Degradation of the material characteristics over the time, resembles fatigue loading effects and influences each failure criterion in a particular way. Failure index foremost hits the limits for delamination and fiber tension fatigue in pressure side of the blade, while matrix compression and delamination fatigue failure modes are prominent modes in suction side. Fiber tension fatigue failure mode is observed to be more sensitive to the cyclic loading with a more steep trend during the time. Other failure modes are almost constant during 80% of the life-time and increase abruptly throughout the last 20%. The method calculates an approximation of the remaining life-cycles for the laminate after undergoing definite load level at any stage of the composite, from damage initiation until its evolution and final failure.

4. Conclusions

Fatigue failure in ply-drop areas of wind turbine blades has been modelled and studied through a progressive and multi-axial fatigue damage modeling in developed Hashin criteria. Damage state has been predicted at different load levels and number of cycles. The damage nucleation by delamination near the resin rich pocket area complies with analytical solutions [18] and available experimental results. The potential of the study in reducing the required number of experimental data for different loading scenarios, improves the design process of new wind turbine blade configurations.

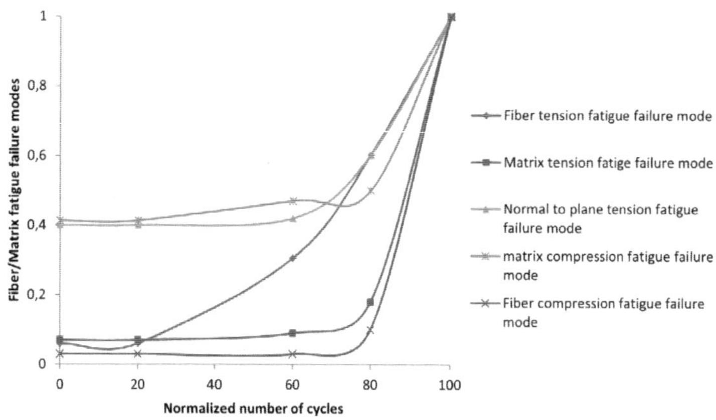

Figure 9. fatigue index with respect to the normalized number of cycles for the fatigue modes of fiber and matrix.

References

[1] Llanos A and Vizzini A 1992 *Journal of Composite Materials* **26** 1968–1983
[2] He K, Hoa S V and Ganesan R 2000 *Composites Science and Technology* **60** 2643–2657
[3] Mandell J and Samborsky D 2010 *Sandia National Laboratories* 1–198
[4] Degrieck J and Van Paepegem W 2001 *Applied Mechanics Reviews* **54** 279–300
[5] Liu Y and Mahadevan S 2013 *Fatigue of Composite Materials* **3** 3
[6] Shokrieh M M and Lessard L B 2000 *Journal of Composite Materials* **34** 1056–1080
[7] MINNETYAN L and CHAMIS C 2013 *Fatigue of Composite Materials* **3** 3
[8] Bak C, Zahle F, Bitsche R, Kim T, Yde A, Henriksen L, Andersen P B, Natarajan A and Hansen M H 2014 *Wind Energy*
[9] Version A 2014 *Analysis Users Guide, Dassault Systems*
[10] Chamis C C 1984 *SAMPE Quarterly* **15** 14–23
[11] Jones R M 1998 *Mechanics of composite materials* (CRC press)
[12] Guideline G W 2010 *Hamburg: Germanischer Lloyd Wind Energie Gmb H*
[13] Dassault Systèmes Simulia 2014 1146
[14] Soden P, Hinton M and Kaddour A 1998 *Composites Science and Technology* **58** 1011–1022
[15] Hashin Z 1980 *Journal of applied mechanics* **47** 329–334
[16] Naderi M and Maligno A 2013 *Journal of Composite Materials* **47** 475–484
[17] Hashin Z 1981 *International Journal of Fracture* **17** 101–109
[18] Pagano N J 2012 *Interlaminar response of composite materials* (Elsevier)

IOP Conf. Series: Materials Science and Engineering **139** (2016) 012023 doi:10.1088/1757-899X/139/1/012023

Influence of the curing cycles on the fatigue performance of unidirectional glass fiber reinforced epoxy composites

Jonas Hüther[1] **and Povl Brøndsted**[2]

[1] Karlsruhe Institute of Technology, Institute for Applied Materials, Department of Hybrid and Lightweight Materials, Karlsruhe, Germany

[2] Technical University of Denmark (Risø campus), Department of Wind Energy, Section for Composites and Materials Mechanics, Roskilde, Denmark

E-mail: Jonas.Huether@kit.edu

Abstract. During the manufacturing process of fiber reinforced polymers the curing reaction of the resin results in shrinkage of the resin and introduces internal stresses in the composites. When curing at higher temperatures in order to shorten up the processing time, higher curing stresses and thermal stresses are built up and frozen, as residual stresses occur. In the present work, a glass fiber reinforced epoxy composite laminate with an unidirectional architecture based on non-crimp fabrics with backing fibers is investigated. Three different curing cycles (time-temperature cycles) are used, leading to different levels of internal stresses. The mechanical properties, static strength and fatigue life time, are measured in three different directions of the material, i.e. the fiber direction, 0°, the 30° off axis direction, and the 90° direction transverse to the fiber direction. It is experimentally demonstrated that the resulting residual stresses barely influences the quasi-static mechanical properties of reinforced glass-fiber composites. It is found that the fatigue performance in the 0° direction is significantly influenced by the internal stresses, whereas the fatigue performance in the off axes directions so is not significantly influenced of these stresses. This is related to the observations that the damage mechanisms in the off axes directions are mainly related to shear failure in the matrix and in the interface between fiber and matrix and different from the damage mechanisms in the fiber direction, where the damage initiates in the transverse backing fibers and is directly related to fiber fractures in the load-carrying axial fiber bundles.

1. Introduction

The introduction is divided into two parts. Firstly, an overview over glass fiber reinforced materials and their manufacturing is given. Secondly, resin curing kinetics are described.

1.1. *Glass Fiber Reinforced Materials*

Epoxy resins are commonly used as matrix materials in glass fiber reinforced composites for rotor blades of wind turbines. To produce a glass fiber reinforced epoxy composite, the dry glass fiber fabrics (non-crimp fabrics) are build up in a mold and sealed in a vacuum bag. Two polymer components, resin and hardener, are mixed and the reinforcing glass fiber based structure is infused using vacuum assisted resin transfer molding (VARTM). During and after the infusion, the resin will undergo a curing reaction and transform to the solid state, the matrix. During this reaction, internal stresses, which heavily depend on the time-temperature cycle during the curing, are built up in the composite.

Lower temperatures demand for longer times to fully cure the material and result in lower stresses whereas higher temperatures and shorter curing times increase these stresses. In manufacturing, it is attractive to reduce the mold time and thus increase the output by increasing the curing temperature. However, this is at the expense of the quality of the laminate.

The question is, how the material's properties are affected and the challenge is to find a compromise which meets the needs of production time and material quality. It is known that higher levels of residuals strains in a composite material reduce the fatigue performance of a glass fiber composite. This is particularly important for wind turbine blades as they are exposed to periodic loads and as repairs during service are difficult. Thus, the correlations between residual stresses and fatigue performance were studied manifoldly, showing that higher residual stresses cause a decrease in fatigue performance for composite materials. [1, 2, 3, 4] For this reason, three different panels are in the focus of this work, demonstrating the nexus between curing cycles, quality and properties.

In a previous study Pereira et al. investigated three unidirectional non-crimp fabric panels cured under three different curing cycles and the magnitude of the stresses was measured [1]. It was experimentally shown how these stresses have a large influence on the fatigue life time of the composite when loaded in the unidirectional fiber direction. It is the objective of the present work to measure if the curing cycles also influence the static properties and the off axes fatigue properties when the same panels are loaded in the off axes directions, that is load directions oriented 30° and 90° to unidirectional fibers.

1.2. Resin Curing Kinetics

The curing of an epoxy resin is the transformation from its liquid reactants to the cured, solid polymer. Thereby, the physical interactions between the initial molecules are partly substituted by covalent bonds. [5, 6, 7]

Before the reaction, each monomer occupies a volume larger than its real size due to physical interaction with its surroundings and other molecules and due to their atomic motion. As the chemical bonds are of a shorter range than the physical interactions, the molecules get packed closer together when transforming to the solid [8]. Macroscopically, this transformation is accompanied by a decrease in volume and thus an increase in density, referred to as shrinkage.[9]

A schematic and two-dimensional representation of this process is depicted in figure 1: In a), a monomer is represented by a hexagon. The inner, grey area is the size of the molecule, the outer contour is the space occupied due to physical interactions. In b), the curing process is sketched. Initially (i) the nineteen monomers can move freely but repulse each other due to physical interactions. During curing, chemical bonds are formed, and the molecules get packed closer together, with physical interactions only relevant on the surface of the formed macro molecule (ii)[8]. Although the total number of represented monomers is the same, the area of this flat macro molecule is reduced, as indicated by the black arrows. Transferred to three dimensions, this model demonstrates volumetric shrinkage during the curing process.

The total shrinkage is defined by the physical properties of the components and its reaction and is neither a function of time nor of temperature. As long as the molecules can move freely, shrinkage can cause a change in shape but will not impair the mechanical properties. However, when the polymer is in contact with stiffer material, such as reinforcing fibers, stresses can built up. This is demonstrated in figure 2: During the curing process, the resin material shrinks and gets attached to the fiber. Being the stiffer material, the fiber will undergo a compressive elastic deformation building up compressive stresses [10]. Further, the difference in the coefficient of thermal expansion of polymers and fibers has an effect on the residual stresses in a composite.

In the study by Pereira et al. a Fiber Bragg Grating (FBG) sensor is placed in a liquid,

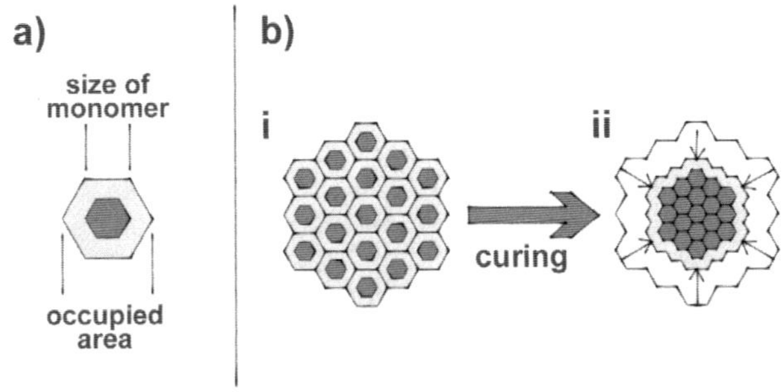

Figure 1. Simplified sketch of curing process. After curing, the molecules are packed closer together and thus the material's density increases (based on Shah and Schubel [8]).

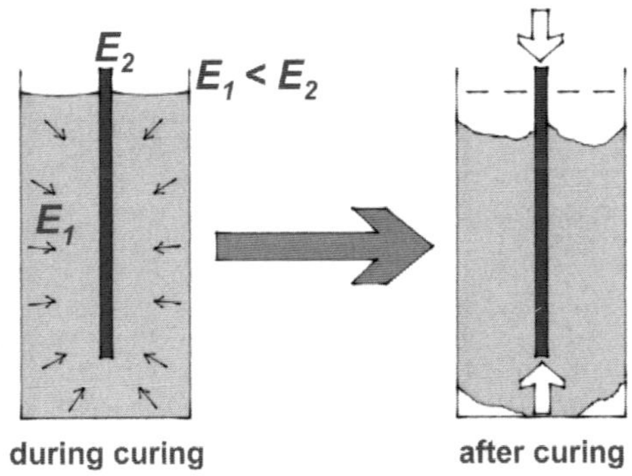

Figure 2. Shrinkage during curing around and obstacle, for instance a glass fiber (black). As Youngs's modulus of the fiber E_2 is significantly higher than Young's modulus of the resin E_1, the fiber will be stressed, as indicated by the white arrows.

freshly mixed resin sample. Such, the evolution of strains can be measured directly and in situ[1]. A typical curve for strains measured by means of FBG sensors is displayed in figure 3. The strain evolution can be tracked over the course of time starting as liquid and then transforming to a gelled and solid state. While the resin system is in its liquid phase, no strains are measured. When the resin transitions into a gelled and later solid state, strains built up and can be detected. After curing is accomplished, the material is cooled to ambient temperature. The final strain value reading then gives a measure of the residual strain existent in the specimen. This value combines the strains due to chemical shrinkage $\Delta\varepsilon_{shr}$ and the strains due to thermal contraction $\Delta\varepsilon_{TC}$. [1, 4, 11, 12, 13]

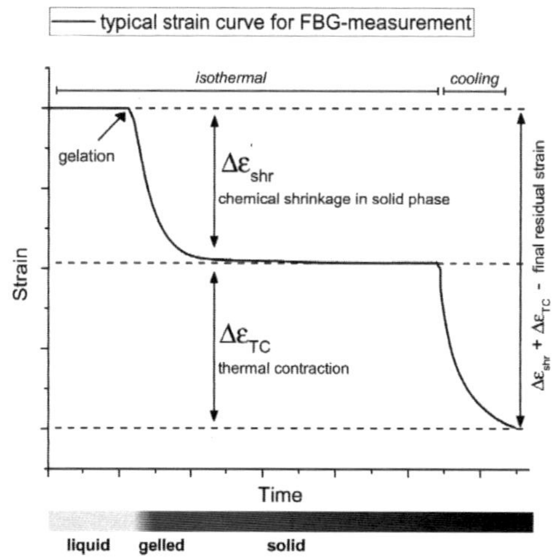

Figure 3. Schematic course of curve of the strain evolution during curing and subsequent cooling. The final strain value is the sum of strains due to chemical shrinkage $\Delta\varepsilon_{shr}$ and strains due to thermal contraction $\Delta\varepsilon_{TC}$.

2. Experiments

In this section, the used materials and experiments will be described.

2.1. Materials

Three panels were manufactured using VARTM under three different projected curing cycles according to figure 4. All cycles lead to a full cure of the material. The one-stage curing cycle at 110 °C (EQA-03) exceeds the maximum glass transition temperature (84 °C) of this system, the curing cycle at 75 °C (EQA-02) remains below it. For the two-stage cycle (EQA-01), temperatures were chosen to be moderate but achieve a full cure.

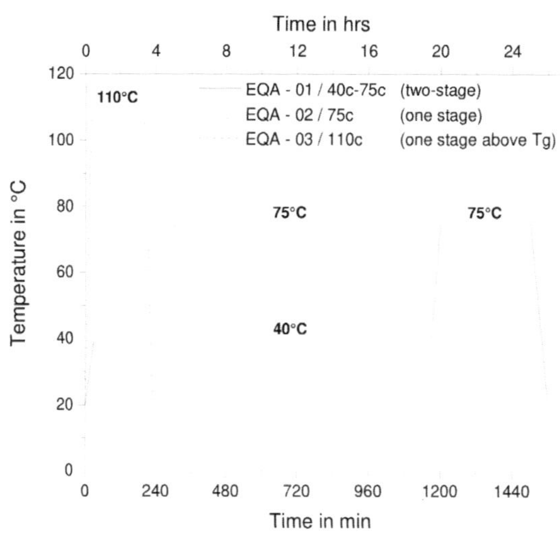

Figure 4. Projected time-temperature profile for the laminates from which specimens were cut.

IOP Conf. Series: Materials Science and Engineering **139** (2016) 012023 doi:10.1088/1757-899X/139/1/012023

Laminate built-up and the test specimen orientation and geometry are found in figures 5 and 6. Figure 5 represents the dry fiber layup of all laminates, figure 6 highlights the cutting directions. Hence in addition to the tests reported by Pereira et al., the new specimen are mainly reinforced with fibers in 30° and 90° to the specimen's longitudinal axis, which is also the axis of the load application.

The fibers were of type *Biax 600* and *UD1150+Biax100* and were vacuum-infused with the resin system *Araldite LY 1564 SP / hardener XB 3486* produced by Huntsman. Prior to the infusion, the resin system was mixed and degassed. The test laminates were cured in three different time-temperature curing cycles as depicted in figure 4.

Figure 5. Layup of the glass fibers.

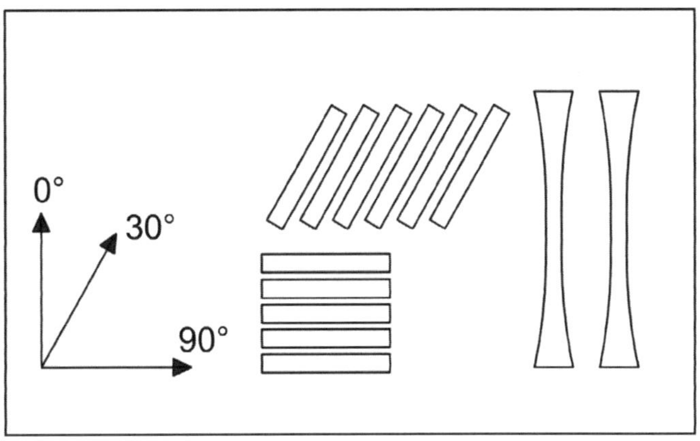

Figure 6. Cutting directions and specimen types.

2.2. Mechanical Test

The following test were performed:

- Static test of the matrix
- 0°-direction – Static, 2 test specimens from panel EQA-01 and EQA-02 and 3 test specimens from panel EQA-03
- 0°-direction – Fatigue, $R = 0.1$, 7 test specimens from panel EQA-01 and 6 test specimens from panel EQA-02 and EQA-03
- 30°-direction – Static, 3 test specimens from each panel
- 30°-direction – Fatigue, $R = 0.1$, 3 test specimens from each panel
- 90°-direction – Static, 3 test specimens from each panel
- 90°-direction – Fatigue, $R = 0.1$, 3 test specimens from each panel

Static tensile tests were performed using an universal servohydraulic test machine. Displacement rate was 2 mm/min. Fatigue tests were run using the same equipment and test were run at 4-5 Hz up to a run out level of $2 \cdot 10^6$ cycles in tensile-tensile mode with $R = 0.1$.

3. Results

In the following, experimental results are provided.

3.1. Residual Strains

As reported by Pereira et al. (refer to figure 3) values for the residual strain yield the results listed in table 1.

Table 1. Residual strains measured using FBG sensors

Material ID	Residual strain in %		
	specimen 1	specimen 2	mean value
EQA-01	0.44	0.44	0.44
EQA-02	1.31	1.44	1.38
EQA-03	1.24	0.82	1.03

3.2. Fiber Content

For quality assessment and to ensure the comparability of the different materials, the fiber volume content for each material was determined using pyrolysis and the results are given in table 2.

Table 2. Fiber volume content of all tested materials

Material ID	Fiber volume content in %
EQA-01	57.8
EQA-02	58.8
EQA-03	57.1

3.3. Quasi-static Test Results

The results of the quasi-static tests for composite specimens and neat resin specimens are displayed graphically in figure 7 (Young's modulus) and figure 8 (tensile strength) and are subsequently listed as tables (table 3 to 5). For composite specimen, the label indicates the cutting direction, that is, the tilting between the unidirectional reinforcement and the specimen axis. Pure resin specimen show no significant variation of quasi-static tensile properties as displayed in the right columns of figure 7 and 8. As for neat resin material, quasi-static properties of composite materials are not significantly influenced by the curing cycle.

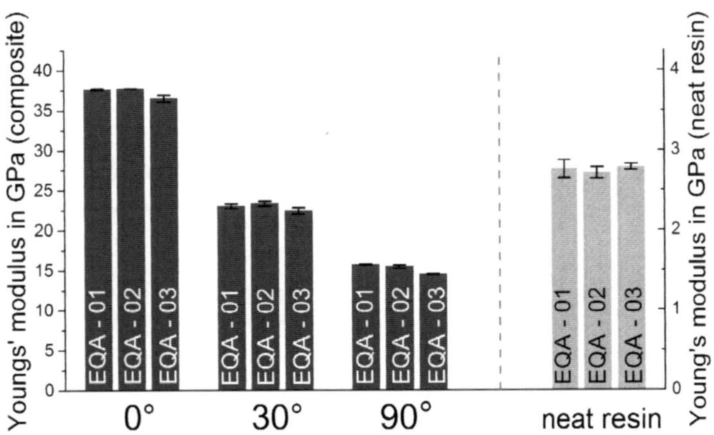

Figure 7. Young's modulus of composite and neat resin material; Right Y-axes (neat resin) adjusted with factor 10.

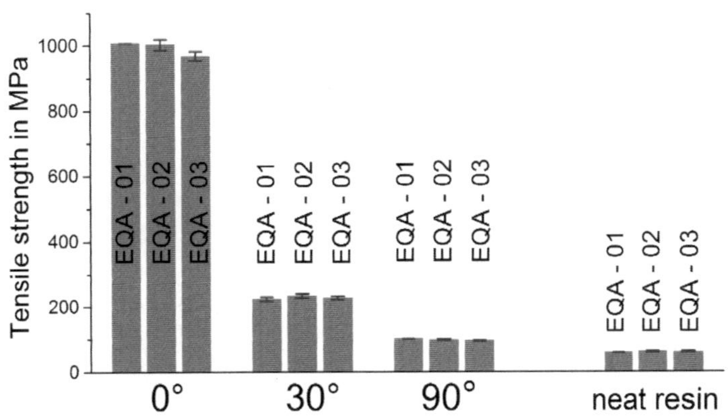

Figure 8. Tensile strength of composite and neat resin material.

Table 3. Young's modulus for composite specimen in GPa

	0°			30°			90°		
ID	EQA-01	EQA-02	EQA-03	EQA-01	EQA-02	EQA-03	EQA-01	EQA-02	EQA-03
mean value	37.6	37.9	36.5	23.0	23.3	22.4	15.7	15.5	14.5
deviation	0.08	0.03	0.41	0.29	0.30	0.39	0.07	0.18	0.06

Table 4. Tensile strength for composite specimen in MPa

	0°			30°			90°		
ID	EQA-01	EQA-02	EQA-03	EQA-01	EQA-02	EQA-03	EQA-01	EQA-02	EQA-03
mean value	1007	1003	968	223	233	227	101	98	98
deviation	0.0	16.1	13.7	6.0	6.8	5.0	0.8	2.8	1.5

Table 5. Quasi-static tensile properties for neat resin samples

ID	Young's modulus in GPa	Tensile strenght in MPa
EQA-01	2.8 ± 0.11	58 ± 0.5
EQA-02	2.7 ± 0.07	61 ± 1.2
EQA-03	2.8 ± 0.09	61 ± 0.4

3.4. Fatigue Test Results

Results for fatigue performance in 0°, 30° and 90° are given in figures 9 to 11. For the specimens in 30°- and 90°-direction, the individual measurements are consolidated in one S-N-curve a time. For the series in 0°-direction (figure 11), three separate S-N-curves have to be distinguished. The fits are calculated using a standard method involving a linear regression with a Basquin-function of type $N = C \times S^m$, with the number of cycles N, the input load S and the constants C and m.

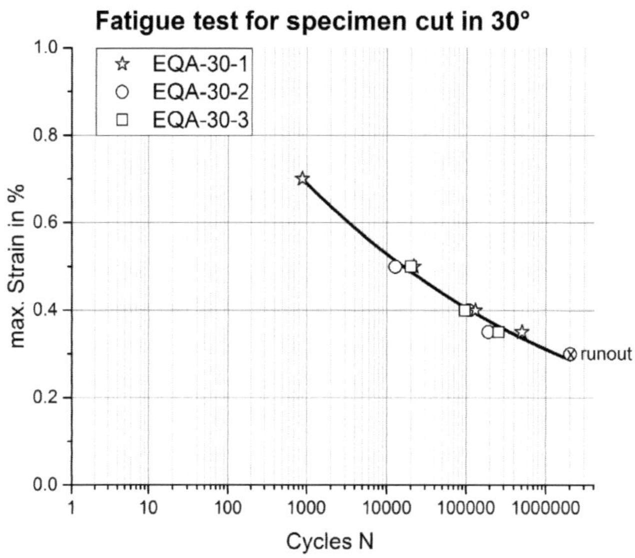

Figure 9. S-N-curve for specimen cut in 30°.

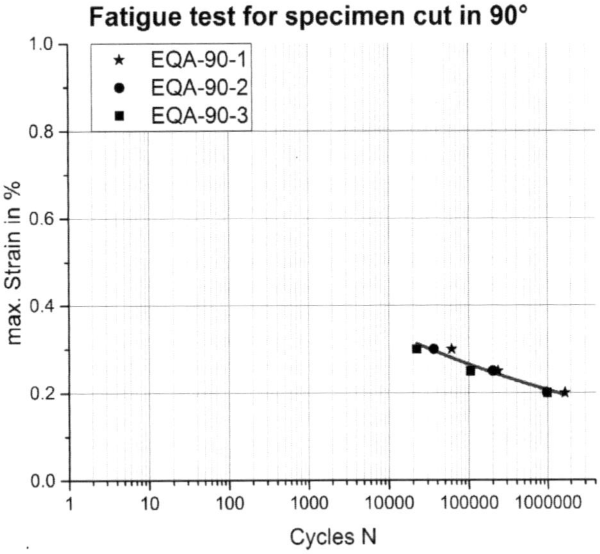

Figure 10. S-N-curve for specimen cut in 90°.

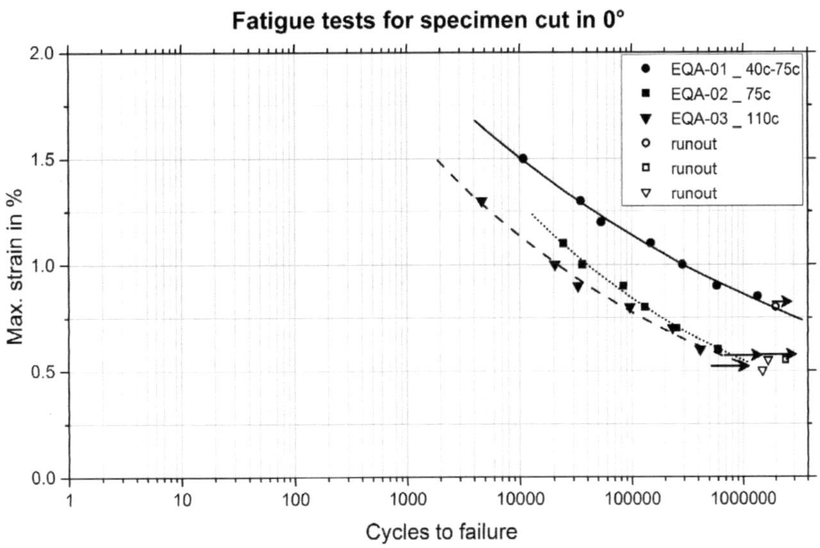

Figure 11. S-N-curve for specimen cut in 0°.

4. Discussion

Variation in the fiber volume contents are low and this is only reflected in a minor differences found in the measured Young's modulus. As the architectures are identical, the major changes in the materials mechanical properties can directly be related to the different curing cycles.

The quasi-static tension tests of the cured resin (the matrix material) clearly show that the mechanical properties of the polymer are not influenced by the curing cycles. The test material has not been constrained and no internal stresses are build up. Also the tensile strengths of the three panels measured in the 0°, 30°, and 90° directions respectively, are not influenced by the internal stresses and the curing cycles. This demonstrates that static properties of fiber reinforced composites are not heavily influenced by the matrix properties, but are only a result of the architecture and the fiber-matrix interface strength. As an interesting and valuable observation, apparently the latter (the interface strength) is not expected to be influenced by the curing temperatures as especially the 90° static strength is the same for all three panels.

The fatigue behavior in the fiber direction is strongly influenced by the internal stresses. The reason is explained and observed by Zangenberg et al. [3] and is due to the fact the fatigue damage is initiated in the transverse backing fibers. The shrinkage stresses in the resin builds up tensile strain in the matrix along the UD fibers as the fibers prevent the resin to shrink freely. This leads to cracks in the backing fibers and these cracks further initiate cracks in the UD fiber bundles. This is demonstrated in figure 12 [3] and further validated in by using CT scanning by Jespersen et al. [14]. Hence the drop in fatigue resistance in the UD fiber direction is directly related to the shrinkage stresses.

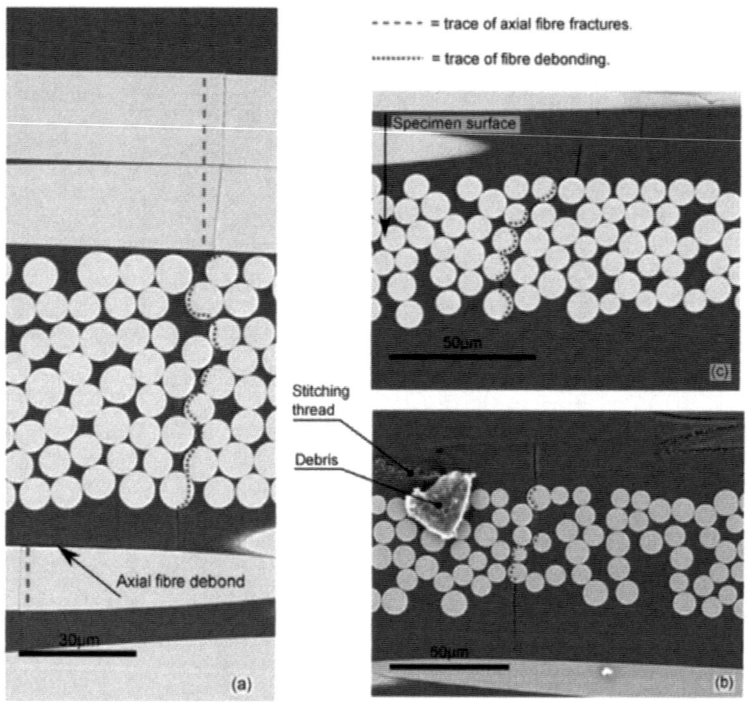

Figure 12. Cracks in UD fiber bundles caused by cracks in the backing fibers [3].

For the off axis fatigue loading in 30° direction we found no influence from the residual stresses. The expected damage mechanism in this loading case is dominated by the shear properties of the laminate and on a micro structural level dominated by the matrix toughness and partly the fiber-matrix interface strength. Hence, apparently similar to the conclusion drawn from the static transverse strength the interfaces are not significantly influenced by the shrinkage stresses.

The damage mechanism expected in the case of 90° loading direction is transverse debonding in the fiber matrix interface. As the UD fibers are constrained in the length directions but free to move and pack in the transverse directions, stresses are not build up in these directions. The only degradation that can be expected might be explained by the assumption that the strength contribution is lower from the backing layers because they are loaded in tension. This can explain the non significant observation, that the fatigue life time might seem to be lower in the panels having undergone a one stage cure.

5. Conclusion

Elevated temperatures during the curing lead to higher residual stresses in unidirectional glass fiber reinforced composites as the resin cannot shrink freely. Neither quasi-static mechanical properties nor the mechanical properties of the pure resin are influenced by the curing cycle.

However, higher residual stresses are related to a decrease in fatigue performance when the direction of loading and reinforcement are the same. Here, cracks in the transversal backing layer initiate cracks in the load carrying UD fibers. For tests with reinforcements not aligned with the load direction (off axes) as demonstrated in this work, no significant influence on the fatigue performance is found.

6. References

[1] Pereira G, Hüther J, Mikkelsen L and Brøndsted P //to be published 2016
[2] Zangenberg J 2013 *PhD-Thesis, DTU Wind Energy*
[3] Zangenberg J H and Brøndsted P a 2014 *Journal of Composite Materials* **48** 2711–2727
[4] Parlevliet P P, Bersee H E and Beukers A 2010 *Polymer Testing* **29** 291–301 ISSN 01429418
[5] Sourour S and Kamal M R 1976 *Thermochimica Acta* **14** 41 – 59
[6] Flammersheim H 1998 *Thermochimica Acta* **310** 153 – 159
[7] Bandyopadhyay A, Valavala P, Clancy T C, Wise K E and Odegard G M 2011 *Polymer* **52** 2445 – 2452
[8] Shah D U and Schubel P J 2010 *Polymer Testing* **29** 629–639 ISSN 01429418
[9] R Cilli A Prakki M A J d A 2010 *Brazilian Dental Science* **6** ISSN 2178-6011
[10] Shokrieh M and Ghanei Mohammadi A 2014 *Residual Stresses in Composite Materials* (Elsevier) ISBN 9780857092700
[11] Barnes J and Byerly G 1994 *Composites Science and Technology* **51** 479–494 ISSN 02663538
[12] Harsch M, Karger-Kocsis J and Herzog F 2007 *Express Polymer Letters* **1** 226–231 ISSN 1788618X
[13] Kosaka T, Osaka K and Sawada Y 2009 *Proccedings of ICCM17 - International Conference on Composite Materials*
[14] Jespersen K and Mikkelsen M 2016 *Proceedings of Risø Simposium 37*

IOP Conf. Series: Materials Science and Engineering **139** (2016) 012024 doi:10.1088/1757-899X/139/1/012024

Fatigue damage observed non-destructively in fibre composite coupon test specimens by X-ray CT

K M Jespersen[1] and L P Mikkelsen

Department of Wind Energy, Section of Composites and Materials Science, Technical University of Denmark, Risø Campus, 4000 Roskilde, DK

E-mail: kmun@dtu.dk[1]

Abstract. This study presents a method for monitoring the 3D fatigue damage progression on a micro-structural level in a glass fibre/polymer coupon test specimen by means of laboratory X-ray Computed Tomography (CT). A modified mount and holder made for the standard test samples to fit into the X-ray CT scanner along with a tension clamp solution is presented. Initially, the same location of the test specimen is inspected by ex-situ X-ray CT during the fatigue loading history, which shows the damage progression on a micro-structural level. The openings of individual uni-directional (UD) fibre fractures are seen to generally increase with the number of cycles, and new regions of UD fibre fractures also appear. There are some UD fibre fractures that are difficult to detect since their opening is small. Therefore, the effect of tension on the crack visibility is examined afterwards using a tension clamp solution. With applied tension some additional cracks become visible and the openings of fibre fractures increases, which shows the importance of applied tension during the scan.

1. Introduction

Wind power turbines are increasingly used in many parts of the world to match the increasing demand for sustainable energy. Among sustainable energy sources wind power is one of the most cost efficient technologies, however they are yet to be competitive to fossil fuels. The cost of energy can be decreased by increasing the blade length, as the power output of the wind turbine scales with the blade length squared. One of the main design factors when increasing the length of the blade is material fatigue, since the blades experience a high degree of repeated loading [1]. The wind variation causes the blades to bend in the direction of the tower (flap-wise bending), and the rotation results in the gravitational loads of the blades repeatedly changing direction causing a repeated edge-wise bending motion. With a life-time in the range of 20-30 years the total number of load cycles sums up to 10^8-10^9, which is significantly higher than for e.g. airplanes and cars [2]. Furthermore, it is the degradation of the stiffness rather than the strength which is a concern for wind turbine blades, since the blades might risk hitting the tower. As the blades mainly experience bending loads, uni-directional glass fibre composites made from non-crimp fabrics (NCFs) are commonly used for the load carrying parts of the blades. However, the fatigue mechanism in these NCF composites is not well understood. If the damage mechanism was properly understood on a micro-structural level it would be possible to decrease safety factors and make more fatigue resistant wind turbine blade materials. Therefore, establishing a method capable of monitoring tension-tension fatigue damage progression in a glass fibre/polymer is the focus of this paper.

Fatigue damage progression in composites is a complex process which includes several interacting damage mechanisms. Many of the damage features are small and difficult to monitor non-destructively. The UD NCF composites used for wind turbine blades commonly include cross-crossed thin supporting bundles (backing bundles) to which the UD bundles are stitched to keep them aligned during manufacturing and handling. For these NCF composites the stiffness degradation during fatigue is mainly caused by UD fibre fractures and is also slightly affected by the initial off-axis cracking in the backing bundles. Furthermore, the location of the UD fibre fractures seems to be highly related to the cross-over points of the off-axis backing bundles [3, 4]. Therefore, the damage appears as local fibre fracture regions at various locations rather than purely homogeneously throughout the material. Furthermore, the fibre fractures and off-axis cracks are features of a few micrometers in size and appear as 3D features such as clusters and chains of fibre fractures [4]. All of this makes it difficult to monitor the damage over time experimentally, since a high resolution non-destructive 3D damage monitoring technique is necessary.

Through time, different methods for in-situ damage monitoring have been used in the attempt to monitor damage progression of fibre composites. On a structural level, several methods such as vibraction analysis, acoustic emission, ultrasonic testing [5], and optical sensors [6] techniques have been used to monitor damage in wind turbine blades, however at a millimeter or even centimeter spatial resolution. 2D imaging methods such as SEM and camera imaging have been used to observe the surface damage progression on a micro-structural level [7, 8], and particularly SEM gives invaluable information on small-scale damage mechanisms due to the high resolution. However, using such methods one does not know what happens below the considered surface. In relation to monitoring the damage progression in fibre composites on a micro-structural level in 3D, X-ray Computed Tomography (CT) is probably the most promising technique. Synchrotron radiation CT has due to the short scan times been used for in-situ studies on fibre composites [9, 10, 11, 12]. However, in relation to fatigue damage progression these studies are limited to a few thousand cycles (e.g. [11, 12]) due to the limited access time. Although the scan time is considerably longer, the obtainable resolution for laboratory X-ray CT imaging has recently become competitive to synchrotron CT. Some studies [13, 14, 15] have considered in-situ damage monitoring in composites by laboratory X-ray CT. However, general for most of the studies is that they consider small sample sizes to obtain a high resolution. As considering in-situ loading of down-sized samples might affect the damage mechanisms in the sample, it is of interest to be able to monitor the damage non-destructively in the regular coupon test specimens, which is scope of the current study.

Our work outlines a method capable of monitoring tension fatigue damage in 3D for a glass fibre/polymer coupon test specimen non-destructively by means of laboratory X-ray CT. Modified solutions for mounting the large samples in the X-ray CT scanner and a method to apply tension to the sample during scanning are explained. Furthermore, the study outlines some limitations and some important factors to take into account when evaluating the results.

2. Composite material and test specimens

The considered material is a glass fibre/epoxy composite made from two types of non-crimp fabrics (NCF); a UD and a biax fabric. In the UD fabric thin supporting off-axis backing bundles are stitched to the UD bundles to keep them aligned. The stacking sequence is $[\pm 45, b/0, b/0]_s$ where "0" is the UD bundle layer and "b" symbolises the backing layer containing both $\pm 45°$ and 90° fibre bundles. This means that the composite contains 4 layers of UD and has a layer of biax at each surface. The material is similar to that considered in [16].

Plates of the UD laminate were cut into butterfly shaped test specimens (for more information on the specimen geometry, see [17]). The specimen geometry with curved edges is specially designed to test UD composites, as the standard plane geometries tend to fail by shear failure in the tabs [17]. The specimen length was 410mm, the width of the gauge section was 15mm, and the thickness of the composite was 4.5mm. 140mm long tapered tabs were attached in both ends of the sample.

3. Experimental methods

To investigate the damage progression in the composite, an ex-situ X-ray CT start-stop fatigue test was performed on a full-size butterfly coupon specimen. Every time the test was interrupted the sample was inspected in the same location using X-ray CT, thereby observing the damage progression in terms of individual UD fibre fractures. The influence of applied tension during X-ray CT scanning was examined by a tension clamp solution.

3.1. Fatigue testing

Fatigue tests were performed on a hydraulic Instron test machine (max 250kN) in load control with a sinusodial waveform and a stress ratio of R=0.1. The fatigue test was carried out with a test frequency of 5Hz and the strain was monitored with two extensometers (one on each side of the sample). Two initial static tests were performed on the sample prior to the fatigue test to determine the initial Young's modulus. This was used to determine the load range where a peak strain of 1% was obtained. The sample used for the ex-situ test was stopped four times (after 47300, 57300, 67300, and 77300 cycles) during the fatigue load history for X-ray CT scanning. Another sample was stopped after 67000 cycles, where damage was expected to be present, for investigation of the effect applied tension on the crack visibility.

3.2. X-ray computed tomography

The X-ray CT experiments were carried out on a Zeiss Xradia Versa 520 scanner, using a 2000x2000 pixel detector with a pixel depth of 16 bit. All scans were performed with a binning of 2 of the pixels on the detector. This means that 2x2 pixels are combined into one giving a higher intensity (shorter scan time) but also a larger pixel size (lower resolution). Three different scan settings were used as shown in Table 1. To locate the damage initially in the ex-situ study, a large field of view (LFOV) scan was performed on the sample. After 47300 cycles, it was possible to see an indication of damage despite the coarse resolution (17.4μm voxel size). A high resolution scan was then performed in that region by means of the "Scout-and-Scan" principle provided by Zeiss and using the "ex-situ" scan settings labeled "ES" in Table 1. For the remaining times the test was interrupted only high resolution scans were performed. For the tension clamp experiments elaborated later, the scan settings "TC" were used and include a higher number of projections to make up for presence of two carbon pins necessary to transfer the load.

Table 1. X-ray CT scan settings used for the large field of view (LFOV), ex-situ (ES), and tension clamp (TC) scans.

Scan name	Source to sample (mm)	Detector to sample (mm)	Exp. time (s)	No. of projections -	Accelerating voltage (keV)	Pixel size (μm)	Optical Mag.	Scan time (h)
LFOV	30	88	2	801	60	17.4	0.4x	2
ES	28	35	7	4601	70	3	4x	11
TC	30	50	12	5201	70	2.5	4x	20

Because of the large sample size mounting in the scanner is not possible with the standard mount and sample holders. Therefore, a short mount was manufactured and a grip (Makro Grip 5-Achs-Spanner from Lang) was modified to fit into the scanner. Fig. 1 a shows the sample placed in the scanner using the short mount and special sample holder. Using this setup it is possible to scan a bit more than the bottom half of the 410mm long sample (there is a limitation of how far down the sample can go). Furthermore, the sample can be mounted in the same way every time with a variation of less than 1mm. A fine tuning of the location was done by comparing the 2D projection images to ensure as little variation as possible between the location of the observed region in the ex-situ study.

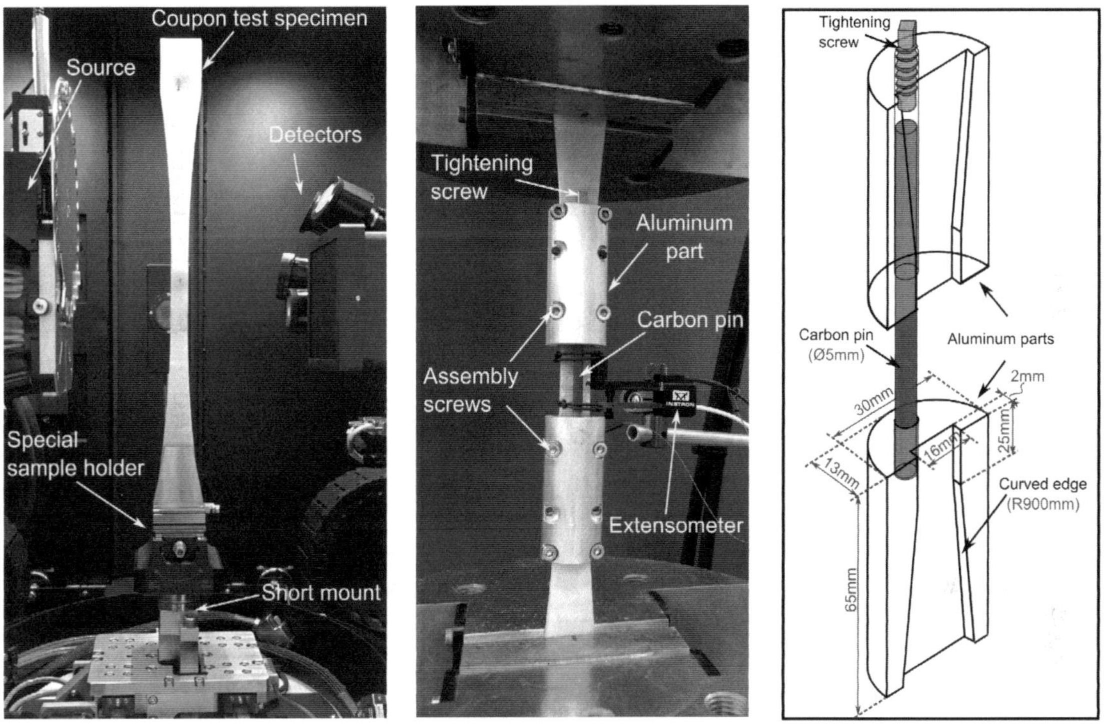

a) Sample mounted in scanner b) Tension clamp on sample c) Principle sketch of clamp tightening mechanism

Figure 1. Specialised experimental setups. a) shows the mounting method for the scanner, b) the tension clamp solution on sample in tension machine, and c) a principle sketch of the tightening mechanism showing half the clamp (scale not exact).

The effect of applied tension on crack visibility was examined using a specially designed tension clamp, which is designed so that the sample can be scanned with the clamp attached. The clamp is shown on the sample in Fig. 1b and Fig. 1c shows a principle sketch of the clamp tightening mechanism along with the approximate dimensions. Fig. 1c only shows half the clamp as it is symmetric, and the holes and assembly screws used to hold the two clamp halves together (seen in Fig. 1b) are excluded for simplicity. While the sample is loaded in a regular tensile test machine, the clamp is mounted on the sample using the assembly screws and the two tightening screws on the top aluminum parts are tightened. By doing so, the carbon pins that transfer the load are also tightened as shown in Fig. 1c. The internal geometry of the aluminium parts of the clamp fits the curvature of the sample, and therefore the aluminum parts are kept in place after the carbon pins are tightened, even after unloading the tensile test machine. That means that when the load is released on the test machine, the clamp will ensure that the central part of the sample is still in tension. The strain in the sample was measured using an extensometer. The sample was loaded up to 7kN corresponding around 0.2-0.3% strain and then the clamp was mounted on the sample. After removing the load, the clamp stabilised at a strain of 0.17% in the gauge section. Obtaining a higher strain using the clamp was attempted, but even if a higher initial load was used the strain in the gauge section went back to the same value. This could be due to damage in the ends of the carbon pins at high loads, and a slightly modified solution is necessary to obtain higher loads. A possible improvement in the future could be to add reinforcement to the ends of the carbon pins. However, in this study a strain of 0.17% was used, which was the highest obtainable strain for the presented clamp solution.

4. Results and discussion

4.1. Fatigue test result

The measured stiffness degradation of the ex-situ fatigue study is shown by the solid line in Fig. 2a where the four interruption points for CT inspection are marked (N_A-N_D). Slight jumps in the data were observed at each of the interruption points, but since this is due to mounting the extensometers a bit differently every time, the curve sections were off-set to give a continuous curve. Fig. 2a also include two other sets of test data (dashed lines) for a slightly higher strain level (1.1%).

By comparison, it is seen that the shape of the ex-situ fatigue curve is similar to those observed for continuous tests to failure. However, the life-time is less for the ex-situ test performed at 1% strain than for the continuous tests carried out at 1.1% strain, which is opposite of what would be expected. However, Fig. 2b shows the S-N data for samples not affected by X-ray relative to the ex-situ test, and even though the life-time seemed less than for the normal tests, it is not significantly different. Whether this slightly shorter life-time was caused by the effect of X-ray or simply that the sample itself was an outlier cannot be said for certain. However, in addition to the slightly shorter life-time the sample was observed to discolor because of X-ray (Fig. 2c) and it is therefore a subject that requires further investigation.

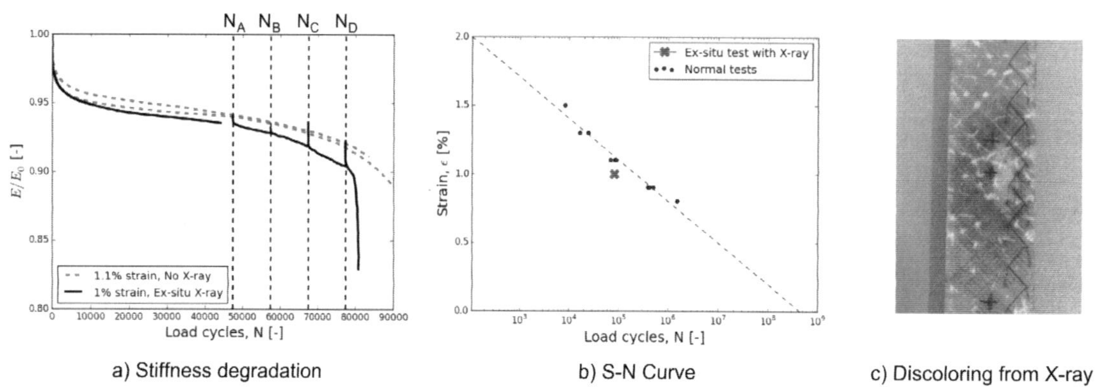

Figure 2. a) shows the stiffness degradation curve for the ex-situ fatigue test (solid line) and examples of continuous fatigue tests (dashed lines) for samples without X-ray exposure, b) the S-N curve data for samples not affected by X-ray compared to the ex-situ sample, and c) the discoloring observed on the sample after N_D cycles (after all the X-ray experiments).

4.2. Ex-situ X-ray CT monitoring of damage progression

Fig. 3 shows virtual 2D slices of the X-ray CT data (no applied tension) for the four different stages of the fatigue test (N_A-N_D in Fig. 2a). There was a slight variation in the scanned location for each scan because of mounting/dismounting the sample in the holder, however as it is seen from Fig. 3 it is quite a small difference for the scans performed after N_A, N_B, and N_C cycles. The reason for the large variation for the scan performed after N_D cycles was that the bottom of the holder had been disassembled between step N_C and N_D, which is important not to do in future studies. As expected, the UD fibre damage is observed to progress with increasing number of cycles. Both new fibre fractures are observed to appear and the opening

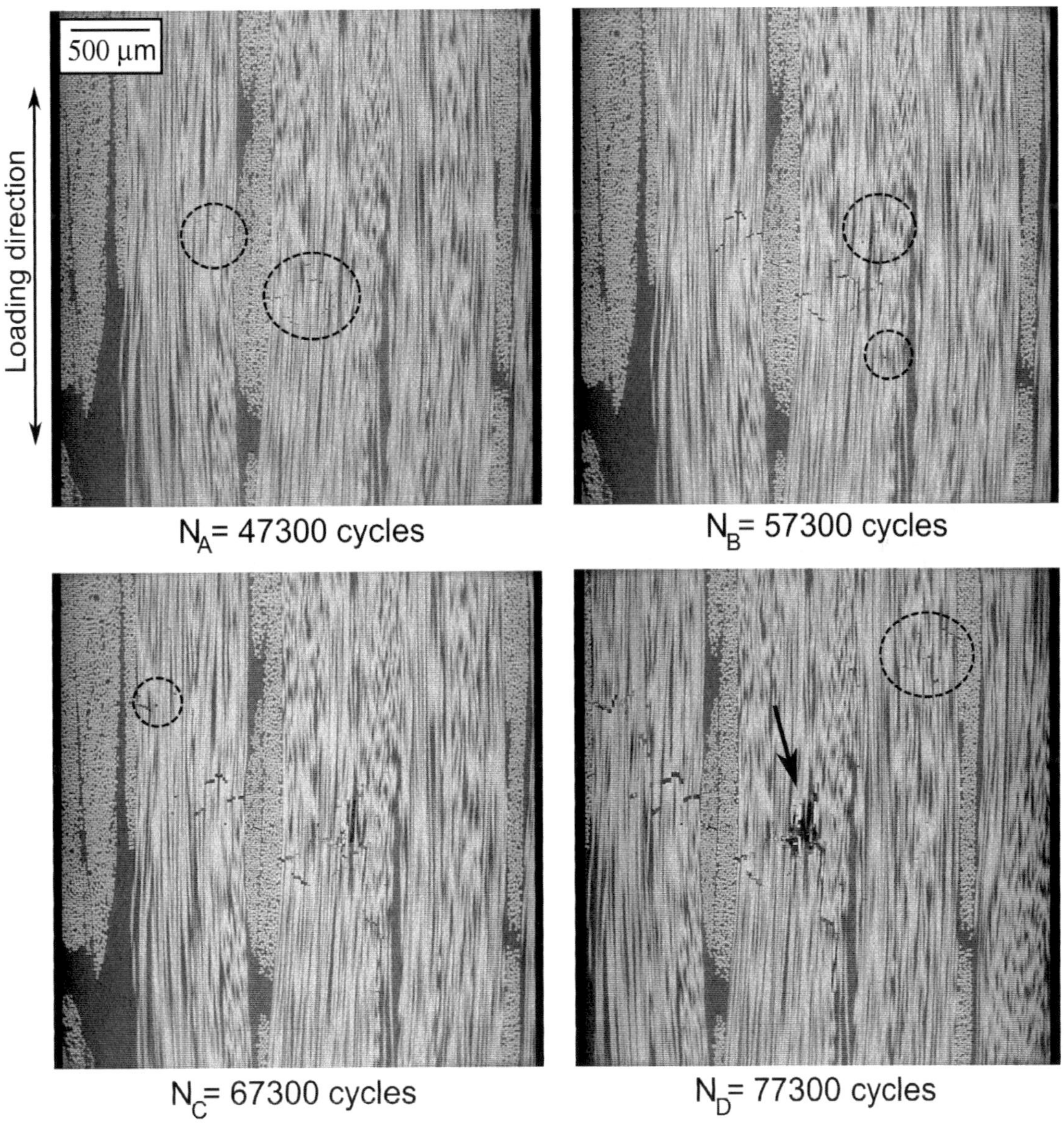

Figure 3. Virtual 2D slice views from 3D X-ray CT scan data showing the damage after each of the four times the fatigue test was interrupted. The ellipses indicate some locations with new damage and the arrow marks a highly damaged region.

of existing fibre fractures is seen to increase for each step. After N_D cycles a damage region with large crack openings compared to other locations was observed (indicated by the arrow in Fig. 3) and the final failure of the sample happened shortly after this (as also seen in Fig. 3). The increasing openings of the UD fibre fractures might be linked of the gradual length increase of the sample during fatigue, however evaluating and comparing the observed damage mechanisms in all the data sets in detail is still ongoing work. In general for all the X-ray data, many fibre fractures were clearly visible, but some had a small opening making them difficult to see. Furthermore, off-axis cracks were not clearly visible because of the small opening. A possible way to increase the visibility of these cracks is to apply tension during scanning in order to open them up, as will be explained in the following section.

4.3. Effect of applied tension during scan

Fig. 4 shows two scans performed in the same damage region of a sample interrupted after 67,000 cycles at 1% strain. Fig. 4a shows a virtual 2D slice from the 3D x-ray data of the unloaded specimen, and Fig. 4b shows the same position with the tension clamp attached keeping the sample at 0.17% strain. It should be noted that 0.17% strain is only around on fifth of the maximum strain during fatigue. Some regions of interest are marked and compared for the loaded sample (region A1-D1) and the unloaded sample (region A2-D2) in Fig. 5.

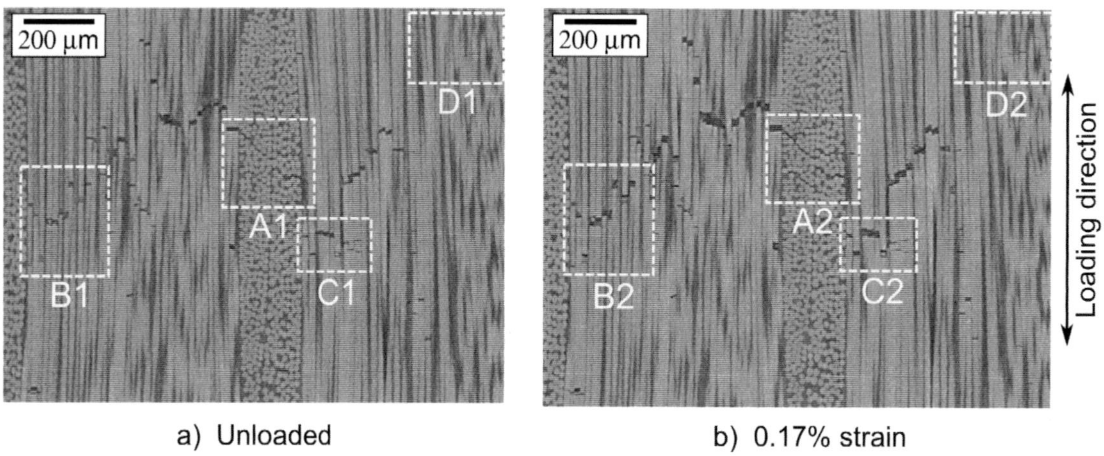

Figure 4. Comparison between observed damage without applied load (a) and with applied load (b). Regions of interest are marked and zoom views can be seen in Fig. 5

By comparing region A1 and A2 in Fig. 5, it is seen that the visibility of the off-axis cracks is only slightly increased at 0.17% strain. The clamp has a larger effect on the visibility of fibre fractures. Some of the fibre fractures are not easy to see without applied tension, whereas they can be seen more clearly with tension (e.g. seen by comparing region B1 and B2). At some locations the opening of the fibre fractures was greatly increased by the tension clamp. In the case of region C1/C2 the opening of some cracks was increased by around a factor two. However, at some locations (e.g. fractures marked by a dashed circle in C2), no change in the opening was observed. There were some locations (e.g. region D1/D2) where the fractures were not visible when the sample was unloaded, whereas they appeared when the sample was strained. The effect of the clamp will be larger if a higher load can be applied, and therefore optimisation of the clamp solution is ongoing work. Furthermore, if an automatic crack counting method was established, it would be possible to quantitatively show the effect of applied tension, and this could also be used to quantify the ex-situ damage. This is currently ongoing work.

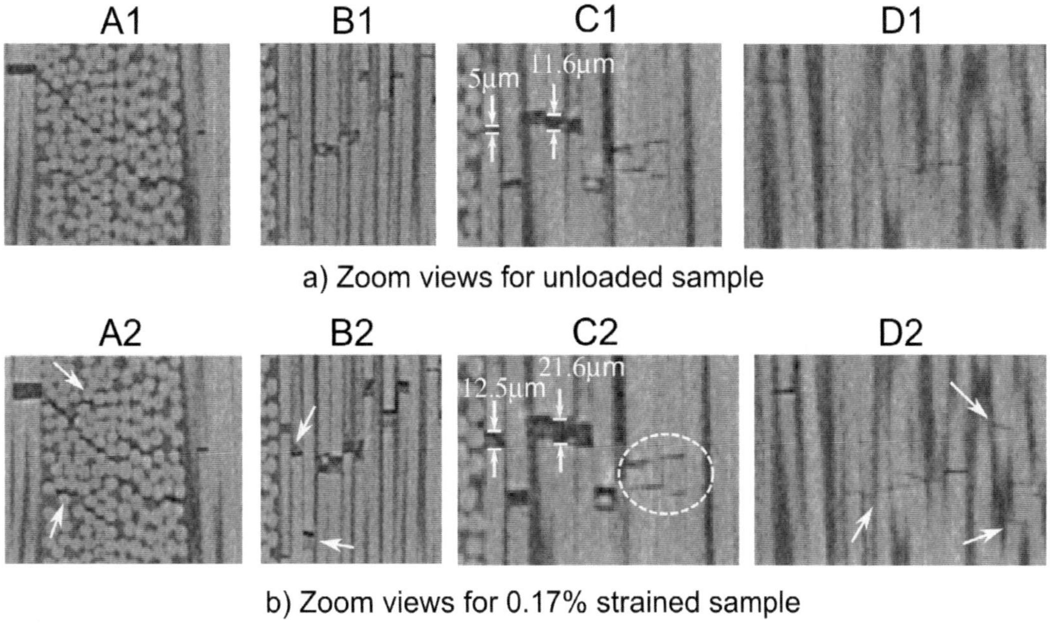

Figure 5. Zoom views in the regions of interest marked in Fig. Fig. 4. a) shows the views for the unloaded sample (A1-D1) and b) the views for loaded sample (A2-D2). The arrows marks some locations with changes when applying load to the sample.

5. Conclusion

In this study, a method for observing fatigue damage progression non-destructively in a glass fibre/polymer coupon test specimens by ex-situ laboratory X-ray CT was explained. Damage was observed in 3D on a micro-structural level including individual fibre fractures and off-axis cracks at four points during the fatigue loading history. The number of fibre fractures along with their opening was seen to increase with increasing number of load cycles. The long time exposure of X-rays on the sample seemed to have a slight negative effect on the tension fatigue life-time, however when scanning four times it did not seem significant. Furthermore, the effect of applying tension during scanning was examined and it was seen that additional fibre fractures and off-axis cracks became visible with load applied. However, it might be necessary to apply a higher strain during scanning to make all the fibre fractures and matrix cracks visible in the X-ray CT images at the considered resolution. Therefore, developing an improved clamp solution is of great interest and is ongoing work.

Acknowledgments

This research was conducted using mechanical testing equipment from DTU Center for Advanced Structural and Material Testing (CASMAT), Grant No. VKR023193 from the Villum Fonden. Financial support from CINEMA: "the allianCe for ImagiNg of Energy MAterials", DSF-grant no. 1305-00032B under "The Danish Council for Strategic Research" is gratefully acknowledged. Additionally, we would like to thank LM Wind Power for manufacturing of test specimens.

References

[1] Nijssen R P L and Brøndsted P 2013 *Advances in Wind Turbine Blade Design and Materials* 175–209
[2] Nijssen R P L 2006 *PhD Thesis, Delt University of Technology*
[3] Zangenberg J, Brøndsted P and Gillespie Jr J W 2014 *Journal of Composite Materials* **48** 2711–2727 ISSN 0021-9983

[4] Jespersen K M, Lowe T, Withers P, Zangenberg J and Mikkelsen L 2016 *Composites Science and Technology* (Submitted)
[5] Pedro F, Márquez G, Mark A, María J, Pérez P and Papaelias M 2012 *Renewable Energy* **46** 169–178
[6] Pereira G F, Mikkelsen L P and McGugan M 2015 *PLOS One* **10** e0141495
[7] Mortell D J, Tanner D A and Mccarthy C T 2014 *Composites Science and Technology* **105** 118–126
[8] Hongwei Z, L M, Brøndsted P, Jinbiao T and Lele G 2010 *Chinese Science Bulletin* **55** 1199–1208
[9] Wright P, Moffat A, Sinclair I and Spearing S M 2010 *Composites Science and Technology* **70** 1444–1452
[10] Scott A E, Mavrigirdati M, Wright P, Sinclair I and Spearing S M 2011 *Composites Science and Technology* **71** 1471–1477
[11] Garcea S C, Macrigirdato M, Scott A E, Sinclair I and Spearing S M 2014 *Composites Science and Technology* **99** 23–30
[12] Garcea S C, Sinclair I and Spearing S M 2015 *Composites Science and Technology* **109** 32–39
[13] Schilling P J, Karedla B P R, Tatiparthi A K, Verges M A and Herrington P D 2005 *Composites Science and Technology* **65** 2071–2078
[14] Sket F, Seltzer R, Molina-Aldareguía J M, Gonzalez C and LLorca J 2012 *Composites Science and Technology* **72** 350–359
[15] Yu B, Blanc R, Soutis C and J W P 2016 *Composites: Part A* **82** 279–290
[16] Pereira G F, Hüther J, Mikkelsen L P and Brøndsted P 2016 *In preparation, DTU Wind Energy*
[17] Zangenberg J 2013 *PhD Thesis, DTU Wind Energy PhD-0018(EN)*

IOP Conf. Series: Materials Science and Engineering **139** (2016) 012025 doi:10.1088/1757-899X/139/1/012025

Determination of mode-I cohesive strength for interfaces

J B Jørgensen[1,#], M D Thouless[2], B F Sørensen[3] and C Kildegaard[1]

[1] LM Wind Power, Østre Alle 1, 6640 Lunderskov, Denmark.
[2] The University of Michigan, 3672 G. G. Brown Addition, Ann Arbor, MI, USA.
[3] The Technical University of Denmark, Dept. of Wind Energy, Frederiksborgvej 399, 4000 Roskilde, Denmark.

E-mail: [#]`jbj@lmwindpower.com`

Abstract. The cohesive strength is one of the governing parameters controlling crack deflection at interfaces, but measuring its magnitude is challenging. In this paper, we demonstrate a novel approach to determine the mode-I cohesive strength of an interface by using a 4-point single-edge-notch beam specimen. The test specimen is made of a glue cast onto a uni-directional, glass-fiber laminate. A crack is cut in the glue, orthogonal to the interface, which creates a high normal stress across the glue/laminate interface during loading. It is observed that a new crack can be initiated along the interface in response to this stress, before the main crack starts to grow. Observations using 2D digital-image correlation showed that an "apparent" strain across the interface initially increases linearly with the applied load, but becomes non-linear upon the initiation of the interface crack. The cohesive strength is determined, using a 2D, linear-elastic, finite-element model of the experiment, as the stress value where the experimental measured "apparent" strain value becomes non-linear across the interface.

1. Introduction

Crack deflection along interfaces is an important failure mechanism in adhesive bonded joints. Several studies on crack deflection have been presented previously, but with a primary focus on modeling. An early work was by Cook and Gordon [1], who used a stress-based approach to model an elliptical notch situated a short distance from a weak interface in a homogeneous substrate. The peak stresses normal to the interface and normal to the notch were compared to show that the interface fails before the substrate (causing crack deflection), if the interface strength is less than about one fifth of the substrate strength.

Later models of crack deflection used an energy-based approach by applying linear-elastic fracture mechanics (LEFM) [2, 3, 4]. These models indicated that, in the absence of a modulus mismatch, the interface toughness should be less than one fourth of the substrate toughness for the crack to deflect. Thus, modeling the deflection of a crack at an interface was, at first, either based on stress [1] or toughness [2]. These two distinct concepts, strength and toughness, are unified in a cohesive law [5]. It was shown that the cohesive strength of the interface is one of the governing parameters that controls crack deflection [5]. This cohesive strength can be measured experimentally using environmental scanning-electron microscopy (ESEM) [6] in conjunction with the J-integral [7, 8]. Unfortunately, this method requires advanced equipment and specialized loading devices.

Experimental studies of crack deflection at interfaces are very limited, and only a few have been found [9, 10]. Kendall [9] derived a deflection criterion for a Griffith crack, and used single-edge-notch-tension (SENT) specimens made from a brittle and transparent ethylene propylene rubber in which crack propagation could be monitored. Unfortunately, details and images of the crack deflection process were not presented in that paper. A subsequent experiment [10] used a wedge to load a single-edge-notch-beam (SENB) to show that an interface crack was initiated before the main crack reached the interface. This competition between growth of the main crack and initiation of an interface crack is similar to the model proposed by Cook and Gordon [1].

In this paper, crack deflection at an interface is studied using a 4-point SENB specimen. The test specimen is manufactured of a brittle vinylester glue cast onto a uni-directional glass-fiber-reinforced polyester laminate. 2D digital-image-correlation (DIC) is used to measure the full displacement field during loading of the specimen. It is found that a new crack initiates in the interface prior to the main crack reaching the interface. This is similar to the experimental observations of Lee and Clegg [10], but we use this failure mode to develop a new approach to measure the cohesive strength of the interface.

2. Approach

A new approach is proposed in this paper for measuring the cohesive strength of the interface, $\hat{\sigma}$. This approach is summarized in figure 1. The strength is measured using 2D DIC (figure 1(a)), in combination with a linear-elastic finite-element (FE) model of the experiment (figure 1(b)). During the test, measurements of an "apparent" strain, ϵ_{yy}, acting normal to the interface allows the point at which the interface crack is initiated to be identified. ϵ_{yy} is obtained from the displacement difference between two points on either side of the interface divided by gauge length, l_g as shown on Figure 2.

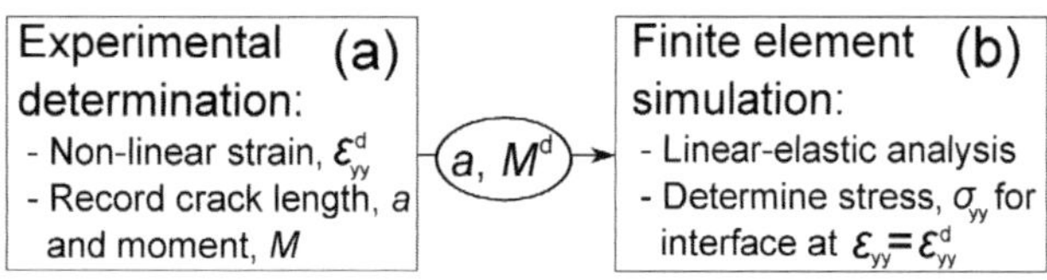

Figure 1. Approach to obtain cohesive strength of the interface.

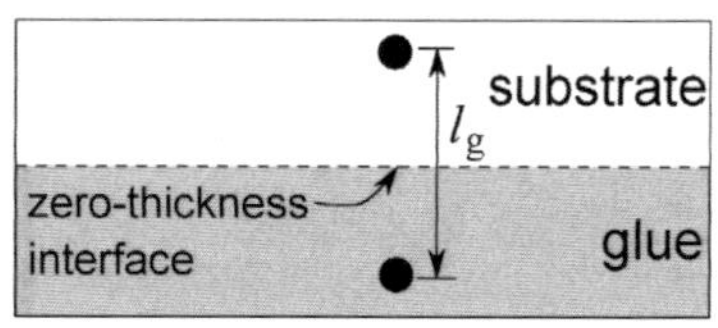

Figure 2. Measuring an "apparent" strain, ϵ_{yy}.

The strain field at the interface increases linearly with load until the point at which the interface crack initiates. The applied bending moment, M^d and the crack length, a at the point when DIC indicates that the "apparent" strain is no longer linear are identified. Linear-elastic materials and a zero-thickness interface are assumed hence the non-linearity in measured "apparent" strain is due to interface separation. Separation of interface is the first step in the crack initiation process and this is the beginning of delamination. These conditions are then used in a linear-elastic finite-element model of the experiment, assuming an orthotropic substrate and an isotropic glue, to calculate the normal stress at the interface. The maximum stress calculated from this numerical analysis is taken to be the cohesive strength of the interface.

3. Methods

The 4-point SENB specimen is illustrated in figure 3, and the dimensions are given in table 1. In this figure and table a_0 is pre-crack length, a is the actual crack length, b is thickness of the glue, c is thickness of the laminate, w is width, and L is the length of the specimen.

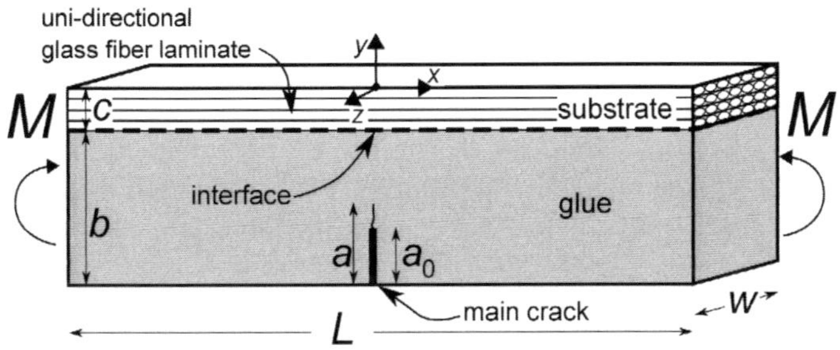

Figure 3. Geometry and nomenclature of the 4 point SENB specimen. The uni-directional fibers are oriented in the x-direction according to the coordinate system.

Table 1. Dimensions of test specimen.

a_0/b [-]	c/b [-]	L/b [-]	w/b [-]
0.6	0.2	6.6	1.1

3.1. Design of experiment by finite element modeling

The purpose of the main crack is to create a high normal stress across the interface. Furthermore, this geometry containing a crack is selected because it can be modeled precisely using FE, allowing the interface stress to be determined accurately. The interface stress is extracted in the symmetry line ($x = 0$, $y = -c$) where the shear stress is zero hence the crack initiation is mode-I. For the approach to work, the interface crack must initiate before the main crack reaches the interface, and, if the main crack starts to grow, it should do so in a stable manner.

The energy-release rate of the main crack, G, shown on figure 3, can be determined for a homogeneous specimen using the results of Tada *et al.* [11]. However, the energy-release rate for the present case of an orthotropic substrate and an isotropic glue depends on the following parameters:

$$a/b, \quad c/b, \quad E_{xx,s}/E_g, \quad E_{yy,s}/E_g, \quad \mu_{xy,s}/E_g, \quad \nu_{xy,s}, \quad \nu_g$$

where E is in-plane stiffness, μ is shear modulus, ν is Poisson's ratio, and the subscripts s and g represent the substrate and glue, respectively.

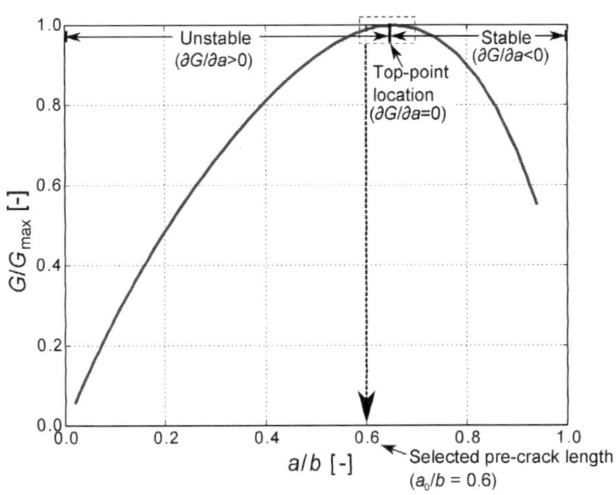

Figure 4. Results from the finite element model of the test specimen geometry, where the normalized energy release rate is determined as a function of relative crack length.

It is important that the main crack grows stably to avoid dynamic effects *i.e.* rapid, unstable crack growth, in the experiment. A 2D plane-strain linear-elastic finite-element model is used to determine a suitable initial length for the main crack to ensure stable crack growth. The normalized energy-release rate of the main crack is determined as a function of the relative crack length, a/b, from the FE calculations and shown in figure 4.

Stable growth of the main crack is achieved if $\partial G/\partial a < 0$. If the pre-crack is very long then the main crack will grow stably, but it will be very close to the interface. Therefore, from figure 4, a normalized pre-crack length of $a_0/b = 0.6$ is selected as the best compromise between length and stability.

3.2. Test specimen and speckle pattern

The test specimen is manufactured from polyester reinforced with fabrics of uni-directional glass fiber, using vacuum-assisted resin-transfer moulding (VARTM). A brittle vinylester glue is subsequently cast onto the glass-fiber laminate creating a zero-thickness interface. The exact material data are confidential, and the results are, therefore, normalized when presented. The pre-crack is cut in the glue using first a thin hack saw, then a standard razor blade, and finally an ultra-thin razor blade of thickness 74 microns. See figure 5 and figure 6 for images of the pre-crack.

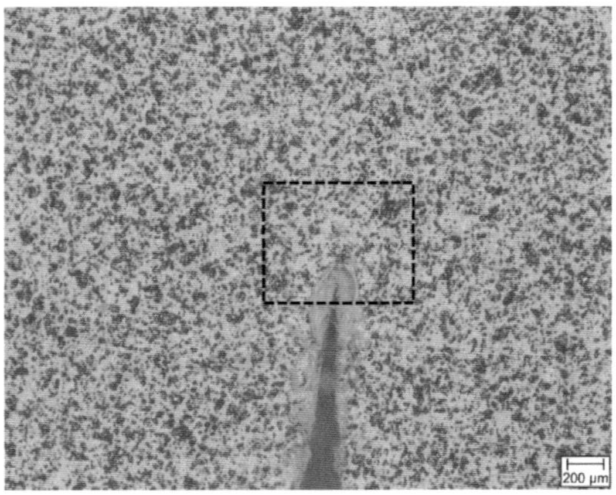

Figure 5. Image taken in optical microscope, of the front of the test specimen showing the pre-crack and speckle pattern.

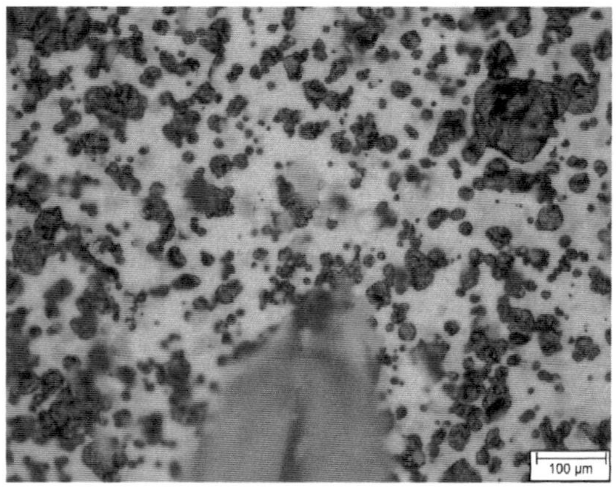

Figure 6. Zoom of the dashed square on figure 5 showing the speckle pattern close to the pre-crack tip.

A speckle pattern is applied to the front surface using an Iwata CM-B airbrush. First, a white baseline paint is applied to cover the front surface of the specimen. Afterwards, a carbon black paint is applied with increased pressure to minimize the speckle sizes. 3-5 pixels across each speckle diameter, and 10 speckles per subset are desired to track displacements accurately, and to maximize the spatial resolution in DIC [12].

It is desirable to have a scaling factor between microns and pixels of about 3 microns/pixel or less to capture the crack initiation accurately. Thus, the speckles should be between 9 microns and 15 microns (3x3 and 5x3). Larger speckles would lower the spatial resolution, since a larger subset should be used to maintain the 10 speckles per subset. After application of the speckle pattern, the speckles are measured in an optical microscope to between 8 and 28 microns, see figure 6. The actual scaling factor is determined, based on a scale bar mounted on the images, to 2.8 microns/px leading to a field of view (FoV) of 6.8 x 5.7 mm (2448x2048 pixels). The pre-crack lengths are measured using a digital vernier caliper with an accuracy of $a_0/b \pm 1\%$, while the crack length is measured during the test using the images, which can be measured with an accuracy of about ± 4 pixels.

3.3. Experimental setup and procedure

A MTS 858 Mini Bionix II servo-hydraulic test machine is used in displacement control to load the specimen at a rate of 0.015 mm/min (0.00025 mm/s) for the cross-head displacement. The specimen is loaded slowly so many images can be captured during the test. The test is conducted at room temperature. Data (time, force, cross head displacement) are collected on a PC at 10 Hz.

Vic-2D DIC system (Vic Snap 8) is used to capture the images with an image frequency of 1 Hz. A CCD sensor of brand Grasshopper GRAS-50S5M and Fujinon CCTV Lens (1:1.8/50mm) are used with the DIC system. A fiber-optic illuminator from Cole-Palmer is used to illuminate the specimen surface. The camera is mounted on a tripod that can be moved in the y-direction and rotated around 3 axes. The lens aperture is set to a medium level of 8, where the minimum is 1.8 and the maximum is 22. This is found to be the best trade-off between capturing surface depth and the amount of light let through the lens.

3.4. Data analysis

Images are correlated with a subset size of 31 pixels and a step size of 3 pixels to obtain the full displacement field using the DIC software, DaVis from LaVision. The subset size is set hence approx. 10 speckles are found in each subset. The step size is set small enough to resolve the fine details of the interface.

The "apparent" strain is determined, by a script, using the displacement at two points across the interface divided by their separation to obtain an average "apparent" strain across the interface. Here a gauge length, l_g of $l_g/b = 0.007$ is used. This method is equivalent to using a virtual strain gauge across the interface in the DIC software. As a check, the normal strain at the interface is also calculated by the DIC software at different points across the interface. The precise location of interface is taken as the point with the largest strain value. The two methods resulted in the same strain value across the interface. The strains are not further post processed since the purpose of the strain measurement is to identify strain non-linearity.

4. Results

Figure 7 illustrates the relative crack length and the normalized moment as a function of time after the start of the experiment. According to figure 7, the moment increases non-linearly with time until about $\sim$ 200 s. This is attributed the establishment of full contact of the rollers on the specimen. Thereafter, the moment increases linearly with time, until the main crack in the glue starts to propagate at $t = 1100$ s. Figure 7 also shows that when the interface crack is fully developed, the main crack grows and reaches the interface to form a doubly-deflected crack at the interface.

The role of the DIC measurements is to identify the "apparent" strain, ϵ_{yy}. These measurements indicate a transition from a linear relationship to a non-linear relationship between the "apparent" strain and the moment at $t = 800$ s (figure 8). The value of the "apparent" strain at which this occurs is designated by ϵ_{yy}^{d}. This is confirmed by observed changes in strain field by contour plots of the vertical strain, similar to those shown in figure 9.

A comparison between figures 9 and 8 shows that both the normal strain across the interface and the applied moment increase linearly with time until $t = 800$ s. In this regime, the "apparent" strain is proportional to the applied moment, as one would expect for a linear-elastic system.

After this point, the moment continues to increase linearly with time until the main crack in the glue grows at $t = 1100$ s, figure 7. However, even while there is still linear elasticity at the macroscopic scale (800 s $< t <$ 1100 s) the strain across the interface increases significantly - this apparent localization of strain is taken to indicate the onset of interfacial delamination. If it is assumed that this is failure of the zero-thickness interface then the level of stress at which the onset of non-linearity occurs can be associated with the cohesive strength of the interface.

IOP Conf. Series: Materials Science and Engineering **139** (2016) 012025 doi:10.1088/1757-899X/139/1/012025

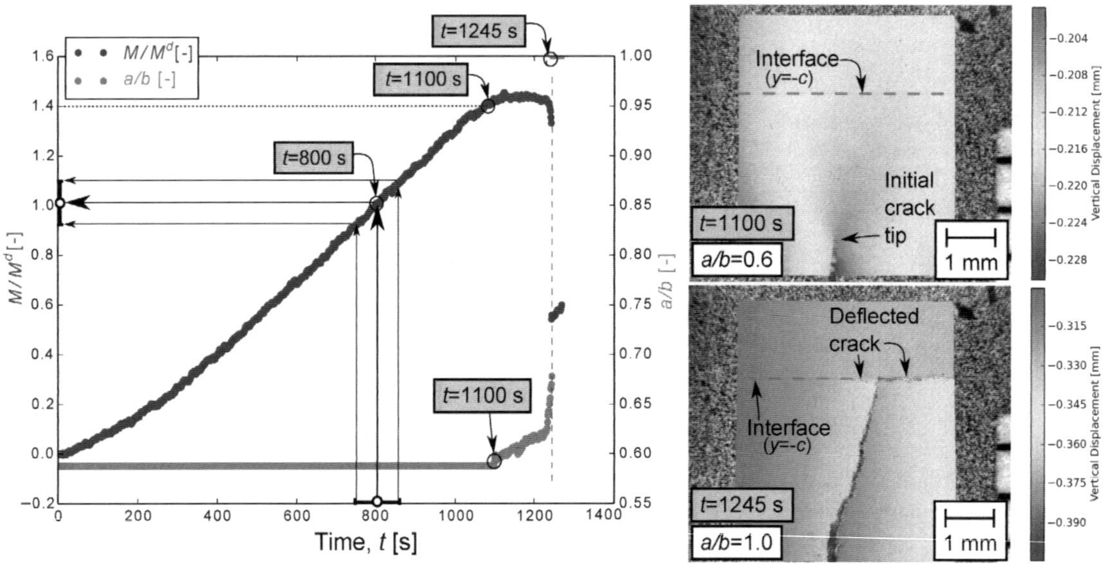

Figure 7. Left: Graph showing normalized moment and relative crack length as a function of elapsed time, t. The vertical blue dashed line indicates the time where the interface crack is fully developed. Right: DIC contour plot of the vertical y-displacement. The top contour plot is at time, t = 1100 s and the bottom contour plot is at time, t = 1245 s, just after the interface crack is fully developed. The red dashed line indicate the interface location.

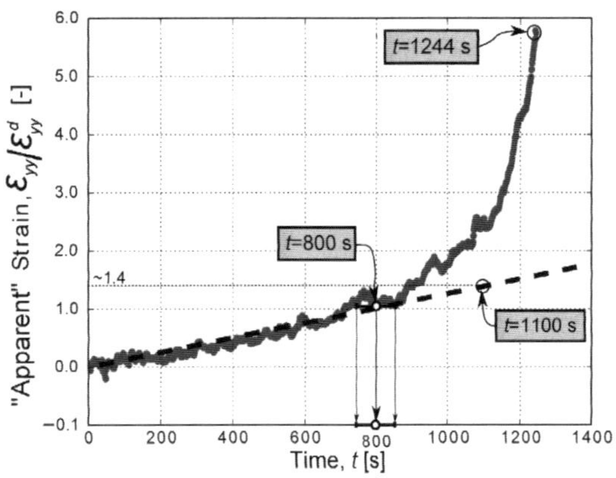

Figure 8. DIC: Normalized "apparent" strain, $\epsilon_{yy}/\epsilon_{yy}^{d}$, shown as a function of time. The "apparent" strain, ϵ_{yy} is normalized by the value of the "apparent" strain at the onset of non-linearity, ϵ_{yy}^{d}. The interface is located at $y = -c$, according to figure 3.

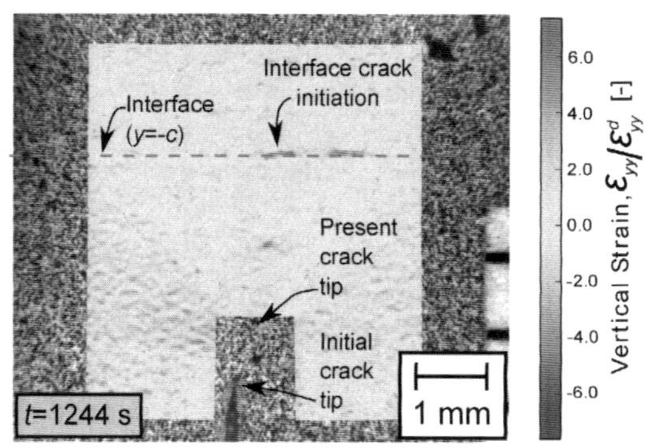

Figure 9. DIC: Contour plot of the vertical strain, normalized by the "apparent" strain at the onset of non-linearity, ϵ_{yy}^{d}, just before the main crack grows to the interface (t = 1244 s). The initiation of the interface crack is clearly identified. The interface is located at $y = -c$ according to figure 3.

The cohesive strength of the interface is determined by using an finite-element model (Section 3.1) to calculate the value of the normal stress across the interface at the conditions under which the onset of a non-linear strain were observed. It is estimated that the time of transition from linear to non-linear strain at the interface can be identified to an accuracy of ± 50 s, the pre-crack length can be measured to an accuracy of $a_0/b \pm 1\%$, and other uncertainties are indicated by error bars in figure 7. As discussed earlier, the material properties that entered into this calculation are confidential information. However, the calculations result in a cohesive strength for the interface of:

$$\hat{\sigma}/\bar{\sigma}_g = 0.081 \pm 0.007 \tag{1}$$

Again, owing to the confidential nature of the system, the cohesive strength has been normalized with the macroscopic strength of the glue, $\bar{\sigma}_g$, which was obtained by a uni-directional tensile test of a dog bone specimen with a gauge length of 115 mm. It would be more appropriate to normalize with the cohesive strength of the glue, but this is not known.

5. Discussion

The cohesive strength can be determined with ESEM using a J-integral based approach [6]. This approach requires manufacturing tiny specimens and using specialized and expensive equipment, such as a special fracture mechanics loading stage for ESEM. A benchmark of the new approach with the ESEM and the J-integral based approach is proposed as a future study.

One of the advantages with the new approach presented in this paper is that there is no need to use advanced scanning-electron microscope equipment, since a standard 4 point bend rig with a DIC camera system can solve the task. During the last decade DIC has become a relatively easy, cheap, and efficient tool for measuring in-plane deformations, and it is available in most labs at universities [13]. The new approach is not limited to the 4-point SENB specimen, but it can be used with any other test specimen, provided that the interface crack initiation can be captured by DIC and the interface stress can be accurately determined using a model (*e.g.* FE or analytical).

The cohesive strength at the interface is extracted in the symmetry line ($x = 0$, $y = -c$) of the SENB specimen where the shear stress is zero and therefore the crack initiation is mode-I. If the interface crack initiates at other locations along the interface, *e.g.* due to a defect, the crack initiation will be mixed mode. This is also confirmed by the FE model.

It is beneficial that the interface crack initiates before the main crack starts to propagate so that the growth of the main crack does not change the strain field. For an interface cohesive law with a high $\hat{\sigma}$ it may not be possible to initiate the interface crack, but maybe the test specimen could be optimized further in the future to enable this. Another design of the test specimen to determine the cohesive strength could be a SENB geometry with an elliptical notch. The advantage of this geometry is that the stress concentration factor would be known, and it will be harder for the main crack to start growing. This might simplify the analysis. An analysis similar to that of Cook and Gordon [1] could then be used to determine the optimum distance from notch to interface.

6. Conclusion

It can be concluded that the mode-I cohesive strength can be determined using a 4-point SENB test specimen in combination with 2D DIC measurements and linear-elastic finite element modeling. For the material system tested, a normalized cohesive strength is determined to $\hat{\sigma}/\bar{\sigma}_g = 0.081 \pm 0.007$, meaning that the interface cohesive strength is about 0.08 of the macroscopic strength of the glue.

IOP Conf. Series: Materials Science and Engineering **139** (2016) 012025 doi:10.1088/1757-899X/139/1/012025

Acknowledgments
Thanks to the LM Wind Power lab for help manufacturing the test specimens and to Fulbright for supporting the research stay at the University of Michigan. Special thanks to James Gorman for his help when conducting the experiments in the lab at the Department of Mechanical Engineering, University of Michigan, Ann Arbor, MI, USA. This research was supported by the Danish Centre for Composite Structure and Materials for Wind Turbines (DCCSM), grant no. 09-067212, from the Danish Strategic Research Council (DSF).

References
[1] Cook J and Gordon J E 1964 *Proc. R. Soc. Lond. A.* **10** 508
[2] He M Y and Hutchinson J W 1989 *Int. J. Solids Structures* **25** 1053
[3] Thouless M D, Cao H C and Mataga P A 1989 *J. Matls. Sci.* **24** 1406
[4] He M Y, Evans A G and Hutchinson J W 1994 *Int. J. Solids Structures* **31** 3443
[5] Parmigiani J P and Thouless M D 2006 *J. Mech. Phys. Solids* **54** 266
[6] Goutianos S, Arevalo R, Sørensen B F and Peijs T 2014 *Appl. Compos. Mater.* **21** 805
[7] Li V C and Ward R J 1989 *Fracture Toughness and Fracture Energy, Mihashi et al. (eds).* 183
[8] Suo Z, Bao G and Fan B 1992 *J. Mech. and Phys. Solids.* **40** 1
[9] Kendall K 1975 *Proc. R. Soc. Lond. A.* **344** 287
[10] Lee W and Clegg W J 1996 *Key Engineering Materials* **116** 193
[11] Tada H, Paris P C and Irwin G R 2000 *The Stress Analysis of Cracks Handbook* 3rd ed (New York: ASME Press)
[12] Crammond G, Boyd S W and Dulieu-Barton J M 2013 *Optics and Lasers in Engineering* **51** 1368
[13] Reedlun B, Daly S, Hector L, Zavattieri P and Shaw J 2013 *Experimental Techniques* **37** 62

IOP Conf. Series: Materials Science and Engineering **139** (2016) 012026 doi:10.1088/1757-899X/139/1/012026

Microstructural analysis using X-ray computed tomography (CT) in flax/epoxy composites

M Kersani[1,2], SV Lomov[2], AW Van Vuure[2], A Bouabdallah[1] and I Verpoest[2]

[1]Faculty of Physics, University of Science and Technology Houari Boumediene, Algiers, Algeria
[2]Department of Metallurgy and Materials Engineering, KU Leuven, Leuven, Belgium

E-mail: kersani. malika@gmail.com

Abstract. Among natural fibres which have recently become attractive to researchers, flax is probably the most commonly used bast-type fibre today. Due to its properties and availability, flax fibre has potential to substitute glass in polymer composites. A flax fibre has a complex structure; it can be classified into elementary fibres, which are grouped into so-called technical fibres. These technical fibres themselves are actually composite structures. Several works [1, 2, 3] were focussed on the study of damage behaviour in unidirectional flax fibres reinforced composites, where materials were subjected to tensile loading. At the microscopic level and at low stress, microcracks arise within the material and by growing they may lead to other forms of damage such as delamination, fibre breakage, interfacial debonding...etc. In order to better understand the damage phenomena and to better control the parameters which lead to the failure, several methods and techniques have been developed on natural fibre reinforced composites [2, 3]. In the present work, X-ray computed tomography (CT) technique has been used to observe damage in flax/epoxy quasi-unidirectional woven laminates, loaded in uniaxial tension. The tensile tests show that these composites offer good mechanical properties. X-ray computed tomography technique allowed us, on the one hand to determine the microstructure parameters of the studied composites and to observe the damage occurring during loading, on the other. The inspection of the several tomography images showed cracks on interface of the yarns and technical fibres.

1. Introduction

Micromechanics damage in laminates commonly initiates with transverse matrix cracking in 90° plies, when loaded in tension. Experimental recent study performed on natural fibre reinforced composites: flax/epoxy [4], didn't reveal such cracks till the late stage of loading. In this previous study, different laminates were subjected to tensile loading then analyzed under an optical microscopy, at different stages of loading. To go deeper in the structure, in the present work samples prepared from similar laminates were subjected to tension then analysed by the X-ray computed tomography (CT). CT analysis is a useful method in finding eventual internal cracks that are not visible by optical microscopy.

2. Materials and methodology

The material used in this work is a quasi-unidirectional woven prepregs, flax/epoxy, with areal density of the reinforcement 170 g/m2, provided by LINEO (Belgium). The woven fabric is pre-impregnated with epoxy resin (prepreg system Araldite LY5150/Aradur 1571/Accelerator 1573/HardenerXB3471). The quasi-unidirectional woven prepregs is made of flax fibres oriented in the warp direction (95.5%) and 4.5% flax fibres in the weft direction (see figure 1). All the studied laminates are constituted of eight layers of the flax prepreg. The orientation in each laminate is made symmetrically to its middle plan. Figure 2 illustrated the sequence of the layers in each laminate. 0° direction is defined as the direction parallel to the loading. The angle in the designation of the laminate means warp orientation of the corresponding fabric layer.

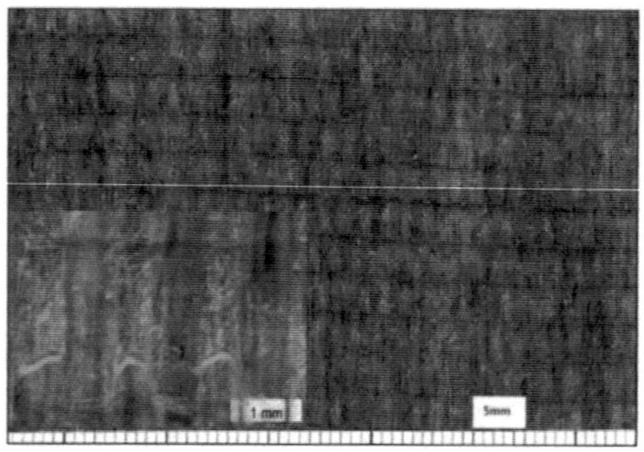

Figure 1. Flax prepreg. Warp direction is vertical. Insert: enlarged image.

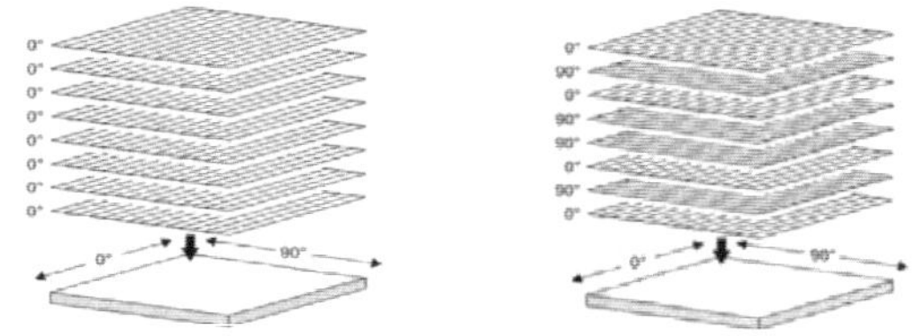

(a) [0°]8 (b) [0°,90]°2s

(c) [0°,90°,+45°,-45]°s (d) [+45°,-45°]2s

Figure 2. The different architectures of the prepared composites.

Several plates with different lay-up configurations: $[0°]_8$,$[0°,90°]_{2s}$,$[0°,90°,+45°,-45°]_s$ and $[+45,-45]_{2s}$ were prepared using vacuum bag and an autoclave. The temperature and the pressure during autoclaving reached 130C and 4 bars, respectively. The total fibre volume fraction of all the laminates is 47 ± 2%.

3. Results and discussion

3.1 Optical microscopy observations

In order to analyze the plate's cross sections specimens of 1x1 cm were cut from the plates' edges; these cross sections, observed under an optical microscope, are shown in figure 3.

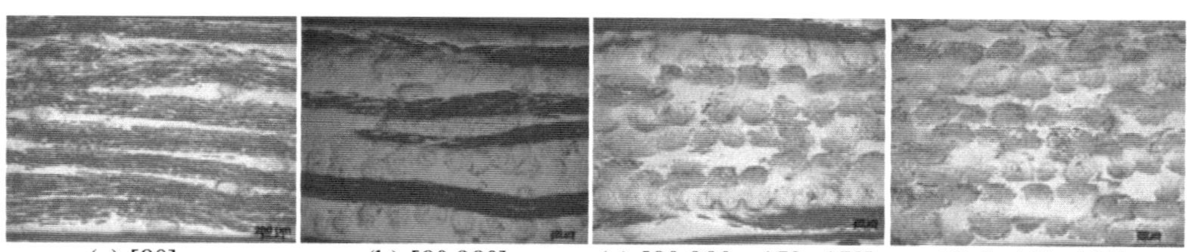

(a) [0°] (b) $[0°,90°]_{2s}$ (c) $[0°,90°,+45°,-45°]_s$ (d) $[+45°,-45°]_{2s}$

Figure 3. Optical microscopy showing the cross sections of the laminates (sections parallel to 0° axis).

These micrographs highlight the different sequences of the laminates and reveal also a good quality of impregnation: no voids were revealed from the optical microscopy observation, in all of the prepared laminates, at the external surface level.

3.2 X-ray computed tomography (CT) of the virgin samples

To go somewhat deeper into the structure and in the attempt to visualize eventual defects, X-ray computed tomography (CT) technique was adopted. The CT method was performed with a Phoenix Nanotom using VGStudioax software. The target material used in the tube was tungsten and the exposure conditions were 69 kV and 150μA. Samples with size 10X10 mm were cut from the prepared plates and then exposed to X-ray beam under different angular positions.

From a large number of the recorded two-dimensional X-ray projection images, the three dimensional (3D) image (figure 4) of the sample is then computationally reconstructed. From the 3D images of the different structures, 2D slices in any plane (x-y, x-z and y-z planes) can be observed. An example of these structures can be seen in figure 5.

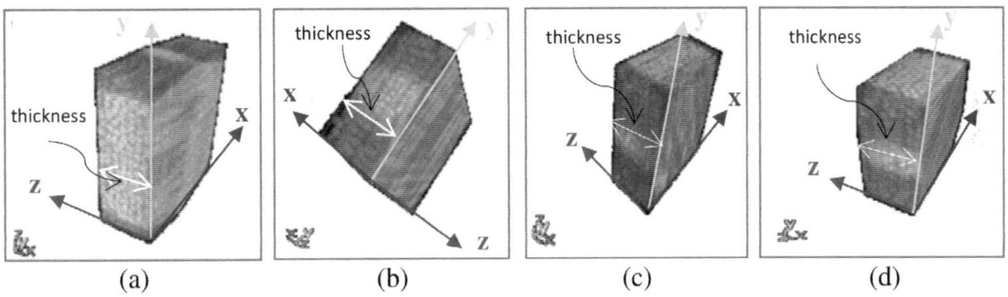

Figure 4. Tomography image illustrating the reconstructed 3D volume of (a) $[0°]_8$, (b) $[0°,90°]_{2s}$, (c) $[0°,90°,+45°,-45°]_s$ and (d) $[+45°,-45°]_{2s}$. (Y axis is parallel to the 0° direction).

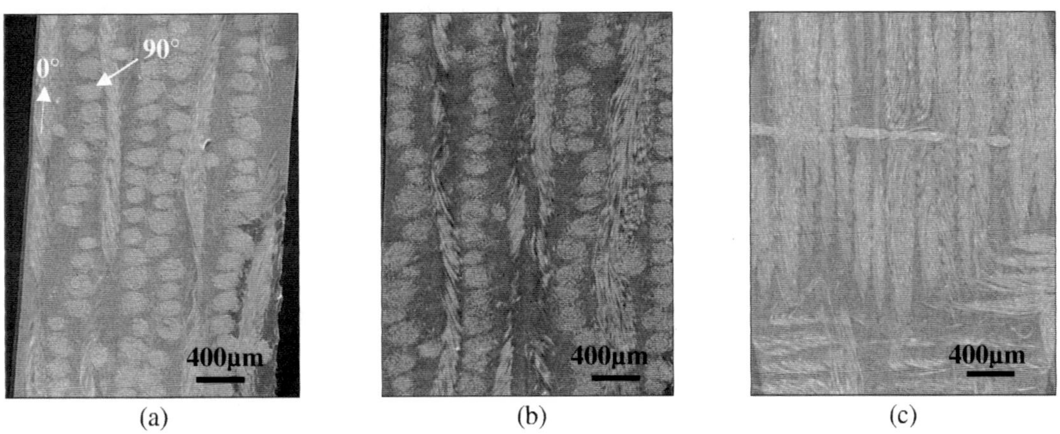

Figure 5. Internal 2D slice in a scanned 3D volume of $[0°,90°]_{2S}$ laminate. (a) In x-y plane, (b) in x-z plane and (c) in y-z plane (y axis is parallel to 0° axis).

The inspection on the several 2D tomographs allowed us to determine the microstructural parameters of the studied composites. Figure 6 shows an example of internal 2D slice in a scanned 3D volume, relative to [0°,90°,+45°,-45°]s structure, in its unloaded state. In this figure the different microstructural characteristics are illustrated. The left image shows 2D slice in x-y plane, where the yarn cross sectional structure can be observed. The yarn about 200-500 μm, is constituted of an assembling of elementary fibres and technical fibres. The technical fibres are themselves composed of ~2 to 5 inseparable elementary fibres. The diameters of technical fibres and elementary fibres can be estimated by 30–300 μm and 15–35 μm, respectively. As it can also be seen in figure 6, elementary fibres are characterised by a lumen; an open channel in their centre. This lumen is observable in few elementary flax fibres. Not all of flax elementary fibres have a visible lumen.

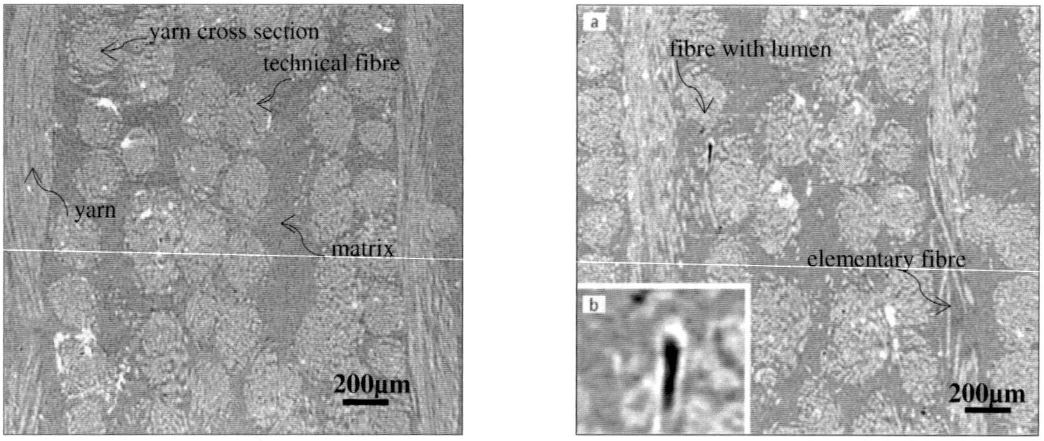

(a) (b)

Figure 6. Figure illustrating the different microstructural parameters in flax epoxy laminate, [0°,90°,+45°,-45°]s structure. (a) 2D slice in x-y plane. (b) 2D slice in x-z plane, the insert is the zoomed lumen.

A more rigorous visual inspection performed on the numerous 2D tomographs in the 3 different planes, revealed also good quality of impregnation. Moreover, matrix porosities which could not be detectable using optical microscopy were revealed by CT analysis. Yet, only very small amount of porosities was detected (figure 7).

(a) (b)

(c)

(d)

Figure 7. Internal 2D slices in x-z plane illustrating porosities in $[45°,-45°]_{2s}$ laminate.

For the purpose of understanding how damage affects the stress-strain response of the material during loading CT method was used. Samples relative to the different studied laminates were subjected to tensile loading and then the overall stress-strain responses were established.

3.3 Tensile tests

The samples relative to each laminate were loaded in uniaxial tension afterwards they were subjected to CT analysis. Tensile tests were performed using an INSTRON 4505 machine with a loading capacity of 100 kN according to ASTM D3039 standard. The crosshead speed, temperature and humidity during the tests were 5 mm/min, 20°C and 50%, respectively.

3.3.1 Stress–strain curves. The stress-strain curves for the samples are represented in figure 8. As it can be seen the tensile test revealed a specific behaviour of the strain-stress curves, characterized by a bilinearity. Three main portions can be observed: the first (1) and the third (3) portions behave linearly and the second (2) one exhibits a non-linear behavior. On the basis of these curves, the tangent modulus (E) degradation, during loading, was evaluated and the variation of E versus strain for one characteristic stress-strain diagram was plotted in the same graph (figure 6). As it can also be seen the non-linear behaviour is manifested at the early stage of loading in all of the studied laminates. The onset of non-linearity is in the range of 0,12%...0,2% strain for $[0°]_8$, $[0°,90°]_{2s}$, $[0°,90°,+45°,-45°]_s$ laminates and 0,5% strain for $[+45°,-45°]_{2s}$ laminate. The onset of the non-linearity of the stress-strain curves is defined as strain for which the tangent slope of the curve is decreased by 10% of its initial value.

Young modulus and the main mechanical properties calculated on the basis of the stress-strain curves: ultimate strength, ultimate strain and Poisson ratio, identified from stress-strain curves, are listed in table 1.

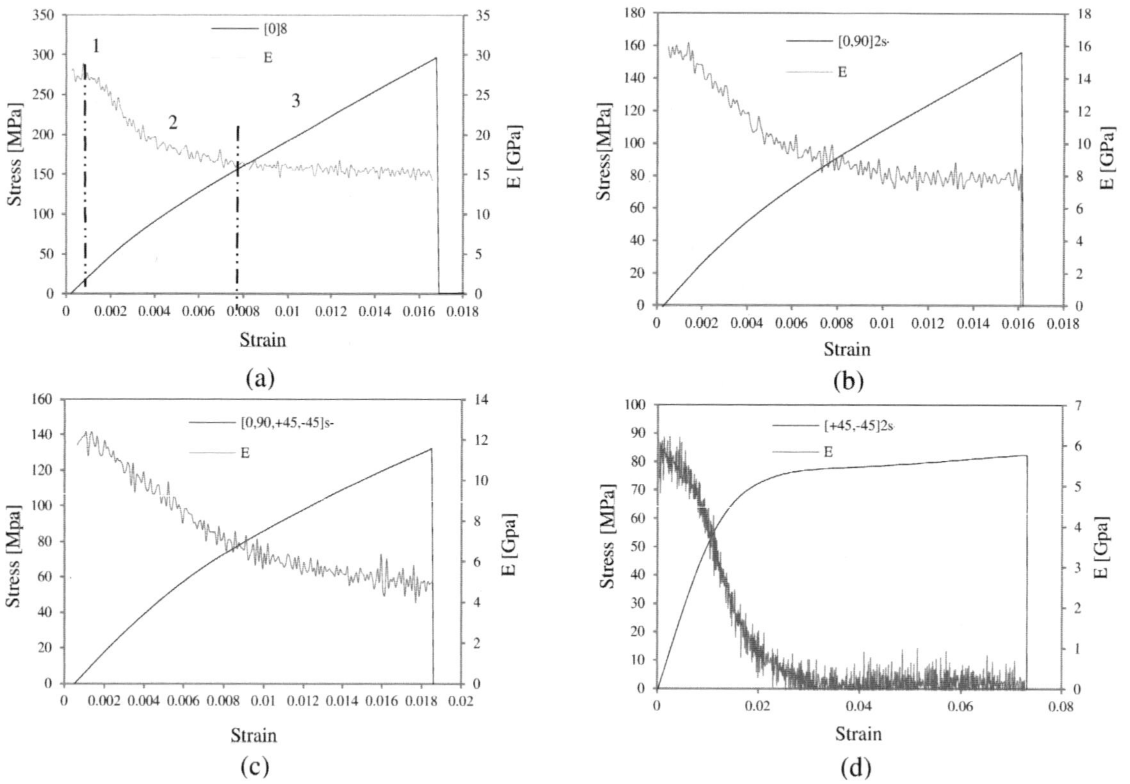

Figure 8. Stress–strain curves and stiffness–strain curve: (a) $[0°]_8$. (b) $[0°, 90°]_{2S}$. (c) $[0°, 90°, 45°, 45°]_S$. (d) $[45°, 45°]_{2S}$.

Table 1. Table summing up the main tensile and AE results. Note: the scatter gives the standard deviation in three tests.

	Onset of non-linearity strain [%]	Initial Young's modulus [GPa]	Poisson's ratio	Ultimate strength [MPa]	Ultimate strain [%]
$[0]_8$	0.12	27.2 ± 0.52	0.409 ± 0.034	296 ± 0.5	1.65 ± 0.055
$[0°, 90°]_{2S}$	0.14	1 5.7 ± 0.15	0.126 ± 0.014	158 ± 2	1.62 ± 0.5
$[0°,90°,45°,45°]_S$	0.2	11.9 ± 0.6	0.319 ± 0.259	126 ± 7.5	1.76 ± 0.15
$[45°,45°]_{2S}$	0.5	5.7 ± 0.11	0.566 ± 0.064	85 ± 4	7.47 ± 0.415

Similar non-linearity behavior of the stress–strain curves relative to the flax fibres and flax fibre reinforced composites was also reported in refs [5,6,7]. The non-linearity phenomena which manifests at the early stage of loading is probably due to intrinsic changes of flax fibres which cause material stiffness decrease. To better understand the non-linearity phenomena and damage mechanisms which occur during loading, tensile tests were performed on another series of samples at different stages of loading. The tensile tested specimens were then subjected to CT analysis.

3.2 X-ray computed tomography (CT) of tensile tested samples at different stages of loading

The 3D images for the tensile tested samples were reconstructed and meticulous visual inspection was performed on ~ 2000 tomographs in each plane of the studied laminates. The analysis revealed that at

IOP Conf. Series: Materials Science and Engineering **139** (2016) 012026 doi:10.1088/1757-899X/139/1/012026

the first stage of loading, corresponding to region2 of the stress-strain curves, microcracks are present on interface of the yarns and technical fibres. In the previous study [4] where damage was monitored by acoustic emission technique, these small cracks started from the applied strain level corresponding to onset of the AE events. Figure 9, figure10 and figure 11 illustrate these small cracks in all the studied laminates.

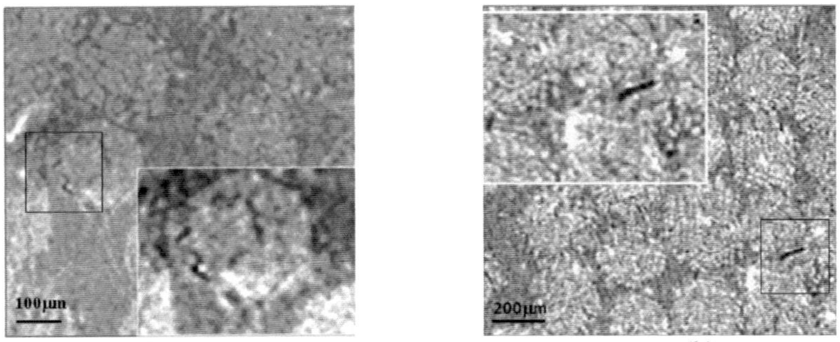

(a) (b)

Figure 9. Internal 2D slice in x-z plane illustrating microcracks on (a) technical fibre interface and (b) yarn/matrix interface, [0°]8 laminate. Insert, the zoomed cracked region.

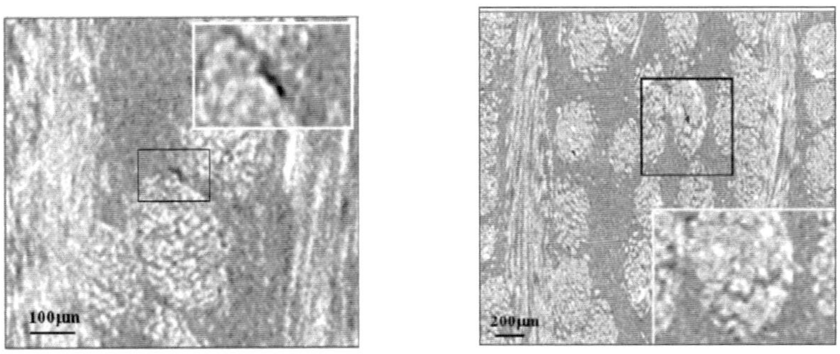

(a) (b)

Figure 10. Internal 2D slice illustrating microcraks on (a) yarn/matrix interface in x-y plane, $[0°,90°]_{2s}$ laminate and (b) technical fibre interface in x-z plane, [0°,90°,+45°,-45°]s laminate. Insert, the zoomed cracked region.

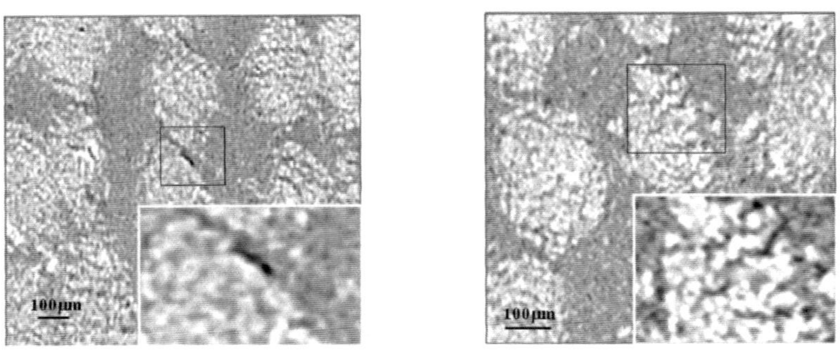

Figure 11. Internal 2D slice illustrating microcracks in x-z plane, [+45°,-45°]2s laminate. (a) on yarn/matrix interface (b) On the technical fibre interface. Insert, the zoomed cracked region.

As the loading increases, these microcracks may growth and developed. Here (figure 12) we present an image where cracks are well illustrated, when load is increased.

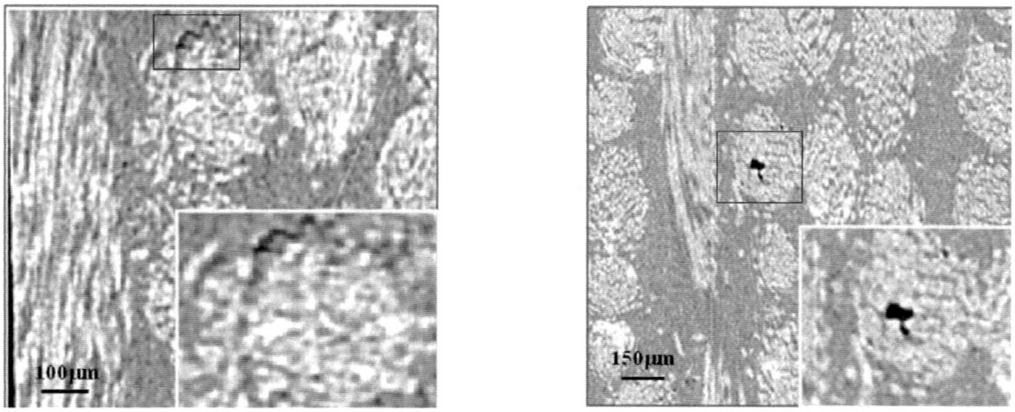

(a) (b)

Figure 12. Internal 2D slice in a scanned 3D volume, [0°,90°,+45°,-45°]s laminate (a) crack is seen at the interface of the yarn, x-y plane (b) shows crack at the technical fibre interface, x-z plane. Insert the zoomed cracked part.

The low adhesion of epoxy resin to flax fibres could favorite the occurrence of cracks on the interfaces of yarns. The presence of a relatively considerable number of lumens in the technical fibre could also stimulate the development of cracks on the technical fibres interfaces [3].

4. Conclusion

The 3D X-ray computed tomography technique allowed us to visualise the external and internal of 3D microstructures. This method was used to characterise four types of laminates. CT analysis complements and corroborates previous study made on similar laminates: The visual inspection performed on numerous tomographs in the different planes (x-y, x-z and z-y) of the damaged samples showed that cracks seem to be located at yarn-matrix and technical fibre interfaces in all stages of loading. Moreover, 3D X-ray tomography revealed also the observed absence of the transverse matrix cracking not only at the external surface level but in all the internal structure of the material. The absence of such cracks can be a major advantage for the use of natural fibre reinforced composites. Even though, CT method enriches the results but how damage affects the stress-strain response of the material remains still not fully understandable. More work should then be also undertaken in future studies.

References

[1] Slattery ATI, 1936, Some observations on the size and shape of the flax fibre ultimates, *J.Text. Inst Trans* **27** T101–T108

[2] CELC, 2012, *Flax and Hemp fibres: a natural solution for the composite industry* (Paris: JEC Composites)

[3] Barley C, Perrot Y, Busnel F, et al, 2006, Transverse tensile behaviour of unidirectional plies reinforced with flax fibres, *Mater Lett* **60** 2984–87

[4] Kersani M, Lomov SV, Van Vuure AW, et al, 2015, Damage in flax/epoxy quasi-unidirectional woven laminates under quasi-static tension *J Compos Mater* **49** 403–13

[5] Baley C, 2002, Analysis of the flax fibres tensile behaviour and analysis of the tensile stiffness increase, *Compos Part A* **33** 939–48

[6] Charlet K, Eve S, Jernot JP, et al, 2009, Tensile deformation of a flax fiber, *Procedia Eng* **1** 233

[7] Hughes M, Carpenter J and Hill C, 2007, Deformation and fracture behaviour of flax fibre reinforced thermosetting polymer matrix composites, *J Mater Sci* **42** 2499–511

Gliding arc surface modification of carrot nanofibre coating – perspective for composite processing

Y Kusano[1], L Berglund[2], Y Aitomäki[2], K Oksman[2] and B Madsen[1]

[1]Department of Wind Energy, Section of Composites and Materials Mechanics, Technical University of Denmark, Risø Campus, DK-4000 Roskilde, Denmark
[2]Division of Materials Science, Composite Centre Sweden, Luleå University of Technology, SE-971 87 Luleå, Sweden

E-mail: yuki@dtu.dk

Abstract. Surfaces of carrot nanofibre coatings were modified by a gliding arc in atmospheric pressure air. The treatment strengthened wetting of deionized water and glycerol, increased an oxygen content, C-O and C=O, and moderately roughened the surfaces. In the perspective of composite materials, these changes to the nanofibres can potentially improve their processability when they are to be impregnated with a polymeric matrix. However, longer exposure to the gliding arc reduced oxidation and roughness of the surface, and thus there exists an optimum condition to achieve good wetting to solvents.

1. Introduction

Non-thermal plasma processing at atmospheric pressure is widely used for surface modification [1, 2]. Its applications include surface cleaning [3], decontamination and sterilization [4,5], deposition of functional coatings [6], and improvement of adhesion, wetting and paintability [7-15].

One of the challenges in non-thermal plasma processing at atmospheric pressure is to achieve high reactivity and high productivity simultaneously [16]. Here, high reactivity is ensured by a high electron temperature in a non-equilibrium plasma so as to change chemical bonding of molecules in the plasma or the plasma-treated surfaces. Meanwhile, high productivity can be demonstrated using high energy densities. However, most plasmas can hardly sustain a non-equilibrium state at high energy densities. One possible approach to overcome this issue is to develop a hybrid plasma with a high energy density in a non-equilibrium state.

A gliding arc is one of the hybrid plasmas [17], generated between diverging electrodes as a low-impedance thermal arc discharge, extended by a gas flow and quenched to a non-thermal condition [18]. It can be operated in atmospheric pressure air and thus advantageously used for large-scale processing [19-26]. An alternating current (AC) gliding arc has a long lifetime extending over hundreds of AC periods without extinction. The plasma column can be elongated to approximately 20 – 30 cm [21-29]. The AC gliding arc is useful for adhesion improvement of glass fibre reinforced polyesters (GFRPs) [22,23], efficiently oxidizing GFRP surfaces when the distance between the edge of the electrodes and the GFRP surface is up to 6 cm in open air. It is also demonstrated that optical techniques are promising for non-instructive diagnostics of the AC gliding arc [24-29]. In particular, the dynamics of the gliding arc is observed by using a high-speed camera. In addition, excited and ground-state hydroxyl radicals are observed in the plasma column and its vicinity, respectively. It is noted that the hydroxyl radical has a high oxidation potential, next to fluorine and higher than ozone and chlorine. Therefore, the hydroxyl radicals generated by the gliding arc can play an important role in surface modification.

Cellulose is renewable, nontoxic, and biodegradable, and is known to be the most abundant biopolymer on earth [30]. It has been extensively studied in terms of biological, chemical, and mechanical properties. In particular, nano-celluloses (NCs) have attracted significant interests due to their high strength, high modulus, high surface area and unique optical properties. Typical lateral dimensions are approximately 5 – 20 nm, while longitudinal dimensions are from tens of nm to several μm [31]. Acid hydrolysis of cellulose will remove the amorphous part of cellulose and lead to highly crystalline and rigid nanoparticles called cellulose nanocrystals or cellulose nanowhiskers [32]. On the other hand, cellulose nanofibres (CNF) can be produced by ultrafine grinding, microfluidization or high pressure homogenization through delamination of fibre cell walls and subsequent liberation of nano-sized fibrils [33].

Promising application for demonstrating mechanical properties of NCs is to mix NCs to impregnate a nanofibre network with a polymer and create strong, light-weight and transparent composite [34]. Its application includes packaging [35,36], vehicles and aeronautical applications. For demonstrating the best performance of NC composites, it is important to achieve substantial dispersion of NCs and strong interfacial interaction between the NC and the polymer matrix. In this respect, surface modification of NCs has attracted significant interests [37,38].

However, surface modification of NCs is not an easy task even using a plasma. For example, when NC is dispersed in a liquid, a plasma as a gas phase does not have a direct contact with NC surfaces. On the other hand, when NCs are dried without a liquid, they are easily aggregated, disturbing uniform surface treatment. There are some studies presented in literature to solve or tackle with these handling problems [1].

In the present work, as a feasibility study, effects of plasma treatment on NC are investigated in a simplified form of NC so as to avoid the above handling problems. That is, nanofibre coatings are treated by a plasma not because of the application for coatings, but because of a simplified specimen feature. More specifically, nanofibres were separated from a carrot residue, dispersed in water, painted on a glass plate, and dried. The subsequent coating surfaces were treated using a gliding arc in atmospheric pressure air. The treated surfaces were characterized by means of contact angle measurements, x-ray photoelectron spectroscopy (XPS), and field emission scanning electron microscopy (FE-SEM) in order to investigate the treatment effect.

2. Experimental methods

2.1 Materials

Carrot nanofibres were isolated from carrot residue supplied by Brämhults Juice AB, Sweden. This raw material is a by-product from carrot juice production. A chemical purification was carried out prior to the fibrillation following the procedure described by Siqueira et al. (2016) [31]. The residue was first washed with distilled water at 85°C, alkali-treated (2% NaOH) at 80°C for 2h, and subsequently bleached with $NaClO_2$ (1.7 %) in an acetic buffer (pH 4.5) at 80°C for 2h for lignin removal. Finally the material was washed until a neutral pH was reached.

This bleached residue was fibrillated by ultrafine grinding using a supermass colloider (MKZA10-20J, Masuko Sangyo, Japan), at consistency of 2%. Prior to the grinding, the suspension was dispersed using a shear mixer (Silverson L4RT Silverson Machine Ltd., England). The grinding was operated at a rotor speed of 1,500 rpm and the grinding stones were gradually adjusted to 100 μm (negative). The total processing time was 40 min. The obtained aqueous carrot nanofibre suspension (2wt% concentration) was used to prepare coatings on glass plates (CORNING® 2947-75x25), and dried at room temperature in atmospheric pressure air for a couple of days. The coating thickness was estimated to be approximately 6 μm.

2.2 Plasma treatment

The gliding arc was generated between two diverging stainless steel tubular electrodes as shown in Figure 1. The outer diameter of the tubular electrodes is 3 mm. Cooling water was fed through the electrodes during operation [22-29]. An air flow was fed between the electrodes with a flow rate of 17.5 standard litres per minutes (SLM). The gliding arc was driven by an AC power supply at a frequency of 31.25 kHz (Generator 6030, SOFTAL Electronic GmbH). The electric power applied to the gliding arc was approximately 700 W. In order to treat the nanofibre coating surface without excess thermal damage from the gliding arc, a poly(methyl-methacrylate) holder was moved forward and backward at a speed of 180 mm/s on which the coated glass plate was fixed. The angle between the gas flow direction and the specimen surface is approximately 90° when the surface is treated. The closest distance between the coating surface and the edges of the electrodes was 15 mm.

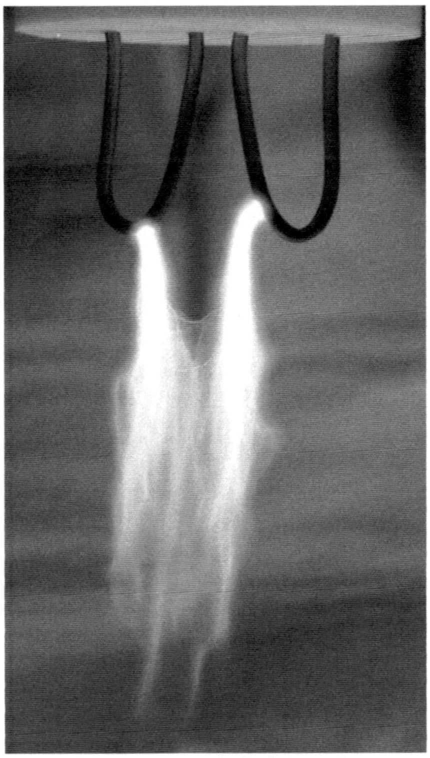

Figure 1. A photo of the gliding arc with water cooled electrodes.

2.3 Surface characterization

Surface characterization is vitally important to understand the surface modification effects and to optimize the process conditions.

Contact angles were measured with deionized water and glycerol in air at room temperature both before and after the gliding arc treatment using a contact angle measurement system (CAM100, Crelab Instruments AB, Sweden). A typical drop size was 1μL. The contact angle was measured within 5 s after the drop was attached onto the surface. Typically 10 measurements were made for each specimen, and their arithmetic mean was calculated.

XPS data were collected using a micro-focused, monochromatic Al Kα X-ray source (1486.6 eV) with a lateral resolution of 30 μm (K-alpha, ThermoFischer Scientific, UK). Atomic concentrations of all elements were calculated by determining the relevant integral peak intensities using the Shirley background. The K-alpha was also used for a high resolution analysis on the carbon 1s (C1s) spectra acquired over 30 scans. The binding energies were referred to the hydrocarbon component (C-H, C-C) at 285 eV. The spectra were de-convoluted through curve fitting, taking purely Gaussian components with linear background subtraction. The full-width at half-maximum (FWHM) for all peaks of C1s was constrained to 1.5 eV.

The surface morphology of the nanofibre coating was observed by using an FE-SEM (Zeiss SUPRA 35, Germany). The NFC surfaces were sputter-coated with approximately 15-nm thick Au before the observation.

3. Results and discussion

The contact angles of the nanofibre coating were measured before and after the gliding arc treatment. Typical images of the deionized water- and glycerol-drops on the nanofibre coatings are shown in Figure 2.

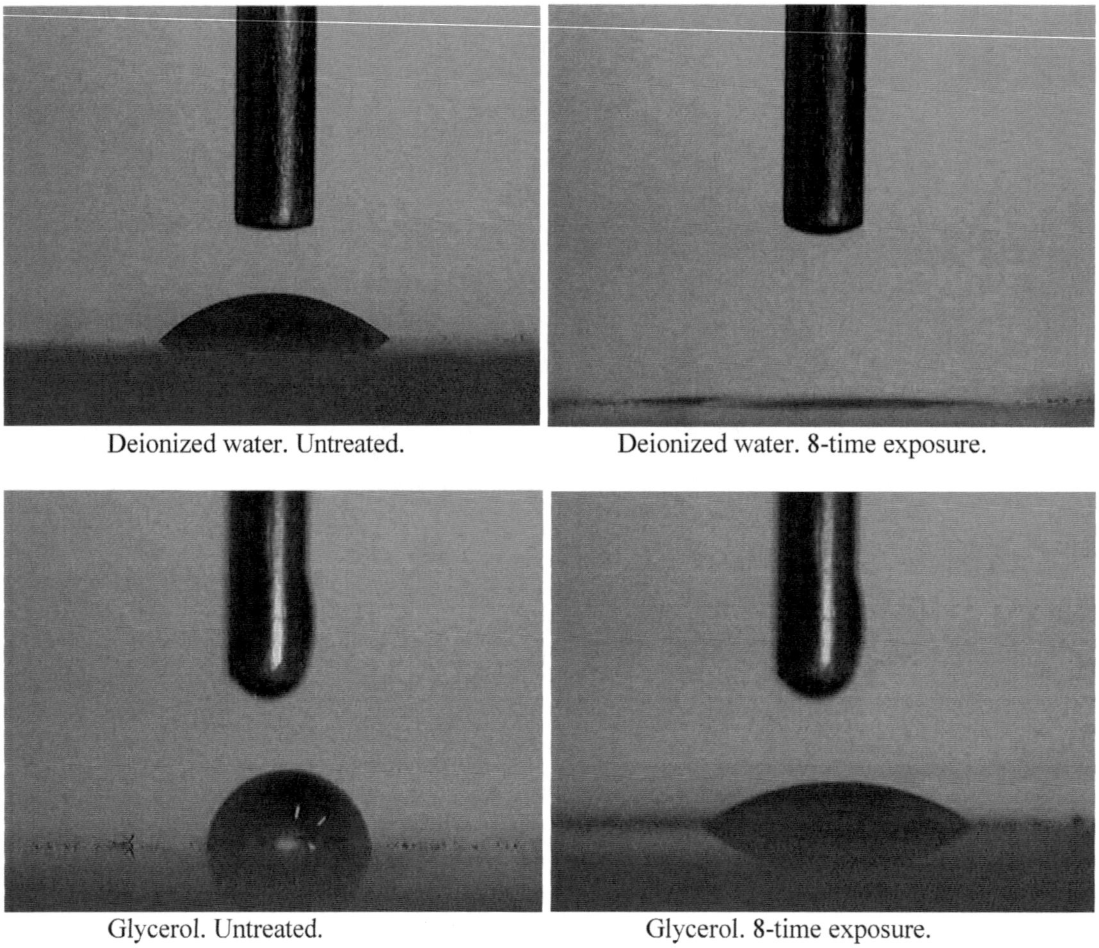

Figure 2. Typical photo images of deionized water-drops and glycerol-drops on the nanofibre coatings.

Figure 3 shows measured contact angles of deionized water and glycerol at different numbers of exposure. The contact angles of deionized water and glycerol before the gliding arc treatment, corresponding to 0-time exposure, were typically between 20° - 35° and 70° - 80°, respectively. The deionized water contact angles tend to decrease as the number of exposure increases, approaching 0°. The glycerol contact angles also tend to decrease as the number of exposure increases, approaching approximately 30°. Enhancement of glycerol wetting is preferable when nanofibres are mixed in a composite, since nanofibres are usually dispersed in an organic solvent for mixing it into a polymeric matrix, potentially improving processability and mechanical properties of the composite. However, it is noted that after 16-time exposure, contact angles of deionized water and glycerol slightly increased.

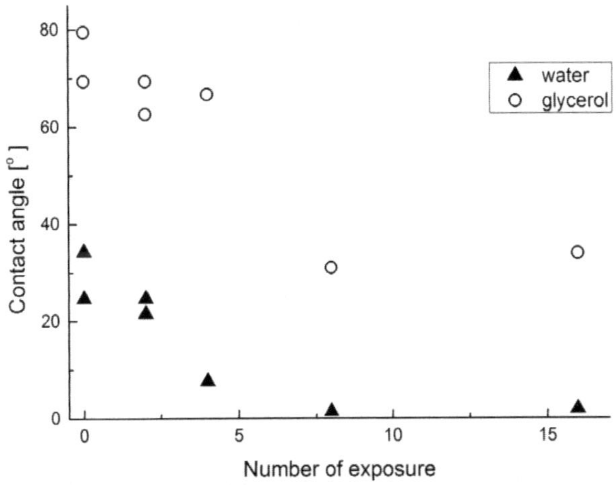

Figure 3. Contact angles of the nanofibre coatings at different times of exposure to the gliding arc. The numbers of exposure are 0 (corresponding to an untreated surface), 2, 4, 8 and 16. Solid triangle: deionized water. Open circle: glycerol.

XPS measurement was carried out to analyze the elemental composition of the nanofibre coatings before and after the gliding arc treatment. The nanofibre surfaces typically contain C, O, Ca, and low concentration impurities. The atomic ratios of O/C and Ca/C calculated using the XPS survey spectra are shown in Figure 4 (a). After 2-time exposure, the O/C ratio increased from 0.59 to approximately 0.72. The O/C ratio may not significantly increase by further exposure, and eventually decreased to approximately 0.64 after 16-time exposure. The Ca/C ratio also increased, after the gliding arc treatment, from 0.02 to 0.03-0.04. The increase in the calcium content can be interpreted as preferential etching of organic components in the coatings. It is indicated that the Ca/C ratio does not monotonically increase as the increase in the exposure to the gliding arc, and that the preferential etching of organic component is not a leading effect of the gliding arc treatment.

Deconvolution of the C1s spectra of the nanofibre coatings before and after the gliding arc treatment was carried out. The result is shown in Figure 4 (b). Peaks at approximately 285 eV, 286.8 eV, and 288.4 eV can be assigned to C-H/C-C, C-O, and C=O (carbonyl). A peak corresponding to carboxy group at approximately 289.5 eV was not detected. Since general plasma treatment can readily create carboxy group on polymer surfaces by oxidation [11], it is suggested that the gliding arc may result in moderate oxidation on polymer surfaces. After 2-time exposure, the peak corresponding to C-H/C-C decreased, while then peaks corresponding to C-O and C=O increased. Similar to the O/C ratio, they may not significantly change by further exposure, and after 16-time exposure, the peak of C-H/C-C increased and the peaks of C-O and C=O decreased.

IOP Conf. Series: Materials Science and Engineering **139** (2016) 012027 doi:10.1088/1757-899X/139/1/012027

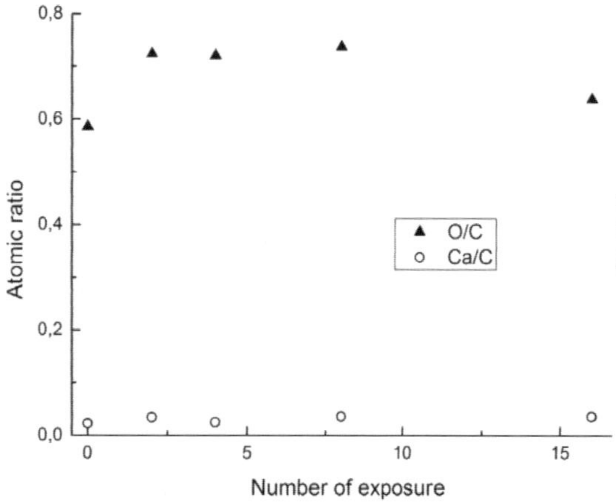

(a)

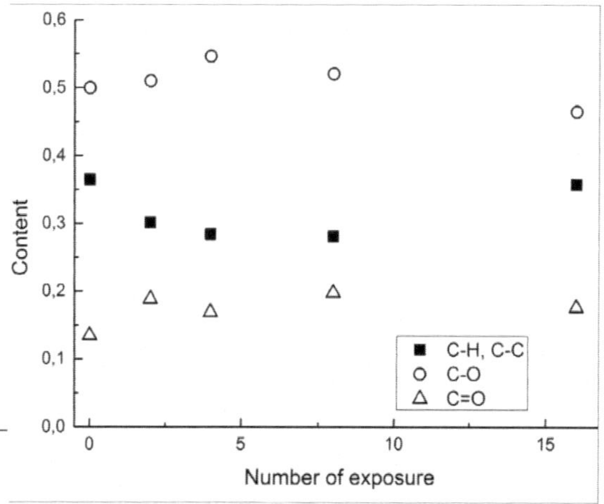

(b)

Figure 4. (a) O/C and Ca/C ratios of the nanofibre coating surfaces at different times of exposure to the gliding arc. (b) Contents of peaks obtained by curve-fitting of C1s spectra of the nanofibre coating surfaces at different times of exposure to the gliding arc.

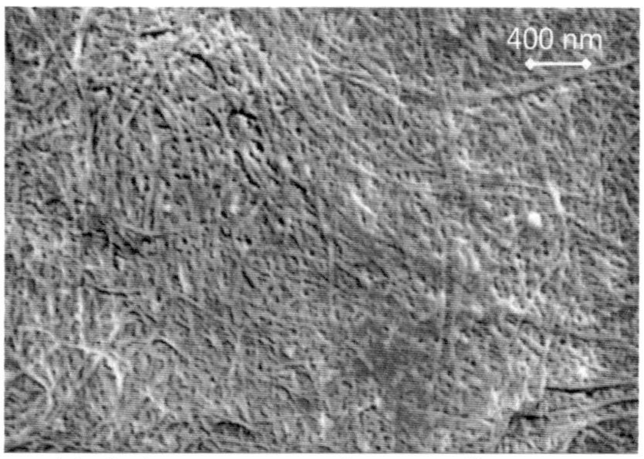

(a) 0-time exposure (untreated).

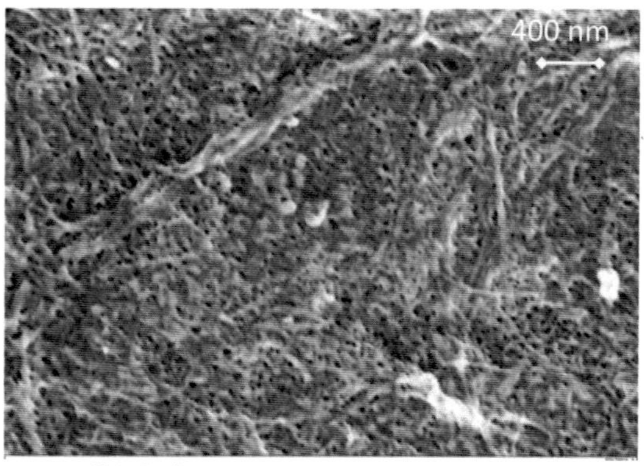

(b) 2-time exposure.

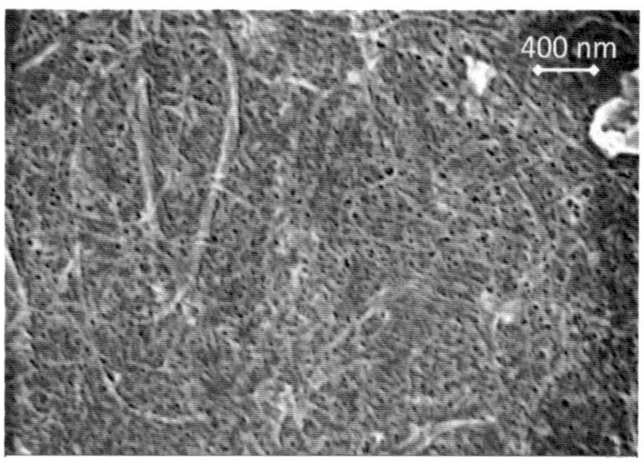

(c) 4-time exposure.

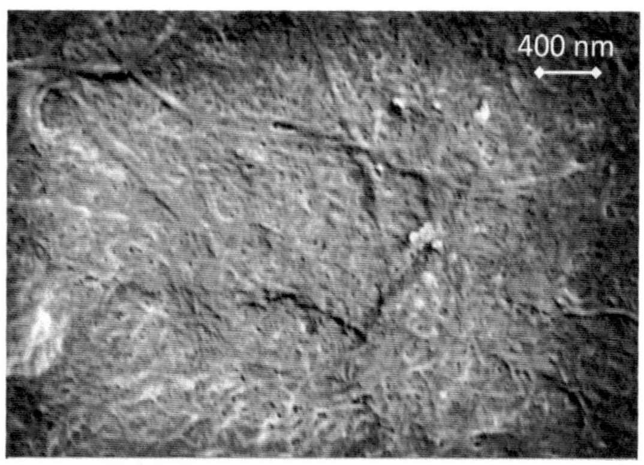

(d) 16-time exposure.

Figure 5. FE-SEM images of the nanofibre coating surfaces.

The morphological changes after the gliding arc treatment is possible only when nano-scale observation is performed, since plasma surface treatment generally changes the top layer up to 10 nm or less [11]. Figure 5 shows FE-SEM images before and after the gliding arc treatment. Figure 5 (a) indicates that the untreated nanofibre coating shows assemblies of nano-cellulose fibrils surrounded by an unstructured phase. 2-time exposure in Figure 5 (b) seems to roughen the surface, most likely due to the preferential etching of the surrounding phase, supported by the XPS result of the increase in the Ca/C ratio. However, further exposure may not significantly roughen the nanofibre coating surfaces in a nano-scale as shown in Figure 5 (c) and (d). Instead, these images indicate that the surface become smoother in nanoscale, while a larger scale deformation might occur (Figure (d)). A possible explanation of this morphological change by the gliding arc treatment is that the nanofibre coating surface could be etched and roughened moderately, and that longer exposure could result in microscale melting at the surface. A similar phenomenon was observed for the gliding arc treatment of glass-fibre reinforced polyester plates [19].

In next steps, treatment optimisation and treatment of larger amounts of nanofibres will be investigated so as to be feasible for production of carrot nanofibre composites with desired properties.

4. Conclusion

Gliding arc treatment can oxidize and roughen the carrot nanofibre surface and strengthen the wetting of deionized water and glycerol, and thus can be potentially useful for surface modification of the carrot nanofibres when they are mixed in a composite since the treatment can potentially improve processability and mechanical properties of the composite.

Acknowledgement

This work was financially supported by the European Commission through the Framework VII programme, INCOM project, contract number 608746.

References

[1] 2011 *Generation and applications of atmospheric pressure plasmas* ed M Kogoma, M Kusano and Y Kusano (New York: NOVA Science Publishers, Inc.) chapters 3-6 pp 45-100
[2] Kogelschatz U 2003 *Plasma Chem. Plasma Proc.* **23(1)** 1-46
[3] Yi CH, Lee YH and Yeom GY 2003 *Surf. Coat. Technol.* **171** 237-240
[4] Leipold F, Kusano Y, Hansen F and Jacobsen T 2010 *Food Control* **21(8)** 1194-1198
[5] Leipold F, Schultz-Jensen N, Kusano Y, Bindslev H and Jacobsen T 2011 *Food Control* **22(8)** 1296-1301
[6] Nwankire CE, Favaro G, Duong Q-H and Dowling DP 2011 *Plasma Proc. Polym.* **8** 305-315.
[7] 1994 *Plasma surface modification of polymers: relevance to adhesion* ed M Strobel, CS Lyons and KL Mittal (Utrecht, VSP)
[8] Kusano Y, Andersen TL and Michelsen PK 2008 *J. Phys. Conf. Series.* **100** 012002
[9] Kusano Y, Teodoru S and Hansen CM 2011 *Surf. Coat. Technol.* **205** 2793-2798
[10] Singh SV, Kusano Y, Morgen P and Michelsen PK 2012 *J. Appl. Phys.* **111** 083501
[11] Kusano Y 2014 *J. Adhesion* **90(9)** 755-777
[12] Teodoru S, Kusano Y, Rozlosnik N and Michelsen PK 2009 *Plasm. Proc. Polym.* **6** S375-S381
[13] Kusano Y, Mortensen H, Stenum B, Kingshott P, Andersen TL, Brøndsted P, Bilde-Sørensen JB, Sørensen BF and Bindslev H 2007 *Plasm. Proc. Polym.* **4(S1)** S455-S459
[14] Kusano Y, Mortensen H, Stenum B, Goutianos S, Mitra B, Ghanbari-Siahkali A, Kingshott P, Sørensen BF and Bindslev H 2007 *Int. J. Adhesion and Adhesives* **27(5)** 402-408
[15] Lee BJ, Kusano Y, Kato N, Naito K, Horiuchi T and Koinuma H 1997 *Jpn. J. Appl. Phys.* **Pt. 1, 36** 2888-2891
[16] Kusano Y 2009 *Surf. Eng.* **25(6)** 415-416
[17] Fridman A, Nester S, Kennedy LA, Saveliev A and Mutaf-Yardimci O 1999 *Prog. Energy*

Combustion Sci. **25** 211-231
[18] Czernichowski A 1994 *Pure Appl. Chem.* **66(6)** 1301-1310
[19] Kusano Y, Teodoru S, Leipold F, Andersen TL, Sørensen BF, Rozlosnik N and Michelsen PK 2008 *Surf. Coat. Technol.* **202** 5579-5582
[20] Kusano Y, Norrman K, Drews J, Morgen P, Leipold F, Singh SV, Bardenshtein A and Krebs N 2011 *Surf. Coat. Technol.* **205** S490-S494
[21] Kusano Y, Andersen TL, Toftegaard HL, Leipold F, Bardenshtein A and Krebs N 2014 *Int. J. Mater. Eng. Innovation* **5(2)** 122-137
[22] Kusano Y, Sørensen BF, Andersen TL and Leipold F 2013 *J. Adhesion* **89(6)** 433-459
[23] Kusano Y, Sørensen BF, Andersen TL, Toftegaard HL, Leipold F, Salewski M, Sun ZW, Zhu JJ, Li ZS and Aldén M 2013 *J. Phys. D Appl. Phys.* **46(13)** 135203
[24] Sun ZW, Zhu JJ, Li ZS, Aldén M, Leipold F, Salewski M and Kusano Y 2013 *Optics Express* **21(5)** 6028-6044
[25] Zhu JJ, Sun ZW, Li ZS, Ehn A, Aldén M, Salewski M, Leipold F and Kusano Y 2014 *J. Phys. D Appl. Phys.* **47**, 295203
[26] Kusano Y, Salewski M, Leipold F, Zhu JJ, Ehn A, Li ZS and Aldén M 2014 *Eur. Phys. J. D* **68** 319
[27] Kusano Y, Zhu JJ, Ehn A, Li ZS, Aldén M, Salewski M, and Leipold F 2015 *Surf. Eng.* **31(4)** 282-288
[28] Zhu JJ, Gao JL, Ehn A, Aldén M, Li ZS, Moseev D, Kusano Y, Salewski M, Alpers A, Gritzmann P and Schwenk M 2015 *Appl. Phys. Let.* **106** 044101
[29] Zhu JJ, Gao JL, Li ZS, Ehn A, Aldén M, Larsson A and Kusano Y 2014 *Appl. Phys. Let.* **105** 234102
[30] Peng BL, Dhar N, Liu HL and Tam KC 2011 *Canadian J. Chem. Eng.* **89(5)** 1191-1206
[31] Siqueira G, Tadokoro S K, Mathew AP and Oksman K 2016 *Comp. Sci. Technol.* **123** 49-56
[32] Bondeson D, Mathew AP and Oksman K 2006 *Cellulose* **13** 171-180
[33] Jonoobi M, Oladi R, Davoudpour Y, Oksman K, Dufresne A and Hamzeh Y 2015 *Cellulose* **22(2)** 935-969
[34] Nakagaito AN and Yano H 2005 *Appl. Phys. A* **80(1)** 155–159
[35] Herrera M, Mathew AP and Oksman K 2014 *Carbohydrate Polym.* **112** 494-501
[36] Herrera M, Sirviö J, Mathew A and Oksman K 2016 *Mater. Design* **93** 19-25
[37] Olsson RT, Azizi Samir MAS, Salazar-Alvarez G, Belova L, Ström V, Berglund LA, Ikkala O, Nogués J and Gedde UW 2010 *Nature Nanotechnol.* **5** 584-588
[38] Siró I, Kusano Y, Norrman K, Goutianos S and Plackett D 2013 *J. Adhesion Sci. Technol.* **27(3)** 294-308

Influence of internal temperature development during manufacturing on thick laminates compression fatigue properties

F Lahuerta and RPL Nijssen

Knowledge Centre WMC. Kluisgat 5, 1771 MV Wieringerwerf, The Netherlands

e-mail: f.lahuerta@wmc.eu, r.p.l.nijssen@wmc.eu

Abstract. Thick laminates (above 6 mm) are increasingly present in large composite structures such as wind turbine blades. Blade designs are based on static and fatigue coupon tests on 1-4mm thick laminates. However, a thickness effect has been observed, showing significantly shorter fatigue life in thick laminates. Different factors are suspected to be involved in the decreased static and dynamic performance of thick laminates. These include the effect of self-heating during fatigue testing, scaling effects, and the influence of residual stresses due to temperature gradients during manufacturing. This work studies the influence of the temperature gradients during resin infusion on the through-thickness fatigue properties in thick laminates. Coupons from thick laminates cured at different curing rates have been tested in fatigue. The work reports the compression fatigue properties of a thick laminate and relates the results with the curing rate. Safety factors between 1.23 and 1.60 regarding the influence of the curing cycles in thick laminates are reported.

1. Introduction

Voids, resin rich areas, dry spots and flaws are inevitably formed during the manufacturing process. Also, the different parameters involved in the manufacturing process such as the curing cycle, and the infusion process are likely to influence the final mechanical properties. This paper focusses on the relationship between curing rate and compression fatigue mechanical properties.

Lee [1] elaborated a series of expressions that can be used in conjunction with cure models to evaluate the effects of the curing cycle on strength values and moduli of the composite after manufacturing. A relation between the heat transfer rates and the interlaminar shear strengths (ILSS) was observed by Davies [2], lately by Liu [3] for epoxy resin in carbon-reinforced composites and recently Esposito [4] reported changes in the ILSS properties with the overheating of a glass fibre reinforced polymer.

While for thin laminates, curing temperatures do not show large gradients through the thickness, in thick laminates strong temperature gradients are observed [5]. Effectively, this is related to a curing cycle variation through the thickness, which can be expected to become more pronounced in thicker laminates. These gradients are caused by the exothermic nature of curing process and thermal diffusion problem dependent on the laminate thickness and thermal conductivities [6].

White and Kim [7] developed and studied the technique of staged curing. In this technique, two or more thin (<5 mm thick) prepreg laminates are manufactured at once with a similar curing cycle per staged laminate. Later, Ciriscioli [8] reported a technique to determine mechanical properties of thick pre-pregs using a porous film in between. This technique was applied to compare the properties of laminates of different thicknesses at the middle thickness position.

With the help of the sub-laminate technique (a technique to extract sub-laminates with the help of peel-ply layers from thick laminates), it was previously reported [9] that mechanical properties vary through the thickness, A strong correlation was found between static strengths and heating rates during the matrix hardening. The influence of the manufacturing and laminate thickness was not reported in detail [10,11].

The present work analyses scaled compression fatigue properties, and compression fatigue properties of composites manufactured at different heating rates using the sub-laminates technique. Where the variation of the fatigue properties through the thickness is related to the variation of the heating rates during the curing.

2. Materials and methods

The present work analyses S-N curves from plates manufactured using the sub-laminate technique [11], as well as S-N curves obtained from thickness scaled compression tests carried out for 4, 10 and 20mm thick coupons [10]. In both cases laminates were vacuum infused with an epoxy resin commonly used for wind energy applications (Hexion RIM135). The core material was a UD glass fibre type E of 600 gr/m^2 non-crimp fabric, and in the case of the scaled compression tests biaxial glass fibre type E of 400 gr/m^2 non-crimp fabric was used for the tabs. Fibre content was measured for each sub-laminate and plate with averaged fibre weight ratios between 68 to 76% (see table 1), void content was below 0.1%, average densities of 1.945 gr/cm^3 and DSC glass transition temperature (T_g) was around 80-85°C.

Table 1. Average cross section and coupons dimensions for each S-N curve. The relation between the coupons series and the measured maximum heating rates during the manufacturing of the coupons.

Name series	Thickness position of sub-laminate	Thickness initial plate	Thickness coupon gauge section	Width gauge section	Cross section area	Number of coupons	Fibre weight ratio, FWR
	[mm]	[mm]	[mm]	[mm]	[mm^2]	[-]	[%]
SubA-C1	4.19	55.65	3.74	14.39	53.8	13	71.3
SubA-C3	28.97	55.65	3.87	14.74	57.0	11	71.0
SubA-C4	39.26	55.65	4.27	14.53	62.0	6	71.3
SubB-C1	4.00	64.00	4.07	14.35	58.4	14	74.7
SubB-C3	26.40	64.00	3.99	15.18	60.5	10	74.9
SubB-C6	60.00	64.00	4.01	15.48	62.0	10	76.2
SubC-C6	60.00	64.00	3.93	15.36	60.3	7	73.2
Th4mm	-	8.00	4.00	10.00	40.0	16	68.9
Th10mm	-	20.00	10.60	25.00	265.0	11	70.7
Th20mm	-	32.00	20.00	55.00	1100.0	9	71.3

Each of the series was tested in fatigue with a stress ratio R=10. Table 1 shows the average dimensions for each of the series as well as the thicknesses of the coupons origin plates. Three 60mm thick laminates were manufactured (A, B and C) with two different curing cycles (see table 1 from [11]). Plate A and B were infused at 25°C, the mould control was setup with an initial soak of 35°C and 380 minutes and post-cured at a mould temperature of 80°C for 800 minutes. In the case of plate C the initial soak mould temperature was 60°C. From such thick laminates, sub-laminates plates were extracted at various thickness positions C1, C3, C4 and C6 (see table 1 and figure 1). Thus, series belonging to the sub-laminate tests are named as SubA, SubB and SubC. Therefore, each sub-laminate is associated with a certain through thickness position and a certain maximum heating rate (see table 2).

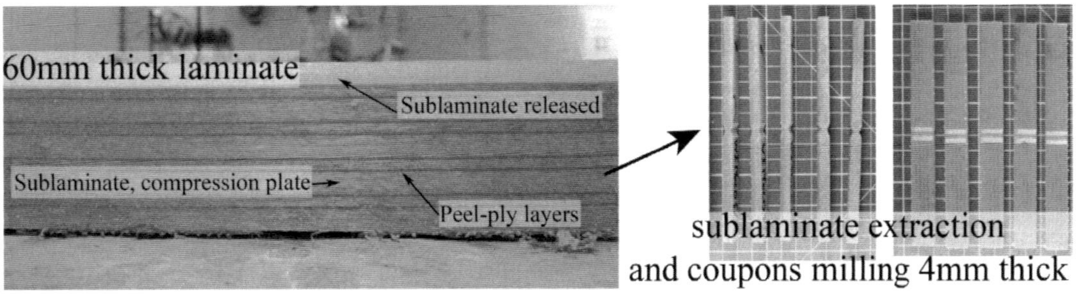

Figure 1. Sub-laminates coupons plates manufacturing and coupons preparation.

Figure 2. Thickness scaled compression tests setups and coupons, for 4, 10 and 20mm.

In addition, three laminates were manufactured to obtain compression specimens with gauge section thicknesses of 4, 10, and 20 mm, respectively. These plates were designated Th4mm, Th10mm, and Th20mm. From these panels, specimens were cut. These are 'scaled' specimens were designed to have maximum similarity between the geometries in terms of stress distribution, as well as maximum stress concentration near the gauge section [12,13].

Both the 4 mm thick coupons sub-laminates and thickness scaled tests were tested on an MTS servo-hydraulic 100 kN test frame, using the standard machine clamping system (see figure 2). Coupons of 10 and 20 mm thickness were tested on 400 kN and 1 MN test frames, respectively. For clamping, these were equipped with an assembly of steel plates and hydraulic clamping (in the case of the 1MN test frame this was hydraulic bolt tightening). Fatigue tests frequencies were limited to 2 Hz for 4 mm thick coupons to 0.5 Hz for 20 mm thick coupons to avoid a reduction of fatigue life due to the self-heating effect [14].

S-N curves were modelled according to ASTM E739 using a least squares regression with an average data sample of 10 coupons per series. A minimum of four different stress levels was tested.

Due to practical reason data was gathered up to 10^7 cycles to failure, and extrapolated to 10^8 and 10^9 cycles to failure using the S-N curves models.

To analyse the scatter, confidence band lines were computed at 95% interval of confidence according to the mentioned standard. As reference stress, design values were obtained using S-N GL curve eq. 5.5.3.3 from GL guidelines [15] with a safety factor $\gamma_{Ma} = 2.21$, $\gamma_{Mb} = 1.63$ and a slope m=10. In addition a compression ultimate strength of 630 MPa and a tension ultimate strength of 849 MPa was used to compute the GL curve. Reserve factors were computed as the ratio between the test stresses values at 95% of confidence (from S-N curves models) and the design stresses from the GL curve at four different cycles to failure (10^6 to 10^9).

Table 2. S-N curve model according to ASTM E739. Slope and offset data, the coefficient of determination and maximum heating rate during the curing per S-N curve. Fatigue compression tests at stress ratio R=10.

Name series	**S-N curve model (ASTM E739)** A Least square offset	B Least square slope	R^2 Least square coef. determination	**Max heating rate during curing**[2]
	[-]	[-]	[-]	[°C/h]
SubA-C1	48.3 ± 12.1	-15.8 ± 4.5	0.787	25.09
SubA-C3	61.5 ± 10.5	-20.8 ± 3.8	0.916	55.87
SubA-C4	35.7 ± 38.7	-11.6 ± 14.2	0.432	64.94
SubB-C1	42.4 ± 11.0	-14.7 ± 4.3	0.753	16.08
SubB-C3	101.2 ± 28.9	-37.6 ± 11.3	0.828	37.43
SubB-C6	60.1 ± 30.3	-21.3 ± 11.9	0.583	48.62
SubC-C6	34.8 ± 116.5	-11.6 ± 46.6	0.048	150.68
Th4mm	65.4 ± 7.4	-23.2 ± 2.8	0.936	9.73
Th10mm	66.4 ± 16.3	-23.6 ± 6.3	0.840	-[1]
Th20mm	91.4 ± 42.0	-34.7 ± 16.8	0.686	18.59

[1] No data available
[2] measured through the thickness during the curing

3. Results and discussion

Two different methodologies were considered to evaluate the influence of the manufacturing in thick laminates fatigue properties: thickness scaled tests and the evaluation of the mechanical properties through the thickness using the sub-laminates technique. To analyse the data, the approach followed was to gather experimental S-N curves with both methodologies for the same material specification. Coupons were manufactured under different conditions and with different thicknesses. The worst scenario in comparison with a design point such as the GL curve was analysed and the reserve factors were computed.

Since the maximum heating rates during the hardening have shown a strong influence in the ultimate strengths [11], the maximum. heating rates were chosen as a manufacturing indicator of the curing cycle. Table 2 shows that the gathered series cover maximum heating rates from 9 to 150 °C/h. Therefore, a large range of possible manufacturing conditions can be found in a thick laminate due to the exothermic reaction and temperature gradients. According to the theory of the weakest link, the laminate component with the lowest fatigue performance will drive the degradation and failure. Therefore from the gathered S-N curves covering the full manufacturing range the worst performance will drive the design if not detailed considerations are made.

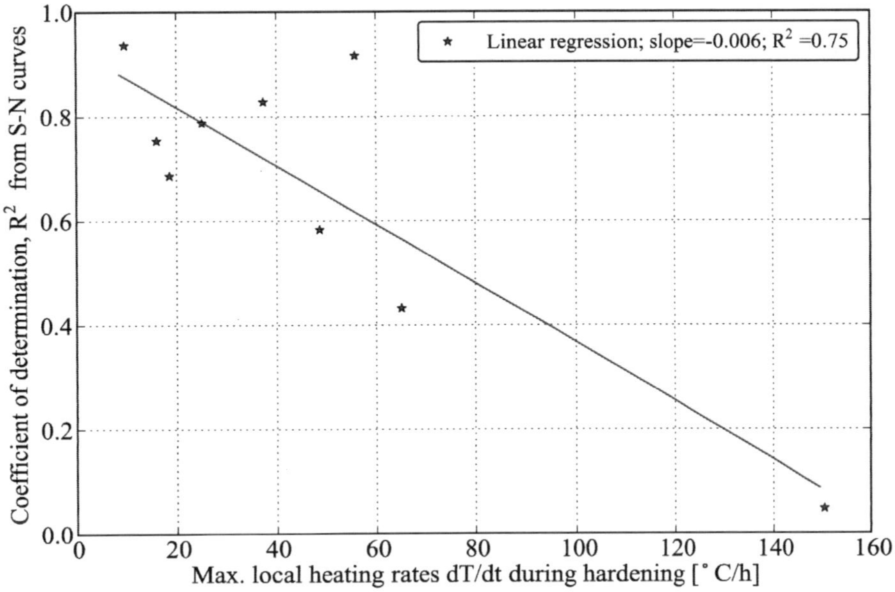

Figure 3. Coefficient of determination R^2 of S-N curves versus maximum local heating rates during the hardening.

Table 3. Stresses and reserve factors computed at different cycles to failure. Stresses computed according S-N curve 95% confidence band line and stresses computed according to GL curve (eq. 5.5.3.3, γ_{Ma} = 2.21, m=10) [15]. Fatigue compression tests at stress ration R=10.

	$\sigma_{95\%}$ Stress at 95% confidence interval				σ_{GL} Stress at GL curve Design stresses				Reserve factor $RF = \frac{\sigma_{95\%}}{\sigma_{GL}}$			
Cycles to failure	10^6	10^7	10^8	10^9	10^6	10^7	10^8	10^9	10^6	10^7	10^8	10^9
Name series	[MPa]				[MPa]				[-]			
SubA-C1	439	379	328	283	138	120	103	89	3.18	3.17	3.17	3.17
SubA-C3	442	396	354	317	138	120	103	89	3.20	3.31	3.43	3.55
SubA-C4	301	247	203	166	138	120	103	89	2.18	2.07	1.96	1.86
SubB-C1	272	232	199	170	138	120	103	89	1.97	1.94	1.92	1.90
SubB-C3	322	303	285	268	138	120	103	89	2.33	2.54	2.76	3.00
SubB-C6	314	281	253	227	138	120	103	89	2.27	2.35	2.44	2.54
SubC-C6	224	184	151	123	138	120	103	89	1.62	1.54	1.46	1.38
Th4mm	345	313	283	256	138	120	103	89	2.50	2.62	2.74	2.87
Th10mm	332	301	273	248	138	120	103	89	2.40	2.52	2.64	2.77
Th20mm	268	250	234	219	138	120	103	89	1.94	2.09	2.27	2.45
								Min values:	*1.62*	*1.54*	*1.46*	*1.38*

Table 2 shows the S-N curves models for the gathered data. For each series a least squares fit was computed as well as the coefficient of determination R^2 of such models. The coefficients of determination are an indicator of the scatter and S-N curves confidence intervals, in other words the uncertainty of the fatigue failure. Table 2 and figure 3 shows that the coefficients of determination correlate with the maximum heating rates. This indicates that large heating rates require of larger confidence intervals and the fatigue failure uncertainty is larger. While significant heating rates during the curing enhance the ultimate strengths, in the case of fatigue, large heating rates increase the scatter and larger intervals of confidence are required.

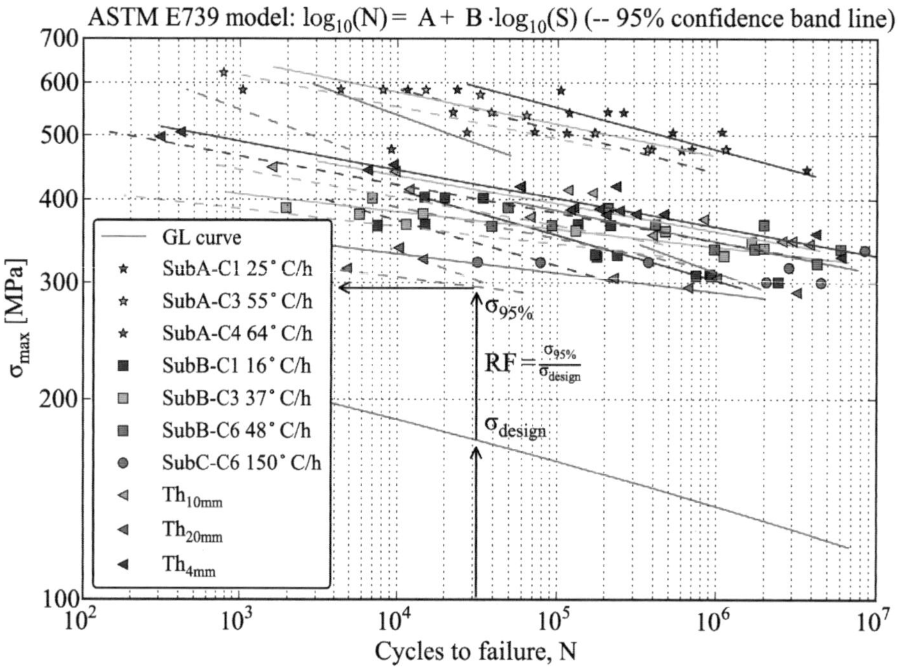

Figure 4. General view of compression S-N curves. Plate A,B and C sub-laminates, curves from scaled compression coupons (4, 10 and 20mm thick) and reference design curve according to GL guidelines (eq. 5.5.3.3, m=10, $\gamma_{Ma} = 2.21$, $\gamma_{Mb} = 1.63$, $\sigma_{compression} = -630MPa$).

Figure 4 shows the compression S-N curves models from the sub-laminates series and the thickness scaled tests. Also, the GL curve is also plotted as a reference of the design level. The S-N curves belonging to plate A show higher average stresses for similar cycles to failure in comparison with plate B due to the influence of the fibre weight ratio FWR. Similar behaviour was observed for the ultimate compression strengths reported in a previous work [11]. Moreover, confidence band lines are also available in figure 4 showing larger intervals of confidence in the case of the series with significant local heating rates (see figure 3).

Moreover, figure 4 contains a graphical example of the reserve factor physical meaning, which is related to the distance between the experimental S-N curve at the 95% confidence band line and the GL curve design level for a given cycles to failure. In this way, the reserves factor were computed for all the series at different cycles to failure, where the lower cases are the most interesting ones since they represent the worst-case scenario. Table 3 shows the stresses at the 95% confidence band line, the GL curve values and the reserves factors. According to the different S-N curves gathered at various manufacturing conditions the lowest reserve factor found at 10^8 cycles is 1.46 and the most optimistic scenario is 3.43. This indicates that the influence of the manufacturing in thick laminates can diminish the reliability of the material fatigue properties more than 200% under certain manufacturing conditions and heating rates. Moreover, taking into account this data when the fatigue performance is

IOP Conf. Series: Materials Science and Engineering **139** (2016) 012028 doi:10.1088/1757-899X/139/1/012028

also reduced by other factors such as ageing or manufacturing defects it might be plausible to reach reserve factors below 1.

The actual experimental data allows evaluating the safety factors to be used in the GL design curve due to the influence of the curing in thick laminates. For such purpose, the experimental data was compared with a GL design curve without any safety factors ($\gamma_{Ma} = 1.$, $\gamma_{Mb} = 1.$) and the safety factor required to reduce the GL design curve below the gathered S-N curves (at 95% band lines) were computed (see figure 5). The computed safety factors oscillate between values of 1.23 to 1.60 at 10^8 cycles to failure for ultimate strengths in compression of 630 and 837 MPa respectively.

It needs to be remarked that GL curve was computed with an ultimate strength of 630 MPa provided by the thickness scaled tests [10] for a 4mm thin laminate. However, according to the sub-laminates quasi-static test results presented in [11], it is possible to achieve ultimate strengths as high as 837 MPa. If the GL design curve is computed with ultimate strengths of 837 MPa different design levels might be obtained levering the expected material performance. Therefore, since the curing cycle influence the thin laminate ultimate strengths, and no clear indications on this respect could be found. Therefore, is unclear which curing cycle should be used to evaluate the ultimate strengths to be used in the GL design curve.

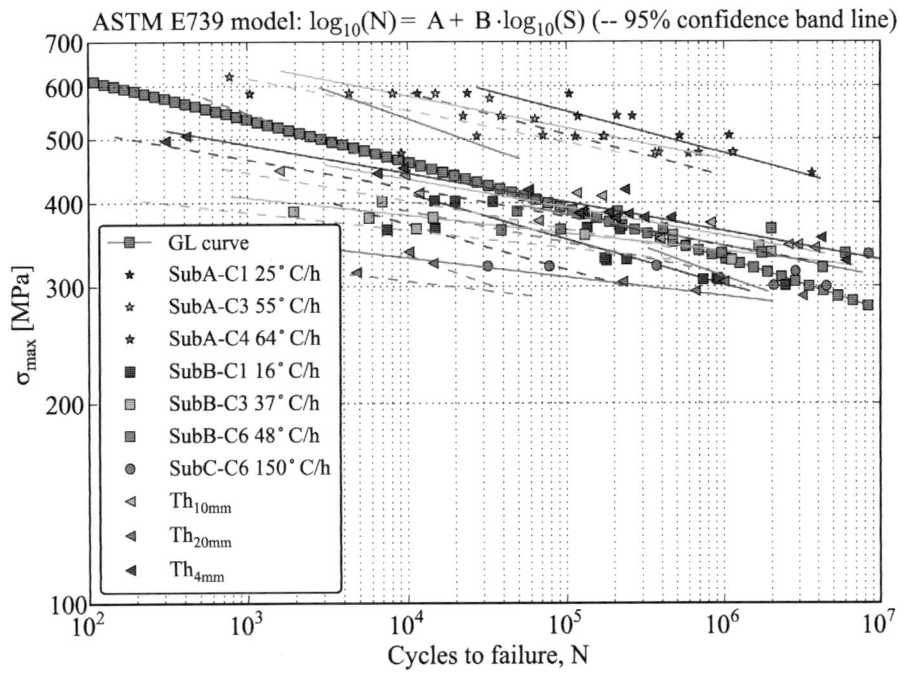

Figure 5. General view of compression S-N curves. Plate A,B and C sub-laminates, curves from scaled compression coupons (4, 10 and 20mm thick) and reference design curve according to GL guidelines (eq. 5.5.3.3, m=10, $\gamma_{Ma} = 1.$, $\gamma_{Mb} = 1.$, $\sigma_{compression} = -837MPa$).

Moreover, similar behaviour might be expected for resin related properties such as the shear, biaxial properties or properties in the transverse direction. Further research is on-going in studying other properties and in understanding the mechanisms that drive the influence of the manufacturing process in thick laminates fatigue properties, where it is intended to investigate the possible differences that might appear in the microstructure with different curing cycles.

4. Conclusions

Thickness scaled tests and sub-laminates tests reveal to be a useful approach to study the influence of internal heat generation during the manufacturing process on fatigue properties. Based on the data provided by thickness scaled tests and sub-laminates tests it was possible to analyse the full operational region in fatigue and determine the worst-case scenarios. The data revealed that reserve factors close to the design curves can be found for certain curing cycles and manufacturing conditions in compression loading. For the tested material, the gathered S-N curves allowed computing the safety factors between 1.23 and 1.60 regarding the influence of the curing cycles in thick laminates.

Acknowledgements

The authors acknowledge the contribution of the IRPWIND and SLOWind projects to motivate and partly fund this research.

References

[1] Lee S and Springer G S 1988 Effects of cure on the mechanical properties of composites *J. Compos. Mater.* **22** 15–29

[2] Davies L, Day R, Bond D, Nesbitt A, Ellis J and Gardon E 2007 Effect of cure cycle heat transfer rates on the physical and mechanical properties of an epoxy matrix composite *Compos. Sci. Technol.* **67** 1892–9

[3] Liu L, Zhang B-M, Wang D-F and Wu Z-J 2006 Effects of cure cycles on void content and mechanical properties of composite laminates *Compos. Struct.* **73** 303–9

[4] Esposito L, Sorrentino L, Penta F and Bellini C 2016 Effect of curing overheating on interlaminar shear strength and its modelling in thick FRP laminates *Int. J. Adv. Manuf. Technol.*

[5] Kumar K V, Safiulla M and Ahmed A An approach to cure thick fiber reinforced composite laminates *Int. J. Adv. Innov. Res.* **1** 158–62

[6] Oh J H and Lee D G 2002 Cure Cycle for Thick Glass / Epoxy Composite Laminates *J. Compos. Mater.* **36** 19–45

[7] White S R and Kim Y K 1996 Staged curing of composite materials *Compos. Part A Appl. Sci. Manuf.* **27** 219–27

[8] Ciriscioli P R, Springer G S and Wang Q 1991 A technique for determining mechanical properties of thick composite laminates *J. Compos. Mater.* **25** 1330

[9] Lahuerta F, Nijssen R P L, van der Meer F P and Sluys L J 2015 Static and dynamic through thickness lamina properties of thick laminates *ICCM20 - 20th International Conference on Composite Materials* (Copenhage: ICCM20 Secretariat) pp 19–24

[10] Lahuerta F, Nijssen R P L, van der Meer F P and Sluys L J 2016 Thickness scaled compression tests in unidirectional glass fibre reinforced composites in static and fatigue loading *Compos. Sci. Technol.* **123** 115–24

[11] Lahuerta F, Nijssen R P L, van der Meer F P and Sluys L J 2016 The influence of curing cycle and through thickness variability of properties in thick laminates *J. Compos. Mater.* Prepublish

[12] Adams D F and Finley G a. 1997 Analysis of Thickness-Tapered Unidirectional Composite Compression Specimens *J. Compos. Mater.* **31** 2283–308

[13] Camponeschi G 1994 The effects of specimen scale on the compression strength of composite materials *Workshop on scaling effects in composite materials and structures NASA-CP-3271* (Langley research center, Hampton (Virginia))

[14] Lahuerta F, Nijssen R P L, van der Meer F P and Sluys L J 2015 Experimental–computational study towards heat generation in thick laminates under fatigue loading *Int. J. Fatigue* **80** 121–7

[15] Anon 2010 *GL guideline for the certification of wind turbines* (Germany)

The sensitivity of the burst performance of impact damaged pressure vessels to material strength properties

K Lasn, N P Vedvik and A T Echtermeyer

Norwegian University of Science and Technology (NTNU),
Composites and Polymers, Department of Engineering Design and Materials,
Richard Birkelandsvei 2B, N-7491 Trondheim, Norway

E-mail: kaspar.lasn@ntnu.no

Abstract. This numerical study is carried out to improve the understanding of short-term residual strength of impacted composite pressure vessels. The relationship between the impact, created damage and residual strength is predicted by finite element (FE) analysis. The burst predictions depend largely on the strength properties used in the material models. However, it is typically not possible to measure all laminate properties on filament wound structures. Reasonable testing efforts are concentrated on critical properties, while obtaining other less sensitive parameters from e.g. literature. A parametric FE model is hereby employed to identify the critical strength properties, focusing on the cylindrical section of the pressure vessel. The model simulates an impactor strike on an empty vessel, which is subsequently pressurized until burst. Monte Carlo Simulations (MCS) are employed to investigate the correlations between strength related material parameters and the burst pressure. The simulations indicate the fracture toughness of the composite, hoop layer tensile strength and the yield stress of the PE liner as the most influential parameters for current vessel and impact configurations. In addition, the conservative variation in strength parameters is shown to have a rather moderate effect (COV ca. 7 %) on residual burst pressures.

1. Introduction

Composite overwrap pressure vessels (COPV) are often used in cars, buses and bulk transport vehicles. They can hold higher pressures and weigh less than steel pressure vessels. COPVs are today mainly used for natural gas storage and transportation. Additional prospective applications include the upcoming hydrogen economy where gaseous hydrogen needs to be stored [1] at internal pressures greater than 700 bars for economic feasibility.

The cylinders used for high pressure applications, and requiring long fatigue life, typically utilize carbon fibre composite overwrap. Liners of *Type IV* cylinders are made from polymers such as HDPE. They are resistant to hydrogen exposure and show good fatigue performance. The pressure vessels are often designed to fail in the cylindrical section. In the following, only this cylindrical section of the pressure vessel is investigated in detail.

The cylinders are often protected from mechanical interactions with the environment. However, accidental impacts are still of concern. Large impact damage, e.g. from a serious car accident or a rollover, is visually obvious. Vessels should thereafter be thoroughly inspected or taken out of service. Low-energy impacts may occur during handling and transport. Such events are: falling of the vessel, dropping of tools and impacts from forklifts. The consequences of low-energy and low-velocity events to residual performance are often unclear. To study the effect of flaws or damage, composite pressure vessels can be experimentally burst tested [2]. Experimental studies however have limitations on the number of vessel and flaw configurations which can be studied. A numerical model can be applied to many different scenarios, but the models are complicated and confidence in the results needs to be established.

Numerical modelling by Finite Element Analysis is used here to study the effect from a given impact event on short-term burst strength. The model calculates first the damage created by the impact and subsequently its influence on the burst behavior. The FE model requires many material properties as input parameters which are not easy to measure. Material properties from flat laminates are not necessarily transferable to filament wound laminates, since fibre architecture and compaction is different. Obtaining a full set of material properties is more demanding for cylindrical specimens than for flat composites due to the lack of standardized methods and the geometry of the curved shapes incompatible with traditional tensile-testing/uniaxial test apparatuses. Consequently, it is desirable to focus testing efforts on a few critical material properties, obtaining other, less relevant parameters from flat laminates and literature.

This work studies the effect from 30 strength related parameters on the predicted burst pressure of the pressure vessel. This helps to determine strength properties which are most influential/critical to the short-term burst behavior.

2. Model description

The 3D FE model is developed in Abaqus 6.14-1, employing the explicit solver. Explicit is typically used for high speed dynamic problems such as impact, material degradation and failure. The description of the numerical model is divided into two sections: general description of the FE model, and discussion of material properties and other input parameters.

2.1. The FE model

The pressure vessel is modelled as an axisymmetric cylinder, attached to end domes, as illustrated in Figure 1. The cylinder consists of a relatively thick PE liner and a thin carbon fibre composite overwrap (Figure 2). The combination of thick PE liner and thin laminate is untypical for commercial pressure vessels. It was motivated by a specific design subjected to experimental studies in our laboratory. The impact and the failure initiation occurs in the middle of the cylindrical part.

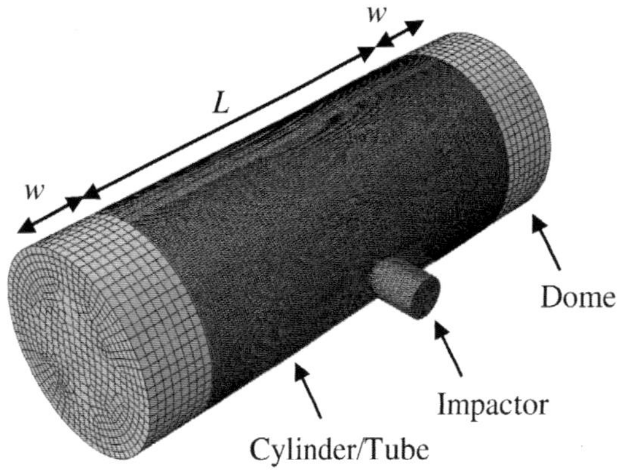

Figure 1. The geometry of the pressure vessel.

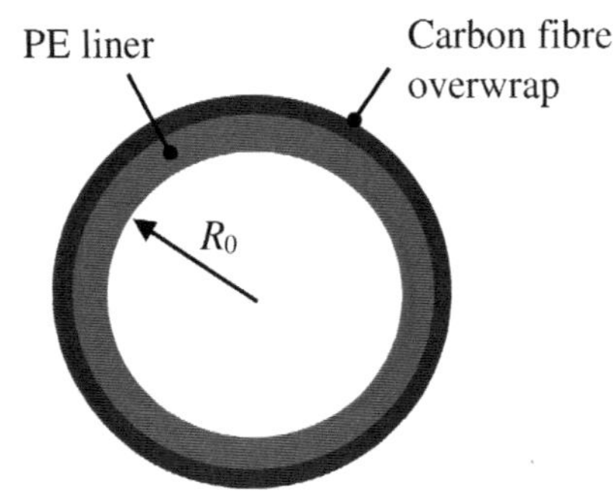

Figure 2. Cross section of the cylinder.

The domes are simplified as two cylinders (i.e. disks) made of linearly elastic isotropic material. The pressure vessel is fixed at the external flat surfaces of the domes. During impact, all displacements of both surfaces are fixed. During pressurizing, axial displacements remain restrained for one end surface, while unconstrained for the opposite surface. The impactor tip is hemispherical and modeled as rigid, positioned to impact the midsection of the tube horizontally. It is given an initial velocity and inertia properties. The impactor is affected by artificial gravity and its vertical position is therefore slightly offset, so the first contact occurs normal to the cylinder surface. Residual vibrations in the vessel are damped after the impact event. During subsequent pressurization, the pressure amplitude is increased according to Abaqus' in-built smooth step definition until P_{max} = 300 bar in course of 0.1 s.

The cylinder of the pressure vessel consists of concentric layers of liner, composite, and cohesive interfaces, modelled with solid and continuum shell elements. The PE liner is isotropic, with elastic-plastic material behavior. The composite layers have orthotropic material symmetry. The stress based Hashin's failure criterion is used for the initiation of intralaminar fibre and matrix damage. Damage evolution is controlled by fracture energies and stabilized by viscosity coefficients. Cohesive layers are implemented between individual layers of composite and the liner. The damage in the cohesive elements is initiated by a quadratic stress criterion and the energy based Benzeggagh-Kenane (BK) formulation is used as the fracture criterion. The cylinder layup [PE / 90° / +15°$_2$ / -15°$_2$ / 90°] is generic, chosen to enforce the failure in hoop layers. The helical layers would be wound as two ±15° pairs in practice — the current layup is a simplified model version.

The connection between individual layers of the cylinder and from the cylinder to the domes is implemented by tie constraints. Contact interactions are prescribed for surfaces in possible contact during the analysis; between the tube elements and the impactor and neighboring composite plies. A hard contact is assumed in the normal direction, and a frictional penalty in the tangential direction.

The base element length for the FE model (see Figure 3) was chosen as δ= 1.7 mm for the composite layers and the impactor, following a mesh convergence study. The tube elements not facing the impactor (backside of the tube) have length 5δ in the circumferential direction. The liner is fairly thick and has three elements in the thickness direction, while the other cylindrical layers have a single element. The solid elements C3D8R in the domes have their linear dimensions set to approx. 4δ.

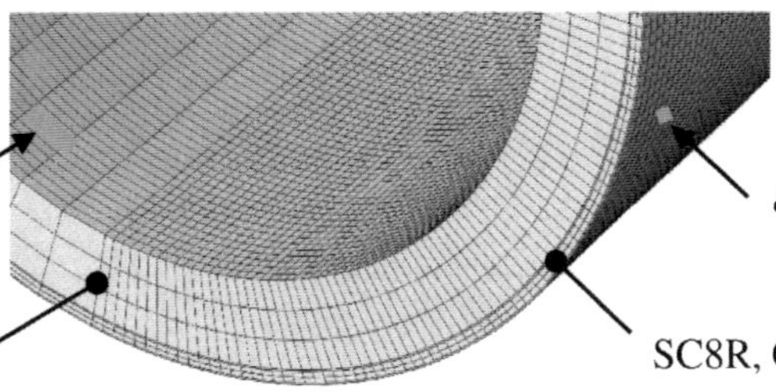

Figure 3. The base mesh length δ, element layout, and element types in the cylinder.

2.2. Material properties

The material properties describe a typical filament wound carbon fibre composite. Specific values are based on the previous work in the Composite Group in NTNU [3-5], as well as information from manufacturers and literature [6-8]. Since this is a numerical study, the chosen material properties are generic. The properties of hoop and helical composite layers are assumed to differ due to different fibre volume fraction V_f, originating from the filament winding process. Helical layers have a lower V_f and therefore larger thickness and lower fibre dominated strengths. Matrix dominated strengths are assumed to be the same for helical and hoop layers.

The material parameters are divided into constant properties and randomly sampled strength related properties. All elastic properties were taken as constant in this study to focus on strength. The sensitivity study for the strength properties is conducted by Monte Carlo Simulations (MCS). Random sampling creates a number of material datasets, with varying strength properties, for FE input. This enables to study the sensitivity of the burst pressure to changes in each individual strength parameter value. A large number of virtual experiments may be simulated to study the behavior of probabilistically sampled material parameters, e.g. as in [9]. Recently, Cai *et al.* [10] used MCS to study the critical buckling pressures for pressure vessels under external hydrostatic pressure. A similar approach is adopted here, even though the number of simulations has been relatively small so far, due to the high computing effort for each simulation. More simulations are in progress.

Table 1 describes the parameters of the model and materials, which are assumed constant in the sampling and the following calculations. These are the geometry, FE mesh size, pressure loading, mass densities, elastic constants and the friction properties in the model. The mass, velocity and shape of the impactor also remains constant in the calculations. It is well known that stiffness and strength of the composite become interconnected during the progressive failure of the material. Hereby, the initial stiffness of the undamaged material is kept constant. Progressive degradation of stiffness can still vary in separate simulations, depending on the sampled strength parameters.

Table 1. The model and material properties with constant values in Monte Carlo Simulations (MCS).

Parameter group	Specifics	Notation	Generic values
General	Layup		[PE / 90° / +15°$_2$ / -15°$_2$ / 90°]
	Maximum pressure and pressurization time with *smooth step*	P_{max}, t_{Pmax}	300 [bar], 10^{-1} [s]
	Base mesh size (composite layers)	δ	1.7 [mm]
	Friction coeff.	μ	0.3 [-]
	Radius of the inner surface of the PE liner	R_0	57.3 [mm]
	Length of the tube	L	0.3 [m]
Dome	Width	w	50 [mm]
	Isotropic properties	ρ, E, ν	1200 [kg/m^3], 100 [GPa], 0.3 [-]
PE liner	Thickness	t_{PE}	12.7 [mm]
	Isotropic properties	ρ, E, ν	960 [kg/m^3], 1.1 [GPa], 0.45 [-]
Composite hoop layers	Thickness	t_{hoop}	0.25 [mm]
	Density	ρ_{hoop}	1550 [kg/m^3]
	Elastic constants	E_1, $E_2 = E_3$	130 [GPa], 11.0 [GPa]
		$G_{12} = G_{13}$, G_{23}	4.7 [GPa], 4.2 [GPa]
		$\nu_{12} = \nu_{13}$, ν_{23}	0.30 [-], 0.50 [-]
Composite helical layers	Thickness (half of the helical pair)	$t_{helical}$	0.40 [mm]
	Density	ρ_{hel}	1400 [kg/m^3]
	Elastic constants	E_1, $E_2 = E_3$	80 [GPa], 7.0 [GPa]
		$G_{12} = G_{13}$, G_{23}	3.0 [GPa], 2.7 [GPa]
		$\nu_{12} = \nu_{13}$, ν_{23}	0.30 [-], 0.50 [-]
Composite (hoop and helical)	Shear contribution coeff. in the Hashin criterion	α	1 [-]
Cohesive layers	Density	ρ_{COH}	1200 [kg/m^3]
	Stiffness	E, $G_1 = G_2$	3.0 [GPa], 1.1 [GPa]
	BK power parameter	η	1.4 [-]
Impactor	Mass	m	1.5 [kg]
	Velocity	v	10 [m/s]
	Radius of the spherical nose	R_{imp}	15 [mm]

Table 2 describes the strength related properties whose sensitivity is investigated by MCS. Each of the parameters in Table 2 is sampled by a normal distribution, defined by the mean and the coefficient of variation (COV). Although the COVs define sufficiently narrow parameter ranges, a small but necessary sampling bias was introduced to avoid the creation of non-physical negative parameter values. Whenever a negative value is sampled, the specific parameter is re-sampled to arrive to a set of all-positive values. The variation of properties in Table 2 is considered as the combined effect of both the lack of experimental data and the scatter in frequently encountered experimental values. Thus it is not aimed to distinguish between various sources and mechanisms of variability.

Table 2. Material properties with values sampled according to normal distributions in Monte Carlo Simulations (MCS).

Parameter group	Specifics	Notation	Mean value in MCS	COV in MCS
PE liner	Elastic-plastic behaviour	True yield stress	23.5 ; 161 [MPa]	10 %
		Plastic strain	0 ; 1.93 [-]	10 %
Composite hoop	Strengths	X_1^T, X_1^C	2700 [MPa], 1400 [MPa]	5 %, 10 %
		X_2^T, X_2^C	80 [MPa], 180 [MPa]	20 %
		$X_{12} = X_{13}$, X_{23}	100 [MPa], 80 [MPa]	20 %
Composite helical	Strengths	X_1^T, X_1^C	1700 [MPa], 800 [MPa]	5 %, 10 %
		X_2^T, X_2^C	80 [MPa], 180 [MPa]	20 %
		$X_{12} = X_{13}$, X_{23}	100 [MPa], 80 [MPa]	20 %
Composite all (hoop and helical)	Fracture toughness	G_{tl}, G_{cl}	150 [N/mm], 75 [N/mm]	30 %
		G_{tt}, G_{ct}	0.83 [N/mm], 0.83 [N/mm]	30 %
	Viscosity, relax. time	η_{ft}, η_{fc}	10^{-8} [s], 10^{-8} [s]	20 % [a]
		η_{mt}, η_{mc}	0.1 [s], 0.1 [s]	20 % [a]
Cohesive layers	Constitutive thickness	t_c	10^{-4} [mm]	15 % [a]
	Strengths	τ_n, τ_s, τ_t	50 [MPa], 50 [MPa], 50 [MPa]	20 %
	Fracture toughness	G_{Ic}, $G_{IIc} = G_{IIIc}$	0.83 [N/mm], 1.8 [N/mm]	30 %

[a] The exponent x in 10^{-x} is sampled.

3. Results and discussion

3.1. General model behaviour

The behavior of 50 virtual specimens was simulated during the Monte Carlo study. These are specimens with constant parameters from Table 1 and randomly sampled parameters as described in Table 2. There are small differences in damage created by the impact and the sequence of events leading to failure in the individual simulations. However, the following characteristic behavior was confirmed from observing individual results from ca. half of the specimens.

At impact (Figure 4), three types of damage are created in the composite: matrix cracking (and fibre/matrix debonding), delamination and fibre failure. In addition, the thick PE liner yields and obtains a dent from the impact. Delaminations occur in three interfaces except between the liner and the composite, see Figure 5. The sizes and shapes of delaminations vary somewhat. The fibre failures occur in the external hoop layer, typically running as a one-element wide crack in the horizontal (axial) direction, as shown in Figure 6. There is at least one case where the fibre failure from the impact is asymmetrical, only on one side. Also, in multiple cases (not always), was a vertical crack of fibre failures created in both helical 15° layers. The internal hoop layer did not acquire fibre failures from the impact. Figures 4-6 describe the impact event for the reference case, with mean values as input parameters.

As the vessel is pressurized, there is very little to no material damage up to a certain point (hereby denoted by 'DI' as damage initiation). After this stage, damage accumulation increases — localized small cohesive failures, isolated fibre failures and some yielding of PE. As the pressure becomes critical, the external and internal hoop layers fail simultaneously by forming a long axial crack (see Figure 7). This event is distinct, and both, the kinetic energy and energy dissipated by material damage increase in approximately a step-like manner. This final failure is clearly recognizable by large scale fibre rupture in hoop layers and it is hereby denoted by 'B' for burst.

The behavior of the model is studied in terms of damage dissipation energy for the whole model (ALLDMD variable in Abaqus) which is accumulated during the pressurizing step. Threshold values of 0.5 J were chosen for damage initiation ('DI') and 100 J for burst ('B'). Figure 8 shows how the internal pressure and the radial displacement of a single point (on the external back-surface of the tube, exactly

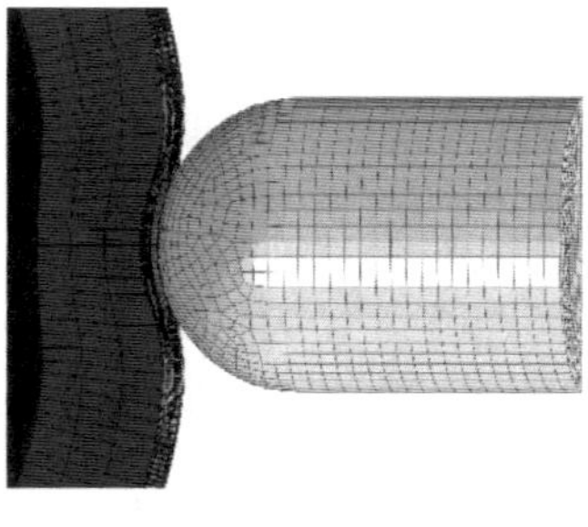

Figure 4. The impactor at maximum depth.

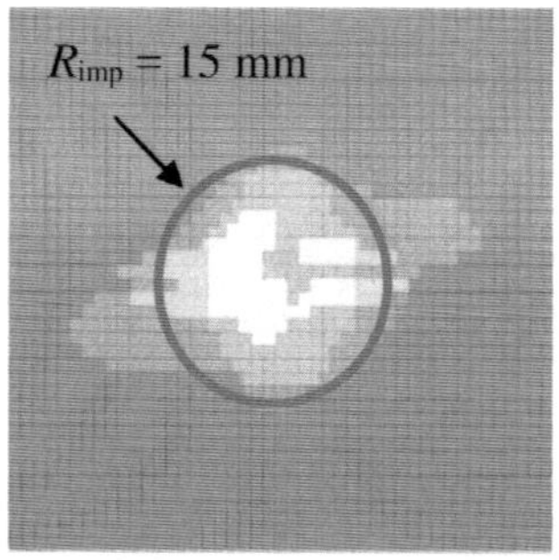

Figure 5. Overlay plot of delaminations from three interfaces.

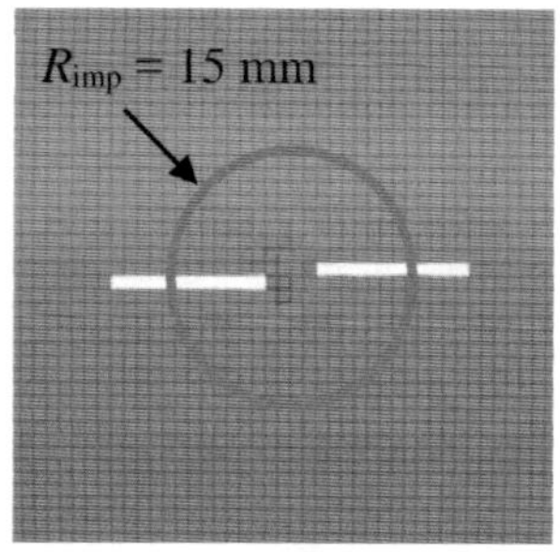

Figure 6. Fibre failures in the external hoop layer.

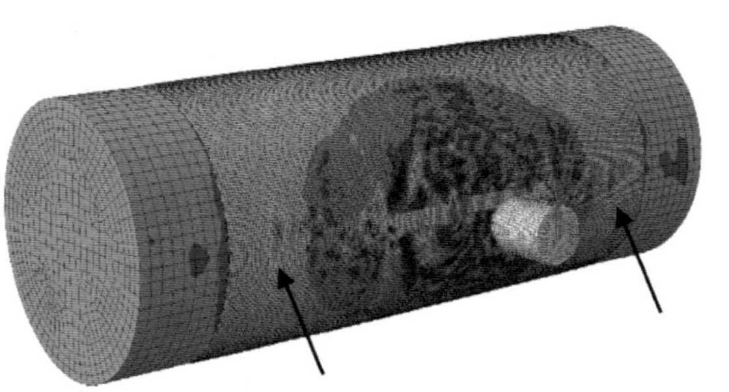

Figure 7. Fibre stresses during the burst event ('B'). The failure of hoop composite layers is in progress. The blue/dark region in the center is unloaded, arrows point to stress concentrations immediately preceding fibre failure.

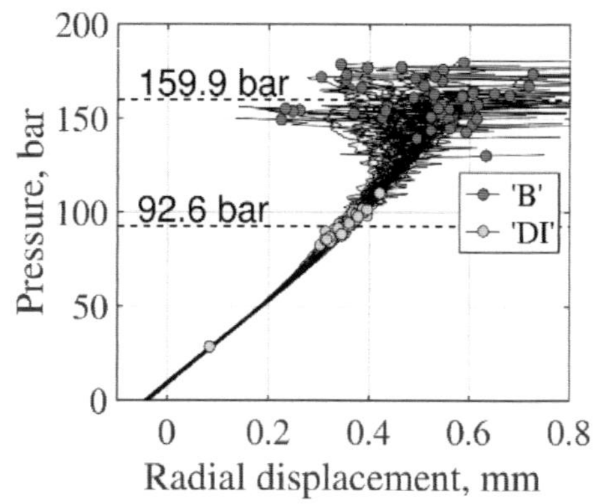

Figure 8. Pressure vs. radial displacement with energy thresholds for damage initiation ('DI') and burst ('B'), mean values are given explicitly.

opposite to the impactor) are related for all 50 calculated samples. The scatter of burst pressures 'B' in Figure 8 has the COV of 7 %. This shows that even as the scatter of strength properties was chosen to be large/conservative (Table 2), the scatter in the calculated burst pressure remains moderate. This is important from practical engineering point of view, indicating that the accuracy of each individual material parameter is not so important.

The pressurization begins with a small negative radial displacement because the damping of residual vibrations was conducted by external pressure in the timestep preceding the internal pressure. The influence on the final burst behaviour is considered negligible. The burst pressure can be estimated for the undamaged vessel as $P \approx X_1^T \cdot 2t_{hoop} / (R_0 + t_{PE}) = 2700 \cdot 0.5 / 70 = 19.3$ MPa = 193 bar (hoop fibre mean $X_1^T = 2700$ MPa). This agrees well with the calculated pressure of 190 bar from non-impacted modification of the FE-analysis. The burst pressures in Figure 8 are 160 ± 22 bar (2 standard deviations), thereby ca. 16 % below the initial value, which is expected due to the damage.
In addition, Figure 9 shows how the damage dissipation energy develops during the final pressurizing step. In principle, the burst event could be defined by a different numerical value or by e.g. kinetic energy altogether. The choice of burst event being equivalent to ALLDMD = 100 J was made after investigating a range of possible kinetic and damage dissipation energies. The details of this substudy are hereby omitted for brevity.

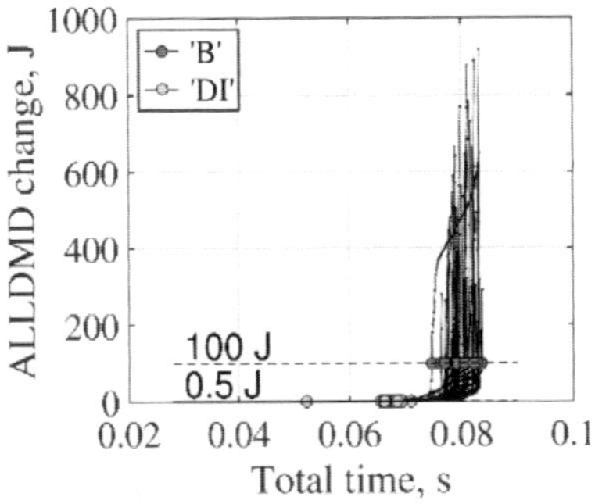

Figure 9. The definition of damage initiation ('DI') and burst ('B') by damage dissipation energy (ALLDMD) during the pressurizing step.

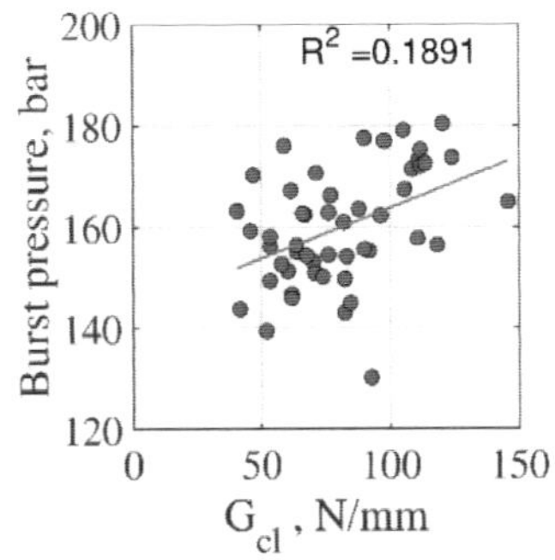

Figure 10. Correlation for the compressive fracture toughness of composite.

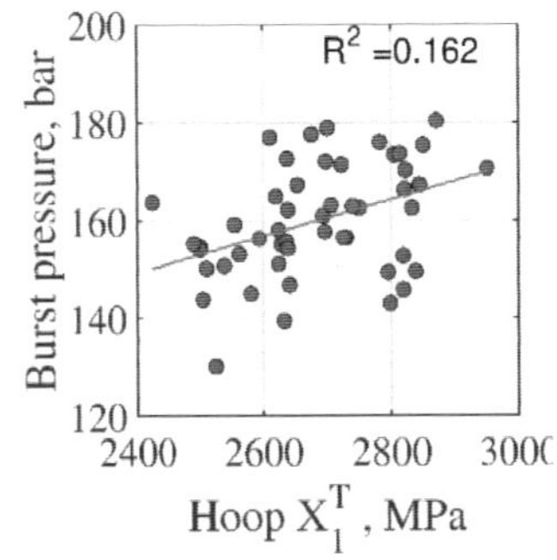

Figure 11. Correlation for hoop layer tensile strength.

3.2. *Correlations*

The correlation between sampled strength variables and burst strengths is evaluated by the coefficient of determination R^2 for linear regression. Values for R^2 can range from 0 to 1 and a higher value implies better correlation between the input parameter and the obtained burst pressure. This correlation does not provide hard evidence for causation, rather an indication which variables can be suspected of affecting the burst pressure more than others.

In general, the obtained correlation coefficients R^2 were fairly weak, as seen from Table 3 where four highest values are given. Figures 10 and 11 show the individual datapoints behind the coefficients which gave the highest R^2 values. This data shows considerable scatter. The scatter in data and the consequently low R^2 indicate that several material parameters affect the impact (therefore the damage) and the burst pressure simultaneously. None of the material parameters however seem to have a clearly superior effect on the burst pressure in current case.

Table 3. The highest correlation coefficients (sample of 50 specimens).

Parameter	Notation	R^2
Composite (hoop and helical) fracture toughness (compressive, longitudinal)	G_{cl}	0.189
Hoop layer tensile strength	X_1^T	0.162
Composite (hoop and helical) fracture toughness (tensile, longitudinal)	G_{tl}	0.141
Yield stress of the PE liner	$\sigma_{y,0}$ PE	0.100

Overall, the results in Table 3 are expected. The greatest influence comes from the compressive fracture toughness of the composite. This is due to impact which creates fibre failures on the external surface of the vessel (Figure 6). The effect from the hoop ply tensile strength and composite tensile fracture toughness are caused by the nature of burst: a hoop fibre tensile failure. The yield stress of the PE liner is an influential material parameter due to the configuration of this specific vessel – a thick liner overwrapped with a thin layer of composite.

It is important to note that since a large number of material parameters are investigated, several of which are influential simultaneously, analyses with small sample sizes can give misleading and strange correlations. This was experienced with this study for small sample sizes of 10-20. To verify that current outcome remains unchanged, more simulations (up to a sample of 100) are still in progress.

4. Conclusions

The following conclusions were drawn based on this numerical study.

- The most influential material parameters for the current vessel and impact configurations were determined (Table 3) as: longitudinal fracture toughness of the composite (both compressive and tensile), hoop layer tensile strength and yield stress of the PE liner.
- The FE prediction of burst pressure for a non-impacted vessel is consistent with hand calculation. The predicted drop in burst pressure from the impact damage was 16 ± 12 %, based on the variations of all parameters modelled. This variation may be acceptable for most engineering calculations.
- The modelling shows that the COV for calculated burst pressures remains fairly low, ca. 7 %, even for the conservative estimates of scatter in the strength properties. The variation of strength parameters has a rather limited effect on the predicted burst pressure, making the exact knowledge of any individual material parameter not so critical. Rather, a group of influential parameters needs to be known fairly well.
- Monte Carlo studies with small sample sizes can lead to erroneous conclusions. The current study and conclusions are based on a sample of 50 simulations. Still, 50 more configurations will be simulated to verify and confirm this outcome.

Acknowledgements

The research leading to these results has received funding from the European Union's Seventh Framework Programme (FP7/2007-2013) for the Fuel Cells and Hydrogen Joint Technology Initiative under grant agreement n° 621194, http://www.hypactor.eu/.

The research was also supported in part with computational resources at NTNU provided by NOTUR, http://www.sigma2.no.

References

[1] Echtermeyer A T and Lasn K 2014 Safety Approach for Composite Pressure Vessels for Road Transport of Hydrogen. Part 2: Safety Factors and Test Requirements *Int. J. Hydrogen Energ.* **39** 14142–52

[2] Makinson J and Newhouse N L 2014 Flaw Testing of Fiber Reinforced Composite Pressure Vessels *J. Press. Vess-T. ASME* **136** 041409-1-5

[3] Shchebetov S, Vedvik N P and Echtermeyer A T 2013 *Typical Static Ply Properties for Composites (IPM-CP-2012-2001 Rev. 01)* NTNU, Trondheim.

[4] Perillo G, Vedvik N P and Echtermeyer A T 2014 Damage Development in Stitch Bonded GFRP Composite Plates under Low Velocity Impact: Experimental and Numerical Results *J. Compos. Mater.* 1-15

[5] Skaar M W 2015 *Modeling and Testing of Impact Damage in Composite Pressure Vessels* MSc Thesis at NTNU https://brage.bibsys.no/xmlui/handle/11250/2350300.

[6] Grafil 34-700 carbon fiber. https://www.rockwestcomposites.com/downloads/34-700.pdf.

[7] Hexion - EPIKOTE Resin MGS RIMR 135 and EPIKURE Curing Agent MGS RIMH 134–RIMH 137. https://www.hexion.com/Products/TechnicalDataSheet.aspx?id=8246.

[8] Daniel I M and Ishai O 2006 *Engineering Mechanics of Composite Materials (Second Edition).* New York: Oxford University Press

[9] Lasn K, Klauson A and Echtermeyer A T 2015 Back-Calculation of Elastic Moduli of a Ply from the Moduli of Cross-Ply Laminates *Mech. Compos. Mater.* **51** 55–68

[10] Cai B, Liu Y, Liu Z, Tian X, Ji R and Zhang Y 2012 Probabilistic Analysis of Composite Pressure Vessel for Subsea Blowout Preventers *Eng. Fail. Anal.* **19** 97–108

IOP Conf. Series: Materials Science and Engineering **139** (2016) 012030 doi:10.1088/1757-899X/139/1/012030

Raman imaging as a tool for assessing the degree of mixing and the interface between polyethylene and cellulose nanocrystals

A E Lewandowska and S J Eichhorn

University of Exeter, College of Engineering, Mathematics and Physical Sciences, Stocker Road, Exeter, EX4 4QL, United Kingdom

E-mail: A.E.Lewandowska@exeter.ac.uk, S.J.Eichhorn@exeter.ac.uk

Abstract. The incorporation of spray-dried cellulose nanocrystals (CNCs) into thermoplastics raises the question about the degree of the mixing, given that typically the former are hydrophilic and the latter hydrophobic. Raman imaging has been used as a non-destructive method for the spatial quantification of the mixing process of CNCs in HDPE. A series of polyethylene based composites was prepared by melt compounding using cellulose nanocrystals (CNCs) as a nano-filler. The cross sectional areas of nanocomposite filaments were investigated using confocal Raman microscopy. The degree of mixing between CNCs and HDPE is quantified using this method. It is shown that CNCs have a strong tendency to agglomerate. However, a certain degree of mixing between CNCs and HDPE is observed from Raman images and chemical images obtained by spectroscopy.

1. Introduction

Recent years have brought an increasing interest in nanofillers based on cellulosic forms such as α-cellulose [1,2], cellulose nanofibres (CNF) [3], microcrystalline cellulose (MCC) [4], and cellulose nanocrystals (CNCs) [5] to reinforce thermoplastic polymers including polyethylene (PE). However, the enhancement of mechanical properties requires good mixing of fillers in the matrix and strong interfacial adhesion between both phases. The hydrophilic nature of cellulose induces the formation of hydrogen bonding between the nanoparticles increasing their tendency to agglomerate in polyethylene matrix (hydrophobic). Composites prepared by a melt compounding process typically use maleic anhydride grafted polyethylene (MAPE) as a compatibilizer in order to improve dispersion within and adhesion between the cellulosic fillers and the matrix. The improvement in cellulose dispersion and the increase in the tensile strength of polyethylene composites were reported after the addition of 2-3 wt% of MAPE. The role of MAPE is to interact between cellulose and polyethylene via an esterification reaction and/or hydrogen bonding between the maleic anhydride groups (-COOH and -C=O) and the hydroxyl groups (-OH) of cellulose [6]. The ratio of ester linkages to hydrogen bonds depends on the ratio of cyclic anhydride to dicarboxylic acid in MAPE [6]. Additionally, a higher dispersion of cellulose was reported after use of polyoxyethylene (PEO) as a dispersion agent [3,5]. However, the improvement in dispersion did not reflect in an increase of the bending strength and bending modulus of the nanocomposites [3].

The physicochemical and mechanical properties of cellulose-polyethylene nanocomposites have been typically assessed using scanning electron microscopy (SEM) [4,5], differential scanning calorimetry (DSC) [4,5], thermogravimetric analysis (TGA) [4], dynamic mechanical analysis (DMA) [3], and mechanical tests [3,4]. These evaluations however are not able to quantify the degree of mixing and the interface between thermoplastics and cellulose nanocrystals. Spectroscopic methods such as Raman and IR provide detailed structural and molecular information on composite structure and their morphologies. Development of spectroscopic imaging has increased its potential to describe the composite morphology. Raman imaging has been used to study different types of materials, including cellulose nanocrystals-polypropylene composites [7], biomaterial degradation *in vivo* [8], the characterization of multilayer films [9] and compatibility of polymer blends [10].

In this paper Raman imaging and chemical imaging are used to characterize cellulose nanocrystals-high density polyethylene (CNCs-HDPE) composites containing spray-dried CNCs as fillers. The quantitative evaluation of the mixed and pure composite components gives an answer on the degree of mixing between polyethylene and cellulose nanocrystals. Furthermore, the role and effectiveness of MAPE as a compatibilizer in the melt compounding process is assessed.

2. Experimental

Spray-dried cellulose nanocrystals were purchased from the University of Maine, Process Development Centre; USA. High density polyethylene (Arboblend HDPE) was supplied by Tecnaro GmbH, while maleic anhydride grafted polyethylene (A-C 575A, MAPE copolymer) was provided by Honeywell. HDPE composites were prepared by melt-compounding in a counter rotating twin-screw extruder (HAAKE Rheomex CTW5, Thermo Fisher Scientific) which formed filaments with a diameter ~2 mm. Prior to extrusion, the fillers and compatibilizer both in powder form were blended in a mortar for 2 minutes. Subsequently, the powder/HDPE matrix was added and the components were mixed in a mortar for a further 8 minutes. The mixture was dried in a vacuum oven at a temperature of 60 °C for 24h to remove humidity. Composites containing CNCs as fillers were extruded at a temperature of 160 °C. The mixing speed was 70 rpm for 7 min. The CNCs/MAPE/HDPE ratio was 0.625/2.50/96.875, 1.25/2.50/96.25, 2.50/2.50/95.00 and 5.00/2.50/92.50 wt%.

The morphology of the cross-section of the composite filaments was studied using a HITACHI S3200N SEM-EDS scanning electron microscope (SEM) operated at an acceleration voltage of 10 kV. The slices of filament were fixed on metal stubs using carbon tape and sputter-coated at ~20 mA with a ~10 nm layer of gold. The magnifications used for the collection of SEM images were 3000× and 6000×. Raman spectroscopy was performed using a confocal Raman microscope, Alpha300 (WITec) equipped with a thermoelectrically cooled CCD detector (down to – 61 °C). A 532 nm wavelength laser was used to collect Raman spectra from cross-sections of composite filaments. During the mapping of the CNCs-HDPE composites the power of the laser was ~13 mW. A 50× objective lens was used for backscattered light collection. The spectrometer grating was 600 g/mm, BLZ = 500 nm. Raman images were recorded in an area of 50×50 μm^2 with a step size of 0.2 µm and an exposure time of 0.1 s and one accumulation. A total of 62500 Raman spectra were measured for each map. The average number of maps per composite was six. WITec Project Plus software was used to analyze Raman images and to convert them into chemical images. These chemical images were subsequently analyzed using Image-J software to estimate the area in μm^2 and the percentage of the area related to each component of the chemical image. The extraction of the objects was performed using an automated threshold with the algorithm IsoData.

3. Results and Discussion

The morphologies of the cross-sectional areas of HDPE composites were investigated using scanning electron microscopy (SEM). Figure 1 illustrates the representative SEM images of HDPE composites containing 1.25% of CNCs (Figure 1A and C) and 5.00% of CNCs (Figure 1B and D) at different magnifications. The surfaces of the cross-sectional areas is smooth with evident cellulose nanocrystals aggregates present within the volume of the matrix. Spray-dried CNCs fillers form spherical-like aggregates in polyethylene, Figure 1. Additionally, SEM images reveal the presence of empty spaces near the spherical CNCs aggregates. These voids probably derive from poor interfacial adhesion between CNCs and HDPE [1]. Another reason for their formation is the process of the cryo-microtome cutting at a temperature of - 22°C, since the orientation of the main axis of the voids is consistent with the direction of cutting. The process of cutting above the glass transition temperature of HDPE can potentially induce a 'drawing out' of the aggregates from the matrix. The existence of some interaction between CNCs and HDPE seems to be supported by the higher magnification images (Figure 1C and D). These images show broken filaments between CNCs aggregates and HDPE matrices. SEM images confirm the formation of aggregates of CNCs in HDPE and the existence of some interaction between nano-fillers and the matrix.

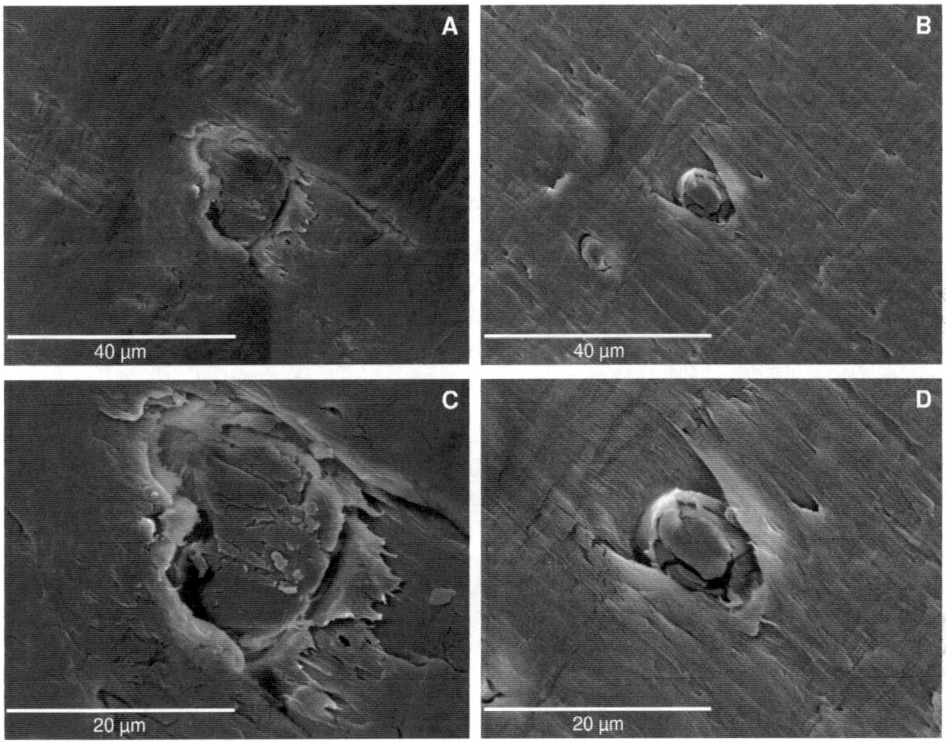

Figure 1. SEM images of cryo-microtome cut CNCs-HDPE composites at a magnification 3000× (A) 1.25% CNCs-HDPE, (B) 5.00% CNCs-HDPE and at a magnification 6000× (C) 1.25% CNCs-HDPE, (D) 5.00% CNCs-HDPE.

The cross-sectional areas of CNCs-HDPE composites were investigated using Raman imaging. The advantage of Raman spectroscopy is its sensitivity to structural and chemical features of the studied materials. The combination of the spectroscopic analysis with optical microscopy facilitates an estimation of the degree of mixing and the interface between the fillers and the matrix.

The HDPE matrix is differentiated from the CNCs by a narrow Raman band located at ~1296 cm^{-1} corresponding to CH_2 twisting modes in the crystalline phase [11,12]. The presence of CNCs is verified by the Raman band located at ~1098 cm^{-1}, corresponding to the C–O ring stretching modes and the β-1,4 glycosidic linkage (C–O–C) stretching modes between the glucose rings of the cellulose chains [13,14]. This feature is absent in HDPE.

Figures 2 and 3 show Raman images of CNCs-HDPE composites. The color scale present on each of the Raman images depicts the relative intensity of selected Raman bands. A yellow color indicates a high intensity of a specific band, whereas a dark color a low intensity. Figure 2 shows Raman images of CNCs-HDPE composites highlighting the intensity of a Raman band located at ~1098 cm^{-1}. The yellow colored fields forming spherical-like shapes correspond to aggregates of CNCs. The shape of the aggregates is similar for all the images and is independent of the loading of CNCs in the composites. This suggests that the fillers' aggregation is related to the spray-drying process used for their preparation. It is noted that the changes in the intensity of the Raman band located at ~1098 cm^{-1} reflect the features of the surface morphology of the aggregates. Raman images presented in Figure 3 show almost the negatives of images in Figure 2. They show the presence of the HDPE matrix using the relative intensity of a Raman band situated at ~1296 cm^{-1}. These images also reveal the surface morphology including features originating possibly from the blade of the knife used for cryo-microtome cutting.

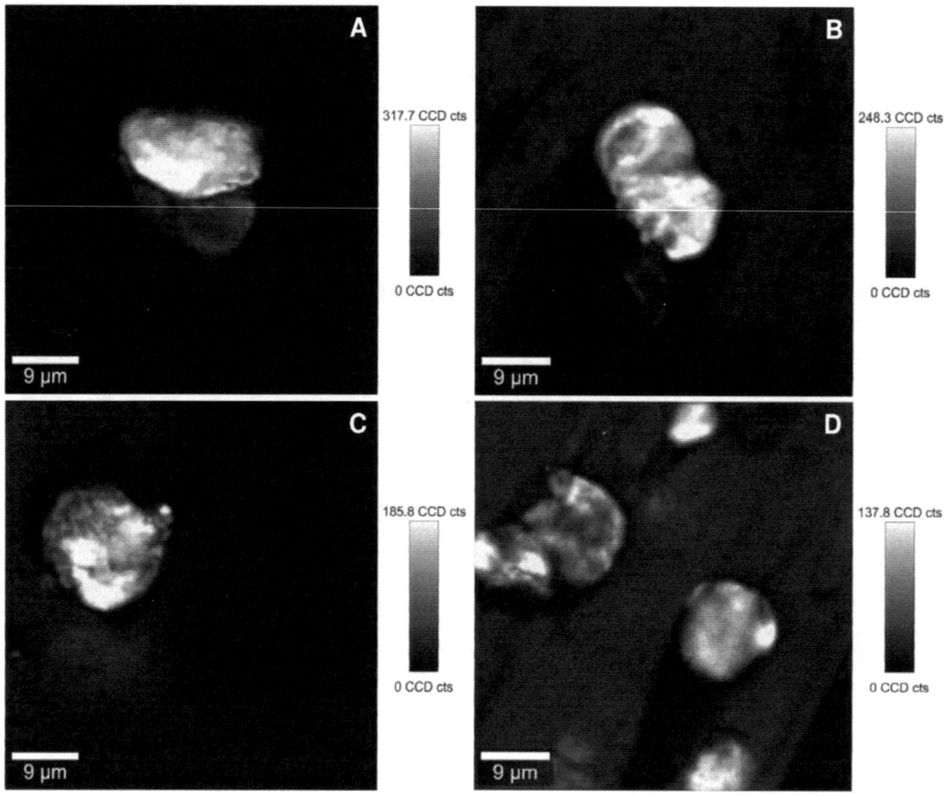

Figure 2. Typical Raman images of CNCs-HDPE composites depicting the intensity of a Raman band located at ~1098 cm^{-1}: (A) 0.625% CNCs-HDPE, (B) 1.25% CNCs-HDPE, (C) 2.50% CNCs-HDPE, (D) 5.00% CNCs-HDPE.

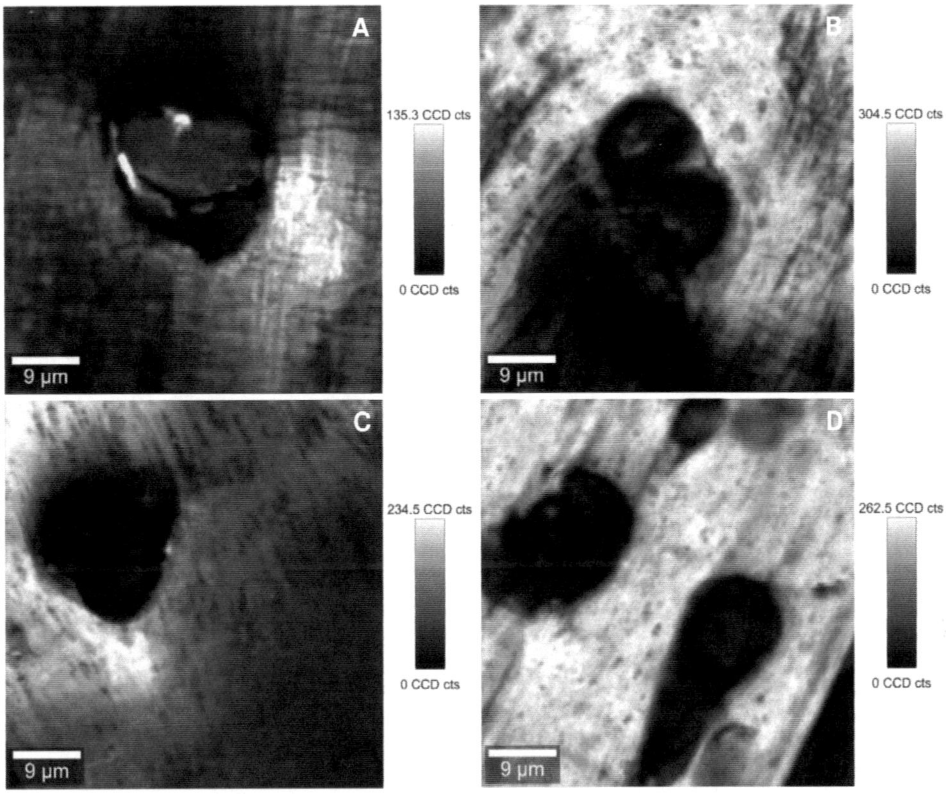

Figure 3. Typical Raman images of CNCs-HDPE composites depicting the intensity of a Raman band located at ~1296 cm^{-1}: (A) 0.625% CNCs-HDPE, (B) 1.25% CNCs-HDPE, (C) 2.50% CNCs-HDPE, (D) 5.00% CNCs-HDPE.

Figure 4 shows typical Raman spectra for the components of the compounded composites. A map of these components transforms them into chemical images showing the regions corresponding to the composition of the composites. It also demonstrates where mixing between the components occurs (Figure 5). The chemical images reveal three major areas present in the CNCs-HDPE composites. Two of them correspond to the pure CNCs (green area) and pure HDPE (red area), which is confirmed by the Raman spectra on Figure 4. Green line Raman spectrum on Figure 4 shows bands characteristic of pure CNCs proving the formation of agglomerates of CNCs in the volume of polyethylene matrix independently of the loading of CNCs in the composites. The areas corresponding to pure HDPE matrix dominate the chemical maps of the composites with different loading (Figure 4 - red line). The most significant for the composite preparation is the blue area on the chemical maps associated with the mixing between CNCs and HDPE. A Raman spectrum corresponding to this region exhibits the bands characteristic of both CNCs and HDPE suggesting a good mixing of the components (Figure 4 - blue line). Despite the presence of the area where mixing between CNCs and HDPE occurring, the chemical images verify once again the tendency of cellulose to aggregate within the HDPE matrix. The aggregates of CNCs consist of both pure CNCs (green area) and mixed CNCs and HDPE (blue area).

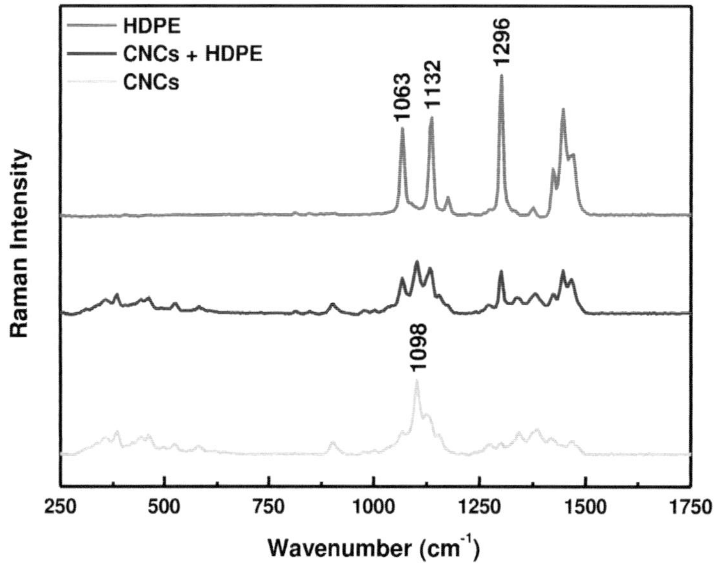

Figure 4 Typical Raman spectra of composite components. The colors used here map onto to the chemical image of 5.0% CNCs(FR)-HDPE in Figure 5C.

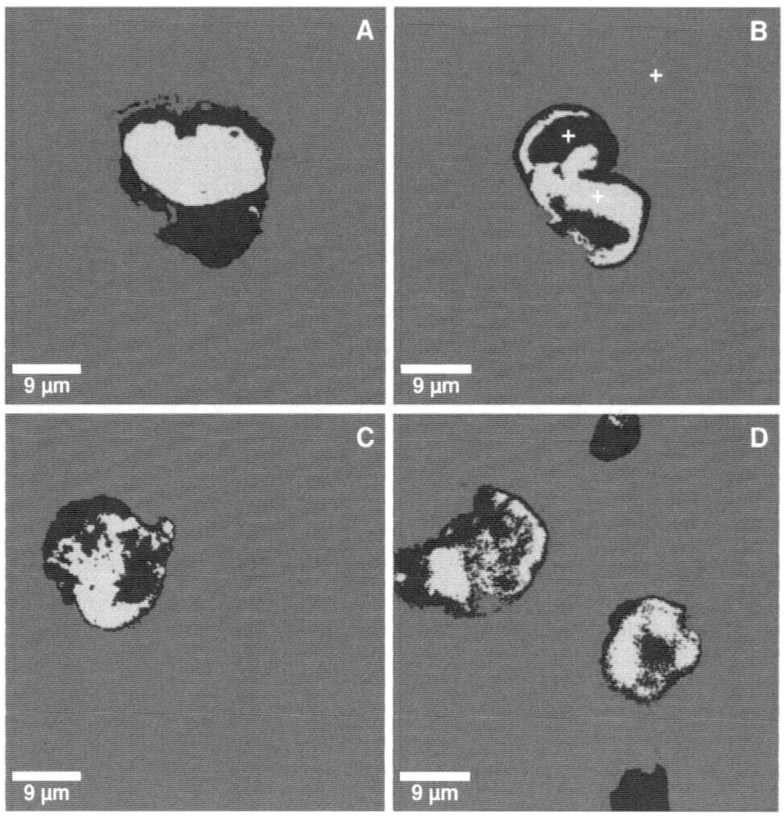

Figure 5. Typical chemical image of 1.25% CNCs-HDPE composite depicting the chemical composition of a mapped cross-section: (A) 0.625% CNCs-HDPE, (B) 1.25% CNCs-HDPE, (C) 2.50% CNCs-HDPE, (D) 5.00% CNCs-HDPE.

The chemical images are used for a statistical assessment of the area corresponding to each of the composite components. Image-J software is used to quantify the area (in μm^2) related to the components (Table 1). The ratio of the fraction of the mixed areas (fillers and matrix) to the areas of pure matrix quantifies the effectiveness of the degree of mixing. The average ratio of the Blue:Red area for CNCs-HDPE composites with different loadings of CNCs is comparable. Since the loading of MAPE compatibilizer is the same for the entire series of composites, it poses an additional limiting factor for increasing the efficiency of mixing between CNCs and HDPE. This suggests that the process of the preparation of CNCs plays an important role in their incorporation into the matrix. The hydrophilic nature of CNCs induces the formation of hydrogen bonding between the nanoparticles increasing their tendency to agglomerate in nonpolar matrices. Additionally, a compact form of the CNCs reduces their ability to mix with the HDPE matrix through their interaction with macromolecules of MAPE and HDPE.

Table 1. Average area fraction of the component of chemical maps quantified using Image J software.

	Area fraction			Ratio of fraction	
Composite	Red[a] [μm^2]	Blue[b] [μm^2]	Green[c] [μm^2]	(Green+Blue)/Red	Blue/Red
0.625% CNCs-HDPE	2282 ± 84	74 ± 26	112 ± 77	0.08 ± 0.04	0.06 ± 0.05
1.25% CNCs-HDPE	2238 ± 102	121 ± 88	107 ± 27	0.10 ± 0.05	0.06 ± 0.04
2.50% CNCs-HDPE	2265 ± 85	105 ± 40	97 ± 81	0.09 ± 0.04	0.05 ± 0.02
5.00% CNCs-HDPE	2220 ± 101	148 ± 54	100 ± 66	0.11 ± 0.05	0.07 ± 0.03

[a] fraction area corresponding to HDPE.
[b] fraction area corresponding to CNCs + HDPE.
[c] fraction area corresponding to CNCs.

4. Conclusions

In recent times there has been a lot of interest in making composites using renewable and biodegradable components. Both bio-polyethylene and cellulose fulfill these criteria. However, high quality composites require good mixing between the filler and the matrix to obtain good interfacial adhesion, and consequently optimize the reinforcing effect of the fillers. Raman microscopy is a non-destructive and useful tool for chemical and spatial quantification of the mixing process of CNCs in HDPE. The data provided by the analysis of Raman images and chemical images includes spatial information regarding the agglomeration of CNCs and the surface morphology features of composites. CNCs are found to form agglomerated 'islands' within the HDPE independent of the filler loading. Raman microscopy has been demonstrated to be able to quantify the degree of mixing between CNCs and HDPE. The quantified degree of mixing for a series of composites shows a comparable degree of mixing for all of the materials. It suggests that for the incorporation process of CNCs into thermoplastics with addition of compatibilizer is equally important as the physicochemical properties and morphology of the filler.

Acknowledgments

The authors would like to thank the EU FP7 funding programme for supporting the work under grant agreement no 604168 (www.newspec.eu).This publication reflects the views only of the author and the European Commission cannot be held responsible for any use which may be made of the information contained therein.

References

[1] Ou R, Xie Y, Wang Q, Sui S, Wolcott MP 2014 *J. Appl. Polym. Sci.* **131** 40331(1-8)

[2] Ou R, Xie Y, Wolcott MP, Sui S, Wang Q 2014 *Mater. Des.* **58** 339-345
[3] Li J, Song Z, Li D, Shang S, Guo Y 2014 *Ind. Crop. Prod.* **59** 318-328
[4] Pöllänen M, Suvanto M, Pakkanen TT 2013 *Compos. Sci. Technol.* **76** 21-28
[5] Ben Azouz K, Ramires EC, Van den Fonteyne W, El Kissi N, Dufresne A 2011 *ACS Macro Lett.* **1** 236-240
[6] Felix JM, Gatenholm P 1991 *J. Appl. Polym. Sci.* **42** 609
[7] Agarwal UP, Sabo R, Reiner RS, Clemons CM, Rudie AW 2012 *Appl.Spec.* **66** 750-756
[8] van Apeldoorn AA, van Manen H-J, Bezemer JM, de Bruijn JD, van Blitterswijk CA, Otto C 2004 *J. Am. Chem. Soc.* **126** 13226-13227
[9] Widjaja E, Garland M 2011 *Mater. Today* **14** 114-117
[10] Huan S, Lin W, Sato H, Yang H, Jiang J, Ozaki Y, Wu H, Shen G, Yu R 2007 *J. Raman Spec.* **38** 260-270
[11] Sato H, Shimoyama M, Kamiya T, Amari T, Šašic S, Ninomiya T, Siesler HW, Ozaki Y 2002 *J. Appl. Polym. Sci.* **86** 443-448
[12] Cherukupalli SS, Ogale AA 2004 *Polym. Eng. Sci.* **44** 1484-1490
[13] Gierlinger N, Schwanninger M, Reinecke A, Burgert I 2006 *Biomacromolecules* **7** 2077-2081
[14] Atalla RH, Agarwal UP 1985 *Science* **227** 636-638

Modelling of volumetric composition and mechanical properties of unidirectional hemp/epoxy composites - Effect of enzymatic fibre treatment

M Liu[1], A Thygesen[1], AS Meyer[1], B Madsen[2]

[1]Center for Bioprocess Engineering, Department of Chemical and Biochemical Engineering, Technical University of Denmark, 2800 Kongens Lyngby, Denmark.
[2]Section of Composites and Materials Mechanics, Department of Wind Energy, Technical University of Denmark, Frederiksborgvej 399, 4000 Roskilde, Denmark.

E-mail: miliu@kt.dtu.dk

Abstract. The objective of the present study is to assess the effect of enzymatic fibre treatments on the fibre performance in unidirectional hemp/epoxy composites by modelling the volumetric composition and mechanical properties of the composites. It is shown that the applied models can well predict the changes in volumetric composition and mechanical properties of the composites when differently treated hemp fibres are used. The decrease in the fibre correlated porosity factor with the enzymatic fibre treatments shows that the removal of pectin by pectinolytic enzymes results in a better fibre impregnation by the epoxy matrix, and the mechanical properties of the composites are thereby increased. The effective fibre stiffness and strength established from the modelling show that the enzymatic removal of pectin also leads to increased mechanical properties of the fibres. Among the investigated samples, the composites with hydrothermally pre-treated and enzymatically treated fibres have the lowest porosity factor of 0.08 and the highest mechanical properties. In these composites, the effective fibre stiffness and strength are determined to be 83 GPa and 667 MPa, respectively, when the porosity efficiency exponent is set equal to 2. Altogether, it is demonstrated that the applied models provide a concept to be used for the evaluation of performance of treated fibres in composites.

1. Introduction

As a result of increasing environmental awareness, research interest has been shifting to use natural fibres as substitute to man-made fibres in fibre reinforced composites due to their unique advantages, such as environmental sustainability, low cost, low density, together with their high stiffness and strength to weight ratio [1,2]. However, their potential use as reinforcement could be considerably reduced by some of their disadvantages including moderate strength of fibres, less controlled processing methods, and seasonal variation in quality [3–5].

In principle, the aim of fibre processing is to obtain more separated and cellulose rich fibres by removing non-cellulosic components, in order to optimize the strength and form of the fibres before being used as reinforcement in composites. Traditional cellulose fibre processing methods for hemp and flax fibres, which have mainly been targeted for yarn production, include field retting and water retting. These retting methods remove non-cellulosic components via spontaneously flourishing microbial activity, and they have been reported to have negative impacts on both fibre properties and the environment [5,6]. As an alternative method, treatment of fibres with pectinolytic enzymes could be a more efficient and controlled method, in addition to overcoming the limitations of the traditional fibre processing methods [7–9].

Porosity is a parameter that is used to assess the quality of composites. Porosity in natural fibre composites is not only created due to insufficient impregnation of the fibres by the matrix, but also due to the air filled cavities inside the fibres, the so-called lumen. Porosity has a direct influence on the volumetric composition (i.e. the volume contents of fibres and matrix) and the mechanical properties of composites [9–11]. A study of the relations between fibre processing routes, and the volumetric composition and mechanical properties of composites is central to the goal of assessing the effect of various fibre treatments.

In the present study, enzymatic treatments with and without hydrothermal pre-treatment were carried out on hemp bast fibres. For comparison, traditional field retting of fibres was also carried out [5]. It is expected that porosity is highest in the composites with the untreated fibres due to the presence of the epidermis layer and parenchyma cells, which consist of a large amount of voids [9,12]. In contrast, it is expected that the enzymatic treatments will produce fibres where the epidermis layer and part of parenchyma cells are removed, in addition to splitting larger fibre bundles into smaller ones [7,9,13]. Those changes are expected to decrease the porosity in the composites, and the mechanical properties will thereby be increased.

The objective of the present study is to use previously developed models for composite volumetric composition and mechanical properties for a quantitative analysis of the effect of enzymatically based fibre processing methods. This approach is shown to provide valuable understanding of the effect of the fibre treatments on the properties of the composites.

2. Model

In the present study, experimental data is modelled by previously reported models [11,14]. Experimental data for composite volumetric composition, density and mechanical properties (i.e. stiffness and strength) were obtained for composites with a fibre weight content (W_f) below the transition value ($W_{f\,trans}$). For the modelling of composites with W_f above $W_{f\,trans}$ only model lines will be shown. A summary of the applied model parameters is shown in Table 1. The parameters, fibre density (ρ_f), matrix density (ρ_m), matrix correlated porosity factor (α_{pm}), maximum obtainable fibre volume content ($V_{f\,max}$), and porosity efficiency exponent for composite stiffness (n_E) and strength (n_σ) are assumed to be constant and independent of the applied fibre treatments. The parameters, transition fibre weight content ($W_{f\,trans}$) and fibre correlated porosity factor (α_{pf}) are assumed to be dependent on the fibre treatment.

Table 1. Summary of model parameters applied in the present study.

Parameter	Meaning	Value	Way of determining the value
ρ_f	fibre density	1.50 g/cm^3	assumed
ρ_m	matrix density	1.14 g/cm^3	measured
α_{pm}	matrix correlated porosity factor	0	assumed
$V_{f\,max}$	maximum obtainable fibre volume content	0.65	assumed
n_E	porosity efficiency exponent for composite stiffness	0 or 2	assumed
n_σ	porosity efficiency exponent for composite strength	0 or 2	assumed
$W_{f\,trans}$	transition fibre weight content	see Table 2	determined from Eq.1
α_{pf}	fibre correlated porosity factor	see Table 2	determined from Eq.2

2.1. *Volumetric composition*

In the selected model [11], the volumetric composition in composites is correlated to the fibre weight content (W_f) in two regions: region A, where W_f is below a transition value ($W_{f\,trans}$), and region B, where W_f is above a transition value ($W_{f\,trans}$), respectively. The transition value ($W_{f\,trans}$) separating region A and B is calculated by Eq.1.

$$W_{f\,trans} = \frac{V_{f\,max}\rho_f(1+\alpha_{pm})}{V_{f\,max}(1+\alpha_{pm}) - V_{f\,max}\rho_m(1+\alpha_{pf}) + \rho_m} \quad (1)$$

In this study, it is assumed that there is no matrix correlated porosity in the composites, and therefore α_{pm} is set to 0. The fibre correlated porosity factor (α_{pf}) is determined by using Eq.2, where the porosity (V_p) is assumed to be a linear function of the fibre volume content (V_f) [11].

$$V_p = \alpha_{pf} \times V_f \quad (2)$$

The volumetric composition in the composites in region A and region B are shown in Eqs.3 – 8.

Region A ($W_f \leq W_{f\,trans}$)

$$V_f = \frac{W_f\rho_m}{W_f\rho_m(1+\alpha_{pf}) + (1-W_f)\rho_f(1+\alpha_{pm})} \quad (3)$$

$$V_m = \frac{(1-W_f)\rho_f}{W_f\rho_m(1+\alpha_{pf}) + (1-W_f)\rho_f(1+\alpha_{pm})} \quad (4)$$

$$V_p = 1 - V_f - V_m = \frac{W_f\rho_m\alpha_{pf} + (1-W_f)\rho_f\alpha_{pm}}{W_f\rho_m(1+\alpha_{pf}) + (1-W_f)\rho_f(1+\alpha_{pm})} \quad (5)$$

Region B ($W_f \geq W_{f\,trans}$)

$$V_f = V_{f\,max} \quad (6)$$

$$V_m = V_{f\,max}\frac{(1-W_f)\rho_f}{W_f\rho_m} \quad (7)$$

$$V_p = 1 - V_f - V_m = 1 - V_{f\,max}\left(1 + \frac{(1-W_f)\rho_f}{W_f\rho_m}\right) \quad (8)$$

2.2. *Composite density*

Equations for composite density, Eqs.10 – 11, can be derived from Eq. 9 by using the expression for volumetric composition in composites.

$$\rho_c = \frac{m_c}{V_c} = \frac{m_f/W_f}{m_f/(\rho_f V_f)} = \frac{\rho_f V_f}{W_f} \quad (9)$$

Region A ($W_f \leq W_{f\,trans}$)

$$\rho_c = \frac{\rho_m\rho_f}{W_f\rho_m(1+\alpha_{pf}) + (1-W_f)\rho_f(1+\alpha_{pm})} \quad (10)$$

Region B ($W_f \geq W_{f\,trans}$)

$$\rho_c = \frac{V_{f\,max}}{W_f}\rho_f \quad (11)$$

In the modelling of composite density, the fibre weight content (W_f) is set as the independent variable, and composite density (ρ_c) is set as the dependent variable. The values of the other parameters are given in Tables 1 and 2. The experimental data of ρ_c vs. W_f is compared with the model predictions (Eqs. 10 – 11).

2.3. *Mechanical properties*

A large number of modified rule of mixtures models for stiffness of composites have been proposed in the literature [14,15]. In one of these models, by including the effect of porosity [14], stiffness (E_c) and strength (σ_{cu}) of composites with a unidirectional fibre orientation and with continuous fibres can be expressed by Eq.12 and Eq.13, respectively.

$$E_c = \left(V_f E_f + V_m E_m\right)\left(1 - V_p\right)^{n_E} \tag{12}$$

$$\sigma_{cu} = \left(V_f \sigma_{fu} + V_m \sigma_m^*\right)\left(1 - V_p\right)^{n_\sigma} \tag{13}$$

where E is the stiffness, V is the volume content, the subscripts c, m, f and p are composite, matrix, fibres and porosity, respectively. σ_{cu} is the composite strength, σ_{fu} is the fibre strength, and σ_m^* is the stress in the matrix at the failure strain of the composite, and n_E and n_σ are the porosity efficiency exponents (PEE) for stiffness and strength, respectively. When PEE = 0, it is assumed that the porosity in the composites has no effect on the mechanical properties of the composites beyond lowering the load bearing volume. When PEE > 0, it is assumed that the porosity in the composites has an effect on the mechanical properties of composites by introducing stress concentrations [14]. Values of PEE equal to 0 and 2 are used in the present study to show these two cases.

By using the models for the composite volumetric composition, equations for the correlation between composite stiffness (E_c) and strength (σ_{cu}) and the fibre weight fraction (W_f) can be established.

Region A ($W_f \leq W_{f\,trans}$)

$$E_c = \frac{\left(W_f \rho_m E_f + (1-W_f)\rho_f E_m\right)\left(W_f \rho_m + (1-W_f)\rho_f\right)^{n_E}}{\left(W_f \rho_m (1+\alpha_{pf}) + (1-W_f)\rho_f (1+\alpha_{pm})\right)^{n_E+1}} \tag{14}$$

$$\sigma_{cu} = \frac{\left(W_f \rho_m \sigma_{fu} + (1-W_f)\rho_f \sigma_m^*\right)\left(W_f \rho_m + (1-W_f)\rho_f\right)^{n_\sigma}}{\left(W_f \rho_m (1+\alpha_{pf}) + (1-W_f)\rho_f (1+\alpha_{pm})\right)^{n_\sigma+1}} \tag{15}$$

Region B ($W_f \geq W_{f\,trans}$)

$$E_c = \frac{\left(W_f \rho_m V_{f\,max} E_f + (1-W_f)\rho_f V_{f\,max} E_m\right)\left(W_f \rho_m V_{f\,max} + (1-W_f)\rho_f V_{f\,max}\right)^{n_E}}{\left(W_f \rho_m\right)^{n_E+1}} \tag{16}$$

$$\sigma_{cu} = \frac{\left(W_f \rho_m V_{f\,max} \sigma_{fu} + (1-W_f)\rho_f V_{f\,max} \sigma_m^*\right)\left(W_f \rho_m V_{f\,max} + (1-W_f)\rho_f V_{f\,max}\right)^{n_\sigma}}{\left(W_f \rho_m\right)^{n_\sigma+1}} \tag{17}$$

In the modelling of composite mechanical properties, the fibre weight content (W_f) is set as the independent variable, and composite stiffness (E_c) and strength (σ_{cu}) are set as the dependent variables. The effective fibre stiffness (E_f) and effective fibre strength (σ_{fu}) are set as derived parameters. The values of the remaining parameters are given in Tables 1 and 2. The models are fitted to the experimental data of E_c vs. W_f and σ_{cu} vs. W_f by using non-linear regression. The goodness of fitting is evaluated by adjusted R-squared values, which has been adjusted for the number of predictors in the model from R-squared values.

3. Materials and methods

3.1. Plant material

Hemp (*Cannabis sativa* L.), variety USO-31, was sown in France (N 48.8526°, E 3.0190°(WGS84)) as described in detail in Liu et al. [5]. The whole hemp stem under the inflorescence base was used as the starting material in the present study. Hemp stem pieces with a length of approx. 15 cm were randomly collected from the stems. The hemp stems were hydrothermally pre-treated at 121 °C for 30 min in an autoclave. Hemp fibre strips were manually peeled off from the pre-treated stems, and then they were enzymatically treated by using pectinases as described in detail by Liu et al. [9]. After enzymatic treatment, the fibre strips were dried at 50 °C for 12 h. For comparison, field retting of the hemp stems was carried out for 20 days after harvest [4].

3.2. Manufacturing of composites

The treated bast fibre strips were manually untangled and aligned to allow the fibres to be processed into unidirectional composites. Bundles of fibre strips were firstly cut to a length of 140 mm, and the fibre strips were then justified to a bunch of fibre strips with masses in the range 0.6 – 2.3 g. Bunches of fibre strips were then put in the mould chambers. Afterwards, a press beam was placed on the top of the fibre strips in each chamber, and two insert beams were used to fix the height of the mould chambers to 2 mm. An epoxy resin (Araldite® LY 1568) and its amine hardener (Aradur® 3489) were mixed at a 100/28 mass ratio, and degassed in a vacuum oven. The setup for the vacuum infusion and moulding process is described in detail in the study by Liu et al. [9]. After curing for 12 h at 80 °C, the composite tensile specimens with dimensions of 140 × 10 × 2 mm were demoulded. Tabs with lengths of 50 mm were mounted on the composite specimens using epoxy resin (DP 460).

3.3. Volumetric composition

Composites with fibre weight contents (W_f) in the range 0 – 0.70 were obtained by varying the amount of fibres (m_f) in the mould chambers during manufacturing of composites. The fibre volume content (V_f) was determined by Eq.18.

$$V_f = \frac{m_f/\rho_f}{m_c/\rho_c} = \frac{m_f}{m_c} \times \frac{\rho_c}{\rho_f} \tag{18}$$

where m_c, ρ_f and ρ_c are composite mass, fibre density and composite density, respectively. When W_f was below 0.30, the composite specimens made by using the above mentioned moulding process were found to have irregular surfaces, and their thickness could not be measured accurately. For those cases, ρ_c was determined by the buoyancy method (Archimedes principle) using water as the displacement medium. When W_f was above 0.30, the composites specimens had flat surfaces, and their density (ρ_c) could be accurately calculated based on their dimensions (length, width and thickness).

The matrix volume content (V_m) was determined using Eq.19.

$$V_m = \frac{m_m/\rho_m}{m_c/\rho_c} = \frac{m_m}{m_c} \times \frac{\rho_c}{\rho_m} \tag{19}$$

where m_m is matrix mass. The porosity (V_p) was then determined using Eq.20.

$$V_p = 1 - V_f - V_m \tag{20}$$

3.4. Tensile properties of composites

For tensile testing of the composite specimens, an Instron Testing Machine 5566 with a load cell of 10 kN was used. Two extensometers were used for strain measurements, and a displacement rate of 1 mm/min (corresponding to a strain rate of 2.5 %/min) was used. Based on the measured stress-strain curves, stiffness (linear regression in the strain interval 0.05 – 0.25%), strength and failure strain was determined. For each type of composite with the differently treated hemp fibres, at least 10 specimens with varied fibre content were tested.

4. Results and discussion

4.1. Composite volumetric composition

The fibre correlated porosity factor (α_{pf}) of the composites manufactured with the differently treated fibres is determined in Figure1a based on the experimental values of V_p and V_f using Eq.2. The values of α_{pf} for the different fibre treatments are shown in Table 2. As shown, composites with field retted and untreated fibres have the highest porosity factor of about 0.16. In contrast, composites with the enzymatically treated fibres have a much lower porosity factor of 0.12, followed by the lowest porosity factor of 0.08 for composites with the hydrothermally pre-treated and enzymatically treated fibres.

The variation of the composite porosity factor with the different fibre treatments can be explained by the changes of the fibre microstructure. When the hemp fibre strips are subjected to pectinases, pectin in the epidermis, in the parenchyma cells, and in the middle lamella regions between fibre cells are partly hydrolysed by enzyme catalysed reactions. This degradation of pectin loosen the bonding between epidermis and cortex, between fibres and parenchyma cells, and between fibres, and consequently, the epidermis and parenchyma cells are partly removed from the hemp fibre strips. With the partly removal of epidermis and parenchyma cells, which consist of a large amount of voids, the fibre bundles are split into smaller fibre bundles [9]. All these changes to the fibre microstructure collectively contribute to the decrease of α_{pf} after the enzymatic treatment.

The transition fibre weight content ($W_{f\ trans}$) for each type of fibre treatment was calculated by using Eq.1 with the obtained fibre correlated porosity factors. The values of $W_{f\ trans}$ are shown in Table 2. It is found that the transition value decreases from 0.77 for composites with field retted and untreated fibres, to 0.75 for composites with enzymatically treated fibres, and finally to 0.74 for composites with hydrothermally pre-treated and enzymatically treated fibres. According to Eq.1, it is evident that the decrease of $W_{f\ trans}$ with fibre processing is directly governed by the decrease of the fibre correlated porosity factor.

Experimental data on the volumetric composition in the composites with the differently treated fibres, and the corresponding model lines are shown in Figure 1b. Besides the difference in the values of $W_{f\ trans}$, it is shown that composites with hydrothermally pre-treated and enzymatically treated fibres have higher fibre volume content (V_f) and matrix volume content (V_m), and lower porosity than the composites with field retted and untreated fibres at any given fibre weight content below $W_{f\ trans}$. According to Eqs.3 – 5, the difference in α_{pf} explains the difference in volumetric composition between the composites. Altogether, the results in Figure 1b reveal that the full volumetric composition in composites with a given type of fibre treatment can be predicted by using the model, and by using a given fibre weight content that is used for the manufacturing of the composites as input.

Table 2. Fibre correlated porosity factor (α_{pf}) and transition fibre weight content ($W_{f\ trans}$) of composites with differently treated hemp fibres.

Fibre sample	Fibre correlated porosity factor (α_{pf})	Transition fibre weight content ($W_{f\ trans}$)
Field retted	0.162	0.772
Untreated	0.157	0.770
Enzymatically treated	0.119	0.753
Hydrothermally pre-treated + enzymatically treated	0.084	0.738

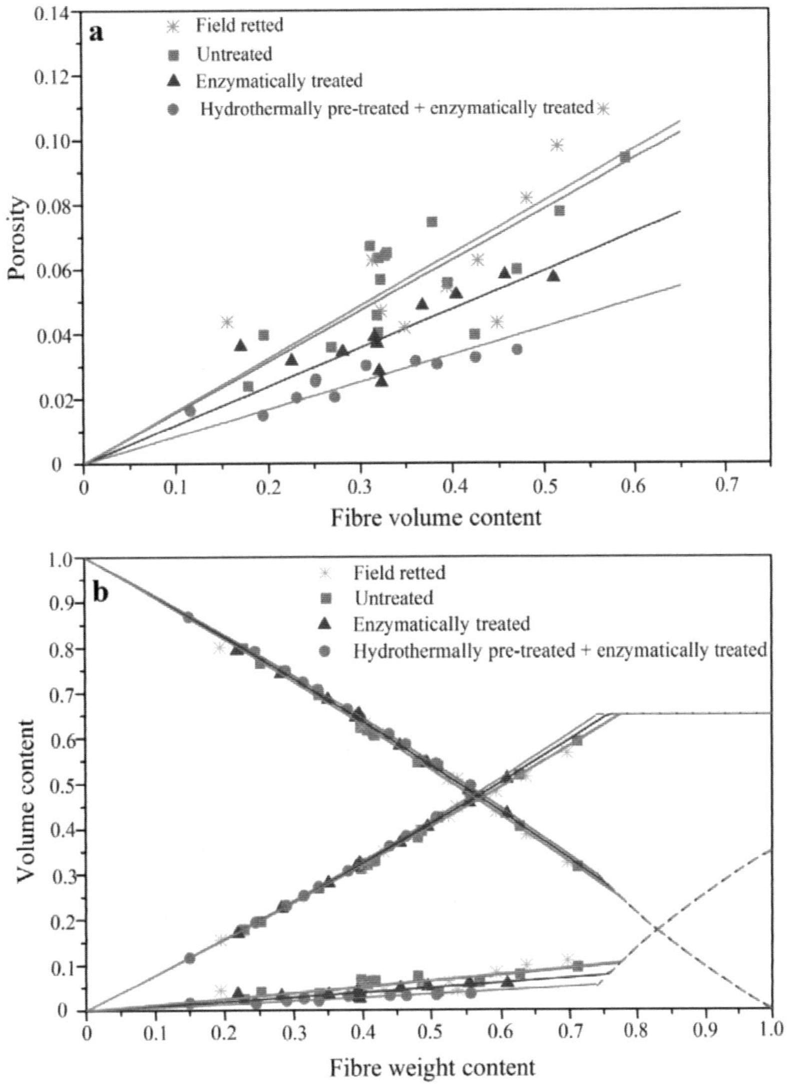

Figure 1. Composite porosity vs. fibre volume content (a), and composite volumetric composition (V_f, V_m and V_p) vs. fibre weight content (b).

4.2. Composite density

Figure 2 shows that the composites with the untreated and field retted fibres have the lowest density, which is a result of the highest porosity content and lowest fibre volume content in these composites (see Figure 1). In contrast, the composites with the enzymatically treated fibres, particularly the hydrothermally pre-treated and enzymatically treated fibres, exhibit clearly higher density due to their lower porosity contents and higher fibre volume contents. The model predictions of composite density in Figure 2 show that the predicted composite density is in good agreement with the experimental data. Therefore, the density of composites with a given type of fibre treatment can also be well predicted as a function of the fibre weight content.

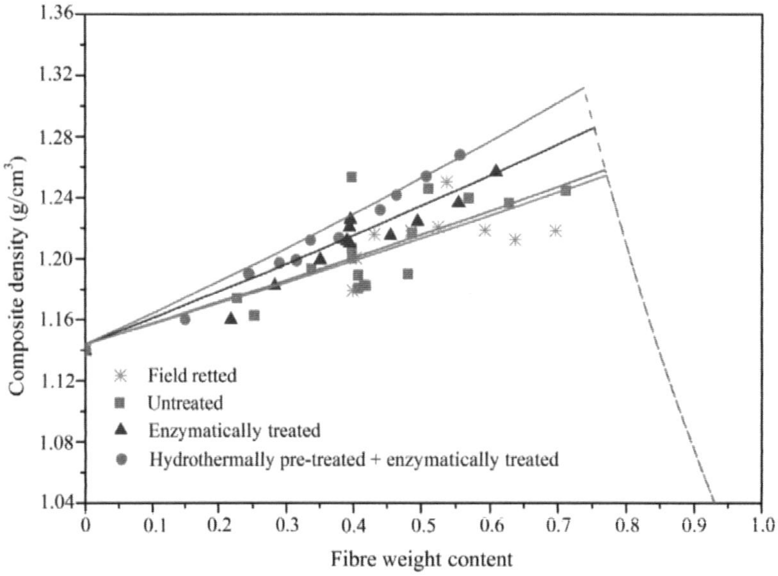

Figure 2. Composite density vs. fibre weight content.

4.3. Composite mechanical properties

In Figures 3 and 4, showing composite stiffness and strength as a function of the fibre weight content, the starting points of the model lines at $W_f = 0$ are equal to the measured epoxy matrix stiffness (E_m) of 2.7 GPa, and the epoxy matrix stress of 27 MPa at the average composite failure strain of 1.0%. Generally, the model lines are in good agreement with the experimental data.

The model lines in Figures 3 and 4 are established by setting the porosity efficiency exponent (n) equal to 0 and 2. For n = 0, it is assumed that all the porosity is located inside the fibres, in the so-called lumen, and this is assumed to have no effect on the mechanical properties of the composites. For n = 2, it is assumed that all the porosity is located outside the fibres, e.g. at the fibre/matrix interface or in the fibre bundles to produce un-impregnated fibres. This is assumed to lead to stress concentrations, which is modelled by setting n equal to 2 [14,15]. When n = 0, as shown in Figures 3a and 4a, composite stiffness and strength are increased non-linearly with W_f with a upward curvature until $W_{f\ trans}$, and thereafter, stiffness and strength are only slightly reduced. When n = 2, as shown in Figures 3b and 4b, composite stiffness and strength are increased non-linearly with W_f with a downward curvature until $W_{f\ trans}$, and thereafter, stiffness and strength are reduced radically. The downward curvature of the model lines is most obvious for the composites with the highest porosity content, such as the composites with the field retted and untreated fibres.

When comparing the model lines in Figures 3 and 4 for composites with the differently treated fibres, it is evident that the composites with hydrothermally pre-treated and enzymatically treated fibres have the highest stiffness and strength, followed by the composites with enzymatically treated and untreated fibres, while the composites with field retted fibres have the lowest stiffness and strength.

The fitted values of the effective fibre stiffness (E_f) and fibre strength (σ_{fu}) established by the model lines in Figures 3 and 4 are shown in Table 3. Field retted samples are found to have the lowest effective fibre stiffness of 52 and 61 GPa, and the lowest effective fibre strength of 474 and 558 MPa, when the porosity efficiency exponent is set equal to 0 and 2, respectively. There is a tendency that the effective fibre stiffness and strength increase from untreated fibres, to enzymatically treated fibres, and finally to hydrothermally pre-treated and enzymatically treated fibres, irrespective of the porosity efficiency exponent values. The hydrothermally pre-treated and enzymatically treated fibres have the highest effective stiffness of 74 GPa and 83 GPa, and the highest effective strength of 625 MPa and 667 MPa, when porosity efficiency exponent is set equal to 0 and 2, respectively.

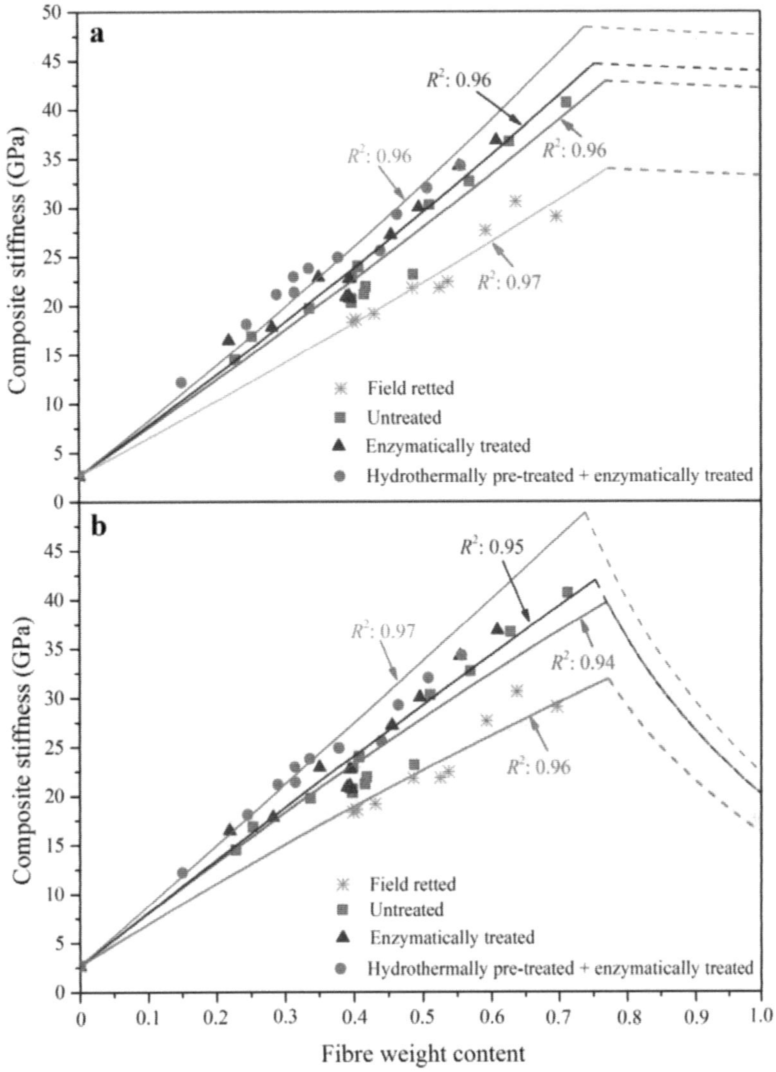

Figure 3. Composite stiffness vs. fibre weight content. Model lines are made using a porosity efficiency exponent of (a) 0 and (b) 2. Values of adjusted R-squared of fitting are shown next to the model lines.

Table 3. Established values of effective fibre stiffness (E_f) and fibre strength (σ_{fu}) in composites with differently treated hemp fibres. The porosity efficiency exponents (n_E and n_σ) are set to be either 0 or 2.

Fibre sample	E_f (GPa)		σ_{fu} (MPa)	
	n_E=0	n_E=2	n_σ=0	n_σ=2
Field retted	52	61	474	558
Untreated	65	75	569	655
Enzymatically treated	68	75	587	644
Hydrothermally pre-treated + enzymatically treated	74	83	625	667

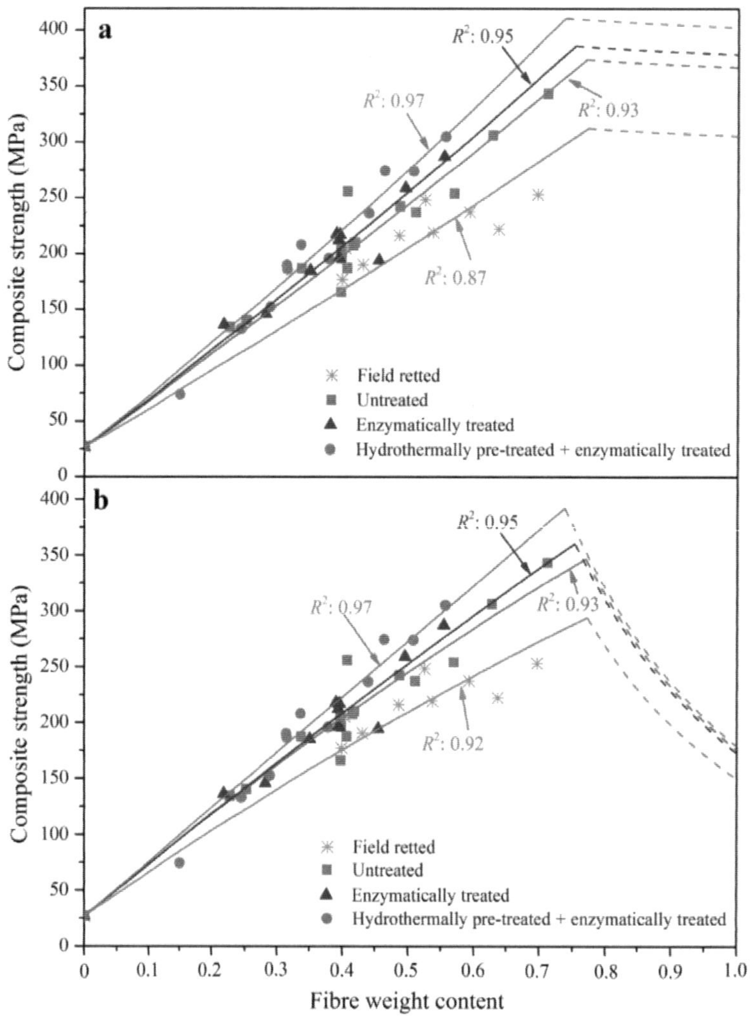

Figure 4. Composite strength vs. fibre weight content. Model lines are made using a porosity efficiency exponent of (a) 0 and (b) 2. Values of adjusted R-squared of fitting are shown next to the model lines.

5. Conclusions

Models for the volumetric composition, density and mechanical properties (i.e. stiffness and strength) of composites with differently treated hemp fibres were applied for evaluating the effect of enzymatic fibre treatments on fibre performance in composites. It is shown that the applied models are in good agreement with the experimental data. The established effective fibre stiffness and strength are used to quantify the effect of the enzymatic fibre treatments on the performance of the fibres in the composites. Altogether, the applied models are shown to be useful tools for the prediction of properties of composites with differently treated hemp fibres.

Acknowledgements

The authors are grateful to the Danish Council for Independent Research supporting the CelFiMat project (No. 0602-02409B: "High quality cellulosic fibres for strong biocomposite materials"). The financial support of China Scholarship Council (CSC, no. 201304910245) for Ming Liu's Ph.D. project is acknowledged. Jonas Kreutzfeldt Heininge and Jan Sjølin from Technical University of Denmark are acknowledged for technical support.

References

[1] Faruk O, Bledzki A K, Fink H P and Sain M 2012 *Prog. Polym. Sci.* **37** 1552
[2] Joshi S V, Drzal L T, Mohanty A K and Arora S 2004 *Compos. Part A Appl. Sci. Manuf.* **35** 371
[3] Arshad M, Kaur M and Ullah A 2016 *ACS Sustain. Chem. Eng.* **4** 1785
[4] Liu M, Fernando D, Meyer A S, Madsen B, Daniel G and Thygesen A 2015 *Ind. Crops Prod.* **76** 880
[5] Liu M, Fernando D, Daniel G, Madsen B, Meyer A S, Ale M and Thygesen A 2015 *Ind. Crops Prod.* **69** 29
[6] Di Candilo M, Bonatti P M, Guidetti C, Focher B, Grippo C, Tamburini E and Mastromei G 2010 *J. Appl. Microbiol.* **108** 194
[7] Li Y and Pickering K L 2008 *Compos. Sci. Technol.* **68** 3293
[8] Dreyer J, Müssig J, Koschke N, Ibenthal W D and Harig H 2002 *J. Ind. Hemp* **7** 43
[9] Liu M, Silva D A S, Fernando D, Meyer A S, Madsen B, Daniel G and Thygesen A 2016 *Compos. Part A Appl. Sci. Manuf.* doi: http://dx.doi.org/10.1016/ j.compositesa.2016.06.003
[10] Shibata S, Bozlur R M, Fukumoto I and Kanda Y 2010 BioResources **5** 2097
[11] Madsen B, Thygesen A and Lilholt H 2007 *Compos. Sci. Technol.* **67** 1584
[12] Shang L, Jiang Z, Liu X, Tian G, Ma J, Yang S 2016 *Bioresources* **11** 2071
[13] Akin D E, Gamble G R, Morrison W H, Rigsby L L, Dodd R B 1996 *J. Sci. Food Agric.* **72** 155
[14] Madsen B, Thygesen A and Lilholt H 2009 *Compos. Sci. Technol.* **69** 1057
[15] Madsen B and Lilholt H 2003 *Compos. Sci. Technol.* **63** 1265

The environmental impact of wind turbine blades

P Liu and C Y Barlow

Institute for Manufacturing, University of Cambridge, 17 Charles Babbage Road, Cambridge, CB3 0FS, United Kingdom

Email: pl384@cam.ac.uk

Abstract. The first generation of wind turbine (WT) blades are now reaching their end of life, signalling the beginning of a large problem for the future. Currently most waste is sent to landfill, which is not an environmentally desirable solution. Awareness of this issue is rising, but no studies have fully assessed the eco impact of WT blades. The present study aims to provide a macroscopic quantitative assessment of the lifetime environmental impact of WT blades. The first stage has been to analyse global data to calculate the amount of WT blade materials consumed in the past. The life cycle environmental impact of a single WT blade has then been estimated using eco data for raw materials, manufacturing processes, transportation, and operation and maintenance processes. For a typical 45.2 meter 1.5 MW blade this is 795 GJ (CO_2 footprint 42.1 tonnes), dominated by manufacturing processes and raw materials (96% of the total. Based on the 2014 installed capacity, the total mass of WTB is 78 kt, their energy consumption is 82 TJ and the carbon dioxide footprint is 4.35 Mt. These figures will provide a basis for suggesting possible solutions to reduce WTB environmental impact.

1. Introduction

Wind energy has developed rapidly over the last two decades to become one of the most promising economical and green sources of renewable energy, responding to concerns about use of fossil fuels and increasing demand for energy. The wind turbine industry is reaching maturity and is still growing steadily [1,2]. The first generation of commercial turbines are reaching the end of their design life, and attention is just starting to turn to the problem of what will happen to the waste as the generators are decommissioned. The environmental implications are significant, but at present there are no estimates of the potential magnitude of the problem. We are addressing one aspect of this, focusing on the blades. A large part of these high-value components is fibre composite, for which there is currently no satisfactory recycling route. The composites recycling industry will grow in the coming years, and one of its requirements will be estimates of the environmental impact the composites waste may cause.

A few life cycle assessments on wind turbines have been published. In 1999, Gürzenich [3] estimated the cumulative energy demand of a 1.5 MW 33 metre blade wind turbine to be around 14,000 GJ, varying slightly when the turbine is installed onshore, close-coastal or coastal. Subsequent studies have covered the environmental impact both of a single wind turbine and of a tens of megawatts wind farm [4–6]. There is a big range of results due to the different data sources and assumptions adopted. The scope of the studies is also limited to a couple of specific turbine models and so cannot provide a full picture of wind power ecological problems. In order to consider the technology development effect, Tremeac[7] compared the difference between 250 kW and 4.5 MW turbines. He found the energy consumption of the emerging turbines (4.5 MW) is much higher than that of an early stage model (250 kW), but the unit energy consumption is similar. Crawford [7] evaluated the energy consumptions of 850 kW and 3 MW turbines. He found the energy consumption to be significantly higher than previous studies and that the size of wind turbines did not appear to be an important factor in optimising their life cycle energy performance. In all these studies the environmental impact of the blades has considered only the major materials (fibre and resin), with rated power used to estimate turbine size, so a number of contributing factors have been omitted. The present study looks at the environmental impact of all stages of the wind turbine blade lifecycle with a spectrum of blade models and a range of industry development scenarios. Analysis of the contributing factors provides insight into ways in which the environmental impact can be reduced.

2. Approach

Bills of materials (BOM) were obtained for 20 different blades from three different wind turbine blade manufacturers. Eco-data from the 2015 CES Selector database [8] was used to calculate the energy consumption, carbon dioxide emissions and water consumption during the life cycle (manufacture, transport, operation and maintenance stages) for different sizes, materials and regions. Visual disturbance, noise and ecological disturbance are not considered in this study. The system boundary was limited to the blade factory, transport route and wind farm.

2.1. Bill of materials

The analysis included all the materials used in blade manufacture categorised as major materials, supporting materials and consumables as shown in table 1. Personal protection equipment (e.g. gloves and masks) and reusable manufacturing equipment (e.g. scissors and moulds) are not included.

Table 1. Materials listed in the bill of materials.

Major	Supporting	Consumable	
Carbon fibre UD	Steel accessories	Continuous filament mat	Resin flow pipes
Glass fibre UD	Copper accessories	Peel-ply/release film	T-fitting and infusion valve
Glass fibre multi-axial fabric	Aluminium accessories	Vacuum bag film	Mould cleaner and releasing agent
Resin	Balsa	Porous membrane	Hand Spray adhesives
Resin Curing agent	PVC	Flow mesh layer	Gel coat
Structural adhesives	Paint	Breather bleeder	
Structural adhesive curing agent	Putty	Vacuum bagging sealant tape	

2.2. Embodied energy

Data from the 2015 CES Selector [8] was used to calculate the embodied energy of the material in the blades. The CES data for glass fibre reinforced plastic (GFRP) and carbon fibre reinforced plastic (CFRP) was found to be in error, being much higher than the sum of fibre fabric and resin, so our calculations for these materials are based on data from the literature. The results of unit embodied energy, carbon dioxide emissions and water consumption are shown in figure 1.

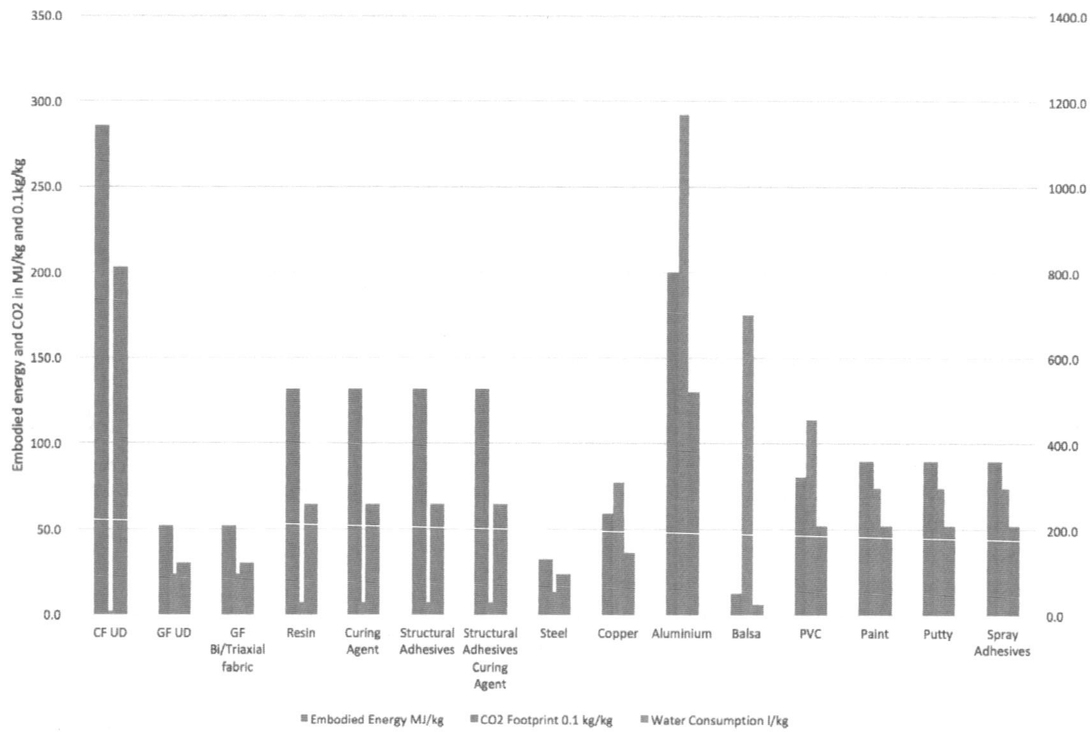

Figure 1. Eco-audit data.

2.3. *Life cycle energy*

2.3.1. Manufacturing. The material unit embodied energy multiplied by the material usage in the BOM was used to calculate the primary energy, carbon emissions and water consumption from the manufacture stage.

2.3.2. Blade size and transportation.

Table 2. Blade size class.

Class	Typical length (m)	Typical weight (tonnes)	Rated power (MW)
Early blade	Less than 40	6.50	1.5
Contemporary blade I	45	10.00	2-3
Contemporary blade II	55	14.00	3-4
Emerging blade	Over 60	25.00	6+

The large size of blades mean that transport options are limited to sea freight and road. Standard options for road transport are trucks capable of taking 14 or 32 tonne loads. Table 2 shows how blade weights have changed over time, divided into four classes. Although most blades are lighter than 14 tonnes and so could in theory be transported using the smaller 14 tonne trucks, due to their large size they normally need a large 32 tonne truck, and a truck can only carry only one blade a time. Transport energy is therefore calculated using data per kilometre for 32 tonne trucks, noting that they are not carrying their full payload. There is some difference in truck fuel efficiency when they carry full load or half load, but it is hard to quantify in this case. We assume the fuel consumption is the same as when the truck fully loaded. The CO2 footprint conversion is 0.071 kg/MJ in transportation process.

Table 3. Transportation eco data from CES 2015.

	Transport energy (MJ/tonne/km)	Transport energy per truck (MJ/ /km)
Sea freight	0.16	n/a
32 tonne truck	0.46	14.72
14 tonne truck	0.85	11.90

We set four scenarios to understand the range of energy consumption during the transportation stage. Generally, blade manufacturers try to make the blades close to regions where wind resources are abundant, where turbines are expected to be installed, to reduce the cost of transportation. For example, a Beijing blade manufacturer transports the blades 200 km by road to Zhangjiakou. However, local manufacture is not always possible so sometimes the blades must be transported over long distances, such as Beijing to Hami, around 2600 km. Sea freight is also commonly used. A Danish blade manufacturer will send blades from Esbjerg to London for installation, 600 nautical miles (one nautical mile is approximately 1.852 kilometres). In one extreme case a China manufacturer won a tender in Brazil; the blade factory is in Lianyungang, the sea freight will be around 13500 nautical miles. The results are shown in table 4.

Table 4. Transport energy consumption and CO_2 footprint.

	Energy (GJ)	CO_2 (t)
Transport Case road 200 km	2.94	0.21
Transport Case road 2593 km	38.17	2.71
Transport Case sea 600 nautical miles	1.78	0.13
Transport Case sea 13471 nautical miles	39.92	2.83

2.3.3. Operation and Maintenance. For operation and maintenance (O&M), we assume the blade life time is the 20 year design life. Routine maintenance and accidental damage are the two major waste sources in the O&M stage. Routine maintenance includes cleaning and minor and major repairs. During the initial few months of operation of new blades, some excess adhesive used in the blade manufacture becomes detached from the interiors of the blades and must be removed during blade maintenance periods. This is typically tens of kilograms for a 1.5 MW blade. Repainting and repair of small defects or stone damage are very common for most blades. Generally, 15 kg fibre and resin is enough for each of these minor repairs [9]. Major repairs only happen on specific blade batches and are usually caused by manufacturing defects or design defects. Such repairs typically involve re-strengthening work on major structures such as shell bonding, shear web bonding or the blade root. Each major repair job consumes tens to hundreds of kilograms of fibre, resin and adhesives [10].

Quite a few blades break in accidents due to extreme weather, with reports indicating that this causes failure in 1-3% of blades each year. Some failures need major repairs and some of them require replacement of the blade. These blades are treated as accidentally damaged blade waste.

Overall, the material usage in the maintenance stage is between 1.5% to 4.5% of the blade weight. Major materials used in this stage are fibre fabric and resin and we assume the usage ratio is 60% fibre and 40% resin. The equivalent energy consumption and CO_2 emission are presented in table 5.

Table 5. O&M energy consumption and CO_2 emission for a 10 tonne blade.

	Energy(GJ)	CO_2(t)
O&M@1.5%	12.6	0.66
O&M@3.0%	25.2	1.32
O&M@4.5%	37.8	1.97

2.3.4. End-of-life. A few possible recycling methods for fibre reinforced plastics have been identified in the literature, but the technology is not mature and most blades are currently sent to landfill. The environmental impact is assumed to be zero.

2.4. Energy payback time

Wind turbine energy generation (E_{gen}) is a function of rated power (P, in MW), capacity factor, and life span (t_s). Capacity factor is the ratio of actual energy production to full power theoretical production at continuous operation. The typical onshore wind turbine capacity factor (γ) is between 20-35%; we assume 30%. The rated power depends on the turbine model. The designed lifespan of a wind turbine blade is assumed to be 20 years. The annual energy production (AEP) is calculated as follows (24h in day, 365 days in year):

$$E_{gen} = P\gamma t_s \times 24 \times 365 = AEP \times t_s$$

Cumulative energy demand (CED) is an indicator for the overall life cycle environmental impact of many non-agricultural goods which is equivalent to energy consumption in this study. Energy payback time (EPBT) is a metric for the time taken for the system to generate the amount of energy required for its own manufacture, transport and installation, operation and disposal [11]. It is defined as:

$$EPBT = \frac{CED}{E_{gen}}$$

3. Results and discussion

The material usage and their proportion of the total blade energy of a typical small blade are shown in table 6.

Table 6. Material usage and energy consumption ratio of a 1.5 MW blade.

	Material by weight	Energy consumption
CF/GF fabric	60.4%	38.6%
Resin and adhesives	32.3%	56.7%
Steel	1.1%	6.0%
Copper	0.3%	2.5%
Aluminium	0.0%	0.6%
Balsa	2.3%	0.3%
PVC	1.7%	0.1%
Paint	0.9%	0.3%
Putty	0.7%	1.3%
Spray Adhesives	0.0%	1.3%

The total fibre content is around 60% of the blade weight and the resin with curing agent and adhesives is around 32%. As previous wind turbine LCA studies assume the materials in blades are 60% fibre and 40% resin [7,12], our result indicates that there is 8% difference between these studies and reality. Resin, curing agent and adhesives dominate the energy consumption (up to 57%). The unit primary energy of the fabric is lower, so despite the higher volume fraction of fibre its contribution to the whole blade energy is around 38%. Primary energies arising from other materials are low in comparison. The energy consumption of manufacture, transport and O&M are around 96.1%, 1.6% and 1.7% respectively.

Table 7. Eco results of 20 blade models from three manufacturers. Energy consumption, CO_2 footprint and water consumption are the results of one single blade. The energy payback time is calculated as the time taken for the turbine to generate energy equal to that required to produce three blades.

	Manufacturer A			
Model	40.2A-1.5	40.3A-1.5-IIIA	42.2A-1.5-IIIB	42.8B-1.5-IIIB
Rated power (MW)	1.5	1.5	1.5	1.5
Energy consumption (GJ)	627	606	699	689
CO_2 footprint (tonnes)	33.3	31.9	36.9	36.3
Water consumption (tonnes)	744	578	898	877
Total energy production (GJ)	283,824	283,824	283,824	283,824
Energy payback time (months)	1.59	1.54	1.77	1.75
	Manufacturer A			
Model	45.2A-1.5-IVB	45.2B-1.5-IV (Bolt embedded)	51.9-1.8-IV	56.8-2.0-IV
Rated power (MW)	1.5	1.5	1.8	2
Energy consumption (GJ)	795	775	924	1238
CO_2 footprint (tonnes)	42.1	41.1	48.7	65.1
Water consumption (tonnes)	989	959	1155	1458
Total energy production (GJ)	283,824	283,824	340,589	378,432
Energy payback time (months)	2.02	1.96	1.95	2.36
	Manufacturer B			
Model	DW93 (Carbon)	43.5-1.5	48.4-2	54.4-2
Rated power (MW)	2	2	2	2
Energy consumption (GJ)	1194	746	944	1129
CO_2 footprint (tonnes)	67.7	39.3	49.9	59.8
Water consumption (tonnes)	1079	1001	1227	1415
Total energy production (GJ)	378,432	378,432	378,432	378,432
Energy payback time (months)	2.27	1.42	1.80	2.15
	Manufacturer B			
Model	55-3.0	56.4-3.6	59.5-2.0	37.5-1.5-IIA+
Rated power (MW)	3	3.6	2	1.5
Energy consumption (GJ)	1299	1556	1461	624
CO_2 footprint (tonnes)	68.6	82.1	77.2	33.0
Water consumption (tonnes)	1599	1897	1845	741
Total energy production (GJ)	567,648	681,178	378,432	283,824
Energy payback time (months)	1.65	1.64	2.78	1.58
	Manufacturer C			
Model	40.25-1.5-IIIA+	42-1.5-IIIB-	47-1.5-IVB	51.5-1.5-IVB
Rated power (MW)	1.5	1.5	1.5	2
Energy consumption (GJ)	640	650	862	1049
CO_2 footprint (tonnes)	33.9	34.4	45.6	55.6
Water consumption (tonnes)	754	776	1038	1363
Total energy production (GJ)	283,824	283,824	283,824	378,432
Energy payback time (months)	1.62	1.65	2.19	2.00

The detailed blade energy consumption data for all models is presented in table 7. We find that longer blades generally have higher energy consumption, because they need more material to manufacture. As a secondary factor, the in-house waste level also varies between models and manufacturers which may be caused by inaccurate control of fabric and resin usage. Better control would reduce material usage, energy consumption and cost. Use of carbon fibre allows a lighter blade; this weight saving benefits the whole wind turbine system. However, the unit energy consumption of CFRP is 5.5 times to that of GFRP. When the full GFRP blade is compared with a similar size partial CFRP blade (CFRP spar cap, rest of the blade is GFRP), the partial CFRP blade energy consumption is around 50% higher than for a GFRP blade and the energy payback time is 15% longer (table. 8). NEEDS [13] predicts that carbon fibre is expected to account for up to 50% of fibres in blades by 2025, which would lead to a more serious environmental impact in the future. Putting this in context, however, we note that the blade primary energy accounts for 6-10% of the wind turbine energy [4,7], so this increment in blade energy consumption will not hugely affect the overall environmental impact of wind turbines. Carbon fibre is a high value material, so there is incentive for developing recycling routes which may provide future benefits.

Table 8. GFRP and CFRP blade comparison.

Model	45.2A-1.5-IVB (full glass fibre, GFRP)	45.3-DW93 (carbon fibre spar, GFRP+CFRP)	% increase of CFRP over GFRP
Total energy consumption (GJ)	795	1194	+50.3%
Total CO_2 footprint (tonnes)	42.1	67.7	+60.9%
Total water consumption (tonnes)	989	1,079	+9.1%
Energy payback time (months)	2.02	2.27	+12.7%

Our results differ from earlier results in the literature. Wagner [4] calculates that the energy consumption of an early (and therefore comparatively small) 1.5 MW-33 metre turbine rotor (3 blades plus cap) is around 1140 GJ and 8.4% of turbine energy consumption with turbine energy payback time at 6 months. For a 3 MW turbine, Crawford [7] estimates these numbers as 5050 GJ and 6% with nearly 12 months turbine payback time, but this result is questionable because it uses a glass fibre fabric embodied energy of 168 GJ/t which is double the number used in most literature. Our present study is expected to have greater accuracy because of the large sample size and up-to-date data.

4. Conclusions

- The energy payback time of blades is between 1.54 and 2.78 months, increasing with blade scale-up. Large wind turbines reduce the levelized cost of electricity (LCOE), but the blade environmental impact is growing rapidly.
- The same rated power blades can have environmental impacts differing by up to 46% (40.3 compared to 47-IVB). The reason is that the early stage blades are installed in higher wind speed sites, so the blades are relatively short. Newer blades are commonly installed in medium and low wind speed regions. Although the rated power is the same, the new blades are longer than the old ones with more material usage, and the environmental impact is increased.
- The energy consumption and carbon dioxide emissions of CFRP blades are much higher than GFRP blades. If blades contain up to 50% CFRP in 2025 as predicted, the environmental impact of WTBs will significantly increase compare to current calculations.
- Manufacturing waste varies in-house and between manufacturers. For example, if manufacturer A can reduce the manufacturing waste level to their best level, it could reduce the material usage between 6% and 23%. But the average manufacturing waste of manufacturer B is lower than others, so there is clearly scope for improving industry norms.
- The manufacture stage accounts for more than 96% of the whole blade life cycle energy consumption, with transport and O&M accounting for 1.6% and 1.7% respectively.

Acknowledgement
The authors would like to thank the Industrial Sustainability Research Group at the University of Cambridge and the industrial cooperation partners for advice and support. This work was supported, in part, by China Scholarship Council (CSC).

References

[1] IEA 2013 *Technology Roadmap: Wind Energy*
[2] GWEC 2014 *Global Wind Energy Outlook 2014*
[3] Gürzenich D, Mathur J, Bansal N K and Wagner H-J 1999 Cumulative energy demand for selected renewable energy technologies *Int. J. Life Cycle Assess.* **4** 143–9
[4] Wagner H J and Pick E 2004 Energy yield ratio and cumulative energy demand for wind energy converters *Energy* **29** 2289–95
[5] Vestas 2006 *Life cycle assessment of offshore and on shore sited wind power plants based on Vestas V90-3.0 MW turbines*
[6] Rashedi A, Sridhar I and Tseng K J 2013 Life cycle assessment of 50MW wind firms and strategies for impact reduction *Renew. Sustain. Energy Rev.* **21** 89–101
[7] Crawford R H 2009 Life cycle energy and greenhouse emissions analysis of wind turbines and the effect of size on energy yield *Renew. Sustain. Energy Rev.* **13** 2653–60
[8] Granta Design 2015 CES Selector 2015
[9] Zhang Z 2016 Private communication with technical director of KhanWind
[10] Sinoma Tech and LZFRP 2015 Private communication with blade aftersales manager of Sinomatech and LZFRP
[11] Merugula L, Khanna V and Bakshi B R 2012 Reinforced Wind Turbine Blades - An Environmental Life Cycle Evaluation *Environ. Sci. Technol.* **46** 9785–92
[12] Guezuraga B, Zauner R and Pölz W 2012 Life cycle assessment of two different 2 MW class wind turbines *Renew. Energy* **37** 37–44
[13] New Energy Externalities Developents for Sustainability (NEEDS) 2008 *RS 1a: Life cycle approaches to assess emerging energy technologies: Final report on off-shore wind technology*

IOP Conf. Series: Materials Science and Engineering **139** (2016) 012033 doi:10.1088/1757-899X/139/1/012033

Control and design of volumetric composition in pultruded hybrid fibre composites

Bo Madsen[1], Fariborz Hashemi[2] and Paridah Tahir[2]

[1] Section of Composites and Materials Mechanics, Department of Wind Energy, Technical University of Denmark, Denmark
[2] Institute of Tropical Forestry and Forest Products (INTROP), Universiti Putra Malaysia, Malaysia

E-mail: boma@dtu.dk

Abstract. Hybrid composites consist of two of more fibre phases in a common matrix phase. This is a challenge for the control and design of the volumetric composition and microstructural uniformity of such composites. In the present study, a model is presented for the prediction of the complete volumetric composition (i.e. volume fractions of fibres, matrix and porosity) in hybrid fibre composites. The model is based on a constant local fibre volume fraction criterion. Good agreement is found between model predictions and experimental data of pultruded hybrid kenaf/glass fibre composites with variable hybrid fibre weight mixing ratios. To demonstrate the suitability of the model, simulations are performed for four different cases of volumetric composition in hybrid kenaf/glass composites.

1. Introduction

The use of more than one type of reinforcement fibres in a common matrix is denoted as hybrid fibre composites [1]. Such composites have attracted the interest of researchers and industry due to the potential synergistic effect of having two (or more) types of fibres with different properties [2]. Furthermore, it makes it possible to allow for more optimal design of the properties of composites to meet the required specifications.

The properties of composites depends on a number of materials parameters, such as the fibre and matrix properties, the fibre orientation, the fibre/matrix interface properties, and the volumetric composition (volume fractions of fibres, matrix and porosity). The latter parameter is vital for reliable predictions of mechanical, physical, and thermal properties of composites [3]. In many cases, the volumetric composition of composites can be accurately controlled during manufacturing of composites, and as such, it can be used as a direct controllable parameter for the performance of composites.

Composites are heterogeneous materials consisting of a fibre phase and a matrix phase. In hybrid composites, the degree of heterogeneity is further increased by the addition of an extra fibre phase. Therefore, control and design of the volumetric composition in hybrid composite is one the main challenges in the manufacturing of such composites. Similarly, the higher degree of heterogeneity of hybrid composites is challenging with respect to controlling the microstructural uniformity. To the best of the author's knowledge there is just one previous study [2] on the analysis of the volumetric composition and microstructural uniformity of hybrid composites. In that study, hybrid glass/carbon fibre composites were manufactured by a vacuum infusion technique. The concept of so-called local fibre volume fractions was proposed, which is based on the volumes of the local non-hybrid composite regions in the hybrid composites. It was found that the local fibre volume fractions in the hybrid glass/carbon composites with different hybrid fibre mixing ratios were constant, and equal to the fibre volume fractions in the related non-hybrid composites. A model for the prediction of the fibre and matrix volume fractions in the composites was developed [2]. The model, however, was limited in assuming zero porosity in the composites.

In the present study, firstly, a generic model for the complete volumetric composition in hybrid fibre composites is developed. Secondly, the constant local fibre volume fraction criterion is used for controlling the volumetric composition and microstructural uniformity of pultruded hybrid kenaf/glass fibre composites. The experimental volumetric composition in the pultruded hybrid composites is analysed by the developed model. Finally, model simulations are made to demonstrate the suitability of the model for designing the volumetric composition in hybrid composites.

2. Theory

2.1. Development of model for volumetric composition in hybrid composites

In conventional non-hybrid composites, one type of fibres is embedded in a matrix. This is illustrated in Figure 1a showing an example of a unidirectional fibre composite where the single fibres are uniformly dispersed in the matrix. In hybrid composites, in principle, several different types of fibres can be combined, however, in practice, the combination of two fibres is the most common [4]. Since the two types of fibres are mixed and embedded in a common matrix, different fibre intermixing configurations exist. Figure 1b shows three different fibre intermixing configurations in hybrid composites: layer/layer mixing, bundle/bundle mixing, and fibre/fibre mixing. This exemplifies the range of possible dispersions from a non-uniform dispersion (layer/layer) to a uniform dispersion (fibre/fibre). The actual fibre intermixing configuration in hybrid composites depends on the configuration of the fibre performs, together with the applied composite manufacturing technique.

In both non-hybrid and hybrid composites, regardless of the fibre intermixing configurations, the volumetric composition is conventionally given by the volume fractions of fibres, matrix and porosity which are related to the total volume of the composite. This so-called *global volumetric composition* is illustrated in Figure 1c. However, as indicated in Figure 1b, and exemplified in details in Figure 2, the two fibre types in hybrid composites can be thought to form local and discrete non-hybrid composite regions consisting of only one fibre type. This leads to the concept of *local volumetric composition*, where the volume fractions of fibres, matrix and porosity are related to the volume of the local non-hybrid composite region. This is readily acceptable for the cases of layer/layer and bundle/bundle mixing configurations, where the distinction between the non-hybrid composite regions is clear; it has however a more conceptual meaning for the case of fibre/fibre mixing configuration. Altogether, as illustrated in Figures 1c and 1d, the global volumetric composition in hybrid fibre composites can be separated into two local volumetric compositions.

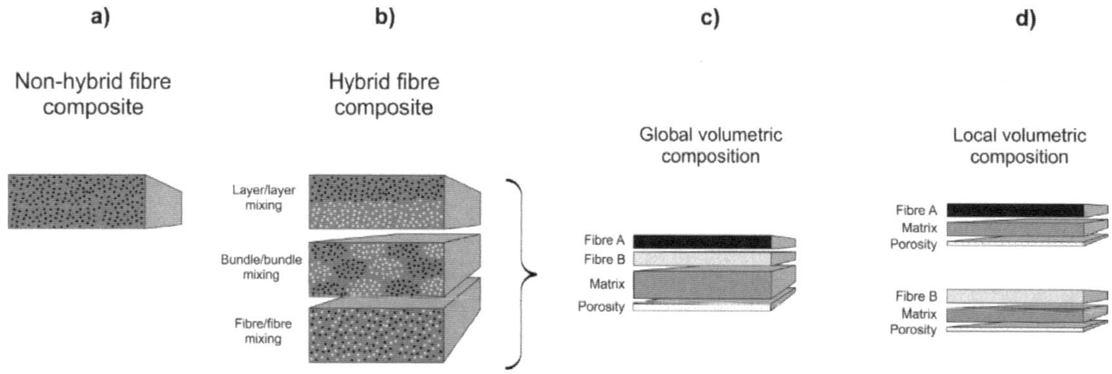

Figure 1. Schematised drawing of the typical fibre configurations in unidirectional (a) non-hybrid and (b) hybrid fibre composites. The concept of (c) global volumetric composition, and (d) local volumetric composition in hybrid fibre composites is illustrated by slab models.

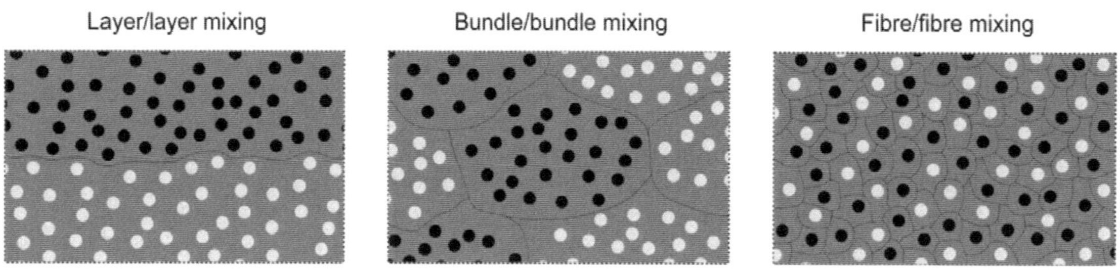

Figure 2. Schematised drawing of the two non-hybrid composite regions (belonging to the black and light gray fibres, respectively) in hybrid fibre composites with different fibre intermixing configurations.

Here follows a presentation of the developed model of the volumetric composition in hybrid composites. The volumetric composition is calculated as a function of the so-called *hybrid fibre mixing ratio*, and the *local fibre volume fractions are used as input parameters.*

The hybrid fibre mixing ratio is the ratio between the two fibre types in the hybrid composite, and it can be quantified by weight (β) or by volume (γ) of the fibres.

$$\beta = \frac{m_{fB}}{m_{fB}+m_{fA}} \quad (1)$$

$$\gamma = \frac{v_{fB}}{v_{fB}+v_{fA}} \quad (2)$$

where m_f is the fibre mass, v_f is the fibre volume, and the subscripts *A* and *B* denote the two fibre types. For β and γ equal to 0, the composite consists only of type A fibres, and for β and γ equal to 1, the composite consists only of type B fibres. Thus, β and γ equal to 0 or 1 define the situation of the two non-hybrid composites. Values of β and γ between 0 and 1 define the situation of hybrid composites with different fibre mixing ratios.

An equation for the relation between β and γ can be formed as following [2]:

$$\gamma = \frac{v_{fB}}{v_{fB}+v_{fA}} = \frac{\frac{m_{fB}}{\rho_{fB}}}{\frac{m_{fB}}{\rho_{fB}}+\frac{m_{fA}}{\rho_{fA}}} = \frac{\rho_{fA}\, m_{fB}}{\rho_{fA}\, m_{fB}+\rho_{fB}\, m_{fA}} \rightarrow \gamma = \frac{\rho_{fA}\,\beta}{\rho_{fA}\,\beta+\rho_{fB}\,(1-\beta)} \tag{3}$$

where ρ_f is the fibre density.

Before derivation of the model equations, some helpful support equations can be established as follows:

$$\gamma = \frac{v_{fB}}{v_{fB}+v_{fA}} = \frac{1}{1+\frac{v_{fA}}{v_{fB}}} \rightarrow \frac{v_{fA}}{v_{fB}} = \frac{1-\gamma}{\gamma} \tag{4}$$

$$\gamma = \frac{v_{fB}}{v_{fB}+v_{fA}} = \frac{V_{fB(\mathrm{II})}v_{c(\mathrm{II})}}{V_{fB(II)}v_{c(II)}+V_{fA(\mathrm{I})}v_{c(\mathrm{I})}} \rightarrow \frac{1}{\gamma} = \frac{V_{fA(\mathrm{I})}v_{c(\mathrm{I})}}{V_{fB(\mathrm{II})}v_{c(\mathrm{II})}}+1 \rightarrow \frac{v_{c(\mathrm{II})}}{v_{c(\mathrm{I})}} = \frac{V_{fA(\mathrm{I})}\gamma}{V_{fB(\mathrm{II})}(1-\gamma)} \tag{5}$$

where v is the absolute volume, V is the volume fraction, the subscript c is the composite, and the subscripts I and II designate the two non-hybrid composite regions for type A fibres and type B fibres, respectively.

Hereafter follows derivation of the model equations. The equations for the global volume fractions of the two types of fibres (V_{fA} and V_{fB}) are derived as follows:

$$V_{fA} = \frac{v_{fA}}{v_c} = \frac{v_{fA(\mathrm{I})}}{v_{c(\mathrm{I})}+v_{c(\mathrm{II})}} = \frac{v_{fA(\mathrm{I})}}{\frac{v_{fA(\mathrm{I})}}{V_{fA(\mathrm{I})}}+\frac{v_{fB(\mathrm{II})}}{V_{fB(\mathrm{II})}}} = \frac{1}{\frac{1}{V_{fA(\mathrm{I})}}+\frac{\frac{\gamma}{1-\gamma}}{V_{fB(\mathrm{II})}}} = \frac{V_{fA(\mathrm{I})}V_{fB(\mathrm{II})}(1-\gamma)}{V_{fA(\mathrm{I})}\gamma+V_{fB(\mathrm{II})}(1-\gamma)} \tag{6}$$

$$V_{fB} = \frac{v_{fB}}{v_c} = \frac{v_{fB(\mathrm{II})}}{v_{c(\mathrm{I})}+v_{c(\mathrm{II})}} = \frac{v_{fB(\mathrm{II})}}{\frac{v_{fA(\mathrm{I})}}{V_{fA(\mathrm{I})}}+\frac{v_{fB(\mathrm{II})}}{V_{fB(\mathrm{II})}}} = \frac{1}{\frac{\frac{1-\gamma}{\gamma}}{V_{fA(\mathrm{I})}}+\frac{1}{V_{fB(\mathrm{II})}}} = \frac{V_{fA(\mathrm{I})}V_{fB(\mathrm{II})}\gamma}{V_{fA(\mathrm{I})}\gamma+V_{fB(\mathrm{II})}(1-\gamma)} \tag{7}$$

The equations for the global volume fractions of matrix and porosity (V_m and V_p) are derived as follows:

$$V_m = \frac{v_m}{v_c} = \frac{v_{m(\mathrm{I})}+v_{m(\mathrm{II})}}{v_{c(\mathrm{I})}+v_{c(\mathrm{II})}} = \frac{v_{m(\mathrm{I})}+v_{m(\mathrm{II})}}{\frac{v_{m(\mathrm{I})}}{V_{m(\mathrm{I})}}+\frac{v_{m(\mathrm{II})}}{V_{m(\mathrm{II})}}} = \frac{1+\frac{V_{m(\mathrm{II})}v_{c(\mathrm{II})}}{V_{m(\mathrm{I})}v_{c(\mathrm{I})}}}{\frac{1}{V_{m(\mathrm{I})}}+\frac{\frac{V_{m(\mathrm{II})}v_{c(\mathrm{II})}}{V_{m(\mathrm{I})}v_{c(\mathrm{I})}}}{V_{m(\mathrm{II})}}} = \frac{1+\frac{V_{m(\mathrm{II})}V_{fA(\mathrm{I})}\gamma}{V_{m(\mathrm{I})}V_{fB(\mathrm{II})}(1-\gamma)}}{\frac{1}{V_{m(\mathrm{I})}}+\frac{\frac{V_{m(\mathrm{II})}V_{fA(\mathrm{I})}\gamma}{V_{m(\mathrm{I})}V_{fB(\mathrm{II})}(1-\gamma)}}{V_{m(\mathrm{II})}}} = \frac{V_{m(\mathrm{II})}V_{fA(\mathrm{I})}\gamma+V_{m(\mathrm{I})}V_{fB(\mathrm{II})}(1-\gamma)}{V_{fA(\mathrm{I})}\gamma+V_{fB(\mathrm{II})}(1-\gamma)} \tag{8}$$

$$V_p = \frac{v_p}{v_c} = \frac{v_{p(\mathrm{I})}+v_{p(\mathrm{II})}}{v_{c(\mathrm{I})}+v_{c(\mathrm{II})}} = \frac{v_{p(\mathrm{I})}+v_{p(\mathrm{II})}}{\frac{v_{p(\mathrm{I})}}{V_{p(\mathrm{I})}}+\frac{v_{p(\mathrm{II})}}{V_{p(\mathrm{II})}}} = \frac{1+\frac{v_{p(\mathrm{II})}}{v_{p(\mathrm{I})}}}{\frac{1}{V_{p(\mathrm{I})}}+\frac{\frac{v_{p(\mathrm{II})}}{v_{p(\mathrm{I})}}}{V_{p(\mathrm{II})}}} = \frac{1+\frac{V_{p(\mathrm{II})}v_{c(\mathrm{II})}}{V_{p(\mathrm{I})}v_{c(\mathrm{I})}}}{\frac{1}{V_{p(\mathrm{I})}}+\frac{\frac{V_{p(\mathrm{II})}v_{c(\mathrm{II})}}{V_{p(\mathrm{I})}v_{c(\mathrm{I})}}}{V_{p(\mathrm{II})}}} = \frac{1+\frac{V_{p(\mathrm{II})}V_{fA(\mathrm{I})}\gamma}{V_{p(\mathrm{I})}V_{fB(\mathrm{II})}(1-\gamma)}}{\frac{1}{V_{p(\mathrm{I})}}+\frac{\frac{V_{p(\mathrm{II})}V_{fA(\mathrm{I})}\gamma}{V_{p(\mathrm{I})}V_{fB(\mathrm{II})}(1-\gamma)}}{V_{p(\mathrm{II})}}} = \frac{V_{p(\mathrm{II})}V_{fA(\mathrm{I})}\gamma+V_{p(\mathrm{I})}V_{fB(\mathrm{II})}(1-\gamma)}{V_{fA(\mathrm{I})}\gamma+V_{fB(\mathrm{II})}(1-\gamma)} \tag{9}$$

In the equations above, γ is applied as the independent variable. In some cases, however, it might be more suitable to use β as the independent variable, since the two types of fibres usually are mixed and controlled by their weights in the hybrid composites. The relation between γ and β is given by Equation (3).

Having derived equations for the four global volume fractions, it can be verified that their summation is equal to one, as required:

$$V_{fA} + V_{fB} + V_m + V_p = 1 \rightarrow$$

$$\frac{V_{fA(\mathrm{I})}V_{fB(\mathrm{II})}(1-\gamma)+V_{fA(\mathrm{I})}V_{fB(\mathrm{II})}\gamma+V_{m(\mathrm{I})}V_{fB(\mathrm{II})}(1-\gamma)+V_{m(\mathrm{II})}V_{fA(\mathrm{I})}\gamma+V_{p(\mathrm{I})}V_{fB(\mathrm{II})}(1-\gamma)+V_{p(\mathrm{II})}V_{fA(\mathrm{I})}\gamma}{V_{fA(\mathrm{I})}\gamma+V_{fB(\mathrm{II})}(1-\gamma)}$$

$$= \frac{V_{fA(\mathrm{I})}\gamma\left(V_{fB(\mathrm{II})}+V_{m(\mathrm{II})}+V_{p(\mathrm{II})}\right)+V_{fB(\mathrm{II})}(1-\gamma)\left(V_{fA(\mathrm{I})}+V_{m(\mathrm{I})}+V_{p(\mathrm{I})}\right)}{V_{fA(\mathrm{I})}\gamma+V_{fB(\mathrm{II})}(1-\gamma)} = \frac{V_{fA(\mathrm{I})}\gamma+V_{fB(\mathrm{II})}(1-\gamma)}{V_{fA(\mathrm{I})}\gamma+V_{fB(\mathrm{II})}(1-\gamma)} = 1 \quad (10)$$

The concept of non-hybrid composite regions in the hybrid fibre composites allows for a definition of the volumetric composition of the hybrid fibre composites with respect to the non-hybrid composite regions. Here follows derivation of the equations for the volume fractions in the two non-hybrid composites:

$$V_{c(\mathrm{I})} = \frac{v_{c(\mathrm{I})}}{v_c} = \frac{v_{c(\mathrm{I})}}{v_{c(\mathrm{I})}+v_{c(\mathrm{II})}} = \frac{1}{1+\frac{v_{c(\mathrm{II})}}{v_{c(\mathrm{I})}}} = \frac{1}{1+\frac{V_{fA(\mathrm{I})}\gamma}{V_{fB(\mathrm{II})}(1-\gamma)}} = \frac{V_{fB(\mathrm{II})}(1-\gamma)}{V_{fA(\mathrm{I})}\gamma+V_{fB(\mathrm{II})}(1-\gamma)} \quad (11)$$

$$V_{c(\mathrm{II})} = 1 - V_{c(\mathrm{I})} \quad (12)$$

In the developed model of the volumetric composition in hybrid composites, Equations (6) – (9), the global volume fractions (V_{fA}, V_{fB}, V_m and V_p) are calculated as a function of the hybrid fibre mixing ratio (γ or β). The input parameters in the model are the local fibre, matrix and porosity volume fractions ($V_{fA(\mathrm{I})}$, $V_{m(\mathrm{I})}$, $V_{p(\mathrm{I})}$ and $V_{fB(\mathrm{II})}$, $V_{m(\mathrm{II})}$, $V_{p(\mathrm{II})}$). For hybrid composites manufactured by certain manufacturing techniques, *the local fibre, matrix and porosity volume fractions are constant for all hybrid fibre mixing ratios, and they can thereby be set equal to the fibre, matrix and porosity volume fractions in the two non-hybrid composites*. This was demonstrated in study by Beauson et al. [2] to be the case for hybrid glass/carbon composites manufactured by vacuum infusion. In the present study, it will be demonstrated also to be the case for hybrid kenaf/glass composites manufactured by pultrusion.

3. Materials and methods

3.1. Materials

Glass fibre rovings with a linear density of 4800 tex (g/1000 m) were purchased from China National Building Materials (Group) Corporation. Kenaf fibre rovings (spun yarn) with a linear density of 1000 tex were provided by KEFI (Malaysia) Sdn.Bhd. Unsaturated polyester resin (Reversol P-9771) was obtained from Revertex (Malaysia) Sdn.Bhd. Calcium magnesium carbonate (Dolomite) was used as filler (7 wt%), and methyl ethyl ketone peroxide (Butanox M-50) was applied as catalyst (1 wt%). In the further analysis and modelling of the manufactured composites, the matrix phase of the composites is taken to be consisting of both the polyester and the filler.

3.2. Design and manufacturing of composites

Six pultruded composites were manufactured: two non-hybrid kenaf and glass composites, and four hybrid kenaf/glass composites. In the above presented model equations, the glass fibres correspond to the A fibre type, and the kenaf fibres correspond to the B fibre type.

The composites were manufactured in the form of cylindrical rods with a diameter of 10 mm by using an industrial pultrusion machine (Pultrex, PX-500-6 T) at Innovative Pultrusion Sdn.Bhd, Malaysia.

To determine the number of rovings to be applied for the manufacturing of the non-hybrid kenaf and glass composites, the wanted fibre volume fraction in the composites was set to be 0.40 and 0.60 respectively. The number of rovings was then calculated by using the following equation:

$$V_f = \frac{v_f}{v_c} = \frac{A_f}{A_c} = \frac{n_r\ T_r}{\rho_f\ \frac{\pi}{4} d^2} \quad \rightarrow \quad n_r = \frac{V_f\ \rho_f\ \frac{\pi}{4} d^2}{T_r} \tag{13}$$

where A_f is the absolute fibre area, n_r is the number of rovings, T_r is the roving linear density, ρ_f is the fibre density, and d is the die diameter.

In the selection of number of rovings to be applied for the manufacturing of the hybrid kenaf/glass composites, it was assured that *the local fibre volume fractions in the hybrid composites was kept identical to the fibre volume fractions in the non-hybrid composites.* The derivation of Equation (19) for the calculation of the required roving numbers in the hybrid composites is shown next.

First, equations for the local fibre volume fractions in the hybrid composites are established:

$$V_{fA(I)} = \frac{v_{fA}}{v_{c\,(I)}} = \frac{A_{fA}}{A_{c\,(I)}} = \frac{n_{rA}\ T_{rA}}{\rho_{fA}\ A_{c\,(I)}}; \quad V_{fB(II)} = \frac{v_{fB}}{v_{c\,(II)}} = \frac{A_{fB}}{A_{c\,(II)}} = \frac{n_{rB}\ T_{rB}}{\rho_{fB}\ A_{c\,(II)}} \tag{14}$$

Then, equations for the fibre volume fractions in the non-hybrid composites are established:

$$V_{fA} = \frac{v_{fA}}{v_c} = \frac{A_{fA}}{A_c} = \frac{n_{rA}^0\ T_{rA}}{\rho_{fA}\ A_c}; \quad V_{fB} = \frac{v_{fB}}{v_c} = \frac{A_{fB}}{A_c} = \frac{n_{rB}^0\ T_{rB}}{\rho_{fB}\ A_c} \tag{15}$$

where the superscript 0 is applied to define the situation for the non-hybrid composites.

Finally, the governing Equation (19) can be derived:

$$V_{fA(I)} = V_{fA} \ \rightarrow \ \frac{n_{rA}\ T_{rA}}{\rho_{fA}\ A_{c\,(I)}} = \frac{n_{rA}^0\ T_{rA}}{\rho_{fA}\ A_c} \ \rightarrow \ \frac{n_{rA}}{n_{rA}^0} = \frac{A_{c\,(I)}}{A_c} \tag{16}$$

$$V_{fB(II)} = V_{fB} \ \rightarrow \ \frac{n_{rB}\ T_{rB}}{\rho_{fB}\ A_{c\,(II)}} = \frac{n_{rB}^0\ T_{rB}}{\rho_{fB}\ A_c} \ \rightarrow \ \frac{n_{rB}}{n_{rB}^0} = \frac{A_{c\,(II)}}{A_c} \tag{17}$$

Since by definition

$$A_{c\,(I)} + A_{c\,(II)} = A_c \rightarrow \frac{A_{c\,(I)}}{A_c} + \frac{A_{c\,(II)}}{A_c} = 1 \tag{18}$$

it follows that

$$\frac{n_{rA}}{n_{rA}^0} + \frac{n_{rB}}{n_{rB}^0} = 1 \tag{19}$$

Hence, by selecting the number of rovings in the hybrid composites (n_{rA} and n_{rB}) to meet the condition of Equation (19), it is assured that the local fibre volume fractions in the hybrid composites, for all hybrid fibre mixing ratios, are kept indentical to the fibre volume fractions in the non-hybrid composites. This will allow to have composites with high degree of structural uniformity, where each type of fibre will have a constant effect on the mechanical properties of the composite. The hybrid effect can be more easily evaluated, which is useful for the design of properties of hybrid composites.

Two series of composites were produced (composites a1-f1, and a2-f2) with various process conditions. Table 1 shows the specifications of the manufactured pultruded hybrid kenaf/glass fibre composites.

Table 1. Specifications of the manufactured pultruded kenaf/glass composites: number of fibre rovings, hybrid fibre mixing ratio (β), die temperature and speed of machine.

Composite code	No of fibre rovings		Hybrid fibre weight mixing ratio, β	Die temperature (°C)		Speed of machine (m/min)
	Kenaf n_{rB}	Glass n_{rA}		1st section	2nd section	
a1 / a2	0	25	0.00	120 / 90	150 / 140	0.68 / 0.40
b1 / b2	10	20	0.09	120 / 90	150 / 140	0.68 / 0.40
c1 / c2	20	15	0.22	120 / 100	150 / 145	0.68 / 0.40
d1 / d2	30	10	0.38	120 / 90	150 / 150	0.68 / 0.40
e1 / e2	40	5	0.63	130 / 100	150 / 155	0.68 / 0.40
f1 / f2	50	0	1.00	140 / 140	150 / 150	0.40 / 0.40

3.3. Determination of volumetric composition in composites

The fibre weight fractions in the composites was calculated based on the applied number of fibre rovings and their linear densities, together with the measured linear density of the pultruded composite profile. The density of the fibres, the cured matrix, and the composites were measured experimentally. The gas pycnometry method was used for density measurement of the kenaf fibres. The buoyancy method was used for density measurement of the glass fibres, the cured matrix and the composites. Based on the found values for fibre weight fractions and materials densities, the global volumetric composition of the composites was determined by using standard composite equations. More details can be found in the study by Hashemi *et al.* [5].

4. Results and discussion

4.1. Experimental volumetric composition in pultruded hybrid kenaf/glass composites

The average density (± stdv.) of kenaf fibres, glass fibres and cured polyester matrix (with fillers) were measured to be 1.57 (± 0.01), 2.60 (± 0.03) and 1.30 (± 0.01) g/cm^3, respectively. Table 1 shows the determined fibre weight fractions, density and volumetric composition of the manufactured two series of pultruded hybrid kenaf/glass composites. It can be seen that the measured kenaf and glass fibre volume fractions on 0.43 and 0.59 in the non-hybrid composites (at $\beta = 1$ and $\beta = 0$, respectively) are almost identical to the predefined kenaf and glass fibre volume fractions on 0.40 and 0.60 (see Materials and methods section). This shows that the creation of porosity in the composites has no effect on the resulting fibre volume fractions, since these are directly controlled by the applied number of rovings in the pultrusion process. This is due to the constant composite volume (as controlled by the die dimensions) in the pultrusion process. Furthermore, in Table 1, it can be seen that the porosity volume fraction in the non-hybrid glass composite is zero, and the porosity is then increased when the kenaf fibre volume fraction is increased. It has previously been shown that there is a linear relationship (R^2=0.96) between the porosity volume fraction and the kenaf fibre volume fraction [5]. The slope of the regression line was found to be 0.66. This slope is equal to the so-called fibre porosity factor (α_{pf}), which is an important parameter for controlling the porosity volume in composites [3]. As was shown in the study by Hashemi et al. [5], the porosity volume fraction in composite type f1 on 0.18 is noticeably lower than the established linear relationship, and therefore, in the further modelling of the experimental results, the results for f1 are not used as model input.

Table 2. Fibre weight fractions, density and volumetric composition of pultruded hybrid kenaf/glass fibre composites.

Composite code	Fibre weight fractions		Density (g/cm^3)	Fibre volume fractions		Matrix volume fraction	Porosity volume fraction
	Kenaf	Glass		Kenaf V_{fB}	Glass V_{fA}	V_m	V_p
a1	0.00	0.74	2.05	0.00	0.59	0.41	0.00
a2	0.00	0.74	2.05	0.00	0.58	0.41	0.00
b1	0.07	0.65	1.86	0.08	0.47	0.40	0.05
b2	0.07	0.65	1.85	0.08	0.47	0.40	0.06
c1	0.15	0.56	1.62	0.16	0.35	0.36	0.13
c2	0.16	0.56	1.65	0.16	0.36	0.36	0.12
d1	0.27	0.43	1.43	0.24	0.24	0.33	0.19
d2	0.25	0.40	1.52	0.24	0.24	0.41	0.12
e1	0.41	0.25	1.24	0.33	0.12	0.32	0.23
e2	0.42	0.25	1.26	0.34	0.12	0.32	0.22
f1	0.57	0.00	1.18	0.42	0.00	0.40	0.18
f2	0.62	0.00	1.08	0.43	0.00	0.32	0.25

4.2. Modelling of volumetric composition in pultruded hybrid kenaf/glass composites

Figure 3 displays the experimental volumetric composition as a function of the hybrid fibre weight mixing ratio (β) for the two series of composites. The model lines in Figure 3 are calculated by the presented volumetric composition model, Equations (6) – (9). The used input parameters are shown in Table 3.

Table 3. Input parameters for the modelling of experimental and simulated cases of volumetric composition in hybrid kenaf/glass fibre composites.

Case	Local volume fractions in glass composite region			Local volume fractions in kenaf composite region		
	Glass $V_{fA(I)}$	Matrix $V_{m(I)}$	Porosity $V_{p(I)}$	Kenaf $V_{fB(II)}$	Matrix $V_{m(II)}$	Porosity $V_{p(II)}$
Experimental case	0.59	0.41	0.00	0.43	0.32	0.25
Simulated case A	0.60	0.40	0.00	0.40	0.60	0.00
Simulated case B	0.60	0.40	0.00	0.50	0.50	0.00
Simulated case C	0.60	0.40	0.00	0.50	0.40	0.10
Simulated case D	0.60	0.35	0.05	0.50	0.40	0.10

As can be seen in Figure 3, there is an excellent agreement between the experimental data points for the kenaf and glass fibre volume fractions, and the model lines. This is due to the good control of fibre volume fractions in the pultrusion process, as also mentioned above. The fibre volume fractions in the hybrid composites are accurately given by the number of fibre rovings calculated by Equation (19). The experimental data points for the matrix and porosity volume fractions, however, deviate slightly from the model lines. The reason for these deviations can be explained by the typical scatter of porosity in composites. The model assumes that the local porosity volume fraction in the non-hybrid kenaf composite region is constant at 0.25 for all hybrid fibre mixing ratios (i.e. equal to the porosity volume fraction in composite f2, at β = 1). However, due to the stochastic element in the creation of porosity in composites, and the influence of many factors, porosity in composites cannot be accurately foreseen. Thus, the likely scatter of the local porosity volume fraction in the hybrid composites explains the slight deviations of the experimental matrix and porosity volume fractions from the model lines in Figure 3. The matrix and porosity volume fractions in the pultruded hybrid composites are directly correlated with each other; if one of them is increased (e.g. the porosity), the other one is decreased (e.g. the matrix).

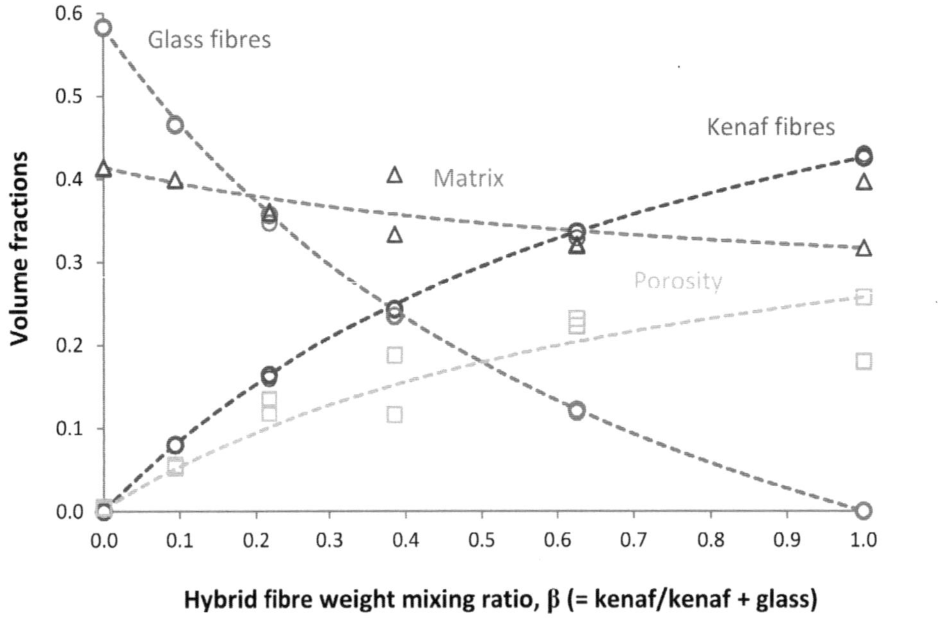

Figure 3. Volumetric composition in pultruded hybrid kenaf/glass composites as a function of the hybrid fibre weight mixing ratio. Shown are experimental data points and model lines.

4.3. Model simulations for different cases of volumetric composition in hybrid kenaf/glass composites
In this section, simulation of the volumetric compositions in hybrid kenaf/glass composites will be performed for four different cases: A, B, C and D. Table 3 shows the model input parameters for these cases. The plots with the model lines are shown in Figures 4-7.

In the simulated case A, the local kenaf and glass fibre volume fractions are set equal to the predefined values of 0.40 and 0.60, respectively, which are almost identical to the experimental values, and then, the local porosity in the kenaf composite is set equal to zero. Thus, case A reflects the situation where the process conditions have been optimised towards avoiding porosity in the kenaf composite region. The model lines are shown in Figure 4. In comparison to the experimental case (Figure 3), the matrix volume fraction in case A is not decreased as a function of β, but instead the matrix volume fraction is increased as a function of β .

In the simulated case B, the fibre volume fraction in the kenaf composite is set equal to 0.50. Thus, case B reflects the situation where the process conditions have been further optimised towards being able to increase the fibre volume fraction in the kenaf composites. The model lines are shown in Figure 5. The maximum obtainable fibre volume fraction in natural fibre composites is typical below the one for glass fibre composites due to the lower packing ability of natural fibres [3].

In the simulated case C, the porosity volume fraction in the kenaf composite is set equal to 0.10 to reflect the realistic situation of not being able to fully avoid porosity in these composites. One type of porosity in natural fibre composites is the one located inside the fibres, in the lumen [3]. The model lines for case C are shown in Figure 6. It can be observed that for these settings, the matrix volume fraction in the hybrid composites is constant at 0.40.

In the simulated case D, the porosity volume fraction in the glass composite is set equal to 0.05 to reflect the likely situation where some porosity is also present in the glass fibre composite. The model lines are shown in Figure 7. The model line for porosity does not start at zero at $\beta = 0$, but it starts at 0.05, and it is increased to 0.10 at $\beta = 1$. Thus, case D reflects the situation of two composites with different (non-zero) porosity contents.

Altogether, the four cases of model simulations, together with the good agreement with the experimental data, demonstrate that the model is suitable for controlling and designing the volumetric composition in pultruded hybrid composites. In further work, the model should be used together with micromechanical models for composites (e.g. see [6]), in order to calculate mechanical properties of hybrid composites as a function of the hybrid fibre weight mixing ratio. In this way, hybrid composites can be accurately designed to meet the mechanical specifications of given applications.

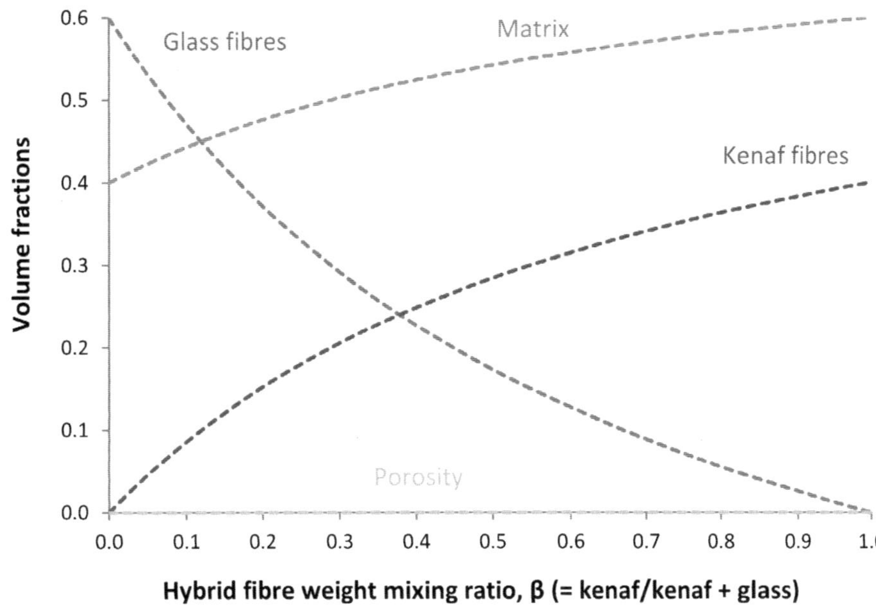

Figure 4. Simulated case A for volumetric composition in hybrid kenaf/glass composites.

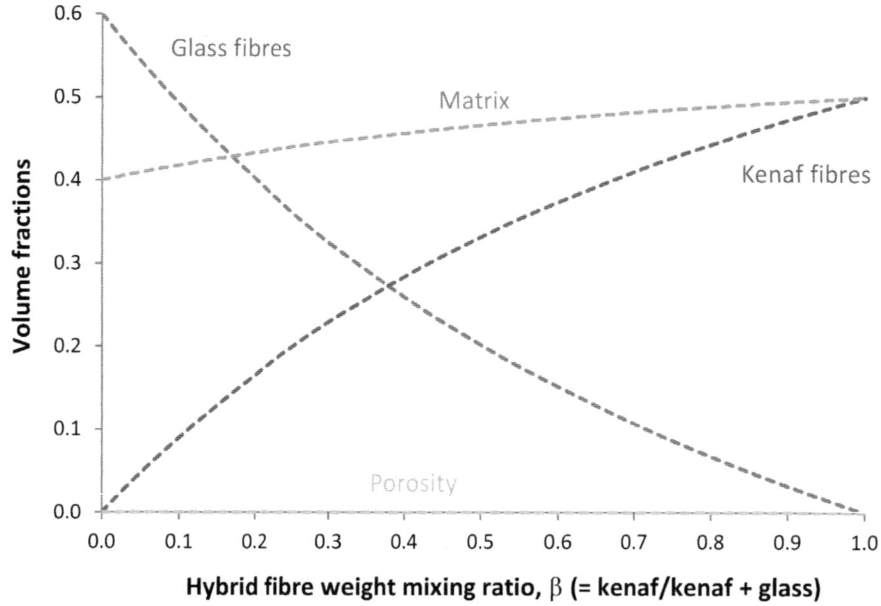

Figure 5. Simulated case B for volumetric composition in hybrid kenaf/glass composites.

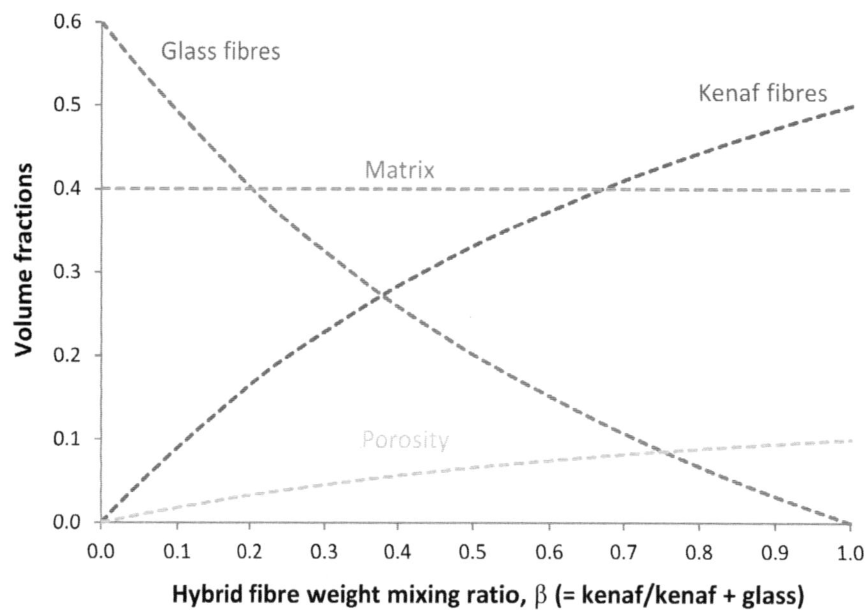

Figure 6. Simulated case C for volumetric composition in hybrid kenaf/glass composites.

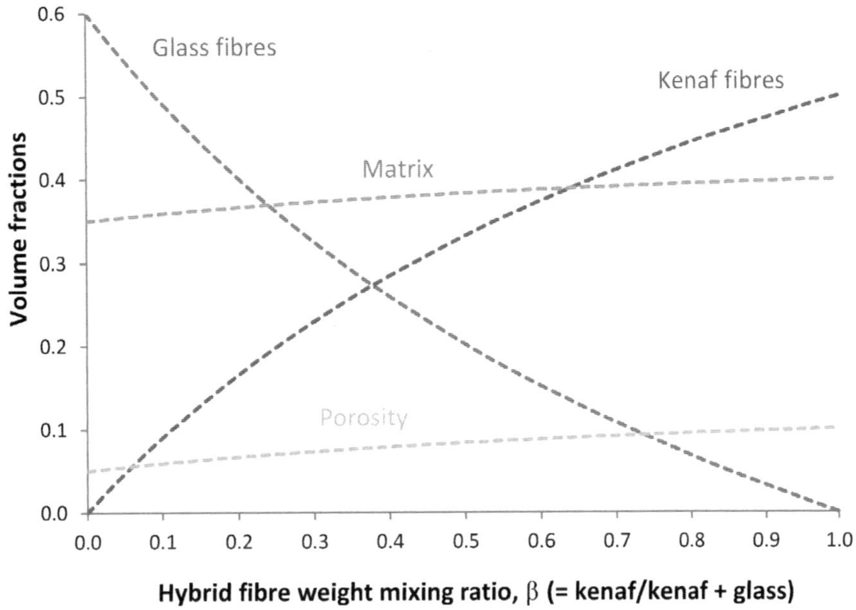

Figure 7. Simulated case D for volumetric composition in hybrid kenaf/glass composites.

5. Conclusions

In the present study, the constant local fibre volume fraction criterion is used for controlling the volumetric composition in pultruded hybrid kenaf/glass composites with five different hybrid fibre weight mixing ratios. Gravimetrical methods are used for the determination of the volumetric composition (i.e. volume fraction of fibres, matrix and porosity) in the composites. A generic model is presented for the prediction of the volumetric composition in hybrid composites, and good agreement with the experimental data is found. It is demonstrated that the model is a suitable tool for control and design of the volumetric composition in pultruded hybrid composites.

References

[1] Hayashi T 1972 On the improvement of mechanical properties of composites by hybrid composition. In: *Proc 8th Intl Reinforced Plastics Conference*, pp149-152

[2] Beauson J, Markussen C M and Madsen B 2013 Analysis of composition and microstructural uniformity of hybrid glass/carbon fibre composites. In: *34th Risø International Symposium on Materials Science*, pp 177-91

[3] Madsen B, Thygesen A and Lilholt H 2007 Plant fibre composites – porosity and volumetric interaction *Composites Science and Technology* **67** 1584-00

[4] Bunsell A R and Harris B 1974 Hybrid carbon and glass fibre composites *Composites* **5** 157-64

[5] Hashemi F, Tahir P M, Madsen B, Jawaid M, Majid D, Brancheriau L and Juliana A 2015 Volumetric composition and shear strength evaluation of pultruded hybrid kenaf/glass fiber composites *Journal of Composite Materials* **0(0)** 1-13

[6] Madsen B, Thygesen A and Lilholt H 2009 Plant fibre composites – porosity and stiffness *Composites Science and Technology* **69** 1057-69

IOP Conf. Series: Materials Science and Engineering **139** (2016) 012034 doi:10.1088/1757-899X/139/1/012034

Study on parameters affecting the mechanical properties of dry fiber bundles during continuous composite manufacturing processes

A Maier[1] and R Schledjewski[1, 2]

[1]Processing of Composites, Department Polymer Engineering and Science, Montanuniversitaet Leoben, Otto-Gloeckel Straße 2, A-8700 Leoben, Austria

[2]Corresponding author: ralf.schledjewski@unileoben.ac.at
Phone: +43 3842 402 2700, Web page: http://www.kunststofftechnik.at/composites/

Abstract. For continuous manufacturing processes mechanical preloading of the fibers occurs during the delivery of the fibers from the spool creel to the actual manufacturing process step. Moreover preloading of the dry roving bundles might be mandatory, e.g. during winding, to be able to produce high quality components. On the one hand too high tensile loads within dry roving bundles might result in a catastrophic failure and on the other hand the part produced under too low pre-tension might have low quality and mechanical properties. In this work, load conditions influencing mechanical properties of dry glass fiber bundles during continuous composite manufacturing processes were analyzed. Load conditions, i.e. fiber delivery speed, necessary pre-tension and other effects of the delivery system during continuous fiber winding, were chosen in process typical ranges. First, the strain rate dependency under static tensile load conditions was investigated. Furthermore different free gauge lengths up to 1.2 m, interactions between fiber points of contact regarding influence of sizing as well as impregnation were tested and the effect of twisting on the mechanical behavior of dry glass fiber bundles during the fiber delivery was studied.

1. Introduction

Composites are used in high amounts in different fields of application such as automotive, aerospace, marine, buildings, civil infrastructure, and sporting. High performance mechanical properties in combination with low weight are most often the reason to use composite materials. Although carbon fibers deliver best performance, glass fibers (GF) are due to lower prices the most widely used reinforcement in polymer matrix composites (PMCs). Glass fibers and glass fiber-reinforced plastics appear to behave as brittle materials. Already 1920 Griffith introduced the concept that glass contains pre-existing flaws, so that the fracture process is one of crack propagation rather than crack initiation. Weibull extended this concept by reasoning that the flaws are randomly distributed throughout the body and are of random severity. The "weakest-link" approach was adopted as a criterion of failure or a brittle material fails when the stress at any one flaw becomes larger than the ability of the surrounding material to resist the local stresses. Applying the statistical laws of probability, Weibull assumed a reasonable distribution function and derived an according expression [1,2,3].

In composite manufacturing processes the dry, i.e. not impregnated fibers are delivered, e.g. from a spool creel, to the point where the final component is formed. So, the dry fibers have to withstand tensile and bending forces due to friction, when in contact with guiding elements, or break systems, installed to receive defined pre-tensioning of the roving. For process optimization a good understanding about dry roving behavior is necessary. Therefore free gauge lengths of glass fiber bundles in dimensions relevant for continuous manufacturing processes in combination with influencing factors such as sizing, impregnation and shear stresses and strains on the maximum tensile loads of glass fibers were investigated in detail in this work.

2. Theoretical Background

2.1. Length effect

Based on the theory of the statistical probabilities for critical voids within the tested volume the tensile strength or the maximum tensile load is linked directly to the statistical failure probability, which in consequence means that the theoretical strength is the ideal strength and based on the increasing test volume (from short single fiber to bulk material) a decreasing strength will occur. The chance of the presence of a critical void inside a tested volume is increasing when the test volume is increasing. The amount of voids or defects, their shape and their distribution inside the test volumes are based on several facts as for example the manufacturing process of the test samples or the type of material itself but are in every tested volume statistically identical. Therefore the theory of the statistical failure distribution is only addressing the presence of one critical defect. That one critical defect leads, based on its higher stress concentration caused by the shape of the void in combination with the void size and position inside test volume, to a higher local stress σ_{local} than the global stress $\sigma_{nominal}$ and with a reduced cross section to an earlier catastrophic failure at lower applied forces F [1,2,3,4,5,6].

2.2. Interaction points

The interaction between single filaments within a fiber bundle and between fiber bundles is always a contact point between the fiber sizing. Therefore the interaction behavior is based on the mechanical and chemical interactions between the types of sizing.

The main purpose for sizing and finisher is to act as both protection of the raw glass fibers and bonding the organic functional groups to the matrix. These functional groups (R) have to assure chemical compatibility with the later used matrix resin [7,8,9]. Silane coupling agents are especially developed for glass fiber surfaces and drastically improve the bonding between glass fibers and epoxy matrices. Furthermore silane groups are perfectly protecting the raw glass fibers against water absorption. Therefore the coupling agents are performing a chemical bond between the glass fiber and the polymerized silane in a multilayer structure [10,11].

2.2.1. Friction. The friction force F_f, which is the fundamental characteristic of friction of two surfaces, is determined by the interaction of the two surfaces, whose real contact area will be denoted by S. Usually, the friction force is a function of the pressure p, the sliding velocity v, the temperature T, the contact time, and other external friction parameters [12,13]. According to prevailing ideas, the friction force can be divided, as shown in equation 1:

$$F_f = F_a + F_d \tag{1}$$

Where F_a is the adhesion component, which is based on the van der Waals interaction between the friction members, and F_d is the deformation component, which is associated in particular with the deformation of asperities induced in the counter body by the harder member of the friction pair [12,13].

2.3. Twisting

Whether using yarns, characterized by a slight twist when delivered, or using rovings, with an inside payout from the bobbin, more or less pronounced twist will result in compaction and shear effects between the filaments. Due to the fact that experimental shear tests are only possible for UD fiber-matrix laminates by using the indirect route via in-plane shear tests, one can imagine that testing the shear behavior of dry fiber rovings, which are not fixed into a matrix system, is even more complex and standardized tests are not available. The experimental testing of shear stress - shear strain behavior of composite laminates can become rather complex. For multi-directional fiber lay-ups and woven or braided fabrics, the V-notched rail shear test is a very common testing method [14]. However, this test cannot be recommended for measuring the shear stress-shear strain behavior of unidirectional laminates because the measured results are not reliable [14]. This results mainly from the fact that, depending on the position of the UD laminate within the testing device, either the matrix or the fibers are highly loaded by shear which is not representative for the shear behavior of the entire laminate. In order to be able to measure the shear stresses and strains of UD laminates nevertheless, the so called "in-plane shear test" can be used [15]. The test's name results from the test procedure, where the UD fibers are arranged under ±45° in a laminate and not a real shear load, but a tensile load is applied during the experimental test. The angle between the tensile load (0°) and the ±45° results in a stress state within the laminate similar to a shear test.

To analyze the effect of twisting/shear on dry glass fibers under tension the tests under twisting were conducted under close circumstances to the "in-plane shear test".

3. Materials and Sample Preparation

3.1. Materials

Glass fiber rovings used are standard StarRov® 086 4800tex E-glass fiber rovings from Johns Manville (Denver, United States) [16].The diameter of each filament was approximately 23 μm according to the manufacturer data sheet. A typical epoxy resin system was used to study the effect of impregnation. EPIKOTE™ Resin MGS LR 160 and EPIKURE™ Curing Agent MGS LH 502 by Hexion Specialty Chemicals B.V. (Rotterdam, Hoogvliet, Netherlands) were used.

3.2. Roving bundle sample preparation

To ensure a consistent load introduction into the specimens and an equal strain behavior of the roving, the specimens had to be prepared before inserting them into the tensile testing rig (Figure 1). This was done by using two-component epoxy glue and roughened aluminum tabs. To ensure that all single filaments were glued together low viscosity glue was used. To ensure that the applied test forces will be held by the epoxy matrix a special treatment was applied. During the bonding process a pressure of 1.5 bar and a temperature of 100 °C were applied for 100 seconds. Due to that process a bonding force of 2500 N/cm² was reached based on the two component epoxy data sheet. Specimens with free testing length between the plates of 50 mm, 150 mm, 300 mm, 500 mm, 800 mm, 1000 mm, and 1200 mm were produced.

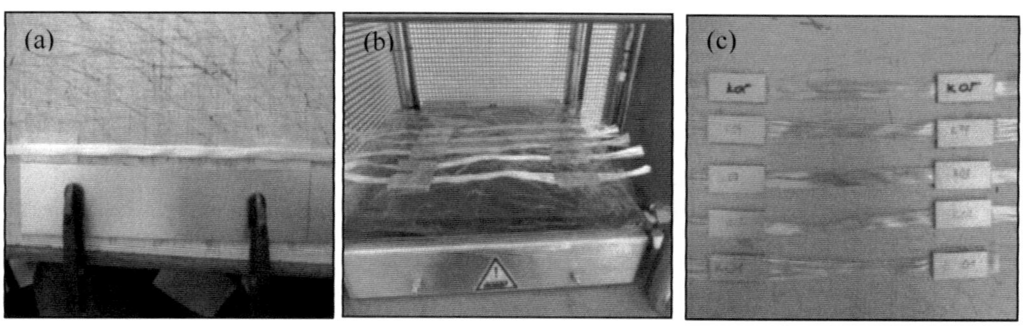

Figure 1. (a) Aligned E-glass roving and aluminum tabs (b) special tabs treatment with heat and pressure (c) resulting testing samples.

4. Experimental results

4.1. Effect of the strain rate on the mechanical properties

The strain rate dependency can be interpreted as a nearly linear increase of the maximum tensile load with an increasing strain rate as expected from the literature [17,18,19]. Based on preliminary tests and the information of the effect of strain rate dependency on the tensile load found in literature, a strain rate of $6*10^{-3}$ s^{-1} was chosen for all tests presented in this paper. This decision was affected mainly by the possible and controllable parameters given by the tensile test machines.

4.2. Effect of free gauge length on the mechanical properties

In Figure 2 the effect of different free gauge lengths on the maximum tensile loads and in consequence the survivability based on the statistically failure distribution of the rovings tested is shown in detail. Included in this figure are the measured data from 50 mm up to 1200 mm free gauge length, which is a scale reasonable to represent the fiber delivery in continuous manufacturing processes. Figure 2 clearly shows that with an increasing free gauge length the maximum tensile load decreases. Even at the very long specimens with more than 1000 mm free gauge length, this trend continues. The top of Figure 2 indicates a 100 % survivability of the fiber bundle which means that during all of the tests no fatal fiber failure was detected.

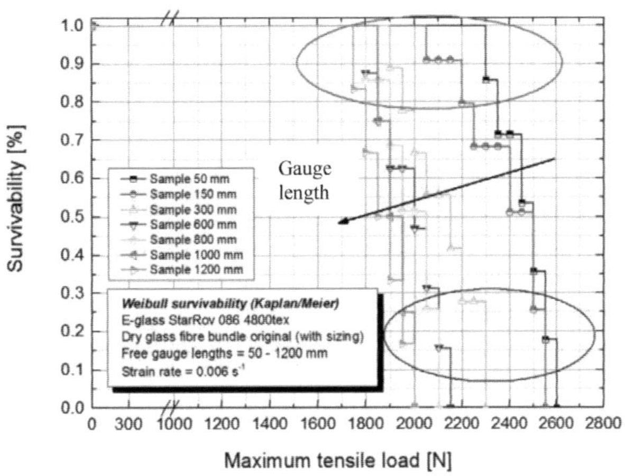

Figure 2. Survivability and maximum tensile load as a function of free sample length.

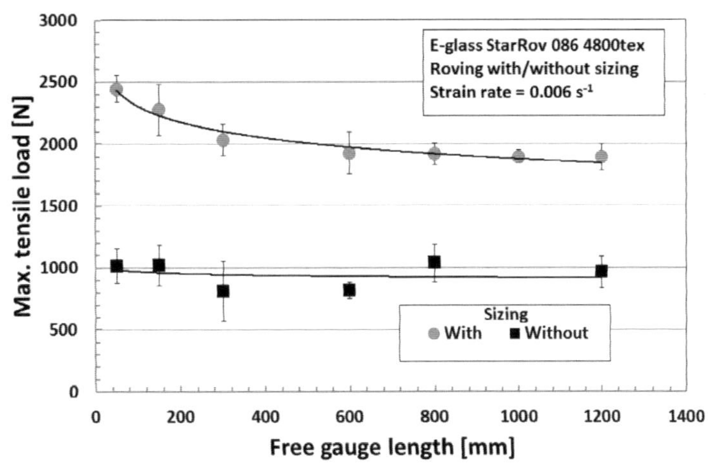

Figure 3. Comparison of glass fiber with and without sizing.

4.3. Effect of sizing
In order to investigate the effect of sizing, the sizing on the fibers as delivered was chemically removed by using piranha solution. Piranha solution acts on organic materials and it is expected any fiber sizing, finish or any other surface treatment done during fiber manufacturing is removed. The glass fiber itself is expected to be inert during such a treatment. Both, fibers as delivered and without sizing, were mechanically tested with different gauge length. The load-strain curves of the chemical treated fibers (without sizing) clearly showed that the missing sizing resulted in a decreased maximum tensile load from approximately 2000 N to less than 1000 N (Figure 3). An explanation for this effect is a more pronounced adhesive interaction of sized glass fiber surfaces compared to the glass fiber surfaces without sizing. Due to friction such adhesive interaction results in a failure behavior of the roving quite comparable to a composite material. The sizing acts here like a matrix and in case of failure of single filaments in the roving the load transfer into surrounding placed filaments is given.

4.4. Effect of twisting on the mechanical properties
During the tests to analyze the influence of twisting an interesting trend was found, showing that the maximum tensile load first increase when low twist is given and later on decrease with an increasing number of twists (Figure 4).

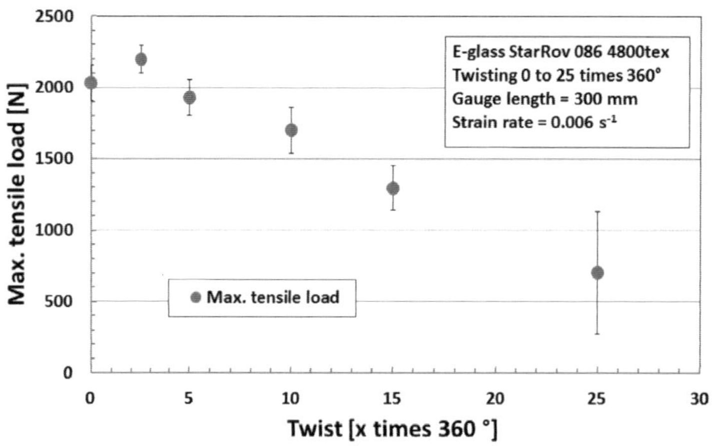

Figure 4. Analysis of the maximum tensile load with standard deviation over number of twist.

The increasing load capability is explained by a compaction of the fiber bundle in the twisted zone. Due to the compaction the adhesive inter-filament load transfer mechanisms are intensified. An increased tolerance against single filament failure due to load transfer into surrounding filaments leads to an increased maximum tensile load during testing. Since the rovings have been fixed by the aluminum tabs before twisting and the roving was glued having a distinct width, filaments located in maximum distance to the rotation axis during twisting will reduce the distance between the end-tabs (Figure 5). Consequently, filaments located more close to the rotation axis will contribute to load bearing during tensile test earliest when an according elongation of the specimen is given. Tensile testing of twisted rovings, as it has been done in this study, results in a stepwise filament load bearing and filament failure which results in an increase of elongation before final fatal failure of the roving (Figure 5). Due to this failure behavior only a part of the filaments is under load and these filaments do reach their maximum load capability already before further filaments do contribute to load bearing. The stiffness of highly twisted samples is significantly lower and the final maximum load reached during testing is significantly lower (Figure 5).

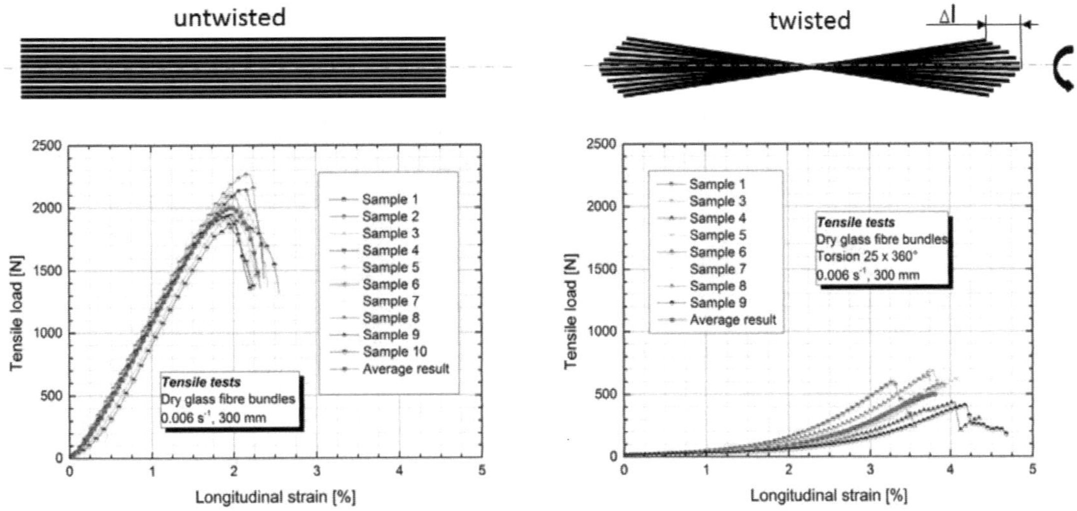

Figure 5. Effect of twist on failure behavior during tensile testing.

4.5. Effect of impregnation

Since the inter-filament load transfer is already identified to govern the failure behavior under tensile load a further comparison was made by using resin impregnated rovings. The resin impregnation was done directly before doing mechanical testing. No resin curing effects are assumed to affect the experimental results. Compared to the dry fiber bundles the impregnated ones result in significantly higher maximum tensile load (Figure 6). These results are quite in line with the results elaborated before (Figure 7). The ability to transfer the mechanical load from one filament to surrounding ones reduces the brittle failure behavior of the fiber bundles. Dry fibers without any surfaces covers (sizing) do only have distinct point contacts having only low adhesion effects. If the filaments are covered by polymeric sizing material, the adhesion improves and load transfer mechanisms are enhanced. Low roving twist results in higher compaction and as a result more pronounced intimate contact and further enhanced inter-filament load transfer is given. Further improvement is reached by impregnation. The resin will enhance the load transfer especially in the region of no direct contact between the filaments. Additionally, the resin might support relocation of the fiber position due to lubrication effects. This could result in more direct contact points or even line contact between the filaments.

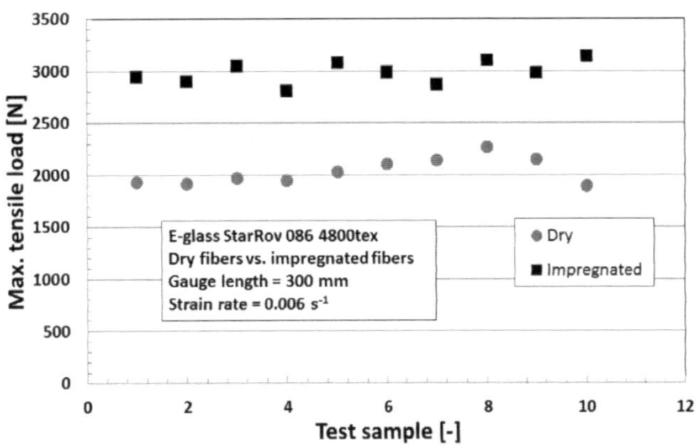

Figure 6. Maximum tensile load for dry and resin impregnated fiber bundles.

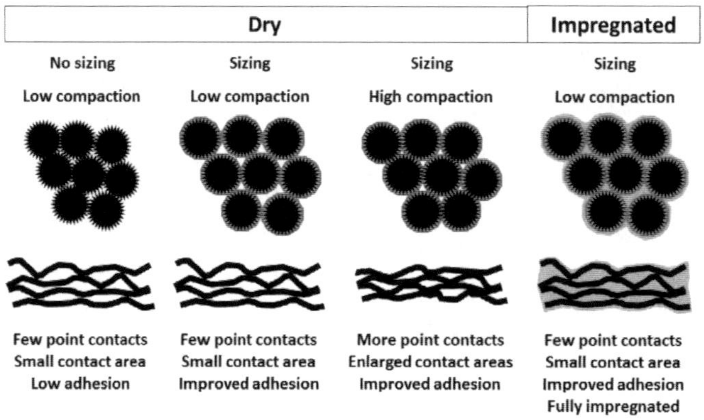

Figure 7. Effects of the variation of load transfer conditions

5. Conclusion

During fiber delivery in continuously running processes several different effects might become relevant. The load bearing behavior of the fiber bundles is influenced by (a) the inter-filament adhesion, (b) the compaction of the fiber bundle, (c) the amount and shape of interaction areas, (d) the free length the load is acting, and (e) whether the fiber bundle is already impregnated, or not.

The resulting maximum (tensile) load the roving can withstand varies in a wide range. For most processing configurations the fiber tension typically do not reach the shown limits, exemplarily elaborated for one roving grade in this paper. In some cases, e.g. if pre-tensioning during winding is given, these limits have to be known very well to ensure save fiber delivery. As shown by the results presented, in the configuration delivering the maximum tensile loads, i.e. impregnated fibers, the maximum tensile load of about 3000 N is equivalent to a tensile strength of about 1500 MPa if all filaments are similar loaded. This is less than half the level of typical tensile strength known for e-glass!

References

[1] Griffith A A 1920 *The phenomena of rupture and flow in solids (Philosophical transactions / Royal Society of London* ser. A, v. 221*)* (London)

[2] Weibull W 1939 A Statistical Theory of the Strength of Materials *Ceneralstabens Litografiska Anstalts Förlag* **1939**

[3] Lund M D 2010 Tensile Strength of Glass fibres *Dissertation* Section of Chemistry Department of Biotechnology, Chemistry and Environmental Engineering, Aalborg University

[4] Soric Z, Galic J and Rukavina T 2008 Determination of tensile strength of glass fiber straps *Mater Struct* **41** 879

[5] Andersons J, Joffe R, Hojo M and Ochiai S 2002 Glass fibre strength distribution determined by common experimental methods *Composites Science and Technology* **62** 131

[6] Neuber H 1985 *Kerbspannungslehre: Theorie der Spannungskonzentration ; genaue Berechnung der Festigkeit* 3rd edn (Berlin: Springer)

[7] Campbell F C 2010 *Structural composite materials* (Materials Park, Ohio: ASM International)

[8] Brown E N, Davis A K, Jonnalagadda K D and Sottos N R 2005 Effect of surface treatment on the hydrolytic stability of E-glass fiber bundle tensile strength *Composites Science and Technology* **65** 129

[9] Jones F R 1996 Interphase Formation and Control in Fibre Composite Materials *KEM* **116-117** 41

[10] Feuillade V, Bergeret A, Quantin J-C and Crespy A 2006 Relationships between the glass fibre sizing composition and the surface quality of sheet moulding compounds (SMC) body panels *Composites Science and Technology* **66** 115

[11] Feuillade V, Bergeret A, Quantin J-C and Crespy A 2006 Characterisation of glass fibres used in automotive industry for SMC body panels *Composites Part A: Applied Science and Manufacturing* **37** 1536–44

[12] Uetz H and Wiedemeyer J op. 1985 *Tribologie der Polymere: Grundlagen und Anwendung in der Technik: Reibung, Verschleiss, Schmierung* (München, Wien: Carl Hanser)

[13] Bartenev G M, Lavrent'ev V V, Lee L-H and Ludema K C 1981 *Friction and wear of polymers (Tribology series* v. 6*)* (Amsterdam, New York, New York: Elsevier Scientific Pub. Co.; Distributors for the U.S. and Canada, Elsevier/North-Holland)

[14] ASTM International 2005 Standard Test Method for Shear Properties of Composite Materials by V-Notched Rail Shear Method *(*ASTM D7078/D7078M-05*)* (West Conshohocken, PA, United States: ASTM International)

[15] ASTM International 2007 Standard Test Method for In-Plane Shear Response of Polymer Matrix Composite Materials by Tensile Test of a ±45° Laminate *(*ASTM D3518/D3518M-94*)* (West Conshohocken, PA, United States: ASTM International)

[16] JohnsManville StarRov: Direct Roving 086 (accessed 7 Dec 2015)

[17] Barré S, Chotard T and Benzeggagh M L 1996 Comparative study of strain rate effects on mechanical properties of glass fibre-reinforced thermoset matrix composite *Composites Part A: Applied Science and Manufacturing* **27** 1169–81

[18] Arao Y, Taniguchi N, Nishiwaki T, Hirayama N and Kawada H 2012 Strain-rate dependence of the tensile strength of glass fibers *J Mater Sci* **47** 4895–903

[19] Elanchezhian C, Ramnath B V and Hemalatha J 2014 Mechanical Behaviour of Glass and Carbon Fibre Reinforced Composites at Varying Strain Rates and Temperatures *3rd International Conference on Materials Processing and Characterisation (ICMPC 2014)* **6** 1405–18

IOP Conf. Series: Materials Science and Engineering **139** (2016) 012035 doi:10.1088/1757-899X/139/1/012035

Nonlinear time dependent behaviour of epoxy resins

C Marotzke and T Feldmann

BAM, Division 5.3, Mechanics of Polymers, Unter den Eichen 87, 12205 Berlin, Germany

christian.marotzke@bam.de

Abstract. The nonlinear behaviour of epoxy resins is studied on standard tensile tests. A strain field measurement system is applied (Aramis) in order to monitor local strains. The residual strain is measured by recovering the specimens for up to 68 hours after unloading. The time span the specimen is exposed to load has a large influence on the creeping process and the residual strain after recovering. This is studied by comparison of instantaneous unloading with keeping the specimen under permanent load for thirty minutes. It is shown that moderate differences in the initial strain can lead to large differences in the creep behaviour as well as in the residual strain.

1. Introduction

The failure initiation of fiber reinforced composites is strongly influenced by the mechanical behaviour of the matrix. Failure of plies with a high loading transverse to the fiber direction is initiated on microscale either by interface or matrix failure. For the simulation of microfailure a district description of the matrix material behaviour is needed. Since epoxy resins are widely used in high strength materials applications e.g. in aircraft, wind energy and automotive industry the present study is focused on them.

Epoxy resins possess a complex mechanical behaviour [1-5]. They exhibit a strong nonlinearity, they are highly time dependent and they can undergo large deformations. They behave as viscoelastic as well as viscoplastic solids. The superposition of time dependence and plasticity makes the determination of the respective material properties very elaborate. A further problem arising in the determination of material properties is that the deformation in tensile specimens is not homogeneous. The occurrence of large deformations is limited to a small zone. As a result, in one region the material may behave predominantly viscoelastic while in another region it behaves viscoplastic. This gives the classical tests such as relaxation and creep test a different meaning. The typical values of strain to failure found in literature range from four to eight percent. These values presumingly are not taking into account the local nature of plasticising. Tensile tests show that at a certain load large deformations develop leading to local strains of more than 50%. The contribution of these high deformations to the total change in length of the specimen, however, is small due to the short length of the zone. It must be doubted that all strain to failure values given in literature are indeed true local strains but rather averages over the specimen length.

The stress/strain behaviour depends on various parameters. That means, for example, that the residual strain left after unloading depends on the level of the strain applied, the duration of the loading process and the time the material could recover after unloading. When we classify the material behaviour as viscoelastic or viscoplastic we have to keep in mind that these material models are still idealised models of the actual mechanical behaviour. The present study does not deal with the development of constitutive equations but is restricted to a phenomenological level. The tests performed so far do not represent a full matrix of parameter variations. The study aims to identify the governing parameters in order to design a test program in an appropriate manner.

2. Specimen preparation

The resin used for the present study is a L135/H137 epoxy system manufactured by Hexion. The tests are performed using standard dogbone tensile specimens with a total length of 150 mm, a width of 10 mm and a thickness of 4 mm. The specimens are cut out of plates by a milling process. The tests are performed by displacement control with a loading speed of 1 mm/min. Since the material can undergo large viscoplastic deformation the formulation utilises Hencky strains (logarithmic strains). In order to allow local strain measurement a strain field measurement method (digital image correlation) is applied. In the present study the Aramis system is used.

The specimens are prepared for the strain field measurement by applying a black/white patterns on the specimens. The patterns cover the full width and have a length of 65 mm. The distance of the grid lines of the Aramis digital imaging system is 1 mm in either direction. The strains are evaluated at specified lines in longitudinal and in transverse direction (figure 1). Line 1 is located in the center of the zone where large deformations appear. In this region the material behaves viscoplastic. If strain values are used in the following for the stress/strain curves the averages over the cross sections are taken. Line 4 and 5 are located on and near the edge of the viscoplastic zone. Line 0 is located in a region where the material behaves viscoelastically. In order to visualise the strain distribution in longitudinal direction of the specimen one line is located in the middle of the specimen (line 6). In this case the strain on every point is taken, this is, no averaging is done.

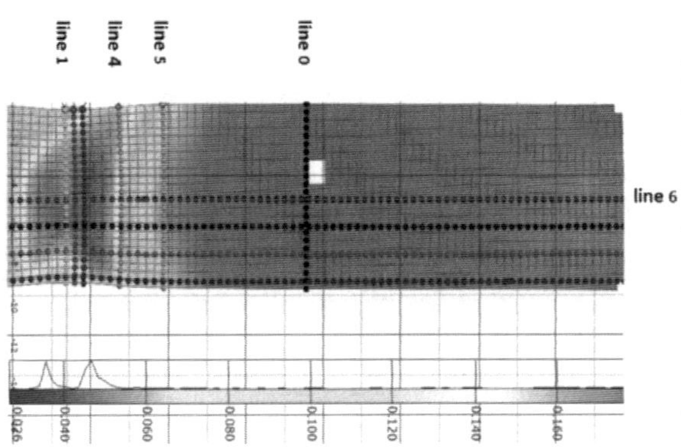

Figure 1: Positions of Aramis strain data lines

3. Stress / displacement and stress / strain curves

The typical stress/displacement behaviour at a constant loading rate of 1 mm/min is shown in figure 2. The specimen is loaded by displacement control with a rate of 1 mm/min. The engineering stress (red line) linearly increases in the first phase of the loading process which corresponds to a basically linear elastic behaviour with evanescent small time dependent deformation. Then the slope decreases and the curve becomes strongly nonlinear leading to a substantial decrease of the engineering stress. The nonlinear behaviour is caused by primarily three effects. First, the time dependence of the mechanical properties grows with the strain level. Second, additional viscoplastic deformations develop at a certain strain level. Third, the cross section decreases leading to an increase of the compliance. The decrease of the cross section necessitates the discrimination between true stresses and engineering stresses. The lateral strain is also measured by the Aramis system allowing the calculation of the true stresses in the high strain zone. Since the material is isotropic the change of the thickness can be calculated directly from the change of the width. The deviation of the true stresses (figure 2a, blue line) from the engineering stresses (red line) becomes significant after 150 seconds loading time which corresponds to an average strain of roughly 2%. When the engineering stresses drop down the true stresses also decrease rapidly but only for a short time. Then the true stresses increase again until specimen failure where the stresses slightly supersede the prior maximum. At failure the difference between the true stresses and the engineering stresses is maximal reaching about 30%.

The different material behaviour in the viscoelastic and in the viscoplastic zone is visualised by the stress/strain curves in the respective zones (figure 2b). The engineering and the true stresses are plotted against the strain in the viscoelastic as well as in the viscoplastic zone. Since the maximum longitudinal strain in the viscoelastic zone is 4.7% the true stress is about 9% higher than the engineering stress. After the stress maximum is reached the strain in the elastic zone even decreases. This is caused by a strain redistribution in the specimen due to large deformations in the viscoplastic zone which is addressed in a later section. The longitudinal strain in the viscoplastic zone grows very rapidly and reaches more than 40% (figure 2b).

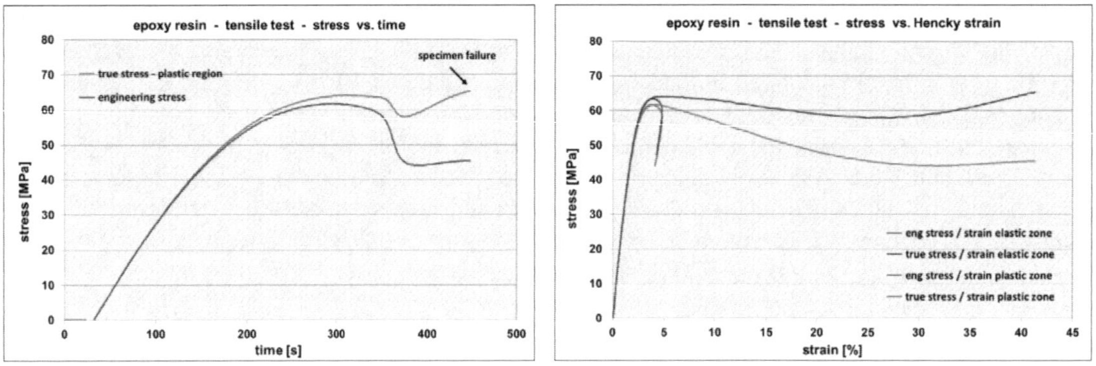

Figure 2 a): Stresses vs. displacement
engineering stress (red line),
true stress (blue line)

b) Engineering stress / true stress vs. strain
viscoelastic zone (blue line, red line),
viscoplastic zone (green line, violet line)

4. Strain distribution in the specimen

The strain distributions delivered by Aramis in and around the high strain region are shown in figure 3. The plots correspond to the last step before ultimate specimen failure occurs. On the left hand side the strains in longitudinal direction (loading direction) are shown. The necking is very pronounced (true scale, no graphical enlargement). The strains decrease slightly towards the edges of the specimen. This is caused by the change from plane strain to plane stress conditions. The transverse strains are negative, this is, the lowest strains appear as red and vice versa. As expected, the transverse strain

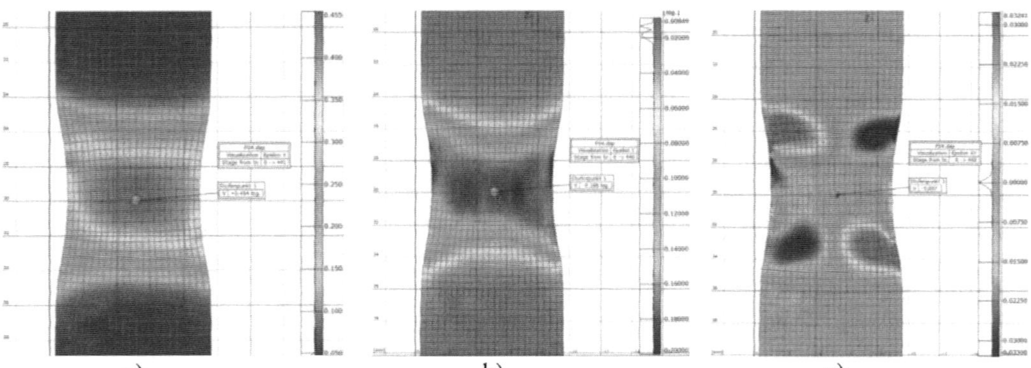

a) b) c)

Figure 3: Strain distribution in the center part of specimen: a) longitudinal strain; b) transverse strain; c) shear strain distribution in general is very similar to the longitudinal strain. On the rims of the viscoplastic zone (e.g. yellow region), however, the strains decrease a little faster towards the edges of the specimen. Substantial shear strains arise at the rims of the viscoplastic zone in the vicinity of the specimen edges which is due to the change of the width of the specimen. Even though the shear strains are much lower compared to the maximum longitudinal and transverse strains they are of the order of the longitudinal strains in the viscoelastic region.

4.1 Strain distribution along the specimen during loading
In order to get an overview over the development of the longitudinal strains during the loading process the strain along the center line (line 1) is examined. Figure 4 shows the strains at different time steps. One has to keep in mind that the coordinate system Aramid refers to is the actual coordinate system, that is, the Eulerian coordinate system. This means that during the loading process a point fixed on the specimen, this is, in terms of continuum mechanics a material point, is shifted. The shift of the position however is moderate. The center point of the zone where large deformations occur shifts within the relevant time span (300s to 448s) by 1.6 mm.

During the increase of the load (applied displacement) a zone with large deformations forms at a certain load. The zone is very limited and virtually does not extend during the further loading process. A typical experiment is shown in figure 4. In the first phase of loading (strain below 4%) the longitudinal strain is distributed basically homogeneously in the specimen. The strain grows proportional to the applied displacement. When the strain exceeds 4% (time step 300, green line) the strain starts to grow faster in a limited zone of about 10 mm in length. With further increase of the applied displacement the strain rapidly grows and reaches a maximum of 46% just before specimen failure. This corresponds to a linear strain of more than 58%. The high strain zone does not substantially extend with the time. Outside this zone the strain changes only moderately. This shows that the epoxy resin has the ability to locally undergo large local deformations. The question is which portion of the deformation is viscoelastic, this is, will recover with the time and how big is the viscoplastic portion. These questions will be addressed later.

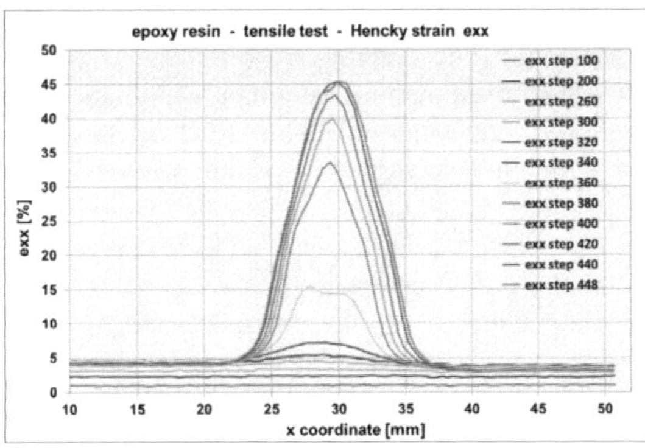

Figure 4: Strain along central line at different time steps (load steps)

4.2 Time dependence of strains at different locations
The course of the strain versus time is analysed at different cross sections along the specimen (figure 5). The data represent the averages over the respective cross sections. Over a long time span the strains measured at any position increase linearly with the time. Since the tests are displacement controlled the time is equivalent to the applied displacement, e.g. to the load. At about 4% strain the curves begin to differ. The strains at two cross sections start to increase very rapidly. These lines correspond to the center (red line) and the rim (yellow line) of the viscoplastic zone. Both strains grow simultaneously up to about 12% which is reached at 350s. At this point the strains at the rim of the viscoplastic zone

start to grow slower. The strains outside the high strain zone even decelerate while the strain in the viscoplastic zone still rapidly grows.

The strain in the viscoelastic region (light blue line) only slightly increases after the 4% mark. When the viscoplastic strains start to grow very fast the strains in the viscoelastic region even slightly decrease. The strain in the cross section closer to the viscoplastic zone grows a little faster and reaches 10% in the moment of ultimate failure. The decrease of the strain in the viscoelastic region is caused by the over proportional growth of the strain in the high strain zone. Since the external displacement is prescribed a redistribution of the strain field has to take place.

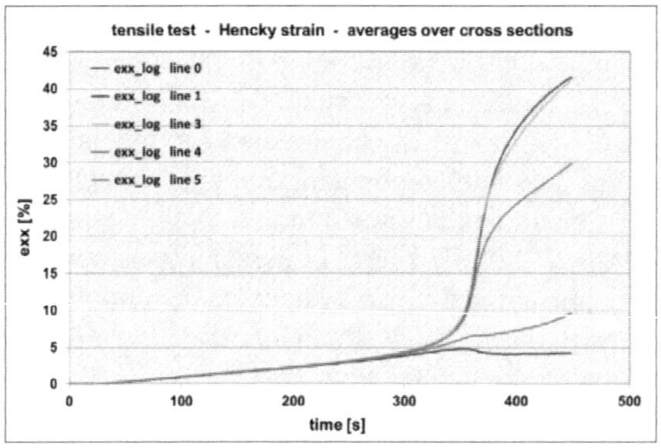

Figure 5: Strain vs. time; average over cross sections at different positions along the specimen

4.3 Strain recovery after unloading

The strain in the high strain zone is a superposition viscoelastic and viscoplastic strain. In order to determine the viscoplastic portion a series of tests was performed where the specimens were unloaded and the strain recovery was monitored. Either the specimens were immediately unloaded or they were exposed to a permanent load before unloading. It's a matter of fact that the strain distribution along the specimen monitored by the digital imaging system is not known during the test. Accordingly it is not possible to reach a specified strain in the respective zones. As a result, the strain levels of different tests cannot be predefined and accordingly underlie some fluctuation. This reduces the accuracy in the estimation of the influence of parameters such as the duration of loading.

4.3.1 Immediate unloading, high initial strain, small residual strain First, the strain recovery is monitored for instantaneous unloading. At a certain displacement, which corresponds to a maximum strain of 11.2% (see below) the specimen is rapidly unloaded with a rate of 85mm/min. Subsequently the strain is monitored over 68 hours. The strain versus time is evaluated for two different zones in the specimen (figure 6a), the viscoelastic (green line) and the viscoplastic zone (red line). For low strains the curves more or less coincide. After the strains have reached 2% the strain in the viscoplastic zone grows increasingly faster while the growth of the strain in the viscoelastic zone is decelerated. Just before unloading the strain in the elastic zone even decreases. This is due to the fact that the strain in the viscoplastic zone grows faster than the external displacement. As already mentioned, this causes a redistribution of the strain in the specimen

In order to improve the visibility of the loading process the strains are displayed on two different time scales in figure 6b. The dotted lines refer to the enlarged time scale representing the first 40 minutes of the recovery phase, the continuous lines correspond to the full time span of 68 hours. Just before unloading the strain in the viscoplastic zone is 11.2% while in the viscoelastic zone it is 4.1%. When the specimen is unloaded both strains instantaneously decrease. Within the first 40 minutes the strain in the viscoelastic region vanishes. The viscoplastic strain decreases to 3.4% which is a loss of 70%. At the end of the recovery process the strain in the viscoplastic zone is reduced to 2,7%. This shows that more than 75% of the initial strain is viscoelastic and the plastic portion of the strain is rather small. A completely different behaviour is encountered if the specimen is further exposed to load which is studied in the next section.

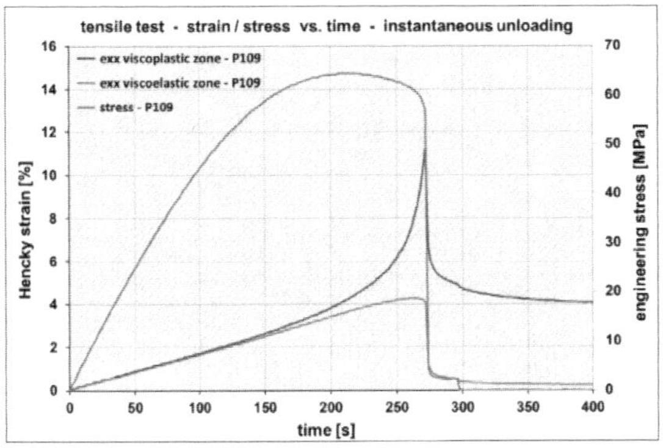

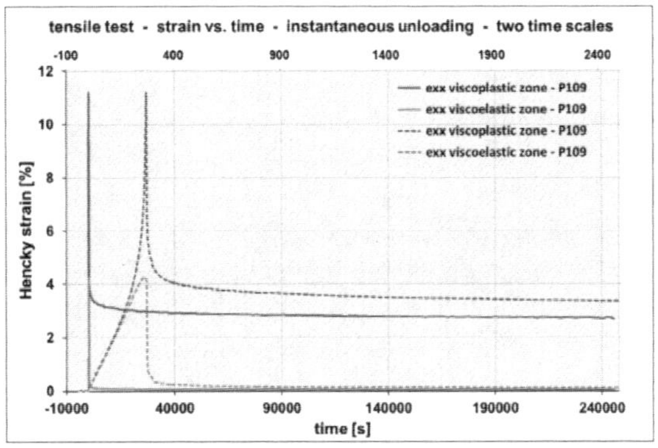

Figure 6a): Instantaneous unloading: stress/strain vs. time in viscoplastic and and viscoelastic zone; initial phase **b)** Instantaneous unloading: strain vs. time in viscoplastic and viscoelastic zone; displayed on two different time scales

4.3.2 Permanent load, higher initial strain, high residual strain In a second test the influence of the time the specimen is exposed to load is studied. To this end the applied displacement is kept constant at a certain level for a specified time span. Keeping the displacement fixed does not give a classical relaxation test since the strain in the specimen is not homogeneous. As a result, it is not possible to control the strain of the total specimen. Alternatively the strain in a certain region could be controlled by using an additional strain measurement tool such as a strain gauge. However, different strains cause a different creep behaviour. Keeping the strain constant, e.g. in the viscoelastic region, does not prevent creeping in the viscoplastic zone. As a result, the stress relaxation is not only due to the relaxation in the region of interest, i.e. where the strain is kept constant but an additional decrease of the stress occurs due to the loss of stiffness caused by creeping in the viscoplastic region. Another alternative is to control the external displacement which is equivalent to prescribe the average strain in the specimen. In this case again a redistribution of the strains occurs in the specimen also resulting in different creep and relaxation processes in the respective regions.

The specimen is loaded up to a specified displacement (4.2 mm) corresponding to a stress of 62.1 MPa (figure 7a, blue curve). The displacement is held constant for 34 minutes. During this time span the stress relaxes down to 36.1 MPa which is a loss of more than 40%. In the moment when the displacement is kept constant the strain in the viscoplastic zone is 6.7% while it is 4.3% in the viscoelastic zone. While the strain in the viscoelastic zone decreases to 3.0%, which is a loss of 30% the strain in the viscoplastic zone enormously grows due to creeping and reaches 29.2% at the end of this phase. This means that the strain grows by 330%. The creeping of the viscoplastic zone enforces a decrease of the strains in the viscoelastic zone since the average strain is constant. This means that three processes are active simultaneously: relaxation, creep and strain recovery.

After the end of the load phase, this is, the increase of the external displacement together with the holding phase, the specimen is unloaded and the strain is monitored for another 12 hours. The strains in the viscoplastic and in the viscoelastic zone decrease instantaneously by 1.7 and 2.5%, respectively (figure 7a). The complete recovery process is shown in figure 7b. While the strain in the viscoelastic region levels off during the recovery phase the strain in the viscoplastic zone recovers by only 5% yielding to a residual strain of 22%. This means that during the permanent loading phase the material has experienced large plastic deformations.

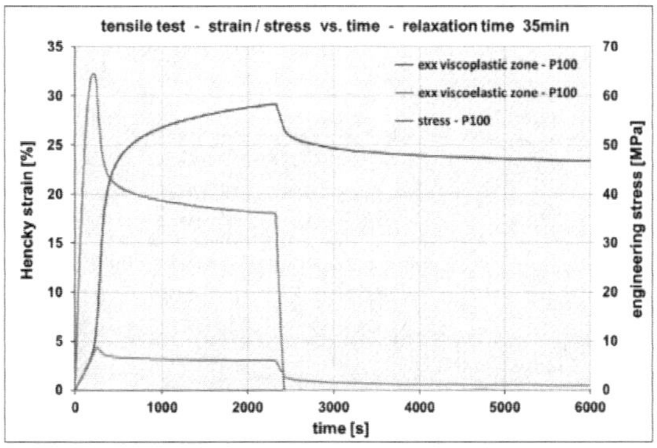

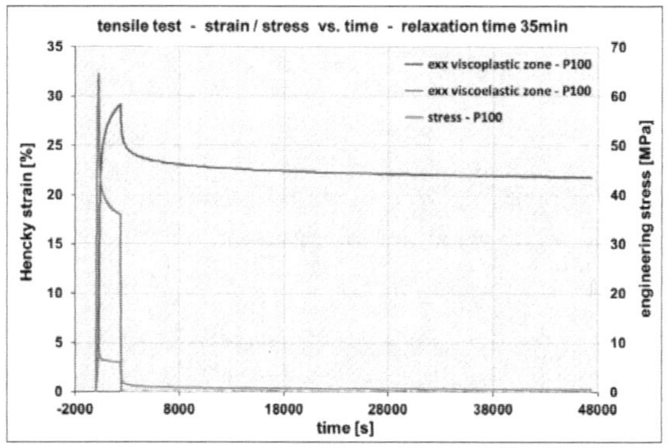

Figure 7a): Holding time 30min: stress/strain vs. time in viscoplastic and viscoelastic zone; initial phase

b) Holding time 30 min: stress/strain vs. time in viscoplastic and viscoelastic zone; initial and recovery phase

4.3.3 Permanent load, medium initial strain, small residual strain A similar test is performed with a slightly lower external displacement of 3.55 mm. Even though the reduction of the applied displacement is just 15% it has major consequences for the strain in the viscoplastic zone. The corresponding stress is even a little higher (62.3 MPa) since in either tests the stress is already beyond the maximum (figure 8a). The displacement is held constant for 30 minutes. During this time span the stress relaxes down to 40.7 MPa which is a loss of 35%. At the end of the loading process when the phase of constant displacement begins the strain in the viscoplastic zone is 4.2% while it is 3.6%.in the viscoelastic zone. While the strain in the viscoelastic zone slightly decreases to 3.3%, the strain in the viscoplastic zone increases substantially and reaches 8.7% which is a growth of about 110%.

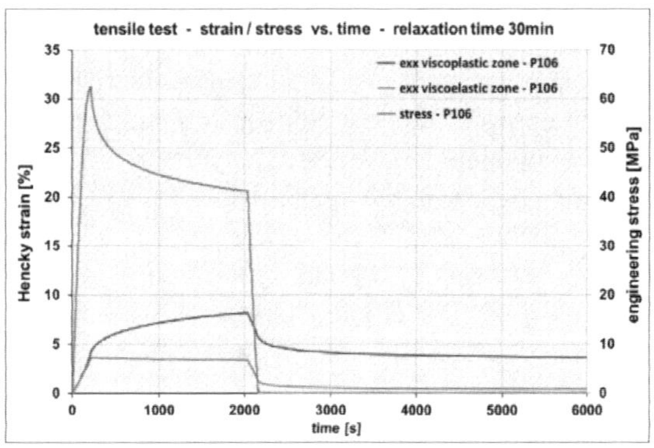

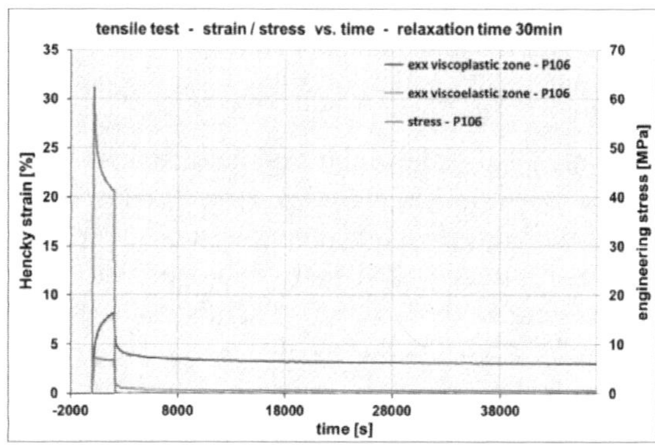

Figure 8a): Holding time 30min: stress/strain vs. time in viscoplastic and viscoelastic zone; initial phase

b) Holding time 30 min: stress/strain vs. time in viscoplastic and viscoelastic zone; initial and recovery phase

Again, the specimen then is unloaded and the strain is monitored for 12 hours. The strains in the viscoplastic and in the viscoelastic zone decrease instantaneously by 1.1 and 2.7%, respectively. The creeping process is shown in figure 8b. After 12 hours recovery time the residual strain in the viscoplastic zone is only 3% while in the viscoelastic zone it has vanished. In the first experiment with permanent loading (section 4.3.3) the initial strain before permanent loading is 6.7% while in the actual test it is 4.2% . During the phase of permanent loading the strain in the actual experiment reaches 8.7% while the other specimen ends up with 29.2% strain. After unloading and strain recovery the residual strain is 3% and 22%, respectively. This shows that a moderate increase of the initial strains has a large influence on the creep and recovery behaviour in the viscoplastic zone. Obviously the material is very sensitive against small changes of the strain level.

IOP Conf. Series: Materials Science and Engineering **139** (2016) 012035 doi:10.1088/1757-899X/139/1/012035

4.3.4 Immediate unloading, medium initial strain, small residual strain In a fourth test, a specimen loaded up to a maximum strain of 6.7% (figure 9a) which equalises the strain in the first test at the end of the loading process. It should be noticed that the scale of the strain is strongly enlarged in figure 9a. The corresponding minimum strain is 4.2%. In contrast to the former test the specimen is rapidly unloaded with a loading rate of 85 mm/min. The strains instantaneously drop by approximately 3%. After 12 hours recovery time the maximum and the minimum strain are decreased to 0.4% and 0.1%, respectively (figure 9b). In figure 9b for the sake of visibility the stresses are not displayed. The strain at the end of the loading process is the same as for the specimen studied in the preceding section where the loading process was followed by phase of permanent load. The instantaneous unloading obviously prevents the specimen to develop noteworthy viscoplastic strains. The specimen behaves predominantly viscoelastic. Comparison with the first test (4.3.1) shows that of course the residual strain increases with the magnitude of the initial strain. However, the influence of the time the specimen is exposed to load is much higher.

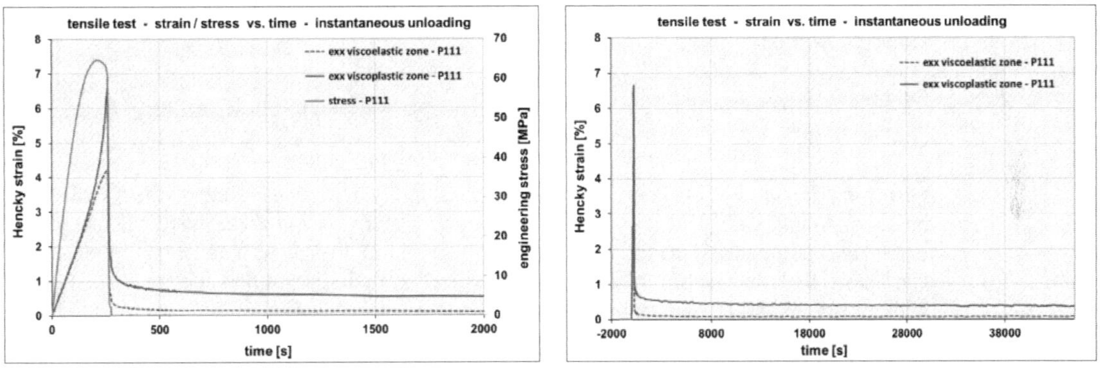

Figure 9a): Instantaneous unloading: stress/strain vs. time in viscoplastic and viscoelastic zone; initial phase

b) Instantaneous unloading: strain vs. time in viscoplastic and viscoelastic zone; initial and recovery phase

Conclusions

The viscoelastic/viscoplastic behaviour of epoxy resin is very sensitive against the level of strain and especially the time the specimen is exposed to load. Extensive creeping can occur with local strains higher than 50%. In order to develop a constitutive equation taking into account the essential features of the material requires an extended test program.

References

[1] Knauss W G, Emri I and Lu H 2008 Mechanics of Polymers: Viscoelasticity *Handbook of Experimental Mechanics*, Ed Sharpe W N (New York: Springer)

[2] Haupt P, Lion A and Backhaus E 2000 On the dynamic behaviour of polymers under finite strains: constitutive modelling and identification of parameters *International Journal of Solids and Structures* vol 37 (Amsterdam: Elsevier) pp 3633-3646

[3] Haupt P and Lion A 1995 Experimental identification and mathematical modelling of viscoplastic material behaviour *Continuum Mechanics and Thermodynamics* vol 7 (Springer: Berlin-Heidelberg) pp149-167

[4] Ward I M and Sweeney J 2012 *Mechanical properties of solid polymers* (London: Wiley & Sons Ltd)

[5] Christensen R M 1997 Yield functions/failure criteria for isotropic materials *Proc. R. Soc. London* part A (London) pp 1473-1491

IOP Conf. Series: Materials Science and Engineering **139** (2016) 012036 doi:10.1088/1757-899X/139/1/012036

A molecular dynamics study on the interaction between epoxy and functionalized graphene sheets

L S Melro[1], R Pyrz, L R Jensen

Department of Mechanical and Manufacturing Engineering, Aalborg University, Aalborg, Denmark

Corresponding author: liliana@m-tech.aau.dk[1]

Abstract. The interaction between graphene and epoxy resin was studied using molecular dynamics simulations. The interfacial shear strength and pull out force were calculated for functionalised graphene layers (carboxyl, carbonyl, and hydroxyl) and epoxy composites interfaces. The influence of functional groups, as well as their distribution and coverage density on the graphene sheets were also analysed through the determination of the Young's modulus. Functionalisation proved to be detrimental to the mechanical properties, nonetheless according to interfacial studies the interaction between graphene and epoxy resin increases.

1. Introduction

Engineering structures are increasingly comprised of carbon fibre reinforced polymer composites due to their low weight to stiffness ratio, resistance to environmental impact, and processability. Understanding the interaction between the fibres and the polymer matrix at different length scales is crucial for developing improved carbon fibre reinforced polymers. At the molecular level carbon fibres consist of stacked layers of graphene. This is a two dimensional layer with carbon-carbon sp^2 hybridized covalent bonds that result in outstanding mechanical properties. However, the crystalline and smooth surface of the graphene hinders the wetting by the polymer matrix. Recent reports suggest that the covalent bonding of functional groups to graphene can overcome this problem [1]. Nevertheless, it has an impact on the mechanical properties of the graphene due to a change in hybridization and introduction of defects. The chemisorption changes the hybridization of the grafted carbon atom from sp^2 to sp^3 resulting in an elongation of the bonds between the sp^3 carbon and its neighbouring sp^2 hybridized atoms. This elongation shortens the angles between them, and the grafted carbon buckles off plane in course of pyramidalization [2]. The mechanical properties of graphene and in particular the influence of functional groups have been addressed by several authors applying Molecular Dynamics (MD), in particular grafting density, orientation and/or location of hydroxyl, methyl, and carboxyl groups [3-5].

On the other hand the adhesion between polymer and graphene is related to the chemistry of both surfaces. As mentioned previously, by covalently bonding of chemical groups on the graphene the interaction between graphene and the polymer matrix can be increased improving the efficiency of load transfer [6]. Hence, a compromise needs to be made between the increase in load transfer capability and the decrease of the mechanical properties of the graphene due to functionalisation. MD simulations are advantageous in addressing nanoscale interfacial issues as it calculates microscopic interaction between molecules that can be scaled up to the macro level through statistical mechanics, obtaining predictions of bulk properties. Interfacial properties of CNT and polymer composites have been examined through

MD by applying the pull out method [4,7-9]. More recently graphite and graphene have been added to this circle as reported by Jang et al. [10] using vinyl ester resin as matrix, and Liu et al. [11] who compared two different matrices, polyethylene and poly(methyl methacrylate).

In this work we employed MD to calculate the interfacial shear strength (ISS) between an epoxy resin and functionalised graphene layers. Each layer has a single type of chemical group – carboxyl, carbonyl, and hydroxyl – and are dispersed on the graphene with a 10 % coverage density. Their mechanical and structural properties, the influence of coverage density (3.0 and 5.0 %) and placement of hydroxyl groups (10 % grafting as a large single cluster and as smaller clusters) were also assessed.

2. Methodology

2.1. Molecular models

2.1.1. Epoxy. Epoxy resin was built using the 3D atomistic tool from Materials Studio 6.0 [12]. The polymer was based on a 50 % crosslinked bisphenol A diglycidyl ether (DGEBA) with triethylenetetramine (TETA) consisting of 744 atoms. The crosslinked molecule was replicated resulting in a periodic cell with 70 packed molecules and 1 configuration which originated 12 250 atoms. Cell dimensions are 5.2 x 4.0 x 4.7 nm and cell angles $\alpha = \beta = \gamma = 90^{\circ}$. The density was set to 1.15 g.cm^{-3}, an average value for epoxy resin density as reported in literature [13].

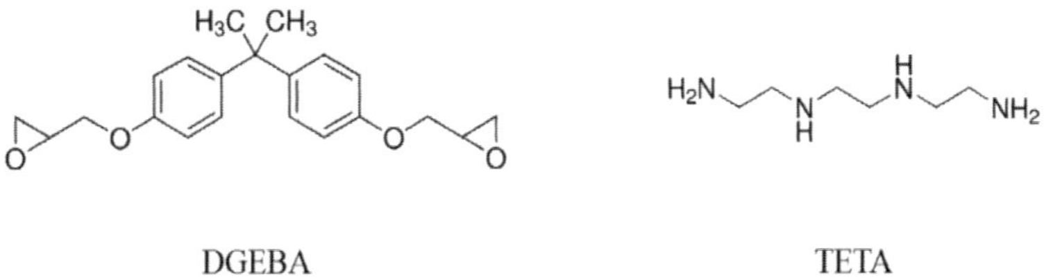

Figure 1. Molecules Bisphenol A diglycidyl ether (DGEBA) and Triethylenetetramine (TETA) used to build the crosslinked epoxy resin

2.1.2. Graphene. A graphene layer simulation cell with 960 atoms and periodic boundary conditions was created with the final dimensions 4.9 x 5.1 x 2.0 nm and angles $\alpha = \beta = \gamma = 90^{\circ}$. The final structure has a density of 2.24 g.cm^{-3}. Functional groups (COOH, C=O, and OH) were individually and randomly grafted onto the 96 carbon atoms from the graphene layer resulting in 10 % functionalization.

2.2. Simulation Setup

The COMPASS force field [14] was used to model the interactions between atoms. In calculating the non-bonded interactions, van der Waals and electrostatic interactions are described by Lennard Jones 6-9 and Coulombic functions, respectively. The atom-based summation method with cut-off radii of 9 Å and long range corrections was used to calculate non-bond interactions.

Dynamics were run with time step of 1 fs and at 298K in every case, unless stated otherwise.

To equilibrate the epoxy system successive MD simulations were run. First, dynamic simulations with the canonical ensemble (NVT) and Andersen thermostat were conducted for 70 ps before annealing. Next, simulated annealing was applied in order to gradually minimise the energy of the structure to prevent it from being trapped in a conformation that represents a local minimum. Thus higher temperatures allow the simulation to overcome energy barriers to move into other low energy areas hence, the temperature is periodically increased from an initial temperature to a mid-cycle temperature and back. At the end of each complete temperature cycle, the lowest energy structure produced by that cycle is output to the trajectory file. The annealing schedule was performed with an initial temperature (300 K), a mid-cycle temperature (600 K), and four heating ramps per cycle and with 500 simulation steps in each dynamics step. The annealing was performed in an isothermal-isobaric ensemble (NPT) with Andersen thermostat and Parrinello-Rahman barostat, respectively. This allowed the shape and

volume of the cell to change during annealing. Following annealing a final equilibration dynamic simulation was performed for 110 ps using the NPT ensemble with Andersen thermostat and Parrinello-Rahman barostat. After equilibrium the angles of the simulation cell were corrected again to 90° as they have been changed during NPT. This step is necessary in order to combine the epoxy system with the graphene molecular system.
For graphene systems dynamics were performed with a NVT ensemble using Berendsen thermostat during 100 ps to ensure equilibration.

2.2.1. Interaction at the interface. The interface between the epoxy and graphene was prepared by layering the crosslinked epoxy cell on top of the energy minimized graphene cell. For calculations only non-covalent bond interactions were considered. Epoxy was set to match the graphene layer and follow lattice orientation from the latter as the former is more flexible and adjusts better to the more rigid graphene. The layer was gradually compressed in the Z coordinate until it reached a density close to the theoretical value of epoxy, 1.15 g.cm^{-3}. Each compression cycle was followed by a dynamics NVT for 50 ps. To avoid large energy increases to the system, this compression was performed stepwise [15]. Finally, NPT dynamics was performed with Berendsen as both thermostat and barostat for 100 ps in order to equilibrate the final structure.

2.2.2. Mechanical properties of graphene. Each molecular model was energy minimized using a combination of the steepest decent and the conjugate gradient methods. The graphene was strained in the x-axis direction with steps of 0.5 %. A dynamics run using NVT ensemble was performed after each step with Berendsen thermostat for 100 ps. Atomistic stress calculations were based on the virial expression [16]. The stress-strain plots were used to calculate the Young's modulus. The influence of grafting density on the mechanical properties of graphene was studied by randomly grafting hydroxyl groups to the graphene structure in different percentages, 3.0 % and 5.0 %. The effect of the distribution of hydroxyl groups as a single cluster or several small clusters was assessed by comparing these models with dispersed hydroxyl groups on the graphene. This latter study was performed for 10 % functionalised layers.

2.3 Pull out simulations
For the calculation of the interaction energies and the interfacial shear strength the periodic boundary conditions were removed and hydrogen atoms were added to the graphene edges and fixed. The graphene was pulled in the x-axis direction with increments of 5 Å until complete separation of graphene and epoxy, as shown in figure 2. Geometry optimization was performed after each pull out.

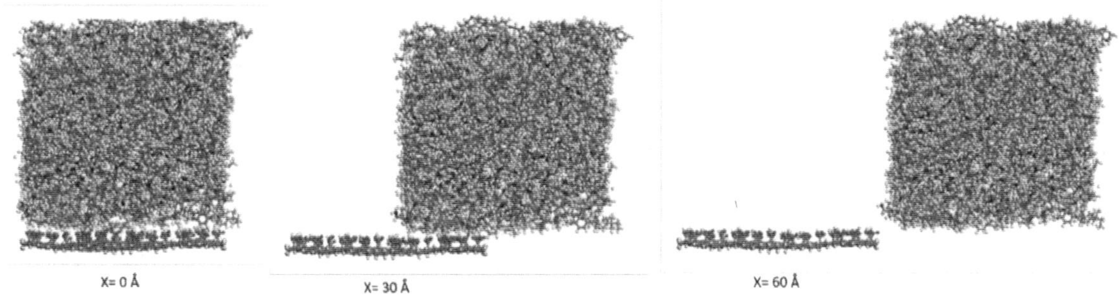

Figure 2. Molecular model of the pull out of graphene-COOH from a crosslinked epoxy resin.

3. Results

3.1. Graphene mechanical properties

The Young's modulus for pristine and functionalised graphene are presented on figure 3. Close to what is reported in the literature for MD simulations [16-18], the pristine graphene obtained the highest Young's modulus, 1.35 TPa. This is attributed to its sp^2 hybridised structure. A decrease on the Young's modulus is verified with the grafting of functional groups onto the graphene layer. Both –COOH and – OH groups decrease the Young's modulus of graphene 42 % and 47 %, respectively. The largest drop occurs for carbonylated graphene, 53 %. With the grafting of –COOH and –OH groups the carbon-carbon bond hybridisation changes to sp^3 in order to promote the chemical bond (figures 5 and 6). This change of in-plane sp^2-hybridisation to out-of-plane sp^3 allows a larger mobility of the carbon due to unsupported sp^3 bonds [19,20]. Several simulations of functionalised graphene layers have reported a decrease on the mechanical properties of graphene [3-5]. With a different type of grafting –C=O promotes the complete breaking of the sp^2 bond between one neighbour atom of the grafted carbon creating a vacancy defect, which has been previously studied with AIREBO potential in MD simulations [21], and proved a decrease in strength of graphene with 2.0 % vacancy concentration.

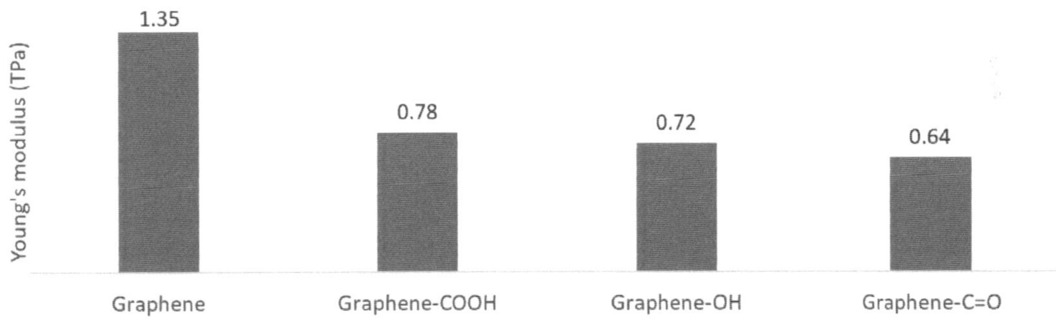

Figure 3. Young's modulus obtained from molecular dynamics simulations for the different layers.

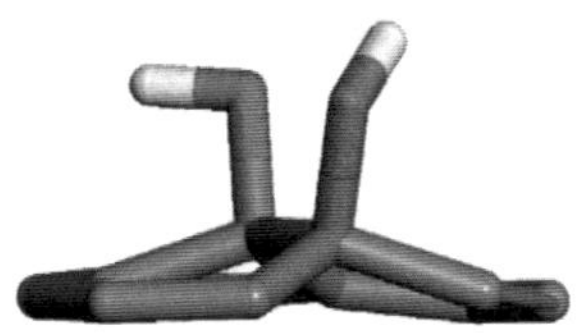

Figure 4. Fragment from graphene-OH after equilibration showing the pull-off of the grafted carbon.

The structure of graphene was studied by calculating the bond length distribution for systems before and after equilibration. Pristine graphene has a homogeneous length distribution of 1.42 Å prior to equilibration [16]. Some minor out-of-plane displacements occurred after equilibration as a consequence of ring strain relief and thermal vibrations. This leads to a small change in bond length distribution of about 0.20 Å (figure 5). The influence of functional groups is however more noticeable right before equilibration. The grafting of carboxylic groups on the graphene and consequent conversion of local carbon displacements from in plane sp^2 to out-of-plane sp^3 hybridisation, while at the same time the grafted carbon is pulled off plane in a pyramidal structure (figure 4), increases the bond length between grafted carbon and structural carbons to approximately 1.53 Å (figure 5). Thus decreasing the surrounding bonds to 1.40 Å. After equilibration the bond length distribution broadens. Out-of-plane displacements are even clearer. Hydroxyl groups have a similar behaviour as carboxylated surfaces except for out-of-plane displacements after equilibration which are less severe (figure 6).

Unlike the previous functional groups, carbonyl promote the breaking of a sp^2 hybridised bond between the grafted carbon and one neighbouring carbon in order to covalently bond an oxygen atom by sp^2 hybridisation. This creates a defect on the graphene surface, resulting in an out-of-plane displacement similar to carboxyl groups with broader bond length distribution. After equilibrium the bond length distribution increases (figure 6). Such behaviour has been reported in the literature [2,19].

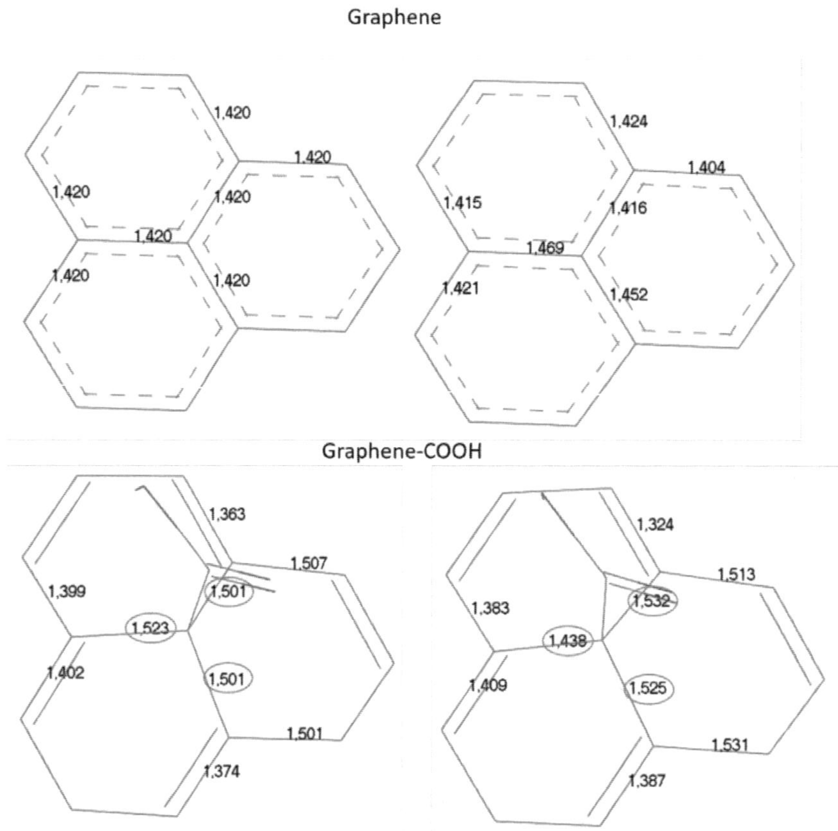

Figure 5. Fragment of a 960 atom graphene layer without functionalisation (top) and functionalised with carboxyl groups (bottom), before (left) and after equilibrium (right), showing the bond lengths in Å near functionalised sites. Circled are the stretched bonds resulting from the pull-off of functionalised carbons.

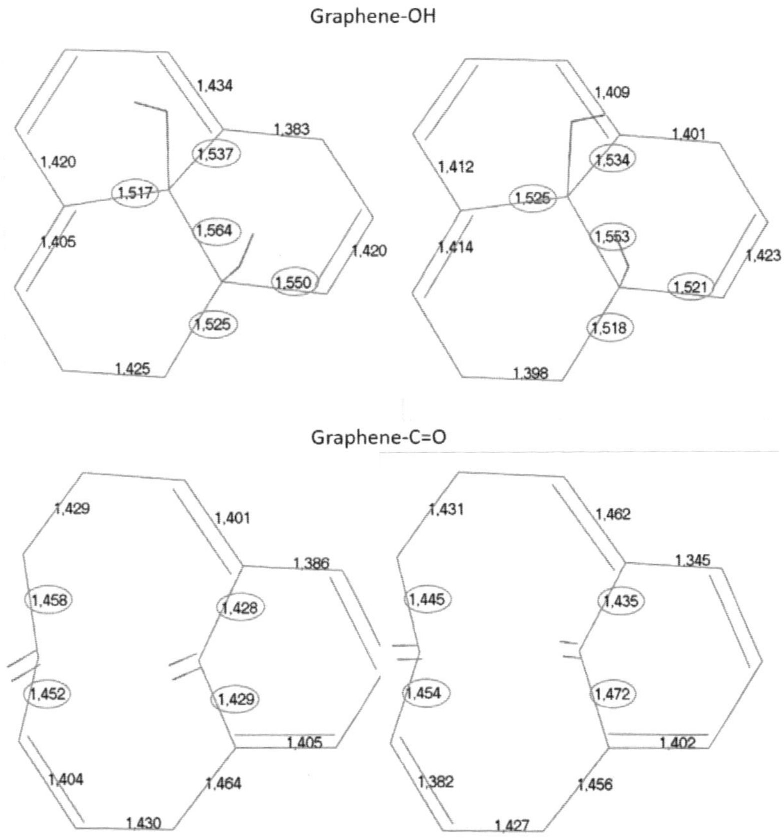

Figure 6. Fragment of a 960 atom graphene layer functionalised with hydroxyl (top) and carbonyl (bottom) groups, before (left) and after equilibrium (right), showing the bond lengths in Å near functionalised sites. Circled are the stretched bonds resulting from the pull-off of functionalised carbons.

Studies performed for the 3.0 % and 5.0 % hydroxyl functionalised graphene proved a slight increase in the out-of-plane dislocations of carbon atoms with increasing coverage density (figure 7). Calculations showed a decrease of the Young's modulus with increasing functionalisation, from 0.77 to 0.74 to 0.72 TPa for 3.0 %, 5.0%, and 10 % grafting, respectively. The major effect is however verified for the smallest degree of functionalisation as with increasing grafting percentages the influence on the Young's modulus is not so noticeable. This seems to suggest that a smaller degree of funtionalisation is enough to decrease the mechanical properties of graphene, and an increase of that percentage will not have a much larger effect. These results come in agreement with a study by Qin and Buehler [3] who verified the influence of functional groups density on the fluctuations of the graphene sheets and consequently their mechanical properties.

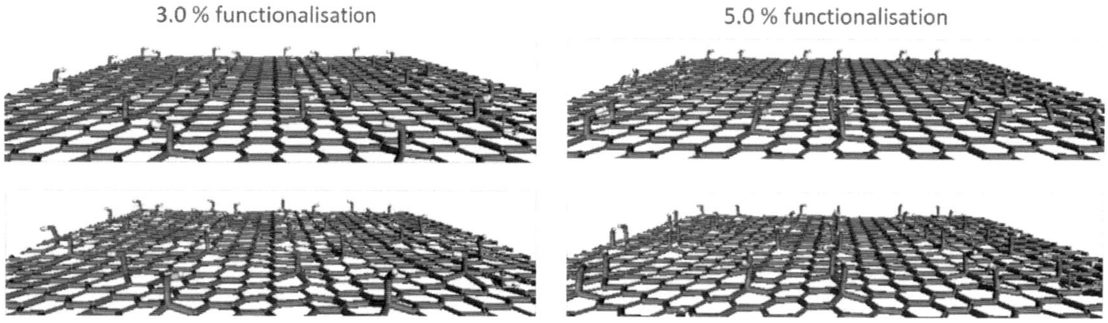

Figure 7. Effect of group density on the graphene conformation for 3.0 % and 5.0 % grafting before (top) and after (bottom) equilibration.

Comparisons between 10 % grafted groups distributed randomly, as a single cluster, and as several clusters (figure 8) were made. The single cluster obtained the lowest Young's modulus, 0.68 TPa, which can be understood by a larger interaction between concentrated hydroxyl groups together with a higher amount of sp^3 bonds confined in a reduced area. The smaller clusters model on the other hand have less impact on the elastic modulus, comparable to randomly dispersed groups, between 0.72 – 0.75 TPa. This study proves that density and distribution of functional groups influence the mechanical properties of graphene, as shown by Pei et al. [4].

Regarding their structural properties, graphene with single cluster and several clusters have an average bond length distribution of 1.42 Å. Out-of-plane displacements are more concentrated for the single cluster, whilst for small clusters and randomly distributed groups they are dispersed and less pronounced.

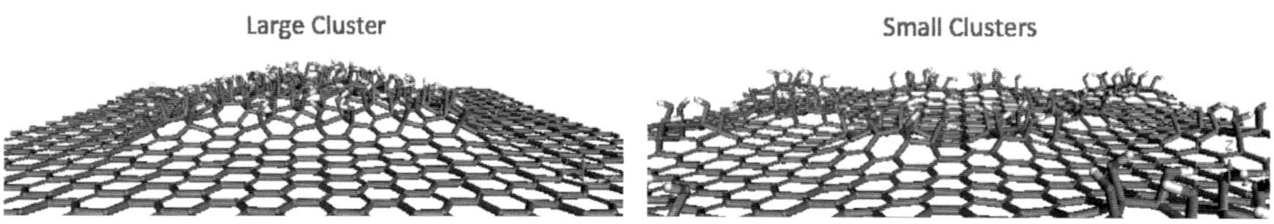

Figure 8. Effect of hydroxyl groups (10 % functionalisation) organised as a large single cluster (left) and several small clusters (right) after equilibration.

3.2. Interfacial characterisation

The interaction energy was estimated from the energy difference (ΔE) between the total potential energy of the composite and the sum of the potential energies of individual molecules as shown in equation (1):

$$\Delta E = E_{total} - \left(E_{graphene} + E_{epoxy}\right) \quad (1)$$

Where E_{total} is the total potential energy of the system, $E_{graphene}$ the potential energy of the graphene, and E_{epoxy} the potential energy of the epoxy resin.

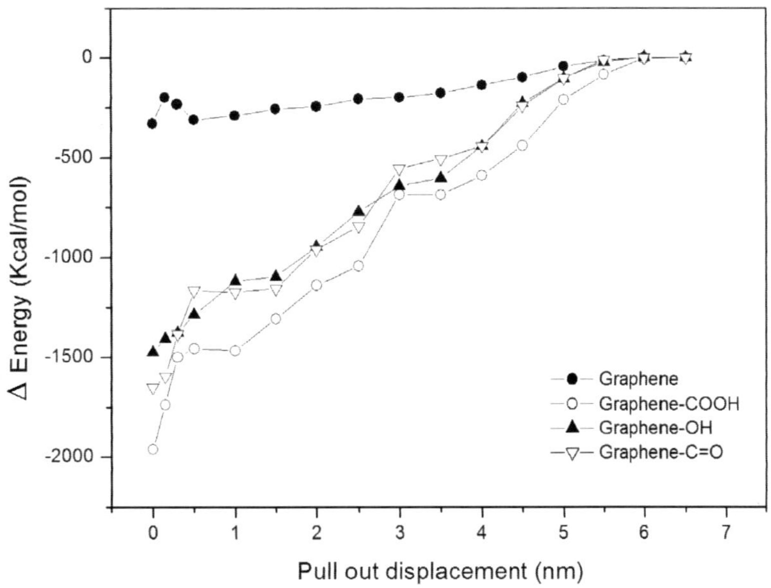

Figure 9. Interaction energies during pull out between epoxy resin and the functionalised graphene layers.

The interaction energy is negative during each stage of the pull-out showing the existence of attractive interaction between polymer and reinforcement. The ISS, through van der Waals interactions, increases linearly with applied strains, and tend to stabilize to a constant value close to zero after complete separation of the surfaces as a result of the lack of interaction.

The interfacial shear strength can be evaluated by analysing the changes in potential energy during the pull out of the graphene from the epoxy and following the relations on equations (2) and (3):

$$\Delta E = \int_0^L w(L-x)\tau_i dx = w\tau_i Lx - \frac{1}{2}\tau_i wx^2 = \frac{1}{2}wL^2\tau_i \quad (2)$$

$$\tau_i = \frac{2\Delta E}{WL^2} \quad (3)$$

Where W is the width of the graphene (W = 49 Å) and L the length (L = 51 Å). As there are no periodic boundary conditions applied and we have a two-layer system, the graphene only interacts with one side of the epoxy, thus we have simply w. Furthermore, even though L is assumed, figure 9 shows that a distance of 60 to 65 Å (depending on the system) is a good value taking into account that after that length there are no considerable interactions.

The pull out force was obtained through differentiation of the potential energy curve and based on equation (4):

$$F_{pull\ out} = \frac{\Delta E}{\Delta x} \quad (4)$$

Where ΔX is the total displacement.

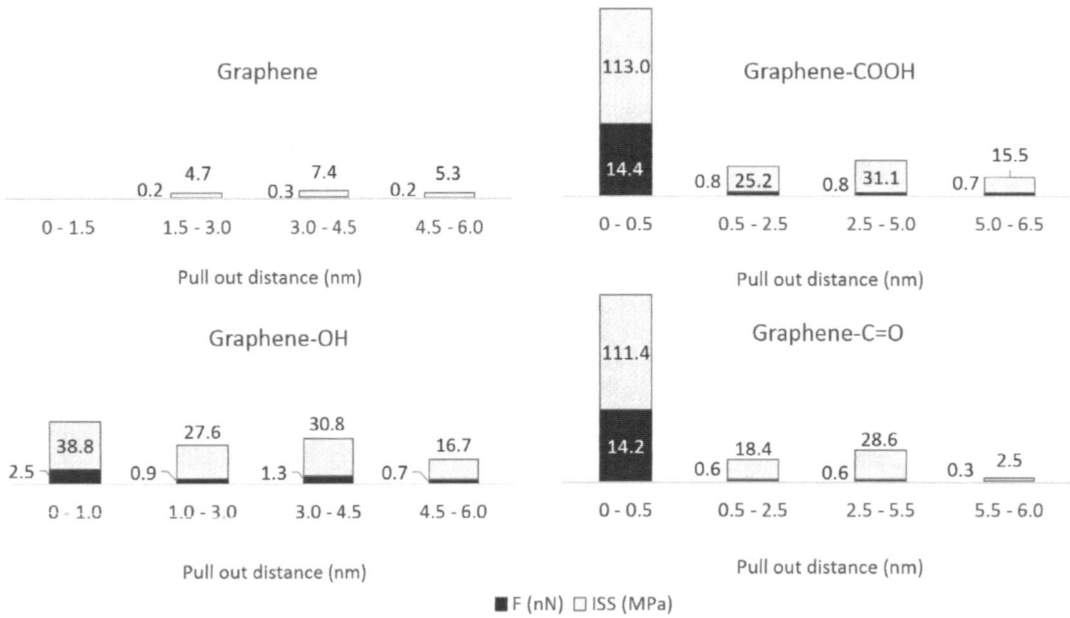

Figure 10. Interfacial shear strength and force required for pulling the graphene layers from the epoxy.

The results displayed on figure 10 show that the force required for pulling the pristine graphene from the polymer is extremely lower than for functionalised layers thus suggesting very little interaction between the two. On opposite sides, carboxylated and carbonylated graphene seem to interact very well with the epoxy has the forces required for pulling the graphene layers are the highest of the four models studied (14 nN), despite lowering considerably after the initial stage. Even the ISS are similar for both, 113 and 111 MPa, respectively. Graphene-OH is the one among the functionalised layers who presents less interaction with the epoxy. It reaches maximum values of 38.8 MPa for ISS and 2.5 nN of force required for the pull out, during the initial stages.

4. Conclusion

The results obtained show that functionalisation of graphene sheets improve considerably the ISS and pull out force for epoxy resin composites. Pristine graphene seems to have almost no interaction with the polymer whereas carboxyl and carbonyl groups increase considerably this interaction. Hydroxylated graphene shows an improvement of the pull out force and ISS compared to pristine graphene however these values are much lower than the other functionalised graphene models studied. Despite an increase in interface interactions, the determination of the Young's modulus showed a decrease of the mechanical properties of functionalised layers, with the smallest drop occurring for carboxylated graphene, 42 %, and the highest for carbonylated, 57 %. A higher coverage density also proved a decrease of the Young's modulus. This is expected due to the change of hybridisation of grafted carbons, which also implies a broadening of the bond length distribution. The organisation of these groups in a single cluster also means a greater negative impact on the mechanical properties. Overall the results suggest a good commitment between loss of mechanical properties and improvement of interfacial strength should be achieved with carboxyl functional groups.

References

[1] Frankland S J, Caglar A, Brenner, D W and Griebel, M 2002 *J. Phys. Chem. B* **106** 3046-3048
[2] Lin Y, Ding F and Yakobson B I 2008 *Phys. Rev. B* **78**, 041402(R)
[3] Qin Z and Buehler M 2012 *Molecular Simulation* **38** 695-703
[4] Pei Q X, Zhang Y W and Shenoy V B 2010 *Nanotechnology* **21** 115709
[5] Zheng Q, Geng Y, Wang S, Li Z and Kim J 2010 *Carbon* **48** 4315-4322

[6] Guru K, Mishra S B and Shukla K K 2015 *Applied Surface Science* **349** 59-65
[7] Gou J, Minaie B, Wang B, Liang Z and Zhang C 2004 *Computational Materials Science* **31** 225-236
[8] Zheng Q, Xia D, Xue Q, Yan K, Gao X and Li Q 2009 *Applied Surface Science* **255** 3534-3543
[9] Wong M, Paramsothy M, Xu X J, Ren Y, Li S and Liao K 2003 *Polymer* **44** 7757-7764
[10] Jang C, Lacy T E, Gwaltney S R, Toghiani H and Pittman Jr C U 2013 *Polymer* **54** 3282-3289
[11] Liu F, Hu N, Ning H, Liu Y, Li Y and Wu L 2015 *Computational Materials Science* **108** 160-167
[12] Accelrys Software Inc 2011 *Materials Studio v6.0.0.*
[13] Brandrup J, Immergut E H and Grulke E A 2003 *Polymer Handbook* (John Wiley & Sons, Inc)
[14] Sun H 1998 *J. Phys. Chem. B* **102** 7338-7364
[15] Hadden C M, Klimek-McDonald D R, Pineda E J, King J A, Reichanadter A M, Miskioglu I, Gowtham S and Odegard G M 2015 *Carbon* **95** 100-112
[16] Tsai J and Tu J 2010 *Materials & Design* **31** 194-199
[17] Zhou L, Wang Y F and Cao G 2013 *J. Phys. Condens. Matter* **25** 1361-648
[18] Jing N, Xue Q, Ling C, Shan M, Zhang T, Zhou X and Jiao Z 2012 *RSC Advances* **2** 9124-9129
[19] Montazeri A, Ebrahimi S and Rafii-Tabar H 2015 *Molecular Simulation* **41** 1212–1218
[20] Pei Q X, Zhang Y W and Shenoy V B 2010 *Carbon* **48** 898-904
[21] Dewapriya M A N and Rajapakse R K N D 2014 *Journal of Applied Mechanics* **81** 081010

IOP Conf. Series: Materials Science and Engineering **139** (2016) 012037 doi:10.1088/1757-899X/139/1/012037

Time-resolved X-ray microtomographic measurement of water transport in wood-fibre reinforced composite material

Arttu Miettinen[1], Tero Harjupatana[1], Markku Kataja[1], Stefania Fortino[2], Kirsi Immonen[2]

[1] University of Jyväskylä, Department of Physics, P.O. Box 35 (YFL) FI-40014 Jyväskylä, Finland
[2] VTT Technical Research Centre of Finland, P.O. Box 1000, FI-02044 VTT, Finland

E-mail: arttu.miettinen@phys.jyu.fi

Abstract. Natural fibre composites are prone to absorb moisture from the environment which may lead to dimensional changes, mold growth, degradation of mechanical properties or other adverse effects. In this work we develop a method for direct non-intrusive measurement of local moisture content inside a material sample. The method is based on X-ray microtomography, digital image correlation and image analysis. As a first application of the method we study axial transport of water in a cylindrical polylactic acid/birch pulp composite material sample with one end exposed to water. Based on the results, the method seems to give plausible estimates of water content profiles inside the cylindrical sample. The results may be used, *e.g.*, in developing and validating models of moisture transport in biocomposites.

1. Introduction

Natural fibre composites appear as a promising and environmentally friendlier alternative for, *e.g.*, engineering plastics and glass fibre reinforced composites [1]. A main challenge with many natural-fibre reinforced thermoplastic composites is their inherent tendency to absorb water. This may lead to fungal decay, mold growth, mechanical degradation and dimensional instability, and thereby to, *e.g.*, decreased service life [1, 2]. A key topic in development of improved natural fibre composites is thus understanding the influence of moisture on the properties of the material, in particular if the material is to be used in variable environmental conditions.

Measuring time-dependent three-dimensional moisture content field inside material is a non-trivial task, especially for heterogeneous materials that contain multiple phases. In this work we present a method for measuring one-dimensional time-dependent water content profile based on three-dimensional X-ray microtomographic images. The method is based on first using 3D digital image correlation for measuring the displacement field of the wetted and possibly swollen sample. The local water content is then found by measuring the change in the values of the effective X-ray attenuation coefficient for each material point (3D image voxel) between the dry state and the wetted and deformed state.

Previously, several authors have used various methods based on X-ray tomography to determine local moisture content in different kinds of materials like stone, clothing and wood [3–6]. In many of those works the effects of swelling deformation of the material has not been considered, or is included by making a specific assumption on the deformation characteristics of the material. Other possible methods for measuring local water concentration, based on sensing local amount of hydrogen nuclei in the material, are neutron radiography [7, 8] and magnetic resonance imaging [9].

A method similar to the one introduced in this paper was recently published and used for measuring water content in an approximately 2 cm wide bentonite clay sample [10]. Application of the method to smaller and less-absorbing composite material samples poses several problems in sample preparation, imaging and image analysis. Here, we demonstrate its practical application on a cylindrical polylactic acid/birch pulp composite material sample of diameter 2 mm. The experimental set-up was designed to yield one-dimensional axial wetting. Finally, the feasibility of the method is assessed based on the results from this first trial experiment.

2. Method for analysing moisture content distribution using X-ray tomography

We start by considering a composite material sample at known low water content, called the 'dry state' of the sample, in what follows. An X-ray tomographic image of the dry sample, denoted by $\mu^0_{\text{eff}}(\vec{x})$, $\vec{x} \in C \subset \mathbb{R}^3$, represents a three-dimensional map of the effective X-ray attenuation coefficient indicating the original reference configuration of the sample. Subjecting the sample to a moist environment initiates absorption of water and deformation due to swelling. Assuming that the absorption process is slow enough so that the state of the sample can be assumed approximately stationary within the tomographic scanning time used, we can take another X-ray tomographic image of the partially wetted sample during the wetting process. This image, denoted by $\mu_{\text{eff}}(\vec{x})$ now represents the wetted and deformed configuration of the sample. The task is now to find the distribution of water content of the wet state and express it in the reference frame defined by the dry sample.

To this end, we first estimate the displacement field $\vec{u}(\vec{x}) = (u_1(\vec{x}), u_2(\vec{x}), u_3(\vec{x}))$ between the dry state and the wet/deformed state using the images μ^0_{eff} and μ_{eff} representing these two states. Assuming that the swelling strain is small enables using a simple block-matching algorithm to find the displacement field. The algorithm is based on comparing displaced subregions of the wet state image to non-displaced subregion of the dry state image and minimizing the squared difference with respect to displacement, *i.e.*

$$\vec{u}(\vec{x}) = \text{argmin}_{\vec{u}' \in S} \sum_{\vec{y} \in W} (\mu_{\text{eff}}(\vec{y} + \vec{u}') - \mu^0_{\text{eff}}(\vec{y}))^2, \tag{1}$$

where S is a neighbourhood of origin and W is a neighbourhood of $\vec{x}$. The set S determines the possible values of the displacement and the set W gives the size of the subregion where the two images are compared.

For given displacement field $\vec{u}(\vec{x})$, a back-projected image of the wet state image is defined by

$$\mu^P_{\text{eff}}(\vec{x}) = \mu_{\text{eff}}(\vec{x} + \vec{u}(\vec{x})). \tag{2}$$

The image $\mu^P_{\text{eff}}(\vec{x})$ thus contains the values of attenuation coefficient of wet state material points associated with the original dry state positions of the same material points.

In addition to local change of partial density[1] of water ρ_w, wetting generally induces deformation and can therefore lead to local change of composite material (solid phase) partial density ρ_c, too. As the presence of both water and solid material contribute to local X-ray attenuation, the change in the water content can not be deduced based on the measured change of the local value of the X-ray attenuation coefficient alone. The necessary additional information

[1] Partial density of component α of a mixture is defined as $\rho_\alpha = m_\alpha/V$, where m_α is the mass of component α contained in volume V of the mixture.

can, however, be obtained utilizing the displacement field $\vec{u}(\vec{x})$ and using conservation of mass of the composite material. The change of ρ_c between the dry state and the wet state is thereby found to be

$$\Delta\rho_c(\vec{x}) = -\rho_c^0(\vec{x})\frac{\delta(\vec{x})}{1+\delta(\vec{x})}, \tag{3}$$

where $\rho_c^0(\vec{x})$ is the density of the composite material in the dry state, and

$$\delta(\vec{x}) = \nabla \cdot \vec{u}. \tag{4}$$

The local value of attenuation coefficient of X-rays can be assumed to depend linearly on local densities ρ_c and ρ_w in the wet composite material [11]. We thus write

$$\mu_{\text{eff}} = \alpha_c\rho_c + \alpha_w\rho_w, \tag{5}$$

where α_c and α_w are X-ray mass attenuation coefficients of the composite material and water, respectively (for the composite material, α_c is the mass averaged attenuation coefficient of its component materials). The coefficients α_c and α_w are unknown a priori and must be found by calibration measurements. The difference of the effective X-ray attenuation coefficient between the back-projected wet state and the dry state is given in the dry state reference frame by

$$\Delta\mu_{\text{eff}}(\vec{x}) = \mu_{\text{eff}}^P(\vec{x}) - \mu_{\text{eff}}^0(\vec{x}) \tag{6}$$
$$= \alpha_c(\rho_c^P(\vec{x}) - \rho_c^0(\vec{x})) + \alpha_w(\rho_w^P(\vec{x}) - \rho_w^0(\vec{x})) \tag{7}$$
$$= \alpha_c\Delta\rho_c(\vec{x}) + \alpha_w\Delta\rho_w(\vec{x}), \tag{8}$$

where $\Delta\rho_w(\vec{x})$ is the change of the partial density of water between the reference and the wet states (local change of water content of a composite material point located originally at $\vec{x}$). Solving for $\Delta\rho_w(\vec{x})$ gives

$$\Delta\rho_w(\vec{x}) = \frac{\Delta\mu_{\text{eff}}(\vec{x}) - \alpha_c\Delta\rho_c(\vec{x})}{\alpha_w}. \tag{9}$$

Finally, assuming that the amount of residual water present in the dry state can be neglected, the partial densities of composite material and water in the wet state are given (again, in the dry state reference frame) by

$$\rho_c(\vec{x}) = \rho_c^0(\vec{x}) + \Delta\rho_c(\vec{x}) = \frac{\mu_{\text{eff}}^0(\vec{x})}{\alpha_c} + \Delta\rho_c \tag{10}$$
$$\rho_w(\vec{x}) = \Delta\rho_w(\vec{x}). \tag{11}$$

3. Test case – polylactic acid/birch pulp composite material

The method discussed above was tested with polylactic acid/birch pulp composite material. The polylactic acid was Natureworks Ingeo 3001D polymer with MFI 22 and the birch pulp was provided by Metsä Fibre. The fibres and the matrix were compounded with a twin screw compounder and injection moulded to a dog bone shaped tensile test specimen containing 40 % fibre by mass. A cylindrical sample of 1.9 mm diameter was cut from the central part of the tensile test specimen using a high-precision milling machine. Care was taken not to damage the intact surface of the sample through which the water infiltration was to take place. The sample was glued on the top of a sample holder made of 1.9 mm diameter carbon fibre rod (see figure 1).

To construct a water chamber, the sample was placed and sealed in a long thin-walled polyethylene terephthalate (PET) tube with wall thickness ~6.35 µm and inner diameter 1.98 mm (Vention Medical, United States). The gap between the sample and the tube inner wall was filled with cyanoacrylate glue that was drawn into the gap by capillary action. The

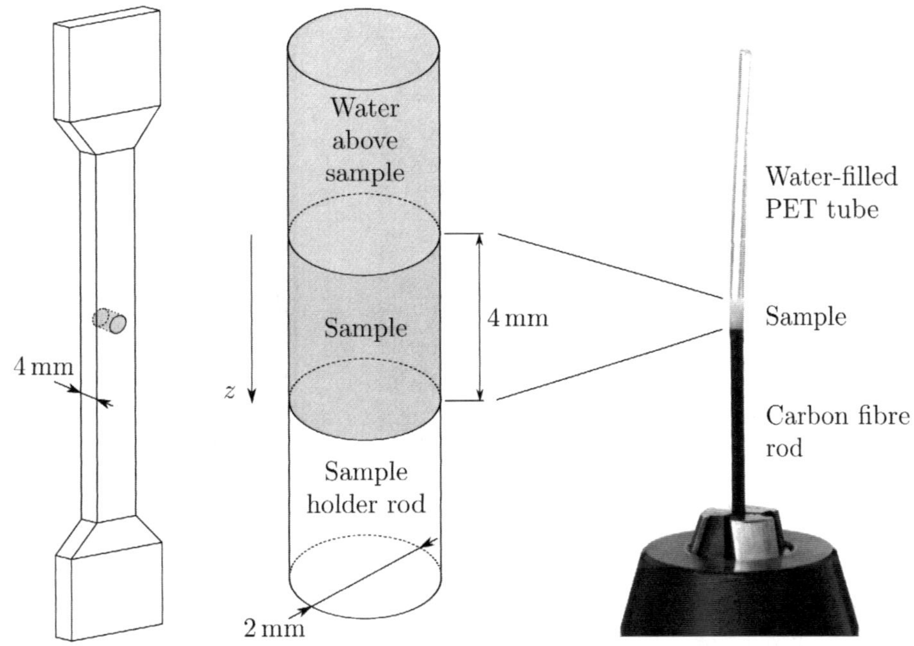

Figure 1. Schematic representation of sample location and orientation with respect to a tensile test specimen (left). The shaded part is the sample. The structure of the sample assembly (middle). Photograph of the realization of the sample assembly (right). The various diagrams are not drawn to scale.

diameter of the final sample, including the PET tube, is approximately 2 mm and its height is approximately 4 mm. The geometry and materials of the sample assembly, shown in figure 1, were selected to allow one-dimensional axial infiltration of water and as free swelling of the sample as possible during the wetting phase of the experiment.

After oven drying in 50°C for 24 h, the sample was imaged using Xradia microCT-400 device with 4.74 µm pixel size, 30 kV acceleration voltage and 3 W electron beam power. Total of 1129 projection images were acquired with 5 s exposure time, totalling in 3.5 h scan time. With the given settings the field-of-view of the tomographic image spans a cylinder of 2.2 mm diameter and height. The entire sample assembly can thus be kept within the field-of-view of the scanner in the radial direction, which is one of the basic requirements for faithful reconstruction of the sample. In the axial direction (z-direction in figure 1) three overlapping sub-scan tomograms had to be acquired to fit the whole 4 mm long sample into the combined field-of-view of the three images. For calibration purposes (see below) the combined image area was selected such that a part of the sample holder rod and a part of the water chamber were visible below and above the sample, respectively. The procedure thus yielded the dry state image $\mu^0_{\text{eff}}(\vec{x})$ for each sub-scan region, see figure 2.

Wetting of the sample was initiated by adding water into the water chamber using a syringe and a needle mounted on a special stage that ensured controlled injection. Immediately after that, acquisition of the three sub-scan images sequentially with predetermined intervals was commenced. The intervals varied from 12 h in the beginning to one week at the end of the wetting process. The total duration of the process was 135 days.

After reconstruction and beam-hardening correction done using the standard utility software of the tomographic scanner, the images were found to contain relatively weak ring artefacts that are typically caused by spatial variation of the X-ray detector properties over its sensitive area.

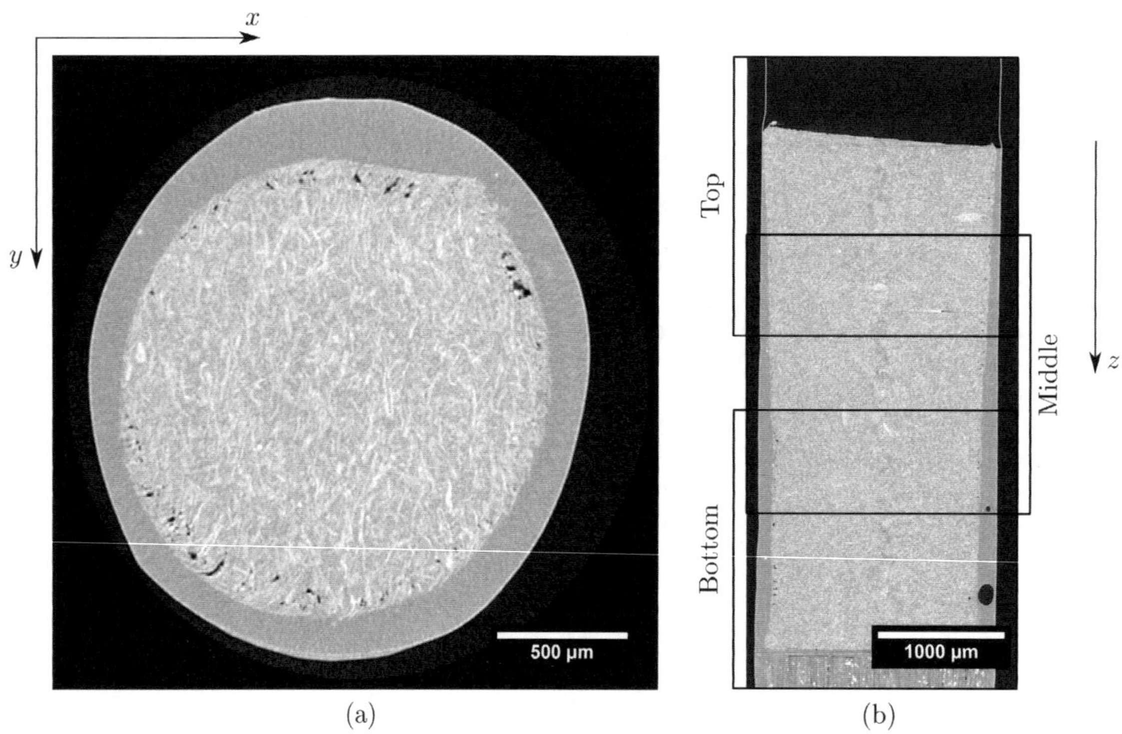

Figure 2. (a) A single horizontal slice of a tomographic image of the dry sample. The thin bright region on the edge of the sample is the PET tube, thick gray part is cyanoacrylate glue, bright small regions are parts of wood fibres and black holes near the edge of the sample are pores caused by milling. (b) A vertical slice of the dry sample showing the three overlapping sub-scan regions. The combined image includes a part of the carbon fibre sample holder rod visible in the bottom of the image and the lower part of the (empty) water chamber.

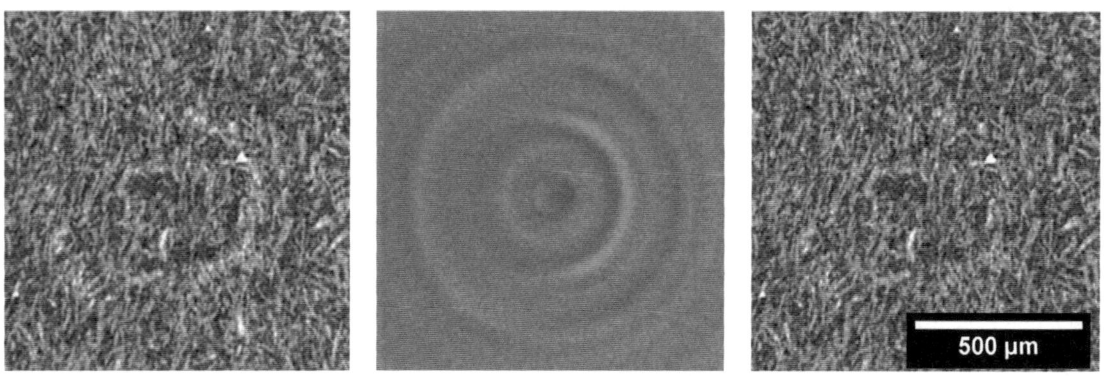

Figure 3. Removal of ring artefacts: A horizontal slice of an original image with notable ring artefact (left), of the normalization image produced as an average of 20 images of a homogeneous phantom sample (middle) and of a final corrected image (right). The scale bar applies to all the three images.

In order to alleviate the effect of ring artefacts in the final results, a phantom sample was made of highly homogeneous poly(methyl methacrylate) rod of diameter 2 mm, and imaged 20 times using the same settings as for the composite sample. The average image of these 20 images, containing virtually but the ring artefacts, was then used to normalize the original tomographic images thereby removing most of the ring artefacts (see figure 3).

We denote the sequence of final corrected images for a selected sub-scan region by $\mu_{\text{eff}}(j,\vec{x})$, where $j = 1,\dots,N$ is the scan index such that $\mu_{\text{eff}}(1,\vec{x})$ is the image taken immediately after water injection at time $t = 0$ and $\mu_{\text{eff}}(N,\vec{x})$ is the image of the sample in its final state at time $t = 135$ days. The displacement fields $\vec{u}(j,\vec{x})$ between the dry state image $\mu^0_{\text{eff}}(\vec{x})$ and the wet state images $\mu_{\text{eff}}(j,\vec{x})$ were then found using equation (1). All the images were transformed into the reference frame of the corresponding dry state using equation (2) thus yielding images $\mu^P_{\text{eff}}(j,\vec{x})$. Spatial averages of $u_3(j,\vec{x})$ and $\mu^P_{\text{eff}}(j,\vec{x})$ over horizontal (x-y) sections of the sample were calculated for each value of the distance from the upper surface of the sample z, yielding the vertical profile of vertical displacement $u_3(j,z)$, and of attenuation coefficient $\mu^P_{\text{eff}}(j,z)$, respectively. The results from each sub-scan section were then patched together using linearly weighted interpolation within the sub-scan overlap regions, thereby forming the corresponding vertical profiles over the entire sample height.

The calibration constant α_w was obtained by

$$\alpha_w = \frac{\langle \mu_{\text{eff}}(1,z)\rangle_w}{\tilde{\rho}_w}, \tag{12}$$

where $\langle\cdot\rangle_w$ denotes average over part of the data corresponding to the region above the sample, containing solely water. The density of water was taken to be $\tilde{\rho}_w = 1000\,\text{kg/m}^3$. Similarly, the calibration constant α_c was obtained by

$$\alpha_c = \frac{\langle \mu^0_{\text{eff}}(z)\rangle_c}{\tilde{\rho}_c}, \tag{13}$$

where $\langle\cdot\rangle_c$ denotes average over the volume of the dry state sample. The density of the dry composite material was measured using a straightforward gravimetric method resulting in $\tilde{\rho}_c = (1280 \pm 20)\,\text{kg/m}^3$. The initial partial density of the composite material was set to

$$\rho^0_c(z) = \frac{\mu^0_{\text{eff}}(z)}{\langle \mu^0_{\text{eff}}(z)\rangle_c}\tilde{\rho}_c. \tag{14}$$

Finally, the partial densities of composite material and water for each j were calculated by substituting $u_3(j,z)$, $\mu^0_{\text{eff}}(z)$, $\mu^P_{\text{eff}}(j,z)$, α_c, α_w and $\rho^0_c(z)$ into Equations 3 – 11. As the final result, the measured evolution of water content in the sample is shown in figure 4.

4. Discussion and conclusions

A method for non-destructive quantitative measurement of time-dependent water content profile inside a composite material sample set in contact with water was introduced and applied to polylactic acid/birch pulp composite. The experimental set-up was designed to allow one-dimensional axial infiltration of water in a cylindrical composite material sample. As a result, measured water content profiles inside the sample were obtained at various times of wetting.

The method requires careful calibration and correction of various artefacts and sources of error inherent in X-ray tomographic techniques. In general, the measured water content profiles obtained in this preliminary study, shown in figure 4, appear qualitatively plausible and resemble those expected for a diffusive transport process, although not necessarily of simple linear form. In some of the measured profiles, anomalous structures appear indicating local deviation from

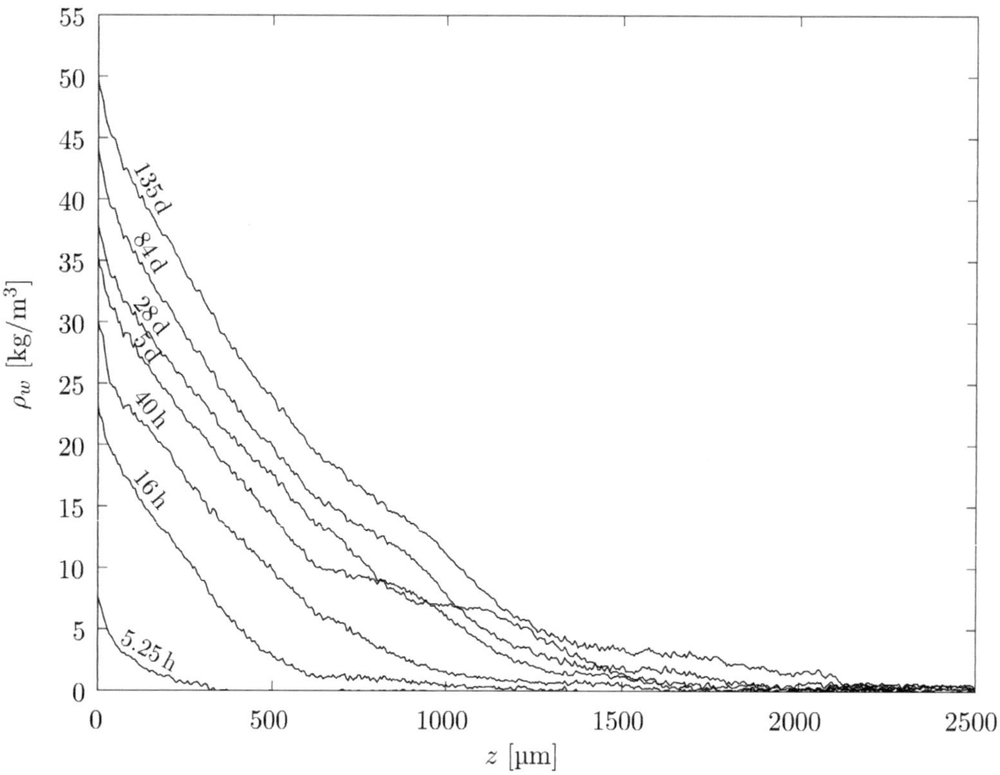

Figure 4. Measured water content profiles in the wood-fibre reinforced composite material sample for various values of wetting time, obtained using X-ray tomographic techniques. The origin of position z is at the surface of the sample exposed to water. For clarity, only a subset of all the measured profiles are shown.

the overall behaviour (see figure 4, near $z = 1000\,\mu m$). Without further investigation, it is not clear whether these anomalies are due to an error, or whether they arise for a physical reason such as presence of inhomogeneities in the sample.

In this first test of the method, a very small sample was used. This prevented valid independent verification of the results by gravimetric measurements, as done in, *e.g.*, [10]. However, based on the measured integrated water content, the total amount of water absorbed by the sample was somewhat less than expected based on independent gravimetric measurements done using larger samples. This might be due to water diffusing and evaporating through the mantle of the cylindrical sample, in spite of the thin layer of sealing glue and plastic tube. In future experiments larger samples and improved outer surface sealing should be used to facilitate quantitative validation of the method.

The detailed results obtained by the present method are useful in validating absorption models (see, *e.g.*, [12, 13]) and in quantifying the necessary material parameters related to water transport and hydromechanical behaviour of the composites. Although the method was applied here in virtually one-dimensional wetting geometry it can, in principle, be used to measure water content maps in a general three-dimensional case. Such extension of the method is left for future work.

IOP Conf. Series: Materials Science and Engineering **139** (2016) 012037 doi:10.1088/1757-899X/139/1/012037

References

[1] Campilho R 2016 *Natural Fiber Composites* (CRC Press)
[2] Joffre T, Wernersson E, Miettinen A, Luengo Hendriks C L and Gamstedt E K 2013 *Composites Science and Technology* **74** 52 – 59
[3] Cai Z 2008 *Forest products journal* **58** 41 – 45
[4] Leonard A, Blacher S, Marchot P, Pirard J P and Crine M 2005 *Canadian journal of chemical engineering* **83**(1) 127 - 131
[5] Ruiz de Argandoña V G, Rodriguez-Rey A, Celorio C, Calleja L and Suárez del Rio L M 2003 *Geological Society Special Publications* **215** 127–134
[6] Weder M, Brühwiler P A and Laib A 2006 *Textile Research Journal* **76** 18–26
[7] Lindsay J T, Matsubayashi M and Islam M N 1994 *Nuclear Instruments and Methods in Physics Research* A **353** 149 – 151
[8] Metwalli E, Hermes H E, Calzada E, Kulozik U, Egelhaaf S U and Muller-Buschbaum P 2016 *Physical Chemistry Chemical Physics* **18**(9) 6458 – 6464
[9] MacMillan M B, Schneider M H, Sharp A R and Balcom B J 2002 *Wood and fiber science* **34** 276 – 286
[10] Harjupatana T, Alaraudanjoki J and Kataja M 2015 *Applied Clay Science* **114** 386 – 394
[11] Stock S R 2009 *Microcomputed tomography* (CRC Press)
[12] Frandsen H L Selected constitutive models for simulating the hygromechanical response of wood
[13] Adhikary K B, Pang S and Staiger M P 2008 *Chemical Engineering Journal* **142** 190 – 198

A simplified model predicting the weight of the load carrying beam in a wind turbine blade

Lars P Mikkelsen

Composites and Materials Mechanics, DTU Wind Energy, Technical University of Denmark, DK-4000 Roskilde, Denmark

lapm@dtu.dk

Abstract. Based on a simplified beam model, the loads, stresses and deflections experienced by a wind turbine blade of a given length is estimated. Due to the simplicity of the model used, the model is well suited for work investigating scaling effects of wind turbine blades. Presently, the model is used to predict the weight of the load carrying beam when using glass fibre reinforced polymers, carbon fibre reinforced polymers or an aluminium alloy as the construction material. Thereby, it is found that the weight of a glass fibre wind turbine blade is increased from 0.5 to 33 tons when the blade length grows from 20 to 90 m. In addition, it can be seen that for a blade using glass fibre reinforced polymers, the design is controlled by the deflection and thereby the material stiffness in order to avoid the blade to hit the tower. On the other hand if using aluminium, the design will be controlled by the fatigue resistance in order to making the material survive the 100 to 500 million load cycles experience of the wind turbine blade throughout the lifetime. The aluminium blade is also found to be considerably heavier compared with the composite blades.

1. Introduction

A wind turbine blade is a long slender structure where the dominating loads are given by the aerodynamics and the gravitation. Throughout the years, the turbines have grown larger requiring longer blades. The longest blade at the moment is the 88.4 m long and 33.7 tons heavy blade from LM Wind Power [1]. The growth of the turbine size has resulted in cost of energy which for onshore installation pricewise can compete with conventional energy sources. Off-shore, this is still not the case and properly even larger turbines are required before this will be the case. During this upscaling, the material in a wind turbine blade is pushed to the limits, where mainly the stiffness and the fatigue resistance are the key material parameters. Wind turbine blades have until now been based on glass fibre composites and only by improving the fibre properties, the matrix fibre interface and the fibre architecture this has been made possible. A pure carbon fibre based solution is still too costly, while aluminium only is relevant for small blades.

Advanced and specialised models and material characterization methods are used in the effort to optimize the material used in the blade. Nevertheless, seen on the large scale, a wind turbine blade is a simple beam structure with well-defined load and boundary conditions. The beam theory describing the blade can be made at different levels of complexity depending on its use. An example of a more complex version is the BECAS model (BEam Cross section Analysis Software), see e.g.[2]. Even though the presented model is much simpler, it is still bridging the gap from the aerodynamic and gravitational loads to an estimate of the material stresses, deflection and blade weight. These weights are then compared with existing wind turbine blades. This approach has been used to estimate the weight of a glass fibre reinforced wind turbine blade for the DTU 10 MW reference turbine [3] with a blade length of 86 m, and it is found that replacing the material with aluminum will increase the blade weight drastically. On the other hand, using a pure carbon fiber based blade will increase the price of the blade significantly.

2. Loads on a wind turbine blade

The wind flow around a turbine rotor can simplified be considered following a 1D momentum theory as illustrated in figure 1. Going from upstream $()_u$ to downstream $()_d$ the wind speed is slowed down from the upstream wind speed u_∞ to the downstream wind speed u_d . Due to the decrease of the wind speed, the corresponding cross sectional area of the wind flow will increase from the upstream area A_u over the known turbine rotor area, A_t to the downstream cross sectional area A_d . Based on this information, we can estimate the aerodynamic loads working on the wind turbine blade, see e.g.[4]. The procedure will be described briefly below.

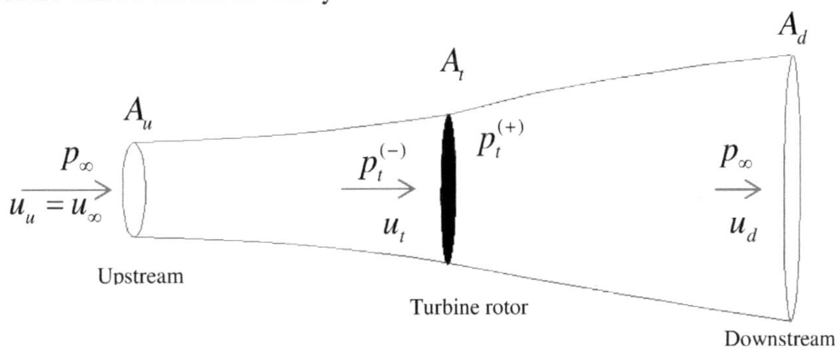

Figure 1. 1D momentum flow

Assuming that the air speed changes continuously across the turbine blades and that the pressure far upstream and far downstream are equal to the pressure of the undisturbed airflow the velocity and the pressure drop at the turbine $()_t$ with the swept rotor area A_t can be found to

$$u_t = \tfrac{1}{2}(u_\infty + u_d) \tag{1}$$

and

$$\Delta p_t = p_t^{(-)} - p_t^{(+)} \tag{2}$$

where the relation between the air speed and the pressure in the upstream and downstream part is given by the Bernoulli equation

$$\begin{aligned} p_\infty + \tfrac{1}{2}\rho_{air} u_\infty^2 &= p_t^{(-)} + \tfrac{1}{2}\rho_{air} u_t^2 \\ p_\infty + \tfrac{1}{2}\rho_{air} u_d^2 &= p_t^{(+)} + \tfrac{1}{2}\rho_{air} u_t^2 \end{aligned} \tag{3}$$

Introducing (3) into (2) , the pressure drop over the turbine can be written as

$$\Delta p_t = \tfrac{1}{2}\rho_{air}(u_\infty^2 - u_d^2) \tag{4}$$

The thrust force working over the turbine rotor, Q_t, can now be found as the swept rotor area, A_t multiplied with the pressure drop Δp_t where equation (4) determine the pressure drop.

$$Q_t = \tfrac{1}{2}\rho_{air} A_t (u_\infty^2 - u_d^2) \tag{5}$$

The only unknown parameter in equation (4) is the downstream wind speed u_d describing how much the turbine is slowing down the wind and can be determined optimizing the power P_t taken out of the wind. The power is given by the difference between the kinetic energy contained in the wind on the upstream to and on the downstream side: $P_t = ½\dot{m}(u_\infty^2 - u_d^2)$. With the mass flow through the turbine given by $\dot{m} = \rho_{air} A_t u_t$ this can be written as

$$P_t = ½\rho_{air} A_t u_t (u_\infty^2 - u_d^2). \tag{6}$$

which corresponds to $P_t = Q_t u_t$. Now introducing the parameter

$$a = (u_\infty - u_t) / u_\infty, \tag{7}$$

a combination of the equations (1), (4) and (6) will results in the following relations

$$P_t = ½\rho_{air} A_t u_\infty^3 4a(1-a)^2 \tag{8}$$

$$Q_t = ½\rho_{air} A_t u_\infty^2 4a(1-a) \tag{9}$$

From equation (8), the so-called Lanchester-Betz limit, $a = 1/3$, describes the maximum obtainable power which corresponds to 59.3% of the reference wind power $P_w = ½\rho_{air} A_t u_\infty^3$. The factor $a = 1/3$ corresponds to a wind speed at the turbine rotor of $u_t = 2/3u_\infty$ and a downstream wind speed of $u_d = 1/3u_\infty$. Most modern wind turbines are pitch regulated. Thereby, it is possible to regulate the position of the wind turbine blades in order to achieve the optimal power output. When the wind turbine power is reaching the generator power $P_t = P_G$, the blade position will be chosen in order to maintain this production and not exceeding it. Examples of wind turbine power curves can be found in [4], [5]. The incoming wind speed, u_∞, where the optimal power, P_t, just reaches the generator power, P_G, is called the rated wind speed, u_R and is a design parameter determining the ratio between the swept rotor area, $A_t = \pi L^2$ where L is the blade length, and the generator power P_G based on the knowledge about the dominating wind speed at a given location. Thus, the parameter a can be written as

$$a = \begin{cases} 1/3 & for \quad u_\infty \le u_R \\ f(P_G, u_\infty) & for \quad u_\infty > u_R \end{cases} \tag{10}$$

where $f(P_G, u_\infty)$ is determined by the condition

$$a(1-a)^2 = \frac{P_G}{2\rho_{air} u_\infty^3 A_t} \tag{11}$$

From (10) and (11), the turbine load (9) can be determined as a function of the remote wind speed u_∞. Thereby, it can be found that the maximum turbine load is found for $u_\infty = u_R$

$$Q_{\max} = \frac{4}{9}\rho_{air} u_R^2 \pi L^2 \tag{12}$$

This load will be distributed with 1/3 to each blade. Along the individual blade, the load intensity (load per unit length), q_x^{flap}, can be approximated by linear varying distribution from a vanishing value at the root to the maximum value at the blade tip given by

$$q_x^{flap} = \frac{8}{27}\rho_{air} u_R^2 \pi x \tag{13}$$

In (13) and later, the terms $()_x$ denote a variable dependent on the axial blade coordinate x measured from the root of the blade. In addition to the aerodynamic load, a gravity load will be present in the wind turbine blade. Assuming now a constant blade material cross section area, A_{blade}, the load intensity in the edge-wise direction is given by

$$q_x^{edge} = \rho_{blade} A_{blade} g \tag{14}$$

with the material density, ρ_{blade}, and the gravitational constant $g \approx 9.81 m/s^2$.

3. Stresses in the blade material

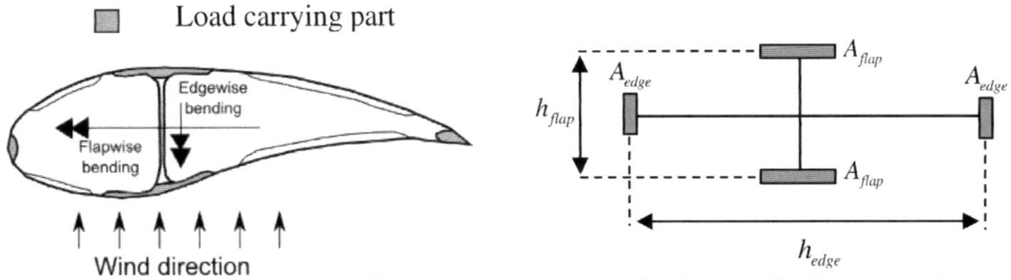

(a) Cross section with loadings (b) A simplified cross-section

Figure 2. Typical wind turbine cross section

In order to calculate the stress values in the blade material, a simplification of the cross section of the blade is used. A typical and the simplified cross section are shown in figure 2. The part marked with green is the load carrying part contributing to the overall stiffness of the blade. Using the simplified cross-section shown in figure 2b together with the load intensities defined earlier in equation (13) and (14), the flap and edge-wise moment in the blade, see e.g. [5] can be found as

$$M_x^{flap} = q_x^{flap}(L-x)\frac{L-x}{2} + \frac{q_L^{flap} - q_x^{flap}}{2}(L-x)\frac{2(L-x)}{3} \tag{15}$$

$$M_x^{edge} = \rho_{blade} A_{blade} g (L-x)\frac{L-x}{2} \tag{16}$$

which after substitution of the load intensity can be written in stresses as

$$\sigma_x^{flap} = \frac{8}{81}\frac{\rho_{air} u_R^2 \pi L^3}{A_{flap} h_x^{flap}}\left(1+\frac{x}{2L}\right)\left(1-\frac{x}{L}\right)^2 \tag{17}$$

$$\sigma_x^{edge} = \frac{1}{2}\frac{\rho_{blade} A_{blade} g L^2}{A_{edge} h_x^{edge}}\left(1-\frac{x}{L}\right)^2 \tag{18}$$

with $A_{blade} = 2(A_{flap} + A_{edge})$. In order to achieve a constant stress level, $\sigma_x^{flap} = \sigma_x^{edge} = \sigma_0$, in the blade material, it can be seen from (17) and (18) that the blade height, h_x^{flap}, and blade width, h_x^{edge}, should vary in the following way

$$\frac{h_x^{flap}}{h_0^{flap}} = \left(1+\frac{x}{2L}\right)\left(1-\frac{x}{L}\right)^2 \quad ; \quad \frac{h_x^{edge}}{h_0^{edge}} = \left(1-\frac{x}{L}\right)^2 \tag{19}$$

Following (19) will result in blade geometries which are solely material strength given. In reality, the blade geometry is given by a combination of structural and aerodynamic requirements. In figure 3, equation (19) is compared with typical blade cross-section variations, where the case $L = 86m$ is taken from [3], while the other two cases are representative blade geometries averaged from a number of known blade geometries. The blade height h_0^{flap} and width h_0^{edge} is specified such that (19) is crossing the representative blade geometry in the largest value points which for the thickness direction is at the blade root while it for the width is at a point located a little away from that, see figure 3. During this, it turns out that h_0^{flap} and h_0^{edge} as a good approximation can be assumed to follow linear functions, with L as the blade length

$$h_0^{flap} = 0.066L - 0.369m \quad ; \quad h_0^{edge} = 0.130L + 0.082m \tag{20}$$

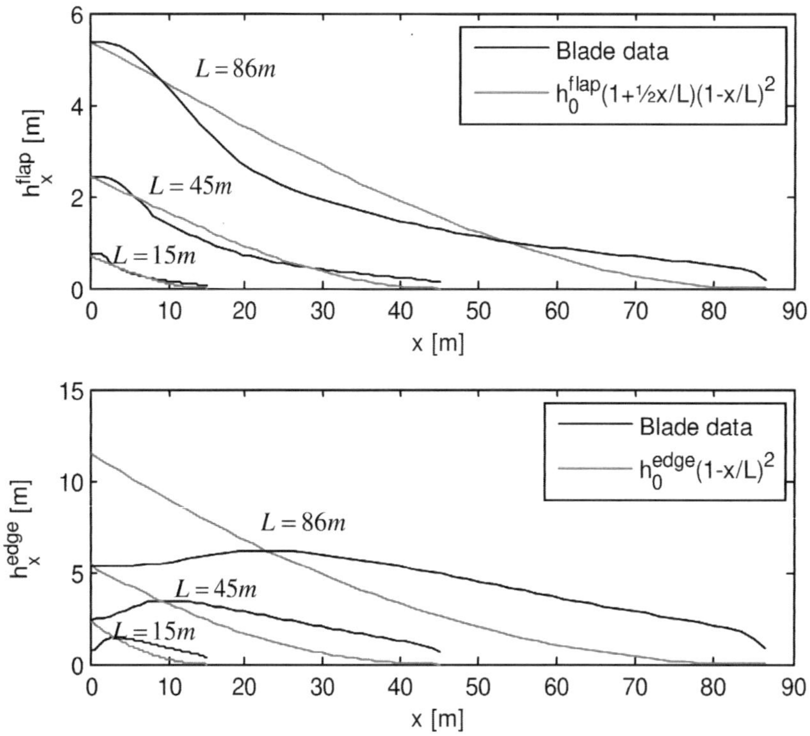

Figure 3 Typical blade geometries compared with the strength determined geometries defined in equation (19).

Table 1. Blade parameters

Type	Parameters
Blade geometry	L ; h_0^{flap} ; h_0^{edge}
Blade material	ρ_{blade} ; σ_0
Rated wind speed	u_R
Density of air	$\rho_{air}=1.2kg/m^3$
Gravitation	$g=9.81m/s^2$

The weight of the load carrying laminates in the blade can be estimated as

$$M_{blade}=2(A_{flap}+A_{edge})L\rho_{blade} \tag{21}$$

with the cross-sections found by combining equation (17), (18) and (19)

$$A_{flap}=\frac{8}{81}\frac{\rho_{air}u_R^2\pi L^3}{\sigma_0 h_0^{flap}} \quad ; \quad A_{edge}=\frac{\rho_{blade}gL^2}{\sigma_0 h_0^{edge}-\rho_{blade}gL^2}A_{flap} \tag{22}$$

Therefore, based on the parameters listed in table 1, the mass of a specific wind turbine blade can be estimated using equation (21) and (22). Keeping in mind the assumption of a constant cross-section along the blade, the height and width variation is obtained aiming for a constant stress value throughout the blade. In addition to a blade design given by the material strength, the blade should also be stiff enough in order to avoid that the blade hits the tower. In [3], the distance from the blade tip to the tower is given as 18m, this is therefore the maximum allowable deflection $w_{\max} = 18\text{m}$ for a $L = 86\text{m}$ turbine. The distance will scale with the turbine size. In the following the transverse deflection, w, of the blade is predicted using Bernoulli-Euler equations

$$\frac{d^2 w}{\partial x^2} = \frac{M_x}{EI_x} \quad \text{with} \quad I_x = \frac{1}{2} A^{flap} \left(h_x^{flap}\right)^2 \tag{23}$$

where M_x is the moment in the beam at the location x and E the material stiffness given later in table 2. Equation (23) is solved using a simple numerical difference scheme shown below where the blade is split up into n parts where, $\theta_0 = 0$; $w_0 = 0$, is the rotation and deflection at the root section.

$$\begin{aligned} \theta_i &= \theta_{i-1} + \frac{1}{2}\left(\frac{M_i}{EI_i} + \frac{M_{i-1}}{EI_{i-1}}\right)\left(x_i - x_{i-1}\right) \quad ; \quad i = 1,....n \\ w_i &= w_{i-1} + \frac{\theta_i + \theta_{i-1}}{2}\left(x_i - x_{i-1}\right) \quad ; \quad i = 1,....n \end{aligned} \tag{24}$$

4. Comparing materials selections for wind turbine blades

Three materials system are considered in the following, see table 2. Due to the high number of load cycles a blade experiences throughout its life time, the strength values are taken as the fatigue failure stress after $5 \cdot 10^8$ cycles.

Table 2. Representative materials properties for glass fibre composites (GFRP), carbon fibre composites (CFRP) and aluminium.

Material	Stiffness	Fatigue strength	Density
GFRP	$E = 44GPa$	$\sigma_0 = 160MPa$	$\rho = 1900kg/m^3$
CFRP	$E = 120GPa$	$\sigma_0 = 300MPa$	$\rho = 1600kg/m^3$
Aluminium	$E = 70GPa$	$\sigma_0 = 100MPa$	$\rho = 2700kg/m^3$

The curves called "fatigue design" in figure 4 show the resulting blade weight as a function of the blade length in the case the material overall is loaded to the fatigue strength, σ_0. In figure 4, this weight variation is compared with the weight of existing blade designs and the 10 MW DTU wind energy reference turbine [3]. In figure 5, the maximum deflections found using the difference scheme (24) are shown. It can be seen that the fatigue given blade design for the glass fibre composite case exceeds the maximum allowed tip deflections given in [3]. Therefore, an additional case is shown in figure 4 and 5 called "deflection design". In that case, the flap area A_{flap} is increase with 50% making it possible to stay below $w_{\max} = 18\text{m}$ for $L = 86\text{m}$. As nearly all the existing wind turbine blades are made of glass fibre composite, this case should be compared with the existing blades. The estimation is found to be 20-40% below the existing turbine blade, which may be considered realistic as only the weight of the load carrying beam in the blade is taken into consideration. In all the estimates, the material cross section, A_{blade}, is for simplicity assumed constant. Nevertheless, in realty, the blade material is often tapered down approaching the end of the blade. During this, a local larger blade height is required in order to make a fatigue strength given design.

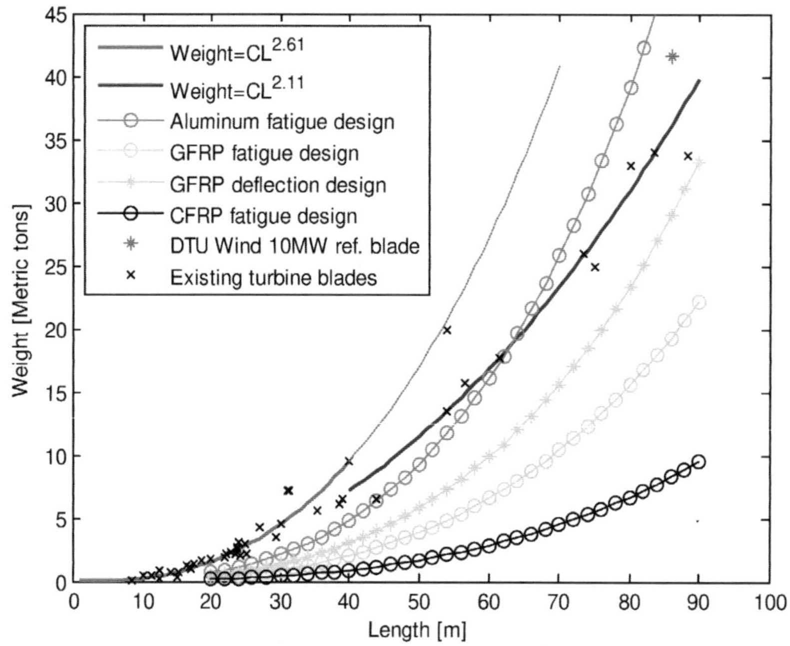

Figure 4. Blade weight predictions and comparison with existing designs

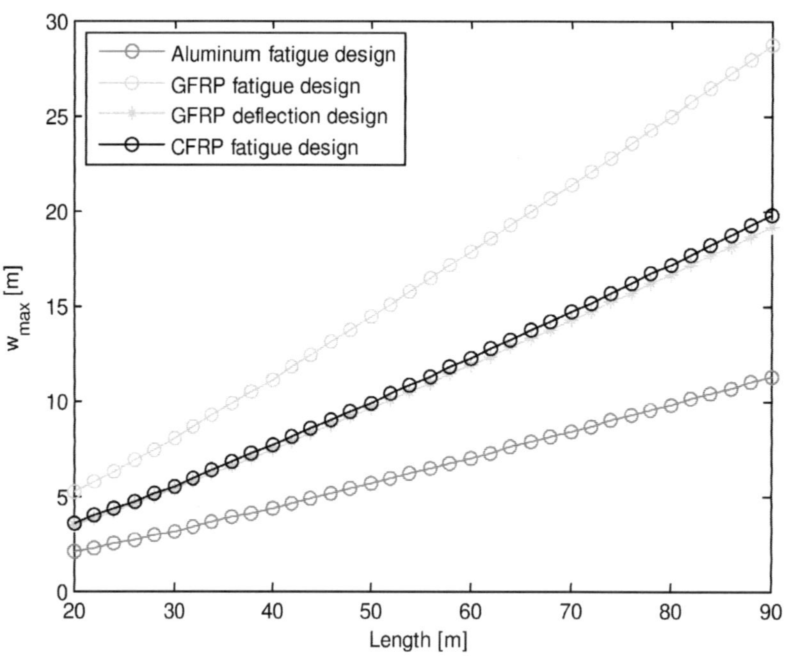

Figure 5. Maximum deflection of the different blades

5. Conclusion

Based on a simplified beam model of a wind turbine blade, is has been shown possible to make a realistic estimate of the weight of the load carrying laminates in a wind turbine blade. Due to its simplicity, the model is well suited for course work investigating scaling effects of wind turbine blades. This approach has been used as a central element in the materials part of the cousera.org course: "Introduction to Wind Energy" [5]. Even though it turns out, that the design of a glass fibre blade is stiffness given, locally the material can still be loaded to its fatigue strength limits. Therefore, optimizing materials for wind turbine blades, large research activities are going on improving the fatigue resistance, see e.g. [6], [7].

Acknowledgments

The author wishes to acknowledge financial support from DCCSM: the Danish Centre for Composite Structures and Materials for Wind Turbines, DSF-grant nr. 09-067212 and from CINEMA: the allianCe for ImagiNg of Energy MAterials, DSF-grant no. 1305-00032B both under the Danish Council for Strategic Research. In addition, senior scientist Kim Branner from DTU Wind Energy is acknowledged for sharing the typical blade geometry data shown in figure 3.

References

[1] Wittrup S, 2016, LM Windpower bygger verdens længste vindmøllevinge *(in Danish)*, *Ingeniøren* [Online], available: https://ing.dk/artikel/lm-windpower-bygger-verdens-laengste vindmoellevinge-185073
[2] Blasques JP and Stolpe M, 2012, Multi-material topology optimization of laminated composite beam cross sections, *Composite Structures* **94** 3278-3289
[3] Bak C, Zahle F, Bitsche R, Yde A, Henriksen LC, Nataraja A and Hansen MH, 2013, Description of the DTU 10 MW Reference Wind Turbine, *DTU Wind Energy Report-I-0092*, Roskilde, Denmark
[4] Tong W, 2010, Fundamentals of wind energy, *WIT Transactions on State of the Art in Science and Engineering, Chapter 1 in Wind Power Generation and Wind Turbine Design* **44** 3–42
[5] Mikkelsen LP, Danielsen HK and Badger M, 2016, Introduction to Wind Energy - Week 4 Materials, *Coursera.org* [Online], available: https://www.coursera.org/learn/wind-energy
[6] Hüther J and Brøndsted P, 2016, Influence of the curing cycles on the fatigue performance of unidirectional glass fiber reinforced epoxy composites, In *IOP Conference Series: Materials Science and Engineering*, to be published
[7] Jespersen KM and Mikkelsen LP, 2016, Fatigue damage observed non-destructively in fibre composite coupon test specimens by X-ray CT, In *IOP Conference Series: Materials Science and Engineering*, to be published

IOP Conf. Series: Materials Science and Engineering **139** (2016) 012039 doi:10.1088/1757-899X/139/1/012039

Hierarchical nanoreinforced composites: Computational analysis of damage mechanisms

Leon Mishnaevsky Jr. [1]**, Alessandro Pontefisso**[1] **and Gaoming Dai** [2]

[1]Technical University of Denmark, Risø Campus, Department of Wind Energy, Frederiksborgvej 399, 4000, Roskilde, Denmark

[2]Siemens Wind Power A/S, Assensvej 11, 9220 Aalborg, Denmark

E-mail: lemi@dtu.dk

Abstract. The potential of hierarchical composites with secondary nanoreinforcement is discussed and analysed on the basis of the computational modelling. The concept of nanostructuring of interfaces as an important reserve of the improvement of the composite properties is discussed. The influence of distribution, shape, orientation of nanoparticles (carbon nanotube, graphene) in unidirectional polymer matrix composites on the strength and damage resistance of the composites is studied in computational studies. The possible directions of the improvement of nanoreinforced composites by controlling shapes, localization and other parameters of nanoreinforcements are reviewed.

1. Introduction

Lightweight strong materials are sought for in many areas of industry. Most often, polymer based composites are considered as most promising lightweight strong materials. However, further enhancement of the composite properties is required for many applications, including better lifetime and fatigue resistance, multiple applications, and so on. In order to achieve the enhancement of the composite properties, improvement of constituents (fiber, polymer) and manufacturing technologies is used. The development of hybrid composites, with nanoeingineered phases, is a very promising direction to design lightweight materials with improved properties [1].

The fiber reinforced composites with nanoengineered matrix show often much better performances than the composites without any nanoreinforcements. For instance, 80% improvement of fracture toughness of carbon fiber reinforced epoxy composites achieved as a result of 0.5 wt.% CNT addition of carbon nanotubes (CNTs) [2]. 30% enhancement of the interlaminar shear strength of woven carbon fabric in epoxy matrix due to the deposition of multi and single walled CNT on fibers [3].

The potential advantages of hierarchical composites over neat composites include the synergy between structural elements at several scale levels, possibility to improve competing properties of materials [4], potential to improve the interface controlled properties, like compression and fatigue strength, by placing nanoparticles on interfaces or fiber sizing, combining the advantages of nanomaterials and composites.

In this paper, the studies of the potential of the improvement of composite service properties based on hierarchical materials design and secondary nanoscale reinforcement (carbone nanotubes, graphene), carried out at the Department of Wind Energy, Technical University of Denmark, are summarized.

2. Nanostructured interfaces as a source of the improvement of composite properties

Figure 1 shows several examples of hierarchical structures of materials with exceptional properties, observed in experiments: nacre structure (brick mortar structure with polymer layers bridged by mineral nanoparticles), ultrafine grained metal (titanium) with atomistic size particles of grain boundaries [7]-[8] and fiber reinforced composites with secondary nanoreinforcement [9]-[11].

So, nacre of mollusks contains 95% of aragonite platelets and only 5% of biopolymer. Thin layers of biopolymer between the platelets can be considered as interfaces in this material [5]. The biopolymer layers represent organic macromolecules, containing polysacharides and protein fibers. Furthermore, the biopolymer thin layers of nacre contain nanopores and also inorganic mineral bridges, linking the aragonite platelets [5]-[6], see Figure 1. Song and Bai [5] evaluated the fracture toughness in the "brick bridge mortar" structure of nacre and showed that the availability of nanostructures in the nacre interfaces (i.e., mineral bridges between aragonite platelets located in the bioolymer layers) is one of the reasons for the high toughness of nacre. The mineral bridges reinforce the weak interface, and control the crack propagation in the interfaces (biopolymer layers). Layers of constrained disturbed polymers surrounding nanoparticles in polymer ensure the drastic enhancement of nanocomposite properties observed in experiments (like 200% increase in stiffness or strength achieved at 0.5% nanoparticle content) [12]. Nanoparticles (nanoclay or carbon nanotubes) located in fiber/matrix interfaces (fiber sizing) of unidirectional composites allow to increase the fatigue lifetime of the composites drastically, as observed experimentally in [10], [11], [14]. One of reasons of extraordinary strength and toughness of nanocrystalline ultrafine grained metals is the high content of grain boundary phases in these materials [15]. Their properties can be further enhanced if the grain boundaries are non-equilibrium, with high density of dislocations and especially with foreign atoms/precipitates [7]. The availability of precipitates strongly delays the damage growth and ensures 83% increase in the critical strains due to the precipitates, and around 300% increase due to the precipitates located in grain boundaries.

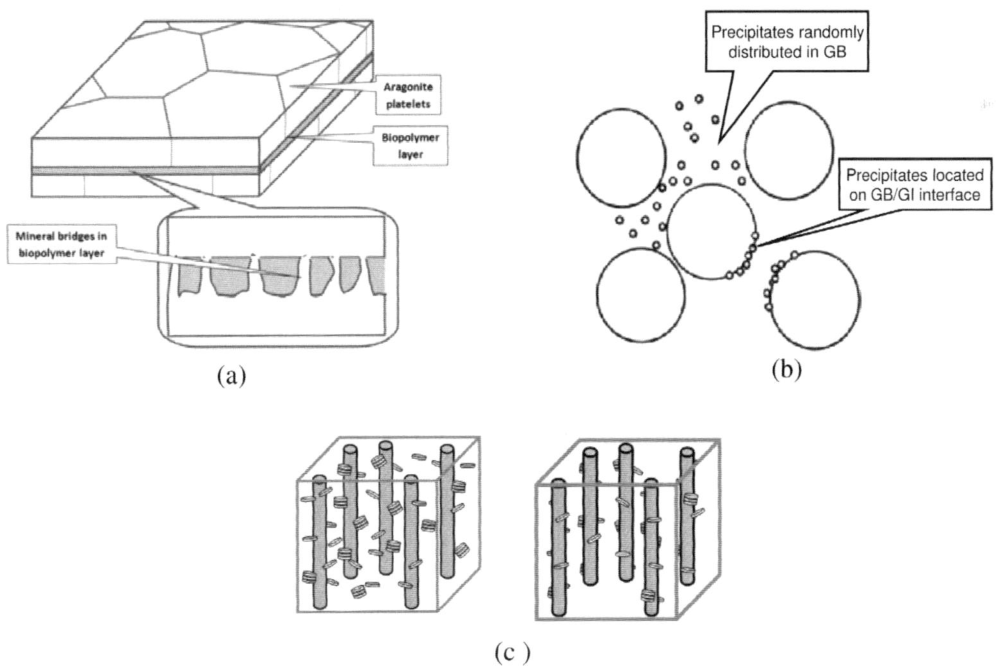

Figure 1. Examples of materials with nanostructures interfaces and grain boundaries: (a) Nacre structure (brick mortar structure with polymer layers bridged by mineral nanoparticles), (b) ultrafine grained metal (titanium) with atomistic size particles of grain boundaries [6]-[7] (here, GB means "grain boundary" of the metal) and (c) fiber reinforced composites with secondary nanoreinforcement [8]-[10]. Reprinted from [16], [8] and [11] with kind permission of Elsevier

On the basis of generalization of these observations, it was demonstrated in [16], that the purposeful nanostructuring of interfaces and grain boundaries represents an important reserve of the improvement of the materials properties [16]. Since the material deformation is often localized in and around defects (interfaces and grain boundaries), the structuring of these regions (adding specially arranged and oriented nanoreinforcements, or adding nanoscale defects, changing the local properties) allows to control the deformation and fracture behavior of these weak areas, thus, determining the degradation process in the whole material. Quite often, the interphase layers with low stiffness lead to the localization of deformation, while the internal structures of the interphase layers (like mineral bridges in nacre, or nanoplatelets in sizing of fiber reinforced composites) allow to control the deformation, damage initiation and fracture processes locally.

3. Computational modelling of hierarchical composites

In order to analyze the effect of the distribution of secondary nanoparticles in the matrix and in the interface on the damage behaviour of hierarchical composites, a computational multiscale model was developed, which includes the fiber/matrix interaction at the higher scale level (microlevel) and nanoparticles/epoxy matrix interaction on nanolevel. A set of programs for the automatic generation of 3D multiscale models of composites was developed [11]-[17]. The programs generate command files for the commercial FE software ABAQUS. The unit cell structures are divided into two levels. The high and lower level models are linked by submodelling techniques. The macro (upper level) unit cells contain three phases: the matrix, fibers and "third phase" interphase layers (which characterizes the interface roughness, interphases [1], [18]. Both matrix and the interphase layer might contain nanoreinforcements. A number of 3D multiscale unit cell models of nanoparticles (carbon nanotubes, nanosheets) and fiber reinforced composites have been generated and used for the computational testing of structures of composites. Each unit cell model (upper or lower level) has an average of 100…300k elements.Figure 2 shows an example of a unit cell model. In this model, polymer composites with glass fibers (GFRC), carbon fibers (CFRC) and hybrid fibers (HC) are reinforced with carbon nanotubes. The aspect ratio of CNTs was 1000, and they were distributed in the sizing of the fibers.

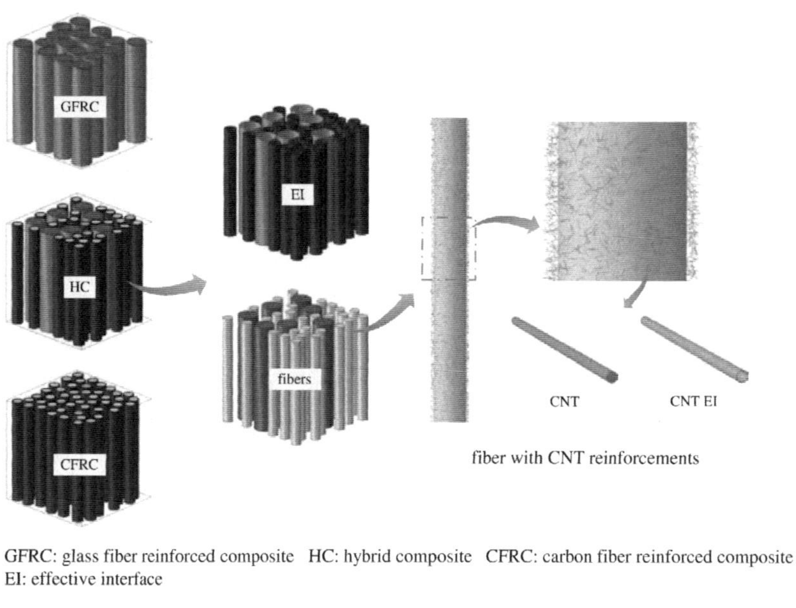

Figure 2. Example of unit cell model of CNT reinforced hybrid composite.Reprinted from [10] with kind permission of Elsevier

The volume fraction of the CNT reinforcement is 0.46%. The model was subject to cyclic compression-tension loading [10]. The crack propagation analysis is carried out in the framework of the linear elastic fracture mechanics (LEFM) approach. The materials are supposed to be elastic. The criteria of damage modelling, materials properties and other details of simulations are described in details in [10].

As a result of simulations, it was shown that the CNT enhances the fatigue performance (stress of the S-N curves corresponding to the same cycles, and lifetime) in all considered composites [10]. For the very low cycle loading, the CNT reinforcement leads to 25%...43% increase in the stress corresponding to a given cycles number (here, taken as 1.72 x 10^7 cycles), while for the millions of cycles, the CNT effect increases the stress by 64…120%. Similarly, the simulations results show that the nanoclay particles located in the matrix or in the fiber sizing, enhance fatigue lifetimes of composites [11].

Figure 3 shows 5 a schema of model of hierarchical fiber composite with nanoclay reinforcements (a) and simulated crack paths in in-between platelets observed in the simulations (aligned and randomly oriented platelets, b and c) [11]. In the simulations, the composites with nanoclay reinforcement achieve the same fatigue life (taken exemplarily at $5.68*10^7$ cycles) as neat composites, while subject to 2…3.5 times higher loadings. As observed in the simulations, composites with the nanoplatelets localized in the fiber/matrix interphase layer (fiber sizing) lead to much higher fatigue lifetime than those with the nanoplatelets in the matrix: 43…49% higher applied stress corresponding to the selected lifetime of $5.68*10^7$ cycles.

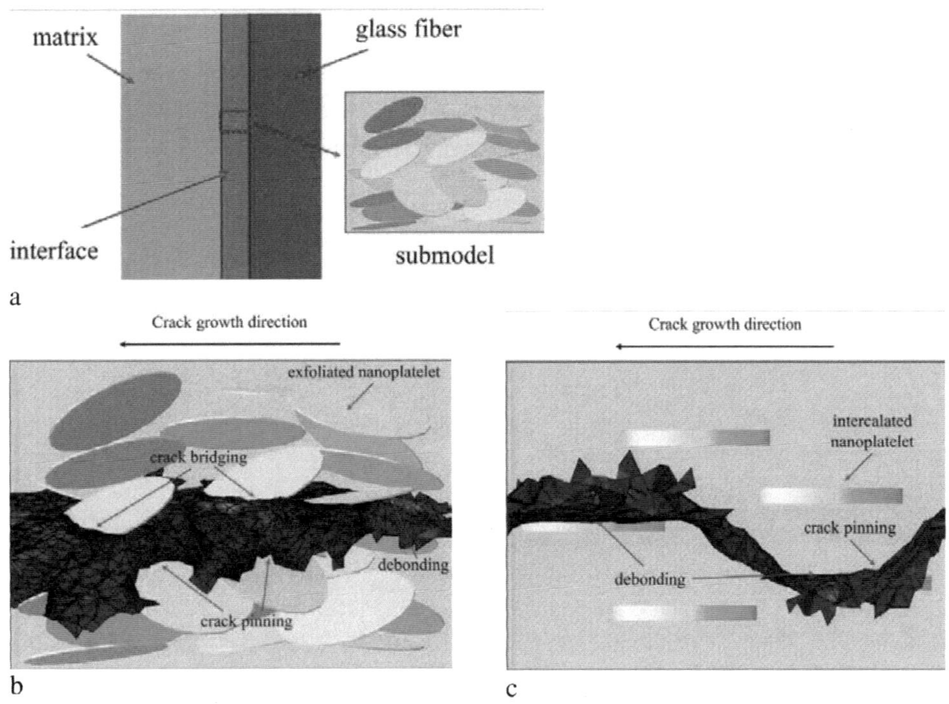

Figure 3. Schema of model of hierarchical fiber composite with nanoclay reinforcements (a) and simulated crack paths in submodels (aligned and randomly oriented platelets, b and c). Reprinted from [[11]] with kind permission of Elsevier

4. Which nanoscale factors control the efficiency of secondary reinforcement?

Nanoscale structure, distribution, shapes and arrangement of secondary nanoreinforcement play a very important role for the optimization of the service properties of hierarchical composites. The effect of the shape, distribution and orientation of nanoparticles (graphene and carbon nanotubes) has been studied in a number of works [13][17]. As demonstrated in the computational studies in [13], the factors influencing the elastic properties and strength of graphene reinforced polymer nanocomposites can be ranked as follows (from stronger to weaker effects): crumpled graphene shape (only for strength: 50% reduction of strength only due to the shape crumpling, but only 10% reduced stiffness) -> graphene sheet misalignment (~50% reduction of stiffness and strength because of misalignment) -> volume content of graphene (50% higher stiffness and 2.5 times higher strength at 4 times higher content of graphene) -> interphase layer strength (7 times higher interphase strength gives only 2.5 times higher composite) and clustering degree (only 25% difference between highly clustered and exfoliated) -> the aspect ratio of sheets (~60% lower strength and 26% higher stiffness due to 20 times higher aspect ratio).

The effect of the shape of carbon nanotube (CNT) reinforcements and hybrid CNT/graphene reinforcements on the strength and damage behavior of nanocomposites has been studied numerically in [17]. Further, the real snake-like shape of carbon nanotubes [19] was taken into account, instead of oversimplified cylindrical models used most often. In order to generate the real snake-like shape of carbon nanotubes (CNT), particles are represented as a chain of hexahedral segments, which is built sequentially, while letting each segment to misalign with respect of the one which precedes it. These segments are joined in the solid modeler module of ABAQUS [17]. The crack growth was modeled using the finite element weakening techniques, implemented in the subroutine User Defined Field. The critical damage value and other simulations details are given in [17].

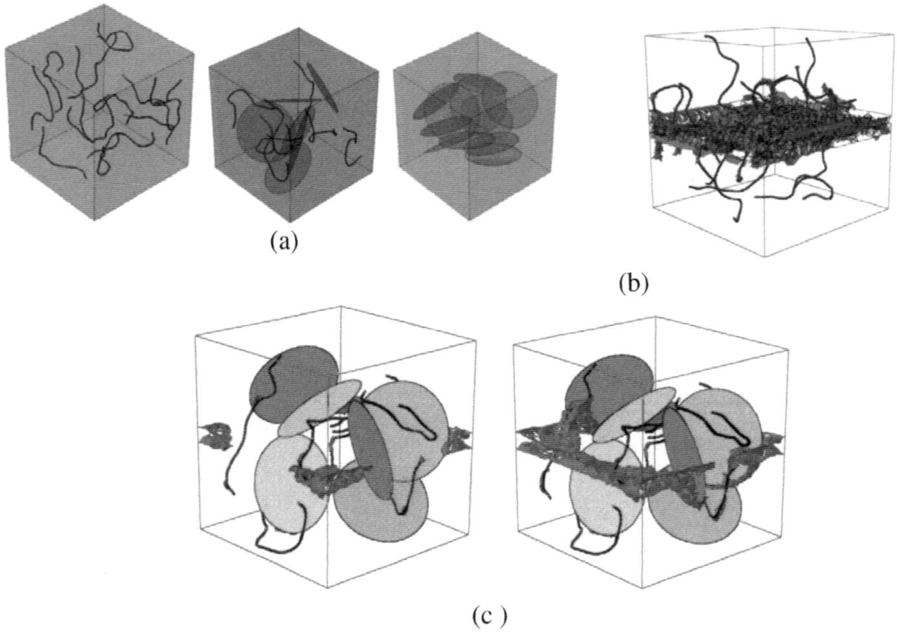

Figure 4. Schema of model of hierarchical fiber composite with nanoclay reinforcements (a) and simulated crack paths in submodels (aligned and randomly oriented platelets, b and c). Reprinted from [11] with kind permission of Elsevier

Figure 4a shows the examples of the unit cells with snake-like CNT particles, pure graphene and hybrid CNT/graphene reinforcement. Figure 4bc show the crack path in the unit cell with pure CNT reinforcement and hybrid CNT/graphene reinforcement. As a result of the computational simulation of damage evolution in the unit cells, it was demonstrated that that the synergy effect of hybrid CNT and graphene nanoreinforcement is caused by different interaction of nanoparticles with cracks at different stages of fracture. While graphene particles cause the crack deviation at the early stages of fracture, the CNTs ensure the debonding and fiber bridging mechanisms after the main crack is formed.

In the simulations it was observed that the peak stress for the composite with idealized, cylindrical reinforcement is slightly higher than that for the composite with more realistic, snake-like nanotube reinforcement. Further, it was shown that the real, snake like reinforcements induce higher local stresses and also lead to the earlier failure as compared to the idealized cylinder like models.

5. Conclusions

In this paper, an overview of recent studies of the potential of improving the strength and damage resistance of composites by using hierarchical structure and secondary nanoscale reinforcement, carried out over last years at the Section of Composites and Mechanics of Materials, Department of Wind Energy, Technical University of Denmark, is presented. It is demonstrated that the hierarchical structures of composites represent a promising way for the enhancement of the materials properties. An interesting and promising way to improve the composite strength is nanostructuring of interfaces and other deformable areas, ensuring the control of damage mechanisms. The shapes, orientation and localization of secondary nanoparticle reinforcement play a very important role for the optimization of the materials performances.

Acknowledgement

The author gratefully acknowledges the financial support of the Danish Innovation Foundation via the project "High reliability of large wind turbines via computational micromechanics based enhancement of materials performances" (Ref. no. 10-094539).

References

[1] Mishnaevsky Jr L, 2007 *Computational Mesomechanics of Composites,* John Wiley, 280 pp.
[2] Godara A, Mezzo L, Luizi F, Warrier A, Lomov S, van Vuure AW, Gorbatikh L, Moldenaers P, Verpoest I 2009 *Carbon,* Vol.47, 12, pp. 2914-2923
[3] Bekyarova E, Thostenson ET, Yu A, Kim H, Gao J, Tang J, Hahn HT, Chou TW, Itkis ME, Haddon RC 2007 *Langmuir* 23 (7) pp. 3970–3974
[4] Kanzaki S, Shimada M, Komeya K and Tsuge A (1999), *Key Engineering Materials,* 161-163, pp. 437-442
[5] Song F and Bai Y L 2003 *J. Mater. Res.,* Vol. 18, No. 8
[6] Song F, Soh AK, Bai YL *Biomaterials* 24, pp 3623–3631
[7] Mishnaevsky Jr. L al. 2014 *Materials Science & Engin* R.Vol. 81, 2014, pp. 1–19
[8] Liu HS, Mishnaevsky Jr L 2014 *Acta materialia,* Vol. 71, pp. 220-233
[9] H.W. Zhou HW, Mishnaevsky Jr. L, Yi HY, Liu YQ, Hu X, Warrier A, Dai GM, *Composites B,* 2016, 88, 201-211
[10] Dai GM, Mishnaevsky Jr L 2015 *Composites B* (2015), pp. 349-360
[11] Dai GM, Mishnaevsky Jr L 2014 *Composites Science and Technol,* Vol. 91, pp. 71-81
[12] Peng RD, Zhou HW, Wang HW, Mishnaevsky Jr L 2012. *Computational Materials Science,* 60 pp 19–31
[13] Dai GM, Mishnaevsky Jr L 2014 *Computational Materials Science,* Vol. 95, pp 684–692
[14] Mishnaevsky Jr L, Dai GM, 2014 *Computational Materials Science,* Vol. 81, pp 630-640
[15] Valiev RZ, Alexandrov IV, Enikeev NA, Murashkin MY 2010 *Rev.Adv.Mater.Sci.* 25, pp. 1-10
[16] Mishnaevsky Jr L, 2015 *Composites B,* Vol. 68, pp. 75–84
[17] Potenfisso A, Mishnaevsky Jr L, 2016, *Composite B,* pp. 338-349
[18] Mishnaevsky Jr L and P. Brøndsted P 2009, *Composites Sci and Technol,* 69/7-8, 2009, pp. 1036-1044
[19] Sidorenko D, Mishnaevsky Jr. L, Levashov E, Loginov P, Petrzhik M, 2015, *Materials and Design,* 83, pp. 536-544

IOP Conf. Series: Materials Science and Engineering **139** (2016) 012040 doi:10.1088/1757-899X/139/1/012040

The influence of removing sizing on strength and stiffness of conventional and high modulus E-glass fibres

Helga Nørgaard Petersen[1], Yukihiro Kusano[2], Povl Brøndsted[2] and Kristoffer Almdal[1]

[1] Department of Micro- and Nanotechnology, Technical University of Denmark, Ørsteds Plads Building 345C, DK-2800 Kgs. Lyngby, Denmark
[2] Department of Wind Energy, Section of Composites and Materials Mechanics, Technical University of Denmark, Risø Campus, Frederiksborgvej 399, DK-4000 Roskilde, Denmark

E-mail: hnpe@dtu.dk

Abstract. Two types of E-glass fibres, a conventional and a high modulus where the last one in the following will be denoted as ECR-glass fibre, were investigated regarding density, diameter, stiffness and strength. The fibres were analysed as pristine and after sizing removal treatments. The sizing was removed by either burning at 565°C or soxhlet extraction with acetone. It was found that the density and the stiffness increased after removing the sizing by the two removal treatments whereas the diameter did not change significantly. The strength of the fibres decreased after burning as the sizing, protecting against water and fibre-fibre damage, had been removed. The strength of the fibres after extraction was not significantly different from the strength of the pristine fibres despite removing the sizing. This indicates that the bonded part of sizing is still protecting the glass fibre surface.

1. Introduction

There is a number of different types of glass fibres all based on silica but with different composition of additives. The silica network creates a three dimensional structure which is amorphous when the melt is rapidly cooled during manufacturing of glass fibres. Some of the additives are mixed in the melt to lower the melting temperature to around 1000°C by spoiling the native structure of silica, others are added to obtain certain bulk properties for specific usage of the fibres [1]. Aluminium, titanium, and zirconium oxides can be included in the silica tetrahedron network increasing the stability but creating charged area derived from lack of oxygen. Other oxides additives such as calcium and magnesium oxides will even out charges and hence modify the network as they are larger and have only one or two bonds to the silica network [1, 2].

E-glass fibres contain mainly aluminium and boron oxides as additives and were initially made for electical applications. They are the most used in fibre reinforced composites because of their high strength and low cost. ECR-glass fibres are a more chemically resistant version of the E-glass, but without boron oxide. Other types of glass fibres include C-, AR-, R- and S-glass fibres with different abilities and usages: corrosive acid resistance, alkali resistance used in cement and concrete, high mechanical strength and acid corrosive resistance, and high mechanical strength and stability in environments with high temperatures and corrosion, respectively [1–3].

Manufacturing of glass fibres is almost the same for all types of fibres. The constituents are mixed and melted at high temperatures in a vessel from which the fibres are drawn through bushings. The fibres are cooled by water spray immediately after and subsequently the sizing is applied, often by passing a roll immersed in the sizing mixture. Then the fibres are gathered in strands and winded on bobbins [1, 2]. The diameter is determined by the size of the bushing holes, the melt temperature, and the draw speed [2]. The strength of the fibres is mainly determined by the glass composition but also by how quickly the melt is quenched, and surface flaws. The flaws can be divided into two types. The first type develops during manufacturing as a result of impurities or stresses induced by the quenching. The second type ocurs because of the subsequent handling, mainly fibre-fibre damages. Existing flaws will extend during tensile stress especially in moist environments ultimately leading to breakage [4, 5].

The sizing applied during manufacturing is a multicomponent surface coating with multiple functions. It is an aqueous suspension with two main elements making up around 90 wt.% of the dry components: a coupling agent and a film former [6]. The coupling agent is considered the most important as it is believed that it controls the adhesion between fibre and matrix. The film former is the protector of the surface against water attack and fibre-fibre damages [7, 8]. Furthermore the sizing is suggested to heal flaws in the glass fibre surface and also reduce the risk of new flaws to arise [9, 10]. Sized glass fibres exhibited higher strength compared to unsized fibres [11]. It has previously been shown that removing the sizing by heating increases the stiffness but at the cost of a decrease in the fibre strength [3, 12, 13]. The heating increases the dimensions and the amount of flaws in addition to the removal of the healing sizing [14].

The sizing applied during manufacturing is needed to be able to handle the fibres until usage. Removing the sizing should therefore be done right before using them e.g. in composites. Burning the fibres at high temperatures will remove all the organic material from the fibres. By extraction it is possible to remove the part of sizing that is physically adsorbed to the surface but not the bonded part. The degree of extraction depends however highly on the solvent used [6, 15–17]. The sizing removed by heat treatment and/or solvent extraction and the glass fibres before and after the sizing removal were intensively characterised [18].

In this paper commercial E-glass and ECR-glass fibres were investigated. Density, diameter, stiffness and strength were measured on pristine fibres and fibres where the sizing had been removed.

2. Methods and Materials

Two types of fibres were analysed: E-glass and ECR-glass fibres. The E-glass fibres were produced by Jushi (Chengdu, China) and the following properties was given by the manufacturer: tex number of 1200, 17 μm diameter of monofilament, moisture content $\leq 0.10\%$ and sizing content in the range of 0.45-0.70 wt.%. The ECR-glass fibres were produced by Owens Corning (Brussels, Belgium) with a tex number of 1200 and a diameter of 17 μm. Soxhlet extraction with acetone of sizing from pristine glass fibres and burning pristine fibres at 565°C were reported previously [18].

Density measurements were performed using a gas pycnometer (Quantachrome Ultrapyc 1200e, England) with nitrogen gas as the displacement medium. Prior to analysis the fibres were dried at room temperature in a vacuum chamber overnight. The fibres were then cut into small pieces to fit in the sample cup and placed in the pycnometer. 2-5 grams of fibres were used. Each measurement was repeated at least eight times.

Tensile testing of single glass fibres were performed on a Favimat+ with a Robot2 from Textechno, Germany. The linear density was measured by recording the resonance oscillation of fibre vibration induced by acoustic waves, according to ASTM D 1577. The individual fibre diameter and cross section were calculated from the linear density and the density measured by the pycnometer. The tensile test was conducted subsequently on the same fibre just used to determine the cross sectional area. 75-100 fibres were tested for each type (E-glass and ECR-glass fibres) and treatment (pristine, fibres after soxhlet extraction, and burning) with same gauge length of 40 mm and test speed of 1 mm/min.

3. Results and Discussion

The measurement of density was done three to eight times by the pycnometer until a satisfying standard deviation of less than 0.02% was obtained within three consecutive measurements. If this was not reached after the eighth measurement the data was discarded. This average density determination was repeated at least eight times. The density was determined for both glass types also after soxhlet extraction and after burning. The measured densities are shown in figure 1.

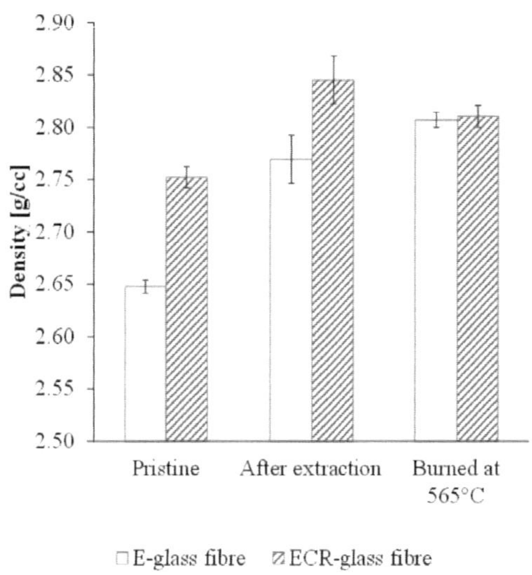

Figure 1. The densities of E-glass and ECR-glass fibres pristine and after removal of sizing by soxhlet extraction or burning at 565°C, all determined by use of a pycnometer.

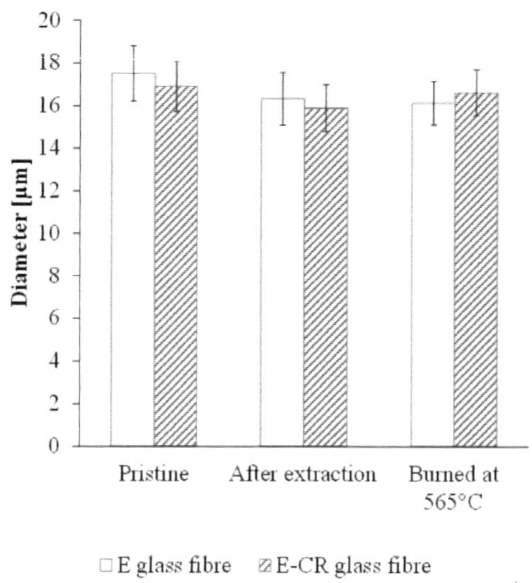

Figure 2. Diameter of the tested fibres (pristine fibres, fibres after soxhlet extraction and fibres after burning at 565°C) calculated from the linear density measured with vibroscopy.

The obtained densities of pristine E-glass and ECR-glass fibres were close to literature values of 2.58 and 2.72 g/cc, respectively [3]. The density of the glass fibres has increased after removal of sizing both by soxhlet extraction and by burning at 565°C. When estimating a density of sizing of 1 g/cc based on typical sizing components and thickness, it is reasonable that the removal of sizing is the cause of the higher density of fibres after soxhlet extraction. Extraction of the sizing is not believed to change the glass structure.

It is noticeable that the density of both E-glass and ECR-glass fibres after burning reaches the same level around 2.81 g/cc. The compositions of E-glass and ECR-glass fibres are very much alike. The main difference is that the ECR-glass is without boron oxide. Instead it contains zirconium oxide [2, 19]. Structurally they are both a random silicon oxide network with aluminium cations as a part of the network. Boron oxide is a part of the network on equal terms with silica where zirconium oxide is a intermedia along with aluminium [1]. Heating the glass fibres to a temperature near annealing temperature could change the bulk structure of the glass fibres making them even more alike. The burned E-glass fibres have a even higher density than the fibres after soxhlet extraction. This can be explained by glass compactment where internal stresses in the glass structure are released. As a result the fibres contract yielding a higher density [13]. It was noted that the fibres after extraction were very electostatic since the antistatic agent in the sizing had been removed, but no other visible changes. The fibres after burning were much stiffer, behaved less flexible and appeared to have small bends from lying on top of other fibres while being heated, all supporting that the glass structure had changed to a more rigid network as a part of the relaxation.

The diameters of the E-glass and ECR-glass fibres were calculated in connection to the single fibre tensile testing from the linear density. The diameters are shown in figure 2. It was expected that the burned fibres had a slightly smaller diameter after the heat treatment. However, the diameter does not change significantly after heat treatment nor after soxhlet extraction. The large variation in diameter of all the fibres obstructs the possibility to observe an alteration of size and shape of the glass fibre.

Between 100 and 125 fibres of each type and treatment were tested but not all were used for the calculation of the average stiffness and strength of the fibres. Some of the data were excluded due to breakage during the linear density measurement or shortly after initiation of the tensile test, for example, if two fibres were tested simultaneously by mistake and if the fibre was sliding in the clamps during testing. Especially the burned fibres were troublesome to test as they were very fragile, so as few as 25 data points were left, whereas up to 73 data points were obtained when testing the fibres after soxhlet extraction. Stiffness was calculated based on the initial slope of the stress-strain curve in the range of 0.05-0.25% strain. The strength was calculated from the maximum force measured. The obtained values of stiffness and strength of single glass fibres as pristine fibres and after treatments are shown in figure 3 and 4, respectively.

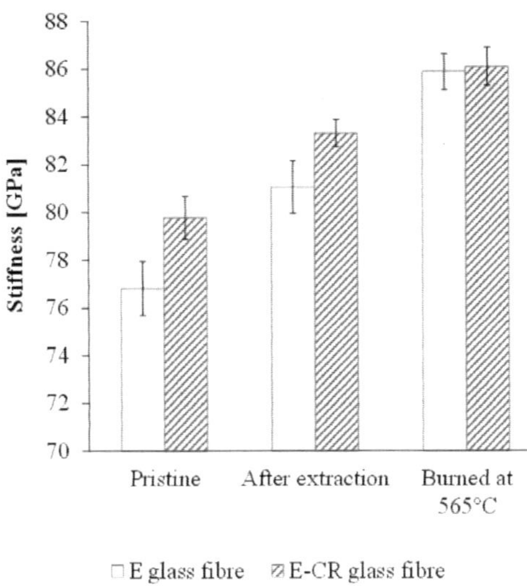

Figure 3. The stiffness of E-glass and ECR-glass fibres pristine and after removal of sizing by soxhlet extraction or burning at 565°C, obtained by single fibre tensile testing.

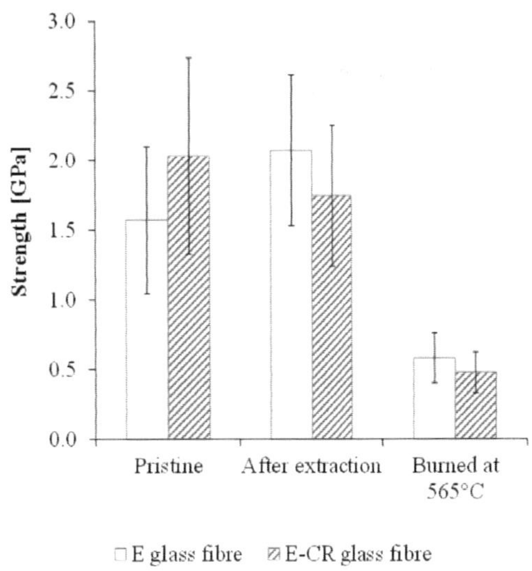

Figure 4. The strength of E-glass and ECR-glass fibres pristine and after removal of sizing by soxhlet extraction or burning at 565°C, obtained by single fibre tensile testing.

The pristine E- glass and ECR-glass fibres were found to have a stiffness of 77 GPa and 80 GPa, respectively, which are close to literature values [2, 3]. The stiffness of the fibres was found to increase after both soxhlet extraction and heat treatment with 3-9 GPa. Similar results for heat treated glass fibres have been found in literature with the explanation that stresses induced by the rapid cooling during manufacturing are released [3, 13]. The strength of the pristine fibres are much lower than values given in literature [2, 3, 19] from which a strength of around 3 GPa was expected. The burned fibres do however still follow the decreasing tendency and have similar values as found in literature [11–13, 20]. The low strength is related to enlargement of existing surface flaws that become vulnable to attack from water [20]. The fibres after soxhlet extraction exhibit strengths close to the one obtained for the pristine fibres with no significant difference. They are however close to strengths measured on bare E-glass fibres [11]. It should be noted that a compliance correction have not been executed since only one gauge length was used. This could result in slightly lower values caused by compliance in the testing equipment [21].

4. Conclusion

The densities of E-glass and ECR-glass fibres were found to increase by 5% and 3%, respectively, when the sizing had been removed by soxhlet extraction using acetone. When removing the sizing by burning at 565°C the increase was 6% and 1%, respectively. The changes in density were related to removal of the less dense sizing and to compactment of the glass structure.

For E-glass fibres the stiffness increased from 77 GPa to 81 GPa after extracting the sizing by soxhlet and even to 86 GPa when pristine fibres were burned. Opposite was found with the strength of the fibre after burning with a decrease from 1.6 GPa to 0.6 GPa, whereas soxhlet extraction gave an increase to 2.1 GPa. ECR-glass fibres exhibit the same tendency regarding stiffness as the E-glass fibres where the pristine fibres were found to have a stiffness of 80 GPa, increasing to 83 GPa after soxhlet extraction and to 86 GPa when burned. The strength changed differently with the treatment for the ECR-glasss fibres with a large decrease after soxhlet extraction from 2.0 GPa to 1.8 GPa where burning yield a strength of 0.48 GPa.

The increase in stiffness after burning was related to relaxation of the glass structure where the decrease in strength was associated with enlargement of flaws and water attack after removal of the protective sizing. The processes behind the increase of strength when removing the sizing by extraction has not been clarified but could be related to the bonded sizing maintaining the healing effect of the surface flaws. The two types of glass fibres behaved very much alike.

Acknowledgements

This work was supported by the Danish Centre for Composite Structures and Materials for Wind Turbines (DCCSM) from the Danish Research Council for Strategic Research (grant number: 09-067212). In connection to the density measurements and tensile testing the authors would like to thank Justine Beauson and Anthony Fraisse.

References

[1] Dwight D W 2000 *Comprehensive composite materials* vol 1 ed Kelly A and Zweben C (Amsterdam: Elsevier) pp 231–61

[2] Hearle J W S 2001 *High-performance fibres* (Cambridge: Woodhead Publishing Ltd.)

[3] Hartman D, Greenwood M and Miller D 1996 High strength glass fibers Tech. rep. Owens Corning Corp.

[4] Hara M 1988 *Glastechnology Bericht* **61** 191–6

[5] Dibenedetto A T and Lex P J 1989 *Polymer Engineering and Science* **29** 543–55

[6] Gorowara R L, Kosik W E, McKnight S H and McCullough R L 2010 *Composites Part A* **32** 323–9

[7] Thomason J L and Adzima L J 2001 *Composites Part A* **32** 313–21

[8] Jones F R 2010 *Journal of Adhesion Science and Technology* **24** 171–202

[9] Zinck P, Mäder E and Gerard J F 2001 *Journal of Materials Science* **36** 5245–52

[10] Ershad–Langroudi A, Mai C, Vigier G and Vassoille R 1997 *Journal of Applied Polymer Science* **65** 2387–93

[11] Jenkins P G, Yang L, Liggat J J and Thomason J L 2015 *J Mater Sci* **50** 1050–7

[12] Brearley W and Holloway D G 1963 *Physics and Chemistry of Glasses* **4** 69–75

[13] Otto W H 1961 *Journal of The American Ceramic Society* **44** 68–72

[14] Piggott M R and Yokom J C 1968 *Glass Technology* **9** 172–5

[15] Tanoglu M, Ziaee S, Mcknight S H, Palmses G R and Gillespie J W 2001 *Journal of Materials Science* **36** 3041–53

[16] Feih S, Wei J, Kingshott P and Sørensen B F 2005 *Composites Part A* **36** 245–55

[17] Thomason J L 1995 *Composites* **26** 487–98

[18] Petersen H N, Kusano Y, Brøndsted P and Almdal K 2013 *Proc. of the 34th Risø Int. Symp. on Materials Science: Processing of fibre composites - challenges for maximum materials performance. 2-5 September, Roskilde, Denmark* ed Madsen B and *et al* (Technical University of Denmark) pp 333–40

[19] Brøndsted P, Lilholt H and Lystrup A 2005 *Annual Review of Materials Research* **35** 505–38

[20] Feih S, Boiocchi E, Mathys G, Mathys Z, Gibson A G and Mouritz A P 2011 *Composites Part B* **42** 350–8

[21] Yang L and Thomason J L 2012 *Journal of Materials Science* **48** 1947–54

Size scale dependence of compressive instabilities in layered composites in the presence of stress gradients

K Poulios and C F Niordson

Department of Mechanical Engineering, Solid Mechanics, Technical University of Denmark, Nils Koppels Allé, DK-2800 Kgs. Lyngby, Denmark

E-mail: kopo@mek.dtu.dk

Abstract. The compressive strength of unidirectionally or layer-wise reinforced composite materials in direction parallel to their reinforcement is limited by micro-buckling instabilities. Although the inherent compressive strength of a given material micro-structure can easily be determined by assessing its stability under a uniform compressive load, this is often not sufficient for predicting failure initiation within a larger structure. In cases, where the composite material micro-structure is locally subjected to strongly non-uniform loadings, compressive instabilities depend not only on the maximum compressive stress but also on spatial stress or strain gradients, rendering failure initiation size scale dependent. The present work demonstrates and investigates the aforementioned effect through numerical simulations of periodically layered structures with notches and holes under bending and compressive loads, respectively. The presented results emphasize the importance of the reinforcing layer thickness on the load carrying capacity of the investigated structures, at a constant volumetric fraction of the reinforcement. The observed strengthening at higher values of the relative layer thickness is attributed to the bending stiffness of the reinforcing layers.

1. Introduction

The load carrying capacity of structures made of strongly anisotropic composite materials is often determined by failure under compressive loads parallel to the reinforcement. Due to its vast practical importance, this kind of failure has been subject of extensive research including both experimental and numerical investigations. Two major effects related to compressive failure of unidirectionally reinforced composites are micro-buckling and formation of kink-bands, [1].

Historically, significant research effort went into understanding the early observed discrepancies between the predictions of the classical limit of Rosen, [2], for elastic micro-buckling and the significantly lower experimentally determined compressive strengths. Argon suggested a different compressive instability limit, based on an assumed local misalignment of the reinforcement and plastic yielding of the matrix material, [3]. Later, Budiansky proposed a combination of Rosen's and Argon's limits into a single expression, [4], and Budiansky and Fleck extended this limit to the case of a matrix material with isotropic hardening of Ramberg-Osgood type, [5].

These very fundamental investigations, determined the inherent compressive strength of a considered composite under a spatially constant and longitudinal compressive stress. In a real structure though, the material micro-structure is generally subjected to arbitrary stress states, possibly including significant spatial gradients. The load carrying capacity of a structure is then determined as a combination of the material micro-structure properties and the macro-structure design and loads. Waas proposed a semi-analytical model for predicting elastic micro-buckling initiation close to a free surface and in the presence of stress gradients, [6]. Fleck and Shu developed a numerical model based on a homogenized representation of the material micro-structure for studying elasto-plastic micro-buckling initiation for different kinds of fiber misalignments, [7] . Kyriakides et al. compared numerical and experimental results of kink-band formation, based on a finite-element model with explicitly discretized micro-structure, [8]. Following the same approach, Wind et al. determined the bending load capacity of a longitudinally reinforced V-notched beam both experimentally and numerically, [9].

Based on numerical examples, the present work demonstrates that at the presence of stress gradients, prediction of compressive failure initiation is not trivial and not easy to capture by analytic models.

2. Numerical Model

This section provides a brief summary of the homogenized finite-element model, introduced in [10], which is used for all simulations, presented in the next section. The main characteristic of the model, is that it involves three unknown fields to solve for: the displacements field $\boldsymbol{u}$, the internal kinematic variable field $\boldsymbol{d}$ and the plastic multiplier field ξ_m. The physical meaning of the vector variable $\boldsymbol{d}$ is illustrated in figure 1, which also defines the unit vectors $\boldsymbol{T}$ and $\boldsymbol{N}$, respectively tangential and normal to the fiber direction.

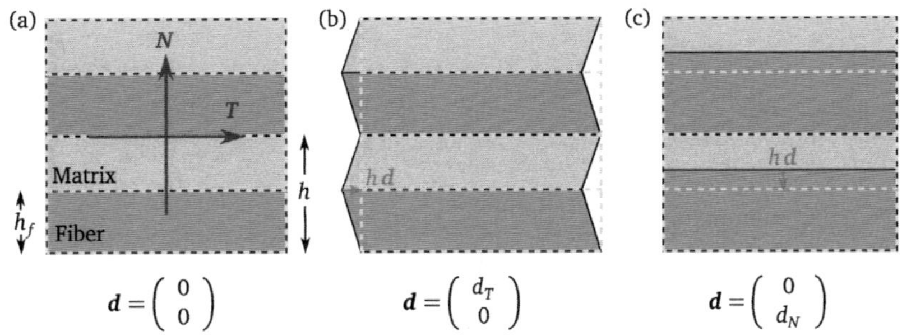

Figure 1. Undeformed and arbitrary deformed configurations of the micro-structure.

A kinematic analysis of the considered micro-structure leads to average deformation gradient tensors for the fiber and the matrix materials respectively defined as

$$\mathbf{F}_f = \boldsymbol{I} + \nabla \boldsymbol{u} - \frac{1}{c_f} \boldsymbol{d} \boldsymbol{N}^T \tag{1}$$

and

$$\mathbf{F}_m = \boldsymbol{I} + \nabla \boldsymbol{u} + \frac{1}{c_m} \boldsymbol{d} \boldsymbol{N}^T \ , \tag{2}$$

in terms of the unknown homogenized fields $\boldsymbol{u}$ and $\boldsymbol{d}$

A careful averaging of the virtual work in the material micro-structure with respect to variations of variables $\boldsymbol{u}$ and $\boldsymbol{d}$, leads to the following expression:

$$\begin{aligned} \delta W = \quad & (c_m \mathbf{P}_m + c_f \mathbf{P}_f) : \nabla \delta \boldsymbol{u} + (\mathbf{P}_m - \mathbf{P}_f) : (\delta \boldsymbol{d} \ \boldsymbol{N}^T) \\ & + \frac{h_f^2}{12 \, c_f} \mathcal{D} \mathbf{P}_f(\mathbf{F}_f) \left[\nabla \boldsymbol{d} \boldsymbol{T} \boldsymbol{T}^T \right] : \left(\nabla \delta \boldsymbol{d} \boldsymbol{T} \boldsymbol{T}^T \right) \ , \end{aligned} \tag{3}$$

where c_f and $c_m = 1 - c_f$ are the volume fractions of the fiber and matrix respectively, $\boldsymbol{P}_f$ and $\boldsymbol{P}_m$ are the corresponding 1st Piola-Kirchhoff stress tensors, $\mathcal{D}\boldsymbol{P}_f$ is the directional derivative of the mapping $\boldsymbol{F}_f \mapsto \boldsymbol{P}_f$ and h_f is the fiber layer height. The first term in Eq. (3) expresses the classical force equilibrium and the second term expresses an internal equilibrium in the sense of the Reuss model. The last term corresponds to the part of the fiber bending energy that is not captured by the homogenized displacements field $\boldsymbol{u}$.

The 1st Piola-Kirchhoff stress tensors contained in Eq. (3) are based on hyper-elastic and hyper-elasto-plastic constitutive laws for the fiber and matrix materials respectively. Under consideration of Eqs. (1) and (2), these stress tensors can be written as closed form functions $\boldsymbol{P}_f(\nabla\boldsymbol{u}, \boldsymbol{d})$ and $\boldsymbol{P}_m(\nabla\boldsymbol{u}, \boldsymbol{d}, \xi_m)$ of the unknown variables and their gradients.

Eq. (3) is the basis of a corresponding finite-element formulation, which results to a discretized system with sufficient equations for solving for $\boldsymbol{u}$ and $\boldsymbol{d}$. One additional equation is necessary for solving for the plastic multiplier ξ_m. This is the plastic yield condition, expressed in weak form as

$$\int_{\Omega_0} r(\nabla\boldsymbol{u}, \boldsymbol{d}, \xi_m)\, \delta\xi_m \; dX = 0 \qquad \forall \;\; \delta\xi_m \;\; , \tag{4}$$

where Ω_0 is the considered solid domain in the undeformed configuration and $r(\nabla\boldsymbol{u}, \boldsymbol{d}, \xi_m)$ an appropriately chosen non-smooth complementarity function, which incorporates the plastic yield surface along with the corresponding Karush–Kuhn–Tucker conditions. A specific choice of the complementarity function r is provided in [10]. It should also be noted that the accumulated plastic strain in the matrix material γ_m at the current time instant depends on the corresponding value in the previous time instant and the unknown multiplier ξ_m for the current load step.

3. Numerical Examples

This section presents numerical results corresponding to two distinct examples under plane strain conditions. In the first example, a V-notched beam is subjected to pure bending load, until micro-buckling initiates at the notch tip due to the developed compressive stresses. The second example demonstrates micro-buckling initiation on the circumference of a circular void in a unidirectionally reinforced material, subjected to far-field longitudinal compression.

3.1. Bending of a V-notched beam

The geometry of the considered V-notched beam for this numerical example is defined through the overall beam height $H = 4.64$ mm and length $L = 20$ mm, the reduced beam height $H_0 = 3.867$ mm at the notch location, the notch opening angle $\theta = 67°$ and the notch tip radius $\rho = 0.196$ mm. All simulation results refer to a unit thickness, $t = 1$ mm. The beam is reinforced in its longitudinal direction with the reinforcing fiber layers exhibiting a sinusoidal misalignment in the region close to the notch tip. The wavelength of the misalignment is $p = 0.156$ mm and its amplitude $a = 0.52$ µm, resulting in a maximum misalignment angle of 0.6°.

The reinforcement has an initial Young's modulus $E_f = 27600$ MPa and Poisson's ratio $\nu_f = 0.3$, while the corresponding elastic properties for the matrix material are $E_m = 1478.3$ MPa and $\nu_m = 0.38$. The matrix material is assumed to exhibit J_2-flow plastic behavior of Ramberg-Osgood type hardening, with the initial yield limit $\sigma_{ym} = 90$ MPa and parameters $n_m = 4$ and $\alpha_m = 3/7$. The fiber volume fraction is $c_f = 0.587$.

In order to determine the bending load capacity of the considered beam numerically, a prescribed bending angle at the beam ends is incremented by means of a numerical continuation procedure. The continuation procedure detects the critical load where a micro-buckling material instability initiates due to compressive stresses in the region close to the notch tip. Figure 2 shows snapshots of a 3.3 mm×1.8 mm region around the notch tip at the critical load for different values of the fiber height h_f.

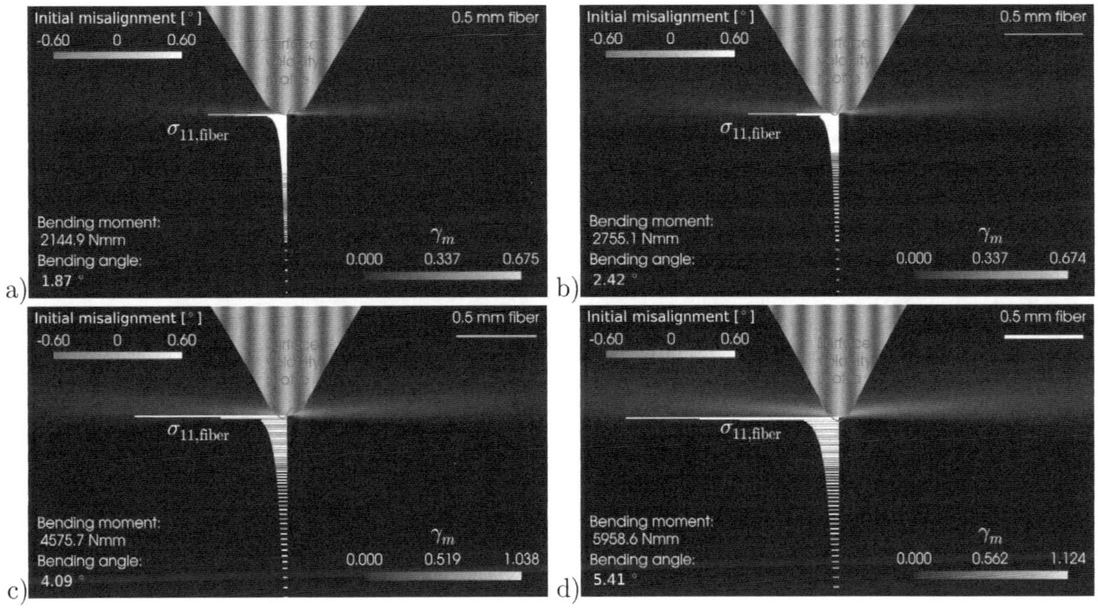

Figure 2. Snapshots of beam bending simulations at critical load for four different reinforcing fiber heights: a) 2.6 µm b) 5.2 µm c) 10.4 µm d) 20.8 µm.

The results presented in figure 2 make clear that the fiber height h_f has a large impact on the bending load capacity of the considered beam. In the case of the highest value of $h_f = 20.8$ µm, an almost triple bending angle and moment can be sustained compared to the case with the lowest $h_f = 2.6$ µm. The difference is also obvious in the longitudinal stress profiles on the cross-section through the notch, included in the snapshots. The micro-buckling modes for the four considered values of h_f are visualized through the surface velocity profile at the notch tip. The increased bending stiffness at higher beam heights results to smoothened velocity profiles and corresponding buckling modes. The color map in figure 2 illustrates the accumulated plastic strain in the matrix material γ_m, while the gray scale map in the background illustrates the variation of the fiber misalignment angle along the horizontal direction.

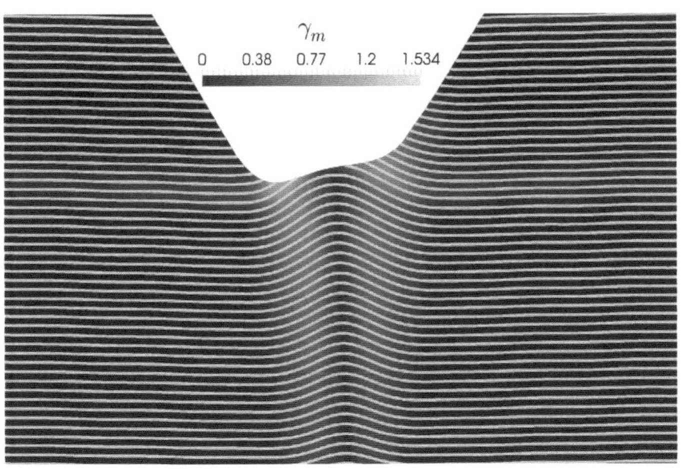

Figure 3. Snapshot of beam bending simulation in the post-buckling regime for reinforcing fiber height $h_f = 5.2$ µm.

For the case with $h_f = 5.2$ μm, figure 3 illustrates the deformed reinforcing fibers at approximately the same bending angle as the critical point shown in figure 2b, but posterior to the initiation of micro-buckling and the formation of a pair of kink-bands. The visualized streamlines were generated accounting for the direction of the deformed fibers but they are not supposed to capture the real density or thickness of the fiber layers.

3.2. Longitudinal compression of a specimen with a circular hole

The modeled geometry in the second numerical example corresponds to a circular disc of outer radius $R = 20$ mm with a concentric circular hole of radius $r = 1$ mm. The reinforcing material elasticity parameters are $E_f = 2.6 \cdot 10^5$ MPa and $G_f = 10^5$ MPa and the relevant matrix material parameters are $E_m = 2.6 \cdot 10^3$ MPa, $G_f = 10^3$ MPa, $\sigma_{ym} = 90$ MPa, $n_m = 4$ and $\alpha_m = 3/7$. As in the previous example, all presented results refer to a unit thickness $t = 1$ mm and plane strain conditions.

Far-field uniaxial compression is applied at the outer boundary with respect to the horizontal displacements, while two different load cases are considered regarding the vertical displacements, which are either constrained to zero or left free. The two cases are respectively referred to as transversely constrained ($\varepsilon_{22} = 0$) and transversely free ($\sigma_{22} = 0$).

The reinforcement is parallel to the horizontal direction and imperfection-free. Investigations with an imposed sinusoidal waviness imperfection of the reinforcement have shown that the effect of such imperfections is rather low, compared to the effect of the circular void itself and they are omitted here for the sake of simplicity.

Due to the available symmetries in the geometry, the reinforcement and the boundary conditions, it is sufficient to model only one quarter of the disc. Additional boundary conditions are necessary along the horizontal and vertical symmetry axes. The vertical and horizontal components of $\boldsymbol{u}$ are respectively fixed to zero along the two aforementioned axes and the horizontal component of $\boldsymbol{d}$ is fixed to zero on both symmetry axes.

As the far-field longitudinal strain ε_{11} is incremented within a numerical continuation scheme, a limit point is reached, where micro-buckling initiates near the top of the void circumference. Figure 4 shows snapshots corresponding to this limit point for two different values of the fiber height. Moreover the figure includes results both for the transversely constrained case, left from the vertical symmetry axis as well as for the transversely free case, right from the symmetry axis. The solid black line corresponds to the undeformed circumference, while the velocity profile on the void circumference is visualized by means of the red vectors.

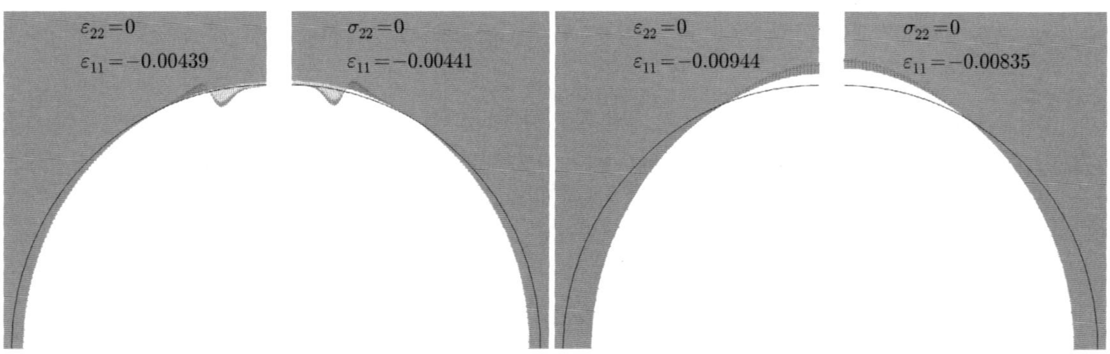

(a) (b)

Figure 4. Snapshots at initiation of micro-buckling close to the circumference of a void in a longitudinal compressed reinforced composite for two different fiber heights: a) $h_f = 5$ μm and b) $h_f = 60$ μm.

For the highest value of h_f, micro-buckling initiates at a higher far-field strain, which is reflected in the considerably larger deformations observed in figure 4b. The critical far-field strain for the lowest value of h_f is approximately 0.44%, independent of the condition in the transverse direction. On the contrary, for the highest value of h_f, different critical strains are attained for the transversely constrained and transversely free cases, respectively equal to 0.94% and 0.84%.

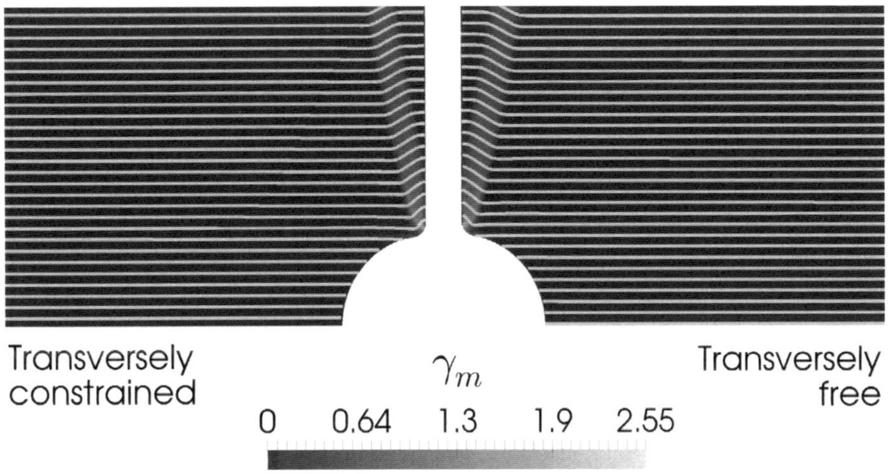

Figure 5. Snapshot in the post-buckling regime for reinforcing fiber height $h_f = 5$ µm.

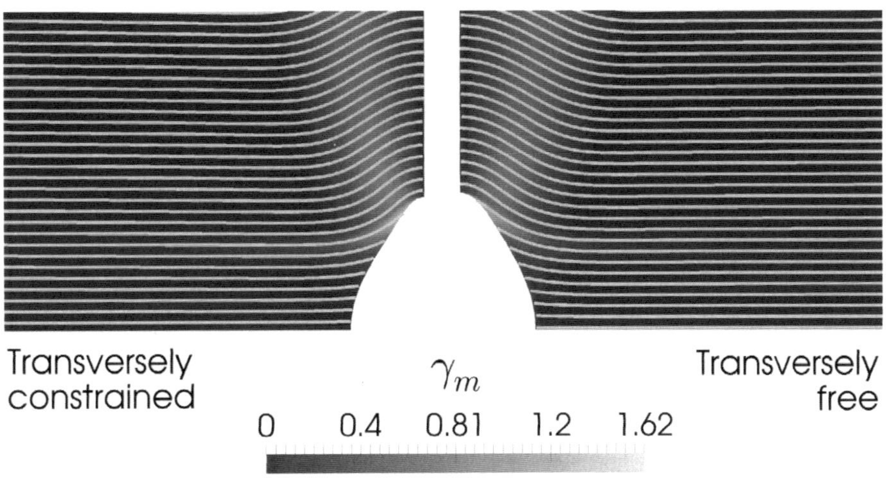

Figure 6. Snapshot in the post-buckling regime for reinforcing fiber height $h_f = 60$ µm.

Figures 5 and 6 show the post-buckling deformed solids in the region near the considered void, for the two different values of the fiber layer height, both for the transversely constrained and the transversely free cases. For the lowest fiber height value, the snapshots presented in figure 5 correspond to approximately the same far-field longitudinal strain as at the buckling initiation point shown in figure 4a, i.e. $\varepsilon_{11} = -0.0044$, but after the formation of the kink-bands that are visible in the region above the void. The differences between the two cases with respect to the boundary condition in the transverse direction are minimal. Figure 6 shows corresponding snapshots in the post-buckling regime with regard to the highest fiber height value. It should be noted that the transversely free case shown in figure 6 is at a significantly lower far-field strain, $\varepsilon_{11} = -0.0084$, than the transversely constrained one, $\varepsilon_{11} = -0.0094$. Despite the lower applied strain, the observed deformations are similar, demonstrating the significant impact of the boundary condition in the transverse direction, for this example. Compared to figure 5, the formed kink-bands shown in figure 6 exhibit a significantly smoother transition, attributed to the higher bending stiffness of the fibers in this case.

4. Conclusions

The two numerical examples included in this work, demonstrate the crucial role of the fiber bending stiffness in predicting micro-buckling initiation in unidirectionally reinforced composite materials. At locations close to notches or holes, where significant stress gradients are present, compressive material instabilities were observed to initiate at very distinct levels of the maximum longitudinal compressive stress, depending on the assumed reinforcing fiber height. The numerical results show that the interplay between micro-structural effects and the overall geometry of the structure introduces a size scale effect which is not possible to capture by analytical models, unless a length scale parameter such as the reinforcing fiber layer height is considered.

Acknowledgments

This work was supported by the Danish Council for Independent Research, through the Sapere Aude project Higher Order Theories in Solid Mechanics.

References

[1] Fleck N 1997 *Adv. Appl. Mech.* **33** pp 43–117
[2] Rosen B W 1965 Mechanics of composite strengthening, in *Fibre Composite Materials* (Materials Park, OH: ASM) chap 3 pp 37–75
[3] Argon A S 1972 *Treatise on Materials Science and Technology* vol 1 pp 79–114
[4] Budiansky B 1983 *Comput. Struct.* **16** pp 3–12
[5] Budiansky B and Fleck N 1993 *J. Mech. Phys. Solids* **41** pp 183–211
[6] Waas A M, Babcock C and Knauss W 1990 *J. Appl. Mech.-T. ASME* **57** pp 138–149
[7] Fleck N and Shu J 1995 *J. Mech. Phys. Solids* **43** pp 1887–1918
[8] Kyriakides S, Perry E J and Liechti K M 1994 *Appl. Mech. Rev.* **47** pp S262–S268
[9] Wind J L, Waas A M and Jensen H M 2015 *Compos. Struct.* **122** pp 51–56
[10] Poulios K and Niordson C F 2016 *J. Mech. Phys. Solids* **94** pp 433–452

IOP Conf. Series: Materials Science and Engineering **139** (2016) 012042 doi:10.1088/1757-899X/139/1/012042

Interlaminar interaction in paper thermoplastic laminate composites

M Prambauer[1,2], C Paulik[2] and C Burgstaller[1,3]

[1] Transfercenter für Kunststofftechnik GmbH, Franz-Fritsch-Straße 11, 4600 Wels, Austria

[2] Institute for Chemical Technology of Organic Materials, Johannes Kepler University, Altenbergerstr. 69, 4040 Linz, Austria

[3] christoph.burgstaller@tckt.at

Abstract. Bio-based composites are a research topic since several decades, which aims for sustainable and durable materials. In the scope of this research, many different sources for bio-based reinforcements have been investigated. Typical issues associated with the use of such are property variations due to cultivation area and climate, besides the influences of the type, pre-treatment and fibre geometry. Another issue can be the availability of such natural fibres. Due to these reasons, we started using paper sheets as reinforcements in laminate composites with thermoplastic materials. In preliminary studies with polypropylene composites, we found good mechanical properties, even higher than could be expected by estimating the composite properties from the constituents by applying simple rule of mixtures type models. We suspect, besides some effect of paper compaction, interlaminar effects to be the reason for this. Therefore, the aim of this work is to investigate the effects of the interfacial interaction on the different paper laminate properties due to different matrix polymers. For this work, we used polypropylene, polyamide 6 and 12 as well as polystyrene. Composites were produced via compression moulding and samples for mechanical testing and density evaluation were cut from the moulded plates. The results from mechanical tests show, that there is a reinforcing effect, regardless of matrix polymer used. Simple rule of mixtures evaluations show, that the different matrices exhibit different degrees of interaction, based on their chemical structure. In addition, also influences due to processing were found.

1. Introduction

Bio-based composites are in the focus of researchers since several decades. Different reinforcements, like natural fibres from stems, leaves or seeds of different plants as well as agricultural by-products like rice husks or wheat straw have been investigated for their potential to reinforce plastic matrices. [1, 2] The reinforcements have been investigated either as short cut fibres or powders or as semi-finished products, like fabrics or fleeces. Although natural reinforcements are bio-based, require less energy input than man-made fibres and are widespread, there are some problems associated to them due to their natural basis. These are property variations due to cultivation area and climate, besides the influences of the type, pre-treatment and fibre geometry [3]. Another factor is the availability, also throughout the year, of specific fibre types and qualities, which limits their applicability.

Therefore, we started using paper sheets as reinforcements in laminate composites with thermoplastic materials, i.e. polypropylene [4]. In the course of this research we found interesting properties for polypropylene paper laminates with a paper volume fraction of approximately 50 vol.-% of 6 GPa elastic modulus and 70 MPa tensile strength. Furthermore, when modelling these composite properties from the constituent properties via simple rule of mixture type models, we found an increase in constituent properties in the composite, which cannot be attributed solely to the paper compaction [5].

Therefore, the aim of this work is to investigate the effects of the interfacial interaction on the different paper laminate properties due to different matrix polymers. This approach was chosen because the characterisation of the interaction between paper fibres and a matrix is rather complicated. Typical micromechanical approaches, like single fibre pull-out or microdebond tests, are inapplicable due to the short fibre length (below several millimetres) of the single paper fibres, and tests like determining interlaminar shear strength with short beam bending tests or peel off tests are not applicable, as these do not show the required failure modes due to the thermoplastic matrix.

2. Materials and Methods

In this work a standard copy paper (Xerox Transit 80 g/m² from Kraft pulp fibres from softwood and apparent density of 0.816 g cm^{-3}) with a thickness of 98 μm was used. A SEM micrograph (taken with a Vega Tescan scanning electron microscope at the University of Applied Sciences in Wels, after sputtering the samples with gold) depicting the paper surface, showing the fibre orientation as well as some inorganic filler can be found in figure 1. Any further information about the constituents is proprietary information with the paper producer and is not available for publication.

For laminate fabrication, the different polymers (table 1), i.e. polypropylene (PP), polyamide 6 (PA6), polyamide 12 (PA12) and polystyrene (PS), were produced into films by means of a flat film extrusion line (P/M Plastic Maschinenbau, 35 mm screw diameter, screw length over screw diameter (L/D) ratio of 18). Typical film thicknesses produced were 140 and 100 μm, and also 250 μm for the PS films. In case of the PP, a compatibilizer (5 wt.-% Scona TPPP 8112) was used in all the formulations to increase the interaction between the matrix and the paper fibres and is therefore not noted further.

Figure 1. SEM micrograph of paper surface of the copy paper used in this study (scale bar represents 200 μm)

Table 1. Polymers used in this study and their properties

Polymer	Density (gcm^{-3})	$T_{(Processing)}$ (°C)	Melt volume flow rate ($cm^3 10^{-1} min^{-1}$)
PP	0.905	210	9 (230°C, 2.16 kg)
PA6	1.140	230	40 (250°C, 2.16 kg)
PA12	1.010	210	--
PS	1.040	210	12 (200°C, 5 kg)

The laminates were produced by stacking the pre-cut films and paper sheets, the latter all oriented in machine direction, in a pre-heated steel mould with 10x10 cm² free volume and subjecting the whole setup to 50 bar for 5 min at the respective processing temperature for each matrix polymer, given in table 1. For composite build up, the outer layers always were polymer layers, so the stacking sequence starts and ends with a film sheet. To achieve the different volume fractions, films with different film thickness, i.e. 100, 140 and 250 µm were used. Additionally, for volume fractions of 20 vol.-% and below, the films were used double to increase the polymer volume fraction. The calculated paper content was checked by optical microscopy and was used for all the comparisons in this paper. An example for such a cross-section is given in figure 2, where the polymer layers can be distinguished from the paper layers, as well as the remaining porosity in the paper can be seen.

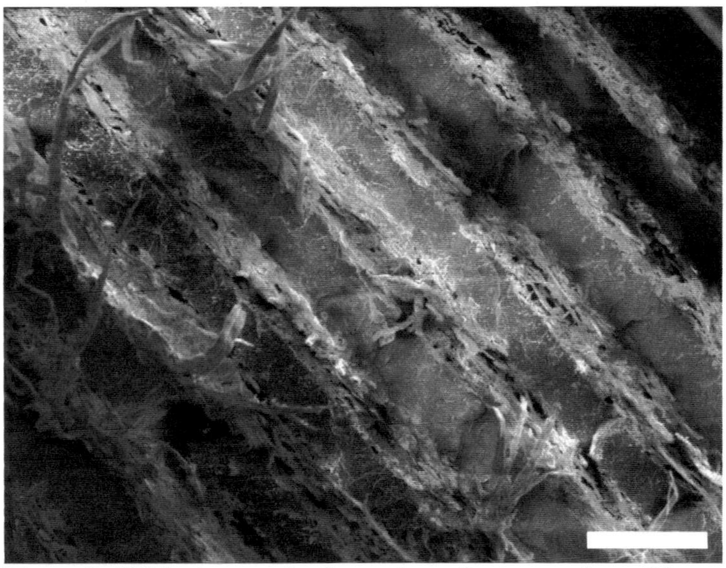

Figure 2. SEM micrograph of cross-layer fracture surface of the composite consisting from 50 vol.-% of copy paper in polypropylene (scale bar represents 200 µm)

Different specimens were milled from the laminates by means of a Coesfeld CPM 3020 mould cutter. Tensile testing was carried out on a Zwick-Roell Z020 20 kN universal testing machine, in accordance with ISO 527-2 (on specimens type 1BA) with three replicates per series with a loading rate of 1 mm min^{-1} for determining elastic modulus and after that 5 mm min^{-1} until the break of the samples. Flexural testing was performed by a Zwick-Roell ZMART.PRO 10 kN universal testing machine, according to ISO 178, on three replicates with sample dimensions 75 x 10 x 3.5 mm³, and a loading rate of 1 mm min^{-1} for determining elastic modulus and after that 2 mm min^{-1} until the break of the samples. The interlaminar shear strength (ILSS) was determined by an unnotched Charpy flatwise impact strength measurement, carried out by a Zwick-Roell 5113.300 impact pendulum, according to ISO 179-1/3fn. Five replicates of type 3 specimens with dimensions of 45.5 x 10 x 3.5 mm^3 and a pendulum with 5 J impact energy were applied. The density of the samples was measured with a Sartorius YDK 01LP density determination kit on an analytical balance (Sartorius AX 224), according to ISO 1183-1, where ethanol was the immersion liquid. The density of the latter was determined before the measurement with the help of a standardised volume glass immersion body.

3. Results and Discussion

Reinforcing polymers with paper leads to increased mechanical tensile and flexural properties. An example for the stress strain curves of paper, polymer and the composite are given in figure 3, where a synergistic reinforcement effect can be seen at the first glance. In our opinion this is due to the fact that the paper strength accounts for paper fibre interaction mainly, while in the case of the composites, the fibre properties itself are – at least partially – harnessed, thus resulting in increased composite properties.

As shown in figure 4, the elastic modulus of the different composites is increasing with increasing paper content. The initial decrease of the elastic modulus in case of the PA6 based composites is suspected to arise from moisture uptake of the PA6 from the ambient air, which therefore reduces the elastic modulus of PA6 [6]. Nevertheless, there is a reinforcement effect visible for all investigated polymer matrices. The same is true for the tensile strength (figure 4), where also an increase with increasing paper content can be found. In case of the PS based composites, the sharp decrease through the addition of the paper is due to the fact that the pure polystyrene strains to about 1.4 % for the tensile strength, but the composite with 10 vol.-% of paper strains only to 0.86 %. Taking this into account for the neat PS, i.e. using the stress value at the reduced strain, one yields approx. 24 MPa, which is in good accordance with the trend of tensile strengths of the PS based composites.

In case of the flexural tests, we can find similar results for flexural modulus and flexural strength (figure 5). For the moduli, the data are slightly different from the ones from tensile testing, but exhibit the same trends. The flexural strength data also show the same trends, with the exception that besides the neat PS also the neat PA6 shows higher flexural strength than the respective composites with 10 vol.-%, which again can be explained by the limited strain applied in the composites in comparison to the unreinforced materials.

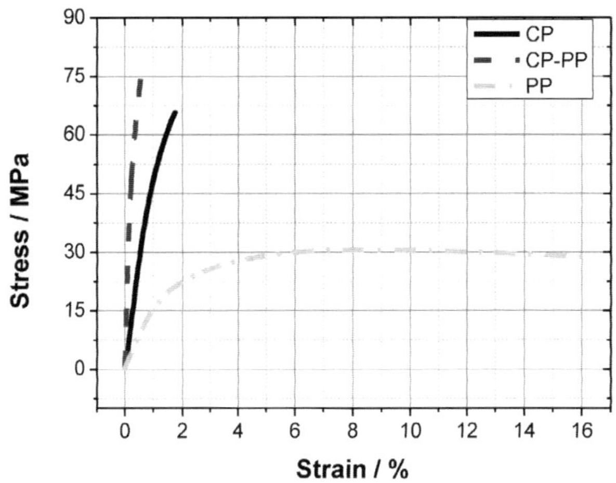

Figure 3. Stress strain curves for polypropylene film (PP), copy paper (CP) and the composite with 50 vol.-% of copy paper in polypropylene (CP-PP)

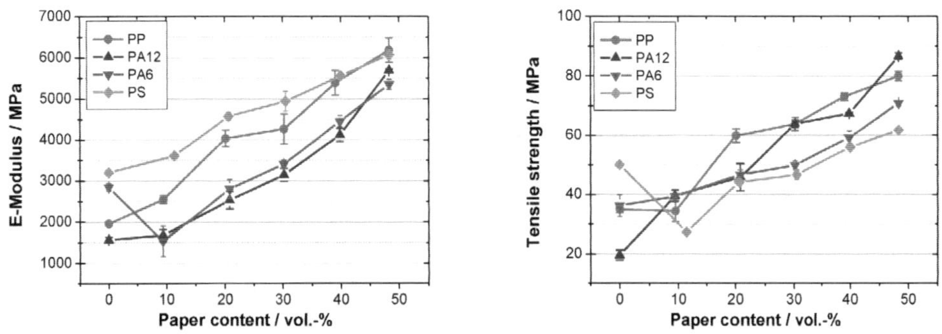

Figure 4. Elastic modulus from tensile tests (left) and tensile strength (right) vs. paper content of the composites with paper and different matrix polymers

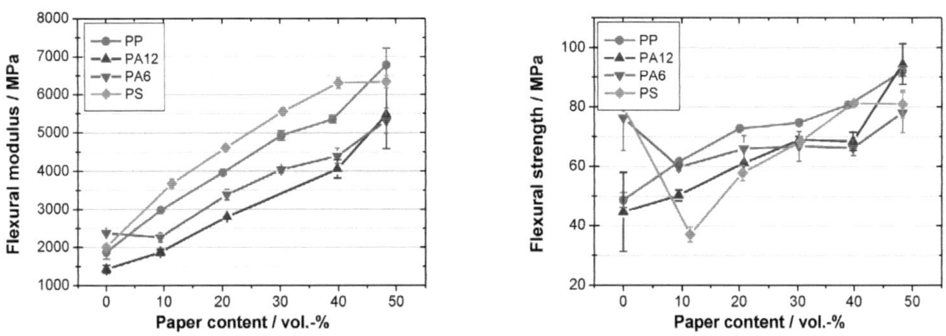

Figure 5. Flexural modulus (left) and flexural strength (right) vs. paper content of the composites with paper and different matrix polymers

The composites density (figure 6) shows an increase in density with an increase in paper content. This may not seem surprising at a first glance, but the paper itself exhibits a density of 1.03 g cm^{-3}, which is surpassed rather quickly by the different composites. There are different reasons for this behaviour. First, the density of the paper is product from the density of the cellulose fibres and the pores in the paper (neglecting here the influences of inorganic fillers, which are also present in most paper grades). The pores also are either within the cellulose fibres (the lumen) or between, which are the pores of the paper. These pores can be compacted (the lumen) or filled with the matrix, which then gives a higher apparent density of the paper in the composite. The effectiveness of the compaction is depending mainly on the applied pressure in the fabrication (besides the compressibility of the matrix), while the pore filling is a parameter which can be related to the interlaminar interaction between the different matrices and the paper. As a first evaluation for this, a linear fit for extrapolating the paper density was applied. The authors are aware that this is only a very crude evaluation, but due to the aforementioned problematics of determining paper fibre interactions with polymer matrices, this seems to be a first indication of how well a matrix can interact with a given paper. We find values (all of them in g cm^{-3}) of 1.29 (PA6), 1.27 (PA12), 1.23 (PP) and 1.18 (PS). Comparing these to the data from ILSS testing (the short beam Charpy impact tests, where a fractured sample from these tests is shown in figure 7), these seem to be in reasonable accordance due to the similar ordering of the values found in ILSS. In case of the ILSS, also PA6 exhibits the highest values and PS the lowest. The difference in the order between PP and PA12 can be due to the fact, that the density evaluation is a simplified method giving averaged value. Also, there is some influence of the impact test itself on the ILSS data, as can be seen from the fractured specimen in figure 7. There is a large delamination visible, but also some laminate fractures in the middle of the specimen. This means that with the ILSS data, always some influence of the matrix impact strength can also be found.

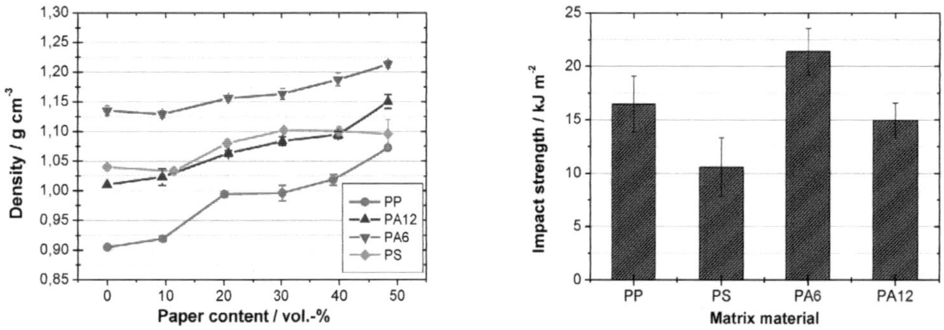

Figure 6. Composite density vs. paper content (left) and impact strength from ILSS testing of composites with 40 vol.-% paper vs. matrix material (left) of the composites with paper and different matrix polymers

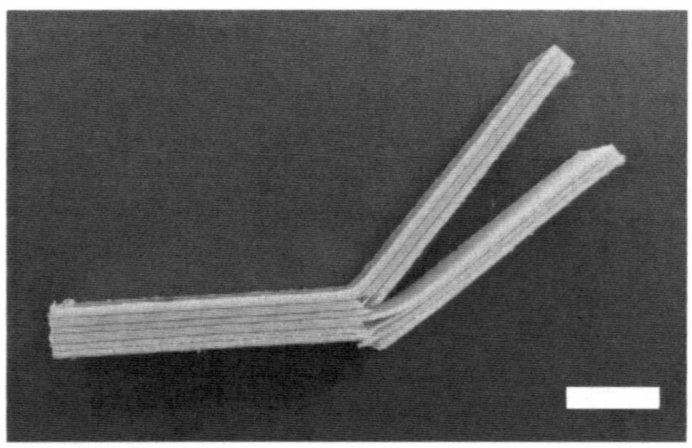

Figure 7. Exemplary fractured test specimen from ILSS impact testing exhibiting laminar failure (scale bar represent 5 mm)

Putting these data into perspective with the results from tensile strength (figure 1), where interlaminar interaction should play an important role, one finds some differences. Using again a linear fitting approach for the tensile strength, thus being a simplified application of a rule-of-mixtures, where the constituents properties add up to the composites properties in relation to their volume fractions, one can use the slope of these fits as a measure for the reinforcing effect. Again, we decided here to not take porosity into account as a separate factor, due to a current lack of proper analysis of that in the investigated composite system, but leave that summarized in the paper contribution. The slopes calculated from the tensile strength vs. the paper volume fraction show, that PA12 exhibits the greatest reinforcing effect, followed by PP, PS and PA6 in that order. With the exception of PA6, one could expect that from the ILSS and density data. Also the chemical structures of the constituents (figure 8) can give some explanation for that. Putting PA6 aside, the chemical structure of PA12, with the amide groups in the chain, will be capable of forming chemical bonds (hydrogen and covalent bonds at the chain ends), as can the PP with the help of the maleic anhydride grafted chains added into the material due to the compatibilizer. The chemical structure of PS can only form polar bonds due to the nucleophilic nature of the benzene rings in the structure.

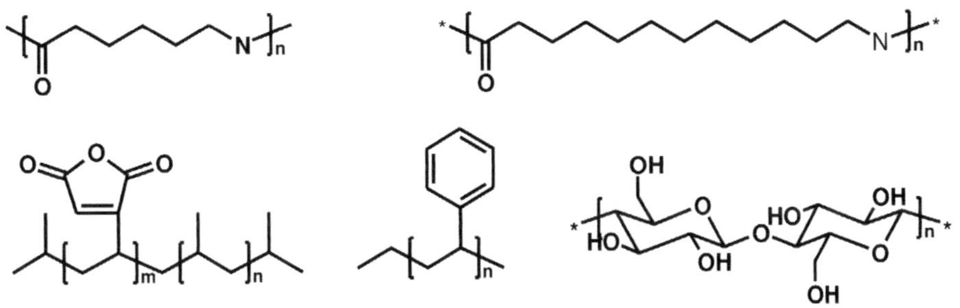

Figure 8. Schematic chemical structures of the used polymers and cellulose; top: left – PA6, right – PA12; bottom: left – PP grafted with maleic anhydride, middle – PS, right - cellulose

Following this logic, PA6 should be even better, due to the fact that it exhibits roughly double as much amide bonds as PA12 does for a comparable chain length. On the other hand, PA6 is more susceptible to water uptake – and therefore hydrolysis – as PA12 [6]. Another factor is, that the melting point for PA6 is about 220 °C, and a processing temperature at 230 °C means that the material is relatively viscous. The ILSS results show a good impregnation of the paper, so another factor can be the degradation of the paper itself. Cellulose degradation starts around 250 °C, but other paper components, like starch-based coatings (which are applied to increase printability [7]), degrade earlier. Therefore, the significantly higher processing temperature of PA6 can play a role, which would explain our findings to some extent.

A last simple use of our rule-of-mixtures fitting, i.e. using the volume fractions weighed constituents properties in order to estimate the respective composite property, is to predict the tensile strength of the paper by extrapolating to 100 vol.-% with the linear fits. Again, this is only a rough estimation, but should give us some more insight on the mechanisms effective in the composites. Here we yield 139 and 130 MPa for PA12 and PP, respectively, while about 106 MPa for the other two matrix polymers. The copy paper itself exhibits about 62 MPa from mechanical testing. Similar predictions for the elastic modulus of the paper yield around 10,400 MPa for all matrices except PA6, where about 9,200 MPa are yielded, while the neat paper shows an elastic modulus of about 5,400 MPa. The fact that the estimated data here is in a much closer range than for the tensile strength shows two things – first, at the lower strain levels where the elastic modulus is measured, the interlaminar adhesion plays a lesser role than at the higher strains where tensile strength is determined, and second, with the exception of PA6, all the matrices exhibit very comparable data, therefore this interaction can be attributed to the paper and its surface structure. The fact that PA6 shows lower values here may again be due to the higher processing temperature and therefore some degradation of the paper surface, as discussed above. Also, all of this shows, that the composites are utilizing not only the properties of the constituents, but also, due to the interaction at the interfaces, utilizing the properties of at least a portion the single fibres.

4. Conclusions

In the scope of this work, we investigated the properties of composites built up as laminates from paper and different thermoplastic polymer matrices. We found, that all the composites show increasing properties with increasing paper content, but also the properties of the matrix play an important role. The chemical structure as well as the processing exhibits influences on the composite properties. Using simple rule of mixture type models to evaluate the composites show, that the pores in the paper are filled or compacted, and that the apparent paper properties in the composites are significantly higher, i.e. about double, than the properties of the neat paper. These results show, that due to the interaction between the constituents, the paper fibres themselves and not only the paper layers are – at least to some extent – utilized for their properties.

IOP Conf. Series: Materials Science and Engineering **139** (2016) 012042 doi:10.1088/1757-899X/139/1/012042

Further investigations will deal, besides other polymer matrices to widen the view for the simple modelling applied here, with detailed investigations on the interaction between the matrix and the fibres, to further understand the mechanisms in such layered composites, and therefore getting the tools to improve their performance further towards industrial applicability.

References

[1] Gauthier R, Joly C, Coupas AC, Gauthier H, Escoubes M 1998 *Polymer Composites* **19** 287
[2] Gassan J 2002 *Composites: Part A* **33** 369
[3] Wambua P, Ivens J, Verpoes I 2003 *Composites Science and Technology* **63** 1259
[4] Prambauer M, Paulik C, Burgstaller C 2015 *Composites: Part A* **74** 107
[5] Prambauer M, Paulik C, Burgstaller C 2016 *Composites: Part A* **88** 59
[6] Oberbach K 2001 *Saechtling Kunststoff Taschenbuch* (Munich: Hanser) p 490
[7] Autio K, Suortti T, Hamunen A, Poutanen K 1992 *Starch* **44** 393

IOP Conf. Series: Materials Science and Engineering **139** (2016) 012043 doi:10.1088/1757-899X/139/1/012043

Properties of drawn W wire used as high performance fibre in tungsten fibre-reinforced tungsten composite

J Riesch[1], J Almanstötter[2], J W Coenen[3], M Fuhr[1,4], H Gietl[1,5], Y Han[1,5], T. Höschen[1], Ch Linsmeier[3], N Travitzky[4], P Zhao[1], R Neu[1,5]

[1] Max-Planck-Institut für Plasmaphysik, 85748 Garching, Germany

[2] OSRAM GmbH, SP PRE PLM DMET, 86830 Schwabmünchen, Germany

[3] Forschungszentrum Jülich GmbH, Institut für Energie- und Klimaforschung - Plasmaphysik, Partner of the Trilateral Euregio Cluster (TEC), 52425 Jülich, Germany

[4] Department of Material Science, Glass and Ceramics, University of Erlangen-Nürnberg, Martensstr. 5, Erlangen, Germany

[5] Technische Universität München, 85748 Garching, Germany

E-mail: johann.riesch@ipp.mpg.de

Abstract. High strength and creep resistance also at high temperature, combined with a high thermal conductivity and high melting point make tungsten (W) an ideal material for highly loaded areas in future fusion reactors. However, as a typical bcc metal tungsten features an intrinsic brittleness up to very high temperature and is prone to operational embrittlement. Tungsten fibre-reinforced tungsten composite (W_f/W) utilizes extrinsic toughening mechanisms similar to ceramic fibre-reinforced ceramics and therefore overcomes the brittleness problem. The properties of the composite are to a large extend determined by the properties of the drawn tungsten wire used as reinforcement fibres. W wire exhibits a superior strength and shows ductile behaviour with exceptional local plasticity. Beside the typical mechanisms observed for ceramic composites the ductile deformation of the fibres is therefore an additional very effective toughening mechanism. Tension tests were used to investigate this phenomenon in more detail. Results show that there is a region of enhanced localized plastic deformation. The specific energy consumption in this region was estimated and used to suggest optimisation options for W_f/W composites.

1. Introduction

The development of capable materials is essential for sophisticated future energy systems like a fusion reactor. Due to its high strength and creep resistance also at high temperature, combined with a high thermal conductivity and melting point tungsten is a very promising candidate for highly loaded areas in such a reactor [1]. However, as a typical bcc metal tungsten features an intrinsic brittleness up to a temperature of typically 500 K - 600 K [2] and up to above 1200 K for high temperature annealed material [3]. In addition, W is prone to operational embrittlement e.g. by grain coarsening [4] or/and neutron irradiation [5].

Tungsten fibre-reinforced tungsten composites (W_f/W) overcome the brittleness problem of tungsten by utilizing extrinsic toughening mechanisms [6]. Therefore, an application of W as a plasma facing material under thermal transients and neutron bombardment becomes feasible [7]. Hence W_f/W was chosen as risk mitigation plasma facing component and high heat flux material in the EU fusion roadmap [8, 9].

W_f/W composites consist of tungsten fibres made of commercial drawn tungsten wire embedded in a tungsten matrix produced either by a chemical process [10, 7] or by powder metallurgy [11]. The composite structure is determined and maintained (during fabrication and operation) by engineered interface systems between fibre and matrix [12]. The principle of extrinsic toughening in such a composite system has been shown in the past years at the Max-Planck-Institute for Plasma Physics, Garching (IPP) [13, 14]. Various active extrinsic toughening mechanisms, e.g. fibre bridging or fibre pull-out, have been shown in model systems consisting of a single W fibre embedded in a chemically deposited tungsten matrix in the as-fabricated case, as well as after embrittlement of the fibre by recrystallization and grain growth (called "embrittled" in the following) [15, 16]. In a further development step a fabrication method for bulk material was developed based on the chemical deposition process and first samples were produced. Mechanical tests on these samples revealed a very large toughness at room temperature and active toughening mechanisms in embrittled conditions [10, 17].

The idea of extrinsic toughening has been developed for ceramic materials and has been successfully used for many years [18]. Besides the typical mechanisms observed in ceramic composites the ductile deformation of the tungsten wire has been shown as a very effective additional mechanism in as-fabricated W_f/W [15]. A significant local plasticity combined with a large local strain of up to 1.36 for a length of 66 µm has been observed. The specific energy consumption in this region is therefore very high, being the reason for the high efficiency of the mechanism. As the specific deformation energy is highest in the necking region the formation, growth and extension of this zone are important for the optimisation of W_f/W composites.

In this contribution we give an overview of the tensile properties of tungsten wire and focus on the ductile deformation and fracture behaviour. Special focus is laid on the necking region. Results of standard tension tests and manual stepwise tests in combination with 3D observation in a confocal microscope are presented. The link between fibre properties and composite properties is discussed in detail.

2. Tensile properties of tungsten wire

Commercial drawn tungsten wire is used as fibres in W_f/W composites. The development of tungsten wire is strongly related with the illumination industry (more details in [19]). Starting with pure tungsten and a powder metallurgical process Coolidge realized the possibility to produce ductile tungsten wire by strong mechanical deformation at elevated temperature (more details e.g. in [20]). Later it was realized that W wire doped with several 10 ppm potassium (K) shows a very beneficial high temperature creep stability. During the drawing process K forms small bubble rows which are aligned on the grain boundaries and lead to a beneficial evolution of the grain structure at high temperature (more details e.g. in [21]).

As a consequence of the drawing process, tungsten wire features a unique microstructure consisting of elongated intertwined grains. The small grain size perpendicular to the wire axis together with the high dislocation density bestows high ductility (high density of mobile dislocations) and also considerable strength on the drawn wire (Hall-Petch strengthening). In figure 1 typical stress-strain curves of an as-fabricated doped tungsten wire tested at room temperature are shown (similar to the curves shown in [22]). After elastic loading a phase of significant work hardening occurs before the maximum stress is reached. The maximum stress is embedded in a region with only small variations in stress. A rapid stress decrease precedes the final fracture of the wire.

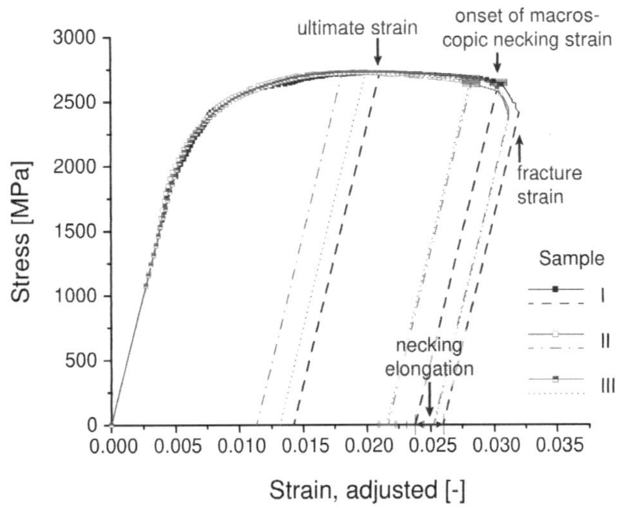

Figure 1. Typical tensile curves of three K doped W wire samples (diameter: 150 µm) (similar to [22]). The onset of apparent necking strain is shown in comparison to the ultimate strain and the fracture strain for each sample.

In figure 2 scanning electron microscopy pictures of a typical fracture surface of an as-fabricated pure tungsten wire after tensile testing is shown. The shown fibre has a diameter of 150 µm and was tested with a tensile speed of $5\,\mu m\,s^{-1}$ for a measuring length of 18 mm The strong necking as an indication of the wire's ductility is clearly visible in the images (figure 2, upper left corner). Besides some larger cracks with different sizes a typical knife-edge necking of individual grains leads to a fibrous structure (bottom row). There is no significant difference to fracture surfaces obtained for doped tungsten wire in previous studies (compare e.g. [7, 23]). Leber et al. [23] describes this behaviour as a consequence of the microstructure. During tensile loading the fine elongated grains produced in the drawing process debond at longitudinal grain boundaries due to the Poisson's contraction. The resulting free-standing individual grains afterwards neck down in a knife-edge fracture mode leading to a typical fibrous surface.

Tensile test have been used to investigate the mechanical properties of both pure [24] and potassium doped tungsten wire [22]. As-fabricated and annealed samples with a diameter of 150 µm have been tested at room temperature for both grades. In the as-fabricated case and after annealing at 1273 K (3 h) pure W wire shows ductile behaviour and an ultimate strength of around 2900 MPa and 2000 MPa, respectively. A heat treatment of 1900 K (30 min) leads to embrittlement and a mean strength of approximately 900 MPa. Potassium doped W wire shows an ultimate strength of about 2700 MPa in the as-fabricated state and a ductile behaviour up to an annealing temperature of 2173 K (30 min). The tensile strength stays up to about 2000 MPa for this temperature. At an annealing temperature of 2500 K (30 min) rapid grain growth leads to embrittlement and a decrease of strength to 1300 MPa. For both grades the embrittlement is correlated with the loss of the fine elongated grain structure and not necessarily with the occurrence of (primary) recrystallisation. For pure tungsten wire this transition takes place at 1273 K, for potassium doped wire at 2173 K.

3. Experimental

Drawn W wire doped with 60-75 ppm potassium was tested in standard tension tests and in manual tests in combination with in-situ microscopy. In a first step the wire was cut into appropriate long pieces. For the standard test mechanically straightened wire was used (details

Figure 2. Fracture surface of a pure tungsten wire sample, fractured in a tensile test at room temperature (details of the testing procedure in [24]). In the side view (top row) the necking (left) and the fibrous fracture surface (right) are visible. The top views (bottom row) reveal the differently sized cracks and the typical knife-edge fracture of individual grains (right).

in [22]). To concentrate the fracture in the centre both ends of a wire sample were attached to paper by glueing (UHU Plus endfest 300) for the standard test or covered by the glue only for the manual test. The uncovered part was approximately 20 mm for the standard test and 2 mm for the manual test.

The standard tests were conducted similar to the procedure presented in [22]. The tests were performed with a universal testing machine (TIRA Test 2820) at room temperature. The load was measured by a 200 N range load cell. The displacement was measured contact-less by a laser speckle extensometer (LSE-4000 DE). The measuring length was in the range of 8 mm for 8 samples (short) and 18 mm for 6 samples (long). The tests were conducted in a displacement controlled mode with a displacement rate of $5\,\mu m\,s^{-1}$. More details can be found in [22]. Both pieces of each fractured sample were investigated in a confocal optical microscope (LEXT OLS 4000 3D Laser Measuring Microscope). In figure 3 the microscopic picture of one piece (left) together with a schematic drawing of the corresponding second piece (right) is shown for a typical fractured fibre. The necking diameter as well as the size of the zone of macroscopic necking were determined together with the necking elongation. Macroscopic necking was defined as the region where the diameter is smaller than 99 % of the original diameter. This corresponds with the accuracy of the measurement in this investigation. The measurements were done for each side of a fractured sample separately and then summed up to determine the total value. The necking diameter was evaluated at the respective fracture surface. To determine the size of

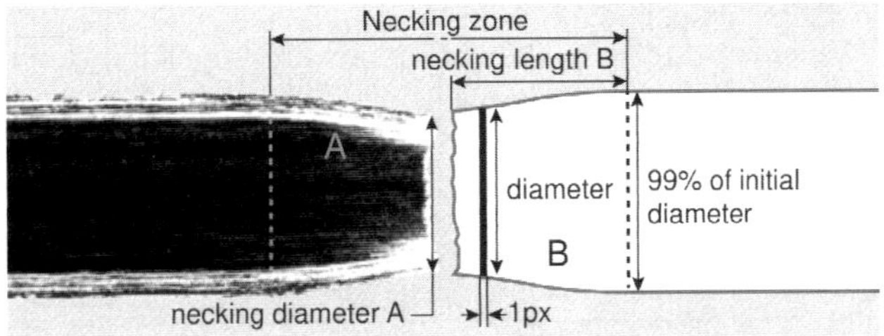

Figure 3. Parts of a typically fractured tungsten fibre. A microscopice picture of the necking region (left) and a schematic drawing (right) are shown.

the macroscopic necking zone the diameter for each pixel along the necking zone was determined as a first step. This zone starts at the 99 % limit and ends at the tip of the fibre. The tip was defined as the point where the diameter shows a strong variation due to the roughness of the fracture surface. The necking length was then determined by counting the pixels between the 99 % limit and the tip of the fibre. Assuming volume conservation the elongation during fracture was determined. For each pixel along the necking length a cylinder volume was calculated with the height of one pixel and the corresponding diameter (see figure 3). These cylinder slices were then summed up to calculate the total volume. By assuming a perfect cylinder for the undeformed sample the original size of the necking zone was obtained. The necking elongation was then calculated by subtracting the original size from the beforehand obtained size of the fractured sample.

The device used for the manual tests was small enough to be placed in the confocal microscope mentioned before allowing the detailed observation of the deformation process. The device consisted of two Teflon coated (to minimize friction) plates at which the wire was attached in the centre. At one side the plates were connected in a rotatable way and on the other side a screw with a fine thread (250 µm pitch) allowed to move the plates apart. The attached fibre was thereby strained. The screw was moved by hand in 24 steps per turn. Due to the very small opening compared to the dimensions of the whole setup the fibre was mainly strained in tension. A microscopic observation was performed every 4-5 steps. If a change of the wire diameter was recognized, observations were performed every step or half-step. The microscopy pictures were used to determine the specific elongation for different regions of the wire. Surface features were used to divide the measuring length in distinct regions and a picture-to-picture comparison between the different steps was used to determine the specific elongation.

4. Results and Discussion

The size of the necking zone was determined to be around 250 µm with an elongation of around 50 µm. It has to be noted that there is a significant variation in the determined necking zone size (187 µm - 369 µm) and necking elongation (41 µm - 69 µm). The necking diameter was around 120 µm which corresponds with to reduction in area of $q = 0.36 \pm 0.02$ and a zero gauge length elongation of $\epsilon_0 = 0.6$. The detailed values are shown in table 1. The rough fracture surface has an impact on the accuracy of the obtained values. However, as surface features on a typical fracture surface of a W wire (see figure 2) are small compared to the obtained values the influence is expected to be low. The reduction in area is very similar to tension test results for pure and doped W wire in previous studies [24, 23, 25]. This suggest that the observed mechanisms and behaviour seem to represent a typical behaviour of W wire. In comparison to the constraint wire in W_f/W the reduction in area is similar to results for moderate testing speeds [15], but larger than for samples tested at very high speeds [7]. The size of the necking zone is much larger than in the constraint case (250 µm compared to 66 µm in [15]), however, the elongation

Table 1. Overview of tension test results. The uncertainties are calculated as standard deviation of the mean.

	Size necking zone [µm]	*Necking elongation* [µm]	*Necking diameter* [µm]	*reduction in area q* [-]	*zero gauge length elongation ϵ_0* [-]
short	243 ± 12	50 ± 2	122 ± 3	0.34 ± 0.02	0.52 ± 0.06
long	259 ± 27	52 ± 5	118 ± 2	0.39 ± 0.02	0.63 ± 0.06
total	250 ± 13	51 ± 2	120 ± 3	0.36 ± 0.04	0.57 ± 0.09

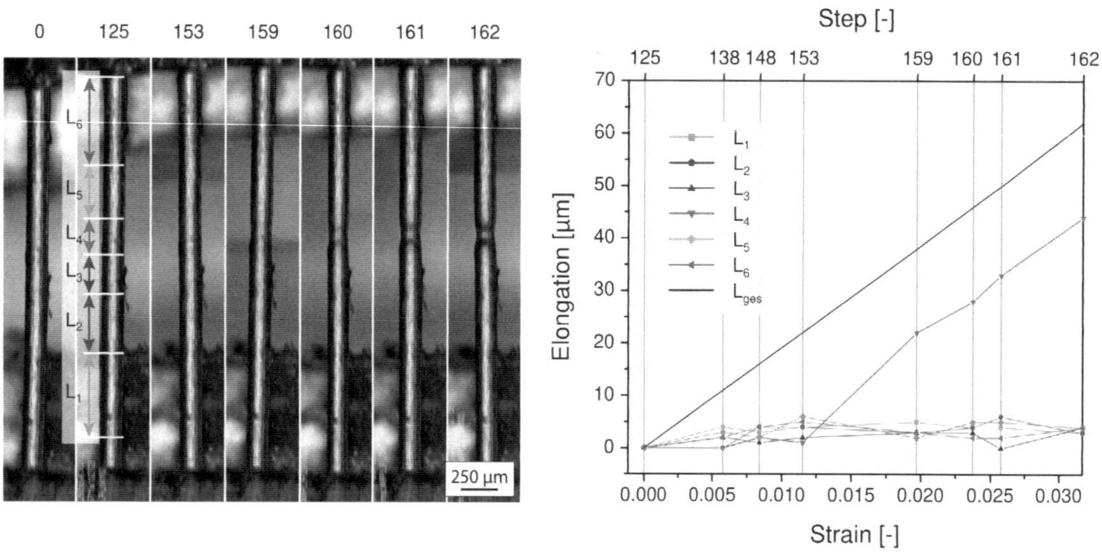

Figure 4. Microscopy pictures of tungsten wire at different displacement steps during a manual tension test (left). Elongation over strain in a manual stepwise tension test (right). The total values as well as the values for distinct regions are given. Most of the elongation is concentrated in region L_4 which represents the region where apparent necking occurs.

is smaller (50 µm compared to 92 µm in [15]). The constraint seems to have a positive impact on plasticity, but results for constraint wire are few and the issue needs further investigation.

The results of the stepwise manual test are summarized in figure 4. In step 159 the diameter reduction exceeds 1 % corresponding with the onset of macroscopic necking. The reduction is also visible in the corresponding picture. To determine the local elongations the wire was divided in regions as shown in figure 4. Region L_4 was chosen in a way to include the region of macroscopic necking. The specific elongation of the different regions is shown in comparison to the total elongation over rising strain (see figure 4 right). Before necking was detectable the elongation was uniformly distributed along the regions (although the regions have a different length). When macroscopic necking occurs almost all the elongation is concentrated in region L_4 whereas the elongation in the other regions seems to stay constant. In step 159 already around 60 % of the total elongation is generated within L_4 which increases up to 70 % in step 162. Due to the limited measurement accuracy a clear mapping of the specific elongation before step 159 and beyond L_4 is difficult. However, the results indicate clearly that with the onset of macroscopic necking most of the elongation -which is a measure of deformation- is concentrated in the region where this necking occurs. Up to this moment the elongation of the other measuring regions contribute significantly to the total elongation.

According to the presented results the plastic deformation in the examined potassium doped tungsten wire can be classified in 3 regions (the boundaries of the different regions are according to figure 1):

- Between the yield point and the ultimate strain: plastic deformation without macroscopic necking and rising load (Ris)
- Between ultimate strain and the onset of macroscopic necking: plastic deformation without macroscopic necking and decreasing load (Dec)
- Between onset of macroscopic necking and final fracture: plastic deformation with macroscopic necking and localized deformation (MaN)

In the following we present an interpretation based on the microstructure of drawn tungsten wire. As it was mentioned before (chapter 2), the grain structure of tungsten wire consists of long elongated grains with a high aspect ratio [23, 25]. Hence the majority of grain boundaries lies parallel to the wire axis and only a small fraction is oriented perpendicular to the fibre axis. The weakness of grain boundaries in tungsten materials has been reported elsewhere [2]. Considering the highly stressed perpendicular grain boundaries and assuming a high concentration of defects in a particular wire region, it is likely that this perpendicular grain boundaries fail in this region by delamination at stresses around the ultimate strength. Even if such boundaries debond they are bridged by the adjacent long grains leading only to a slight decrease in stress and a preservation of to a ductile behaviour (region: Dec). Due to the perpendicular direction the impact of this debonding on the diameter reduction is expected to be low and therefore below the measuring limit in this investigation. However, the delamination would lead to a weakening in the particular area and thus promote the occurrence of macroscopic necking in this area (region: MaN). After macroscopic necking has occured, the deformation of the drawn tungsten wire could follow the procedure suggested by other researchers [23, 25] and fail in the typical knife-edge fracture of individual grains. Although this model coincides with the experimental results of this study, the deformation of tungsten wire needs further investigation.

The macroscopic necking is correlated with the stress-strain curves for the three samples shown in figure 1. At first the necking elongation is normalized with the measuring length. Following the elastic line the onset of the apparent necking is found at the stress-strain curve (see figure 1). It is remarkable that the onset of the macroscopic necking deviates significantly from the ultimate strain which is typically seen as the starting point of necking [26]. The macroscopic necking seems to be correlated with the crook at the end of the tensile curve.

The above presented results suggest that the plastic deformation is very large in the region of macroscopic necking. For a first estimation the specific plastic deformation energy is determined by integrating the stress-strain curve from the onset of the macroscopic necking until the fracture and by normalizing the energy with the necking length rather than the measuring length. For comparison, the specific energy until ultimate strain is also calculated by integrating the stress-strain curve (the elastic fraction is subtracted).

- $\tilde{w}_{\mathrm{f,pl,Ris}} = (0.032 \pm 0.002)\,\mathrm{J\,m^{-3}}$
- $\tilde{w}_{\mathrm{f,pl,MaN}} = (0.56 \pm 0.07)\,\mathrm{J\,m^{-3}}$

The error in this estimation especially for the necking region is not negligible. Nevertheless, it is significant that the specific energy consumption in this area is more than one order of magnitude higher in comparison to the part where the deformation is distributed uniformly along the whole fibre (uniform deformation). The region between uniform strain and the onset of the macroscopic necking (region: Dec) probably shows a mixture of localized deformation and evenly distributed elongation. Further research is necessary.

If the analysed wire is used in W_f/W composites the plastic deformation plays a significant role for the toughness of the material. The amount of energy dissipation is determined by the actual length being deformed and the specific plastic deformation energy [15]. Due to the large local plasticity in the zone of macroscopic necking the size of this zone is a good guiding value for the optimum debonding length. Based on the obtained results a debonding length in the range of 250 µm would be ideal. A crack opening of 50 µm would be sufficient to allow a full plastic deformation and therefore a maximum energy consumption.

5. Summary and Outlook

The large toughness of W_f/W composites is strongly related to the large specific deformation energy and high strength of the tungsten wire used as reinforcing fibre. Tension tests were used to investigate the deformation of potassium doped tungsten wire in detail. The necking diameter as well as the size of macroscopic necking and the corresponding elongation were determined. Tests with in-situ observation revealed that the elongation preceding fracture predominately occurs in a small region as soon as macroscopic necking occurs but not necessarily when ultimate strain is reached. It was shown that the specific deformation energy in this region is much higher than in the other parts of the wire. Based on the size of this area an optimum debonding length in W_f/W composites was suggested.

The investigations showed that the deformation behaviour is strongly related to the microstructure and its evolution. Further research is essential to relate the macroscopic response to the microscopic behaviour. An optical strain measurement setup allowing the determination of local strains will be used together with a detailed microscopic analysis in a first step. Furthermore, tension tests with in-situ tomographic observation are planned.

Acknowledgements

The authors want to thank G. Matern, S. Elgeti and M. Balden for their assistance in microscopy. This work has been carried out within the framework of the EUROfusion Consortium and has received funding from the Euratom research and training programme 2014-2018 under grant agreement No 633053. The views and opinions expressed herein do not necessarily reflect those of the European Commission.

References

[1] Coenen J W et al 2016 *Physica Scripta* **T167** 014002
[2] Lassner E and Schubert W D 1999 *Tungsten - Properties, Chemistry, Technology of the Element, Alloys, and Chemical Compounds* (Kluwer Academic / Plenum Publishers)
[3] Reiser J, Rieth M, Dafferner B and Hoffmann A 2013 *Journal of Nuclear Materials* **442** S204
[4] Yih S and Wang C 1979 *Tungsten: Source, Metallurgy, Properties, and Applications* (Springer Science+Business Media New York)
[5] Barabash V, Federici G, Rödig M, Snead L and Wu C 2000 *Journal of Nuclear Materials* **283–287** 138
[6] Linsmeier C et al 2016 *Nuclear Fusion* Submitted
[7] Riesch J, Aumann M, Coenen J W, Gietl H, Holzner G, Höschen T, Huber P, Li M, Linsmeier Ch and Neu R 2016 *Nuclear Materials and Energy* DOI: http://dx.doi.org/10.1016/j.nme.2016.03.005
[8] Romanelli F 2012 *Fusion Electricity – A roadmap to the realisation of fusin energy* (European Fusion Development Agreement, EFDA) URL `https://www.euro-fusion.org/wpcms/wp-content/uploads/2013/`
[9] Stork D et al 2014 *Journal of Nuclear Materials* **455** 277
[10] Riesch J, Höschen T, Linsmeier Ch, Wurster S and You J H 2014 *Physica Scripta* **T159** 014031
[11] Jasper B, Coenen J W, Riesch J, Höschen T, Bram M and Linsmeier Ch 2015 *Material Science Forum* **825–826** 125
[12] Du J, Höschen T, Rasinski M, Wurster S, Grosinger W and You J H 2010 *Composites Science and Technology* **70** 1482

[13] Du J 2011 *A Feasibility Study of Tungsten-Fiber-Reinforced Tungsten Composites with Engineered interfaces* Ph.D. thesis Technische Universität München URL `http://mediatum.ub.tum.de/node?id=998317`
[14] Riesch J 2012 *Entwicklung und Charakterisierung eines wolframfaserverstärkten Wolfram-Verbundwerkstoffs* Ph.D. thesis Technische Universität München URL `http://mediatum.ub.tum.de/node?id=1106428`
[15] Riesch J, Buffiere J Y, Höschen T, di Michiel M, Scheel M, Linsmeier Ch and You J H 2013 *Acta Materialia* **61** 7060
[16] Riesch J, Buffiere J Y, Höschen T, Scheel M, Linsmeier Ch and You J H 2016 *Composites Part A: Applied Science and Manufacturing* Submitted
[17] Neu R et al 2016 *Fusion Engineering and Design* DOI: http://dx.doi.org/10.1016/j.fusengdes.2016.01.027
[18] Evans A 1990 *Journal of the American Ceramic Society* **73** 187
[19] Schade P 2006 *International Journal of Refractory Metals and Hard Materials* **24** 332
[20] Briant C and Bewlay B 1995 *MRS BULLETIN* **20** 67
[21] Briant C L and Hall E L 1989 *Metallurgical Transactions A* **20A** 1669
[22] Riesch J, Han Y, Almanstötter J, Coenen J W, Höschen T, Jasper B, Zhao P, Linsmeier Ch and Neu R 2016 *Physica Scripta* **T167** 014006
[23] Leber J, Tavernelli J, White D and Hehemann R 1976 *Journal of Less-Common Metals* **48** 119
[24] Zhao P, Riesch J, Höschen T, Almannstötter J, Balden M, Coenen J, Himml R, Pantleon W, von Toussaint U and Neu R 2016 *International Journal of Refractory Metals and Hard Materials* Submitted
[25] Lee D 1975 *Metallurgical Transactions A* **6** 2083
[26] Davis J R 2004 *Tensile Testing* vol 2 (ASM International)

Experimental/numerical study of anisotropic water diffusion in glass/epoxy composites

I B C M Rocha[1,2], S Raijmaekers[1], R P L Nijssen[1], F P van der Meer[2] and L J Sluys[2]

[1]Knowledge Centre WMC, Kluisgat 5, 1771MV Wieringerwerf, The Netherlands
[2]Delft University of Technology, P.O. Box 5048, 2600GA Delft, The Netherlands

E-mail: i.barcelos@wmc.eu, s.raijmaekers@wmc.eu

Abstract. In this work a glass/epoxy composite commonly used in wind turbine blades is exposed to a humid environment at an elevated temperature. To research the anisotropic diffusion behaviour observed in unidirectional composite specimens, experimental results of slices cut along the three directional planes of the laminate immersed in demineralised water at 50°C are coupled with numerical modelling. The weight of the slices was measured at regular intervals, from which the uptake behaviour could be deduced. The process was modelled using a 3-dimensional RVE of the material, where diffusion is modelled as steady-state and the diffusivity in each direction was measured by applying concentration gradients to the model. The experimental data shows similar water uptake behaviour for samples in both transverse directions, while the water uptake in the fibre direction was significantly faster. A proper fit according to Fick's law was obtained for the transverse direction, while this was not possible for the samples in fibre direction, suggesting a strong dependency of the diffusion behaviour on the fibre orientation. Results from the proposed numerical models show that the geometric effect of fibres acting as barriers for the water movement is indeed responsible for part of the observed anisotropy.

1. Introduction

Research on composite material usage optimization for wind turbine blades has been on the rise in the past few years, and particular focus is being given to reducing design uncertainty through a better understanding of material interaction with extreme service environments. Among many types of environmental attack, the combined influence of temperature and moisture ingression is often regarded as the most critical one [1]. Upon exposure to humid environments, composite structures tend to absorb water molecules through a diffusion process driven by pressure gradients. Experimental observations in glass/epoxy composites suggest that these water molecules are only capable of moving through and settling in the resin but not in the fibres [2]. This brings complexity to the diffusion phenomenon, since the water molecules have to go around the fibres for diffusion in the transverse direction [3], making the diffusion anisotropic. Furthermore, chemical interaction between the epoxy and fibre sizing creates an interphase region around the fibres where water diffusivity can be different from the one in bulk resin regions [4], which also promotes an anisotropic diffusion behaviour. In this work, an attempt is made to elucidate the anisotropic diffusion process using a combination of experimental measurements and microscopic numerical analysis.

In order to quantify the effect of fibre orientation on the diffusivity of water inside the material, it is important to measure the water uptake for isolated directions. To do so, a thick glass/epoxy composite panel was manufactured, from which equal sized specimens were made in the three directional planes of the laminate. The panel was made using a resin system commonly used in wind turbine manufacturing. After cutting, the samples were polished to avoid erroneous measurements due to rough damaged surfaces. Before starting the immersion, the specimens were dried at 50°C until a stable moisture-free state was achieved. The samples were subsequently immersed in demineralized water at 50C and weighed at intervals. The resulting water uptake curves were then used in order to investigate the anisotropic nature of the diffusion process.

In order to investigate the cause of the anisotropy, the diffusion process was simulated on the microscale. A 3-dimensional Representative Volume Element (RVE) of the material was used and the diffusivity in each direction was computed by imposing periodic concentration boundary conditions and subjecting the RVE to a unit flux. In order to account for phenomena such as diffusion at the interphase, a level set field with constant radius was built around each fibre of the RVE and the diffusivity at each material point could be computed as a function of its distance to the nearest fibre. By fitting the experimental data using the numerical models, the relative contribution of each diffusion mechanism is assessed.

2. Experiments

2.1. Materials

The resin system is the Momentive EPIKOTE RIMR 135 / EPIKURE RIMH 1366, consisting of a monomer (70-100% 4,4-Isopropylidenediphenol-Epichlorohydrin Copolymer and 0-30% 1,6-Hexanediol Diglycidyl Ether) and a hardener (25-50% Alkyletheramine, 20-25% Isophoronediamine and up to 20% Aminoethylpiperazine) [5], mixed in a 100:30 ratio between monomer and hardener. For the UD laminate, the Saertex PPG 2002 2400tex (963 g/m^2) unidirectional fabric was used, featuring a silanic coupling agent in order to improve interface adhesion.

A thick composite panel with 50 plies (approximately 30mm) was manufactured through vacuum infusion moulding. The panel was cured for 3 hours at 30°C and 5 hours at 50°C, and subsequently post-cured for 10 hours at 70°C. Fibres were oriented in a single direction (0°) for 95% of the total areal weight of the fabric, with 90° stabilization roving (sr) being used, accounting for 5% of the total areal weight. Since a pure UD panel was required to enable a direct link with the numerical results, the stability roving of the fabric was removed from each ply, except for a boundary region of about 5 cm at the top and bottom side. The fabric is stacked with the side of the stability roving being alternated, resulting in a balanced and symmetric laminate $[(sr\ /\ 0°) - (0°\ /\ sr)]_{25}$.

Specimens of about 30 x 30mm were cut from the panels using a water cooled diamond saw. After cutting slices in the three directional planes of the laminate, they were polished using a polishing machine with grid size up to 1200 in order to avoid erroneous measurements due to rough damaged surfaces. The slices were polished until a thickness of approximately 0.8 to 0.9 mm was reached. This uniform thickness was chosen to conform a width-to-thickness ratio of at least 30, and to minimise possible edge effects.

2.2. Conditioning

After polishing the slices, they were first dried in an exsiccator at 50°C in order to remove moisture content accumulated during manufacturing and storage. The samples were dried until a stable moisture-free state was achieved, while vacuum was applied to the exsiccator in order to accelerate the drying process and avoid the occurrence of oxidative reactions. Subsequently the samples were weighed, measured and photographed, which was considered as reference state, followed by immersion in demineralised water at 50°C. At intervals, the samples were taken out of the water and weighed. The interval at which the samples were weighed decreased from hourly at the first day to twice on the second day and from there on, daily to every other day. Only one sample was taken out of the water at once, quickly wiped dry, weighed on an analytical balance and returned into the bucket of water, all within a timespan of about one minute to minimize the influence on the diffusion process.

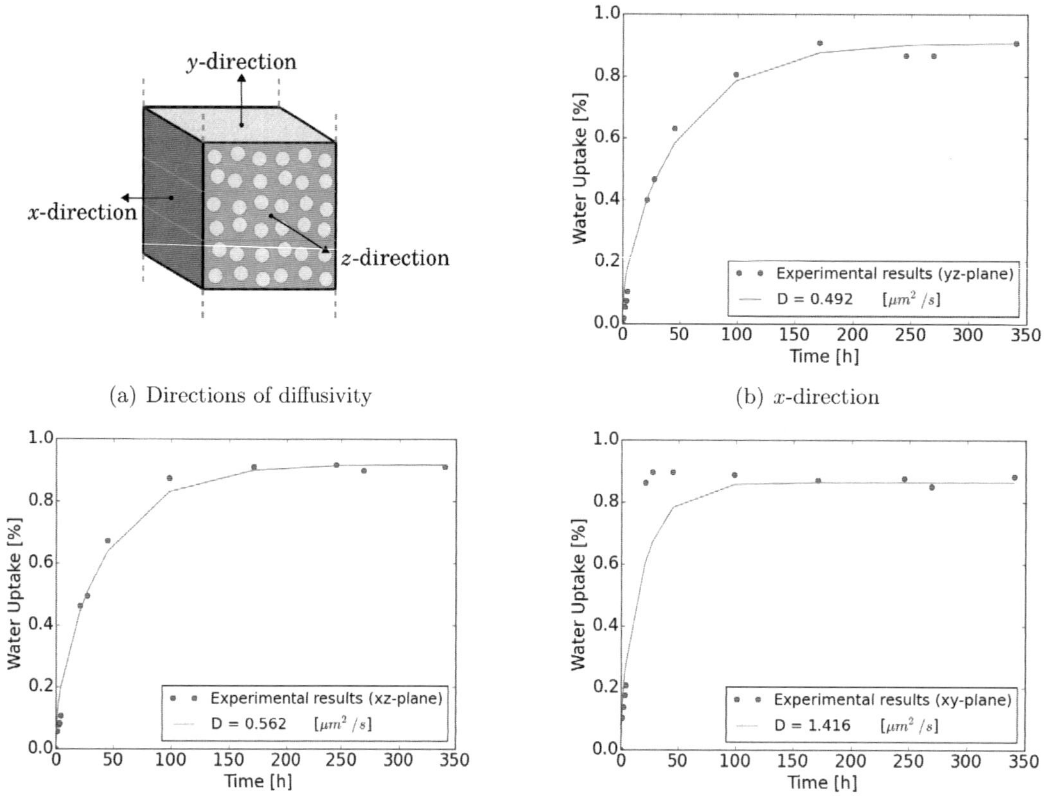

(a) Directions of diffusivity (b) x-direction

(c) y-direction (d) z-direction

Figure 1. Experimental results for water uptake and diffusivity in x, y & z-direction.

2.3. Results and Discussion

Specimens were cut in the three directional planes of the laminate (Figure 1(a)), where the x direction is the transverse plane in the through-thickness direction of the laminate, the y direction is the transverse plane in the in-plane direction of the laminate and the z direction is the fibre direction of the laminate. In Figure 1, for each direction a representative water uptake curve is shown, together with a fitted diffusivity curve according to Fick's law. Because of the high width-to-thickness ratio of the samples, we can neglect the water diffusion through the edges and assume the diffusivity to be one-dimensional. The red dots represent the individual weight measurements performed at intervals, while the red line is a fit for the diffusivity described by the following equation [6]:

$$D = \pi \left(\frac{h}{4w_\infty} \right)^2 \left(\frac{w_2 - w_1}{\sqrt{t_2} - \sqrt{t_1}} \right)^2 \quad (1)$$

where D is the diffusivity parameter, h being the specimen thickness, w_∞ the water uptake at saturation and w_1 and w_2 are two different uptake levels chosen from experimental measurements at the corresponding time values t_1 and t_2. The best fit is found by performing a least squares procedure on each possible combination of w_1 and w_2.

In Figure 1(b) and 1(c) with respectively the x and y-direction, a similar uptake behaviour was found, which could be explained by a similar arrangement of the fibres inside the sample. For both directions the fibres are oriented in the plane of the sample, with the only difference being the through-thickness versus in-plane orientation of the sample plane. For these directions a close fit is obtained with the experimental data, suggesting that the assumption of Fickian uptake behaviour is valid. The z-direction (Figure 1(d)) on the other hand shows different uptake behaviour, where the sample reaches saturation in a shorter period of time. The initial part of the experimental data for this direction shows a poor fit with the Fickian diffusivity curve, suggesting other mechanisms being at play during the diffusion process requiring further research. For the z-direction, saturation is reached after approximately 27 hours of immersion, while the x and y samples are saturated after approximately 170 hours.

The water uptake percentage at saturation is consistent for each direction and lies around 0.9% with respect to the dry (reference) state of the samples. The fibre weight percentage for the 50-ply panel was determined at 71.0% and with 2.4% of sizing and stitches (same fabric as used in [6]), the resin accounts for 26.6% in weight of the laminate. Assuming that glass fibres are impervious to water and water diffuses in the resin material only, and taken that for neat resin a water uptake percentage of 3.4% was found at saturation in [6], the expected uptake for the composite samples would be $0.266 \cdot 3.4\% = 0.90\%$, similar to the experimental values.

3. Numerical Modelling

3.1. Motivation

In order to further understand the mechanisms that cause the experimentally measured diffusivity to show a markedly anisotropic behaviour, a numerical modelling approach is adopted in this work. In this work, it is considered that the diffusion can be modelled as orthotropic, being fully described by a diagonal diffusivity matrix $\mathbf{D}$ with three independent diffusivity values, D_x, D_y and D_z, with the latter being the diffusivity along the fibres.

Here, it is hypothesised that such orthotropic behaviour is a consequence of the inhomogeneous microscopic structure of the composite material. Therefore, the diffusivity values are obtained through a steady-state diffusion analysis at the microscale:

$$D\nabla^2 c = 0 \tag{2}$$

where ∇^2 is the Laplace operator, c being the water concentration and D is now a scalar since the microstructural components (fibres and resin) are isotropic.

3.2. RVE Model

The microscopic Representative Volume Element (RVE) considered in this work is a three-dimensional representation of unidirectional fibres and surrounding matrix, with the fibres aligned in the z direction. Geometrically periodic RVEs with pseudo-random fibre distributions are generated using the discrete element software Hades. The two-dimensional geometry is meshed with 3-node triangles using Gmsh. Examples of RVEs with 1, 16 and 49 fibres can be seen in Figure 2. Lastly, since the RVE must be three-dimensional in order to consider the diffusion process along the fibres (z direction), the 2D RVEs used as input are extruded at runtime and the 3-node triangles are replaced by 6-node wedge elements.

In order for the RVE to consistently represent a macroscopic material point, periodic water concentration boundary conditions are enforced in a similar way to what was done in [7] for

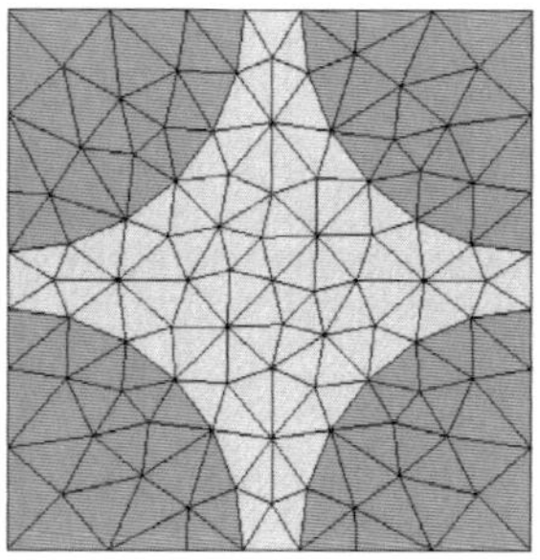
(a) 1 fibre

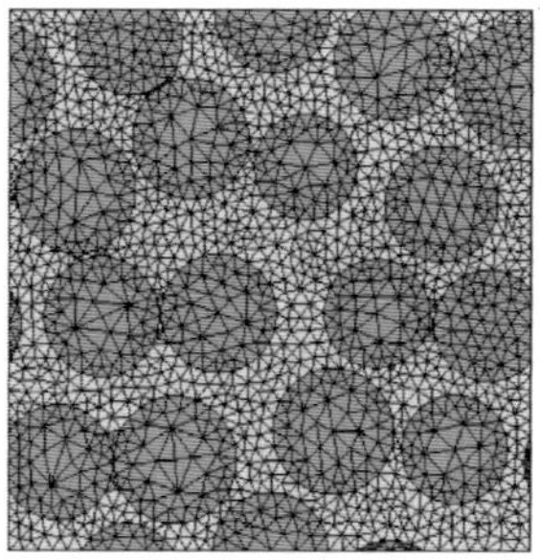
(b) 16 fibres

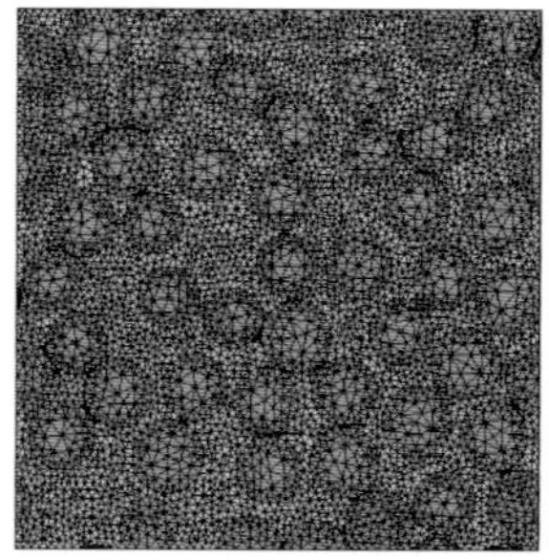
(c) 49 fibres

Figure 2. Examples of generated RVEs with different number of fibres

heat conduction. For each direction (x, y and z), the difference in concentration for nodes on opposing surfaces is equal to a prescribed concentration gradient value. Additionally, in order to guarantee that the steady-state problem will have a unique solution, the volume average of the concentration in the whole RVE is equal to a prescribed value, which represents the concentration of the macroscopic material point represented by the micromodel. Figure 3(a) shows the concentration gradients which can be prescribed and Figure 3(b) shows an example of the obtained concentration field when unitary gradients are applied in every direction and the average concentration is fixed at zero. It is important to note that the model is only executed on the resin elements, as it is considered that water does not diffuse into glass fibres.

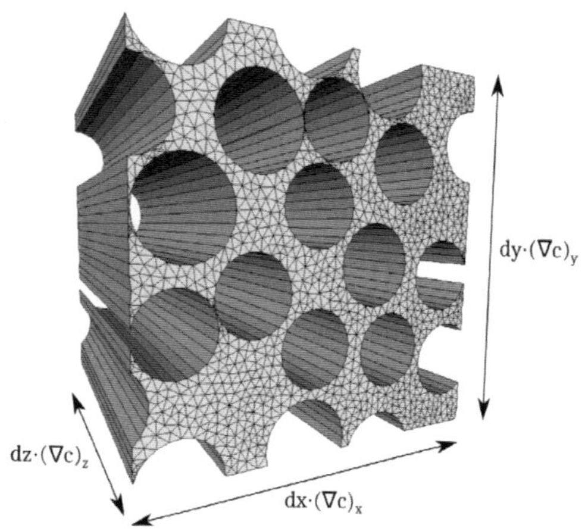

(a) Prescribed concentration gradients

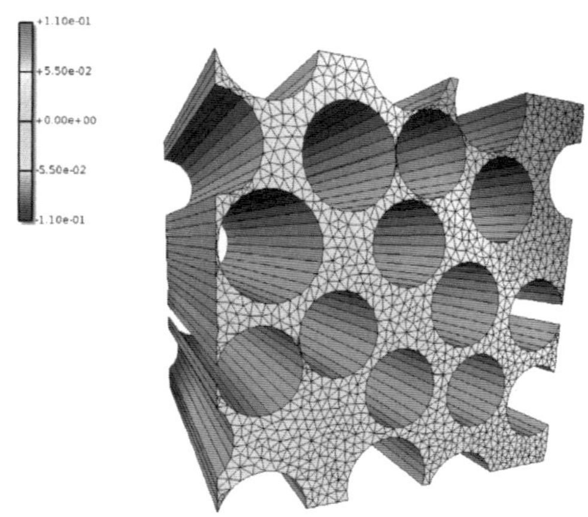

(b) Resultant concentration field

Figure 3. Concentration gradients applied as boundary conditions and resultant concentration field

After the steady-state solution is found, the volume average in the RVE of each flux component is computed, which would effectively represent the resultant macroscopic flux corresponding to the applied concentration gradients. Since the micro analysis is linear, applying unit concentration gradients in every direction results in flux values equal to the desired homogenised diffusivities.

Such formulation accounts for the diffusivity anisotropy caused by the geometric inhomogeneity (i.e. glass fibres acting as barriers for the water) [3]. However, literature evidence [4] suggests that diffusion may happen faster around the fibres when compared to bulk resin regions. In order to take this effect into account, a level set field is generated in order to describe the distance of any point in the mesh to the surface of a fibre. This distance is then used to determine the diffusivity of a point by linearly interpolating between a lower bound representing the diffusivity in the bulk (unmodified) resin (D_{bulk}) and an upper bound representing the diffusivity at the fibre-matrix interface (D_{max}). An example of the resultant diffusivity field can be seen in Figure 4 for an interphase width of $2\mu m$.

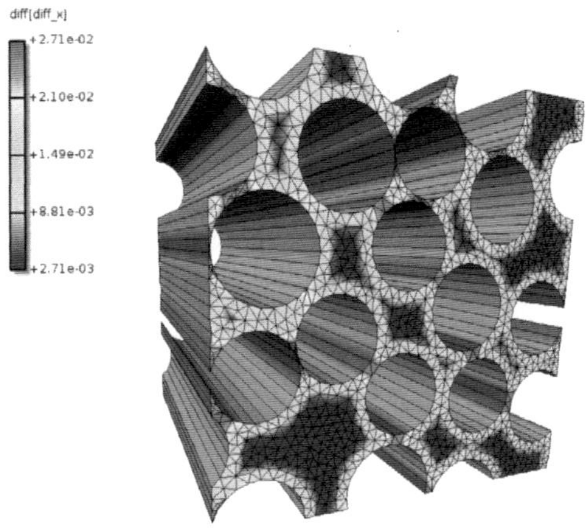

Figure 4. Diffusivity field considering an interphase width of $2\mu m$.

3.3. Numerical Results

The numerical strategy discussed in the previous sections was used in an attempt to reproduce the obtained experimental results. The RVE used in the analyses had 36 fibres with a diameter of $15\mu m$, a fibre volume fraction of 0.54 and a total cell size of $0.11 x 0.11 x 0.11\, mm^3$. The mesh consisted of wedge finite elements with characteristic length of $0.5\, \mu m$ in the x-y plane, so that an interphase layer of $1\, \mu m$ thickness would correspond to two element layers. Furthermore, in order to allow for a smoother diffusivity variation as the distance to the nearest fibre increases, a total of 81 integration points was used per finite element. The input bulk diffusivity value of $0.724\ \mu m^2/s$ for neat resin was obtained from previous experiments [6].

For the first model (*No interphase*), the level set fields were not used and the bulk diffusivity was used at every material point. Therefore, only the effect of geometric inhomogeneity was taken into account in this case. As expected, the homogenised transverse diffusivities were lower than the neat resin value, due to the barrier effect caused by the fibres. On the other hand, in the z direction, no obstacles are present and the homogenised value is equal to the neat resin one. Compared with experimentally obtained results, the diffusivities provided by the model are approximately 37% lower for D_x and D_y and 44% lower for D_z, suggesting the presence of additional mechanisms that accelerate the water movement inside the composite material.

The next two models included interphase modelling by using level set fields, with the interphase thickness t_{int} and maximum diffusivity D_{max} values being obtained from the work of Joliff et al [4], where the authors propose that water moves approximately 5 times faster in the interphase region. The thick interphase model considers a region of highly modified material properties of $4\, \mu m$ thickness based on MTA measurements, while the thin interphase model considers a thickness of $1\, \mu m$ based on AFM results reported in [4]. As expected, the inclusion of an interphase yields higher homogenised diffusivity values. For both cases, however, the obtained values were higher than the experimental ones. Such disagreement may be due to differences in experimental aspects such as resin system, commercial fibre sizing, manufacturing

Table 1. Comparison between experiments and numerical simulations.

	D_{bulk} $[\mu m^2/s]$	D_{max} $[\mu m^2/s]$	$\frac{D_{\text{max}}}{D_{\text{bulk}}}$ [-]	t_{int} $[\mu m]$	D_x $[\mu m^2/s]$	D_y $[\mu m^2/s]$	D_z $[\mu m^2/s]$
Experimental	-	-	-	-	**0.500**	**0.531**	**1.303**
No interphase	0.724	-	-	-	0.316	0.329	0.724
Thick interphase [4]	0.724	3.619	5.0	4.0	1.221	1.223	2.269
Thin interphase [4]	0.724	3.619	5.0	1.0	0.732	0.728	1.209
Fit w/ thick interphase	0.724	1.375	1.9	4.0	0.526	0.531	1.071
Fit w/ thin interphase	0.724	2.111	2.9	1.0	0.528	0.531	0.957

and curing processes, as well as in modelling choices such as fibre distribution and mesh size, among others.

For the two final models, the interphase thickness values obtained by Joliff et al [4] were kept and D_{max} was adjusted to fit the experimental values. As the experimentally measured ratio between D_x and D_y differs from the one obtained for this particular RVE, D_y was chosen as the one to be fit exactly, while a difference still remains in D_x. For the thick interphase model, the value obtained for D_{max} was approximately twice as high as the bulk value, while for a thin interphase a value 3 times higher than D_{bulk} was obtained. It is worth noting that neither model was able to correctly predict D_z, with an error of 26% for a thin interphase and of 18% for a thick interphase. This suggests that a combination of geometric inhomogeneity and linearly increasing diffusivity at the interphase are not enough to explain the observed differences in diffusivity. Further investigation is needed to elucidate the matter.

4. Conclusion

This work investigated the anisotropic diffusion behaviour observed in unidirectional glass/epoxy composite specimens through the coupling of experimental observations and numerical modelling. The experimental procedure consisted of manufacturing a 50-ply composite panel and cutting slices along its three planes of orthotropy. The slices were polished, dried to a moisture-free state and subsequently immersed in demineralised water at 50°C while having their water uptake measured at regular intervals. The process was modelled using a 3-dimensional RVE of the material representing a material point of a specimen during the transient phase of the diffusion phenomenon. Periodic concentration boundary conditions and a mass balance constraint were applied to the RVE and the diffusivity in each direction was measured by applied unit concentration gradients to the cell.

For experimental data, similar water uptake behaviour was observed for samples in the x and y direction due to a comparable fibre arrangement. A proper fit according to Fick's law was obtained for these directions, while this was not possible for the samples in z direction. Further research is necessary into the additional mechanisms playing a role for diffusion along the z direction. For each direction the water uptake percentage at saturation reached up to 0.90%, which is in line with the expected uptake when considering the fibre weight percentage of the panel. The z samples were fully saturated after immersion for about 27 hours, while the x and y samples required about 170 hours to reach full saturation, suggesting a strong dependency of the diffusion behaviour on the fibre orientation.

Results from the proposed numerical models show that the geometric effect of fibres acting as barriers for the water movement is indeed responsible for part of the observed anisotropy, with the homogenised diffusivities in the x and y direction being approximately 2.2 times slower than in the z direction. However, it was not possible to fit the experimental values by using only the bulk epoxy diffusivity, as the model still provided values approximately 40% lower than the ones observed experimentally. Lastly, level set fields are used to allow faster diffusion along the fibre-matrix interphase. Using the interphase thickness values measured by Joliff [4], a fit for the transverse diffusivity values was obtained. However, a fit for all three directions could not be obtained with the current micromodel.

Acknowledgements
The authors acknowledge the contribution of the TKI-WoZ and IRPWIND projects for motivating and partly funding this research and of the TKI-MIMIC project industrial partners for providing useful feedback.

References
[1] Maxwell A S, Broughton W R, Dean G and Sims G D 2005 Review of accelerated ageing methods and lifetime prediction techniques for polymeric materials Tech. rep. National Physical Laboratory (NPL)
[2] Gautier L, Mortaigne B and Bellenger V 1999 *Composites Science and Technology* **59** 2329
[3] Choi H S, Ahn K J, Nam J D and Chun H J 2001 *Composites Part A: Applied Science and Manufacturing* **32** 709
[4] Joliff Y, Rekik W, Belec L and Chailan J F 2014 *Composite Structures* **108** 876
[5] 2006 Technical data sheet - EPIKOTE resin MGS RIMR 135 and EPIKURE curing agent MGS RIMH 134-RIMH 137 Tech. rep. Momentive
[6] Rocha I B C M, Raijmaekers S, Nijssen R P L and van der Meer F 2015 *20th International Conference on Composite Materials.*
[7] Ozdemir I, Brekelmans W A M and Geers M G D 2008 *International Journal for Numerical Methods in Engineering* **73** 185

High performance SMC matrix for structural applications

T Salard[1,2], F Lortie[1], J F Gérard[1] and C Peyre[2]

[1]Ingénierie des Matériaux Polymères, UMR 5223, Institut National des Sciences Appliquées de Lyon, Bât 403, 69621 Villeurbanne Cedex, France
[2]Plastic Omnium-MCR, ZI des Iles Feray, 07300 Tournon s/ Rhône, France

E-mail : thomas.salard@insa-lyon.fr

Abstract. Mechanical properties of a common SMC (Sheet Molding Compound) matrix constituted of a vinylester resin and a Low-Profile Additive (LPA) were compared to those of vinylester modified with core-shell rubber (CSR) particles. Valuable properties are brought by CSR, especially high impact strength, high fracture toughness with little loss in stiffness, in spite of the presence of CSR agglomerates in blends.

1. Introduction

The automotive industry currently knows an important energetic transition with the development of vehicles which must respond to drastic environmental constraints. In this context, the reduction of overall weight of cars is a key point and a way to perform it is to substitute heavy steel parts by lightweight composite materials.

Sheet molding compounds (SMCs) are a class of composite materials used for a long time in automotive industry because of their low weight, low cost and good mechanical properties [1], [2]. Generally they are constituted of a polymer matrix reinforced by short glass fibers and mineral fillers [1]. The matrix is usually constituted of unsaturated polyester or vinylester resins which own interesting intrinsic properties (high modulus, good solvent resistance, low cost) and which are blended with a thermoplastic low profile additive (LPA) to compensate for polymerization shrinkage.

Up to now SMCs have been mainly found in semi-structural applications like rear closures or rear floors but the possibility to extend their range of use in structural parts is studied. However, it requires to significantly improve their mechanical properties and especially their impact strength with no loss in stiffness. Many techniques have been studied to produce composites with better performances. Among them, several studies describe toughening strategies of thermoset resins (soft toughening agents like reactive liquid polymers, core-shell particles, block copolymers or different kind of inorganic fillers) but most of them are related to epoxy resins or remain at the matrix scale [3], [4].

The present work focuses on the modification of a vinylester resin by core-shell rubber (CSR) particles. Usually, CSR as powder are difficult to disperse in thermosetting precursors and a high level of shear is required to favor a good dispersion. Also, the resulting mixtures are often very viscous and are hardly compatible with the requirements of SMC industrial process for which a low initial viscosity is sought to promote a good impregnation of fibers [5]. Hence, to avoid problems of dispersion and viscosity, the present work investigates the toughening of thermosetting resins by CSR initially pre-dispersed in a carrier resin.

2. Experiments

2.1 Materials

A vinylester resin VE (in solution in 35 wt. % of styrene) was used for this work. Two grades of CSR, named M1 and M2 were employed. Both of them are liquid masterbatches composed of 40 wt. % of CSR particles (CSR1 for M1 and CSR2 for M2) dispersed in a carrier resin (C1 for M1 and C2 for M2). C1 and C2 are constituted of a vinylester or unsaturated polyester diluted in HPMA (Hydroxypropyl Methacrylate). The properties given by the supplier of the two masterbatches are listed in Table 1. A classic PVAc (Poly(Vinyl Acetate))-based additive (polymer in solution in 57 wt. % of styrene) employed in SMC to compensate for polymerization shrinkage was mixed with vinylester to make a standard SMC matrix, which was compared to VE-CSR blends. MEKP (Methyl Ethyl Ketone Peroxide) and Cobalt Octoate allowed the initiation of the polymerization.

To prepare the formulations, resin and additive were mixed for a few minutes in a planetary mixer which prevents from air bubbles. This step was repeated twice by adding first Cobalt Octoate and secondly MEKP. Then, the mixture was poured into a vertical mold (300 x 220 x 4 mm^3) before gelation occurs. It was first polymerized for 24h at room temperature, and then post-cured during 2h at 90°C, 2h at 120°C and 2h at 150°C. DSC measurements were made to assure that the reaction was complete. By adding the masterbatches M1 or M2 to the vinylester resin, the properties can be modified by the CSR particles and/or by the carrier resin. To discriminate the influence of the carrier resin from that of the CSR particles, blends with carrier resin but without CSR particles were also prepared. The respective compositions of the prepared formulations are given in Table 2.

Table 1. Comparison of the CSR grades.

	CSR particles		Carrier resin	
Masterbatch	Core type	CSR size(nm)	Reactive diluent	Nature of the prepolymer
M1	Styrene-butadiene rubber (SBr)	100	HPMA	vinylester
M2	Polybutadiene (PBd)	200	HPMA	polyester

Table 2. Prepared formulations.

		M1		M2		PVAc-based additive	
Formulation name	VE amount (parts)	CSR1 amount (parts)	C1 amount (parts)	CSR2 amount (parts)	C2 amount (parts)	PVAc (parts)	Styrene (parts)
ref	100	0	0	0	0	0	0
5M1	87.5	5	7.5	0	0	0	0
5M2	87.5	0	0	5	7.5	0	0
8M2	80	0	0	8	12	0	0
10M2	75	0	0	10	15	0	0
15M2	62.5	0	0	15	22.5	0	0
5C1	87.5	0	7.5	0	0	0	0
5C2	87.5	0	0	0	7.5	0	0
8C2	80	0	0	0	12	0	0
15C2	62.5	0	0	0	22.5	0	0
8PVAc	80	0	0	0	0	8.5	12.5

2.2 Methods

The glass transitions of neat resin and CSR-modified vinylester networks were determined using dynamic mechanical analysis on cured samples in three-point bending mode at 1Hz from -120°C to 230°C with a heating rate of 3°C/min.

Compressive tests were performed according to ISO 604 at a testing speed of 1 mm/min on samples with dimensions of 10 x 10 x 4 mm^3 and the compressive yield strength σ_y was recorded.

Three-point bending test were carried out on samples with dimensions of 80 x 10 x 4 mm^3. The support span and cross-head speed were set at 60 mm and 2 mm/min. The flexural modulus E_f, flexural strength σ_f and flexural strain ε_f were determined in accordance to ISO 178.

The Charpy impact resistance of the blends was measured according to IS0 179-1. The dimensions of the specimens were 80 x 10 x 4 mm^3 and the support span was 62 mm. The hammer employed had an energy of 0,5 J or 1 J according to the energy required to break the samples.

Fracture toughness measurements were made according to ISO 13586 on three-point bending mode with samples with dimensions of 60 x 10 x 4 mm^3 (Single-Edge Notched samples). Specimens were first notched with a saw. Then a razor blade is put in the notch and is tapped until a sharp crack is created. The value of the crack length shall be around the half of the sample width. The plain strain critical stress intensity factor K_{1C} is calculated using the equation:

$$K_{1C} = \frac{P}{BW^{\frac{1}{2}}} f\left(\frac{a}{W}\right) \quad (1)$$

Where P is the critical load, B the sample thickness, W the sample width and f(a/W) a shape factor which depends of the crack length to sample width ratio.

The fracture surfaces of specimens used of K_{1C} measurements were observed with a scanning electron microscope under an acceleration voltage of 10 kV in order to study the dispersion of CSR particles in the matrix and to analyze the toughening mechanisms encountered in the presence of CSR.

3. Results and discussion

Storage modulus G' and tan(δ) are plotted from DMTA results (Figure 1.a). Values of $T\alpha$ of materials were chosen from the maximum of the tan(δ) peak attributed to the α relaxation. A secondary transition at around -90°C (γ relaxation) for the neat resin was attributed to motions of local groups of the polymer chains as reported by others authors [6], [7]. The addition of CSR materbatch to resin leads to a decrease of $T\alpha$ from 165°C for the neat resin to 100°C for the 15-M2 blend. For CSR blends, the α relaxation of CSR core is also visible at -65°C for M1 blends (SBr core), and at around -90°C for M2 blends (PBd core), superposed with the γ transition of the vinylester network. Results from Figure 1.b shows that drop of $T\alpha$ was completely due to the carrier resin which acts as a plasticizer for the matrix since the decrease of $T\alpha$ was almost identical whether CSR particles are present or not. These observations were confirmed by a decrease of the rubbery plateau modulus when M1 or M2 is added, showing a reduction of networks' crosslinking density. 8PVAc system is bi-phasic with a PVAc-rich phase (α transition near 30°C) and a VE-rich phase with a α relaxation lower than that of the neat resin (145°C instead of 165°C) due to a part of solubilized PVAc which plasticizes the phase. This depletion can also be partly attributed to a modification of the ratio of double bonds $C{=}C_{VE}/C{=}C_{styrene}$ by the introduction of styrene brought by the PVAc solution.

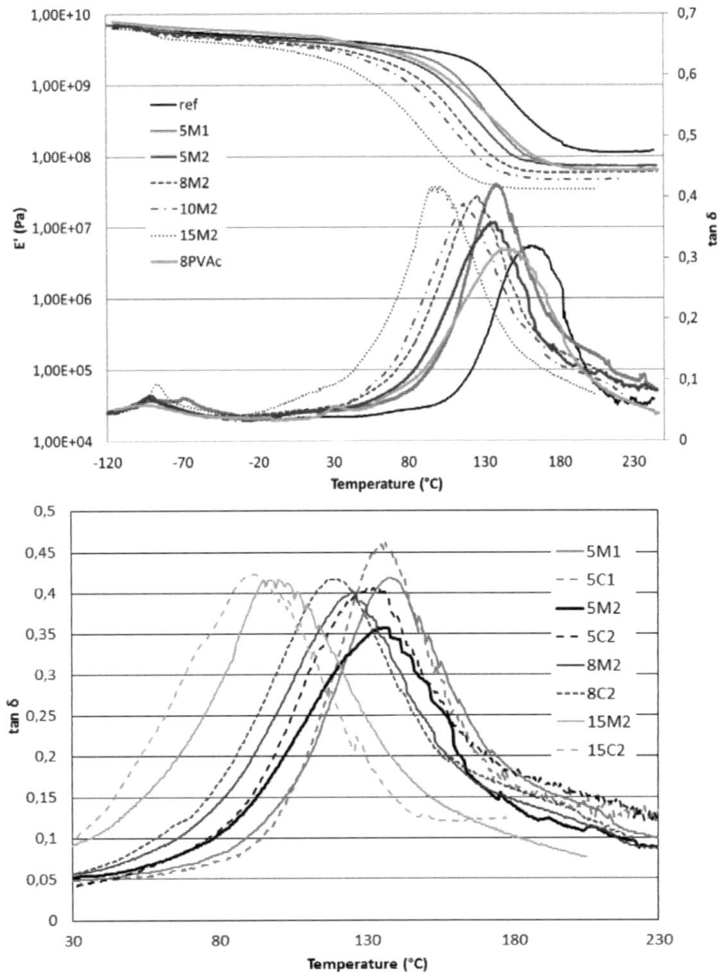

(a)

(b)

Figure 1. DMTA curves : (a) storage modulus and tan δ Vs temperature curves of CSR-blend systems ; (b) tan δ Vs temperature curves of formulations with and without CSR particles.

As expected, the compressive yield strength of blends decreases with an increase of the CSR content (Figure 2). This result can be explained by the fact that a part of rigid VE matrix was replaced by low-modulus rubber particles which promote plastic deformation of the material. Others authors reported these observations with thermosets in presence of CSR particles [8], [9]. In our case, the value of σ_y was almost the same for 5M1 and 5M2 proving that rubber fraction may be the main factor determining this value instead of the nature of CSR particles (size and nature of the core).

Stress-strain curves obtained for the materials are shown in Figure 3.a. The flexural modulus decreases with CSR content because of the introduction of a low-modulus phase in the material. Indeed, the modulus is reduced from 3.3 GPa for the neat resin to 1.6 GPa for 15M2 blend. Keeping a high modulus is a key factor for structural materials and the drop in modulus noticed for 10M2 and 15M2 (respectively 2.4 GPa and 1.6 GPa) was considered too high to employ these blends for structural applications. Increasing the amount of CSR leads to blends more ductile because the presence of a soft phase reduces the flexural stress whereas flexural strain keeps rising. Both carrier resin and CSR particles influence flexural properties of materials by improving ductility (Figure 3.b). However, the neat resin possesses low flexural stress and strain (Figure 3.a) because of its important brittleness and its high sensitivity to the presence of flaws. It is noted that there is no difference between M1 and M2 blends, confirming the predominance of rubber content on others parameters for mechanical reinforcement. As for the 5 wt. % CSR blends, the addition of a soft phase PVAc (8.5 wt. %) improves the ductility of the material and delays the failure of the materials if compared to the neat resin.

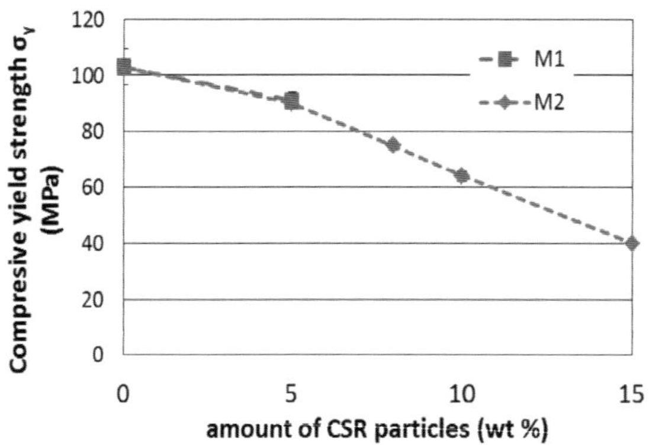

Figure 2. Compressive yield strength of CSR blends.

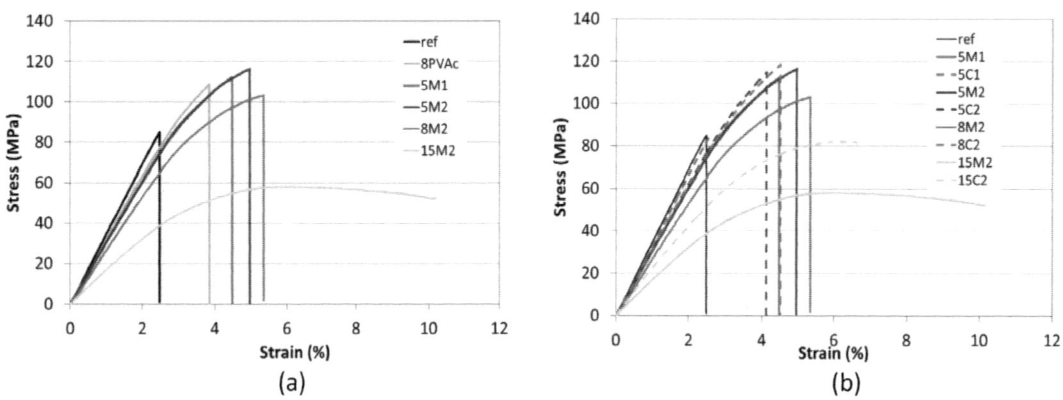

Figure 3. (a) Flexural stress-strain curves of different blends; (b) Comparison of stress strain curves of blends with and without CSR particles.

The effect of masterbatches M1 and M2 on impact strength of materials was studied (Figure 4.a). Adding M1 or M2 to the neat resin leads to an outstanding increase of the impact strength from 4.9 kJ/m² for the neat resin to 15.4 kJ/m² for 15M2 blend, i.e. an improvement higher than 200%. 5 wt. % of CSR is high enough to reach performances surpassing the ones of standard SMC matrix 8PVAc. As for static tests, differences between 5M1 and 5M2 are small considering standard deviations. Even if toughening may be partly brought by the carrier resin, the major part of the improvement seems to come from CSR particles or from a synergistic effect between carrier resin and CSR (Figure 4.b). Toughening mechanisms will be discussed later but an increase in impact strength could originate from the absence of large defects which could promote the initiation of cracks in materials. It can be deduced that CSR particles are probably macroscopically well-dispersed in the matrix.

Fracture toughness is also significantly improved in presence of masterbatches (Figure 5.a). K_{IC} values go from 0.45 MPa.m$^{1/2}$ to 1.2 MPa.m$^{1/2}$ for 10M2 and 15M2 blends. It appears that M2 masterbatch is more efficient than M1 comparing 5M1 and 5M2 even if deviations must be taken into account. Like for impact strength measurements, it was noticed that carrier resins contribute in improving K_{IC} values of matrices but the main toughening effect is noticed when CSR particles are present. Many authors described the role played by CSR particles in thermoset toughening [4], [10],

[11]. Around a crack tip, the cavitation of the rubbery core of CSR particles leads to a reduction of the hydrostatic component of the stress field and allows the transition from a triaxial stress state (plane-strain) to a plane-stress state, initiating deformation of the matrix by shear-banding which contributes to toughen materials. The reduction of compressive yield strength due to CSR particles can also facilitate plastic deformation of the matrix which is able to dissipate more energy.

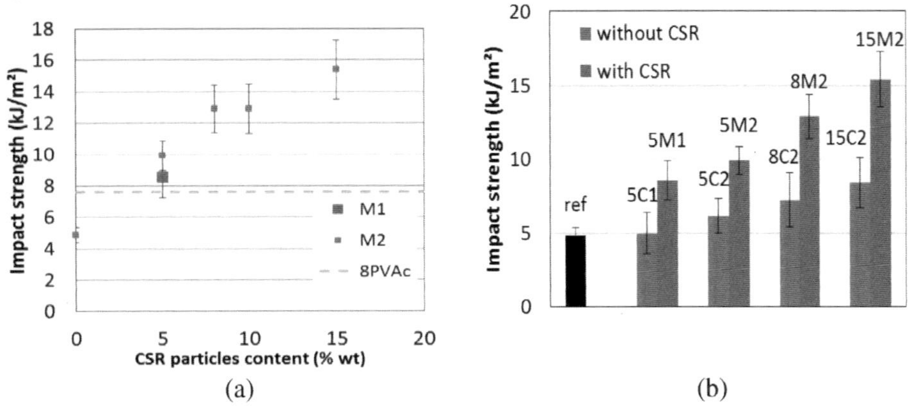

Figure 4. (a) Impact strength of different formulations with M1, M2 and PVAc ; (b) Influence of both CSR particles and carrier resin on the increase of the impact strength.

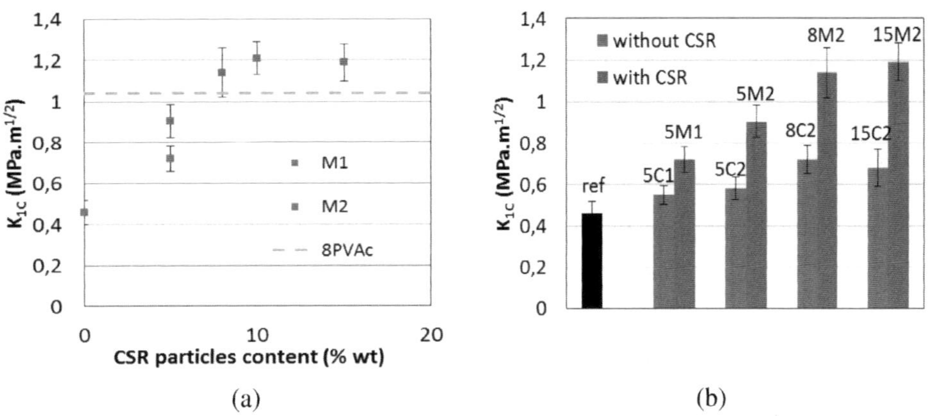

Figure 5. (a) K1C values of different formulations with M1, M2 and PVAc ; (b) Influence of both CSR particles and carrier resin on the fracture toughness.

SEM observations on fracture surfaces after fracture toughness measurements were performed on different formulations (Figure 6). Fracture surface of neat resin is smooth and reveals the brittle nature of thermosetting polymers like vinylester (Figure 6.a). Feather like marks are visible, associated with fast crack growth showing that cracks propagated on several planes. For CSR blends, SEM pictures macroscopically show that, CSR are well dispersed (Figure 6.b,d). But at higher magnification (Figure 6.c,e), SEM pictures present areas more or less rich in CSR particles, like a co-continuous microstructure, proving a CSR reorganization probably taking place during polymerization of the matrix. Even if CSR particles were not well dispersed, the resulting morphology seems to enhance mechanical properties. CSR blends exhibit rougher surfaces if compared to neat resin, indicating that more energy was absorbed during the propagation of the crack. Many voids are visible and could result from core cavitation of CSR or CSR debonding from the matrix. Further investigations like TEM or OM analyses will be carried out to deeper identify toughening mechanisms. SEM observations (not shown here) were also realized on CSR-free blends (with carrier resin) and fracture surfaces look like neat resin one. Thus, roughness of fracture surfaces of CSR blends is solely attributed to the presence of CSR particles.

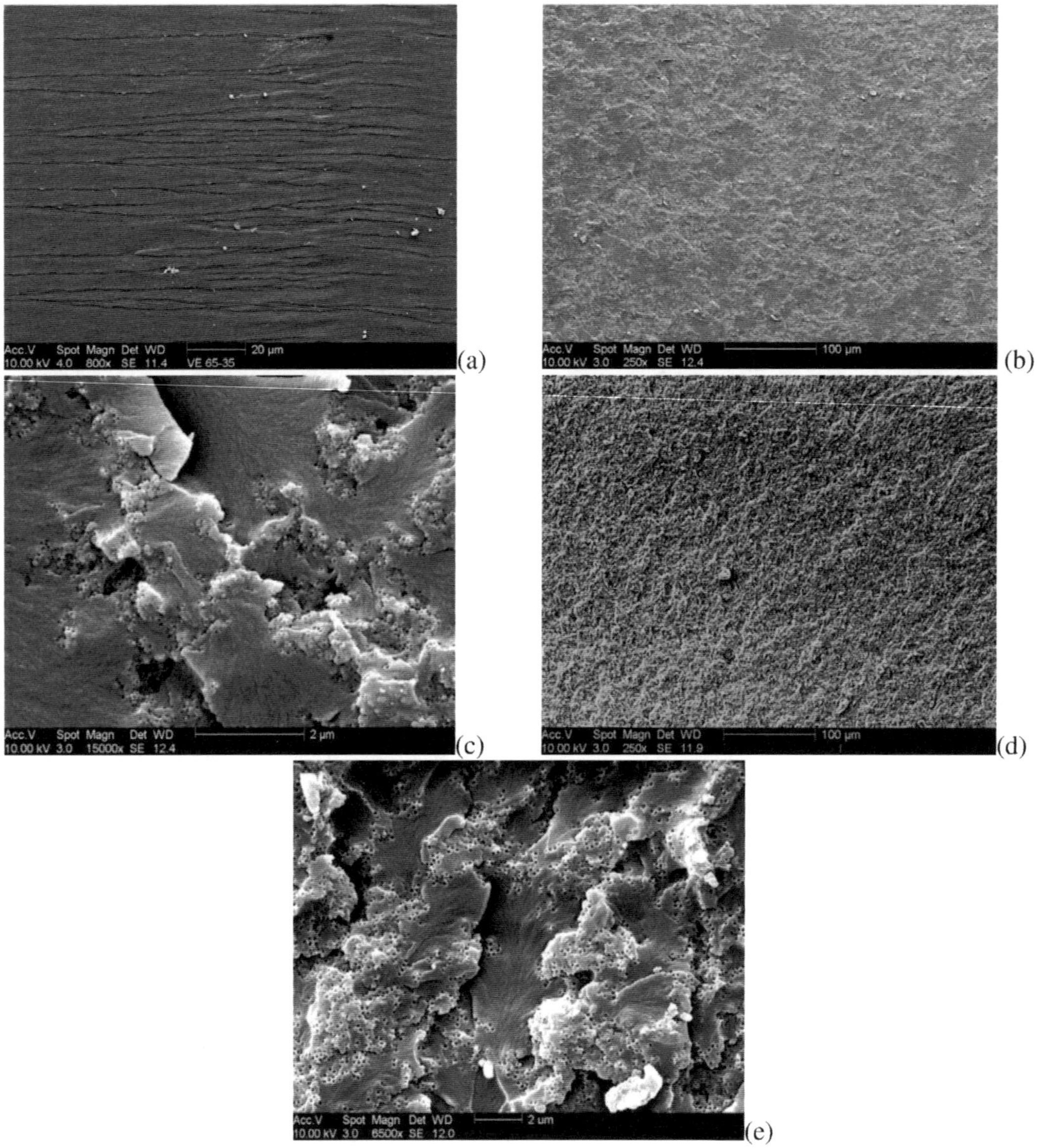

Figure 6. SEM pictures of : (a) neat resin ; (b),(c) 5M1 system ; (d),(e) 10M2 system.

4. Conclusion

A vinylester resin was modified with masterbatches of CSR dispersed in carrier resins and blends were compared to a common SMC matrix. Results demonstrated an interesting improvement in impact strength and fracture toughness in presence of CSR. In order to give access to structural applications, it is important to get a matrix which exhibits a good compromise between stiffness and toughness. That is why 10M2 and 15M2 systems are not suitable candidates because of the too important loss in modulus. However, 5M1, 5M2 and 8M2 systems are really interesting especially thanks to their impact strength significantly higher than that of standard SMC matrix 8PVAc, and their rigidity not sacrificed. These blends will be tested in SMC formulations and the effect on mechanical properties will be determined. Complementary analyses like TEM and OM experiments on DN-4PB samples particularly described by Sue [12], will be led on matrix systems in order to identify toughening mechanisms related to CSR particles.

References

[1] Orgeas L and Dumont P J J 2012 Sheet Molding Compounds *Wiley Encyclopedia of composites* ed Nicolais L, Borzachiello A and al. (Wiley) pp 2683-2718
[2] Nomura T, Yokoi T and Mishima Y 1991 *Int. J. Mater. Prod. Tec.* **6** 143
[3] Pascault J P, Sautereau H, Verdu J and Williams R J J 2002 Yielding and Fracture of Polymer Networks *Thermosetting Polymers* (New York : Marcel Dekker) pp 350-388.
[4] Bagheri R, Marouf B T and Pearson R A 2009 *J. Macromol. Sci. Rev.* **49** 201
[5] Hayes B S and Seferis J C 2001 *Polym. Composite* **22** 451
[6] Cook W D and Delatycki O 1975 *J. Polym. Sci. Pol. Phys.* **13**, 1049
[7] Melot D, Escaig B, Lefebvre J M, Eustache R P and Laupretre F 1994 *J. Polym. Sci. Pol. Phys.* **32** 249
[8] Ahmadi M, Moghbeli M R and Shokrieh M M 2012 *Iran. Polym. J.* **21** 855
[9] Quan D and Ivankovic A 2015 *Polymer* **66** 16
[10] Huang Y J, Wu J, Liang J G, Hsu M W and Ma J K 2008 *J. App. Polym. Sci.* **107** 939
[11] Ahmadi M, Moghbeli M R and Shokrieh M M 2012 *Polym. Eng. Sci.* **52** 1928
[12] Sue H J 1992 *J. Mater. Sci.***27** 3098

IOP Conf. Series: Materials Science and Engineering **139** (2016) 012046 doi:10.1088/1757-899X/139/1/012046

Influence of microwave plasma treatment on the surface properties of carbon fibers and their adhesion in a polypropylene matrix

C Scheffler,[1,3] E Wölfel[1], T Förster [1], C Poitzsch[1], L Kotte[2] and G Mäder[2]

[1]Leibniz Institut für Polymerforschung e.V. (IPF), Department Composite Materials, Hohe Straße 6, 01069 Dresden, Germany

[2]Fraunhofer Institut für Werkstoff- und Strahltechnik (IWS), Department Chemical Surface and Reaction Technology, Winterbergstraße 28, 01277 Dresden, Germany

[3]Corresponding author: scheffler@ipfdd.de

Abstract. A commercially available carbon fiber (CF) with an epoxy-based sizing (EP-sized CF) and an unsized CF have been plasma treated to study the effect on the fiber-matrix adhesion towards a polypropylene matrix. The EP-sized fiber was chosen because of its predictable low adhesion in a polypropylene (PP) matrix. The fibers have been modified using a microwave low-pressure $O_2/CO_2/N_2$-gas plasma source (Cyrannus®) developed at IWS in a batch process. One aim of this study was the evaluation of parameters using high energies and short time periods in the plasma chamber to see the effect on mechanical performance of CF. These results will be the fundamental work for a planned continuous plasma modification line. The CF surface was characterized by determining the surface energies, single fiber tensile strength and XPS analysis. The adhesion behavior before and after plasma treatment was studied by single fiber pull-out test (SFPO) and scanning electron microscopy (SEM). It was shown that the CO_2- and O_2-plasma increases the number of functional groups on the fiber surface during short time plasma treatment of 30 s. Carboxylic groups on the unsized CF surface resulting from O_2-plasma treatment lead to an enhanced fiber-matrix adhesion, whereas the fiber strength was merely reduced.

1. Introduction

The excellent properties of carbon fibers (CF), such as high specific strength and stiffness, performance to weight ratio, high thermal and chemical resistance and also conductivity, make them applicable in various fields of mechanical engineering, automotive and air transport industries [1][2][3]. However, the poor interfacial adhesion between reinforcing CF surface and the polymer matrix limits the crucial properties of the composites toughness as well as longitudinal and transverse strength. To solve these critical issue researchers followed different ways that are summarized by Sharma et al. [4] to two main approaches: "wet" and "dry" surface modification methods. The wet chemical approaches include: (i) the application of polymer finish to CF surface (sizing), (ii) acidic medication, (iii) electrochemical modification and (iv) electro-polymer coating. Surface modification of CF by dry methods are (i) high energy irradiation modification, (ii) nickel surface coating, (iii) thermal modifications, (iv) miscellaneous dry treatments and (v) surface modification in "multi-scales" by nano particle modification and nanotube coating using for example chemical vapour deposition or electrophoretic deposition. Plasma surface modification, also a dry method, changes the fiber surface layer physicochemically by introducing functional groups without affecting the layers underneath. For plasma modification inert (Ar, N_2) [13]-[17] and reactive gases (O_2, CO_2, air, CF_4) [18]-[23] or mixtures of both are applied. Summarizing the literature, it can be concluded that all gases and process parameters during plasma treatment lead to more or less improved wetting behavior. The damage of the filament surface is more severe by application of reactive compared to inert gases and increases with treatment time, especially in the case of O_2. Also a higher surface roughness arises under all conditions. Based on XPS measurements it is stated [18][19][22], that using O_2-plasma, as pure gas as well as in a gas mixture, increases the [O]:[C]-ratio significantly. Mainly OH^--groups are generated, which lead to higher polarity and surface energy. In the most studies very long treatment times up to 30 min combined with low energies were chosen for the modification. In this work, the focus was set on low treatment times of 30 s, since short treatment times are a precondition for potential continuous modification processes. However, in this study the surface tension was determined over a time period up to 180 s to gain basic process knowledge. A maleic anhydride grafted PP was used as a matrix to investigate changes of the adhesion behavior.

2. Experimental

2.1. Materials

A commercially available PAN-based CF roving (50 k) with an epoxy-(EP)-based sizing (sizing content 1 wt.-% according to manufacture specification) and an unsized (electrochemically activated) PAN-based CF roving (50 k) were selected for the treatment by microwave plasma. The filament diameter was 7.2 ± 0.3 µm for EP-sized CF and 6.9 ± 0.4 µm for unsized CF, respectively. The EP-sized fiber was chosen because of its predictable low adhesion in a polypropylene matrix since EP-sizing is applicable in epoxy-based matrix systems. The EP-sized fiber serves as control fiber since it is expected that the EP-sizing will be removed during plasma treatment and this effect will be clearly seen during all testing methods.

As a matrix system for single fiber pull-out test (SFPO) an isotactic polypropylene (PP, HG455FB, Borealis) was used being modified by 2 wt.-% of maleic anhydride grafted PP. Before using the PP for manufacturing the single fiber model composites the PP was dried for four hours in the vacuum oven at 80°C to avoid any effect of moisture, although the moisture uptake of PP is known to be small.

2.2 Methods

2.2.1. Plasma modification. For surface modification of CF the microwave plasma equipment (developed at Fraunhofer IWS Dresden) was used, which works discontinuously in a batch process. The plasma chamber is a cylindrical 6 inch Cyrannus® plasma source and single 10 cm pieces of a 50K filament yarn have been introduced for each plasma treatment. The following parameters were used during treatment of CF: purging the chamber with argon for 5 min, pressure 5-210 mbar, O_2- and CO_2-plasma gas with 1000-5000 sccm.

2.2.2. X-ray photoelectron spectroscopy (XPS). X-ray photoelectron spectroscopy (XPS) studies were carried out by means of an Axis Ultra photoelectron spectrometer (Kratos Analytical, Manchester, UK) equipped with a monochromatic Al Kα (hν = 1486.6 eV) X-ray source of 300 W at 20 mA. The kinetic energy of the photoelectrons was determined with a hemispheric analyzer set to pass energy of 160 eV for wide scan spectra and 20 eV for high-resolution spectra, respectively. Quantitative elemental compositions were determined from peak areas using experimentally determined sensitivity factors and the spectrometer transmission function.

2.2.3. Single fiber tensile test. The tensile tests on the carbon filaments before and after plasma treatment were carried out on the testing machine FAVIGRAPH ME (Textechno, Germany) with a 20 cN force cell. The testing velocity was 25 mm/min at a gauge length of 50 mm. The fineness of each single filament was measured using a vibroscopical testing method according to ASTM D 1577. The filament strength and strain were measured in compliance with DIN ISO 5079. The distribution of failure stresses for each treatment condition is based on 50 measurements. The failure stresses were evaluated by Weibull distribution function as described in [12]. The scale parameter σ_o represents the stress at which 63.2 % of the filaments break ($P(\sigma_o) = 0{,}6321$). The shape parameter m (also Weibull modulus) describes the distribution of the failure stress. A higher value of m indicates that the filaments fail in a narrower range of failure stresses. Weibull parameter are determined by plotting $\ln(-\ln(1-P))$ against $\ln(\sigma)$. Thereby, m is equal to slope of generated curve and σ_0 equates to the interception with $\ln(-\ln(1-P)) = 0$.

2.2.4. Wetting behavior. The surface energy measurement was made using a Krüss K14 tensiometer (Krüss GmbH, Germany) in water, ethylene glycol and n-hexane. One end of a single fiber was attached to the hook of an electro micro-balance (sensitivity ± 0.1 µg). The free end of the fiber was immersed into the liquid (final depth of immersion 2 mm, rate 6 mm/min, mean of 20 measurements) [5]. From the test, the force versus position loop was recorded and the force was converted to the contact angle (θ) using the Wilhelmy equation. The surface tension (γ_F) and its two components (non-polar or dispersive part, γ^d and the polar part, γ^p) of the fiber were then calculated to evaluate the wetting behavior of the fiber surfaces.

2.2.5. Single fiber pull out. The interfacial adhesion strength was evaluated by single fiber pull out test [6][7]. Using an embedding equipment designed and constructed at the Leibniz Institute, the model micro-composites were prepared by accurately embedding one end of the single fiber in the matrix perpendicularly with a pre-selected embedding length l_e (l_e = 300 µm) and an embedding temperature of 210 °C at controlled atmosphere and temperature. The pull out test was carried out on a self-made pull-out apparatus with force accuracy of 1 mN, displacement accuracy of 0.07 µm and a loading rate of 0.01 µm/s at ambient conditions.

The force- displacement curves were detected and the maximum force, F_{max}, required for pulling the fiber out of the matrix was measured. After testing, the fiber diameter, d_f was measured by optical microscopy. The adhesion bond strength between the fiber and the matrix was characterized by the values of the apparent interfacial shear strength ($\tau_{app} = F_{max}/2\pi * d_f * l_e$.). Each fiber/matrix combination was evaluated in about 15-20 single tests.

3. Results

3.1. Surface energy

The surface energies of CF depending on the time of plasma treatment are summarized in Fig. 1. Comparing the results of O_2- and CO_2-plasma treatment shows that the O_2-plasma leads to a decreased polar fraction for EP-sized CF after 30 s. Additional testing by FTIR (not included in this paper) proved that this was due to the removal of the sizing. By further plasma treatment the surface tension and polar fraction increases again leading to a value of γ_F=28.4 mN/m after 60 s in O_2-plasma. In contrast, the highest polar fraction during this investigation was found for CO_2-plasma treated unsized CF after 60 s.

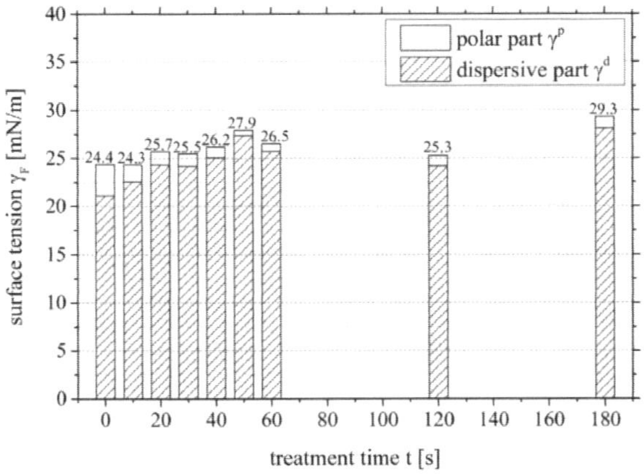

a) CO_2-plasma treated EP-sized CF

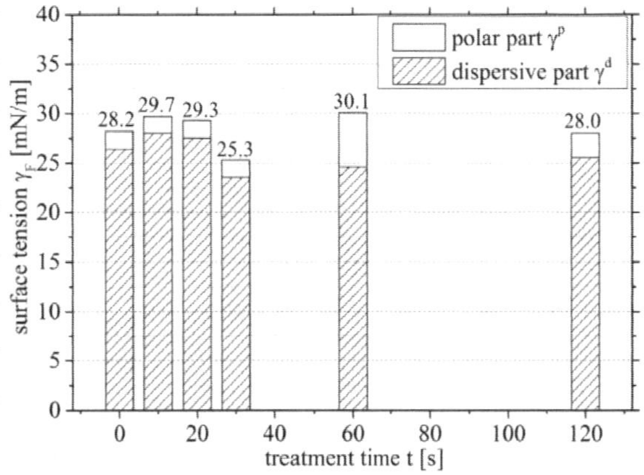

b) CO_2-plasma treated unsized CF

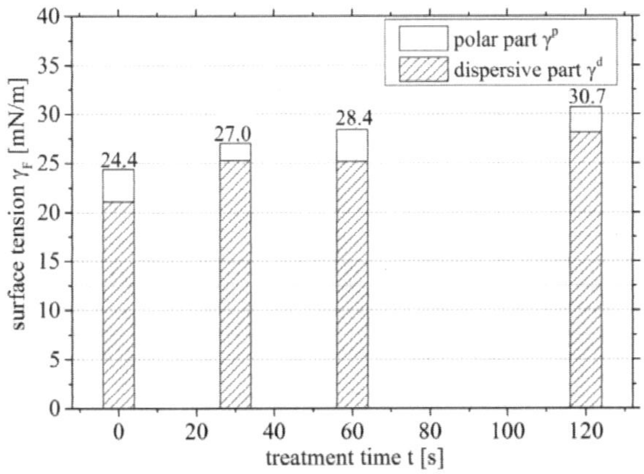

c) O_2-plasma treated EP-sized CF

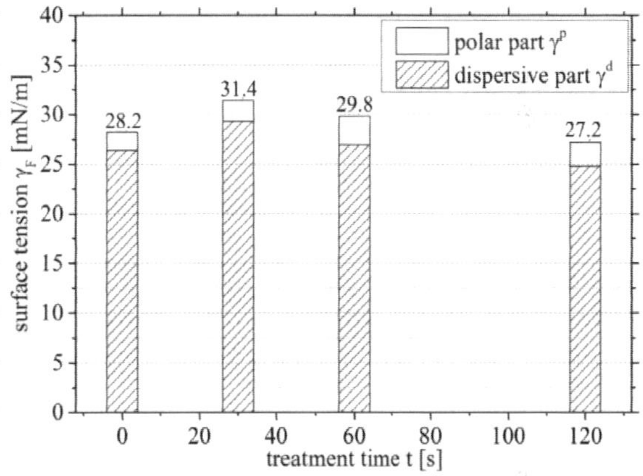

d) O_2-plasma treated unsized CF

Fig. 1. Surface tension of CO_2-plasma treated a) EP-sized CF, b) unsized CF and O_2-plasma treated c) EP-sized CF, d) unsized CF vs. treatment time.

3.2. XPS

In the literature the carboxylic groups –COOH and –C=O have been discussed for the improvement of fiber-matrix adhesion as well as hydroxyl groups –C-OH for increasing the polarity of the fiber surface [8]. To avoid additional effects by sizing only unsized CF were analyzed by XPS after a plasma treatment time of 30 s. During the plasma treatment the [O]:[C]-ratio was increased by CO_2-plasma from 0.061 to 0.281, but also O_2-plasma treatment lead to a slightly increased [O]:[C]-ratio from 0.061 to 0.096.

By deconvolution of C1s peaks the bonding characteristics and oxidation state, respectively, can be concluded by resolving in component peaks and their functionalities [11], see Table 1. Compared to the initial state of unsized CF, a slightly increased number of OH-groups after CO_2-plasma treatment and of HO-C=O-/⁻O-C=O-groups after O_2-plasma treatment was detected.

Table 1. XPS bonding characteristics of unsized CF compared to plasma-treated CF.

Peak	Unsized CF	CO_2, 30 s	O_2, 30 s
		Component [at.-%]	
Unsaturated –C=C- or =C-C= bondings	66.78	63.19	64.66
Saturated hydrocarbons C_xH_z	7.28	8.60	7.08
C-N-bondings of amines, C=N-bondings of imines, C≡N-bondings of nitrils	3.41	2.28	2.59
C-O-bondings of alcoholic and phenolic groups C-OH-groups	2.61	3.87	2.56
Carbonyl carbon atoms (C=O) of ketons or quinones	1.79	1.47	1.57
Carbonyl carbon atoms of carbonic acid groups (HO-C=O) or their carboxylates ($^-$O-C=O)	0.07	0.37	0.82

3.3. Filament tensile strength

The tensile failure stresses for CF in the unsized and sized state are subjected to high scatter. The heterogeneity of CF seems to be based on a not homogeneous degree of graphitization of the single fibers [10]. Additionally, the distribution homogeneity of sizings contributes to the extent of scatter for failure stresses. The Weibull distribution function was used to derive the probability of failure. As shown in Fig. 2 and Table 2 the Weibull-modulus of EP-sized CF is decreased during 30 s plasma treatment with CO_2- and O_2-plasma. As already arises from the determination of the surface energies, the sizing is removed during the plasma treatment. This leads to reduced values of the Weibull-modulus, changing from m = 7.22 to m = 2.07 for O_2-plasma treatment. If EP-sized CF is exposed to longer treatment times of 120 s the failure stress is significantly reduced. Compared to the EP-sized CF no reduced failure stresses due to plasma treatment could be detected for unsized CF, nor for 30 s and neither after 120 s of treatment. It seems that the removal of sizing during plasma treatment is accompanied by effects that lead to severe damage of the CF surface, which do not occur on the unsized carbon surface.

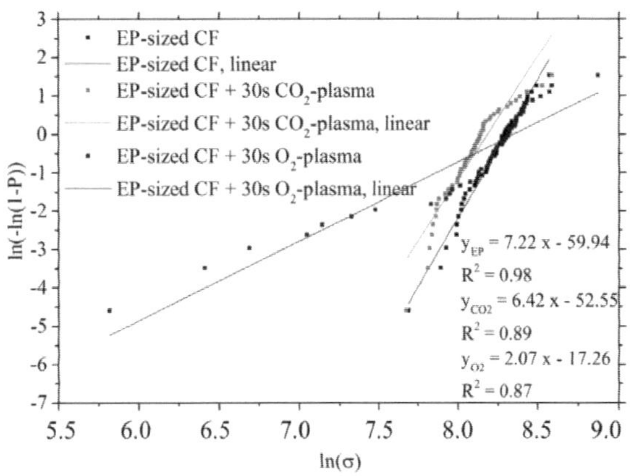

a) 30s plasma treated EP-sized CF

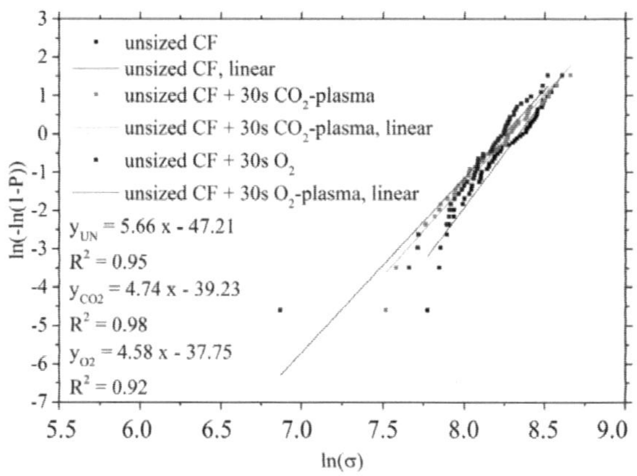

b) 30s plasma treated unsized CF

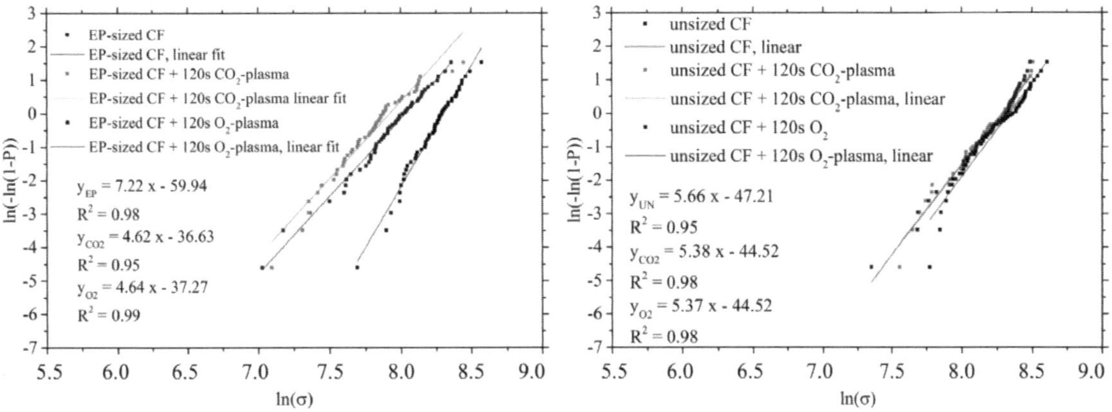

c) 120s plasma treated EP-sized CF d) 120s plasma treated unsized CF

Fig. 2. Weibull plot of the 30 s plasma treated a) EP-sized CF, b) unsized CF and 120 s plasma treated c) EP-sized CF, d) unsized CF.

Table 2. Weibull modulus and characteristic strength of untreated and plasma treated CF.

	scale parameter σ_0 [MPa]	**Weibull modulus m**	**scale parameter σ_0 [MPa]**	**Weibull modulus m**
EP-sized CF	4033	7.22	4033	7.22
unsized CF	4175	5.66	4175	5.66
treatment	**30s**		**120s**	
EP-sized CF + CO_2-plasma	3582	6.42	2753	4.62
EP-sized CF + O_2-plasma	4234	2.07	3079	4.64
unsized CF + CO_2-plasma	3948	4.74	3949	5.38
unsized CF + O_2-plasma	3814	4.58	3982	5.37

3.4. Fiber-matrix adhesion

The SFPO is a very sensitive method related to changes due to fiber surface modification. For the EP-sized CF the results of SFPO (Table 3) mirror the results that have been received by testing the wetting behavior and the filament tensile strength. It was shown that the EP-sizing is removed during plasma-modification. Since the EP-sizing is not appropriate for fiber-matrix interaction because of its chemical and physical properties, the lowest adhesion strength is measured for EP-sized CF. The removal of the sizing exposes the unsized carbon surface that enables an improved interaction with the MAH-g-PP matrix due to the functional groups on the surface as was determined by XPS. However, also the shear strength of the unsized CF is on the level of EP-sized CF after removing of the sizing. After treatment in CO_2- and O_2-plasma over 30 s no increase of the shear strength was detected, though a longer treatment time of 60s leads to enhanced pull-out force and the highest shear strength in this work. Based on the XPS data after 30 s plasma treatment time it can be assumed that after 60 s the amount of HO-C=O-/$^-$O-C=O-groups might have increased and therefore causes improved fiber-matrix interaction.

Table 3. Apparent shear strength and pull-out work for EP-sized and unsized CF embedded in PP matrix before and after plasma treatment.

sample	treatment	t_{app} [MPa]	SD [MPa]
EP-sized CF	none	19.6	9.5
	CO_2, 30 s	25.4	9.7
	O_2, 30 s	27.8	5.6
unsized CF	none	26.5	5.4
	CO_2, 30 s	22.2	4.7
	O_2, 30 s	27.5	9.3
	O_2, 60 s	30.5	5.5

4. Conclusion

Based on this data it can be concluded that primarily the O_2-plasma modification of CF surfaces reveals potential for the application of functional groups on the carbon surface and therefore an improved fiber-matrix adhesion. However, a crucial change of the carbon fiber surface properties due to the carefully chosen parameters for plasma treatment used in this study is not achieved yet. The nearly constant mechanical properties of the unsized CF imply the possibility of using wider limits for process parameters that come closer to the requirements for a continuously plasma modification concept (short treatment times at high energies).

References

[1] Soutis C 2005 Fiber reinforced composites in aircraft construction *Prog Aerospace Sci* **41** 143–51

[2] Paiva J M F, Santos A D N and Rezende M C 2009 Mechanical and morphological characterizations of carbon fiber fabric reinforced epoxy composites used in aeronautical field *Mater Res*, **12 (3)** 367–74

[3] Edie D D and McHugh J J 1999 High performance carbon fibers *Carbon materials for advanced technologies* ed T D Burchell (Pergamon) chapter 4 119–38.

[4] Sharma M, Gao S L, Mäder E, Sharma H, Wei L Y and Bijwe J 2014 Carbon fiber surfaces and composite interphases *Compos Sci* **102** 35-50

[5] Kamath Y K, Dansizer C J and Weigmann H D 1984 Surface wettability of human hair. I. effect of deposition of polymers and surfactants *J Appl Polym Sci* **29** 1011-26

[6] Mäder E, Grundke K, Jacobasch H J and Wachinger G 1994 Surface, interphase and composite property relations in fibre-reinforced polymers *Compos* **25 (7)** 739-44

[7] Fu S Y, Lauke B, Mäder E, Yue C Y and Hu X 2000 Tensile properties of short-glass-fiber-and short-carbon-fiber-reinforced polypropylene composites *Composites Part A* **31 (10)** 1117-25

[8] Dilsiz N and Wightman J P 2000 Effect of acid-base properties of unsized and sized carbon fibers on fiber/epoxy matrix adhesion *Colloids and Surf* A **164** (2000) 325-36

[9] Tiwari S, Sharma M, Panier S, Mutel B, Mitschang P and Bijwe J 2011 Influence of cold remote oxygen plasma treatment on carbon fabric and its composites with specialty polymers *J Mater Sci* **46** 964-74

[10] Huson M.G, Church J S, Kafi A A, Woodhead A L, Khoo J, Kiran M S R N and Fox B L 2014 Heterogenity of carbon fibre *Carbon* **68** 240-49

[11] Moulder J F, Stickle W F, Sobol P E and Bomben K D 1992 *Handbook of X-ray photoelectron spectroscopy* ed Chastain J (Perkin-Elmer Corporation) Physical Electronics Division

[12] Scheffler C, Förster T, Mäder E, Heinrich G, Hempel S and Mechtcherine V 2009 Aging of alkali-resistant glass and basalt fibers in alkaline solutions: Evaluation of the failure stress by Weibull distribution function *J Non-Cryst Solids* **355 (52)** 2588-95

[13] Bogoeva-Gaceva G, Mäder E, Häussler L and Dekanski A 1997 Characterization of the surface and interphase of plasma-treated HM carbon fibres *Compos Part A* **28 (5)** 445-52

[14] Tiwari S, Sharma M, Panier S, Mutel B, Mitschang P and Bijwe J 2011 Influence of cold remote nitrogen oxygen plasma treatment on carbon fabric and its composites with specialty polymers *J Mater Sci* **46 (4)** 964-74

[15] Han S H, Oh H J and Kim S S 2014 Evaluation of fiber surface treatment on the interfacial behavior of carbon fiber-reinforced polypropylene composites *Compos Part B* **60** 98-105

[16] Farrow G J and Jones C 1994 The effect of low power nitrogen plasma treatment of carbon fibres on the interfacial shear strength of carbon fibre/epoxy composites *J Adhes* **45 (1-4)** 29-42

[17] Yuan L Y, Shyu S S and Lai J Y 1991 Plasma surface treatments on carbon fibers. II. Mechanical property and interfacial shear strength *J Appl Polym Sci* **42 (9)** 2525-34

[18] Park S J, Chang Y H, Moon C W, Suh D H, Im S S and Kim Y C 2010 A study of atmospheric plasma treatment on surface energetics of carbon fibers *Bull Korean Chem Soc* **31 (2)** 335-38

[19] Li J 2009 The research on the interfacial compatibility of polypropylene composite filled with surface treated carbon fiber *Appl Surf Sci* **255 (20)** 8682-84

[20] Bismarck A, Kumru M E and Springer J 1999 Influence of oxygen plasma treatment of PAN-based carbon fibers on their electrokinetic and wetting properties *J Colloid Interface Sci* **210 (1)** 60-72

[21] Allred R E and Schimpf W C 1994 CO_2 plasma modification of high-modulus carbon fibers and their adhesion to epoxy resins *J Adhes Sci Technol* **8 (4)** 383-94

[22] Jang J and Yang H 2000 The effect of surface treatment on the performance improvement of carbon fiber/polybenzoxazine composites *J Mater Sci* **35 (9)** 2297-2303

[23] Tsutsumi K, Ban K, Shibata K, Okazaki S and Kogoma M 1996 Wettability and adhesion characteristics of plasma-treated carbon fibers *J Adhes* **57 (1-4)** 45-53

[24] Ho K K, Lamoriniere S, Kalinka G, Schulz E and Bismarck A 2007 Interfacial behavior between atmospheric-plasma-fluorinated carbon fibers and poly (vinylidene fluoride) *J Colloid Interface Sci* **313(2)** 476-484

[25] Xie J, Xin D, Cao H, Wang C, Zhao Y, Yao L and Qiu Y 2011 Improving carbon fiber adhesion to polyimide with atmospheric pressure plasma treatment *Surf Coat Technol* **206 (2)** 191-201

[26] Erden S, Ho K K, Lamoriniere S, Lee A F, Yildiz H and Bismarck A 2010 Continuous atmospheric plasma oxidation of carbon fibres: influence on the fibre surface and bulk properties and adhesion to polyamide 12 *Plasma Chem Plasma Process* **30 (4)** 471-87

IOP Conf. Series: Materials Science and Engineering **139** (2016) 012047 doi:10.1088/1757-899X/139/1/012047

Flexural testing on carbon fibre laminates taking into account their different behaviour under tension and compression

M C Serna Moreno[1], A Romero Gutiérrez and J L Martínez Vicente

Universidad de Castilla-La Mancha, Instituto Investigaciones Energéticas y Aplicaciones Industriales & ETS Ingenieros Industriales, Av. Camilo José Cela s/n, 13071 Ciudad Real, Spain

[1]E-mail: mariacarmen.serna@uclm.es

Abstract. An analytical model has been derived for describing the results of three-point-bending tests in materials with different behaviour under tension and compression. The shift of the neutral plane and the damage initiation mode and its location have been defined. The validity of the equations has been reviewed by testing carbon fibre-reinforced polymers (CFRP), typically employed in different weight-critical applications. Both unidirectional and cross-ply laminates have been studied. The initial failure mode produced depends directly on the beam span-thickness relation. Therefore, specimens with different thicknesses have been analysed for examining the damage initiation due to either the bending moment or the out-of-plane shear load. The experimental description of the damage initiation and evolution has been shown by means of optical microscopy. The good agreement between the analytical estimations and the experimental results shows the validity of the analytical model exposed.

1. Introduction

Three-point-bending tests are used extensively in mechanical characterization due to the simplicity of the specimen preparation and testing. It is an interesting way to subject a specimen to tension, compression and out-of-plane shear simultaneously for checking its structural integrity. Well documented formulas are available for analysing materials with equal tension and compression properties submitted to flexural testing [1,2]. However several structural materials have different tensile and compressive response and, so that, an analytical model which takes into account this aspect would be desirable.

Composites constructed from carbon fibres pre-impregnated by an epoxy polymer (pre-preg) generally present unconventional structural behaviour which needs thorough examination. In particular, the elastic responses under tension and compression in the principal material directions are both different. This fact was utilized by Jones [3] and, more recently, by other authors [4,5] for obtaining the elastic moduli and the maximum normal stresses in the tensile and compressive regions generated on unidirectional specimens submitted to flexural testing in parallel to the fibres direction. But none of the various analytical works found in the literature regarding cross-ply laminates in flexure took into account the dissimilar material behaviour in tension and compression, since these were mainly focused on different failure criteria, the displacements field or the flexural strength and modulus [6-13]. Nevertheless, the effect of this characteristic on the results of three point bending tests is notable. The deviation of the neutral axis from the mid-height plane and the non-symmetric strain and stress distributions are some of the consequences.

The analytical model stated in [14,15], based on the technique of the homogenised section, describes the mentioned features related to the tensile and compressive material response and predicts the first failure mode produced depending on the beam span-thickness relation. In this work we unify the nomenclature of both studies in order to highline the predictive capacity of the model in any composite with no force coupling. Two unidirectional and three symmetric cross-ply laminates comprised by continuous carbon fibre-reinforced epoxy laminae were analysed in order to assess experimentally the mentioned theory. $[0]_5$, $[0]_8$, $[90_2\ 0_2]_S$, $[0_2\ 90_2]_S$ and $[0_3\ 90_2]_S$ composites were fabricated and tested, being the experimental observables the applied load and the maximum tensile strain in the most loaded section. Microscopy analysis was developed for reviewing whether flexural or interlaminar shear failure modes were obtained in the tests. The analytical estimation of the neutral axis position as well as the mode and region of damage initiation were in good agreement with the experimental observations and measurements.

2. Material

Composite plates were fabricated stacking several plies of pre-preg by means of a hand lay-up process and applying the curing cycle with a hot platen press. Afterwards, these plates were machined with the form of parallelepiped specimens in order to perform the experiments using a three-point bending tool with a span $L = 25$ mm installed in a testing machine. Each ply consisted of unidirectional carbon-fibres IMA-12K in a resin matrix M21E from Hexcel™. The thickness of each ply after the curing process was $\Delta = 0.25$ mm and these presented an orthotropic linear elastic behaviour whose properties in the principal directions under traction (super-index t) and compression (super-index c) are listed in table 1. The elastic moduli of both loading cases in the fibres direction (sub-index f) and in perpendicular to the fibres direction (sub-index m) are respectively E_f^t, E_f^c, E_m^t and E_m^c. In parallel, f_f^t, f_f^c, f_m^t and f_m^c are the normal strengths in the principal directions under tensile and compressive loading. Lastly, the interlaminar shear strength of the material is defined as f_s. Except the values of E_m^c and f_m^c, the material properties have been determined in previous research by means of uniaxial [16] and flexural testing [14]. Four significant digits have been taken into account due to the accuracy of the measurement equipment. The value of E_m^c was calculated in [15] from the experimental results by means of the orthotropic relations and f_m^c was obtained from references [11-13].

Table 1. Properties in principal directions depending on the loading case.

E_f^t (GPa)	E_f^c (GPa)	E_m^t (GPa)	E_m^c (GPa)	f_f^t (MPa)	f_f^c (MPa)	f_m^t (MPa)	f_m^c (MPa)	f_s (MPa)
177.6 ± 5.7	97.03 ± 3.03	11.84 ± 1.05	7.46*	2601. ± 121.	1356. ± 80.	56.07 ± 2.12	200**	46.65 ± 1.12

*Calculated [15]. **Estimation from [11-13].

3. Analytical model

A three point bending test consist in a bar of rectangular cross section resting on two supports, which is loaded and deflected by means of a loading nose acting in the midway between the supports [1,2]. Therefore a pinned–pinned specimen submitted to a central load P is studied, considering a span L and a rectangular cross section of width b and total thickness h. It is assumed that no extensional-shear-bending–twisting coupling exists and then only normal stress σ (given in this work in absolute value) and out-of-plane shear stress τ are produced. The neutral axis of the laminate is presumed to be deviated a distance d from the mid-height of the section due to the different tensile and compressive material behaviour. By way of example, figure 1a shows the cross section of a $[0_2\ 90_2]_S$ laminate with total thickness $h = 8\Delta$.

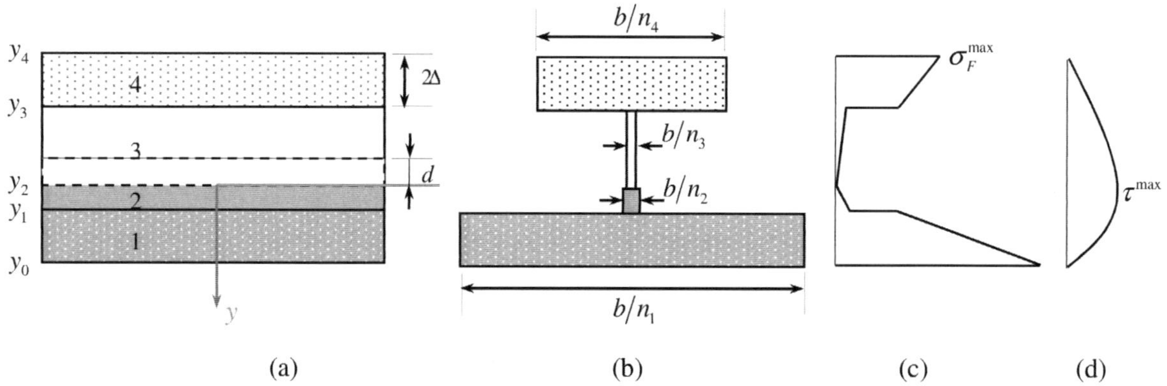

Figure 1. Laminate $[0_2\ 90_2]_S$. (a) Limiting y-coordinates of each region in the actual section. (b) Corresponding width modifications in the homogenised section. (c) Normal stress and (d) out-of-plane shear stress distributions. Not to scale.

To overcome the analytical difficulty generated by the different response to tensile and compressive loading, the cross section is homogenised for unifying its elastic response [14,15] and handling with the problem by means of the linear classical bending theory of slender beams [17]. The rectangular cross section has to be examined for distinguishing the different regions submitted to tension and compression, which have different material response. Let us consider that the total section is divided in an even number of regions N with different answer under loading. The variable i (defined from 1 to N) numbers the regions from bottom to top, being y_{i-1} and y_i the limiting coordinates of each i-region and h_i the maximum y-coordinate of each region in absolute value. Then, the cross section is modified in order to analyse it under the assumption that its whole material response is unique and equal to a reference behaviour. The original width of each region has to be reduced so as to maintain the equilibrium of the resultant forces, while the thickness is kept for conserving the original strain distribution. The highest elastic modulus in the loading direction is considered the one of reference. This value of reference divided by the elastic modulus of the i-region is defined as the moduli ratio n_i, being the modified width of the i-region defined as $b_i = b/n_i$. Once the geometrical modification has been developed (figure 1b), the static moment of the whole homogenised section about the neutral axis has to be zero. This fact is used for deducing the distance d between the mid-height plane and the neutral axis using the relation given by equation (1). As well, the static moment of "half" of the modified area, S_z, can be obtained by means of equation (2).

$$\sum_{k=1}^{N} \frac{\left(y_{k-1}^2 - y_k^2\right)}{n_k} = 0 \tag{1}$$

$$S_z = \sum_{k=1}^{N/2} b_k \frac{\left(y_{k-1}^2 - y_k^2\right)}{2} \tag{2}$$

In order to determine the first failure mode and its location, f_i is assumed to be the normal strength of each region in the loading direction and f_s the interlaminar shear strength of the material. The top and bottom furthest fibres to the neutral plane are the most deformed due to the bending moment and, in case of initial flexural failure, the most likely to fail due to the normal stress distribution generated. From the two furthest regions to the neutral axis the less resistant is the one with lower normal failure strength, which is called the flexural failure region "F" and it has a normal failure strength f_F, an elastic moduli ratio n_F and a maximum coordinate h_F. Meanwhile, in case of inter-laminar opening, the failure mode is expected to be initialized in the position of the neutral axis where the maximum out-of-plane shear stress is produced. Equations (3) and (4) show the threshold between both failure modes, obtained taking into account the relation $\tau^{\max}/\sigma_F^{\max}$ (figure 1c and figure 1d) between the maximum shear stress produced in the neutral axis and the maximum normal stress developed in the flexural failure region.

First flexural failure: $$\frac{2n_F S_z}{bLh_F} < \frac{f_s}{f_F} \quad (3)$$

First out-of-plane shear failure: $$\frac{2n_F S_z}{bLh_F} > \frac{f_s}{f_F} \quad (4)$$

4. Experimental observations and discussion

Three point-bending experiments were conducted by means of a standard fixture [1,2] installed in an universal testing machine. These tests were performed on specimens with the shape of rectangular prisms with slightly different dimensions due to the cutting process. The main limitations were related to their length and width, which were chosen for fulfilling the space requirements imposed by the three-point bending tool and for achieving an homogenous shear opening all along the width of the cross section. Regarding the thickness, the principal restriction was to employ a sufficient number of 0º-layers in order to obtain specimens rigid enough to avoid big displacements that could lead to non-linear effects. Five lay-ups were chosen to reproduce experimentally the different failure modes. Two unidirectional composites, $[0]_5$ and $[0]_8$, and three cross-ply laminates, $[90_2\,0_2]_S$, $[0_2\,90_2]_S$ and $[0_3\,90_2]_S$, were fabricated and tested. A strain gage was located on the lower surface of the mid-span section (the cross section of the beam where the central load is applied) for registering the maximum tensile strain ε_t. Meanwhile, the applied load P was acquired by means of the loading cell of the testing machine. The location of the failure regions was analysed by optical microscopy using a LEICA DR IRM microscope. Images of the damage initiation, taken in perpendicular direction to the most loaded section, are shown in figure 2 and figure 3 at magnification of 50x.

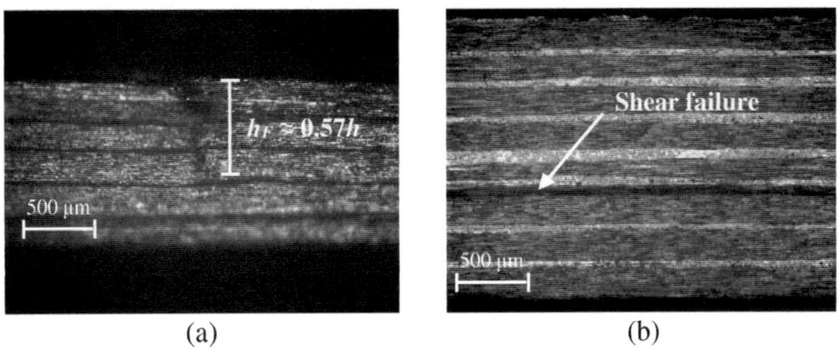

Figure 2. Damage initiation [14]: (a) $[0]_5$, (b) $[0]_8$. Magnified images 50x.

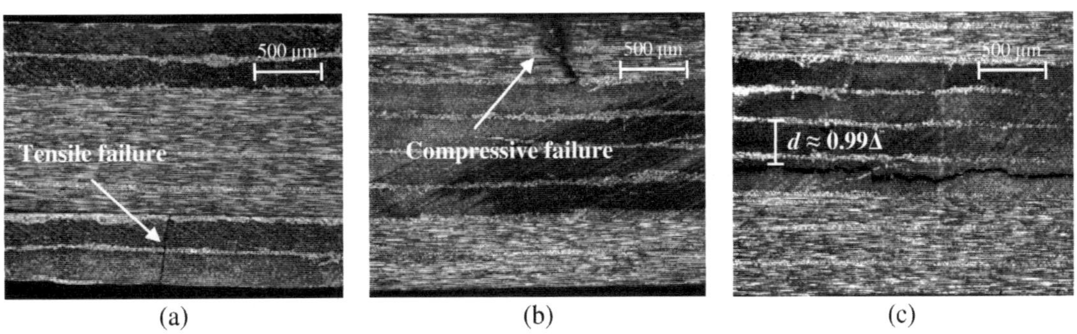

Figure 3. Damage initiation [15]: (a) $[90_2\,0_2]_S$, (b) $[0_2\,90_2]_S$, (c) $[0_3\,90_2]_S$. Magnified images 50x.

Due to the repetitively of the results, the experimental information at the instant of first failure for five representative specimens are listed in table 2 together with the evaluation of the analytical estimations. The comparison between the analytical estimations and the experimental observations is addressed in the next sub-sections.

Table 2. Experimental and analytical definitions in the moment of first failure [14,15].

	h (mm)	b (mm)	P (kN)	$m\varepsilon_t$	d (mm)	f_F (MPa)	n_F	h_F (mm)	S_z (mm^3)	First damage*
[0]$_5$	1.25	9.80	0.60	10.00	0.37Δ	1356.	1.83	$h/2 + d = 0.57h$	$0.09bh^2$	C
[0]$_8$	2.00	9.86	1.20	8.86	0.60Δ	1356.	1.83	$h/2 + d = 0.57h$	$0.09bh^2$	S
[90$_2$ 0$_2$]$_S$	2.00	11.25	0.13	4.28	0.33Δ	56.07	15.00	$h/2 - d = 0.46h$	$0.03bh^2$	T
[0$_2$ 90$_2$]$_S$	2.00	10.35	1.13	6.52	0.84Δ	1356.	1.83	$h/2 + d = 0.60h$	$0.07bh^2$	C
[0$_3$ 90$_2$]$_S$	2.50	10.80	1.72	7.71	0.99Δ	1356.	1.83	$h/2 + d = 0.60h$	$0.08bh^2$	S

*C: Flexural failure (compression), T: Flexural failure (tension), S: Out-of-plane shear failure.

4.1. Unidirectional laminates

The existing normative for developing three point-bending tests on unidirectional composites [1,2] considers that the response of the material is the same under tension and compression and, in opposition to our study, it analyses a problem with symmetric normal and shear stress distributions. Nevertheless, the non-symmetrical failure modes found experimentally suggest the importance of the different loading response (figure 2a and figure 2b). Tests on unidirectional laminates were developed to compare the results of the standards with the theory derived in [14,15] which takes into account the difference in behaviour under tensile and compressive loading. The proposed laminates consists of five and eight plies (h = 1.25 mm and h = 2 mm, respectively), whose section contains two regions with different behaviour under tensile and compressive loading (N = 2). The highest elastic modulus in the loading direction was present in the region submitted to tension, which was taken as the reference for defining the moduli ratio of each region and developing the geometrical modifications. The lower normal strength in the loading direction was present in the region submitted to compression, so it was considered as the region where the first flexural failure could be initiated. The geometrical threshold given by equations (3) and (4) predicts that the thickness of the specimen has to be lower than 1.5 mm for obtaining flexural failure and higher than 1.5 mm for developing shear opening-modes. Therefore, the two different laminates should show experimentally opposite failure modes. It is important to highlight that, in the studied case, the value of this threshold (h = 1.5 mm) is 13% smaller than the classical limit ($h = 2Lf_s/f_F$ = 1.7 mm) [18] obtained considering that the material presents the same response under tension and compression.

4.1.1. Laminate [0]$_5$. The 5-plies specimen presented initial flexural failure in the region submitted to compression due to the lower compressive strength in the fibres direction. The values of the experimental observables listed in table 2 correspond to the measurements when the loss of load-carrying capacity was observed. Typical compressive failure modes could be observed using optical microscopy (figure 2a). The dissimilar lengths of the regions submitted to tension and compression could be appreciated. Then the simplification of the response developed by the normative, which predicts equal dimensions for the tensile and compressive regions, does not reproduce completely the experimental observations. The length of the damaged region estimated analytically was $h_F = 0.57h$ = 712.5 μm (table 2), which differs a 14% from the position of the mid-height plane of the section. After the moment of the load-carrying capacity loss, during the last stages of the loading process, the maximum shear stress could be able to achieve the shear strength of the material and then marginal shear failure modes could also be observed.

4.1.2. Laminate $[0]_8$. The 8-plies specimens exhibited shear failure modes at the position of the neutral axis, at an average depth equal to 0.57 times the total thickness h. This measurement coincides with the analytical estimation given in table 2, what confirms that the model developed reproduced the experimental results. The not-centred position of the interlaminar opening-mode can be appreciated in figure 2b for a representative specimen. The shear damage was initiated when the maximum shear stress achieved the interlaminar shear strength. Meanwhile, the maximum tensile and compressive normal stresses were lower than their correspondent normal strengths, thus only shear failure was induced. Once the experiment went on, the initial crack was propagated in the longitudinal direction of the specimen. Nevertheless, after the onset of the shear opening-mode, the loading continued over the upper part of the specimen delimited by the shear failure plane. Then the resistant layer had a lower effective thickness, in average smaller than the established limit ($h_F = 0.57h = 1.14 < 1.5$ mm). This condition could be enough to trigger unexpected flexural failure modes after the stage of shear damage initiation.

4.2. Cross-ply laminates

In order to examine the correctness of the analytical technique in other symmetric and balanced composites, three different cross-ply laminates, $[90_2\ 0_2]_S$ $[0_2\ 90_2]_S$ and $[0_3\ 90_2]_S$, were experimentally tested. Different linear force-strain response and first failure load were appreciated. In particular, the strong influence of the stacking sequence is observed in the loading progression of the $[90_2\ 0_2]_S$ and $[0_2\ 90_2]_S$ laminates, being the force-strain slope and the first failure load of the $[0_2\ 90_2]_S$ specimen 5.62 times and 8.69 times higher than the ones of the $[90_2\ 0_2]_S$ lay-up (table 2), respectively. According to the threshold given by equations (3) and (4), $[90_2\ 0_2]_S$ and $[0_2\ 90_2]_S$ composites should present flexural failure in the areas submitted to tension and compression respectively. Meanwhile, $[0_3\ 90_2]_S$ laminates should develop interlaminar shear failure. In summary, these configurations present completely opposite failure modes that are detailed in the next paragraphs.

4.2.1. Laminate $[90_2\ 0_2]_S$. The top and bottom 90°-plies were the most deformed due to the bending moment. Among them, the bottom 90°-region submitted to tension was the furthest to the neutral axis with lower normal strength and it was considered as the region where the first flexural failure could be initiated. Using the analytical limitations given by equations (3) and (4), in this laminate first bending failure would be achieved if $h < 11.6$ mm, while first interlaminar shear failure would be produced if $h > 11.6$ mm. As the total thickness was $h = 2$ mm, this composite should fail experimentally due to the bending moment in the 90°-lower layers submitted to tensile loading. This result was confirmed experimentally and figure 3a shows the damage initiation in the so-called flexural failure region under tension. Not only one main crack could appear in the lower 90°-region of the mid-span section, but also several micro-cracks could emerge in its close vicinity at the same time. Observing the load-strain evolution, it could be inferred that the damage initiation is governed by the response of the first failure region. After the damage initiation, the bottom region lost its load-carrying capacity and the remaining lay-up [90 90 0 0 0 0] supported the increasing external loading. Then the top 90°-region turned to be the most deformed with lower value of the normal strength of the material, being the following to fail when the normal compressive strength of the matrix was achieved. Afterwards the lasting core of the specimen, formed by four 0°-layers, was submitted to flexural loading. Due to the fact that the compressive normal strength in the fibres direction is lower than their tensile normal strength, the damage evolution continued in the upper 0°-plies submitted to compression and this led to the total failure of the specimen at a final load close to 0.96 kN.

4.2.2. Laminate $[0_2\ 90_2]_S$. Similar procedure was followed for studying the symmetric cross-ply $[0_2\ 90_2]_S$ (figure 1), which had the same number of plies and orientations than the laminate of the previous subsection. However, the different stacking sequence provoked the first failure due to the bending moment in the top 0°-layers submitted to compressive loading. In this case, the two top and bottom 0°-regions were the most deformed during the loading process. The level of the maximum normal stresses in both regions is sketched in figure 1c and a clear initial damage was produced in the top 0°-region (figure 3b). The reason is that the tensile strength in the fibres direction is two times higher than their compressive strength, being the top 0º-layers the so called flexural failure region. The experimental observation was in agreement with the limitations given by equations (3) and (4), which stated that the conditions for developing first bending or interlaminar shear failure modes were $h < 2.08$ mm and $h > 2.08$ mm respectively. The total thickness of the laminate was $h = 2$ mm, so the analytical prediction was that the composite should fail experimentally due to the bending moment under compression in the location of the flexural failure region. In the moment of first failure, the maximum shear stress τ^{max} was just 14% lower than the interlaminar shear strength of the material f_s. Then, after the loss of the load-carrying capacity of the top 0º-plies, the remaining [90 90 90 90 0 0] laminate developed a fast shear failure in the 90º-region due to the closeness to the shear strength.

4.2.3. Laminate $[0_3\,90_2]_S$. Including to the previous laminate an extra 0º-layer on its top and its bottom should cause interlaminar shear failure in the position of the neutral axis, because according to equation (4) the condition for obtaining first delamination $h > 1.9$ mm was satisfied. The addition of both 0º-plies provoked the reduction of the maximum normal stresses in all the regions, while the maximum shear stress was increased. This growth was enough for achieving the value of the interlaminar shear strength f_s while $\sigma_F^{max} < f_f^c$. The first opening-crack was initiated in the neutral axis (figure 3c), whose location has been analytically estimated at a distance $d = 0.99\Delta$ from the mid-height plane. The analytical position fitted perfectly with the experimental evidence, proving that the neutral axis was not centred in the mid-height of the section due to the different tensile and compressive material response. After the damage initiation, the experimental observations confirmed that the initial damage quickly expanded between the two lower 90º-plies. In figure 3e the las top 0º-ply does not appear due to the magnification used (50x).

5. Conclusions

Composites generally exhibit different stress–strain behaviour under tension and compression. The effect of this characteristic was examined in terms of three-point bending tests on unidirectional and cross-ply CFRP laminates. Using the technique of the homogenised section, the problem could be studied considering a homogeneous response of the material. Taking into account the different tensile and compressive material response, the full analytical descriptions of the normal and shear stress distributions allowed to describe the thickness condition which determined if the specimen would present either flexural or shear failure modes. The observations by optical microscopy confirmed the non-symmetric stress states produced in the central cross section and the region where the damage initiation was localised. This demonstrated the great capability of the model exposed for defining laminates whose flexural response and first failure can be beforehand controlled.

6. Acknowledgements

This work was financially supported by the University of Castilla-La Mancha.

References

[1] UNE-EN 2562:1997: *Aerospace series. Carbon fibre reinforced plastics. Unidirectional laminates. Flexural test parallel to the fibre direction*; 1997.
[2] UNE-EN ISO 14130:1999: *Fibre-reinforced plastic composites. Determination of apparent interlaminar shear strength by short beam method*; 1999.
[3] Jones R M 1976 *J. Compos. Mater.* **10** 342–54.
[4] Mujika F, Carbajal N, Arrese A and Mondragón I 2006 *Polym. Test.* **25** 766–71.
[5] Carbajal N and Mujika F 2009 *Polym. Test.* **28** 150–6.
[6] Pagano N J 1976 *J. Compos. Mater.* **1** 336–342.
[7] Turvey G J 1980 *Int. J. Solids Struct.* **16(5)** 451–63.
[8] Turvey G J 1982 *Fibre Sci. Technol.* **16(1)** 1–10.
[9] Kam T Y and Sher H F 1995 *J. Compos. Mater.* **29(4)** 463–82.

[10] Mujika F 2012 *J. Compos. Mater.* **46(3)** 259–74.
[11] *Mechanical Properties of Carbon Fiber Composite Materials* 2014, Fiber/Epoxy resin (120°C Cure), 1-800-811-2009, ACP Composites.
[12] Marín L, Trias D, Badalló P, Rus G and Mayugo J A 2012 *Compos. Struct.* **94** 3321–6.
[13] Bunsell A R and Renard J 2005 *Fundamentals of fibre reinforced composite materials* CRC Press.
[14] Serna Moreno M C, Romero Gutiérrez A and Martínez Vicente J L 2016 *Compos. Struct.* **136** 706–11.
[15] Serna Moreno M C, Romero Gutiérrez A and Martínez Vicente J L 2016 *Compos. Struct.* **146** 62-8.
[16] Martínez Vicente J L, Serna Moreno M C, Caminero Torija M A and López Cela J J 2014, ECCM16, Seville (Spain).
[17] Garrido J A and Foces A 1994 *Resistencia de Materiales* Secretariado de Publicaciones, Universidad de Valladolid.
[18] Sideridis E and Papadopoulos G A 2004 *J. Appl. Polym. Sci.* **93** 63–74.

IOP Conf. Series: Materials Science and Engineering **139** (2016) 012048 doi:10.1088/1757-899X/139/1/012048

An exploration of the relationship of chemical and physical parameters in the micromechanical characterisation of the apparent interfacial strength in glass fibre epoxy systems

J L Thomason, L Yang, D Bryce and R Minty

University of Strathclyde, Department of Mechanical and Aerospace Engineering, 75 Montrose Street, Glasgow G1 1XJ, United Kingdom

Email: james.thomason@strath.ac.uk

Abstract. This paper focuses on the cure shrinkage and the thermomechanical properties of an amine cured epoxy resin system and its adhesion to glass fibre. The fibre-matrix interfacial shear strength (IFSS) was characterized using the microbond test over a range of test temperatures and a range of amine:epoxy ratios. The apparent IFSS in this glass-epoxy system was shown to be strongly dependent on the testing temperature and the matrix stoichiometry. High levels of cure shrinkage were measured in the IFSS microdroplets which resulted in internal stresses causing significant levels of droplet deformation. The results presented here can be interpreted as providing further support for the hypothesis that a significant fraction of the interfacial stress transfer capability in epoxy composites can be attributed to a combination of residual radial compressive stress and static friction at the fibre-matrix interface.

1. Introduction

The performance of fibre-reinforced polymer composites is defined by a combination of the fibre and matrix properties and the ability to transfer stresses across the fibre–matrix interface. It is well recognised that optimization of the stress transfer capability of the fibre-matrix interface region is critical to achieving the required composite performance level. The ability to transfer stress across this interface is often reduced to a discussion of 'adhesion' that is a simple term to describe a combination of complex phenomena on which there is still significant debate as to their relative significance and their characterisation. Certainly, one of the generally accepted manifestations of 'adhesion' is the mechanically measured value of interfacial shear strength (IFSS).

Despite the high level of attention commonly focused on chemical influences on glass fibre-epoxy adhesion, a number of authors have also commented on the role of shrinkage stresses contributing to the stress transfer capability at the fibre-matrix interface [1-7]. Recent work has shown that the level of apparent adhesion in glass fibre reinforced composites may be strongly influenced by the glass transition temperature (Tg), thermal expansion coefficient, and modulus-temperature relationship of the composite matrix (or interphase) [3-7]. In epoxy polymers these factors are known to be dependent on the ratio of curing agent to epoxy resin and the thermal history of the system. In addition to the thermal stresses at the interface caused by the heating-cooling cycle during composite production thermosetting matrices also undergo a volume change during their chemical reaction due to the phenomenon of cure-shrinkage. Since the epoxy matrix is shrinking while its Tg is simultaneously increasing due to the polymerisation it is possible for a proportion of this cure shrinkage to be frozen into the system causing additional stress at the fibre-matrix interface.

It is also well known that the performance of the fibre-matrix interface is critically related to the nature of the surface coating (or size) which has been applied to the fibres during their manufacture. The sizings used on glass fibres used to reinforce epoxy resin polymers will nearly always contain epoxy molecules [8,9] although the exact nature of such glass fibre sizes is often kept an industrial secret. Consequently, there is a strong possibility of the stoichiometry of a composite matrix or, more likely a local interphase region around the fibres being different from that of the initial resin system preparation [10]. Clearly all of the above factors can be tied together with the local stoichiometry and the thermal history of the epoxy system. In this work a study was undertaken to characterise the effects of variation of the epoxy-curing agent ratio on the thermal and mechanical performance of an epoxy polymer. Furthermore, the effect of these chemical changes on the level of cure shrinkage and the apparent level of adhesion in glass fibre-epoxy microbond specimens was evaluated.

2. Materials and methods

Boron free E-glass fibres (OC-APS) with average diameter of 17.5μm and coated with γ-aminopropyltriethoxysilane were supplied by Owens Corning. Araldite 506 epoxy resin and Triethylenetetramine (TETA) curing agent were purchased from Sigma-Aldrich. The stoichiometric ratio (amine:epoxy ratio r=1) for this system was calculated at 12.0% by weight of TETA. The epoxy and TETA were thoroughly mixed in appropriate proportions and degassed under vacuum for 12 minutes. Droplets were then deposited on a single glass fibre using a thin wire. Approximately 40 droplets were placed on individual fibres before these samples were transferred into a convection oven, where they were heated first to 60 °C and held isothermally for 1 hour followed by another 2 hours heating at 120°C. After heating, the samples were left in the oven to cool down. The configuration and development of the microbond test (MBT) and TMA-Microbond test (TMA-MBT) rigs has been reported previously [7,11]. Differential scanning calorimetry (DSC) was carried out in a TA Instruments Q2000 DSC under a nitrogen gas flow of 50 ml/minute. The pre-cured epoxy material was cut down to specimen sizes with a mass in the range of 15-20 mg. These samples were subjected to a DSC heat-cool-heat cycle from -10 °C to +150 °C at 10 °C/minute. The Tg values reported are mid-point values from the second heating ramp.

The change in dimensions of the typical epoxy microdroplets during curing was observed in a Mettler Toledo FP90 hot-stage placed under a Olympus BX51 microscope. The hot-stage heating schedule was similar to the microbond samples oven curing, heat from 20 °C to 60 °C at 2 °C per minute, isothermal at 60 °C for 60 minutes, then further heating to 120 °C. Images of the samples were recorded at one minute intervals. These droplet images were approximated as ellipses and the major and minor dimensions of each was measured to obtain an estimate of the droplet volume. Ten droplets were measured at each of five amine:epoxy group ratios investigated.

3. Results and discussion

3.1. IFSS dependence on measurement temperature

The results for IFSS obtained for an r=1 epoxy matrix, measured at test temperatures in the range 20 °C to 150 °C, are summarized in Figure 1 that shows the average values with 95% confidence limits (between 10-20 samples per temperature). It can be clearly seen that there exists a significant temperature dependence of measured IFSS in this thermosetting system. The IFSS falls from 54 MPa measured at 20 °C to just 2 MPa when measured at 150°C. It is noticeable that the highest rate of change of IFSS with temperature is also in the region of the Tg of this epoxy system which occurs in the range 70-90 °C [7]. The magnitude of the IFSS with this fibre and an epoxy matrix (GF-EP) is much greater than for the same fibre in polypropylene (GF-PP) at any particular temperature [5]. This supports the general expectation that the stress transfer capability of the GF-EP interface is much greater than that of GF-PP. However, there is also a striking similarity in the form of the IFSS versus temperature dependence for these two very different composite systems. Both systems exhibit a significant step-change in the IFSS around the associated matrix Tg [5,7].

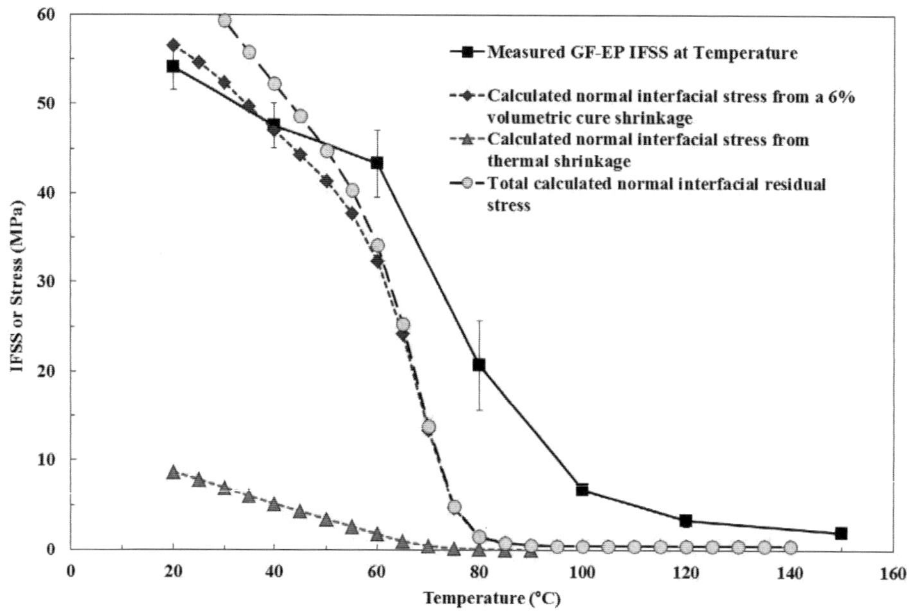

Figure 1. Glass Fibre-Epoxy (r=1) IFSS versus temperature compared to residual radial compressive stress at the interface [7]

Figure 1 also shows the maximum potential contribution to the interfacial radial compressive stress from thermal and cure shrinkage where a 6% volumetric cure shrinkage had been employed [7]. These are maximum values since it has been assumed that no relaxation of the stresses occurs below the matrix Tg. Furthermore the buildup of the stress values have been calculated using a temperature step of 5 °C reducing from a Tg value of 80 °C and using temperature dependent measured input values of the epoxy matrix modulus (by DMA) and expansion coefficient (by TMA) [7]. It can be seen that the residual radial interfacial stress obtained from such a level of cure shrinkage is significantly greater than the residual thermal stress. This result appears to be well aligned with the results of other researchers [12]. It is clear from the data in Figure 1 that the sum of thermal and cure shrinkage related residual radial interfacial stress is of an appropriate magnitude in order for acceptable values (<1) of the coefficient of static friction to deliver an interfacial stress transfer contribution of similar magnitude to the measured IFSS. Previous work has indicated that a large fraction of the IFSS in thermoplastic composites can be attributed to residual radial compressive stresses at the interface [5,6]. It appears from the current results that it is also possible to make a case for residual stress combined with static friction being a major contributor to the apparent IFSS in a thermoset system.

However, a major challenge to this hypothesis is the weight of opinion that the chemistry and chemical reactions in the system must in some way play a role in the stress transfer capability of the interface. In order to maintain the primary hypothesis of residual compressive stress being a major contribution to the apparent IFSS it becomes necessary to investigate how the chemistry of the polymerising matrix system could affect the major drivers of this residual stress. Hence it is pertinent to investigate parameters such as IFSS, matrix Tg, expansion coefficient, modulus, and cure shrinkage as a function of temperature and the stoichiometry of the epoxy matrix.

3.2. IFSS room temperature dependence on matrix stoichiometry

Figure 2 presents the IFSS results (from two different operators) obtained for the OC-APS fibres as a function of the matrix r value (amine/epoxy group ratio). Each data-point represents an average of approximately 40 individual successful microbond measurements. The line in this Figure is added purely as a guide to the apparent trend. The primary observation in this Figure is that the IFSS exhibits a broad peak about the stoichiometric value (r=1) for this epoxy matrix system. Consequently, a reasonable conclusion could be that small local variations in the matrix composition around the fibre (in the interphase if one exists) due to epoxy or amine groups present in the fibre sizing may not lead to significant variations of the IFSS. It appears that local variations in the value of r of greater than ±0.5 from the stoichiometric ratio would be necessary to significantly change the level of IFSS. Nevertheless, a second important observation from Figure 2 is the extremely steep drop in IFSS observed when the r ratio drops below approximately 0.5. It can be seen that if for any physical reason r<0.5 it would have serious consequences for the apparent adhesion in this system and presumably also for the further mechanical performance of a composite. It will be interesting to see if these trends are maintained when the IFSS is measured at higher temperatures.

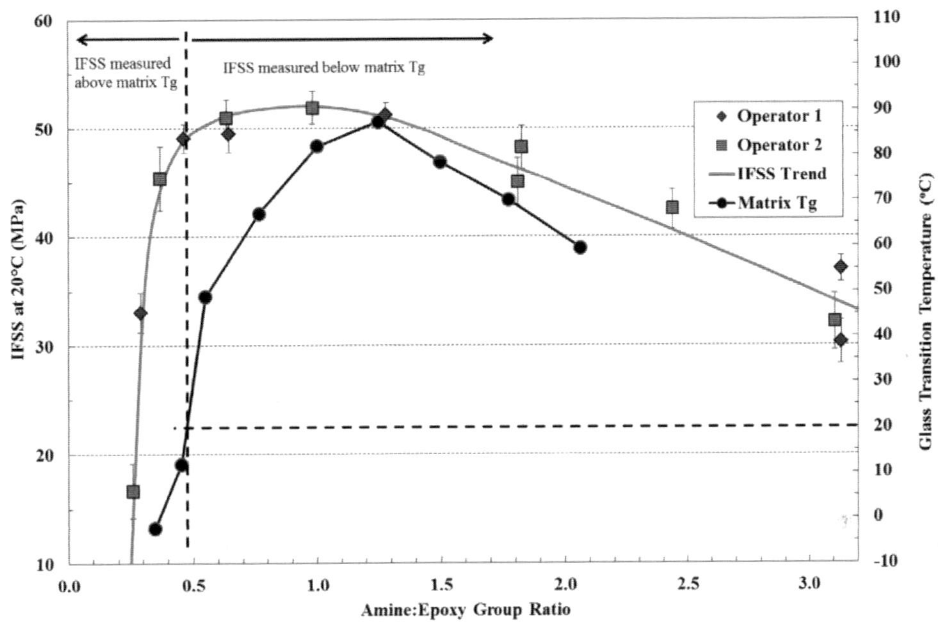

Figure 2. IFSS and matrix Tg versus amine:epoxy group ratio in matrix

In terms of the modelling of the residual compressive stresses (from both thermal and cure shrinkage) at the fibre-matrix interface, the temperature at which such stresses are "frozen-in" and cannot easily relax away is an important parameter. In epoxy based systems it is accepted that the characteristic temperature at which this occurs is the matrix Tg. The results for the DSC determined cured polymer Tg are also shown in Figure 2 as a function of the system stoichiometry. The maximum Tg for this epoxy system appears to be in the range 1<r<1.2 close to the calculated stoichiometric ratio. Samples with either an excess of epoxy or an excess of hardener gave a much lower Tg value. This Tg dependence of stoichiometry has been observed in epoxy resins by other researchers [13–15].

Theoretically, for r=1, all amine hydrogens react with all epoxide groups, giving a network of interconnected rings [15]. These rings consist of two or three epoxidic chains, which have reacted with the corresponding amine group. In the epoxy-rich systems (r<1) a fraction of the available epoxy groups remain unreacted. This results in a network with some less restrained, bulky groups situated at the end of the chains. This increase of the number of chain ends increases the free volume of the polymer, decreasing the Tg. In the systems with an excess of amine (r>1), the epoxy groups react initially with the primary amine groups, but there are unreacted secondary amine groups remaining. Under these conditions, the epoxy tends to give branched structures [16], and occasionally may give rings consisting of the partially reacted tri-amine molecules and the epoxide chain. These structures contribute to an increase of the free volume of the system and consequently exhibit a lower Tg [15]. It is observed that the presence of unreacted epoxy groups (r<1) has a more noted effect on the polymer Tg than the same amount of unreacted amine groups (r>1).

Comparison of the general trends observed in Figure 2 for IFSS and matrix Tg as a function of the matrix stoichiometry reveal some similarity with a maximum value occurring around r=1 and both IFSS and Tg reducing as r moves away from unity (in either direction). A particularly interesting correlation is the position of the large step down in IFSS when r<0.5. It is also at this approximate value of r that the matrix Tg drops below room temperature. Consequently, the very low values of IFSS observed for r<0.5 in Figure 2 were measured at a temperature above the matrix Tg when presumably there is little build-up of residual radial interfacial stress to contribute to the apparent measured IFSS. This appears to be the same phenomenon observed in Figure 1 where the experiment design involved a fixed matrix Tg (≈80 °C for r=1) and measurement of IFSS at temperatures above and below this value. In Figure 2 the experiment is designed with a constant IFSS measurement temperature but a changing epoxy matrix Tg. In both cases a similar effect can be observed of reduced IFSS when the measurement temperature is above the Tg of the matrix polymer. Further work to characterise the system IFSS as a function of both r value and temperature is planned.

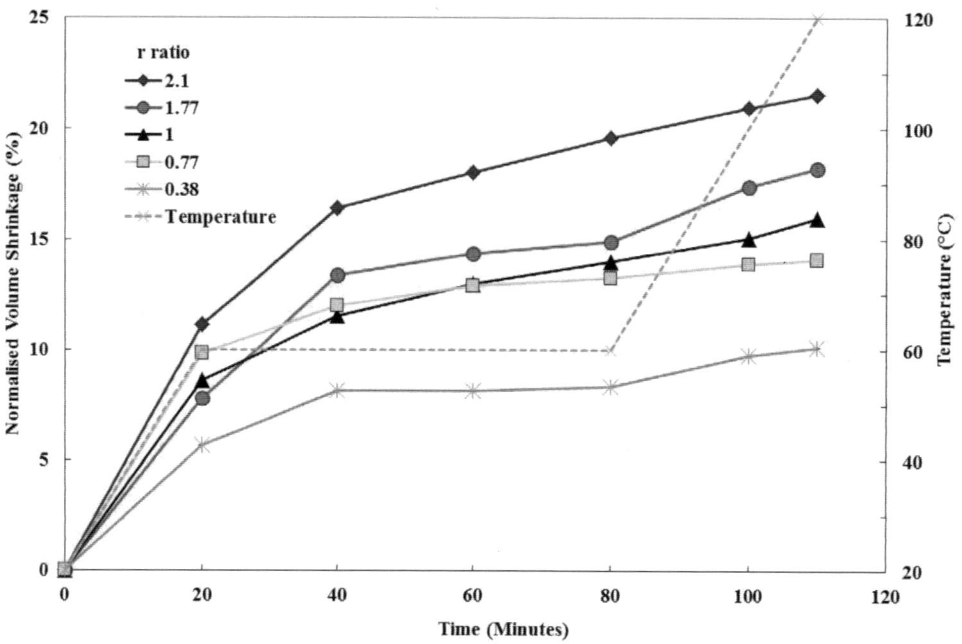

Figure 3. Normalised volumetric shrinkage of microbond droplets during curing

3.3. Droplet shrinkage during curing

The results for the volumetric shrinkage normalised to the original droplet volume for different r ratio droplets are shown in Figure 3. Each data point is the average of the dimension change of ten different droplets. It is interesting to note that the over-riding change in volume is shrinkage, even during the first heating period (0-20 min) the cure shrinkage is greater than the increase in droplet volume due to thermal expansion with a volumetric shrinkage of 6-11% in this period depending on the droplet stoichiometry. The droplets continue to shrink during the isothermal period (20-80 min) at 60 °C and shrink even further during the final heating step (80-110 min) to 120 °C (again overcoming any thermal expansion during heating). The overall volumetric cure shrinkage during the droplet curing is in the range of 11-21% with a clear trend towards increasing cure shrinkage with increasing r ratio.

These values are well in excess of the 6% volumetric cure shrinkage which was used in the modeling of the residual radial compressive stress at the fibre-matrix interface shown in Figure 1. There is a great deal more work required to fully quantify the effect of this cure shrinkage; however these results clearly indicate that there is potentially sufficient volume change in the preparation of microdroplets for IFSS measurement to explain the values of apparent adhesion using residual stresses alone.

3.4. Residual stress and microdroplet deformation

An interesting result of the continual monitoring of the microdroplets during the epoxy curing was the observation of sudden deformations appearing in the droplet shape, frequently during the final heating step to 120 °C. Figure 4 shows an example of this phenomenon. Figure 4a shows a well formed droplet (typical of what is often seen in other microbond publications) before curing. A large deformation of the droplet after cure can be observed in Figure 4b.

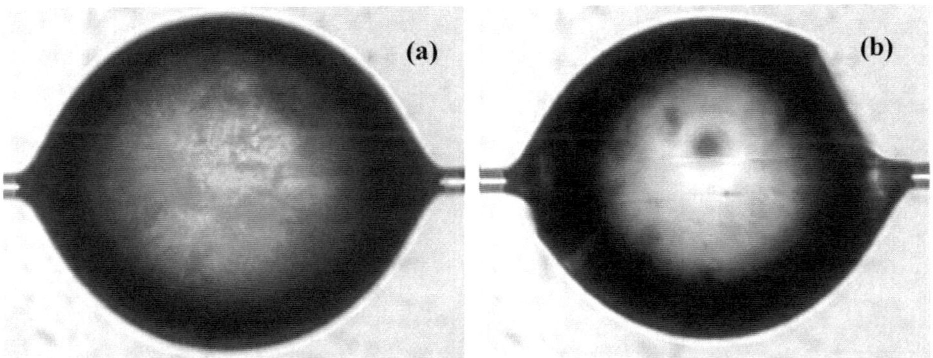

Figure 4. Optical micrographs of epoxy droplet on glass fibre. (a) before cure. (b) after cure

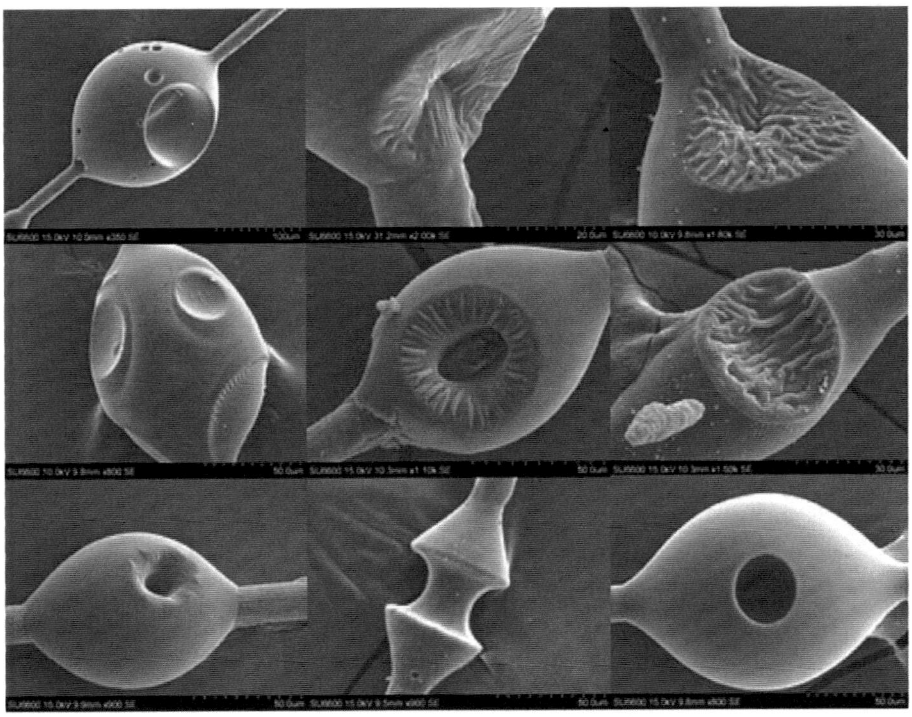

Figure 5. SEM micrographs showing various deformations of cured microbond samples

The phenomenon was also captured on video in certain cases and it appears to be the result of a buildup, during the isothermal cure at 60 °C, of residual cure-shrinkage stress in the droplet. This residual stress becomes frozen in as the Tg of the reacting droplet rises above the isothermal cure temperature. However, on further heating, the droplet is again placed in the rubbery state and the local relaxation of some of the built up residual stress can apparently cause these deformations. These droplet deformations are not well resolved in optical microscopy however a broad survey of cured microdroplets in the scanning electron microscope (SEM) revealed the wide range of deformation types and sizes. A few examples are shown in Figure 5. A further attempt was made to semi-quantify this deformation effect. Every droplet produced for the microbond testing in this work had been photographed to estimate its dimensions. It appears that many of these droplet deformations are not clearly visible in optical microscopy – unless you are specifically searching for them. It is certainly the case that there does not appear to be any reference to these deformations in the existing literature on microbond testing in epoxy systems.

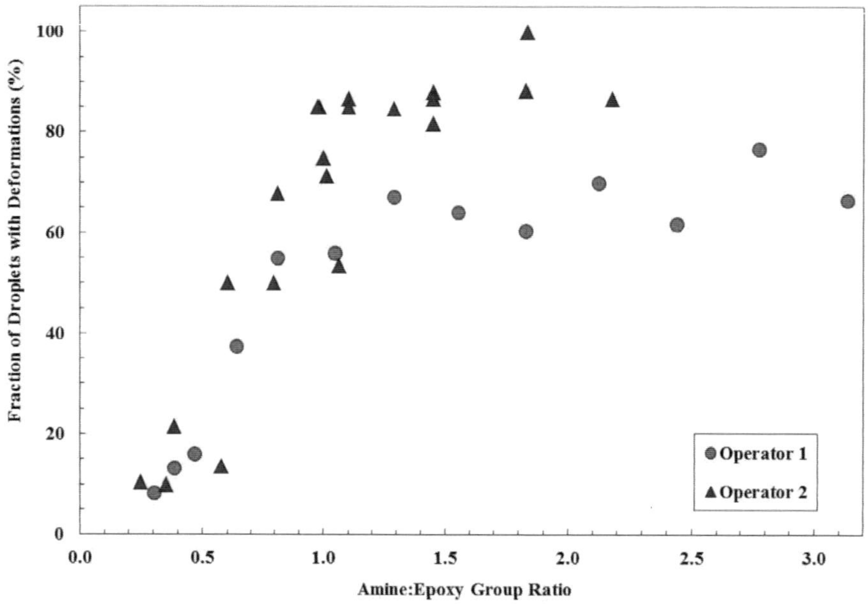

Figure 6. Observation of droplet deformations versus amine:epoxy group ratio in matrix

Figure 6 shows the results of a review of all these droplet micrographs in terms of the fraction of droplets which exhibited deformations. As stated above, the data can only be considered semi-quantitative at this time as there is clear evidence of an operator bias. However both datasets clearly indicate the same trend. At very low r ratios there is little evidence of deformations in the microdroplets. However, the fraction of deformed droplets rises rapidly as the r value is raised above 0.5. As discussed above it is at this r value that the Tg of the droplet rises above room temperature and residual stress can start to build in the droplet. Once again this phenomenon clearly requires further investigation, not least as a source for the variability often obtained in microdroplet evaluation of IFSS in epoxy systems. However, the apparent correlation between the lack of residual stress in the droplets with $r<0.5$ (as evidenced by the low fraction exhibiting deformations) with a very low value of IFSS is very interesting in terms of the hypothesis that apparent adhesion may by explained by residual radial compressive stresses at the fibre-matrix interface alone – even in reactive thermosetting systems.

4. Concluding remarks

The results presented here can be interpreted as providing further support for the hypothesis that a significant fraction of apparent interfacial shear strength (IFSS) in fibre-reinforced epoxy composites can be attributed to a combination of residual radial compressive stress and static friction at the fibre-matrix interface. The temperature dependence of apparent IFSS of a glass fibre-epoxy system in the range 20°C up to 150°C showed a highly significant inverse dependence on testing temperature, with a major step change in the glass transition region of the epoxy matrix. This temperature dependence of the glass fibre-epoxy IFSS was compared to the change in residual radial compressive stresses at the interface as the test temperature is changed. The analysis indicated that the magnitude of the thermal residual stress due to mismatch in the thermal expansion coefficients of fibre and matrix was insufficient to explain the magnitude the of system IFSS. However, when the additional potential residual stress generated by a 6% cure shrinkage of the epoxy matrix was considered, then the magnitude of the residual stress at the interface could be found to be of the same order of magnitude as the measured IFSS. Results on the room temperature dependence of the IFSS in the same system as a function of the amine:epoxy group ratio (r ratio) in the matrix revealed a strong dependence of the IFSS and the matrix thermal and mechanical properties on the r ratio. At $r<0.5$ a correlation was observed between large reduction of IFSS with the lowering of the matrix Tg below room temperature. Direct optical measurement of the cure shrinkage of IFSS microdroplets revealed volumetric shrinkage in the range 11-21% during curing depending on the matrix r ratio. It was also observed that the resultant residual stress in the microdroplets was sufficient to cause severe deformations of the droplets during curing and the occurrence of these deformations was also shown to be dependent on the matrix r value and most likely on the matrix Tg. These observations are important to all researchers actively using the microbond technique to characterise fibre-matrix IFSS.

Acknowledgments

One of the authors (RM) would like to acknowledge the financial support from Engineering and Physical Sciences Research Council under the EPSRC DTP programme. The authors would also like to acknowledge the assistance of the Advanced Materials Research Laboratory of the University of Strathclyde with the mechanical testing and electron microscopy.

References

[1] Nairn JA 1985 *Polym. Compos.* **6** 123
[2] Wagner HD and Nairn JA 1997 *Compos. Sci. Technol.* **57** 1289
[3] Thomason JL 2002 *Compos. Part A* **33** 1283
[4] Thomason JL 2010 *Polym. Compos.* **31** 1525
[5] Thomason JL and Yang L 2011 *Compos. Sci. Technol.* **71** 1600
[6] WenBo L, Shu Z, LiFeng H, WeiCheng J, Fan Y, XiaoFei L and RongGuo W 2013 *Polym. Compos.* **34** 1921
[7] Thomason JL and Yang L 2014 *Compos. Sci. Technol.* **96** 7
[8] Thomason JL Glass fibre sizings: A review of the scientific literature. Blurb Inc. 2012. ISBN: 798-0-9573814-1-4. URL: http://www.blurb.co.uk/b/6244677-glass-fibre-sizings
[9] Thomason JL. Glass fibre sizing: A review of size formulation patents. Blurb Inc. 2015. ISBN: 798-0-9573814-3-8. URL: http://www.blurb.co.uk/b/6244662-glass-fibre-sizing
[10] Thomason JL 1995 *Composites* **26** 487
[11] Yang L and Thomason JL 2012 *Polymer Testing* **31** 895
[12] Jakobsen J, Jensen M and Andreasen JH 2013 *Polymer Testing* **32** 1350
[13] Meyer F, Sanz G, Eceiza A, Mondragon I and Mijović J 1995 *Polymer* **36** 1407
[14] Garcia F, Soares B, Pita V, Sánchez R and Rieumont J 2007 *J. Appl. Polym. Sci.* **106** 2047
[15] Calventus Y, Montserrat S and Hutchinson JM 2001 *Polymer* **42** 7081
[16] Morgan RJ, Kong FM and Walkup CN 1984 *Polymer* **25** 375

Curing kinetics and thermomechanical properties of latent epoxy/carbon fiber composites

S Dalle Vacche, V Michaud, M Demierre, P-E Bourban and J-A E Månson
Laboratoire de Technologie des Composites et Polymères, Ecole Polytechnique Fédérale de Lausanne, Station 12, CH-1015 Lausanne, Switzerland

e-mail : sara.dallevacche@epfl.ch

Abstract. In this work, resins based on diglycidyl ether of bisphenol A (DGEBA) epoxy and a latent hardener, dicyandiamide (DICY), as well as carbon fiber (CF) composites based on them, were prepared with three commercial accelerators: a methylene bis (phenyl dimethyl urea), a cycloaliphatic substituted urea, and a modified polyamine. The curing kinetics of the three DGEBA/DICY/accelerator systems were investigated by chemorheology and differential scanning calorimetry (DSC), in isothermal and over temperature change conditions. Differences in the reaction onset temperature, and in the glass transition temperature (T_g) were highlighted. For curing of thick resin samples, a slow curing cycle at the lowest possible temperature was used, followed by high temperature (160 – 180 °C) post-curing. Indeed, fast curing at higher temperatures caused the formation of hot spots and led to local burning of the samples. The obtained thermomechanical properties, assessed by ultimate tensile testing and dynamic mechanical analysis (DMA) in single cantilever configuration, were all in the expected range for epoxy resins, with tensile moduli close to 3 GPa and T_g > 140 °C. The long-term stability of these resins at room temperature was verified by DSC. Composite samples were prepared by hand lay-up by manually impregnating four layers of 5-harness satin CF textile, and curing in vacuum bag. Impregnation quality and void content were assessed by optical microscopy. The flexural properties of the post-cured composites were assessed by three-point bending test at room temperature and showed no relevant differences, all composites having bending moduli of 45 - 50 GPa. Finally, composites cured with a faster high temperature curing cycle (20 min at 140 °C) were prepared with the DGEBA/DICY/ methylene bis (phenyl dimethyl urea) system, obtaining similar properties as with the slower curing cycle, showing that the prepreg system allowed more flexibility in terms of curing cycle than the bulk resin samples.

1. Introduction

Epoxy/carbon fiber prepregs are used in a wide range of applications, including e.g aerospace, boat building, and automotive. Due to their reactivity at room temperature, commonly used epoxy prepregs need to be stored at -18 °C and usually have a short "out-life". For application such as on site repair, where cold storage is not practical, the use of dry fibers and two-component epoxy resins, which are mixed at the time of use, is preferred. Latent curing systems based on dicyandiamide (DICY) are only reactive above a certain temperature; nevertheless this comes at the expense of a high curing temperature (> 180 °C) and time [1 - 3]. Several types of accelerators, which enhance the reactivity of the resin, still guaranteeing a certain extent of room temperature stability, have been developed and studied [4, 5]. The use of an epoxy/DICY based resin as a matrix for carbon fiber prepregs would allow their long-term storage at room temperature, reducing storage costs and simplifying logistics.

However prepregs based on such systems, and related literature, are still not widely available. Some studies recently appeared on glass and carbon fiber prepregs, either house made or commercial [6 - 9]. This work investigates the possibility of producing out of autoclave carbon fiber composites based on epoxy/DICY latent resin formulations containing different commercial accelerators, comparing their room temperature stability and properties of the cured material.

2. Experimental

A diglycidyl ether of bisphenol A (DGEBA) epoxy resin with EEW = 184-190 (Epikote™ Resin 828 by Momentive Performance Materials Inc., Waterford, NY, USA) was used, in combination with different latent hardener/accelerator systems. Two similar grades of the latent hardener, dicyandiamide (DICY), were obtained from two suppliers: Omicure™ DDA 10 by CVC Thermoset Specialties (Moorestown, NJ, USA), mean particle size 12 μm, and Amicure® CG1200G by Air Products Nederlands B.V. (Utrecht, The Netherlands), particle size 11 μm. Two accelerators were used in combination with DDA10, namely Omicure™ U-52M, a methylene bis(phenyl dimethyl urea), and Omicure™ U-35M, a cycloaliphatic substituted urea, both by CVC Thermoset Specialties (Moorestown, NJ, USA). Ancamine® 2441 (Air Products Nederlands B.V., Utrecht, The Netherlands), a modified polyamine that can be used both as hardener and as accelerator for DICY, was used in combination with Amicure® CG1200G.

Table 1. Compositions (phr) of the epoxy resin formulations used in this study, with the corresponding time to double viscosity (t_{shelf}) and glass transition temperatures of the cured resins (T_g) determined on DSC second scan, (* first DSC scan to 275°C at 20°C per minute, **first DSC scan at 10 °C/min) reported by the suppliers).

Code	Epikote 828	Omicure DDA 10	Amicure CG1200G	Omicure U-52M	Omicure U-35M	Ancamine 2441	t_{shelf} (weeks)	T_g (°C)
ED_U52	100	8	-	1	-	-	57	133*
ED_U35	100	8	-	-	1	-	> 70	129*
EA_A	100	-	6	-	-	5	> 12	135**

In order to prepare the resin formulations, the DGEBA epoxy resin was heated to 40 °C to decrease viscosity and the hardener and accelerator were mixed into it with a mechanical stirrer for 15 min. The mixture was then degassed in a vacuum oven at 40 °C prior to use. The compositions were chosen according to the indications in the datasheets, and are summarized in Table 1, together with their shelf lives (t_{shelf}) at room temperature and the glass transition temperatures (T_g) of the cured resins indicated by the suppliers.

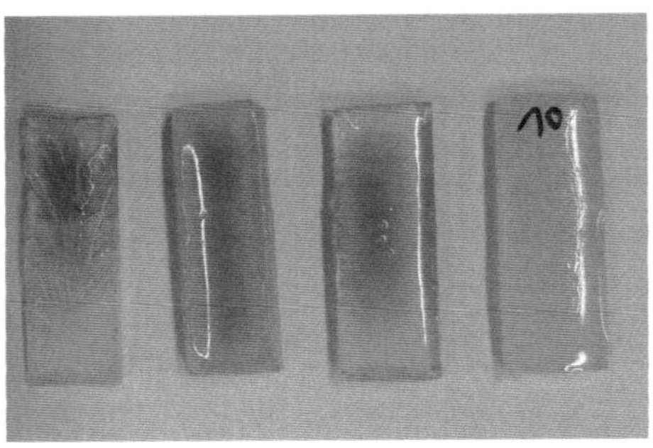

Figure 1. Three resin samples (left side) cured at high temperature, showing local burning, and one sample (last on the right) cured at low temperature.

Resin plates with a thickness of 3 - 4 mm for dynamic mechanical analysis and tensile tests were cured in open silicon molds. As preliminary curing tests at temperatures above 135 °C led to inhomogeneous curing or local burning (Figure 1), the samples were initially slowly cured at low temperature until solid, then the temperature was increased to 160 °C during 20 min. Finally they were demolded, and part of them were post-cured for 10 min at 180 °C. The temperature and duration of the initial curing were adapted according to the resin formulation, as summarized in Table 2.

In order to fabricate composite plates, four layers of 5-harness satin carbon fiber fabric were weighed and then impregnated by hand with the resin formulations. The four superposed layers of impregnated fabric were then sealed in a vacuum bag, which was placed either in an oven or in an autoclave. Vacuum was applied at room temperature for 1 hour for debulking, and then the selected curing cycle was carried out. After demolding, the composite was weighed and the mass of resin was determined by subtracting the mass of the fibers. From the mass of the fibers and resin, and their densities, the volume fraction of fibers (on a void free basis) was calculated to be $V_f = 0.52 \pm 0.02$. Specimens were then cut with a diamond saw and post-cured. Curing and post-curing cycles are detailed in Table 3.

Table 2. Cured and post-cured resin samples

Code	Resin formulation	Cure cycle	Post-cure cycle
ED-U52-C	EDU52	60 min at 100 °C 30 min at 115°C 20 min at 160 °C	no
EDU52-PC			10 min at 180 °C
ED-U35-C	EDU35	30 min at 115 °C 30 min at 120 °C 50 min at 125 °C 20 min at 160 °C	no
ED-U35-PC			10 min at 180 °C
EA-A-C	EAA	60 min at 100 °C 20 min at 160 °C	no
EA-A-PC			10 min at 180 °C

Table 3. Carbon fiber composite samples

Code	Resin formulation	Cure cycle	Post-cure cycle
ED-U52_F	ED-U52	60 min @ 120 °C	20 min @ 160 °C + 10 min ramp to 180 °C + 10 min @ 180 °C
ED-U52_F_140	ED-U52	20 min @ 140 °C	
ED-U52_F_140_A	ED-U52	20 min @ 140 °C in autoclave (P = 0.5 MPa)	
ED-U35_F	ED-U35	60 min @ 120 °C	
EA-A_F	EA-A	60 min @ 120 °C	

Differential scanning calorimetry was performed with a Q100 DSC (TA Instruments, USA) under a constant N_2 flow of 50 mL/min. Non-isothermal scans on fresh and aged uncured resin consisted of a heating cycle from 40 °C and 250 °C at 10 °C min^{-1}. Isothermal scans on uncured resins were performed by rapidly equilibrating to the desired temperature, and holding the temperature for a fixed time. After each non-isothermal or isothermal scan the sample was left in the DSC cell where it was rapidly equilibrated to -40 °C, and a heating cycle was performed up to 250 °C at 10 °C min^{-1} to assess the glass transition temperature and the residual reactivity.

Chemorheology was performed with an AR2000 rheometer (TA Instruments, USA), in oscillatory mode, holding at 25 °C for 10 min, then increasing temperature from 25 °C to 180 °C at 10 °C min^{-1} and finally holding at 180 °C until the complex viscosity reached a plateau. The frequency was 1Hz and the strain was controlled between 0.01% and 20% in order to follow the viscosity evolution.

Dynamic mechanical analysis (Q800 DMA, TA Instruments, USA) was performed in single cantilever (resin specimens) or in three-point bending (composite specimens) configurations, with an excitation frequency of 1 Hz. The temperature was raised from 25 °C to 240 °C at a heating rate of

3 °C min^{-1}. For single cantilever test the applied amplitude was 15 μm, while for 3-point bending a strain of 0.05 % was applied.

The mechanical properties in tension (resin specimens) and three-point bending (composite specimens) were assessed at room temperature with a Universal Testing Machine (UTM, LFM-125kN, Walter & Bai) equipped with a 1kN load cell (HBM Test and Measurement, Darmstadt, Germany). The tensile measurements were carried out on dog-bone shaped specimens at a speed of 1.2 mm min^{-1}, while for 3-point bending the speed was set at 6 mm min^{-1} and the span at 40 mm.

Optical micrographs were taken with an Olympus BX60 microscope equipped with a camera. The samples were embedded in clear epoxy resin and wet ground with SiC paper down to 4000 grit. Porosity was estimated from the micrographs by image analysis with ImageJ software.

3. Results and discussion

3.1. Curing kinetics and aging

The DSC temperature ramps performed on fresh uncured resin formulations revealed different curing kinetics depending on the hardener/accelerator system, as shown in Figure 1. ED-U52 showed an onset temperature of 130 °C and ED-U35 of 140 °C. Both exhibited a relatively narrow exotherm peak, with maxima at 158 °C and 166 °C, and reaction enthalpies equal to 375 ± 52 J/g and 352 ± 52 J/g, respectively. No residual reactivity appeared above 190 °C. EA-A showed a broader exotherm peak, with an onset of reaction already at 65 °C and a peak maximum at 145 °C. A second small peak with a maximum at 214 °C, close to the melting temperature of DICY, seems to indicate that some residual reaction occurred after melting of undissolved DICY. The reaction enthalpy was 346 ± 19 J/g. ED-U52 and ED-U35 were tested after aging at room temperature for 8 months. Their reaction enthalpy and their T_g were not significantly different from those measured on fresh resins, however aged ED-U52 reacted at slightly lower temperature than the fresh one, with an onset of reaction at 125 °C and a peak maximum at 152 °C. On the other hand, although the reaction enthalpy of EA-A after 6 months aging was not found to have dramatically decreased (324 J/g), and the obtained T_g was equal to that of the fresh resin, after 8 months the resin had become a hard and brittle reticulated solid, and DSC was not carried out on it. ED resins thus seem better adapted as the resin of a room-temperature storage prepreg, as already indicated from Table 1 for the pure resin.

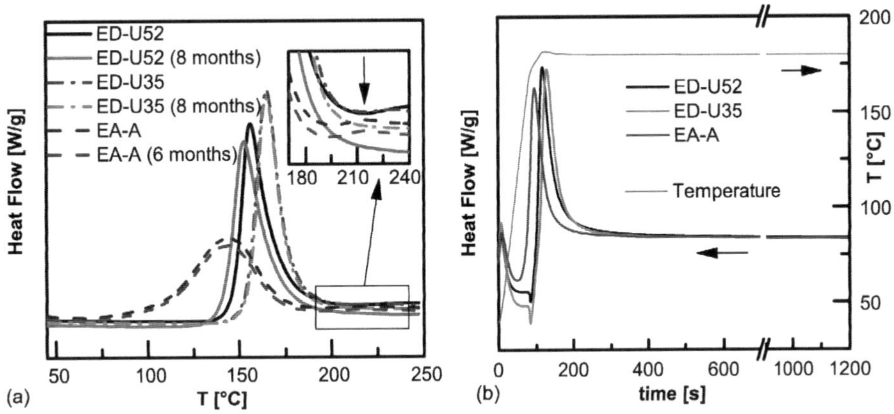

Figure 2. (a) Non-isothermal DSC scans of fresh and aged uncured resins, the inset highlights the reaction peak at 214 °C for EA-A resins, and (b) heat flow as a function of time during isothermal curing of fresh resins at 180 °C.

Isothermal curing at 180 °C during 20 min (Figure 2b) led to fully cured resins for all the three accelerated formulations, as demonstrated by the absence of residual reactivity during the second DSC scan. Increasing the curing time up to 1 h did not further increase T_g, which was 141 °C for ED-U52, 137 °C for ED-U35 and 138 °C for EA-A. As shown in Figure 2b, EA-A started reacting well before the DSC cell could reach the isothermal temperature, considerably faster than ED-U52 and ED-U35. The reaction of ED-U52 was only somewhat more rapid than that of ED-U35, as shown by the maximum of the reaction peak being delayed of about 10 s for the latter.

The complex viscosity at 25 °C was found to be similar for the three accelerated formulations, i.e 15 Pa s for ED-U52 and ED-U35, and 18 Pa s for EA-A. The viscosity initially decreased with increasing temperature, and then increased due to the curing reaction (Figure 3). The chemorheology test confirmed the DSC results. Preliminary tests performed on resin/hardener mixtures, with no accelerators, had shown that, with DICY alone, the viscosity increase started only after about 1 – 2 min at 180 °C. All accelerated formulations, ED-U52, ED-U35, and EA-A, showed a faster viscosity build-up than with DICY alone. The viscosity of ED-U52 started to increase at a lower temperature and more rapidly than that of ED-U35, while the plateau value was similar for the two systems. For EA-A, the viscosity increase started already at 130 °C and proceeded very rapidly to reach already at 160 °C a viscosity close to that of the plateau value of the other two resins, However for this resin the instrument was not able to follow the viscosity evolution up to 180 °C and a plateau value at this temperature could not be attained to compare with the other resins.

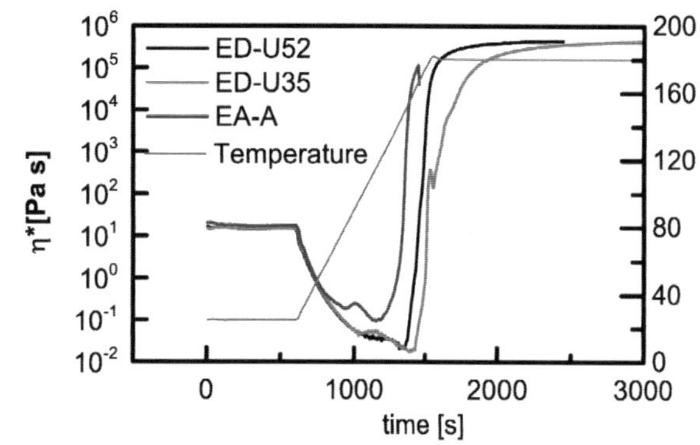

Figure 3. Complex viscosity η* of fresh resins and temperature during cure in the rheometer.

3.2. Properties of cured resin

Despite attempting to optimize the slow cure cycle for ED-U35 formulation, it was not possible to avoid a certain degree of settling of the hardener and accelerator, and the specimens showed a slight cure gradient through the thickness. This phenomenon was not observed for the other two resins, possibly due to the faster viscosity increase that hindered the settling of hardener/accelerator particles.

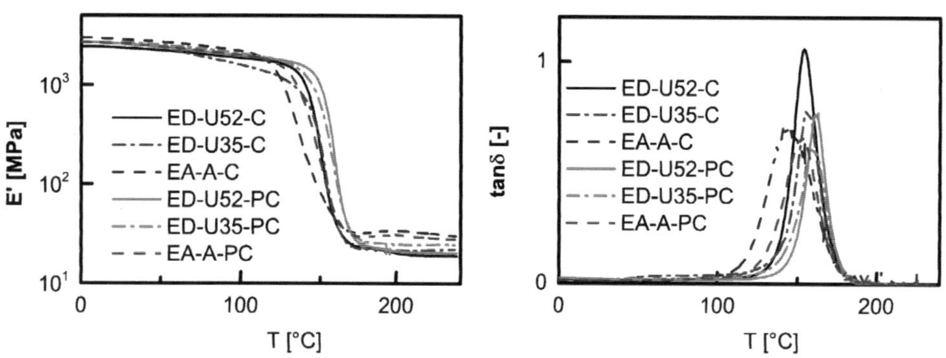

Figure 4. Storage modulus and tanδ of cured and post-cured resin specimens

The cured resins ED-U52-C, ED-U35-C and EA-A-C, although having similar properties at room temperature, exhibited a different evolution of E' and tanδ with temperature (Figure 4). ED-U52-C showed the highest T_g, with a narrow tanδ peak centered around 154 °C. ED-U35-C on the other hand, showed a slight drop of the storage modulus and a shoulder on the tanδ curve above 40 °C, followed by a steeper E' drop and a narrow tanδ peak at 156 °C. Finally EA-A-C had a wide glass transition region, indicated by a broad tanδ peak, at 145°C. All the three cured resins exhibited some degree of residual reactivity after T_g, as indicated by a very slight increase of their storage moduli around 180 – 200 °C. ED-U52-PC and ED-U35-PC had similar storage moduli over the entire range of temperatures, and a narrow tanδ peak centered around a T_g of 160 °C. On the other hand EA-A-PC had similar storage modulus as the other post-cured resins before and after the glass transition region, however the glass transition occurred at somewhat lower temperature (145 °C) with a wider tanδ peak. The storage moduli at room temperature were similar for all cured and post-cured resins, and were in the range of 2.5 ± 0.3 GPa.

The tensile moduli of post-cured specimens of all the three investigated resin formulations did not show statistically significant differences, with values in the 2.6 – 3.1 GPa range, as shown in Figure 5. ED-U52-PC and ED-U35-PC specimens had similar values of stress and strain at break, close to 60 MPa and just above 3.5 %, respectively. All the values obtained for the tensile properties fall in the typical range for a DGEBA epoxy.

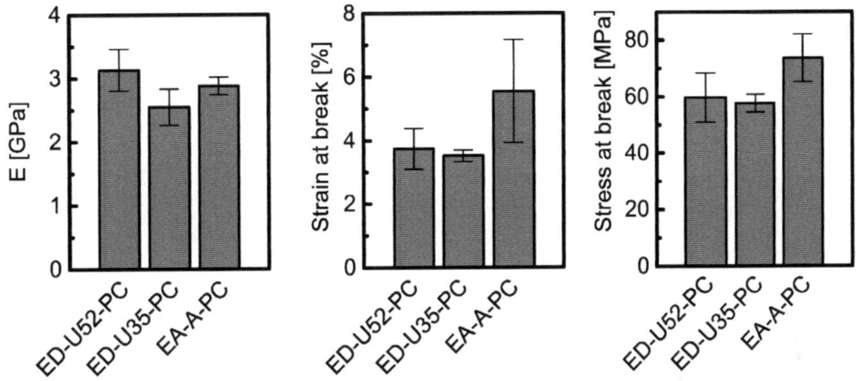

Figure 5. Tensile properties of post-cured resins

3.3. Properties of carbon fiber composites

Optical micrographs of cross-sections of composite specimens are given in Figure 6. The composites ED-U52-F, ED-U35-F and EA-A-F, which went through the same curing cycle, and ED-U52-F-140, which was cured at higher temperature and shorter time than his homologous ED-U52-F (see Table 3), showed similar impregnation and a porosity of about 6 %. The voids were located mostly at the interface between fabric layers. On the other hand ED-U52-F-140-A, for which the curing cycle, with the same temperature and time than ED-U52-F-140, was carried out in an autoclave, had lower porosity, i.e. about 4 %. These values are rather high even for a hand lay-up composite, but could be attributed to the high viscosity of the DGEBA resin, combined with insufficient degassing steps.

The storage moduli (Figure 7) of the composite specimens ED-U52-F, ED-U35-F and EA-A-F had very similar values at temperatures below 100 °C, which was to be expected on the basis of similar matrix properties, fiber volume fraction, and porosity. Finally, comparing ED-U52-F-140 with ED-U52-F-140-A, a difference in the E' before glass transition was observed, that can be attributed to the lower porosity obtained with autoclave curing. On the other hand, glass transition occurred at similar temperature for all the specimens, as shown by the similar tanδ plots (Figure 7).

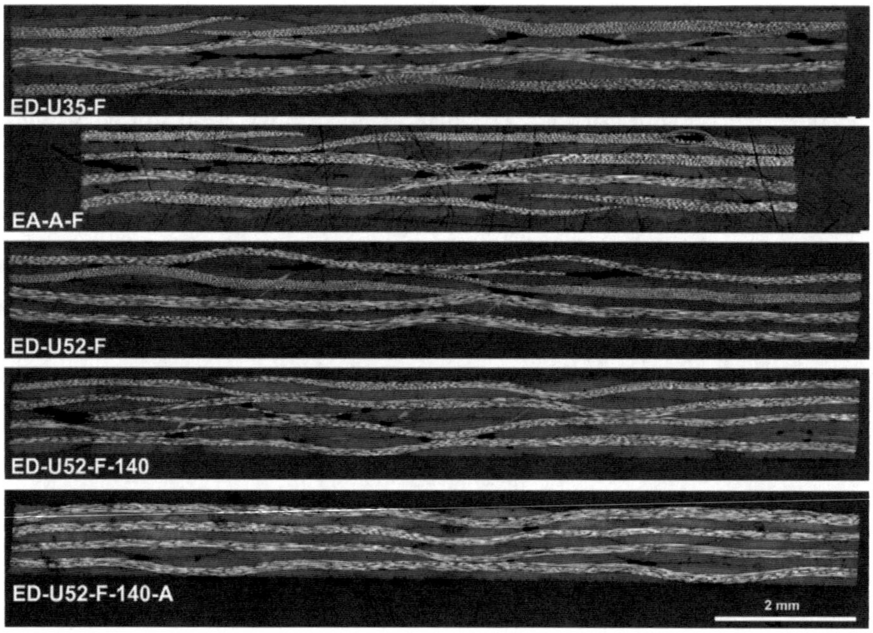

Figure 6. Optical micrographs of cross-sections of carbon fiber composites. Examples of voids at the interfaces between CF layers are indicated by arrows, and voids inside the layers by ellipses.

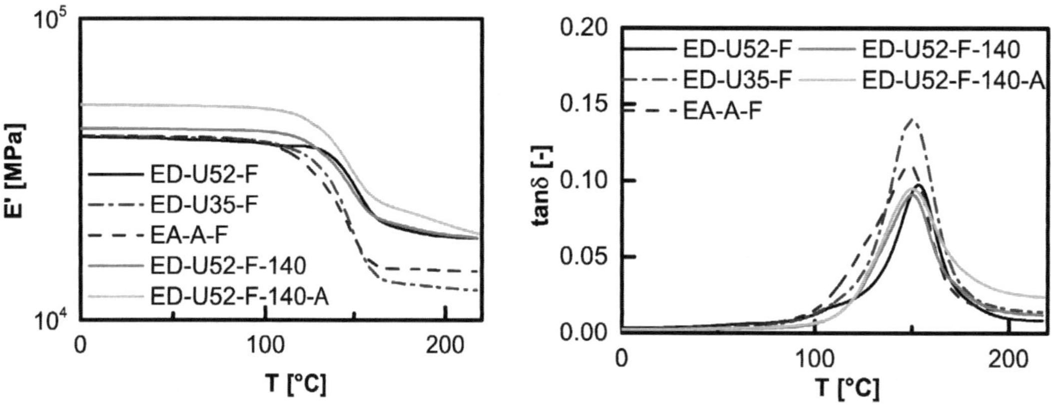

Figure 7. Storage moduli and tanδ as a function of temperature for composite specimens

The flexural properties of vacuum bag cured composites (Figure 8) did not show relevant differences due to the type of hardener/accelerator system, or to the curing cycle, with the only exception of the lower flexural strength and marginally lower strain at break showed by EA-A-F specimens. An upper limit for the modulus of a composite containing a volume fraction V_f of carbon fibers in a 5-harness satin configuration may be estimated by the rule of mixtures $E_{long} = V_f \cdot E_f + (1 - V_f) \cdot E_m$. Introducing the values of the tensile moduli of the fibers and of the matrix (assumed to be E_f = 230 GPa and E_m = 3 GPa), and considering as a first approximation that half of the fibers contribute to the longitudinal modulus in a 5-harness satin fabric, a value of E_{long} of about 60 GPa is obtained. The flexural moduli obtained experimentally are thus not far from the theoretical upper value of the longitudinal modulus, considering the fiber crimp of the satin fabric, and the shear contribution to the flexural modulus value.

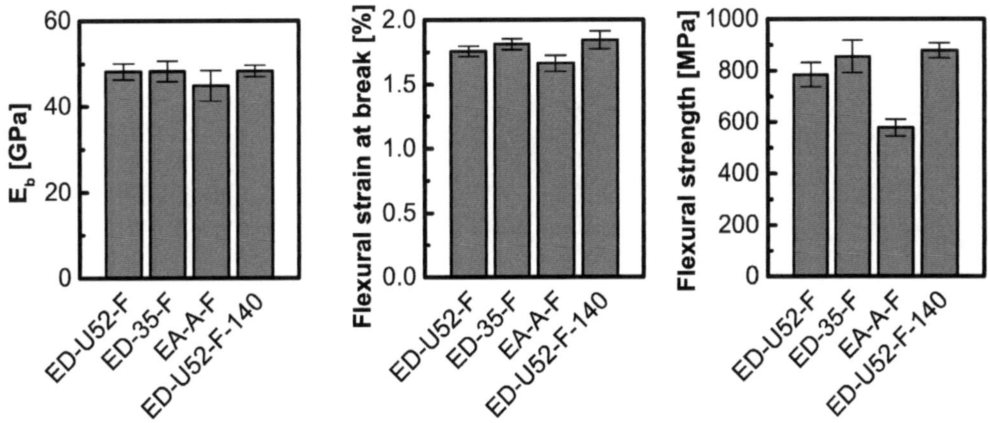

Figure 8. Flexural modulus (E_b), flexural strain at break and flexural strength of post-cured composites.

4. Conclusions

Among the three epoxy/DICY/accelerator systems considered in this work, ED-U52 and ED-U35 demonstrated no significant aging after 8 months storage of the resin at room temperature. EA-A, although still having retained a substantial degree of reactivity after 6 months, was found to have reticulated after 8 months, confirming its lower storage stability at room temperature. Nevertheless, when fully cured, both the bulk resin and the composite samples obtained from the three different accelerated resin formulations showed similar thermomechanical properties. Furthermore, when the resin systems were used as a matrix for carbon fiber composites it was demonstrated that the impregnation and the void content of the composite specimens cured in a vacuum bag were similar for all resin formulations, as expected on the basis of their similar viscosity at room temperature, resulting in similar mechanical properties. The void content decreased, and the storage modulus increased, as expected, when cure was performed in an autoclave.

This study indicated that vacuum bag curing of carbon fiber composites based on all the epoxy/DICY/accelerator systems considered is possible and leads to acceptable thermomechanical properties, although it would be desirable to decrease the void content, by e.g. using an epoxy resin with lower viscosity. Furthermore the three formulations led to similar properties of the composite specimens, and therefore the choice of a specific accelerator can be based on other considerations, as e.g. shelf life, desired curing temperature and time, or cost. These types of resin formulations could therefore be of interest for on site repair application that could not afford refrigerated storage of prepregs or the use of an autoclave.

References

[1] Sacher E 1973 *Polymer* **14** 91
[2] Amdouni N, Sautereau H, Gérard J-F, and Pascault J-P 1990 *Polymer* **31** 1245
[3] Pfitzmann A, Fischer A, Fryauf A, and Fedtke M 1992 *Polym. Bull.* **27** 557
[4] Poisson N, Maazouz A, Sautereau H, Taha M, Gambert X 1998 *J. Appl. Polym. Sci.* **69** 2487
[5] Liu X D, Kimura M, Sudo A, and Endo T 2010 *J. Polym. Sci. Part Polym. Chem.* **48** 5298
[6] Hayaty M, Beheshty M H, and Esfandeh M 2011 *Polym. Adv. Technol.* **22** 1001
[7] Hayaty M, Beheshty M H, and Esfandeh M 2011 *J. Appl. Polym. Sci.* **120** 62
[8] Hayaty M, Beheshty M H, and Esfandeh M 2011 *J. Appl. Polym. Sci.* **120** 1483
[9] Ozdemir N G 2016 *Express Polym. Lett.* **10** 394

IOP Conf. Series: Materials Science and Engineering **139** (2016) 012050 doi:10.1088/1757-899X/139/1/012050

Microdamage analysis in thermally aged CF/polyimide laminates

J Varna[1], H Zrida[1] and P Fernberg[1,2]

[1]Department of Engineering Sciences and Mathematics, Lulea University of Technology, SE 97187 Lulea, Sweden
[2]Swerea SICOMP AB, SE-94126, Pitea, Sweden

E-mail: Janis.Varna@LTU.SE

Abstract. Microdamage in layers of CF Thornel® T650 8-harness satin woven composite with thermosetting polyimide NEXIMID® MHT-R resin was analysed. After cooling to room temperature multiple intra-bundle cracking due to tensile transverse thermal stresses was observed in the studied [(+45/-45)/(90/0)]2s composite. The composite was subjected to thermal cycling quantifying the increase of crack density in layers. Comparison of two ramps with the same lowest temperature shows that the highest temperature in the cycle has a significant detrimental effect. Exposure for 40 days to 288°C caused many new cracks after cooling down to room temperature. Both aged and not aged specimens were tested in uniaxial quasi-static tension. Cracking was analysed using fracture mechanics and probabilistic approaches. Cracking in off-axis layers was predicted based on Weibull analysis of the 90-layer. The thermal treatment degraded the cracking resistance of the surface layer and of the next layer.

1. Introduction

An increasing interest of aerospace industry in composite materials for applications in harsh environments has been driving improvement and development of new composites. Composites designated for aero-engine applications considered in the current paper are subjected to high temperatures. Long exposure to elevated temperatures can cause irreversible changes in the morphology (termed aging in this paper) reducing strength and stiffness [1]. In addition, cyclic variations between high and low temperatures can lead to fatigue resulting in microdamage accumulation [2].

The effect of thermal aging on mechanical performance of CF/RTM6 resin and CF/BMI resin cross-ply laminates was studied in [3] showing that the intralaminar crack density in a following tensile test is a clear function of aging time. The elevated temperature level during thermal cycling with a fixed lowest temperature has an accelerating effect on the microdamage development [4] which currently does not have solid explanation. Apparently irreversible phenomena take place at the highest temperature and microdamage evolves when the lowest temperature in the cycle is reached.

The main objective of the current work is to study intralaminar cracking during thermal cycling and tensile quasi-static loading of "as received" and "aged" quasi-isotropic carbon fibre/polyimide woven composites. Microcracking initiation and growth will be analysed using Weibull initiation strength distribution approach and fracture mechanics, focussing on

- Differences in crack density growth in different layers

- Analysing the effect of high temperature exposure on intralaminar cracking

2. Experiments

2.1. Description of materials.

Quasi-isotropic $[(+45/-45)/(90/0)]_{2s}$ composite was manufactured (RTM) using 8-harness satin Cytec T650 CF and thermosetting polyimide resin denoted NEXIMID® MHT-R (MHT-R). The material was cured at 340°C for 30min followed by 2.5h post cure at 370°C using 12 bar pressure. In Figures the composite is labelled as 432 plate. The volume fraction of fibres is around 60%. More details on manufacturing and properties of the laminate are given in [5]. The average thickness of one "layer" (bundle) is approximately 0.18 mm. The bundle width is about 1.5mm and the aspect ratio partially justifies the used "layer assumption". Both edges of all specimens had fine polishing for damage quantification using optical microscopy. Part of specimens was subjected to aging at 288°C for 960 hours (40 days) in air environment, afterwards cooled down to room temperature (RT) for inspection.

2.2. Experimental procedures.

Two different thermal cycling ramps were used. In the ramp (R1) the temperature was varied between -60°C and RT whereas in ramp (R2) the change was between -60°C and 288°C. The heating and cooling rate was approximately 100 °C for minute and the specimen was exposed to each constant temperature for 10 minutes for up to 150 cycles.

Tensile loading and unloading tests with increasing maximum were performed at RT with displacement rate 2mm/min. Strains were measured using extensometer with gauge length 50mm. The damage state in all specimens was analyzed before any treatment (notation not aged (NA)) and also after aging (A). After certain number of cycles in thermal fatigue and after each loading-unloading step in quasi-static tensile loading the specimen was inspected using optical microscope. The respective number of cracks N_i over distance L (50mm) on the specimen edge was counted and the crack density in each layer was calculated as

$$\rho_i = \frac{N_i}{L \sin \theta_i} \tag{1}$$

where θ_i is the fiber orientation angle in the layer. The void content was measured using image analysis software (NIS-elements BR) together with optical microscope. The void content was defined as the area fraction of voids on the observed edge surface. The average void content was measured over the area of about 100mm^2. From one specimen to another, the average void content varied from 0.3% to 2.4% which is a very high variability.

3. Results and discussion

3.1. Initial damage state.

The initial damage state is caused by thermal stresses at the lowest temperature. According to the Classical Laminate Theory (CLT) thermal transverse stresses in a quasi-isotropic laminate is the same in all layers and the in-plane shear stress is zero. Hence, the number of initiated cracks statistically should be the same in all layers. Still, on the specimen edge the crack density in different layers was different: much higher crack density in the surface +45-layers; lower in the 90-layers but still larger than in the internal ±45-layers, where it was very low. To explain these observations, 3-D FEM thermal stress analysis was performed to study free edge effect on thermal stress distribution: $[(+45/-45)/(90/0)]_{2s}$ specimen was modelled (a specimen cut from the central part of the plate has free edges where the stress state is different than in the same position in the plate). Clear differences between transverse edge stresses in ±45-layers and in 90-layers were found, see details in [6]. In 90-layers the transverse stress is magnified by approximately 20% whereas in all 45-layers it is only 30% of the CLT stress. This means that in the specimen new cracks may have been created in the edge regions of the 90-layer. In 45-layers new cracks were not introduced by making specimens. Therefore, the crack density in the central part of the 90-layer should be similar to the damage state observed on the edge of the 45-layer.

To explain the much higher crack density in the surface layers we have to distinguish two phases in crack evolution: initiation (mainly transverse growth) and propagation along fibers. The propagation of a crack may be described by Linear Elastic fracture mechanics (LEFM) which states that the crack will propagate when the energy release rate (ERR) G reaches or exceeds the critical value G_C . The crack initiation process is stochastic and characterized by coalescence of fibre/matrix debonds leading to "defect" creation and growth in the ply thickness direction. According to LEFM the stress required for central crack growth in an infinite plate is $1.12\sqrt{2}$ times higher than for an edge crack of the same size in a semi-infinite plate [6]. Generalizing this result to laminates we can expect that under thermal loading more cracks are initiated in the surface layer of the quasi-isotropic laminate. The available energy for crack propagation along fibres depends on the ply thickness and on the position of the layer: in the surface layer it is about 2.4 time larger [7]. Hence, the propagation of an initiated crack in the surface layer would start at much lower transverse stresses.

3.2. Thermal fatigue.

Each data point for crack density in Fig. 1 corresponds to the average on two edges of one specimen for all layers of a given orientation. For all layer orientations the R2 (-60°C to 288°C) ramp induced more damage than R1 (-60°C to RT). This means that the temperature interval, $\Delta T = T_{min} - T_{max}$ at fixed T_{min} has effect on the cracking in thermal fatigue. It is consistent with the Paris law for growth of individual crack: ΔG is larger for R2 ramp than for R1 ramp. The crack density in thermal cycling was the highest in surface layers followed by 90- and even lower in internal 45-layers (not shown). Much larger crack density in surface layers was observed also in [8] where thermal cycling at cryogenic temperatures was performed.

In fatigue, certain number of cycles N_{init} is required to initiate a crack and a different number of cycles N_{prop} is required for its propagation along fibres. If $N_{init} \gg N_{prop}$, the number of cycles for full development of the crack is initiation governed. The initiation in the surface layer is accelerated by the free surface effect. The ERR for crack growth along fibres is also much higher in the surface layer than in internal layers and initiated cracks almost instantly grow through the specimen. To investigate whether the edge cracks propagate inside the specimen and vice versa, the edge of one specimen subjected to 150 cycles in R2 was polished to quantify the damage state versus the distance d from the specimen edge: a) the crack density inside the specimen was approximately the same in the 90- and internal 45-layers; b) the crack density on the edge of the surface 45-layer does not change with the distance from the edge, proving that in this layer there is an excess of ERR for propagation.

3.3. Enhanced temperature.

Specimens were exposed to 288°C in ambient air environment (aged) and cooled down to RT to count cracks on specimen edges, see Fig. 2. The thermal stresses at RT introduced new cracks. Surprisingly the crack density on the edge of the 90-layer is higher than in the surface +45-layer reaching about 1.7cr/mm. Edge polishing of a 40 days aged specimen was performed to establish whether the edge data are representative for the bulk of the composite. The results show that in the 90-layer the amount of cracks away from edge is very similar to the number of cracks found on the edge of the same specimen before aging. It indicates that in the bulk of the 90-layer the material degradation (resistance to cracking at room temperature) due to high temperature exposure for 40 days is rather small: cracks initiated on the degraded edge do not propagate inside the layer.

There was no edge "concentration" in crack density for surface layers. During the aging test the whole surface 45-layers are in contact with the environment. Therefore more uniform material degradation can be expected in surface layers leading to more uniform resistance to crack initiation and propagation.

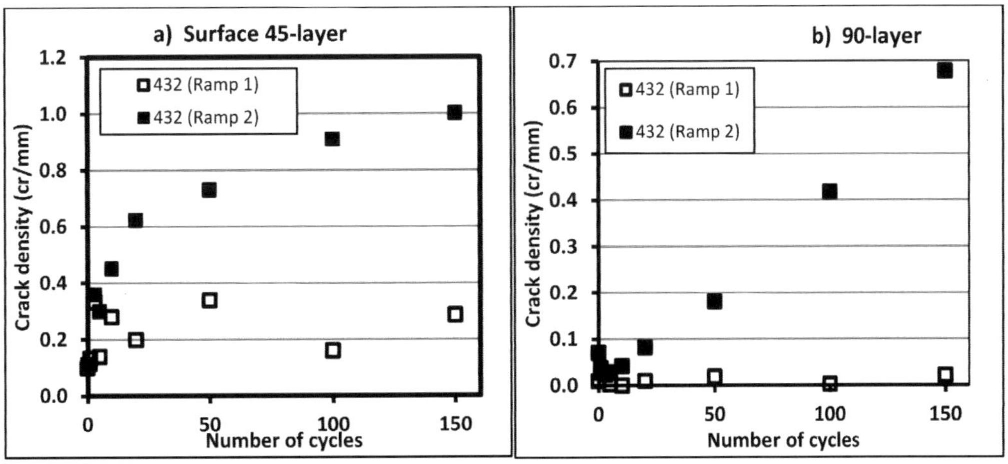

Figure 1. Edge crack density growth in two thermal fatigue ramps: R1 (-60;RT)°C; R2 (-60;+288)°C.

The degraded zone propagates in thickness direction from the specimen surface and the crack density in internal 45-layer depends on the distance of this layer from the surface (the crack density in the 45-layer next to surface layer is much higher than in 45-layers close to the symmetry plane of the laminate where the crack density is very similar as in 90-layers). This means that not only the surface layer but also the next layer has degraded, which is expected because the thickness of one layer (0.18mm) is much smaller than the size of the observed highly degraded zone at the edge (about 1 mm).

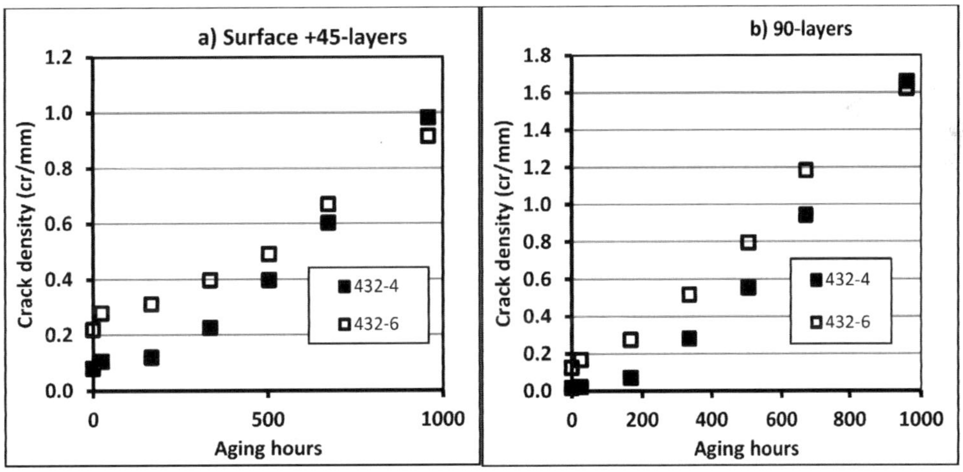

Figure 2. Crack density on the specimen edge versus aging time: a) surface 45-layers; b) 90-layers.

3.4. Damage state in tensile mechanical loading.

The severely damaged edge zone of the aged specimens was removed by polishing before mechanical testing. The effect of aging on cracking resistance was expected to be high in surface layers. The tensile loading introduced new cracks on the edge of aged (A) and not aged (NA) specimens. The cracking data presented below are based on edge observations only. To check that edge data are representative the damage statistics on the edge was compared with the damage state (after polishing) at several distances from the edge finding that statistically the crack density is not reducing.

Apparently most of the initiated cracks had enough of ERR to propagate along the fibre direction and that crack initiation should be the focus for the used composite. These results give sufficient confidence in that edge data are representative for the damage state inside the specimen.

Thermal stresses do not change during the tensile test whereas the mechanical stress is proportional to the applied strain. According to CLT at 1% strain the total transverse stress/shear stress ratio in the 45-layer is $\sigma_{LT}^{45}/\sigma_{T}^{45} \approx 0.45$ and at lower strain the fraction of the shear stress is lower. For this reason, analysing failure in layers we will neglect the shear stress contribution.

The conditions for crack initiation in surface and internal layers are different: in internal layer the surface defect growth is suppressed by constraint from the adjacent layers and defects in the middle of the layer will grow into a crack. In [9] relationship between crack initiation stress in internal and surface layers was suggested

$$\sigma_{T,init}^{internal} = k\sigma_{T,init}^{surface} \quad (2)$$

where $k = 1.12\sqrt{2}$ for surface layer and $k = 1$ for internal layer. The initiation of a crack is a stochastic process with certain probability and with random position. In simulations of crack initiation we assume that any layer in its transverse direction can be considered as consisting of a chain of elements following the two-parameter Weibull strength distribution

$$P_f = 1 - exp\left[-\left(\frac{k\sigma_T}{\sigma_{ino}}\right)^m\right] \quad (3)$$

Parameter k is introduced in (3) to reflect the favourable situation for crack initiation in the surface layer. If the crack density in layer is low enough to allow for large plateau regions in stress distribution, the probability of failure at transverse stress σ_T can be estimated using expression [8]

$$P_f = \frac{\rho(\sigma_T)}{\rho_{max}} \quad (4)$$

In (4) ρ_{max} is the assumed maximum possible crack density in a layer when due to interaction the transverse stress between two neighbouring cracks is too low to lead to new intralaminar crack formation. Based on experimental observations $\rho_{max} \approx 1/t$, where t is the thickness of the layer. Using the experimental crack density dependence on transverse stress to calculate P_f, we plot the $log(-log(1 - P_f))$ versus $log(\sigma_T)$ and use the standard procedure to estimate Weibull parameters. The result is rather sensitive to the used data points on both extremes: data points at high crack density are affected by stress perturbations from neighbouring cracks while the stress at first crack varies a lot from specimen to specimen. The estimates of Weibull parameters can be used in more accurate Monte-Carlo simulations over the whole crack density region including the high crack density region where the stress perturbations caused by cracks are overlapping.

Cracking was analysed in 90-layers converting the strain dependence of crack density to transverse stress dependence using CLT and the estimated thermal stresses. Using (4) the crack density data was transformed to probability of failure and (3) was used to estimate Weibull parameters (referred as P_f-approach). Weibull parameters for both aged and not aged composite are given in Table 1.

3.4.1. Damage accumulation analysis starts with plotting the log-log relationship for 90-layers in NA specimens as shown in Fig. **3a.** The first point on the left corresponds to crack density caused by thermal stresses which has rather uncertain value. The last three points deviate from linearity which is a sign of crack interaction. According to Fig. 3b, where the test data and the simulation using (4) are shown, deviation starts at $\rho_{90} \approx 1.6\ cr/mm$. In general the obtained Weibull parameters together with (3), (4) give good description of the cracking development.

The obtained Weibull parameters were used to predict crack density in internal +45-layers and in the surface +45-layer of the NA laminate. All internal +45-layers behave in a very similar way and the presented data are averages. The data and simulations using the P_f-approach are presented in Fig. 4 which shows good agreement proving that cracking in +45-layers has the same transverse stress dependence as in 90-layers. The crack density in the surface +45- layer, shown in Fig. 4b is significantly higher. The Pf –approach with $k = 1$ gives too low crack density. An attempt to use in simulations equation (3) with $k = 1.12\sqrt{2}$ failed – far too high crack density was predicted, see Fig. 4b. These results show that the role of the free surface on defect growth has been oversimplified in (2).

Table 1 Weibull parameters from 90-layer cracking

Composite	**m**	σ_{in0} **(MPa)**
432-NA	7.84	135.2
432-A	6.88	134.0

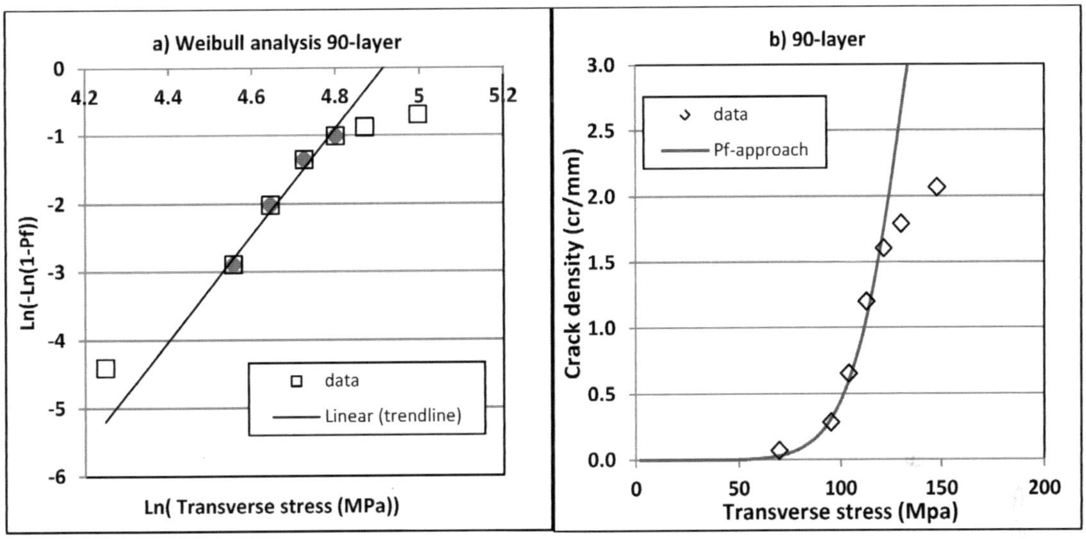

Figure 3. Cracking in 90-layers (NA-432): a) Weibull plot of test data and linear fit; b) crack density versus transverse stress σ_T^{90}, P_f-approach (4).

3.4.2. Effect of aging on damage development in tension was studied obtaining Weibull parameters for 432-A specimens, see Table 1, from aged 90-layers cracking data. They were used in the P_f -approach to predict crack density growth in 45-layers, see Fig. 5. For internal +45-layers predictions are in good agreement with test data. Similarly as for NA specimens, the crack density in the surface layers could not be predicted: it is even outside the previously discussed bounds. The crack density in these layers a) is much higher than in the surface layers of an unaged specimen and b) the difference in crack density with internal layers of the same orientation is much larger than before the aging test. That means that the surface layer has different failure properties than the internal layers: it has lower initiation stress and therefore ρ is much higher than in NA case. The probability of transverse failure in 90-layers of not aged and aged composite is very similar: the high temperature exposure has not affected the damage behaviour of 90-layers.

In Fig. 6 crack density is shown separately in the L3 and in the L7 –layer, both with $-45°$ orientation. The L3 layer is the third layer from the midplane of the $[+45/-45/90/0/+45/-45/90/0]_s$ laminate and L7 is the seventh layer from the midplane. Thus, the L7 is close to the surface. The P_f -approach simulation describing cracking in +45-layers, is shown as solid line. For not aged specimens the crack density in L3 layer is just slightly lower than in the L7 layer. In aged specimens in Fig. 6b the

IOP Conf. Series: Materials Science and Engineering **139** (2016) 012050 doi:10.1088/1757-899X/139/1/012050

difference is striking: much higher crack density in L7 layers. In the L3 layers the cracking is similar as for not aged -45-layers. An obvious explanation for that is aging: layers close to the laminate mid-plane have not degraded. Only the surface layer and the following L7 -45-layer are aged. There are more cracks in -45-layers than in +45-layers which we explain by the effect of damage in adjacent 90-layers.

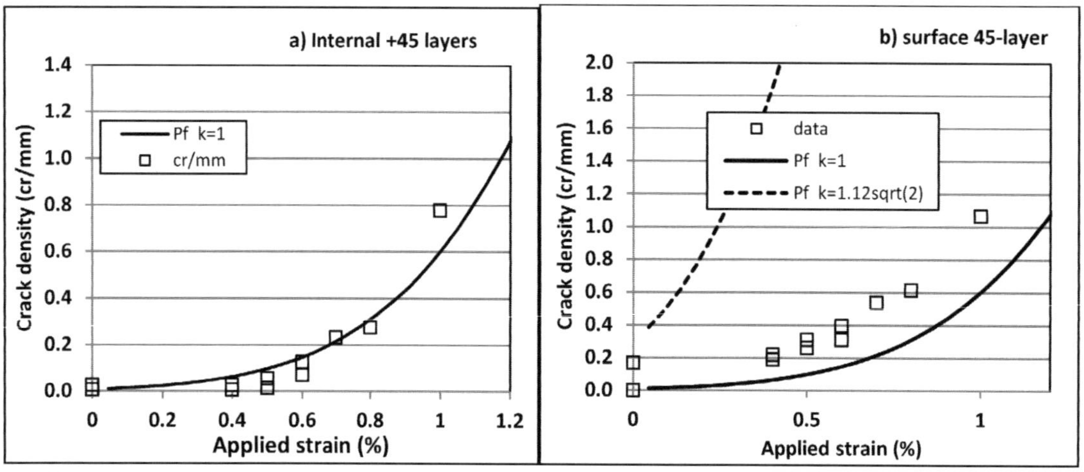

Figure 4. Crack density versus strain in 45-layers of NA-432: data and predictions using Pf-approach.

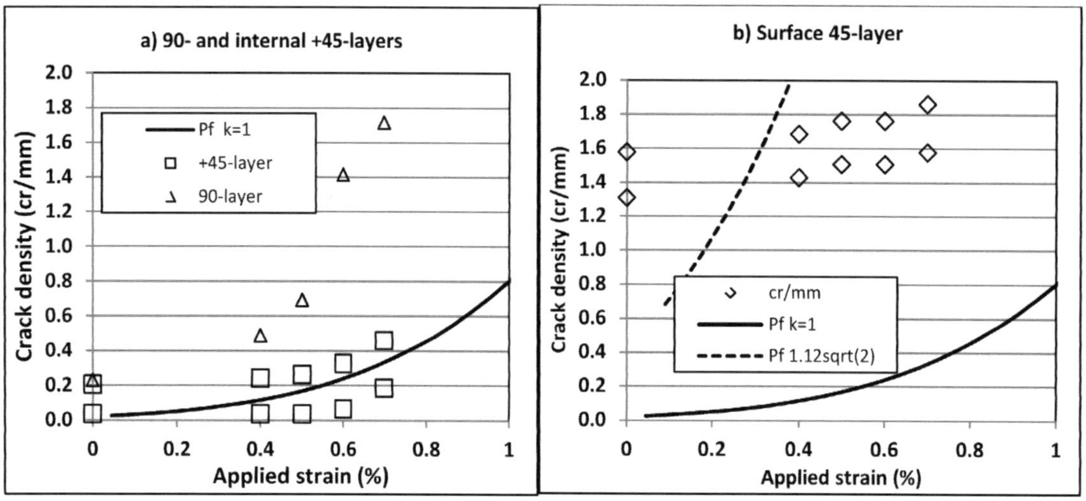

Figure 5. Microcracking in aged A-432 composite.

4. Conclusions

Microscopy was used to analyze the damage state in quasi-isotropic CF/polyimide 8-harness satin composite after manufacturing and its evolution a) due to exposure to 288C for 40days; b) due to thermal cycling; c) in result of tensile mechanical loading. In all cases much higher rate of crack density growth was found in surface layers because of the free surface effect on initiation and also on the propagation of intralaminar cracks. Due to temperature exposure about 1mm of the edge zone was severely degraded causing more cracks whereas damage inside the specimen was not affected.

In tensile loading the crack density in different layers of the laminate was analyzed using for crack initiation Weibull transverse strength distribution. It was found that cracking in internal +45-layers can be predicted based on Weibull properties of 90-layers. The crack density in the surface +45-layer was much higher even for not aged specimens. The thermal treatment significantly reduced the resistance of the surface layer to cracking under mechanical tension. It degraded also the layer next to it, whereas failure resistance of central layers was not affected.

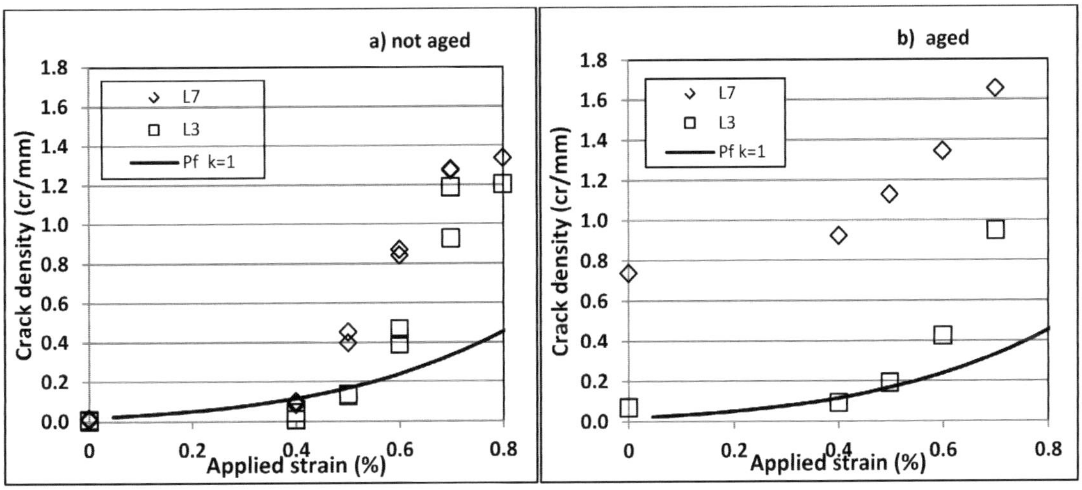

Figure 6. Cracking in -45-layers of the 432 quasi-isotropic laminate (layers are numbered L3, L7 regarding their position with respect to mid-plane): a) NA; b) A.

References

[1] Owens G A and Schofield S E 1988 Composites science and technology **33** 177
[2] Tsotsis T K, Keller S, Lee K, Bardis J, Bish J 2001 Composites science and technology, **61** 75
[3] Giannadakis K, Varna J 2009 IOP Conf. Series: Materials Science and Engineering **5**
[4] Bechel V T, Camping J D, Kim R Y2005 Composites Part B **36** 171
[5] Tsampas S A, Fernberg P S and Joffe R 2015 20th International conference on composite materials
[6] Zrida H, Fernberg P, Varna J and Ayadi Z 2016 Int. Journal of Fatigue submitted.
[7] Broek D 1978 Elementary engineering fracture mechanics, Sijtholf & Noordhoff, Alpena an den Rijn, The Netherlands
[8] Varna J, 2013 Journal of Composite Materials, **47(20-21)** 2443
[9] Bechel V T 2005 46th AIAA/ASME/ASCE/AHS/ASC Structures, Structural Dynamics and Materials conf., Austin Texas
[10] Berglund L A, Varna J, Yuan J 1992 Journal of Reinforced Plastics and Composites **11** 643